U0275597

寰宇文献 Universal Library | SINOLOGY 系列

SELECTED WORKS OF BERTHOLD LAUFER

劳费尔著作集

第二卷

[美] 劳费尔 著

黄曙辉 编

中西书局
ZHONGXI BOOK COMPANY

图书在版编目(CIP)数据

劳费尔著作集 /(美)劳费尔著；黄曙辉编. —上
海：中西书局，2022
(寰宇文献)
ISBN 978-7-5475-2015-4

Ⅰ.①劳… Ⅱ.①劳… ②黄… Ⅲ.①劳费尔–人类
学–文集 Ⅳ.①Q98-53

中国版本图书馆CIP数据核字（2022）第207067号

第 2 卷

024

一部关于苯教的藏文史书

通報

T'oung pao

ARCHIVES

POUR SERVIR À

L'ÉTUDE DE L'HISTOIRE, DES LANGUES, DE LA GÉOGRAPHIE ET DE L'ETHNOGRAPHIE DE L'ASIE ORIENTALE

(CHINE, JAPON, CORÉE, INDO-CHINE, ASIE CENTRALE et MALAISIE).

RÉDIGÉES PAR MM.

GUSTAVE SCHLEGEL

Professeur de Chinois à l'Université de Leide

ET

HENRI CORDIER

Professeur à l'Ecole spéciale des Langues orientales vivantes et à l'Ecole libre des Sciences politiques à Paris.

Série II. Vol. II.

LIBRAIRIE ET IMPRIMERIE
CI-DEVANT
E. J. BRILL.
LEIDE — 1901.

ÜBER EIN TIBETISCHES GESCHICHTSWERK DER BONPO

VON

BERTHOLD LAUFER.

Vor einiger Zeit übersandte mir der indische Pandit Sarat Chandra Dás den in der Presse von Darjeeling hergestellten, 61 Octavseiten umfassenden Abzug eines tibetischen Werkes, das den Titel führt: *rgyal rabs bon gyi ₒbyuṅ gnas* d. h. Entstehung oder Geschichte des Königsgeschlechts nach der Tradition der Bon, im Gegensatz zu andern von buddhistischer Seite verfassten Werken gleichen Titels und Inhalts. Das erste Blatt des Originals ist leider verloren gegangen; der Herausgeber hatte indessen kurz nach Auffindung desselben in Tibet das erste Kapitel ins Englische übersetzt, so dass nach Mitteilung des Anfangs in seiner Übertragung der Text mit dem zweiten Blatte beginnt. Als Verfasser werden im Kolophon am Schlusse (61, 24) der *K῾yuṅ po Blo gros rgyal mts῾an* und der *K῾yuṅ po rGyal ba t῾od dkar* bezeichnet, die im Gedanken an den Nutzen der Lehre und der Wesen das Buch in *Ri k῾rod* verfasst haben. Das Werk handelt von dem Ursprung der Bonreligion, von ihrem Stifter *gŠen rabs*, von der Dauer und Ausbreitung seiner Lehre wie von ihrem endlichen Verfall, der mit dem Tode des Königs *gLaṅ dar ma* seinen Abschluss erreicht hat. Das Ganze zerfällt in

26 nicht numerierte Kapitel, die jedes nach einer eingehenden Disposition wiederum in kleinere Abschnitte zerlegt werden. Bei der Mehrzahl ihrer Angaben bezeichnen die Autoren genau die Quellen, aus denen sie geschöpft haben, und stellen in solchen Fällen, wo die Ansichten widerstreiten, die Meinungen zweier oder mehrerer Gewährsmänner gegenüber. Soweit ich bis jetzt feststellen konnte, scheint sich die Zahl der zur Benutzung herangezogenen Quellen auf etwa 25 zu belaufen. Auch an kritischen Bemerkungen und sogar an Zweifeln in die Richtigkeit mancher Überlieferungen fehlt es nicht. So wird z.B. p. 9, 26 erwähnt, dass *gÑa k'ri btsan po* vor der Geburt Buddha's König von Tibet war, eine Behauptung, die, wie sogleich hinzugefügt wird, in Widerspruch steht mit der Annahme, dass Tibet nach dem Nirvâṇa des Çâkyamuni noch keine Bevölkerung gehabt habe. Bei Citierung der bekannten Ursprungs-sage der Tibeter betreffend ihre Abstammung von einem Affen und einer Felsen-Râkṣasî (*brag srin mo*) wird der Tradition gedacht, dass die Bod einst geschwänzt (*rṅa ma can*) gewesen seien, und die Anmerkung hinzugefügt: »Das ist nicht wahr, denn die Tibeter haben in der That keine Schwänze; auch im Osten, an der Grenze Chinas, soll es geschwänzte Menschen geben, was gleichfalls nicht richtig ist" (p. 9, 20). Mögen uns solche Äusserungen auch naiv vorkommen, so ist nicht zu verkennen, dass sie inmitten eines wundergläubigen Volkes von einer gewissen rationalistischen Denkart Zeugnis ablegen und den Keim einer kritischen Betrachtung der Dinge enthalten. Ja, in einigen Abschnitten lässt sich nicht leugnen, dass wenigstens der Ansatz zu einer Art pragmatischer Geschichtschreibung gemacht ist. So wird das 21. Capitel auf p. 48, das den Titel » Niedergang der Bonreligion unter der Regierung des *K'ri sroṅ*" führt, im Eingang — fast nach der Schablone einer Chrie — in vier Paragraphen zerlegt, die vier Fragen enthalten, die Frage nach der Zeit des Niedergangs, nach dem Lande des Niedergangs, nach dem Herrscher,

unter dessen Regierung derselbe stattgefunden, endlich nach der Art und Weise des Niedergangs selbst. Nachdem die drei ersten Fragen kurz damit beantwortet sind, dass es die Zeit war, in der das Lebensalter der Menschen fünfzig [1]) betrug, dass der Schauplatz Tibet und Žaṅ žuṅ waren, nach der Unterwerfung dieses Landes unter die tibetische Herrschaft und der Ermordung seines Fürsten *Lig mi rgya yab*, und dass der betreffende König von Tibet *K'ri sroṅ ldeu btsan* war, wird der Verlauf des Untergangs der Bonreligion erzählt, indem zuerst von den Ursachen [2]) desselben gehandelt wird, die in den Flüchen eines Bonpo namens *Gra gum btsan po* zu suchen sind, und dann von den die Hauptursache begleitenden Erscheinungen oder Ereignissen [3]). Neben den schriftlich aufgezeichneten Quellen findet auch die mündliche Tradition [4]) Berücksichtigung, wie z.B. wiederholt der Überlieferungen von Žaṅ žuṅ Erwähnung geschieht.

Wie die Composition, so unterscheidet sich auch die Schreibweise dieses Buches wesentlich von der in anderen bisher bekannt gewordenen tibetischen Werken. Die Phraseologie ist oft eigentümlich und bietet nicht geringe Schwierigkeiten; die Perioden sind kurz, der Ausdruck gedrungen, zuweilen bis zur Dunkelheit. Dazu kommt eine teilweise noch gänzlich unbekannte Terminologie, besonders in den mythologischen Partieen. Diese füllen die 25 ersten Seiten des Buches, wozu ich auch die rein sagenhafte Lebensgeschichte des *gŠen rabs* rechne. Interessant ist die auf p. 8 begegnende Anspielung auf das Râmâyana. Es wird erzählt, dass der Schneeberg Tise [5]) und der Manasarovarasee [6]) in ₒOl mo luṅ riṅ [7]) in Persien gewesen seien; von dem in Tibet befindlichen Tise sagt Saskya Paṇḍita, dass der Affenkönig Halumandha von dem Schneeberg

---

1) tib. *ts'e lo lṅa bcu pai dus su.*

2) tib. *rgyu.*   3) tib. *rkyen.*   4) tib. *sñan rgyud.*

5) Sanskrit: Kailâsa, s. Mémoires de la Société Finno-Ougrienne, vol. XI, p. 79.

6) tib. *ma p'am*, Sanskr. Anavatapta, s. ibid. p. 93.

7) *luṅ riṅ* = langes Thal.

Tise im Lande Persien ein Stück abgebrochen, mitgeschleppt, ins Land der Bod geschleudert und so den Tise gemacht habe. Halumandha [8]), dessen Reich als *Halumadhai spreu gliṅ* p. 7, 4 bezeichnet wird, ist Hanumân, der Affenfürst des Râmâyana. Offenbar spielt unsere Stelle auf die Erzählung im 6. Buche (Yuddhakâṇḍa) des indischen Epos an, in der Hanumân, um die vier Heilkräuter zu holen, zum Berge Kailâsa fliegt, und da sich diese verstecken, den ganzen Berg ausreisst und mitbringt [9]). Man erinnere sich, dass auch der Affenfürst, der sagenhafte Stammvater der Tibeter, eine Inkarnation des Avalokiteçvara, wenigstens nach der Version im 34. Capitel des *Maṇi bka ₀bum*, mit Hanumân (in der Form *Hilumandju*) identificiert wird [10]). Diese Ursprungssage wird, wie schon erwähnt, in unserem Werke nur kurz citiert (p. 9, 17—20), aber es ist beachtenswert, dass hier, was in keiner andern Version [11])

---

8) Halumanda im *Padma t'aṅ yig*, s. Grünwedel, Ta-she-sung, Sep.-A. aus Bastian-Festschrift, p. 14, 24.

9) So auch im 6. Akt von Bhavabhûti's Mahâvîracarita. Im 13. Akt des Mahânâtaka ist es der Berg Drohina, den Hanumân zusammen mit der Heilpflanze herbeibringt.

10) Rockhill, The Land of the Lamas, p. 355.

11) Den ersten Bericht über diese Sage hat, soweit ich sehe, Bogle in Verbindung mit der Flutsage gegeben, s. Markham, Narratives of the mission of G. Bogle to Tibet and of the journey of Th. Manning to Lhasa, p. 341. Turner, Gesandtschaftsreise an den Hof des Teshoo Lama, p. 225. Georgi, Alphabetum Tibetanum, p. 280. Pallas, Sammlungen historischer Nachrichten über die mongolischen Völkerschaften, Bd. II p. 406 und Note; mit dieser Stelle ist Julien, Voyages des pèlerins bouddhistes, vol. II, p. 387 zu vergleichen. Klaproth, Fragmens bouddhiques, p. 34 (aus mongolischer Quelle). Schmidt, Forschungen auf dem Gebiete der älteren... Bildungsgeschichte der Völker Mittelasiens, p. 210 (nach dem Bodhimör). E. B. Tylor, Primitive culture, p. 342; 2. ed., p. 376. Bastian, Der Mensch in der Geschichte, Bd. III, p. 347, 349. Rockhill, The life of Buddha, p. 204; derselbe, The land of the Lamas, p. 355—361. Chandra Dás, The origin of the Tibetans, Proceedings of the Asiatic Soc. of Bengal, 1892, No. II, p. 86—88; derselbe, Journal of the Buddhist Text Society, vol. IV, part 2, p. (3); derselbe, ibidem, vol. V, part 1, p. 1—4. Kreitner, Im fernen Osten, p. 834. Spinner, Tibetanisches aus dem britischen Himalaya, in Zeitschrift für Missionskunde und Religionswissenschaft, Bd. VI, 1891, p 132 (aus mündlicher Tradition). Wenzel, The legend of the origin of the Tibetan race, im Festgruss an R. Roth, Stuttgart 1893, p. 170—172 (übersetzt nach dem 7. Capitel des rgyal rabs gsal bai me long). Köppen, Die lamaische

der Fall ist, der Schauplatz, wo der Affe seine Meditationen ver-
richtet, auf einen Felsen des Y a r l u ṅ verlegt wird, des auf den
Karten als Y a l u n g bezeichneten bekannten Nebenflusses des Blauen
Stromes, der nach einer anderen im folgenden besprochenen Tra-

---

Hierarchie und Kirche, p. 44 ff. Köppen's Deutungen der Sage entspringen rein subjektiven
Empfindungen und sind deshalb völlig verfehlt. Er hält sie der Wurzel nach für nicht
buddhistisch, dagegen seien die Heiligen und auch die Affen — denn Tibet hätte keine
Affen — handgreiflich indisch-buddhistische Zuthaten; die hindustanischen geistlichen Väter
hätten dieselben wegen der Aehnlichkeit der mongolischen Physiognomie ihrer gläubigen
Söhne mit der der Affen hinzugedichtet! Zunächst besitzt Tibet wohl Affen, nämlich Macacus tibetanus, s. W a l l a c e, Die geographische Verbreitung der Tiere, deutsche Ausgabe
von A. B. Meyer, Bd. II, p. 197; S a n d b e r g, Handbook of colloquial Tibetan, p. 169;
Rhinopithecus roxellana in Osttibet und Kokonôr-Gebiet, s. A. D a v i d, Journal de mon
troisième voyage d'explorations dans l'empire chinois, vol. II, p. 324; Semnopithecus
schistaceus, s. W a l l a c e, l. c. p. 195, S a n d b e r g l. c.; ferner Macacus vestitus, entdeckt
von G. B o n v a l o t, s. dessen De Paris au Tonkin à travers le Tibet inconnu, p. 487
(s. auch p. 377). Vergl. ferner H. B o w e r, Diary of a journey across Tibet, p. 61, 236
und J ä s c h k e in Zeitschrift der Deutschen Morgenländ. Ges., Bd. 23, p. 553. In den
Liedern des Milaraspa werden wiederholt die munteren Sprünge der Affen beschrieben.
Auch aus der Geschichte ist uns der Affe eine bekannte Erscheinung: Verzeichnisse von
Tributlieferungen, welche die Tibeter den Chinesen schuldeten, führen unter anderen Dingen
Affen und Affenfelle auf, und die chinesischen Annalen erzählen, dass die in eine Menge
kleiner Clans zersplitterten Tibeter der ältesten Zeit alljährlich ihren Häuptlingen einen
Eid leisteten, wobei sie Menschen, Schafe, Hunde und Affen zum Opfer darbrachten; s.
R o c k h i l l, Tibet, in Journal of the Royal Asiatic Soc. 1891, p. 204, und The land of the
Lamas, p. 337, 339. Dass vollends die Inder nicht einer solchen Travestierung fähig wa-
ren, wie sie Köppen nur von seinen abendländischen Vorstellungen suggeriert wurde, bedarf
kaum eines Beweises. Erscheint doch der Affe schon im Rigveda als der Liebling des Indra,
ist er doch der Freund Buddha's, um von der Bedeutung der Affen im Râmâyana ganz zu
schweigen, erscheint doch Buddha selbst in den Erzählungen der Jâtaka sieben Mal als
Affenkönig und einmal als des Affenkönigs Sohn. In Tibet selbst ist die Verehrung des
Affen (Rhinopithecus Roxellana) durch das ausdrückliche Zeugnis des Abbé Armand D a v i d
(s. das obige Citat) erwiesen, der sogar von grosser Achtung und Liebe zu demselben
spricht, und wenn wir bei Osvaldo R o e r o, Ricordi dei viaggi al Cashemir, Piccolo e
Medio Tibet e Turkestan, vol. II, p. 200 von der Stadt Mundi in Kashmir lesen, dass
dort eine unermessliche Schar heiliger unverletzlicher Affen auf Bäumen lebt, die sich ihre
Nahrung aus den Häusern der Einwohner ruhig stehlen dürfen, so werden wir an den
Bericht von A e l i a n, Hist. anim. XVI 10 erinnert, dass in einer indischen Stadt auf
Befehl des Königs Affen von der Grösse hyrkanischer Hunde täglich eine in Reis beste-
hende Mahlzeit erhielten. Weitere Hinweise über Affenverehrung im Gebiet des Buddhismus
sehe man bei H. C o r d i e r, Les voyages en Asie au XIV. siècle du bienheureux frère
Odoric de Pordenone, p. 331, 332, 338—339.

dition als Stammsitz der tibetischen Dynastie gilt. Auf p. 26 unseres
Buches wird nämlich folgende Sage erzählt: In früherer Zeit lebte
Pāṇḍu, der König des Sonnenthrones. Seine Gemahlin hiess Krasna.
Wiewohl sie alle guten Zeichen besass, hatte sie keinen Sohn,
worüber sie sehr niedergeschlagen war. Da fiel die Herrschaft an
seinen Bruder Duḥçāsana [12]), und das Königspaar wanderte ver-
düstert in der Waldeinsamkeit umher. Dort trafen sie mit dem
grossen Einsiedler *Ts'an staṅs bkai bcad* zusammen, dem sie Verehrung bezeigten und Lebensmittel zubrachten. Infolge der grossen
aus dem Opferfeuer strömenden Hitze verbrannte sich das Königspaar die Hände, so dass dem Munde des Pāṇḍu ein Schrei entfuhr,
der den Einsiedler aus seiner Beschauung erweckte. »O König,
Vater und Mutter", rief er, »wie freue ich mich! Was für ein
Begehr führt euch zu mir?" »Ich war der das Jambudvīpa beherr-
schende König; doch da ich sohnlos bin, hat mich mein Bruder
Duḥçāsana der Herrschaft beraubt. Ich bitte, mir die Siddhi eines
Sohnes zu gewähren". Der Einsiedler überreichte ihm ein Gefäss
mit einem Zauberwasser und sagte: »Weihe dies dem Himmel, und
es wird dir ein Sohn erstehen; salbe deinen Leib, und er wird von
Krankheit geheilt". Da brachte er dem Himmel sechs Libationen
dar, und es erschienen die sechs Götter der Welt: im Südwesten
der Svastika-König der Bon, Indra, Sūrya, die Açvin, Kāmeçvara [13])
und Mahākāla [14]) segneten die königliche Gemahlin, die darauf ohne
den Beischlaf des Königs einen Sohn gebar. Dieser hatte eine Svastika
auf den Ohren gezeichnet, die Augen eines Vogels, ein Gehege
muschelweisser Zähne und die Hände wie eine Gans durch Schwimm-
häute verbunden. Er hiess der Göttersohn Karṇa [15]). Der königliche

---

12) tib. *sgra ṅan.*

13) tib. ₒ*dod pai dbaṅ p'yug,* Beiname Kubera's.

14) *legs ldan,* eine besondere Form des Mahākāla nach G r ü n w e d e l, Mythologie des
Buddhismus in Tibet und der Mongolei, p. 177.

15) tib. *rna ba can.*

Vater schämte sich seiner, liess ihn in einen kupfernen Kasten legen und in den Ganges werfen. Der trieb bis zur Stadt Vaiçālī. Alle sahen ihn und erstaunten. Man zog den Knaben auf. Er entwickelte viele Vorzeichen, die auf seine grosse Bestimmung hinwiesen. Als er hörte, dass Duḥçāsana ein Heer zusammenziehe, floh er und stieg von dem Götterberge *Gyang t'o* nach *Mar sog k'a* hinab. Dort gab es eine Prophezeiung, dass der als Königssohn von hoher Bestimmung herabgestiegene Himmelskönig Herrscher von Tibet werden würde. Der Erdgott Bonpo und die andern zwölf Verständigen holten ihn herbei, trugen ihn auf dem Nacken und weihten ihn zum Könige. Sie verliehen ihm den Namen *gÑa k'ri ltsan po*. Dies ist der erste König in Tibet.

Der hier erwähnte Götterberg *Gyang t'o* ist mit dem in der ostmongolischen Version des Sanang Setsen genannten »*ündür küriye-tü teghri aghūla*” identisch, was Schmidt durch die Worte »hochbekränzter Götterberg” übersetzt; denn *gyaṅ* ist nach Desgodins eine aus gestampfter Erde aufgeführte Mauer, dem das mongolische *küriye* »Mauer, Palisade” (s. Kowalewski, Dict. mongol-russe-français 2638 *b*) entspricht; der Name des Berges bedeutet also »hohe Mauer”. Denselben Namen, mit der Orthographie *gyaṅ mt'o*, finde ich auch in einem mir handschriftlich gehörigen Werke, betitelt *bod c'os rgyal gyi gduṅ rabs padma dkar poi p'reṅ ba* (»Weisser-Lotuskranz der Genealogie der tibetischen Dharmarāja”), fol. 63, wo die Lage des Berges in der Provinz dBus, also Centraltibet, angegeben wird. *Mar sog k'a*, das untere *Sog k'a*, ist ein Teil von *Yar luṅ sog k'a*, das nach p. 38, 5 unseres Buches zu den sieben Distrikten des »linken Hornes” (*gyon ru*), d. i. das östliche Tibet, gerechnet wird. Sanang Setsen beschreibt die Art und Weise, wie der Königssohn aus Mittel- nach Osttibet gelangte, durch die Ausführung, dass er über neunfache Gebirgsstufen in das Thal des Yarlung herabstieg. Unter diesem Begriff sind die von den tibeti-

schen Geographen als *rim dgu* bezeichneten neun Bergreihen [16]) zu
verstehen, die sich im Osten von Indien bis Yünnan hinziehen.
Später wird er von Yarlung auf den Schneeberg *Yar lha šam po*
(zwischen Lhasa und der Grenze von Bhutan) gebracht und dort
zum König ausgerufen. Ich kann nicht mit Köppen [17]) übereinstimmen, der in dieser Version des Sanang Setsen und besonders
in der des Bodhimör wegen des grossen Unterschiedes in der geographischen Lage des *Yar lha šam po* und des *Yar lui* einen
Widerspruch constatieren zu müssen glaubt und diesen daraus erklärt, dass die Lamen von Lhasa, als sie daran gingen, die tibetische Geschichte zurecht zu machen (!), in der Tradition das Thal
des *Yar lui* und die ihm benachbarten Gebirgslandschaften als die
Urheimat des Volkes bezeichnet vorfanden, dass sie aber, um die
Einwanderung jenes indischen Fürstensohnes, als des angeblichen
Stammvaters der Könige des Schneelandes, zu ermöglichen oder
doch zu erleichtern, das *Yar lui* Thal nach Südwesten, etwa in die
Mitte zwischen den Himalaya und den späteren Sitz der Herrschaft
(Lhasa), an den Fuss jenes Schneebergs versetzten, der — vielleicht
zur Erinnerung an jenen östlichen Strom — *Yar lha šam po* be-
nannt worden war. Abgesehen davon, ob die von Köppen unter
Berufung auf Klaproth gegebene Identifikation des *Yar lha šam po*
richtig ist, ist in der Sage von einer Verlegung des *Yar lui* in das
Gebiet dieses Berges gar keine Rede, sondern nur von einer Wanderung des Prinzen vom *Yar lui* zu dem Berge oder umgekehrt
wie im Bodhimör. Es ist gar kein Grund vorhanden, hier Köppen's
Legende von der »priesterlichen Mache", die uns meist als ein
recht zweifelhaftes Argument erscheint, ins Feld zu führen. Sage
ist eben Sage, und die vorliegende will nichts anderes als die Über-

---

16) Wasiljew, Географія Тибета переводъ изъ Тибетскаго сочиненія Минь-
чжулъ Хутукты. Pet. 1895, p. 1.

17) Die lamaische Hierarchie und Kirche, p. 50.

lieferung vom Ursprung des tibetischen Königsgeschlechts am *Yar luṅ* und die Verlegung des Schwerpunkts der königlichen Macht nach Centraltibet symbolisch zum Ausdruck bringen. Mit dieser Tradition wird nun die Sage von dem vertriebenen indischen Königssohne verknüpft. Und hier zeigt sich eine eigentümliche Erscheinung in der oben mitgeteilten Version derselben. Während nämlich die bisher aus buddhistischen Werken bekannt gewordenen Versionen den König *gÑa k'ri btsan po* zu einem Angehörigen des Çâkya-Geschlechts, meist zu einem Sohn des Königs Prasenajit von Kosala [18]), machen und sich in der Erzählung auf die bekannte Aussetzungs- und Auffindungssage [19]) beschränken, handelt es sich hier nicht um eine buddhistische, sondern um eine brahmanische Überlieferung, die sich an den Kampf der Pāṇḍu- und Kurusöhne des Mahābhārata anschliesst. Darauf weisen sowohl die aus dem Epos entlehnten Namen Pāṇḍu, Duḥçasana, Karṇa und Krasna, letzteres wahrscheinlich ein Nachklang an Kṛṣṇā, Beiname der Draupadī, wie insbesondere die im Anschluss an die obige Erzählung gemachte Angabe, dass Pāṇḍu fünf Söhne [20]) erzeugt, dass Yuddhiṣṭhira und seine vier Brüder mit Duḥçāsana's zwölf Heeren der 700 Jambudvīpa kämpfen, diesen besiegen und dann die Herrschaft erlangen. Es ist bemerkenswert, dass diese Sage gerade in einem Werke der Bonpo Aufnahme gefunden hat, was nicht zum geringsten dem Antagonismus dieser Sekte gegenüber den Buddhisten entspringt. Aber auch unter diesem Gesichtspunkt betrachtet, wird sie als tibetische Tradition nicht auf

---

18) I. J. Schmidt, Forschungen im Gebiete der älteren... Bildungsgeschichte der Völker Mittel-Asiens, p. 20—27; Geschichte der Ostmongolen, p. 21—23, 316—317; Schlagintweit, Die Könige von Tibet, p. 831—833. Lassen, Indische Altertumskunde, IV, p. 713. Chandra Dâs, Journal of the Asiatic Society of Bengal 1881, p. 213.

19) Zahlreiche Parallelen zu derselben bei Potanin, Очерки сѣверо-западной Монголіи, IV, p. 872—877.

20) Tib. *gyu brtan* = Yuddhiṣṭhira; tib. *srid sgrub* = Arjuna; tib. *„jig ñe* = Bhīma. Die beiden übrigen *spyan gsal* „der Hellängige" und *yzugs mdzes* „der Schöngestaltige" vermag ich nicht mit Nakula und Sahadeva zu identificieren.

Rechnung eines Zufalls zu setzen sein. Denn wenn sich schon im Mahābhārata alte Beziehungen Indiens zu Tibet zeigen [21]), wenn es ferner wahr ist, dass das schon von Herodot [22]) erwähnte Ameisengold *pipīlika* aus Tibet stammte [23]), lässt sich nicht leugnen, dass vielleicht in dieser Sage ein, wenn auch einstweilen nicht näher zu definierender, historischer Kern enthalten sein mag.

Von den acht Genealogieen, die das Buch auf p. 23—31 enthält, will ich hier zwei kurz berühren. Auf p. 31 werden die Könige des Reiches *šar rgya p'ag šaṅ* erwähut. Dieses Land wird auf p. 23, 16 mit etwas veränderter Orthographie *šar rgya yi ₒp'ags ts'aṅ* genannt, woraus schon hervorgeht, dass es sich nicht um einen tibetischen Namen handelt. Über die geographische Lage desselben ist nichts bemerkt, wenn man nicht den Ausdruck *šar rgya* als ‚östliches China’ auffassen will. Dem würde aber die Angabe widersprechen, dass dieses Reich unter den Mongolen gestiftet wird. Der Gründer desselben ist nämlich ein ungenannter chinesischer Richter [24]), der bei einem Aufstand der Mongolen kämpft (wo und gegen wen, ist nicht gesagt), siegt, den Thron besteigt und König *T'as tsoṅ* oder *Tas miṅ* genannt wird. Sein Nachfolger ist *Hūṅ ši*, auf diesen folgt *Kyen ti*, auf diesen *C'aṅ ₒt'uṅ* [25]). Auf p. 23, 16 ist übrigens von sechs Königen dieses Reiches die Rede, in dieser Genealogie werden aber nur jene vier aufgezählt. Die Genealogie der Mongolenkönige beginnt mit dem Ausspruch, dass der Gemahlin des Uigurenkönigs »Eselsohr” (*Boṅ rna*) vom Himmel ein weisser Mann auf weissem Pferde erschienen und ihr beigeschlafen habe, worauf sie einen Sohn,

---

21) Lassen, Indische Altertumskunde, I, p. 848—851.

22) Hist. liber III, cap. 102—105.

23) Vergl. M. Malte-Brun, Mémoire sur l'Inde septentrionale d'Hérodote et de Ctésias comparée au Petit-Tibet des Modernes, Nouvelles Annales des voyages, II, p. 307 —383, und Schiern, Über den Ursprung der Sage von den goldgrabenden Ameisen, 1873.

24) tib. *rgyai k'rims dpon.*

25) Dieser Name könnte tibetisch sein und ‚Biertrinker’ bedeuten.

3

den als Himmelssohn (*gnam gyi bu*) berühmten *Jiṅgir* gebar. Diese
Sage erinnert uns an die conceptio immaculata der *Alung Goa*,
die nach *Sanang Setsen* im Traume von einem Jüngling begattet
wird, der dem *Altan Tobci* zufolge der beim Weggang sich in einen
gelben, kahlen, sein Maul beleckenden Hund verwandelnde Mond
in Jünglingsgestalt war [26]). In der genealogischen Tafel finden sich
mannigfache von den Nachrichten bei *Sanang Setsen* und ₀*Jigs med
nam mk'a* [27]) abweichende Angaben, von denen ich die wichtigeren
hervorhebe. Nach dem mongolischen und tibetischen Historiker ist
Godan der jüngere Bruder des *Guyug* (*Gulug*), während er hier zu
dessen Sohn gemacht ist. *Sanang Setsen* schreibt zwar dem *Tului*
vier Söhne zu, zählt aber nur die drei *Möngke*, *Chubilai* und *Erik
Böke* auf, die auch hier erscheinen, während bei Huth noch
Hwopilai (*Chubilai*) hinzugefügt ist, welcher den Königstitel *Sec'en*
(*Setsen*) führt. Dieser Titel wie die Herrschaft während dieser Zeit
wird in unserem Werke dem *Ariboga* zugeschrieben, der acht Söhne
hat, von deren Namen sich *Jigin* und *Magala* mit *C'iṅgem* und
Maṅgala, zwei von den vier Söhnen des *Chubilai*, identificieren
lassen. Der zweite Sohn des *Jigin*, der von *Sanang Setsen* und
₀*Jigs med nam mk'a Dharmapála* genannt wird, heisst hier tibetisch
lKugs pa, d. i. der Stumme. *Ra k'yi p'ag* [28]), der mit *T'og t'i mur*
kämpft und den Königsnamen *Jayat'u* annimmt, auf unserer Stamm-
tafel ist wohl der *Rijapika* des *Sanang Setsen* und der *Rin c'en* ₀*p'ags*
des ₀*Jigs med nam mk'a*, erscheint aber bei diesem als Sohn des
Haisang. *Toghan temür*, hier der Sohn des *Kušala*, ist nach den
beiden andern Gewährsmännern ein Sohn des *Puyantu Khan*. Ihm
folgt *Kuša* (*Kosala*), dann dessen Sohn *T'o gan t'u mur*, dann

---

26) Vergl. über diese Sage bes Schott, Abulghasi und Sanang Setsen, in Zeitschrift
für Ethnologie 1874, p. 107 und Monatsberichte der Preussischen Akademie 1873, p. 5—7.

27) Übersetzt von G Huth, Geschichte des Buddhismus in der Mongolei, Bd. II.

28) In derselben Schreibung bei Chandra Dás im Journal of the Asiatic Society of
Bengal, vol. LI, part I, p. 75.

Rin c'en dpal, der Sohn des *T'og t'i mur* [29]), und diesem der Sohn des *Toghan temür, Ayubhidara*, den die übrigen Quellen nicht erwähnen. Dann fährt der Text (31, 6) fort: »Darauf ging die Herrschaft an China über. Was die auf tibetischem Gebiet ernannte Königslinie von *Ts'añ* betrifft, so war *La t'i mu mog* der Sohn des *Se c'en*. Er hatte zwei Söhne *C'os dpal* und *bDe Gan*. Die Söhne des älteren *C'os dpal* sind: *bZañ po dpal, T'o ma dpal, Ye šes dpal, Rin c'en dpal*. Die Söhne des jüngeren *bDe Gan* sind: *Ratna dpal* und *T'o ba dpal*, dessen Sohn *dKon mc'og dpal* und *Prajñâ dbañ*. Die Söhne des *Hu le hu* sind *Kamala* und *Dharmapâla*". Die Quelle ist leider bei diesem Abschnitte nicht angegeben.

Ich lasse nun eine Übersetzung von Capitel 21—23 (p. 48.—53) folgen, worin die inneren Kämpfe geschildert werden, die bei der Einführung des Buddhismus unter König *K'ri sroñ ldeu btsan* (740—786 A.D.) stattfanden. Die Fehden der altnationalen und der buddhistischen Parteien werden erzählt, zwischen denen der König als ein schwacher, von Hofintriguen geleiteter Charakter erscheint.

21. Capitel.

Niedergang der Bonreligion unter K'ri sroñ.

Zu jener Zeit traten häufig in der Provinz dBus im Lande Tibet Krankheiten, Würmer, Frost und Hagel auf. Da man durch keine Mittel des Übels Herr werden konnte, warf der loskundige *sPe ne gu* die Lose und verkündete: »In diesem Lande lebt ein Sohn ohne Vater, der die Ursache des Unheils ist". Auf die Frage, wer es sei, erwiderte er: »Es ist ein fünfzehnjähriger Knabe mit rotem Fleisch und Adern, die Augenbrauen oben geteilt, mit einem Gehege muschelweisser Zähne; wenn ihr einen solchen bemerkt habt, der ist es". »Was ist da am besten zu thun?" fragte man. »Zwölf Bonpo",

---

29) Bei Huth: *T'og t'emur*, jüngster Sohn des *Pūyant'u Khan*.

sagte er, »die nicht aus demselben Geschlechte stammen, sollen die Ceremonie der grossen Himmelsreinigung [30]) vollziehen, dann ihn auf einen bräunlichen Ochsen setzen und in das Land verjagen, das mehr als eine Sprachfamilie hat; so wäre es am besten". So thaten sie und verbannten ihn in das Land Kashmir. Ohne unterwegs auf den gefährlichen Brückenpfaden oder durch Raubtiere unterzugehen, traf er mit dem Meister *Padma* (d. i. *Padmasambhava*) zusammen, der nach Kashmir gekommen war, lernte bei ihm die Lehren des Buddhismus und erhielt wegen seiner grossen Weisheit die Bezeichnung *Bodhisattva*.

Da trug man mit Anklagen, die darauf abzielten, die Bon-Priesterinnen von Tibet in die Verbannung zu schicken, dem König Verleumdungen zu. »Diese Bonlehre zu erlernen ist schwer; was das Wort betrifft, so ist es die heilige buddhistische Lehre, die Wahrheiten enthält. Wir begehren daher die Unterdrückung der Bonlehre und die Ausübung des Buddhismus", so lästerten sie. Der Göttersohn [31]) sprach: »Wenn es passend ist, die Bon untergehen zu lassen, ist es denn auch passend, dass Sonne und Mond untergehen?" Obwohl er eine Zeit lang nicht auf sie hörte, sagten *gYu sgra* und die übrigen Minister, welche am Buddhismus Gefallen fanden: »O Herrscher! Um Bier zu trinken, muss man es zuvor in Wasser kochen; um Fleisch zu essen, muss man zuvor ein Tier schlachten. Wenn wir unsere Vorfahren zu übertreffen wünschen, müssen wir zuvor die Bonreligion vernichten und dann die Lehre des Buddhismus einführen. Den Buddhismus müssen wir aus Indien holen. Zu diesem Zweck sind auch geistliche Lehrer erforderlich. Wenn wir dieses Werk ausgeführt haben, dann empfangen erst unsere Vorfahren ihre echte Würde. Dann wird man wohl sagen können: ‚Der väterliche Esel hat als Sohn ein Maultier, der väter-

---

30) tib. *gnam sel c'en po.*

31) tib. *lha sras,* eine Bezeichnung des Königs.

liche Bulle hat als Sohn einen Yakbastard erzeugt'. Zu eben dieser
Zeit ist einer erschienen, der selbst das Licht von Sonne und Mond
übertrifft". Darauf berichteten sie ihm der Reihe nach von einem
Traumbilde, das der König von Mon [32]) zu jener Zeit gesehen habe.
Dieser hätte nämlich geträumt, der silberne Mond sei nur während
der ersten acht Tage des Monats aufgegangen und dann in der
Erde verschwunden, darauf seien dreitausend wärmende goldene
Sonnen aufgegangen, und allen Wesen sei Glückseligkeit zu teil
geworden. Als der König von Tibet das vernahm, dachte er:
»Sollte das der Ausspruch eines Bodhisattva sein?" Die Minister
aber lagen ihm immer und immer wieder beständig in den Ohren,
bis er sich endlich überzeugte, dass der Traum des Königs von
Mon als ein Vorzeichen gekommen sei und gleichsam das Erscheinen der Lehre des Buddhismus nach dem Untergang der Bonreligion
bedeute, und die Unterdrückung der letzteren befahl.

₀Gos k'ri und die übrigen Minister, die Anhänger der Bon waren, richteten Bitten an ihn, doch ohne ihnen Gehör zu schenken,
sprach der König: »Die Könige von Indien, die am Buddhismus
Wohlgefallen finden, sind frei von Krankheiten, langlebig und mit
grossen Reichtümern gesegnet. Wenn sie in der Todesstunde ihr
Gebet gesprochen haben, fahren sie nicht zur Hölle. Deshalb ist
auch uns dergleichen erforderlich. Da nun auf Grund reiflicher
Überlegung mein Gebet dem Buddhismus angehört, woher soll man
den Buddhismus nehmen?" Darauf berief er den *Padmasambhava*
aus Udyāna, den *Bodhisattva, Çrīkūta* [33]), *Nāgadhvaja* [34]) und
Ratnavara [35]) und errichtete nach dem Vorbilde der gsas-Schreine
der Bon dreizehn Kapellen für die Götter von *bSam yas lhun po rtse*,

---

32) Über *Mon* s. Mémoires de la Société Finno-Ougrienne, vol. XI, p. 94—101.

33) tib. *ka ba dpal brtsegs.*

34) tib. *klui rgyal mts'an.*

35) tib. *rin c'en mc'og.*

mT'a ₒdul und *Ru gnon*. Zu Lebzeiten des Königs übte man die Bonreligion im oberen Teile des Landes, im mittleren erhob sich der Buddhismus bald, bald schwand er wieder, im unteren übte man beide Lehren zugleich. Der König befolgte den Buddhismus. Die Unterthanen hingen der Mehrzahl nach der Bonreligion an. Darauf führten ein Bonpo und ein Bande einen Wettstreit in der Magie auf; erweckten Getötete und Verstorbene, liessen Lebende sterben, massen ihre Kräfte und Zaubermacht. Da der Bonpo nicht unterlag, unterdrückte das Volk von Tibet den Buddhismus und hing freudig der Bonlehre an. Der Könige dachte, dass es bei gleichmässiger Ausübung von Bon und Buddhismus nicht passend sei, den Buddhismus zu unterdrücken. Da schloss der König vorher mit den Bande einen eidlichen Vertrag und erliess ein Gesetz, die Bonlehre zu vernichten und den Buddhismus zu pflegen. Der Rumpf der Bonpo war ausgedehnt; der Neid der Bande war gross; des Herrschers Ohren waren fein [36]); der königlichen Gemahlin und der Bande Lippen waren scharf; die Minister waren geriebene Verleumder; das tibetische Volk ein grosser Freund des Neuen [37]). Die verdienstlichen Handlungen der Wesen nahmen ab. Es war die Zeit des Untergangs für die Bonreligion gekommen. Die Bande und buddhistischen Minister streuten Verleumdungen aus: »Gegenwärtig ist des Königs Herz zwischen zwei Parteien geteilt; doch da die Bonlehre auf die Nachkommenschaft der Söhne und Enkel übergehen wird, so ist es jetzt gut, diese Bonpo beiseite zu schaffen". So und auch vieles andere, wie z.B., dass die Bonpo dem König mit Zaubereien nachstellten, sagten sie. Der König nahm es sich zu Herzen, versammelte die *gŠen po* und sprach: »Ihr Bonpo, da ihr zu mächtig seid, so vermute ich, dass ihr mir meine Unterthanen abspenstig machen wollt. Entweder bekehrt euch zum

---

36) d. h. leicht empfänglich für alles, was ihm hinterbracht und geraten wurde.
37) tib. *gsar grogs c'e* rebus novis studere.

Buddhismus und werdet Bande, oder geht und verlasst das Reich von Tibet, oder werdet dienende Unterthanen und zahlt Steuern! So wählt euch denn das angenehmste!" Die Besonnenen erwiderten: »Da die Herrschaft stark und des Königs Lebenszeit nicht kurz ist, so sind wir es zufrieden, in den geistlichen Stand zu treten". So wurden sie Geistliche und bekehrten sich zum Buddhismus. Dann baten sie, da das Walten der Svastikâ-Bonlehre in Zukunft die Menschen befreien würde, auf Grund eines Gelöbnisses aus der Zeit der königlichen Vorfahren den König um die Erlaubnis, alle Bonschätze verbergen zu dürfen, die er ihnen auch gewährte, worauf sie dieselben verbargen.

Die *gŠen po* wussten, dass ein für allemal die Zeit des Untergangs der Bonreligion gekommen sei. Obwohl sie durch schreckliche Thaten und verderbliche Mittel den König mit seiner Umgebung leicht hätten vernichten können, bezwangen sie sich und fügten sich dem Willen des Königs. Da sie wussten, dass König, Minister und Bande den Untergang der Bon beschlossen hatten, flogen die Siddha [38]) zum Himmel empor. Einige gingen auf den Schieferberg [39]) und den Gletscherfelsen [40]).

Einige von denen, die sich verpflichtet hatten, Bande zu werden, trugen im Herzen die Gesinnung der Bon, während nur ihr Mund und ihr Leib den Buddhismus übten. Viele Bonpo wurden in den Brahmaputra [41]) geworfen. Die gsas-Schreine und Stûpa der Bon wurden zerstört. Bei einigen machte man einen neuen Bewurf und verwandelte sie in buddhistische Kapellen, bei andern änderte man den Namen u.s.w. Eine Zusammenfassung dieser Angelegenheiten findet man im *bsGrag byaṅ* und vielen anderen Werken.

---

38) tib. *grub pa t'ob pa rnams.*

39) tib. *gya ri.*   40) tib. *gaṅs brag.*   41) tib. *gtsaṅ po.*

22. Capitel.

Die Verbergung der Schätze.

Im *bsGrag byaṅ* heisst es: Da teilten die neun Männer des grossen Zaubers [42]), die Besonnenen [43]) und die übrigen Bon verschiedene Klassen ein. Einen König der Schätze, vier Minister und einen zweiten Minister, im ganzen Sechs, verbargen sie. Insbesondere fünf grosse Geheimschätze und 1700 kleine Schätze verbargen sie [44]). In allen buddhistischen Klöstern verbargen sie Bonschriften. Ebenso verbargen sie in Bergen und Felsen Opfergaben und viele andere Spenden. So heisst es. Wenn das Wort nicht wahr ist, mögen diese Svastikâ-Bon des schatzbergenden Palastes vollständig zugrunde gehen. Wenn das Wort wahr ist, mögen König, Minister und Bande zugrunde gerichtet werden, weil sie die Ausübung der Bonreligion nicht erlauben! Möge die Königsfamilie in den Dörfern betteln gehen und das Volk um Kleidungsstücke ansprechen! Möge sich die Bonlehre von allen Enden her verbreiten! Solche und viele andere gute und schlechte Wünsche stiessen sie aus. Die anderswohin wandernden *gŠen po* erlangten die Vollendung teils im Feuer, teils im Wasser, teils in der Luft und lebten glückseliger als zuvor.

23. Capitel.

Geschichte der Periode *dbu yog*.

Spe ne gu, Bhe šod kram, Par nu ma šod und *Sum pa mu kʻyud* bestiegen ein Boot, legten ein weisses Lederpolster und eine Kesselpauke hinein und liessen es in die Mitte des Brahmaputra [45]) trei-

---

42) tib. *mtʻu cʻen mi dgu.*   43) tib. *dran pa.*

44) Diese Bergung wird unter dem Namen *bon gter ma* auch im achten Buche des *grub mtʻa šel kyi me loṅ* erzählt. S. Chandra Dás in Journal of the Asiatic Society of Bengal, vol. L, part I, p. 199.

45) Das ist der Tibet durchströmende Oberlauf desselben, tib. *gtsaṅ po* oder *yar cʻab.*

ben. Da stiessen sie einen Fluch aus: »r*Jei lha*, lenke den Lauf des Stromes ab! Treibe das Wasser des Brahmaputra nach aufwärts zurück! Lass im Lande Tibet deine mannigfaltigen Zauberkünste spielen!" Mit diesen Worten flogen die vier Bonpo gen Himmel, stiegen im Lande *gYa gon gyim bu* hernieder und weilten dort im Besitze der Seligkeit.

Als sich nun der Oberlauf des Brahmaputra nach seiner Quelle zurückwandte, wurden viele Begleiter des *Padmasambhava* [46]) von der Strömung fortgerissen. Im Westen war der Berg *Yar lha šam po*, im Süden der *lDon lha*, im Norden der *T'an lha*, im Osten der *sBom ra* sichtbar. Der erhabene Türkisglanz nahm ab, der erhabene See trocknete aus. Epidemieen brachen aus und andres Ungemach verschiedener Art. Der König erkrankte, seine Kraft war geschwächt. Vom Blitz wurden die kleinen Sterne von Lhasa vernichtet. Der Königssohn Siddhârtha wurde vom Blitz getötet. Die sieben Bande wurden vom Blitz erschlagen. Es ist bekannt, dass zu jener Zeit die im grossen Schlossfelsen wohnende Bonfrau *Tse za* entsandt und *Yan k'u bon gnam gšen* gesehen wurde. Nachdem man den *Sum pa mu p'yva* und andre Bon herbeigeholt und zu dem Zeltgott des Herrschers [47]) und zu dem männlichen Gott der Unterthanen [48]) gebetet hatte, floss der Brahmaputra wieder nach unten herab. König und Unterthanen wurden glücklich. Den Bon gaben sie gläubigen Sinnes von dem oberen Teil des Landes drei Bezirke, nämlich das obere *Za gad gser*, das untere *gTsan po gžun* und *C'u bar bre sna*. Von dem unteren Teil des Landes gaben sie *Bon mo lun rin* [49]), ₀*P'an yul* [50]) und *Yar mo t'an*. Von dem oberen

---

46) Der sich damals, der Einladung des Königs *K'ri sron lde btsan* (740—786 A.D.) folgend, auf dem Wege von Udyâna nach Tibet befand.

47) tib. *rjei gur lha*.

48) tib. ₀*bangs kyi p'o lha*. *P'o lha* ist nach Jäschke die Schutzgottheit der rechten Seite eines Mannes.

49) d. h. ‚langes Thal der Bon-frau'.

50) Nach Jäschke Name des nächsten Alpenthales nördlich von Lhasa, dessen Bewohner einen besondern Dialekt sprechen sollen.

Teil gab man als Untergebene *So, So na* und *Glo bo.* Von *dBus* gab man *Dre, Sloñ* und *Koñ. Lha sa t'añ p'u ste* gab man als den erbetenen Boden und *Yar luñ* zum dauernden Aufenthalt. Der Herrscher sprach: »Um mich selbst zu erhalten ist Bon sowohl wie Buddhismus erforderlich; auch um der Unterthanen Leben zu schützen, sind beide erforderlich. Auch um das Heil der Wesen zu bewirken, sind beide erforderlich. Furchtbar [51]) ist die Bonreligion, verehrungswürdig der Buddhismus; deshalb bitte ich jetzt beide gleichzeitig zu üben". Im Werke *bsGrag byañ* heisst es: Zu Lebzeiten des *K'ri sroñ* fand dreimal Ausbreitung und Untergang der Bonreligion statt. Dies ist nur ein Abriss der dort gegebenen Erläuterungen.

Dieser König hatte eine Hauptgemahlin und zwei Nebenfrauen. Diese beiden waren *Jo mo ₒBroñ za* und *P'o yoñ za.* Obwohl sie nicht an der Herrschaft teilnahmen, da sie keine Söhne hatten, besassen sie doch wegen ihrer Anhänglichkeit an den Buddhismus grossen Einfluss. *Ts'e spoñ za* dagegen hatte, wiewohl sie Mutter dreier Söhne war, wegen ihrer Anhänglichkeit an die Bonreligion, nur geringen Einfluss. Der König hatte sie arglistig verlassen. In ihrem Herzeleid hierüber sandte *Ts'e spoñ za* zwei Minister aus, denen sie ihr Geheimnis anvertrauen konnte, und liess aus dem im Norden sieben Pferdetagereisen entfernten Landesbezirk die *sTag pa ra rgya ts'ab* genannten Bonpo zu sich rufen, die mit wundervollen Zauberkräften begabt und dem schnellfüssigen Winde gleich waren. Als die Meister, drei an Zahl, angelangt waren, bat sie dieselben, gegen den König Zauberei zu üben. Sie erklärten, die Zaubermittel müssten dem befleckten Gewande des Königs gelten. Da schickte sie ihren siebzehnjährigen Sohn *Mu ri sgam po* mit dem Auftrag fort: »Geh und fordere den befleckten seidenen Ober-

---

51) Wegen der ihr zur Verfügung stehenden schädigenden Naturgewalten.

rock des Königs als Sühngeldgewand für die Opferceremonie des
Žabs brtan" [52]). So ging er. Der König spielte gerade mit seinen
Dienern auf dem Palaste ein Spiel. Als er auf den als Thorwächter
eingesetzten *Žan bu rin* traf, sagte er ihm, dass er der Königssohn
sGam po sei. Gleichwohl öffnete ihm der Schliesser nicht. Da entspann sich ein Kampf um das Thor, das er schliesslich erbrach.
Er trat in das väterliche Haus ein, und als ihm *Žan bu* zurief:
»Wer ist denn eigentlich dieser Thürerbrecher?", versetzte er ihm
mit dem Schwerte einen Schlag auf den Kopf und tötete ihn. Er
wusch das Blut vom Schwerte ab und drang weiter vor. Der König
fragte ihn: »*sGam po*, weshalb bist du gekommen? Wer hat das
Thor geöffnet? Wohin ist *Žan bu rin* gegangen?" Er erwiderte:
»Ich habe das Thor mit einem Stein geöffnet. *Žan bu rin* ist einen
weiten Weg gegangen [53]). Ich bin gekommen, um dein beflecktes
Gewand, das Oberkleid, für eine Opferceremonie der königlichen
Gemahlin zu fordern". Der König erschrak und händigte ihm sogleich das befleckte Oberkleid aus. Er übergab es seiner Mutter
und ging fort, um zu essen. Als der König sagen hörte, dass man
an jenem Abend Zauberei mit ihm üben wolle, besprach er sich

---

52) *Žabs brtan* bedeutet wörtlich ‚fester Fuss'. Ich vermute indessen, dass *brtan* für
brtad verdruckt ist, womit nach Jäschke eine Art Beschwörung bezeichnet wird, die darin
besteht, dass man das Bild und den Namen eines Feindes im Boden unter einem Idol ver-
birgt und die Gottheit anfleht, denselben zu töten. Um eine verwandte Ceremonie handelt
es sich hier in der That, wie auch aus dem unten gebrauchten Ausdruck *Lingga* hervorgeht,
das nach Jäschke das Bild eines Feindes bedeutet, welches in der Ceremonie des *sbyin*
sreg verbrannt wird, um ihn so durch Zauberei zu töten. Zu dem als Sühngeld gegebenen
Gewand (*glud gos*) ist zu bemerken, dass *glud* im besonderen das Bildnis eines Mannes
bezeichnet, das an seiner Statt beim *gtor ma* (Brandopfer) weggeworfen wird. Desgodins,
Dictionnaire tibétain-latin-français, p. 178, erwähnt *glud ₒbyuṅ* bĕi 4 elementa redemptionis,
species oblationis, in qua aliqua effigies substituitur personae redimendae. In diesem Falle
wird das Bild des Königs auf sein Gewand gezeichnet. Unter der „Befleckung" desselben
ist jedenfalls seine Sündhaftigkeit gegenüber der Bonreligion und die damit zusammenhän-
gende Vernachlässigung seiner Gattin zu verstehen.

53) Vergl Goethe, Faust, II. Teil (Akt I, Sc. 2): „Mein alter Narr ging, fürcht'
ich, weit ins Weite".

darüber mit *Padmasambhava*, der ihm antwortete: »Wenn des Königs
Lebenszeit abgelaufen ist, dann könnte selbst der wirklich vollendete
Buddha, wenn er hier erschiene, die Frist nicht verlängern. Handelt
es sich aber nur um zeitweilige Gefahren, so bin ich der Mann,
sie zu beseitigen". Darauf zeichneten die Bonpo auf das Gewand
ein *Liṅga*, und als durch die schrecklichen Beschwörungen der
Sum pa der Zauber vollendet, war der König in vierzehn Tagen
überwunden. Darauf kehrten die Bonpo schnellfüssig in ihr Land
zurück.

In seiner Todesstunde legte er das Bekenntnis ab, dass er die
Bonreligion unterdrückt habe, und legte in seinem Testamente nieder,
dass die künftigen Geschlechter Bon und Buddhismus in gleicher Weise
pflegen sollten. So lautet der Bericht im *bKa ₒbum*.

Dem *bsGrag byaṅ* zufolge hätte ihm *Tsʿe spoṅ za* Gift gegeben,
wodurch er verschieden sei. Da *sGam po* seiner Mutter dadurch,
dass er sie den Schmutz des Gewandes empfangen liess, eine Gemüts-
krankheit verursacht hatte, vergiftete ihn seine Mutter, worauf er
starb [54]). Darauf verminderten sich die Verdienste der Königsfamilie.
Das ist die Geschichte des *dbu yog* der Unterdrückung der Bonlehre,
so heisst es. Noch heutzutage gibt es in *rMaṅ ₒoṅ* und anderen Orten
viele Leute, welche die Stadt der tibetischen Königsfamilie meiden.

---

54) Das *Grub mtʿa šel kyi me long* nennt ihn *Muni btsan po* und lässt ihn nach dem
Tode des Königs ein Jahr und neun Monate regieren, worauf ihn seine Mutter vergiftete,
um ihren jüngsten Sohn auf den Thron zu setzen. Ebenda wird ihm der Versuch einer
communistischen Staatsverfassung auf Grundlage einer gleichmässigen Verteilung aller Güter
zugeschrieben. S. Journal of the Asiatic Society of Bengal, vol. L, part I, p. 226, 227.

025

关于动物语言的童话

KELETI SZEMLE.

KÖZLEMÉNYEK AZ URAL-ALTAJI NÉP- ÉS NYELVTUDOMÁNY KÖRÉBŐL.

A M. TUD. AKADÉMIA TÁMOGATÁSÁVAL

A MAGYAR NÉPRAJZI TÁRSASÁG KELETI SZAKOSZTÁLYÁNAK ÉS A KELETI KERESKEDELMI AKADÉMIÁNAK ÉRTESITŐJE.

— ◆ —

REVUE ORIENTALE

POUR LES ÉTUDES OURALO-ALTAÏQUES.

SUBVENTIONNÉE PAR L'ACADÉMIE HONGROISE DES SCIENCES.

★

JOURNAL DE LA SECTION ORIENTALE DE LA SOCIÉTÉ ETHNOGRAPHIQUE HONGROISE ET DE L'ACADÉMIE ORIENTALE DE COMMERCE À BUDAPEST.

• • •

SZERKESZTIK
Rédigée par

Dᴿ KÚNOS IGNÁCZ ★ Dᴿ MUNKÁCSI BERNÁT.

KÖZREMŰKÖDŐK:
Collaborateurs

Asbóth Oszkár (Budapest), Willy Bang (Louvain), Luigi Bonelli (Napoli), Karl Foy (Berlin), Friedrich Hirth (München), Clément Huart (Paris), Georg Jacob (Erlangen), Nikolaj Katanoff (Kasan), Katona Lajos (Budapest), Kuun Géza gróf (Maros-Németi), Franz Kühnert (Wien), Mahler Ede (Budapest), L. Mseriantz (Moscou), H. Paasonen (Helsingfors), W. Radloff (St. Petersbourg), Szinnyei József (Budapest), Wilhelm Thomsen (Kopenhague), Vámbéry Ármin (Budapest), Heinrich Winkler (Breslau), Zichy Jenő gróf (Budapest) etc.

TOME **II.** KÖTET.

BUDAPEST.

En commission chez **Otto Harrassowitz,** *Leipsic.*

1901.

rischen Sprachen Einzelne nicht eben Original-Eigentum der kau-
kasischen Sprachen sind, sondern umgekehrt, sie gerieten aus den
finnisch-magyarischen hieher; mit anderen Worten, dass in denselben, wie es bei sprachlicher Berührung in der Regel vorkommt
auch Fälle von Rückwirkung vorhanden sind. Es ist zu ersehen,
dass wir hier so vom sprachwissenschaftlichen, wie vom præ-
historischen Gesichtspunkte aus vor bedeutsamen Problemen stehen, welche dessen würdig sind, dass der Forscher vor den mit
ihrer Lösung verbundenen Schwierigkeiten nicht zurückschrecke,
selbst wenn ihn das Vorgefühl drückt, dass bezüglich einzelner
Partien sein Weg über Irrtümer hinweg zur endlichen Wahrheit
führen werde.

ZUM MÄRCHEN VON DER TIERSPRACHE.

I.

Im 52. Bande der Zeitschrift der Deutschen Morgenländischen
Gesellschaft, S. 287, habe ich eine mongolische Version des bekannten
Märchens von der Tiersprache mitgeteilt, dessen Quelle und Verbreitung Th. BENFEY in Orient und Occident, Bd. II, S. 133—171, eingehend
untersucht hat. Inzwischen ist mir eine Abhandlung von Prof. Katanof mit dem Titel «Zur Frage der Ähnlichkeit osttürkischer Erzählungen
mit slavischen» (Къ вопросу о сходствѣ восточно-тюркскихъ
сказокъ со славянскими), Kasan 1897, Sonderabdruck aus Band
XIV der Nachrichten der Gesellschaft für Archäologie, Geschichte und
Ethnographie an der Kais. Universität Kasan, zugänglich geworden,
deren erster Teil vier neue türkische Versionen dieses Märchens bietet.
Eine deutsche Übersetzung mag daher manchem willkommen sein. In
der ersten hier mitgeteilten Form der Erzählung hat der Mann, der die
Sprache der Tiere versteht, sie von König Salomon erlernt; in der
zweiten findet sich keine Angabe über diesen Punkt, ähnlich wie in der
mongolischen Geschichte nur im allgemeinen von einem ungenannten
«Lehrer» die Rede ist; in der dritten und vierten heisst es, dass ihn
Schlangen darin unterrichtet haben. In der čagataischen Version lacht
der Hahn, in der sagaischen der Hengst; in beiden bringt der Mann,

durch das Gespräch der Tiere veranlasst, sein neugieriges Weib durch eine Tracht Prügel zur Vernunft, in Übereinstimmung mit dem Abschluss des mongolischen Berichts ; in der turfanischen und beltirischen Version dagegen gibt er den Schmeichelein seiner Frau nach und erzählt ihr alles, worauf ihn der Tod ereilt. Die beltirische Version besteht aus der Verschmelzung zweier Märchenmotive, von denen das erstere eine Variante der sehr weit verbreiteten Erzählung mit dem Grundgedanken «Undank ist der Welt Lohn» darstellt. Auch innerhalb der sagaischen Geschichte lassen sich zwei in einander geflossene Bestandteile nach- weisen : dem Märchen von der Tiersprache geht, nur lose damit ver- knüpft, die Unterhaltung zweier Geister des Feuers voraus, die einen Reichen wegen seines Geizes arm und einen Armen reich machen. Die- ses Motiv macht auf mich den Eindruck, dass sein Ursprung bei den Türken selbst, zum mindesten nicht ausserhalb Sibiriens zu suchen ist, weil es sich auf die Ceremonie des Feueropfers gründet, die noch heute von allen sibirischen Völkerschaften beobachtet wird. Die mongolische Variante des Märchens von der Tiersprache stammt aller Wahrschein- lichkeit nach aus einer Quelle, welche den türkischen Versionen nicht zu Grunde gelegen hat; aber auch diese dürften ihre Herkunft verschie- denen Quellen zu verdanken haben. Diese Frage lässt sich aber erst an der Hand eines Materials lösen, das weit umfangreicher sein muss als das bisher gebotene.

1. Čagataische Version nach Rubguzi.*)

Einst kam zu Salomon ein Mann und sagte : «O Abgesandter Gottes, lehre mich die Sprache der Tiere !» Salomon antwortete : «Lerne sie nicht, doch wenn du, nachdem du sie erlernt hast, zu irgend einem darüber sprichst, wirst du sterben.» Jener Mann sagte : «Ich werde nicht sprechen, niemandem werde ich meine Geheimnisse mitteilen !» Salomon lehrte ihn die Sprache der Tiere. Der Mann kam des Abends nach Hause und sagte zu seinem Arbeiter: «Treibe den Ochsen und den Esel in den Stall !» Der Arbeiter that so. Der Esel fragte den Ochsen : «Mein Freund,

---

*) Sein Buch führt den Titel Kysasu'l-anbija und ist in Kasan 1891 gedruckt worden. Die Zeit der Abfassung dieses unter den östlichen sun- nitischen Mohammedanern verbreiteten Werkes fällt in das 15. Jahr- hundert. Katanof hält die daraus mitgeteilte Version möglicherweise für das Vorbild der turfanischen.

wie erging es dir heute?» Der Ochse antwortete: «Was frägst du? Schlecht
geht es mir: ich bin erschöpft.» Der Esel sagte: «Ich will dich eine
List lehren, und du wirst dich morgen vor der Arbeit retten.» Der Ochse
fragte: «Was habe ich zu thun?» Der Esel erwiderte: «Wenn man dir
heute Futter geben wird, dann iss nicht; morgen wird man dich für
krank halten und nicht zur Arbeit mitnehmen!» Da der Mann die
Sprache der Tiere erlernt hatte, so lächelte er und schaute nach, ob ihn
sein Weib nicht sähe. Er erhob sich und ging auf die Strasse hinaus. In
dieser Nacht frass der Ochse kein Heu. Am andern Tage sprach der
Hausherr zu dem Arbeiter: «Lass heute den Ochsen zurück und nimm
den Esel zur Arbeit». So tat auch der Arbeiter. Der Abend nahte
heran; der Arbeiter brachte sie wieder in einen Raum. Abends fragte der
Ochse den Esel: «Mein Freund, bist du nicht abgespannt?» Der Esel
erwiderte: «Freund, wieviel auch immer zu tun wäre, der Mann bemüht sich, es vollständig auszuführen.» Der Ochse fragte: «Hat mein
Herr heute etwas über mich gesagt?» Der Esel antwortete: «Ja!» Der
Ochse fragte: «Was denn?» Der Esel antwortete: «Dein Herr spricht:
morgen werde ich den Ochsen zur Arbeit bestimmen; wenn er nicht gut
gehen und am Abend, wann er nach Hause zurückgekehrt ist, sein Heu
nicht fressen wird, dann werde ich ihn erstechen und für die Bettler eine
wohlthätige Mahlzeit veranstalten». Der Ochse erschrak und begann sogleich sein Heu zu verzehren. Der Herr des Ochsen hatte wieder das
Gespräch des Esels mit dem Ochsen belauscht und lächelte. Sein Weib
sprach: «Du hast gestern gelächelt und lächelst heute! Hast du nicht
eine Geliebte, welche kommt und dir Nachricht von der Strasse bringt?»
Die Frau begann mit ihrem alten Manne zu streiten, indem sie sagte:
«Gib mir die Scheidung oder teile mir das Geheimnis mit, weshalb du
lachst.» Der Mann stand auf und wollte irgend einen Menschen her-
beiholen, um ihm sein Geheimnis mitzuteilen, damit jener es seiner
Frau erzähle. Er ging auf die Strasse hinaus und bemerkte, wie ein
Hahn sich mit einem Hunde unterhielt. Der Hahn warf sich auf die
Henne, und der Hund sprach: «Weshalb freust du dich, wenn heute
dein Herr sterben wird?» Jener Hahn sagte: «Wenn mein Herr sterben
wird, so mag er sterben, weil er keine Männlichkeit besitzt. Ich habe
zwanzig Weiber und werde mit allen fertig, während er nur eine Frau
hat und nicht mit ihr allein fertig werden kann». Der Hund sagte:
«Hahn, gibt es keinen Ausweg aus dieser Klemme?» Der Hahn erwiderte: «Dein Herr möge in den Garten gehen, dort Gerten vom Maul-

beerbaum schneiden und hierher bringen, dann soll er die Thür mit
der Kette verschliessen und sein Weib so lange schlagen, bis sie ihre
Worte bereut und sagt: «Herr, vergib mir!» und sie wird niemals mehr
mit Fragen zudringlich werden.» Derr Mann kehrte zurück und that so,
wie der Hahn gesagt hatte, und sein Weib stellte keine Fragen mehr.

2. Turfanische Version.*)

Zur Zeit des Propheten Salomon lebte ein Mensch, der die Sprache der Tiere erlernt hatte. Dieser Mensch kam in sein Haus. Er besass
einen Stier und einen Esel, die er in den Pflug einspannte. Am Abend
kehrten die Tiere nach Hause zurück. Der erwähnte Esel fragte: «Wie
geht es dir?» Der Stier antwortete: «Man hat mich in den Pflug ge-
spannt, geprügelt und gezwungen zu arbeiten.» Der Esel sagte: «Iss nicht
dein Futter!» und der Stier frass nicht. Jener Mensch stand neben dem
Esel und dem Stier und hörte ihr Gespräch. Andern Tags spannte man
den Esel in den Pflug, schlug ihn und arbeitete auf ihm. Abends kamen
sie nach Hause zurück. Der Stier fragte den Esel: «Mich hatte man ergriffen, geprügelt und gezwungen zu arbeiten, doch jetzt steht es gut mit
mir; geht es dir schlecht? Doch mein Herr sagte: mein Stier ist erkrankt!
und will mich dem Fleischer übergeben; was soll ich jetzt thun?» Der
Esel sprach: «Nunmehr iss und laut — laut schmatze, schnaube und
brülle!» Der Stier begann zu brüllen und wurde gesund. Jener Mensch
hörte das Gespräch des Esels mit dem Stier, und da er es verstand,
lächelte er. Sein Weib fragte ihn: «Was lächelst du? Du lächelst unbedacht.» Diesem Menschen war es bei Todesgefahr verboten worden, mit
einem andern davon zu sprechen, dass er die Tiersprache kenne. Er
wollte nicht davon sprechen, aber sein Weib drang in ihn und sprach:
«Weshalb lachst du unbedacht?» Dieser Mensch besass, so sagt man,
viele Hennen und einen Hahn. Der Hahn sagte: «Ich gehöre zum
Hühnergeschlecht, aber mit welcher Zahl von Hennen kann ich fertig
werden und sie jagen, und da fürchtet sich unser Herr vor einer einzigen Frau!» Darauf fügte er hinzu: «Es wäre gut, wenn er diesem
Weibe Schläge austeilte.» Dieser Mensch besass ferner einen Hund

---

*) Sie wurde von Katanof im April 1892 in der chinesischen Stadt
Urumtschi nach den Worten des Schuhmachers Záit aufgeschrieben, der
türkisch wenig lesen und schreiben konnte, aber geläufig chinesisch sprach.

und eine Katze. Als er seinem Weibe Schläge gab, sprach der Hund zur Katze: «In wieviel Tagen wird mein Herr sterben? Wenn er sterben wird, wird es viel Essen und Trinken geben.» Die Katze sprach: «Wenn der Herr sterben wird, was wirst du machen?» Der Hund erwiderte: «Wer mir Spülwasser geben wird, zu dem werde ich hingehen.» Nachdem jener Mensch den Schmeicheleien seiner Frau nachgegeben hatte, erzählte er ihr alles und starb.

3. Beltirische Version.*)

Ein Mensch rüstete sich, um wilde Tiere und Vögel zu töten Er nahm Flinte und Stock und begab sich auf die Jagd. Als er aufwärts kam, nahte in der Richtung auf ihn zu ein Waldbrand, und vor diesem her kroch eine Schlange. Die Schlange sprach: «Rette mich vor diesem Feuer, und ich werde dir eine dreifache Wohlthat erweisen: ich werde dich drei Sprachen lehren.» Da steckte er die Schlange in seine Busentasche, um sie nach Verlöschen des Brandes wieder herauszulassen. Die Schlange wollte nicht heruntergehen und wand sich um seinen Hals. «Warum hast du mich umschlungen?» «Gutes vergilt man nicht mit Gutem, — lege deine Stirn hierher, und ich werde dich beissen»! «Steht es so, so dauerst du mich nicht, doch wenden wir uns an drei Gerichte! Wenn wir anlangen werden, dann magst du mich beissen.» Nun giengen sie weiter und trafen auf ein Dorf. Auf der Strasse gieng ein altes Pferd, das Heu sammelte. Der Mensch rief dem Pferde zu: «He, Pferd, bleib stehen, richte über uns!» Das Pferd fragte: «Was hat sich mit euch zugetragen?» Der Mann erzählte alles der Reihe nach, und das Ross bemerkte: «Ja, ja, Schlange, wirklich vergilt man Gutes nicht mit Gutem, — es thut mir leid um seine Stirn.» Darauf sprach das Pferd: «Ich habe den Acker bei reichen Hausleuten gepflügt und sie ernährt, doch da ich alt geworden bin, hat mich mein Herr verjagt.» Der Mensch sagte: «Warte nur, wir gehen noch zu zwei Gerichten.» Wieder giengen sie weiter. Sie begegneten hinter dem Dorfe einem Hund. Sie grüssen

---

*) Von Katanof im Dec. 1889 im Minusin'schen Kreise am linken Ufer des Mittellaufes des Abakan, 12 Werst vom Dorfe Askys, nach den Worten des Beltiren Appák Asotschákob aufgeschrieben, der sehr schlecht russisch sprach.

den Hund und sprechen : «Hund, richte über uns !» Der Hund fragte:
«Was hat sich mit euch zugetragen ?» Der Mensch erzählte ihm alles
und der Hund bemerkte: «Es erweist sich, dass man Gutes nicht mit
Gutem vergilt: ich jagte hinter Zobeln und Vögeln her, und wenn ich
sie gefunden hatte, stellte ich sie meinem Herrn zu ; doch jetzt, wo ich
alt geworden bin, hat er mich verjagt, da ich schon den Geruchssinn
verloren habe ; leid tut es mir um seine Stirn.» Doch der Mensch sagte
zu der Schlange : «Warte nur, es bleibt noch ein Gericht übrig». «Nun,
gehen wir noch zu einem Gericht !» sprach die Schlange. So zogen sie
weiter auf einem grossen Wege. Dies war eine Poststrasse, die in der
Steppe lag. Während sie gingen, schritt ein Fuchs über die Steppe.
Der Mensch rief ihn heran und erzählte alles der Reihe nach. Der Fuchs
sagte : «Jetzt, Schlange, geh herunter, und wie dieser Mensch gerade
steht, so stehe auch du !» Die Schlange glitt herab und stand aufrecht.
Der Fuchs fragte : «Du, Mensch, weshalb hast du Flinte und Stock von
Hause mitgenommen?» Der Mensch begriff, dass er die Schlange mit
dem Stock erschlagen müsse, und hieb mit einem Schlag die Schlange
mitten durch. Der Fuchs lief hierhin und dorthin. Der Mensch nahm
seine Flinte von der Schulter, und nachdem er den Fuchs erschossen,
bewies er noch einmal, dass man Gutes nicht mit Gutem vergilt. Der
Mensch nahm den Fuchs auf, ging noch Hause und verstand schon alle
Sprachen. Als er zu Hause anlangte, beklagten sich Vögel, Kühe und
Pferde bei ihm, indem sie sagten : «Deine Frau hat uns nicht gefüttert».
Er schalt deswegen seine Frau aus, doch davon, dass er die verschiedenen Sprachen erlernt hatte, sagte er zu niemandem etwas, denn das
drohte ihm den Tod. Seine Frau fragte ihn täglich, weshalb er um die
Tiere herumgehe, doch er schwieg. Hinter dem Ofen hervor kamen die
Hühner, und der Hahn krähte : «Kikeriki, ich werde mit vierzig Wei-
bern fertig, doch du kannst in Furcht vor dem Tode nicht mit einer fertig werden.» Als er so gerufen hatte, schlug er mit den Flügeln und
trieb seine Hühner hinter den Ofen. Darauf befragte ihn seine Frau darüber, woher er die Tiersprache erlernt hätte. Als er einst bei seiner
Frau lag, konnte er nicht an sich halten und erzählte ihr alles, doch
am Morgen war er bereits tot.

4. Sagaische Version.*)

Ein armer Mensch ging hinter dem Vieh seines Herrn her. Während er dahinging, konnte ein Schlange, die unter einen Stein geraten war, nicht unter demselben hervorkriechen. Die Schlange sprach: «Befreie mich von hier!» und der Mensch half ihr herauszukriechen. Die Schlange sagte: «Sieh, wie man Gutes vergilt!» und biss den Menschen in den Kopf. Darauf begann jener Mensch die Sprache jedes Menschen, des Viehs und der Vögel zu verstehen. Er weidete zehn Hunderte Schafe. Während er sie trieb, blieb hinten ein Lamm zurück. Die Mutter rief ihm zu: «Mein Kind, weshalb bleibst du stehen? Komm schnell hierer.» Das Lamm sprach: «Ich will nicht gehen, weil in mir das Glück aller Schafe ist.» Als die Frist des Dienstes abgelaufen war, empfing der Hirt von seinem Herrn Geld und dieses Lamm. Nachdem er abgerechnet hatte, begab er sich mit dem Lamm zu einem Armen. Als er bei ihm sass, zischte plötzlich das Feuer zweimal, und der Arme warf Stücke seiner Speise ins Feuer.**) Der ehemalige Hirt verstand, dass der Feuergeist des reichen Mannes, bei dem er die Tiere geweidet, und der Feuergeist des Armen sich unterhielten. Der Feuergeist des reichen Mannes sprach: «Dein Wirt pflegt dich gut, obwohl er arm ist, mein Wirt aber nährt mich ungeachtet seines Reichtums nicht». Der Feuergeist verliess den reichen Mann und lebte hinfort bei dem Armen. Darauf verarmte der Reiche, der Arme jedoch wurde reich. Der Hirt begab sich mit seinem Schafe nach Hause zu seiner Frau. Als sie einst die Weideplätze verliessen, bemerkte der Mann zwei sich streitende Meisen. Es stellte sich heraus, dass die eine von ihnen den Käse ihres Herrn ausstreute, damit später etwas zur Nahrung da sei, aber die andre nicht erlauben wollte, dies zu thun. Als der ehemalige Hirt den Sachverhalt erkannt hatte, lachte er. Sein Weib fragte ihn, weshalb er gelacht habe, doch er sagte nichts, denn er fürchtete sich vor dem Tode. Er besass einen Hengst,

---

*) Von Katanof im Januar 1890 am linken Ufer des Baraflusses, 10 Werst vom Dorfe Askys, Kreis Minusinsk, nach den Worten des Sagajers Kebéi vom Stamme Tomnar, 33 Jahre alt, aufgezeichnet.

**) Bevor die Minusin'schen Tataren Speise und Trank zu sich nehmen, werfen sie einen Teil davon und andres ins Feuer als Opfer für den Geist des Feuers, des Hausbeschützers (so Katanof; diese Sitte findet sich indessen bei allen sibirischen Stämmen, ich selbst habe sie zu wiederholten Malen bei Tungusen und Giljaken beobachtet).

4*

der vierzig Stuten regierte. Eine, die beste Stute, hatte ein fettes Fohlen.
Der Herr wollte dasselbe schlachten und verzehren. Das Fohlen drängte
sich an die Mutter und begann, stark an ihr zu saugen. Die Mutter
fragte nach der Ursache davon, und das Fohlen sprach: «Der Herr
wünscht mich zu schlachten, da er mein süsses Fleisch verspeisen will,
doch danach wird er die Neugierde seines Weibes befriedigen und ruhig
sterben!» Als der Hengst dies vernommen hatte, wieherte er laut und
sagte: «Ich werde mit vierzig Stuten fertig, doch er kann mit einer einzigen Frau nicht fertig werden!» Als der ehemalige Hirt dies gehört
hatte, kehrte er nach Hause zurück und peitschte sein Weib ordentlich
aus. Das Weib besserte sich, und sie wurden reich.

B. Laufer.

II.

Zu den von Benfey a. a. O. herangezogenen Parallelen des Märchens von der Tiersprache will ich diesmal nur auf die *ungarische* Variante hinweisen, die im III. Bande des «Nyelvőr» S. 227 mitgeteilt ist.

Ein armer Schafhirt rettet eine Schlange, den Sohn des Schlangen-
königs vom Feuertode. Vor den Schlangenkönig geführt, stellt ihm
dieser die Wahl seiner Belohnung frei. Er heischt das Verständnis
der Tiersprache, das ihm unter der Bedingung verliehen wird, darüber
kein Sterbenswörtchen verlauten zu lassen, weil er dies mit dem Leben
büssen würde.

Unterwegs nachhause, gewahrt er auf einem ausgehöhlten Baume
zwei Elstern, und vernimmt aus ihrem Geschwätz, dass ein grosser
Schatz in der Höhle des Baumes steckt. Der Mann kommt mit einem
Wagen und führt den Schatz heim. Nun heiratet er die Tochter des
Schäfermeisters, die ihn wiederholt nach der Herkunft seines Reichtums befragt. «Frage nicht darnach», sagt er ihr, «Gott hat es uns
gegeben».

Eines Tages sagt die alte Katze zu der Jungen, wie sie beide au
dem Ofengesimse sitzen: «Du bist kleiner, so schlüpfe in die Kammer
und hol' uns Speck heraus». Da musste der Schäfer hell auflachen. Sein
Weib fragt ihn nach dem Grunde seines Lachens, und wie sie denselben
hört, fragt sie weiter, wie er es erfahren? Doch ist ihr Drängen diesmal
noch umsonst.

Ein anderes Mal gehen sie zu Markte. Die Frau reitet die Stute,
der Mann den Hengst. Da die Stute ein wenig zurückbleibt, wiehert ihr

026

关于阿尔泰语的属格的形成

KELETI SZEMLE.

KÖZLEMÉNYEK AZ URAL-ALTAJI NÉP- ÉS NYELVTUDOMÁNY KÖRÉBŐL.

A M. TUD. AKADÉMIA TÁMOGATÁSÁVAL

A MAGYAR NÉPRAJZI TÁRSASÁG KELETI SZAKOSZTÁLYÁNAK ÉS A KELETI KERESKEDELMI AKADÉMIÁNAK ÉRTESITŐJE.

—◆—

REVUE ORIENTALE

POUR LES ÉTUDES OURALO-ALTAÏQUES.

SUBVENTIONNÉE PAR L'ACADÉMIE HONGROISE DES SCIENCES.

*

JOURNAL DE LA SECTION ORIENTALE DE LA SOCIÉTÉ ETHNOGRAPHIQUE HONGROISE ET DE L'ACADÉMIE ORIENTALE DE COMMERCE À BUDAPEST.

···

SZERKESZTIK
Rédigée par

Dᴿ KÚNOS IGNÁCZ ⋆ Dᴿ MUNKÁCSI BERNÁT.

KÖZREMŰKÖDŐK:
Collaborateurs

Asbóth Oszkár (Budapest), Willy Bang (Louvain), Luigi Bonelli (Napoli), Karl Foy (Berlin), Friedrich Hirth (München), Clément Huart (Paris), Georg Jacob (Erlangen), Nikolaj Katanoff (Kasan), Katona Lajos (Budapest), Kuun Géza gróf (Maros-Németi), Franz Kühnert (Wien), Mahler Ede (Budapest), L. Mseriantz (Moscou), H. Paasonen (Helsingfors), W. Radloff (St. Petersbourg), Szinnyei József (Budapest), Wilhelm Thomsen (Kopenhague), Vámbéry Ármin (Budapest), Heinrich Winkler (Breslau), Zichy Jenő gróf (Budapest) etc.

TOME **II.** KÖTET.

BUDAPEST.

En commission chez **Otto Harrassowitz,** *Leipsic.*

1901.

Bitte richtet, dass er all den verschiedenen Gottheiten Bericht erstatten soll *(poro jumon witnosaźe, witnen tapnen kalase küllö jumolan)*.

(Fortsetzung folgt.)

ZUR ENTSTEHUNG DES GENITIVS DER ALTAISCHEN SPRACHEN

von Dr. Berthold Laufer.

Die Gleichförmigkeit, welche das ganze Gebiet der altaischen Sprachengruppe in der Bildung des Genitivs beherrscht, ist trotz mannigfacher im einzelnen auftretender Varianten schon frühe erkannt worden. Bereits Castrén, Grammatik der samojedischen Sprachen § 225, macht auf den allgemein hervortretenden Charakter *n* aufmerksam, und Grunzel, Entwurf einer vergleichenden Grammatik der altaischen Sprachen S. 50, gelangt zu dem Schlusse, dass die Affixe *-ni* und *-nin* die ursprünglichen Genitivformen vorstellen und alle Abweichungen sich aus phonetischen Gesetzen ergeben.

Ich werde kurz zu zeigen versuchen, dass die altaischen, insbesondere die tungusischen und mongolischen Sprachen, in ihrem ältesten Zustand eines einheitlichen Genitivs entbehrt haben, und dass die jetzt als Genitivsuffixe geltenden Silben ursprünglich Pronominalsuffixe der dritten Person gewesen sind.

Im mandschu wird der Genitiv durch Suffigierung von *-i* bei Stämmen mit auslautendem Vokal und *n*, und von *ni* nach den übrigen Consonanten gebildet. Bang (Mandschurica in T'oung Pao, vol. I 331) hält *ni* für die ursprüngliche Bildung. Ohne auf seine Argumente für diese Behauptung des näheren einzugehen, vermag ich zu constatieren, dass thatsächlich in den meisten tungusischen Dialekten *ni* eine vorherrschende Stellung einnimmt, vor allem im goldischen, das ich eingehender kennen gelernt habe. Jedoch unterscheidet sich in dieser Sprache die Auffassung und Bildung des Genitivs wesentlich von der des mandschu und der übrigen tungusischen Sprachformen. Im goldischen tritt nämlich die Silbe *-ni* nicht, wie in den verwandten Idiomen, an den bestimmenden Ausdruck, sondern an den zu bestimmenden : z. B. *loća aja-ni* «der Herrscher der Russen» *takto tora-ni* «die Pfähle des

Vorratshauses,» *Oisa ami-ni* «der Vater des Oisa.» Aus dieser Erscheinung ist mit Notwendigkeit zu schliessen, dass -*ni* in diesem Falle nur ein Pronominalaffix der dritten Person vertreten kann, denn allein diese Auffassung vermag jene Erscheinung zu erklären. *Oisa ami-ni* bedeutet wörtlich : «Oisa Vater-sein», wie etwa im vulgärdeutschen gesagt würde : «dem Oisa sein Vater» ; *buni fokto-ni* «Jenseits Weg-sein», «Jenseits sein Weg», d. i. «der Weg ins Totenland» ; *sole fokto-ni* «stromaufwärts Weg-sein», «der stromaufwärts führende Weg», Bezeichnung des «Telegraphen» am Amur. Dieses Phänomen der Ersetzung des Genitivs durch ein zurückweisendes Possessivsuffix steht auf dem Gebiet der altaischen Sprachen nicht allein da, sondern findet sich auch ganz analog im jakutischen und tschuwaschischen, und FRIEDRICH MÜLLER (Grundriss der Sprachwissenschaft II S. 270) zieht aus dieser Thatsache den Schluss, dass diese genitivische Construktion in den jakutisch-türkischen Sprachen die ursprüngliche sei. Zunächst erhebt sich nun die Frage, ob und in welchem Zusammenhange das goldische Pronominalaffix -*ni* mit den in dieser und verwandten Sprachen gebräuchlichen Formen des persönlichen Pronomens der dritten Person steht. Das goldische vermag auf diese Frage keine direkte Antwort zu geben, da hier nur die sekundäre Bildung *n'oan*, *n'oa-ni* «er» im Gebrauch ist, die indessen sicher auch zu -*ni* in Verwandtschaftsverhältnis steht. Dagegen sehen wir, dass im mandschu das Genitivsuffix -*i* mit der Wurzel des Pronomens der dritten Person *i* thatsächlich zusammenfällt. Daraus ist mit Sicherheit zu folgern, dass das Genitivsuffix -*i* des mandschu ursprünglich auch als ein pronominales Affix der dritten Person anzusehen ist. Ferner hat bereits SCHIEFNER sowohl in der ochotskischen Mundart des tungusischen, wie in der von Anadyr ein reflexives singularisches Suffix -*i* gefunden (Melanges asiatiques III 697, VII 329). Andererseits kennt das ochotskische ein Possessivaffix der dritten Person -*n* (Mél. as III 696), das gleichfalls in den von CASTRÉN erforschten westlichen Dialekten des tungusischen begegnet (s. CASTRÉN, Grundzüge einer tungusischen Sprachlehre § 68, 69): *haga* ‚Schale', *haga-n* ‚seine Schale'. Wir haben also wohl zwei Stämme zu unterscheiden, -*i* und -*n*, die sich dann weiterhin zu -*i + n* und -*n + i* verbinden; -*in* liegt der Deklination von *i* des mandschu zu grunde (-*imbe*, *inde*, -*ini*, -*inci*), -*ni* den genitivischen Possessivaffixen des goldischen und tungusischen überhaupt. Der Unterschied in der Auffassung des mandschu und goldischen besteht darin, dass das mandschu den Ausdruck «der Garten des Vaters» als «Vater-sein Garten»,

«Vater-ihm Garten», «der dem Vater zugehörende Garten», das goldische als «Vater-Garten-sein», «Vater-Garten-ihm», das sich auf den Vater beziehende Gehören, «Eigentum des Gartens». Es ist hinreichend klar, dass in diesem Falle das goldische gegenüber dem mandschu die ältere Redeform repräsentiert; denn in letzterem sind -i und -ni zu reinen Suffixen herabgesunken, deren ursprüngliche Bedeutung gar nicht mehr gefühlt wird, während diese im goldischen noch deutlich ins Bewusstsein fällt. Die Entwicklung der Genetivauffassung des mandschu ist nur aus der des goldischen denkbar, nicht umgekehrt.

Die Genetivendungen der ostmongolischen Schriftsprache sind -u, -ü für Stämme mit auslautendem n; -un, -ün für solche mit den übrigen Consonanten, und -in für Themen mit vocalischem Auslaut· Diese Verhältnisse repräsentieren aber offenbar erst eine spätere Entwicklungsstufe, während die frühere Verfassung im kalmükischen und in den übrigen mongologischen Dialekten zu Tage tritt. In der kalmükischen Schriftsprache, die manche altertümliche Erscheinungen bewahrt hat, finden sich als Genetivsuffixe -i nach n und -in nach den übrigen Consonanten. Dasselbe ist in der ostmongolischen Umgangssprache (Бобровниковъ, Грамматика монгольско-калмыцкаго языка § 171), wie im burjatischen der Fall, wo der Genetiv bei consonantisch auslautenden Wörtern stets durch -i ausgedrückt wird (Castrén, Versuch einer burjätischen Sprachlehre § 35). Der Vokal -i herrscht ebenfalls im Dialekt der Khalkha vor (*Vitale* et *de Sercey*, Grammaire et vocabulaire de la langue mongole, dialecte des Khalkhas, Peking 1897. p. 5, 6). Einige Stämme wie die Ataganzen, Sartagulzen gebrauchen -i auch statt -in Бобровниковъ, l. c. § 172 no.) Die Ursprünglichkeit des Charakters -i gegenüber -u wird ferner vor allem dadurch bewiesen, dass derselbe noch in den alten Werken der ostmongolischen Litteratur wie im Geserepos und in Inschriften auftritt. Vergl. z B. Geser 1, 1 *burchan-i* 1, 12 *ene choton-i biden-i khen*, 1, 15 *ene choto biden-i yaghün-i tostu*, 187, 3 *Geser-i chailju baichu du*. In der mongolischen Inschrift von *Tsaghan Baišin* treffen wir die Genitive *oron-i*, *kümün-i amitan-i*, und sogar *setkil-i*, *tangharik-yi*, *bülük-yi* (Huth, Die Inschriften von Tsaghan Baišin, S. 40), ein deutliches Zeichen, dass in damaliger Zeit der Wandlungsprocess von -i zu -u in der Schriftsprache noch keineswegs überwunden war. Denn auf diese allein scheint sich dieser Wandel zu erstrecken, der seine Ursache vielleicht in dem Bedürfnis hatte, bei den -n-Stämmen den Genetiv von dem gleichlautenden Accusativ zu differenzieren, was seinerseits dann den analogen Übergang von -*in* zu -*un* zur unmittelbaren Folge hatte. Also -*i* und -*in* werden als die ursprünglichen mongolischen Genetivaffixe zu betrachten sein. Auch diese stehen mit dem Pronomen der dritten Person in lebendigem Zusammenhang.

1. Sie sind verwandt mit den Fürwörtern der dritten Person *inu, anu* der alten Büchersprache, und *ni* der Umgangssprache Бобровниковъ § 144). Diese Pronomina folgen stets dem Worte, zu welhem sie gehören. Als ständige Verbindung kommt vor: *beye-ni* er selbst (Бобровиниковъ p. 82). Aber auch schon in der alten Sprache tritt *ni* oder *i* nach auslautendem *n* als Pronominalsuffix dritter Person auf, das z. B. in der Gesersage oft an Postpositionen und Casuspartikeln sich anschmiegt, wie bereits SCHOTT in seinen Altaischen Studien (Abhandlungen der Berliner Akad. 1869, 272), gezeigt hat. Als Beispiele führt derselbe an: *micha-berni* «ex carne eius», *beye-dü-ni* «in corpore eius», *adūn-du-ni* «in stabulo eius», *beye-yi-ni* «corpus (accus.) eius», *ečige-yin-i* «patris eius». Dieselbe Erscheinung haben neuerdings *Vitale* und *de Sercey* (l. c. p. 19) im Dialekt der Khalkha nachgewiesen. Sie sagen wörtlich: «Es ist zu bemerken, dass für die dritte Person des Fürworts noch eine postpositive Form des persönlichen Pronomens vorhanden ist, die dadurch gebildet wird, dass man den Buchstaben *n'* anfügt, dem eine phuonischer Vokal vorausgeht. Diese Form kann sich niemals auf das Subjekt des Verbums beziehen und allen Casus der Deklination, den Adverbien etc. angefügt werden». Beispiele: *têr huni gérin* «das Haus dieses Mannes», *hani morin* «Pferd des Kaisers», *bi bicigin' uzc baina* «ich lese sein Buch». In den beiden ersten Beispielen liegt also eine doppelte Bezeichnung des Genitivs vor: wörtlich: «dieser Mann-sein Haus-sein», «Kaiser-sein Pferd-sein».*) Die Deklination wird folgendermassen veranschaulicht: gen. *gernin'*, loc. *gertin'*, acc. *gerin'*, instr. *gereren'*, elatio *geressen'*.

2. Im burjatischen treffen wir gleichfalls ein Pronominalaffix für die dritte Person des Singulars und Plurals, nämlich *n'i* oder *n* (CASTRÉN, l. c. 90—93), wobei auch an das Verbalsuffix für die dritte Person -*na*, -*ne* zu erinnern ist (ib. § 140). In ersterer Hinsicht bietet demnach das burjatische eine genaue Analogie zum tungusischen. Im Hinblick auf diese

---

*) Vergl. magyarisch: (*az*) *atydnak a ház-a* «des Vaters Haus» und WINKLER in dieser Zeitschrift I, 3, S. 195.

Erscheinungen wäre vielleicht die Frage aufzuwerfen, ob nicht das bewegliche Nominalaffix -*n* (vergl mong. *nidüi* und *nidüin* «Auge», *usu* und *usun* «Wasser», *modo* und *modon* «Baum» u. viele andre) von Hause aus auch nichts anderes als ein Possessivaffix der dritten Person gewesen sei, dessen ursprüngliche Bedeutung im Laufe der Zeit in Vergessenheit geriet. Unter dieser Voraussetzung ist es dann sehr leicht erklärlich, warum an die auf -*n* auslautenden Stämme nur -*i* und nicht -*ni* herantritt, oder man kann vielmehr mit ebenso viel Recht sagen, dass in Fällen wie *nidüini*, *usuni* u. s. w. das aus -*n*+*i* entstandene Affix -*ni* von den Stämmen *nidüi*-, *usu*- abzutrennen ist.

3. In den reflexiven Accusativsuffixen des mongolischen *bēn*, *yēn* ist offenbar das pronominale Element *n* enthalten. Wie alle langen Vokale der altaischen Sprachen das Produkt von Contractionen darstellen, so sind demgemäss auch *bēn* und *yēn* als Verschmelzungen aus zwei Silben entstanden zu denken. Und von vornherein liegt doch die Vermutung nahe, dass diese Suffixe aus Accusativsuffix + Pronominalsuffix entstanden sein müssen. In der That ist in *yēn* das Accusativsuffix *yi* und in *bēn* die mit dem mandschu übereinstimmende Accusativpartikel *be* zu erkennen. Vermutlich wird also *yēn* aus *yi* + *en* und *bēn* aus *be* + *en* entstanden sein*), und dieses *en* steht am nächsten dem mongolischen demonstrativen Pronomen *ene*, bei den Khalkha *en*. Und *en* steht zu *in*, resp. *ni* in engster verwandtschaftlicher Beziehung.

Diesen Thatsachen innerhalb der mongolischen Sprache wie den analogen Erscheinungen der tungusischen Gruppe gegenüber kann die Übereinstimmung der Genetivaffixe -*i*, resp. richtiger · *ni* und -*in* mit dem Pronomen *in-u*, *ni* und dem Pronominalaffix -*ni* (-*i*, -*in*, -*en*) nicht als auf einem Zufall beruhend angesehen werden. Besteht aber zwischen beiden Phänomenen ein innerer Zusammenhang, so geht daraus hervor, dass auch die mongolischen Genetivaffixe ursprünglich Pronominalaffixe der dritten Person gewesen sind. Der Dialekt der Kkalkha vollends

---

*) Diese Untersuchung wirft auch ein Licht auf die Entstehung der mongolischen Instrumentalaffixe *ber*, *yēr*, die gleichfalls von den Accusativaffixen *be*, *yi* + *er* gebildet sein müssen; *är*, *er*, *ör*, *ör* sind noch jetzt die Instrumentalaffixe der kalmükischen nach consonantischem Auslaut und stellen also jedenfalls die primären Affixe dieses Casus dar. Statt *bēn*, *yēn* setzt Боломиновъ, Русско-монголо-бурятскій переводчикъ, второе изданіе, Пет. 1898, p. XV, -*an* und-*en* an.

zeigt eine auffallende Analogie zu den oben berührten Bildungen des goldischen und jakutischen.

Ob und inwieweit meine Darlegung auch für die türkischen und vielleicht die finno-ugrischen Sprachen zutrifft, muss ich der Untersuchung berufenerer Kenner dieser Gebiete überlassen. Von grosser Wichtigkeit für meine Auffassung scheint mir die von RADLOFF, Die alttürkischen Inschriften der Mongolei, neue Folge, Pet. 1897, S. 61, constatierte Thatsache zu sein, dass das seltene Auftreten der Genitivformen in den Inschriften beweise, dieselben seien seit sehr kurzer Zeit im Gebrauch gewesen. Dass im jakutischen der Genitiv vollkommen fehle, scheine darauf hinzudeuten, dass die Türken, die den Kern des türkischen Elementes im Jakutenvolke bildeten, in Anwendung des Genitivs etwa auf dem Standpunkte der alten Türken standen.

DIE LITERATUR DER MAGYARISCHEN VOLKSMÄRCHEN.

— Von LUDWIG KATONA. —

Vor ungefähr achtzig Jahren, also nur um ein Jahrzehnt später als die berühmte Grimm'sche Sammlung, erschien der erste bescheidene Band ungarischer Volksmärchen,*) vorläufig nur eine in deutscher Sprache einem grösseren Leserkreis vorgeführte Auslese jener umfangreicheren Originalaufzeichnung von 80 Märchen, die Georg v. Gaal grösstentheils nach dem Dictate oder auch den eigenhändigen Abschriften ausgedienter Soldaten angelegt haben soll. Mit vollem Rechte wird gegen diese Mittheilung im III. B. (3. Aufl.) der Grimm'schen Kinder- u. Hausmärchen S. 345 der Tadel erhoben, dass die Darstellung «zu gedehnt sei und manchmal an jene falsche Ironie streife, von der sich moderne Erzähler, wie es scheint, nicht leicht losmachen,» was wohl so zu verstehn sein wird, dass Gaal zumeist in der Musäus'schen gezierten Art erzählt und nur sehr selten den echten Volkston trifft. Dies mag auch darin seinen Grund haben, dass seine Gewährsmänner hie und da nachweislich bereits aus gedruckten Quellen, aus Heften der Pfennig-Literatur schöpften.

---

*) Mährchen (sic) der Magyaren, bearbeitet und herausgegeben von *Georg v. Gaal*. Wien. Wallishausser. 1822. (Mit einem Titelkupfer.) X, 454 S. 8°.

027

德雷斯顿皇家图书馆的藏文手稿目录

Zeitschrift

der

Deutschen Morgenländischen Gesellschaft.

Herausgegeben

von den Geschäftsführern,

in Halle Dr. **Pischel,**   in Leipzig Dr. **Fischer,**
Dr. **Praetorius,**   Dr. **Windisch,**

unter der verantwortlichen Redaction

des Prof. Dr. E. Windisch.

Fünfundfünfzigster Band.

Leipzig 1901,

in Commission bei F. A. Brockhaus.

Verzeichnis der tibetischen Handschriften der Königlichen Bibliothek zu Dresden.

Von

Berthold Laufer.

Über die tibetischen Werke der Königlichen Bibliothek zu Dresden lagen bisher keine Mitteilungen vor. Auch der dortige Handschriftenkatalog enthält keine Angaben über dieselben. Wo nichts bemerkt, handelt es sich um Handschriften; Holzdrucke sind nur drei vorhanden, nämlich Nr. 77, 105, 133. Eine sachliche Einteilung liess sich bei der Beschaffenheit des Materials nicht durchführen. Für die Anordnung der im Kanjur befindlichen Schriften war naturgemäss die Reihenfolge derselben im Index des Kanjur massgebend. Unter Aussentitel ist die in die Mitte des ersten Blattes gesetzte Aufschrift zu verstehen, unter Innentitel der oder die das Werk zu Beginn des zweiten Blattes eröffnenden Titel, unter Randtitel die längs des linken Randes jeder Vorderseite quer geschriebenen Titel, unter Schlusstitel die vor dem Kolophon am Schluss des Werkes stehenden Titel. Innen- und Schlusstitel sind in der Regel identisch, Aussen- und Innentitel oft verschieden gefasst; in manchen Büchern weichen alle drei Titel von einander ab, manche besitzen nur Innentitel. Die Randtitel geben den Haupttitel verkürzt in seinen wesentlichen Stichwörtern wieder. Wo im Verzeichnis kein Randtitel angegeben, ist anzunehmen, dass derselbe fehlt. Die Kenntnis der Randtitel ist von grosser Wichtigkeit, da in der Litteratur gewöhnlich nach diesen citiert wird; daher habe ich dieselben auch alphabetisch in einem Index zusammengestellt. Schliesslich sei bemerkt, dass es sich bei diesem Verzeichnis nicht um eine offizielle Katalogisierung handelt, vielmehr die Arbeit meiner persönlichen Initiative entspringt. Die zahlreichen Mängel, die derselben anhaften, mögen ihre Entschuldigung in dem Umstande finden, dass ich nur wenige Tage in Dresden zubringen konnte. Der Verwaltung der Königlichen Bibliothek sei für die ausserordentliche Zuvorkommenheit, mit der mir sowohl ihre tibetischen als mongolischen Handschriftenschätze zur Verfügung gestellt wurden, auch an dieser Stelle der verbindlichste Dank ausgesprochen.

7*

1.

8 fol.

Aussentitel: dpan skoṅ [1]. pʻyag rgya [2]) pa bẑugs so. Rand-titel: dpaṅ skoṅ.

Schluss: bod du dam pai cʻos obyuṅ bai sña·ltas su lha tʻo tʻo ri sñan šal [3]) gyi sku riṅ la pʻo braṅ yum bu bla mkʻar [4]) du nam mkʻa las babs mi, rabs ldan odii don šes pa ooṅ ẑes rmi lam du luṅ bstan te cʻos kyi dbu brñes so || gcig ẑus.

„Als Vorzeichen der Entstehung der heiligen Religion in Tibet fiel zur Zeit des *Lha tʻo tʻo ri sñan šal* im Palaste *Yum bu bla mkʻar* diese Schrift vom Himmel herab, und indem ihm im Traume prophezeit wurde, dass die künftige Generation den Sinn derselben verstehen werde, erlangte der König den Anfang der Religion."

Vgl. über dieses Ereignis I. J. Schmidt, Sanang Setsen, p. 25—27, 319—320; E. Schlagintweit, Die Könige von Tibet, p. 837; Journal Asiatic Society of Bengal, vol. L, part I, 1881, p. 216, und vol. LI, part I, No. I, 1882, p. 2.

Die Schrift ist im Kanjur enthalten, s. K.—I. Nr. 266 (p. 43), obwohl sie in Csoma's Analyse nicht aufgeführt ist, mit über-einstimmendem Schluss. *Lha tʻo tʻo ri* ist angeblich der 27. in der Reihe der tibetischen Könige und soll 441—561 gelebt haben; das Ereignis, auf das oben angespielt wird, soll 521 stattgefunden haben.

2.

15 fol.

Aussentitel: klui spaṅ bskoṅ bẑugs so.

Randtitel: klui spa.

Innentitel: žaṅ žuṅ gi skad du | ta la pa ta ya na ha | sum pai skad du | ‚a ra na pa li ya | rgya gar skad du | nâgarâja-dhaya | bod skad du | klui dpaṅ po koṅ.

Über die Sprache von *Žaṅ žuṅ* s. Sitzungsberichte der Bayerischen Akademie 1898, Heft III, p. 590—592.

Das Land *Sum pa* ist erwähnt im *Grub mtʻa šel kyi me loṅ*, s. Journal Asiatic Soc. of Bengal, vol. LI, part I, No. I, 1882, p. 58, 66. Ebenda, vol. L, part I, p. 189, 196 wird ein Gelehrter *sPuṅs gsaṅ taṅ* aus dem Lande *Sum pa* als Anhänger der Bon-religion genannt. Unter dem ersten mythischen König *gÑa kʻri btsan po* soll aus diesem Lande die Bonreligion in Tibet eingeführt

---

1) Jäschke, Tibetan-English Dictionary, p. 329, liest *spaṅ skoṅ*; Schlagintweit, Könige von Tibet, fol. 15 a: *paṅ koṅ*.

2) K.-I. Nr. 266: *brgya*.

3) Über die verschiedenen Schreibweisen von *sñan šal* s. Schlagintweit, l. c., p. 837 no. 4; Huth, Geschichte des Buddhismus in der Mongolei II, 6: *gñan btsan*; Tāranātha II, 126: *gñan gtsan*.

4) Dies wird wohl die richtige Lesart sein statt *blaṅ gaṅ* des Bodhimör und *bla sgaṅ* des *rGyal rabs*.

worden sein, s. Proceedings of the Asiatic Soc. of Bengal 1892, No. 2, p. 90.

span bskon (Bedeutung nicht sicher) ist wohl mit *dpan skon* (s. Nr. 1) identisch.

In den Grundzügen stimmt dieser Text mit dem 1. und 3. Teil des von mir in den Mémoires de la Société Finno-Ougrienne XI veröffentlichten *Klu ʘbum bsdus pai sñiñ po* überein.

3.

50 fol.

Ohne Titel. fol. 48 b 4: bdud rtsi sñiñ po yan lag brgyad pa gsañ ba man ñag gi rgyud las span blañ mu bžir brtag pai leu ste ñi šu drug pao. Das 26. Kapitel des *Man ñag*, des 3. Teiles des medizinischen Werkes *rGyud bži*. Vgl. Heinrich L a u f e r , Beiträge zur Kenntnis der tibetischen Medicin, 1. Teil, Berlin 1900, S. 12.

4.

69 fol. Unvollständig.

bdud rtsi sñiñ po yan lag brgyad pa gsañ ba man ñag gi rgyud kyi tsʻig don pʻyin ci ma log par ʘgrel pa mes poi žal luñ žes bya ba las dum bu dañ po rtsa bai rgyud kyi rnam bšad bžugs so.

Randtitel: mes poi žal luñ.

„Erstes Stück der Erklärung des Wurzeltraktats (*rtsa bai rgyud*, d. i. der erste Teil des medizinischen Werkes *rGyud bži*) aus dem *Mes poi žal luñ* genannten, in den Wortbedeutungen des Unterweisungstraktates (*man ñag gi rgyud*, der dritte Teil des *rGyud bži*) untrüglichen Kommentars."

Schriften aus dem Kanjur,[1] Nr. 5—105.

5.

11 fol.

Skr. *śatasāhasrikaprajñāpāramitā.* T. oʻpʻags pa šes rab pʻa rol tu pʻyin pa stoñ pʻrag brgya[2] pai don mdor[3] par bsdus pa.

Schluss: de bžin gšegs pa tʻams cad kyi yum cʻen mo šes rab kyi pʻa rol pʻyin pa stoñ pʻrag brgya pai don mdor[3] par bsdus pa rdzogs so.

Verz. 102—117. K.-I. Nr. 8 (p. 2). Kurzer Auszug.

---

1) Verz. = Verzeichnis der tibetischen Handschriften und Holzdrucke im Asiatischen Museum der Akademie der Wissenschaften, von I. J. S c h m i d t und O. B ö h t l i n g k.

K.-I. = Der Index des Kanjur, her. v. d. Akademie der Wissenschaften und bevorwortet von I. J. S c h m i d t.

As. Res. = Asiatic Researches

2) brgyan.

3) mnor.

6.

5 fol.

Skr. *śatasāhasrikaprajñāpāramitā.* T. o p'ags pa šes rab kyi p'a rol tu p'yin pa stoṅ p'rag brgya pai don mdor du bsdus pa. Randtitel: o bum c'uṅ.

Schluss: jo bo rjes bal po ‚*A su la snaṅ* | bal pos šo lo kar bkod pa rdzogs so | jo bo rje dpal ldan ‚*Atišas* lo bcu gñis kyi bar du gsuṅ rab rnams gzigs pas | o bum c'uṅ o di k'o na p'an yon šin tu c'e bar o dug pas | p'yi rabs kyi gaṅ zag rnams kyis kyaṅ | o di la klog don byed pa gal c'e gsuṅs so.

Kurzer Auszug aus K.-I. Nr. 8 (p. 2).

„Der ehrwürdige Herr, der Nepalese *Asula snaṅ* hat diese Schrift in nepalesischen Çloka verfasst. Der ehrwürdige Herr, *śrīmant Atīśa,* hat nach zwölfjähriger Prüfung der heiligen Schriften eben dieses o *Bum c'uṅ* (das kleine Hunderttausend, im Gegensatz zu der grossen Vorlage, o *Bum c'en*) wegen seines ausserordentlichen Segens auch den Menschen der künftigen Geschlechter zur nutzbringenden Lektüre angelegentlichst empfohlen.“

7.

28 fol.

Aussentitel: o p'ags pa sdud pa bžugs so.

Innentitel: Skr. *āryaprajñāpāramitāsañcayagāthā.* T. o p'ags pa šes rab kyi p'a rol tu p'yin pa sdud pa ts'igs su bcad pa.

Schluss: Anführung des Innentitels: šes rab kyi p'a rol tu p'yin pa k'ri brgyad stoṅ pa las p'yuṅ bai o p'ags pa sdud pa ts'ig leur bcad pa o di | slob dpon Seṅge bzaṅ pos žu dag mdzad pas rgya dpe daṅ | bod dpe dag pa la gtsugs nas | ža lu lo tsā ba dge sloṅ *Dharmapâlabhadras* slar yaṅ dag par byas pao.

Innen- und Schlusstitel stimmen mit K.-I. Nr. 13 (p. 3) überein. Das Kolophon giebt indessen an, dass es ein versifizierter aus dem *Aṣṭadaśasāhāsrikaprajñāpāramitānāmamahāyānasūtra* (K—I. Nr. 10) entnommener Auszug ist, der von dem *Ācārya Siṁhabhadra* verbessert und von dem Übersetzer von *Žalu,* dem Bhikṣu *Dharmapālabhadra* mit Zugrundelegung der reinen indischen und tibetischen Bücher noch einmal gereinigt wurde.

8.

31 fol.

Skr. *āryaprajñāpāramitāsañcayagāthā.* T. o p'ags pa šes rab kyi p'a rol tu p'yin pa sdud pa ts'igs su bcad pa. Randtitel: sdud pa.

Schluss: šer gyi p'a rol tu p'yin pa k'ri brgyad stoṅ pa las p'yuṅ bai o p'ags pa sdud pa ts'ig leur bcad pa o di | slob dpon Seṅ ge bzaṅ pos žu dag mdzad pas rgya dpe | bod dpe dag pa la gtsugs nas ‖ ‖ ža lu lo tsā ba dge sloṅ *Dharmapālabhadras* slar yaṅ dag par byas pa.

Dasselbe Werk wie das vorhergehende.

9.

4 fol.

Skr. *āryaprajñāpāramitānāma aṣṭaśatakam.* T. o p'ags pa
p'a rol tu p'yin pai mts'an brgya rtsa brgyad pa gzuṅs sṅags daṅ
bcas pa bẓugs. ·Randtitel: yum mts'an.

K.-I. Nr. 25 (p. 4), Nr. 553 (p. 81).

10.

Fragment: fol. 1 fehlt. fol. 2—5.

Randtitel: kou. ·Schluss: yum c'en mo šes rab kyi p'a rol tu
p'yin pai bšags pa mdo rdzogs so | ẓus dag | *mi̇galam* (sic! statt
maṅgalam) *astu.*

11.

46 fol.

Aussentitel: rdo rje gcod pa bẓugs so.

Innentitel: Skr. *āryavajracchedikāpāramitānāmamahāyāna-
sūtra.* T. o p'ags pa šes rab kyi p'a rol tu p'yin pa rdo rje gcod
pa ẓes bya ba t'eg pa c'en poi mdo.

fol. 45: Skr. *śatasahāsrikaprajñāpāramitāgarbha.* T. šes rab
kyi p'a rol tu p'yin pa stoṅ p'rag brgya pai sñiṅ po. Schluss:
sṅags de brjod pa šes rab kyi p'a rol tu p'yin pa o bum ston pa
daṅ mñam mo -ᴼᴼ- yum c'en mo šes rab kyi p'a rol tu p'yin pai
sñiṅ po rdzogs so.

K.-I. Nr. 16 (p. 3). Vergl. die Einleitung zu Max Müller's
Ausgabe des Sanskrittextes in Anecdota Oxoniensia, Aryan Series,
vol. I, part. I, Oxf. 1881.

12.

Holzdruck. 71 fol. Unvollständig.

Skr. *āryavajracchedikāprajñāpāramitānāmamahāyānasūtra.*
T. a) Aussentitel: rdo rje gcod pa bẓugs so. b) Innentitel: o p'ags
pa šes rab kyi p'a rol tu p'yin pa rdo rje gcod pa ẓes bya ba t'eg
pa c'en poi mdo.

Auf fol. 1 links ein Bild des *T'ub pa dbaṅ po,* d. i. *Śākyamuni,* rechts Darstellung der *Šes rab p'ar p'yin ma,* d. i. *Para-
mitâ,* „Göttin der transcendentalen Weisheit" (s. Pantheon S. 78
Nr. 158). Auf fol. 2 ist links *Kun dya pa,* d. i. *Ananda* und
rechts *gNas brtan rab o byor,* d. i. der *Sthavira Subhūti* (s. Grünwedel, Mythologie des Buddhismus, S. 188—190) abgebildet.

13.

20 fol.

Skr. *āryavajracchedikūprajñāpāramitānāmamahāyānasūtra.*
T. o p'ags pa šes rab kyi p'a rol tu p'yin pa rdo rje gcod pa ẓes
bya ba t'eg pa c'en poi mdo.

Randtitel: rdor gcod.

K.-I. Nr. 16 (p. 3).

14.

38 fol.

Vajracchedikā.

Schluss: rdo rje gcod pai sñiñ po o di lan cig bzlas pas rdo rje gcod pa k'ri dgu stoñ bklags pa dañ mñam par o gyur ro.

Die Namen Buddhas uud buddhistischer Termini wie *šam t'abs, lhuñ bzed, dge sloñ* u...a. sind rot geschrieben.

15.

Fragment. Einzelne Blätter in falscher Reihenfolge geheftet. Schwarzes Papier mit gelber Schrift, die auf den letzten Blättern fast bis zur Unleserlichkeit verblasst ist.

Vajracchedikā.

16.

55 fol. Schwarzes Papier mit weisser, stark verblasster Schrift.

Vajracchedikā.

Nr 17—27: *Vajracchedikā.*

17.

54 fol. fol. 1 fehlt. Blätter nicht numeriert. Schwarzes Papier mit grün bemaltem Rand und Goldschrift.

rdo rje gcod pa bžugs so.

18.

53 fol. Nicht numeriert. Unvollständig.

19.

38 fol. Kleines Format, kleine Schrift.

20.

35 fol. Vorhanden fol. 1—14, 25—39, 44—45, 49, 52—54.

21.

44 fol. Unvollständig. fol. 44 fehlt.

22.

55 fol. Nicht numeriert.

23.

50 fol. Vorhanden fol. 1—37, 44—49.

24.

48 fol. Unvollständig.

25.

47 fol. Mittlere Zeile rot geschrieben.

26.

44 fol.

27.

53 fol. Schluss: rdo rje gcod pai sñiṅ po o di lan cig bzlas pas rdo rje gcod pa k'ri [1] dgu stoṅ bklags pa daṅ mñam par o gyur ro. Vgl. Nr. 14.

28.

3 fol.

Aussentitel: ñi ma daṅ zla bai mdo bžugs so.

Innentitel: Skr. *sūryasūtra*. T. ñi mai mdo.

Schluss: zla bai (!) mdo rdzogs so | paṇḍita c'en po Ânan-da*śrî-i* žal sña nas | maṅ du t'os pai lo tsts'a ba Śâkyai dge sloṅ *Ñi ma rgyal mts'an dpal bzaṅ pos* | skad gñis smra ba rnams kyi gdan sa | gtsug lag k'aṅ c'en po *dpal t'ar pa gliṅ* du bsgyur ciṅ žus te gtan la p'ab pao.

K.-I. Nr. 41 (p. 6): ñi mai mdo, Nr. 42 (p. 6): zla bai mdo.

29—30.

7 fol.

Skr. *āryamaitriyapariprcchadharma aṣṭanāmamahāyānasūtra.* T. o p'ags pa byams pas žus pa c'os brgyad pa žes bya ba t'eg pa c'en poi mdo. Randtitel: byams žus.

Schluss: fol. 6 b 3 o p'ags pa dkon mc'og brtsegs pa c'en poi c'os kyi rnam graṅs stoṅ p'rag brgya pa las | byams pas žus pa c'os brgyad pa žes bya bai leu ste | o duṅ pa bži bcu rtsa gñis pa rdzogs so || || śloka brgyad cu rtsa gsum mc'is | rgya gar gyi mk'an po *Jinamitra* daṅ | *Dânaśila* daṅ | žu c'en gyi lo tsts'a ba *Bande Ye śes sdes* bsgyur ciṅ žus te skad gsar c'ad kyis kyaṅ bcos nas gtan la p'ab pao.

K.-I. Nr. 86 (p. 14) mit übereinstimmendem Kolophon; es fehlt hier aber im Titel *dharma aṣṭa* (= c'os brgyad pa), ein Zusatz, der sich indessen auch bei Csoma (As. Res. XX 411, Nr. 37) findet.

fol. 6 b 6: Skr. *āryasāgaranāgarājapariprcchanāmamahāyā-nasūtra.* T. o p'ags pa klui [rgyal po] rgya mts'os žus pa žes bya ba t'eg pa c'en poi mdo. Randtitel: klui žus.

Schluss fol. 7 b 3: rgya gar gyi mk'an po *Surendrabodhi* daṅ žu c'en gyi lo tsts'a ba *Bande ye śes sdes* bsgyur ciṅ žus te gtan la p'ab pao.

K.-I. Nr. 155 (p. 26).

1) k'ro.

31

2 fol.

Skr. *ārya ātajñānanāmamahāyānasūtra.* T. o p'ags pa o da
ka ye šes žes bya ba t'eg pa c'en poi mdo. Randtitel: mda ka.
K.-I. Nr. 122 (p. 20).

32.

172 fol.

Skr. *āryamahāsaṁnipūtaratnaketudhāraṇīmahāyānasūtra.* T.
o p'ags pa o dus pa c'en po rin po c'e tog gi gzuñs žes bya ba t'eg
pa c'en poi mdo. Randtitel: tog gzuñs.

Schluss: o dus pa c'en po rin po c'e tog gi gzuñs žes bya ba
las | mt'ar p'yin pai leu ste bcu gsum pao ‖ ‖ o p'ags pa o dus
pa c'en po rin po c'e tog gi gzuñs žes bya ba t'egs pa c'en poi
mdo rdzogs so | | rgya gar gyi mk'an po *Silendrabodhi* dañ |
Jinamitra dañ | žu[1]) c'en po gyi lo tsts'a ba *Ban dhe*[2]) *Ye šes
sdes* žus te skad gsar bcad kyis kyañ bcos nas gtan la p'ab pao.
K.-I. Nr. 138 (p. 23), wo im Kolophon *Jinamitra* fehlt.

33.

3 fol.

bdud gžom pa sogs sñags le ts'an brgyad bžugs so. Randtitel: bdud gžom sogs.

Anfang: o p'ags pa blo gros rgya mts'os žus pai mdo las byuñ
bai bdud gžom pai sñags ni. Ohne Kolophon.

o p'ags pa blo gros rgya mts'os žus pa žes bya ba t'eg pa c'en
poi mdo = K.-I. Nr. 152 (p. 25).

34.

12 fol.

Skr. *āryaratnacandrapariprcchanāmamahāyānasūtra.* T.
o p'ags pa (k'yeu) rin c'en zla bas žus pa žes bya ba t'eg pa c'en
poi mdo. Randtitel: rin c'en zla ba.

Schluss: rgya gar gyi mk'an po *Viśuddhasiṁha* dañ lo tsā
ba *Bande dGe dpal* gyis bsgyur | rgya gar gyi mk'an po *Vidyâ-
karasiṁha* dañ | žu c'en gyi lo tsā ba *Bande Devacandras* žus te
gtan la p'ab pa.

K.-I. Nr. 164 (p. 28).

35.

6 fol.

Skr. *āryamahallikāpariprcchanāmamahāyānasūtra.* T. o p'ags
pa bgres mos žus pa žes bya ba t'eg pa c'en poi mdo.

Schluss: rgya gar gyi mk'an po *Jinamitra* dañ | *Dânaśila*
dañ | žu c'en gyi lo ca ba *Bande Ye šes sdes* bsgyur ciñ žus te
gtan la p'ab pa.

K.-I. Nr. 171 (p. 29).

---

1) *bžu.* 2) *sde!*

36.

24 fol.

Skr. *āryamahāmegha.* T. o p'ags pa sprin c'en poo.

Schluss: o p'ags pa sprin c'en po t'eg pa c'en poi mdo las c'ar dbań po rluń gi dkyil o k'or gyi leu żes bya ba | drug cu rtsa bżi pa ts'og dań bcas pa rdzogs so.

K.-I. Nr. 234 (p. 39).

37.

31 fol.

Skr. *ārya ākāśagarbhanāmamahāyānasūtra.* T. o p'ags pa nam mk'ai sñiń po żes bya ba t'eg pa c'en poi mdo. Randtitel: nam mk'ai sñiń po.

K.-I. Nr. 259 (p. 43).

38.

23 fol. fol. 16—19 fehlen.

Skr. *āryadaśadigandhakāravidhvaṃsana[1])nāmamahāyānasū-tra.* T. o p'ags pa p'yogs bcui mun pa rnam par sel ba żes bya ba ni t'eg pa c'en poi mdo.

Schluss: rgya gar gyi mk'an po *Viśuddhasiṃha* dań | lo tsts'a ba *Bande rtsańs de Bendrarakṣita*-s [2]) bsgyur [3]) | żu c'en gyi lo tsts'a ba *Bande Klui rgyal mts'an* gyis skad gsar bcad kyis bcos te [4]) gtan la p'ab pao.

K.-I. Nr. 268 (p. 44).

39.

7 fol.

Skr. *Kūṭāgārasūtra.* T. K'ań bu brtsegs pai mdo. Rand-titel: brtsegs.

K.-I. Nr. 330 (p. 51).

40.

14 fol.

Skr. *āryagośṛṅgavyākaraṇanāmamahāyānasūtra.* T. o p'ags pa glań ru luń bstan żes bya ba t'eg pa c'en poi mdo.

K.-I. Nr. 355 (p. 53).

41.

2 fol.

Skr. *bhagavatīprajñāpāramitāhṛdaya.* T. bcom ldan o das ma śes rab kyi p'a rol tu p'yin pai sñiń po. Randtitel: śer sñiń.

K.-I. Nr. 525 (p. 79), Nr. 21 (p. 4). Verz. 214—216.

---

1) *da gi 'â na dha kâ ra bi dhva na sa na.*

2) *ban dre rakṣa tas.*

3) *bsbyuńs* (sic!).

4) Statt: *skad gsar bcad kyis bcos te* im K.-I.: *żu c'en bgyis te.*

42—46.

10 fol.

Nur Aussentitel: sañs rgyas bcom ldan o das kyi mts'an brgya
rtsa brgyad pa gzuñs sñags dañ bcas pa bžugs so. Randtitel:
ston pai mts'an.

fol. 8 b 5: o p'ags pa šākya t'ub pai sñiñ poi gzuñs rdzogs so.

fol. 9 a 4: o p'ags pa rnam par snañ mdzad kyi sñiñ po žes
bya bai gzuñs rdzogs so.

fol. 9 b 5: zla bai o od kyi mts'an rjes su dran pa rdzogs so.

fol. 10 a 1: sañs rgyas rin c'en gtsug tor can gyi mts'an rjes
su dran pa rdzogs so.

K.-I. Nr. 526—529, 531 (p. 79), Nr. 848, 836, 837, 844,
845 (p. 111).

47.

115 fol.

Skr. *āryasuvarṇaprabhāsottamasūtrendrarājanāmamahāyā-*
nasūtra. T. o p'ags pa gser o od dam pa mdo sdei dbañ poi rgyal
po žes bya ba t'eg pa c'en poi mdo | bam po dañ po. Ohne
Randtitel.

Schluss: o p'ags pa dam pa mdo sdei dbañ poi rgyal po las
bsdus pa leu žes bya ba ste ñi žu gcig pao.

K.-I. Nr. 556 (p. 81), Nr. 557 (p. 82). Verz. Nr. 244—248
(p. 9).

48.

12 fol.

Skr! *āryasuvarṇasatanāmamahāyānasūtra.* T. a) Aussentitel:
gser o od ryañ skyab žes bya ba bžugs so. b) Innentitel: o p'ags
pa gser o od dam pa mdo sdei dbañ poi rgyal po las yañ skyabs
žes bya ba. Vergl. Nr. 47.

K.-I. Nr. 556 (p. 81), mit Kolophon, das hier fehlt, und
Nr. 557 (p. 82).

49.

7 fol.

Aussentitel: nor lha gser o od bžugs só. Innentitel: Skr.
āryaratnadhāraṇī (entspricht nicht den tibetischen Titeln). T.
o p'ags pa gser o od dam pa mdo sdei dbañ poi rgyal po las | nor
p'yugs skyoñ žiñ spel ba žes bya bai gzuñs. Randtitel: nor lha.

Schluss: opags pa gser o od dam pa mdo sdei dbañ poi rgyal
po las | ts'e rabs t'ams cad du yo byad p'un sum ts'ogs pai leu
ste bcu bdun pa rdzogs so. Danach wird es sich wohl um das
17. Kapitel von K.-I. Nr. 556 (bezw. Nr. 557, p. 81, 82) handeln,
nach C s o m a 29 Kapitel umfassend (As. Res. XX, 515).

50.

3 fol.

Skr. *āryamārīcīnāmadhāraṇī.* T. o p'ags pa o od zer can žes
bya ba gzuñs. Randtitel: o od zer.

Schluss: paṇḍita *Amogha*[1])*vajra* daṅ | lo tsâ ba dge sloṅ
Rin c'en grags pas bsgyur bao.
K.-I. Nr. 564 (p. 82), Nr. 961 (p. 124).

51.

8 fol.
Skr. *āryajayavatīnāmadhāraṇī.* T. op'ags pa rgyal ba can
žes bya bai gzuṅs. Randtitel: rgyal ba can.
K.-I. Nr. 567 (p. 83), Nr. 977 (p. 126).

52—57.

13 fol.
Skr. *āryahiraṇyavatīnāmadhāraṇī.* T. op'ags pa dbyig daṅ
ldan pa žes bya bai gzuṅs. Randtitel: dbyig ldan.
Schluss: fol. 4 b 1: rgya gar gyi mk'an po *Jinamitra* daṅ |
Dānaśila daṅ | žu c'en gyi lo tsts'a ba *Bande Ye šes sdes* bsgyur
ciṅ žus te skad gsar bcad kyis kyaṅ bcos nas gtan la p'ab pa.
K.-I. Nr. 570 (p. 83). Nr. 964 (p. 124).

fol. 4 b 2: Skr. *jaṅgulīnāmavidyā.* T. op'ags pa dug sel pa
žes bya bai rig sṅags. Randtitel: dug sel.
K.-I. Nr. 571 (p. 83), Nr. 963 (p. 124).

fol. 6 a: Skr. *siddhapaṭhita*[2])*bhagavatī*[3]) *ārya aṅgulīnāma-
vidyārājñī*[4]). T. bklags pas grub pa bcom ldan o das ma op'ags
ma sor mo can žes bya ba rig pai rgyal mo. Randtitel: bklags grub.
K.-I. Nr. 572 (p. 84), Nr. 966 (p. 125).

fol. 8 b 1: Skr. *āryasarvadharmamātṛkānāmadhāraṇī.* T.
op'ags pa c'os t'ams cad kyi yum žes bya bai gzuṅs. Randtitel:
c'os yum.
K.-I. Nr. 573 (p. 84), Nr. 969 (p. 125).

fol. 9 a 2: *āryacūḍāmaṇīnāmadhāraṇī.* T. op'ags pa gtsug
gis nor bu žes bya bai gzuṅs. Randtitel: gtsug nor. Schluss:
fol. 11 b 7: rgya gar gyi mk'an po *Śilendrabodhi* daṅ | žu c'en
gyi lo tsts'a ba *Bandhe Ye šes sdes* bsgyur ciṅ žus te gtan la
p'ab pa.
K.-I. Nr. 574 (p. 84), Nr. 897 (p. 117).

fol. 12 a 1: Skr. *āryaṣaḍakṣaravidyā.* T. op'ags pa yi ge
drug pa žes bya bai rig sṅags. Randtitel: yi ge drug.
K.-I. Nr. 575 (p. 84), Nr. 892 (p. 117).

58.

19. fol.
Skr. *śatasāhasrikaprajñāpāramitā.* T. Aussentitel fehlt. An-
fang: šes rab p'a rol tu p'yin pa p'yag o ts'al lo | 'oṁ mu ni mu
ni dharma.

---

1) *,a mo sta.*   2) *pyriti.*   3) *bhagavâna.*   4) *rāñjai.*

Schluss: šes rab kyi p'a rol tu p'yin pai gzuñs rdzogs šo || bsod
nams rin po bžin du brtan pa dañ ⌐ rgyud ni ñi zla bžin du gsal
pa dañ | sñan pa nam mk'a bžin du k'yab dañ | gsum po des kyañ
diñ odir bkra šis šog | dgeo | legso | bkra šis par ogyur cig.

Der Skr.-Titel stimmt mit dem tibetischen Schlusstitel nicht
überein. Es wird sich wohl um K.-I. Nr. 578 (p. 84), Nr. 907
(p. 119) handeln.

59.

3 fol.

Aussentitel: gtsug tor rnam rgyal gyi gzuñs mdo bžugs so.
Innentitel: Skr. *sarvatathāgata uṣṇiṣavijayanāmadhāraṇikalpa-*
sahita. T. de bžin gšegs pa t'ams cad kyi gtsug tor rnam par
rgyal ba žes bya bai gzuñs rtog pa dañ bcas pa. Randtitel: rnam
rgyal. Schluss: op'ags pa gtsug tor rnam par rgyal bai gzuñs
rtogs pa dañ bcas pa rdzogs so.

K.-I. Nr. 593—595 (p. 86).

60.

9 fol.

Skr. *āryasarvadurgatipariśodhanī uṣṇiṣavijayanāmadhāraṇī.*
T. ñan ogro t'ams cad yoñs su sbyoñ ba gtsug tor rnam par rgyal
ba žes bya bai gzuñs. Randtitel: ñan sbyoñ gtsug tor.

K.-I. Nr. 596 (p. 86), Nr. 957 (p. 124).

61—62.

12 fol.

Aussentitel: gdugs dkar mc'og grub bžugs so.
Innentitel: Skr. *āryatathāgatoṣṇiṣasitātapatrā aparājitamahā-*
pratyañgiraparamasiddhanāmadhāraṇi. T. op'ags pa de bžin
gšegs pai gtsug tor nas byuñ bai gdugs dkar po can gžan gyis mi
t'ub pa p'yir zlog pa c'en po mc'og tu grub pa žes bya bai gzuñs.
Randtitel: gdugs· dkar. fol. 11 b 7: Schluss wie Innentitel.

K.-I. Nr. 591 (p. 85), Nr. 959 (p. 124).

fol. 12 a: Skr. *āryoṣṇiṣajvalanāmadhāraṇi*. T. op'ags pa
gtsug tor obar ba žes bya bai gzuñs.

K.-I. Nr. 599 (p. 87), Nr. 935 (p. 121).

63—64.

3 fol.

Skr. *āryasarva antarāyaviśodhanināmadhāraṇi*. T. op'ags
pa bar du gcod pa t'ams cad rnam par sbyoñ ba žes bya bai gzuñs.
Randtitel: bar du gcod pa.

K.-I. Nr. 607 (p. 87), Nr. 901 (p. 118).

fol. 3 a 3: Skr.: *āryamaṇibhadranāmadhāraṇi*. T. op'ags pa

nor bu bzañ poi gzuñs žes bya ba. Doch mehr als diese Titel-
angabe nicht vorhanden.

K.-I. Nr. 759 (p. 104), Nr. 943 (p. 122).

65.

4 fol.

Skr. *āryadhvaja agrakeyūranāmadhāraṇī.* T. o p'ags pa rgyal
mts'an rtse moi dpuñ rgyan ces bya bai gzuñs. Randtitel: rgyal
mts'an. Schluss: rgya gar gyi mk'an po *Jinamitra* dañ | *Dâna-
śila* dañ | žu c'en gyi lo tsts'a ba *Bande Ye śes dsdses* bsgyur te |
skad gsar c'ad kyis bcos te gtan la p'ab pao.

K.-I. Nr. 611 (p. 88), Nr. 885 (p. 116).

66—70.

6 fol.

Skr. *āryacakṣuviśodhanināmavidyā-[mantra].* T. o p'ags pa
mig rnam par sbyoñ ba žes bya bai rig sñags. Randtitel: mig
rnam par spyod.

. K.-I. Nr. 618 (p. 89), Nr. 981 (p. 126).

fol. 3 b 6: Skr. *ārya akṣirogapraśamaṇasūtra.* T. o p'ags
pa mig nad rab tu ži bar byed pai mdo.

K.-I. Nr. 619 (p. 89).

fol. 4 b 1: Ohne Skr.-Titel. dkon mc'og gsum la p'yag o ts'al
lo | kṣayai nad sel bai sñags | lus la nad byuñ na | nad k'oñ skems
kyis btab na | bsil yab c'us gtor te | lan drug cu rtsa gcig sñags
nas | dei lus la γyabs na nad med par o gyur ro | kṣayai nad sel
bai gzuñs rdzogs .so.

K.-I. Nr. 796 (p. 106), Nr. 1031 (p. 129).

fol. 4 b 3: Skr. *ārya arśapraśamaṇisūtra.* T. o p'ags pa
gžañ o brum rab tu ži bar byed pai mdo.

fol. 6 a 3: rgya gar gyi mk'an po *Jinamitra* dañ | *Dânaśila*
dañ | žu c'en gyi lo tsts'a ba *Bande Ye śes sdes* bsgyur ciñ gtan
la p'ab pa.

K.-I. Nr. 620 (p. 89), Nr. 993 (p. 127).

fol. 6 a 4: Skr. *āryajvarapraśamaṇināmadhāraṇī.* T. o p'ags
pa rims nad rab tu ži bar byed pa žes bya bai gzuñs.

K.-I. Nr. 624 (p. 89), Nr. 989 (p. 127). Das an diesen beiden
Stellen gegebene Kolophon, identisch mit dem vorhergehenden
fehlt hier.

71.

8 fol.

Skr. *āryavaiśāli[1])praveśamahāsūtra.* T. o p'ags pa yañs pai
groñ k'yer o jug pai mdo c'en po. Randtitel: yañs pa.

Schluss: rgya gar gyi mk'an po *Surendrabodhi* dañ | žu c'en

---

1) *bi bu le!*

gyi lo tsts'a ba *Bandhe Ye šes dsdses* bsgyur ciñ žus te gtan la p'ab pao.

K.-I. Nr. 627 (p. 90) mit übereinstimmendem Kolophon, Nr. 1067 (p. 132).

72—73.

3 fol.

Skr. *āryacauravidhvaṁsananāmadhāraṇī*. T. o p'ags pa mi rgod rnam par o joms pa žes bya ba gzuñs. Randtitel: mi rgod.

K.-I. Nr. 628 (p. 90), Nr. 934 (p. 121).

fol. 2 b 5: Skr. *āryasarva antarasaṁgrāsadhāraṇī*. T. o p'ags pa bar du gcod pa t'ams cad sel bai gzuñs sñags. Am Schluss: gsum žuso.

K.-I. Nr. 629 (p. 90), Nr. 983 (p. 126).

74.

4 fol.

T. byams pai mts'an brgya rtsa brgyad pa gzuñs sñags dañ bcas pa bžugs so. Randtitel: byams pai mts'an.

Schluss: o p'ags pa šes rab kyi p'a rol tu p'yin pai mts'an brgya rtsa brgyad pa rdzogs so.

K.-I. Nr. 634 (p. 91), Nr. 850 (p. 112).

75.

4 fol.

Ohne Skr.-Titel. T. o p'ags pa sai sñiñ po mts'an brgya rtsa brgyad pa gzuñs sñags dañ bcas pa. Randtitel: sa sñiñ mts'an brgya.

K.-I. Nr. 640 (p. 91), Nr. 856 (p. 112).

76.

2 fol.

Skr. *āryamaitripratijñānāmadhāraṇī*. T. o p'ags pa byams pas dam bcas pa žes bya bai gzuñs.

Randtitel: byams pa dam bcas.

Schluss: o p'ags pa o jam dpal gyi šes rab dañ blo o p'el žes bya bai gzuñs rdzogs so.

K.-I. Nr. 642 (p. 91), Nr. 865 (p. 113).

77.

Holzdruck. 6 fol.

Skr. *āryamaitripratijñānāmadhāraṇī*. T. o p'ags pa byams pas dam bcas pa žes bya bai gzuñs bžugs so. Schluss ebenso.

K.-I. Nr. 642 (p. 91).

78.

2 fol.

Skr. *āryavighnavināyakaratādhāraṇī*. T. o p'ags pa bgegs sel bai gzuñs.

K.-I. Nr. 654 (p. 93), Nr. 932 (p. 121).

79.

5 fol.

Skr. *grahamātṛkānāmadhāraṇī.* T. gza rnams kyi yum bžugs so.
Schluss: gza t'ams cad la mc'od pa byas par o gyur ro. Randtitel: gza yum.

K.-I. Nr. 659 (p. 93), Nr. 660, Nr. 970, Nr. 971 (p. 125).

80.

7 fol.

Aussentitel: o p'ags pa nor gyi rgyun žes bya ba k'yim bdag
zla ba bzaṅ pos žus pa bžugs so.
Innentitel: Skr. *āryavasudhāranāmadhāraṇī.* T. o p'ags pa
·nor gyi rgyun žes bya bai gzuṅs.
Randtitel: nor rgyun.

K.-I. Nr. 661 (p. 93), Nr. 980 (p. 126).

81.

3 fol.

Skr. *āryagaṇapatihṛdaya.* T. o p'ags pa ts'ogs kyi bdag poi
gzuṅs bžugs so. Randtitel: ts'ogs bdag. Schluss: gaṇapatii gzuṅs
rdzogs so. — *hṛdaya* müsste tib. *sñiṅ po* entsprechen wie:

K.-I. Nr. 1058 (p. 132), Nr. 664 (p. 94).

82.

9 fol.

Skr. *āryāparimitā āyurjñānanāmamahāyānasūtra.* T. o p'ags
pa ts'e daṅ ye šes dpag tu med pa žes bya ba t'eg pa c'en poi
mdo. Randtitel: ts'e mdo.

Schluss: de la o gyur k'yed o bran bu yoṅ gi o dug (?) na o aṅ |
o dir rje btsun t'ams cad mk'yen pa *Tāranāthai* žal sṅa nas | ts'e
daṅ ye šes dpag tu med pai mdo la ṭīkā mdzad pai dgoṅs pa daṅ
mt'un pa ñid *dpal dga ldan p'un ts'ogs gliṅ* du par tu bsgrubs
pa lags so || o di la brten nas bdag gžan skye dgu mt'a dag o c'i
ba med pa ts'ei dpal la dbaṅ t'ob par gyur cig. Der ehrwürdige
allwissende *Tāranātha* hat zu diesem Sūtra einen Kommentar (*ṭīkā*)
verfasst; in Übereinstimmung mit dessen Auslegung wurde das Werk
in *dpal dGa ldan p'un ts'ogs gliṅ* gedruckt.

K.-I. Nr. 673 (p. 94), Nr. 674 (p. 95), Nr. 825 (p. 109).

83.

4 fol.

Aussentitel: o c'i med rṅa sgra žes bya bai gzuṅs mdo bžugs
so. Innentitel: Skr. *ārya aparimitāyurjñānahṛdayanāmadhāraṇi.*
T. o p'ags pa ts'e daṅ ye šes dpag tu med pai sñin po žes bya bai
gzuṅs. Randtitel: rṅa sgra.

Schluss: Anführung des Innentitels: rgya gar gyi mk'an po

Puṇyasámbhava daṅ | žu c'en gyi lo tsâ ba *Ba ts'ab ñi ma grags* kyis bsgyur bao.

K.-I. Nr. 675 (p. 95), Nr. 826 (p. 109).

84.

4 fol.

fol. 1 fehlt.   fol. 2. Randtitel: don žags.

Schluss: o p'ags pa spyan ras gzigs don yod žags pai sñiṅ po žes bya bai gzuṅs rdzogs so.   Vergl. K.-I. Nr. 682 (p. 96): *ārya amoghapāṣahṛdayaṁ mahāyānanāmadhāraṇī*, die nach Csoma (As. Res. XX 535, Nr. 1) von Avalokiteśvara verkündet wird.

85.

3 fol.

Skr. *ārya avalokiteśvaranāmadhāraṇī.*   T. o p'ags pa spyan ras gzigs dbaṅ p'yug gi gzuṅs.   Randtitel: spyan ras gzigs.

Schluss: fol. 2 a 7.   fol. 2 b 4: o p'ags pa spyan ras gzigs kyi sñiṅ po rdzogs so.   fol. 3 a 1: seṅge sgrai gzuṅs rdzogs so.   Randtitel: seṅge sgra.

Schluss: rgya gar gyi mk'an po *Nag gi dbaṅ p'yug* daṅ | *Klog skya šes rab brtsegs* kyis bsgyur bao.

K.-I. Nr. 692 (p. 97), Nr. 885 (p. 116); Nr. 691 (p. 97), Nr. 886 (p. 116); Nr. 700 (p. 98).

86.

5 fol.

Skr. *āryasamantabhadranāmadhāraṇī.*   T. o p'ags pa kun tu bzaṅ po žes bya bai gzuṅs.   Randtitel: kun tu bzaṅ po.

Schluss: rgya gar gyi mk'an po *Jinamitra* daṅ | *Dānaśīla* daṅ | žu c'en gyi lo tsts'a ba *Bandhe Ye šes sdes* bsgyur ciṅ žus te skad gsar c'ad kyis kyaṅ bcos nas gtan la p'ab pao.

K.-I. Nr. 695 (p. 97), Nr. 879 (p. 115).

87.

4 fol.

Skr. *ārya abhayapradanāma aparājita.*   T. o p'ags pa gžan gyis mi t'ub pa mi o jigs pa sbyin pa žes bya ba.   Randtitel: mi t'ub pa.

Schluss: rgya gar gyi mk'an po *Prajñāvarma* daṅ | žu c'en gyi lo tsts'a ba *Bandhe Ye šes sdes* la sogs pas bsgyur ciṅ žus te gtan la p'ab pa.   Mit kleiner Schrift: stoṅ o gyur ces bya bai gzuṅs rdzogs so.

K.-I. Nr. 704 (p. 98), Nr. 903 (p. 118); Nr. 706 (p. 98), Nr. 905 (p. 118).

88—89.

4 fol.

śloka brgya lobs pa sogs gzuṅs sna ts'ogs bžugs so.   Ohne Innentitel.   Randtitel: śloka.

gdoñ o k'ru bai ts'e c'u k'yor gañ la sñags o di lan gsum mam
bdun bzlas te o t'uñ na | ñin gcig la yi ge śloka brgya lobs par
o gyur te | sñon lobs pa rnams kyañ brjed par mi o gyur ro | śloka
brgya lobs pa rdzogs so. „Wenn man beim Waschen des Gesichts
eine Handvoll Wasser unter drei - oder siebenmaligem Hersagen
dieses Mantra trinkt, wird man an einem Tage hundert geschriebene
Sloka lernen und auch das früher Gelernte nicht vergessen."

K.-I. Nr. 707, 708 (p. 98).

fol. 1, 4: o p'ags pa śes rab kyi p'a rol tu p'yin pa stoñ p'rag
ñi śu lña pai gzuñs. Schluss: fol. 3 a 4: p'a rol tu p'yin pa drug
bzuñ bar o gyur bai gzuñs rdzogs so. Randtitel fol. 2 a: stoñ p'rag
brgya. pa, 3 a: p'an p'yin drug sogs.

fol. 4 a 1: o p'ags pa sdoñ po bkod pai sñiñ po rdzogs so.
fol. 4 a 4: o p'ags pa ma so sor o brañ ma c'en mo bzuñ bar o gyur
bai gzuñs rdzogs so. Randtitel: tiñe o dzin sogs. o p'ags pa lañ
kar gśegs pa la p'yag o ts'al lo. fol. 4 b: o p'ags pa lañ kar gśegs
pai mdo t'ams cad bklags par o gyur bai gzuñs sñags rdzogs so.

K.-I. Nr. 577 (p. 84), Nr. 908 (p. 119); Nr. 585 (p. 85), Nr. 915
(p. 119); Nr. 588 (p. 85), Nr. 917 (p. 119); Nr. 586 (p. 85), Nr. 916
(p. 119); Nr. 589 (p. 85), Nr. 918 (p. 119).

90.

7 fol.

Skr. *āryatārābhaṭṭārakānāma aṣṭaśatakam.* T. rje btsun ma
o p'ags ma sgrol mai mts'an [ma] brgya rtsa brgyad pa źes bya ba.
Randtitel: sgrol mai mts'an brgya.

K.-I. Nr. 723 (p. 100), Nr. 973 (p. 125).

fol. 6—7: Randtitel: târa. Anfang: p'yag o ts'al sgrol ma
myur ma dpa mo | spyan ni skad cig glog dañ o dra ma. Schluss:
rje btsun o p'ags ma sgrol ma la yañ dag par rdzogs pai sañs rgyas
rnam par snañ mdzad kyis bstod pa rdzogs so.

91.

1 fol. grünes Papier.

o p'ags ma sgrol ma gzuñs rdzogs so. Randtitel: târa.
K.-I. Nr. 725 (p. 100), Nr. 974 (p. 126).

92.

4 fol.

Skr. *āryavijayavatīnāmapratyañgirā.* T. o p'ags pa p'yir bzlog
pa rnam rgyal (ba can) źes bya ba bźugs so. Randtitel: p'yir zlog
rnam rgyal.

K.-I. Nr. 730 (p. 101), Nr. 941 (p. 122).

93.

9 fol.

Aussentitel: p'yir zlog pa rnam par rgyal ba źes bya ba.

8*

Innentitel: Skr. *pratyañgiramantrabhirva*(?)*cakranāma*. T.
p'yir zlog pa ñan sñags kyi o k'or lo žes bya ba.
Vgl. 92.

94.

2 fol.

Skr. *āryaparṇaśavarināmadhāraṇī*. T. o p'ags pa ri k'rod lo
ma gyon pai gzuñs. Randtitel: ri k'rod ma.

K.-I. Nr. 732 (p. 101), Nr. 968 (p. 125).

Parṇaśavara ist nach PW. Bezeichnung eines von Blättern
lebenden wilden Volksstamms im Dekkhan. Das tibetische Äqui-
valent bedeutet „die sich mit Blättern kleidenden Bergbewohner".
Vgl. über dieses Volk E. Schlagintweit, Die Lebensbeschreibung
von Padma Sambhava, in Abhandlungen der bayer. Akademie, I. Cl.
XXI. Bd. II. Abt., 1899, p. 438.

95.

8 fol.

Skr. *āryabalavatīnāmapratyañgirā*. T. o p'ags pa p'yir bzlog
pa stobs can žes bya ba bžugs so. Randtitel: p'yir bzlog, von
fol. 7 an: brgyad ɤyul rgyal.

Schluss: o p'ags pa ɤyul las c'a rgyal ba žes bya ba gzuñs
rdzogs so.

K.-I. Nr. 733 (p. 101), Nr. 933 (p. 121).

96.

2 fol.

Skr. *mahāśrī*[1])*sūtra*. T. dpal c'en moi mdo. Randtitel: dpal
c'en mo.

K.-I. Nr. 736 (p. 101), Nr. 978 (p. 126).

97.

7 fol.

Skr. *āryavajra ajita analapramohanīnāmadhāraṇī*. T. o p'ags
pa rdo rje mi p'am pa me ltar rab tu rmoñ byed ces bya bai
gzuñs. Randtitel: rdo rje mi p'am.

Schluss: rgya gar gyi mk'an po *Jinamitra* dañ | žu c'en gyi
lo tsts'a ba *Bandhe Ye šes sdes* bsgyur ciñ skad gsar bcad kyis
kyañ bcos nas gtan la p'ab pa.

K.-I. Nr. 747 (p. 102), Nr. 927 (p. 120), wo den Übersetzer-
namen noch *Dānaśīla* hinzugefügt ist.

98.

4 fol.

Skr. *āryadaśavajrapāṇihṛdaya*. T. o p'ags pa lag na rdo rje
bcui sñiñ po. Randtitel: lag na rdo rje.

K.-I. Nr. 749 (p. 103), Nr. 924 (p. 120).

_____

1) *mahūśa.* K.-I. *-lakṣmiṇī.* Csoma (As. Res. XX 536, Nr. 4) *-śrayā.*

99.

19 fol.

Skr. *āryamahābalanāmamahāyānasūtra*. T. o p'ags pa stobs po c'e źes bya ba t'eg pa c'en poi mdo.

K.-I. Nr. 752 (p. 103), Nr. 920 (p. 119).

100.

4 fol.

Skr. *vajratuṇḍa[1])nāmanāgasamaya*. T. rdo rjei mc'u źes bya bai klui dam ts'ig go. Randtitel: rdo rje mc'u.

K.-I. Nr. 754 (p. 103), Nr. 937 (p. 121).

101.

2 fol.

Skr. *āryavidyārājaśvāsamahānāma*. T. o p'ags pa rig sñags kyi rgyal po dbugs c'en po źes bya ba. Randtitel: dbugs c'eno.

Schluss: rgya gar gyi mk'an po *Prajñāvarma* dañ | źu c'en gyi lo tsts'a ba *Bandhe Ye śes sdes* bsgyur ciñ źus te | gtan la p'ab pao.

K.-I. Nr. 768 (p. 105), Nr. 942 (p. 122), ohne Skr.-Titel.

102.

5 fol.

Skr. *pañcatathāgatamañgalagāthā*. T. de bźin gśegs pa lñai bkra śis kyi ts'igs su bcad pa. Randtitel: bkris.

K.-I. Nr. 816 (p. 108), Nr. 1079 (p. 133).

103—104.

6 fol.

Skr. *āryavajrabhairavadhāraṇīnāma*. T. o p'ags pa rdo rje o jigs byed kyi gzuñs źes bya ba. Randtitel: o jigs byed.

Schluss: fol. 2 b 2: o p'ags pa rdo rje o jigs byed kyi gzuñs źes bya ba | ma ruñs pa p'yir bzlog pa rdzogs so || o p'ags pas gsuñs pai gsuñs rnams rnam mañ yañ || rdo rje o jigs byed źal nas gsuñs pai gzuñs | bsruñ byai las bdun ldan pai gzuñs mc'og o di | kun gyis t'un moñs ma yin rnal o byor dam pai gzuñs | rnal o byor gyi rnal o byor c'en po *Don yod rdo rjei* źal sña nas bsgyur nas | bod kyi *Bandhe sKyo o od obyuñ* la gnañ ño.

K.-I. Nr. 929 (p. 121), wo es im Kolophon rnal o byor gyi dbañ p'yug c'en po heisst. Grünwedel, Mythologie des Buddhismus, S. 101.

fol. 2 b 6: Skr. *āryadrāviḍavidyārājā*. T. o p'ags pa o gro ldiñ bai rig sñags kyi rgyal po. Randtitel: o gro ldiñ.

Schluss: rgya gar gyi mk'an po *Jinamitra* dañ | *Dānaśīla*

1) dunba.

dañ | žu c'en gyi lo tsts'a ba *Bandhe Ye šes sdes* bsgyur ciñ žus
te skad gsar c'ad kyis kyañ bcos nas gtan la p'ab bao.

K.-I. Nr. 609 (p. 88), Nr. 902 (p. 118). Hier ist *dra mi dva*
und *drā mi dā* geschrieben (C s o m a, As. Rès. XX 525, wie oben).

<h2 style="text-align:center">105.</h2>

Holzdruck. Fragment: fol. 6—7 fehlen, 8—10 vorhanden.

Skr. *āryabhadracāryapraṇidhānarāja.* T. o p'ags pa bzañ po
spyod pa smon lam gyi rgyal po.

K.-I. Nr. 1069 (p. 133).

Andere Sanskrit-Tibetische Schriften.

<h2 style="text-align:center">106.</h2>

fol. 22—37. 16 fol.

Skr. *āryamañgalakūṭanāmamahāyānasūtra.* T. o p'ags pa
bkra šis rtsegs pa žes bya ba t'eg pa c'en poi mdo.

Schluss: o p'ags pa bkra šis brtsegs pa žes bya ba t'eg pa
c'en poi mdo las c'o ga dañ bcas pa rdzogs so.

<h2 style="text-align:center">107.</h2>

4 fol.

Skr. *āryatriratnagamananāmamahāyānasūtra.* T. o p'ags pa
dkon mc'og gsum la skyabs su o gro ba žes bya ba t'eg pa c'en
poi mdo. Randtitel: dkon skyabs.

Schluss: rgya gar gyi mk'an po *Sarvajñādeva* dañ | žu c'en
gyi lo tsts'a ba *Bande dpal brtsegs* kyis bsgyur ciñ žus te gtan
la p'ab pa.

<h2 style="text-align:center">108.</h2>

3 fol.

Skr. *āryadrumasūtra.* T. o p'ags pa ljon šiñ gi mdo. Randtitel: ljon šiñ.

<h2 style="text-align:center">109.</h2>

3 fol.

Skr. *bhūmisūtra.* T. sai mdo.

Schluss: rgya gar gyi mk'an po *Padmākaravarma* dañ | žu
c'en gyi lo tsts'a ba dge sloñ *Rin c'en bzañ pos* bsgyur ciñ žus
te gtan la p'ab pao.

<h2 style="text-align:center">110.</h2>

4 fol.

T.[1]) o p'ags pa stag mos žus pa žes bya ba t'eg pa c'en poi
mdo. Randtitel: stag žus.

---

1) Der beigefügte Skr.-Titel: *ārya su-ba-bu-ba-rmi-ti-nāmasūtra* ist mir
unverständlich. Die Rückübersetzung des tibetischen Titels ins Sanskrit müsste
lauten: *āryavyāghrīpariprcchanāmamahāyānasūtra.*

Schluss: byañ c'ub sems dpai rgyud las rtogs pa dañ poi leu rdzogs so | gcig žus.

„Erstes Kapitel der Betrachtungen aus dem *Bodhisattvatantra.*"

111.

10 fol.

Nur Aussentitel: stag mos žus pa žes bya bai mdo bžugs. Randtitel: stag žus.

Schluss: o p'ags pa stag mos žus pa žes bya bai mdo rdzogs so | žus so.

112.

12 fol.

Skr. *āryamañjuśrīnāmasaingīti.* T. o p'ags o jam .dpal gyi mts'an yañ dag par brjod pa. Ohne Randtitel.

Schluss: o di lo c'en *Rin c'en bzañ poi* o gyur la | *Šoñ blo gros brtan pas* bcos pa la don dañ mi o gal žiñ grags c'e ba rnams lo c'en o gyur ñid gžir bžag ‖ o gyur gñis ka la mi bcos su mi ruñ ba rnams dag par rgya gar gyi dpe dañ | rgya o grel c'en mo rnams dañ mt'un par ža lu lo tsts'a ba *Dharmapālabhadra* žes ba gyi bas lhun po spañ du o jam sdud bzañ gsum bco bor gyur pai mdo p'ran grags c'e ba k'uñs ma rags rim žig par du bsgrub[1]) pa dus žus c'en legs par bgyis so gsuñ bai dpe ñe(?) de ñid liñ(?) c'es kyi p'yi mor(?) byas te *sku rab rnam rgyal rtser* bar du bsgrubs pai par ma ñid yid c'es kyi p'yi mor(?) byas te slar yañ *dga ldan p'un ts'ogs gliñ* du par du bsgrubs[2]) pao , dge legs o p'ol ‖.

Soweit ich dieses Kolophon verstehe, ist daraus folgendes zu entnehmen:

Es gab zwei Übersetzungen des vorliegenden Werkes, eine von dem grossen Übersetzer (lo c'en) *Ratnabhadra* (*Rin c'en bzañ po*) und eine andre von *Šoñ blo gros brtan pa*[3]), der die sinngetreuen und hochberühmten Übersetzungen des *Lo c'en* zu Grunde legte. Was in . dieser zweiten Übersetzung noch unkorrekt war, hat der Übersetzer von *Žalu, Dharmapālabhadra* mit Vergleichung der indischen Bücher und der grossen indischen Kommentare gereinigt und einer guten Verbesserung unterzogen, als in *Lhun po spañ* eine grosse Reihe der in o *Jam sdud bzañ gsum bco bo* übersetzten hochberühmten kleinen Sûtra im Original gedruckt wurden. Dann ward das Buch in der Presse von *sKu rab rnam rgyal rtse* und noch einmal in *dGa ldan p'un ts'ogs gliñ* gedruckt.

113.

4 fol.

Skr. *nakṣatramātṛkānāmadhāraṇī.* T. skar mai yum žes bya bai gzuñs bžugs so. Randtitel: skar yum.

---

1) *bsbru!*   2) *bsgyugs.*

3) **Vielleicht identisch mit dem zu Tanjur 117, 3 genannten** *Šoñ blo brtan,* s. H u t h in Sitz. Berl. Akad. 1898, p. 268.

Schluss: dran sron skar ma dga bas žus pai mdo las skar mai
yum žes bya ba | skar ma ñan pa t'ams cad bzlog par byed pai
mdo rdzogs ·so | žus dag.

114.

5 fol.

Skr. *āryakuberaratna.* T. a) Aussentitel: gnod sbyin kubera
nor spel bai gzuñs. b) Innentitel: o p'ags pa nor p'yugs bsruñ žiñ
spel ba žes bya bai gzuñs. Randtitel: kubera.

115.

2 fol.

Aussentitel: o bru spel bai gzuñs.

Innentitel: Skr. *āryagaṇaratnavayūdharanāmadhāraṇī.* T.
o p'ags pa ts'ogs kyi bdag po rin po c'e o brui dkor mdzod dañ |
o bru dañ loñs spyod spel ba žes bya bai gzuñs. Randtitel: o bru spel.

Schluss: o p'ags pa ts'ogs kyi bdag po dkor mdzod žiñ k'ams
t'ams cad spel žiñ bsruñ ba žes bya bai gzuñs leu bcu drug pa
rdzogs so.

116—118.

4 fol.

Aussentitel: rtañ gzuñs bžugs so.

Innentitel: Skr. *āryatathāgataremanta.* T. o p'ags pa remanta
žes bya bai gzuñs.

Randtitel: rtañ gzuñs.

Anfang: rta naḍ t'ams cad rab tu ži bar byed pa yi dkon
mc'og gsum la p'yag o ts'al lo.

Schluss: dpal ârya remanta žes bya bai gzuñs rdzogs so.
fol. 2 b 4: Skr. *āryaśrīmahākāladhāraṇī.* T. o p'ags pa mgon
po nag po rtai gzuñs. Randtitel: rtañ gzuñs. Schluss: mi p'yug
rta dañ bcas pai bsruñ ba rdzogs so. fol. 3 a 3: Skr. *śrīmahāyoginī.*
T. dpal nag po c'en po k'ams gsum la dbañ bsgyur ba. fol. 4 b 5:
ârya remanta rakṣa rakṣa svâhâ. fol. 4 b 7: gnod sbyin kubera
zes bya bai gzuñs rdzogs so. Vgl. Nr. 114.

119.

4 fol.

Skr. *vajravidāraṇanāmadhāraṇī.* T. rdo rje rnam pa o jom
pa žes bya bai gzuñs. Randtitel: rnam o joms.

Schluss: gzuñs mdo o di ni rdo rje o c'añ c'en po rje btsun
Târanâthas žus dag gnañ bai dpe las bris pao.

„Dieses *Dhāraṇī-sûtra* ist nach einem von dem *Mahāvajra-
dhara Bhaṭṭāraka Târanātha* verbesserten Exemplar geschrieben."

120.

7 fol.

Skr. *āryabhadracaryapraṇidhānarāja.* T. o p'ags pa bzañ po
spyod pa smon lam gyi rgyal po. Randtitel: bzañ spyod.

Verz. 394.

121.

3 fol.

Skr. *āryamaitrīpraṇidhānarāja*. T. o p'ags pa byams pai smon lam gyi rgyal po. Randtitel: byaṁ smon.

Verz. 395, 2.

122.

9 fol.

Skr. *śrīvajraratīrā(?)nāmadhāraṇī*. T. dpal rdo rje sder moi gzuṅs bžugs so. Randtitel: rdo rje sder mo.

123.

10 fol.

Skr. *devīmahākālīhasa uṣṇīṣanāmadhāraṇī*. T. lha mo nag mo c'en mo rol bar byed pai gtsug tor žes bya bai gzuṅs. Randtitel: lha mo rol.

124.

3 fol.

Aussentitel: ,a par yaṅ dag šes kyi gzuṅs rdzogs so.

Innentitel: Skr. *yakṣa aparaviśuddhanāmadhāraṇī*. T. gnod sbyin gžan gyis mi t'ub pa yaṅ dag šes kyis gzuṅs. Randtitel: yaṅ dag šes.

125.

15 fol.

Skr. *vajrahṛipaśaśayuga*(?). T. a) Aussentitel: gzaḥ yab gzuṅs bžugs so. b) Innentitel: gzaḥ nad t'ams cad rab tu ži bar byed pai gzuṅs. Randtitel: gza yab. Schluss: gzaḥ yab gzuṅs kyi mdo draṅ sroṅ yab kyi gzuṅs rdzogs so.

„Die alle Planetenkrankheiten beschwichtigende Dhāraṇī."

126.

6 fol.

Skr. *suvarṇabhavasmṛtaṁge(?)nāmadhāraṇī*. T. gser o od rṅa sgra žes bya bai gzuṅs. Randtitel: rṅa sgra.

Tib. *gser o od* pflegt Skr. *suvarṇaprabha*, und tib. *rṅa sgra* Skr. *dundubhisvara* zu entsprechen.

127.

8 fol.

Skr. *ārya anirmita āyurjñāna abhiṣiñcahṛdayanāmadhāraṇī*. T. o p'ags pa ts'e dpag tu med pai sñiṅ po ts'e dbaṅ bskur žes bya bai gzuṅs. Randtitel: ts'e sñiṅ.

128.

3 fol.

Skr. *kāyavākyacittrastambhanavijayadhāraṇī*. T. lus ṅag yid

gsum bciñs pa las rab tu rgyal bar (Aussentitel: grol bar) byed
pa žes bya bai gzuñs. Randtitel: bciñs grol.

Schluss: rgya gar gyi mk'an po paṇḍita *Gayadhara* dañ |
bod kyi lo tsts'a ba *Śākya ye šes* kyis *Mañ yul byams sbran
gyi gtsug lag k'añ* du bsgyur bao.

„Der indische Gelehrte, der *Paṇḍita Gayadhara* und der
tibetische Übersetzer *Śākyajñāna* haben diese Schrift im Kloster
Byams sbran in *Mañ yul* übersetzt."

129.

2 fol.

Skr. *āryabahuputrapratisaranāmadhāraṇi*.   T. o p'ags pa bu
mañ po so sor obrañ pa žes bya bai gzuñs. Randtitel: bu mañ
po. Schluss: rgya gar gyi mk'an po *Jinamitra* dañ | *Dānaśila*
dañ | žu c'en gyi lo tsts'a ba *Bandhe Ye šes sdes* bsgyur ciñ žus
te gtan la p'ab pa.

Schriften ohne Sanskrit-Titel.

130.

2 fol.

T. rgyal poi c'o p'rul ston pa p'yir zlog pa žes bya ba t'eg
pa. c'en poi mdo. Randtitel: rgyal poi c'o o p'rul.

131.

4 fol.

Nur Aussentitel: byañ c'ub sems dpai ltuñ ba bšags pai mdo
bžugs so. Randtitel: ltuñ bšag. Ohne Kolophon.
„Sûtra von der Sühnung der Sünden der Bodhisattva."

132.

7 fol. Unvollständig.

dpal rdo rje ojigs byed kyi bdag ñag odon gyi rim pa bžugs
so. Vergl. Nr. 103.

Über *Śrīvajrabhairava (dpal rdo rje ojigs byed)* s. Grünwedel, Mythologie des Buddhismus, p. 101.

133.

Holzdruck ohne Titel. 8 fol.

Anfang: namo sems can t'ams cad dus rtag par bla ma la
skyabs su mc'io | sañs rgyas la skyabs su mc'io | c'os la skyabs su
mc'io | dge odun la skyabs su mc'io | (Die bekannte Zufluchts-
formel).

Schluss: sdig pa bdag gis bgyi ba ci mc'is pa | de dag t'ams
cad bdag gis so sor bšags | p'yag ots'al ba dañ mc'od ciñ bšags

pa daṅ | rjes su yi raṅ bskul žiṅ gsol ba yis | dge ba cuṅ zad
bdag gis ci bsags pa | t'ams cad rdzogs pai byaṅ c'ub c'en por
bsṅos. [1]).

„Alle von mir begangenen Sünden, welche sie auch sein
mögen, habe ich gesühnt: durch Verehrung und Opferspenden habe
ich sie gesühnt. In der Folge sind die durch Selbstermahnung und
Wohlthätigkeit [2]) ein wenig von mir angesammelten Tugendwerke
auf die ganz vollendete grosse Bodhi gerichtet."

134.

3 fol. fol. 20—22.

T. o p'ags pa snaṅ ba brgyad [3]) ces bya bai gzuṅs. Anfang:
lo mi ruṅ ba daṅ | skar ma mi ruṅ daṅ | gza mi ruṅ ba daṅ |
t'uṅ [4]) c'uṅ mi ruṅ ba daṅ | ñan pa de rnams kyis o dul bsñal o di
yin te | ,a ra na ma ma hâ gra hâ | na ma byin na de | su yu
na svàhâ |

Schluss: ltas ñan pa t'ams cad zlogs cig. „Dhāraṇī, genannt
die acht Erscheinungen."

135.

82 fol. Fragment.

Randtitel: c'os spyod „Religionsübung." Vorhanden sind
fol. 17—19, 35—37, 42—46, 49—55, 60—80, 83—87, 90—113,
125—130, 142—150, 161—162, 171—173, 175, 177, 179, 190.

Ein Werk gleichen Titels erwähnt C s o m a (As. Res. XX, 574)
in der Abteilung mdo (sûtra) des Tanjur.

136.

98 fol.

byaṅ c'ub lam gyi rim pai o k'rid yig o jam pai dbyaṅs kyi
žal luṅ žes bya ba bžugs so. Randtitel: lam rim o k'rid.

„Führer durch die Stationen des Weges zur Bodhi, genannt
Mahnwort des Mañjughoṣa."

Über Titel mit ähnlichen Stichwörtern (*byaṅ c'ub lam gyi
rim*) s. Verz. Nr. 387 (p. 34), Nr. 412 (p. 38), Nr. 435 (p. 48); zu
letzterem vgl. Journal of the Royal Asiatic Society, 1892, p. 141;
H u t h, Geschichte des Buddhismus in der Mongolei, Bd. II, p. 399, 403.

Schluss fol. 97 b 3: byaṅ c'ub lam gyi rim pai o k'rid gžuṅ
o jam pai dbyaṅs kyi žal luṅ žes bya ba o di ni | rgyal bai gsuṅ
rab mt'a dag la gžan driṅ mi o jog pai rtsod dus kyi kun mk'yen
c'en po o k'on ston c'os kyi rgyal poi žal luṅ dri ma med pa o jam
dbyaṅs bla mai drin las legs par t'os šiṅ smra mk'as dag gi dbaṅ
p'yug rje btsun bla ma *dKon cog c'os o p'el bai* druṅ du lam gyi
rim pa c'en mo ts'ig gcig kyaṅ ma lus pai bzabs bšad lan gñis

---

1) *bsñoi.*
2) *gsol ba* Bewirtung der Geistlichkeit mit Speise und Trank.
3) *rgyad.*   4) *tuṅ.*

kyi bar . du mnos pai bka drin las c'os ts'ul o di ñid smra ba la
spobs pai mgrin pa cuñ zad o degs nus pai skal ba can du gyur
pai za hor gyi bande *Nag dbañ blo bzañ rgya mts'o o jigs med
go c'a t'ub bstan lañ ts'oi sde* miñ gžan o jam dbyañs dga bai
bžes gñen du o bod pas | o p'ags pai yul du bhi lambha žes p'yogs
o dir rnam o p'yañ du..o bod ciñ | o jam dbyañs goñ mai rgyal k'ab
tu | wu zui žes pa sa p'o k'yii lo | legs sbyar gyi skad du šra ba
ṇar grags pa bya sboi zla ba | rgya nag pi ts'ä yol žes hor zla
bdun pa | dus o k'or bai mun pa bsal ba o briñ poi dga ba | yon
tan gyi dbyañs ,a lna gsal byed k'i | dbyañs o c'ar bai dmar p'yogs
kyi dga ba gñis pa | dbyans ,a | gsal byed bha | na ts'od byis pa |
k'ams sa | o dod yon dri | ñi ma me bžii o grub sbyor gyi tañka la
señ gei dus sbyor la | bka dañ bstan bcos o gyur ro cog gi ts'al c'en
po *dpal ldan o bras spuñs* c'os kyi sde c'en por sbyar bai yi ge
pa ni *Groñ smad pa o P'rin las rgya mts'os* bris pa.

„Was dieses Werk anbetrifft, so hat es damit folgende Be-
wandtnis: Die fleckenlosen Lehren des in allen heiligen Schriften
des Jina auf andrer Wissen sich nicht verlassenden, grossen All-
wissenden des Dvāparayuga, des die Religionsmüden belehrenden
Dharmarāja hat dank der erhabenen Gnade des *Mañjughoṣa* der
Herr der Beredten, der ehrwürdige Lama *dKon cog c'os o p'el* [1])
vortrefflich studiert. Der bei diesem zweimal bis auf das letzte
Wort gegebene sorgfältige Erklärungen des grossen Werkes der
'Pfadstationen' (*lam gyi rim pa*) empfing und dank solcher Gnade
eben diese Lehrweise predigend die Fähigkeit erlangte, den Nacken
des Mutes ein wenig emporzuheben, der Bande von Zahor, *Nag
dbañ blo bzañ rgya mts'o* [2]), der mit einem andern Namen o *Jigs
med go c'a t'ub bstan lañ ts'oi sde* als Kalyāṇamitra des Mañjughoṣa bezeichnet wird, hat in dem Mañjughoṣa-Palast der früheren
Könige [3]), der nach dem Pralamba [4]) genannten Distrikt in Āryadeśa
der „Herabhängende" (*rnam o p'yañ*) heisst, dieses Werk verfasst,
und zwar in dem auf chinesisch *wu-zui* [5]) genannten männlichen
Erde-Hunde-Jahr [6]), in dem in der Sanskritsprache als Śravaṇa be-

---

1) Doktor der Litteratur und Prediger, wird er für das Jahr 1626 als
Lehrer des elfjährigen späteren *Nag dbañ blo bzañ rgya mts'o* erwähnt.
Huth, Geschichte des Buddhismus in der Mongolei, Bd. II, p. 266.

2) Der fünfte Dalai Lama, 1616—1681. Seine Biographie bei **Huth**,
l. c. p. 265 ff. Ebenda findet sich die Angabe, dass sein Vater aus einer Familie
von *Zahor* stammt.

3) Den auf dem Rotberge (*dmar po ri*) gelegenen, verfallenen Palast der
alten tibetischen Könige hat der fünfte Dalai Lama 1643 mit grosser Pracht
wieder aufbauen lassen. **Köppen**, Die lamaische Hierarchie und Kirche, p. 340.

4) Tib. *op'yañ* = *lambate*, tib. *rab tu op'yañ* = *pralambate* nach
Vyutpatti (Tanj. As. Mus.) fol. 276a, 1. *Pralamba* nach PW., Name einer
Lokalität.

5) *wu* = tib. *sa*, *zui* = tib. *k'yi*. **Csoma**, Grammar of the Tibetan
language, p. 149; **Foucaux**, Grammaire de la langue tibétaine, p. 150.

6) d. i. 1658 A. D.

kannten *Bya-sbo* [1]) Monat, dem auf chinesisch *Pi-tsʻā-yol* genannten siebenten Hor-Monat, an dem die Finsternis des Zeitkreislaufs vertreibenden mittleren *Nandikā*-Tage [2]), [die fünf Guṇa-Vokale (? *yon tan gyi dbyaṅs*) a, die 32 (*kʻi*) Konsonanten] [3]), am zweiten *Nandikā*-Tage des Harmonie hervorbringenden abnehmenden Mondes [4]), [Vokal a, Konsonant bh, an Alter ein Kind, Element Erde, Schmutz der irdischen Güter] [3]), im Bilde der glücklichen Konstellation der Sonne mit dem Vierfeuergestirn [5]), in der Stunde des Löwen [6]). Der die grossen Teile der Übersetzungen des *Kanjur* und *Tanjur* im grossen Kloster *Śrīdhanakaṭaka* [7]) verfasst hat, der Grammatiker *Groṅ smad pa* [8]) *opʻrin las rgya mtsʻo* hat es geschrieben.

Herr Prof. H. Jacobi in Bonn, dem ich den Schlusspassus dieses Kolophons vorlegte, war so liebenswürdig, mir am 24. September folgendes zu schreiben: „Manches in der Datumangabe ist im Dunkel. Es scheint mir das indische Datum zu sein: *Çrāvaṇa ba di 2*. Nach der *pūrṇimānta* Rechnungsweise war es 1658 A. D., Dienstag 6. Juli alten Stiles. Dienstag = *maṅgalavāra* ist offenbar „die Finsternis des Zeitkreislaufes vertreibender mittlerer" (nämlich dritter Wochentag). Die Sonne stand in *Puṣya* (Krebs), der Mond in Konjunktion mit *Dhaniṣṭhā* (Delphin), welches *Nakṣatra* aus vier Sternen besteht. Mit des „Löwen Stunde" ist vielleicht der Löwe als *lagna* bezeichnet; das wäre die dritte oder vierte Stunde nach Sonnenaufgang. Das übrige ist mir ganz dunkel."

137.

3 fol.

T. rdo rje rgyal mtsʻan gyi yoṅs su bsṅo ba bżugs so. Randtitel: bsṅo ba.

138.

4 fol. Ein Stück der rechten Seite des ersten Blattes fehlt.

Nur Aussentitel: dkar cʻag dgos o dod kun o byuṅ bżugs so. Randtitel: dkar cʻag.

---

1) Tibetische Bezeichnung des siebenten Monats, s. Desgodins, Dictionnaire tibétain-latin-français, p. 878, der aber als chinesischen Namen *gau yol* angiebt.

2) Tib. *dga ba* ist vermutlich mit Skr. *nandikā* zu identifizieren.

3) Die in [ ] gesetzten Stellen sind mir unverständlich; vielleicht handelt es sich um astrologische Bestimmungen.

4) *dmar*- oder *nag-pʻyogs* (*kṛṣṇa*). Desgodins, l. c. p. 762; Thibaut, Astronomie, Astrologie und Mathematik p. 12, § 7.

5) tib. *me bżi* (oder *bya ma*) Name des 12. Nakṣatra, Skr. *hasta*.

6) tib. *seṅ gei dus*, ist die Zeit der fünften Doppelstunde, in welcher der Löwe, das fünfte Zeichen des Zodiacus, (tib. *kʻyim gyi o kʻor lo*) die Meridianlinie überschreitet. Vgl. über den tibetischen Zodiacus Chandra Dás in Proc. ASB. 1890, No. I, p. 2—5.

7) tib. *dpal ldan o bras spuṅs*, s. Tāranātha II, p. 142.

8) d. i. der aus der unteren Stadt.

Alphabetisches Verzeichnis der Randtitel [1]).

Die Zahl hinter dem Titel bezeichnet die Nummer der Handschrift.

kou 10.
kubera 114.
*kubera 118.
kun tu bzañ po 86.
klui spa 2.
klui žus 30.
kṣayai nad sel ba 68.
dkar c'ag 138.
dkon skyabs 107.
*bkra šis rtsegs pa 106.
bkris 102.
bklags grub 54.
skar yum 113.

*glañ ru luñ bstan 40.
*bgegs sel 78.
bgres mos žus pa 35.
ogro ldiñ 104.
rgyal poi c'o op'rul 130.
rgyal ba can 51.
rgyal mts'an 65.
sgrol mai mts'an brgya 90.
brgyad ɣyul rgyal 95.

ñan sbyoñ gtsug tor 60.
rña sgra 83.
bsño ba 137.

bciñs grol 128.

c'os spyod 135.
c'os yum 55.
*ojam dpal gyi mts'an 112.
ojigs byed 103.
ljon šiñ 108.

ñi mai mdo 28.

tàra 91.
tiñe odzin sogs 89.
tog gzuñs 32.
rtañ gzuñs 116, 117.
ltuñ bšag 131.

stag žus 110, 111.
stoñ p'rag brgya pa 5, 89.
ston pai mts'an 42.
*stobs po c'e 99.

don žags 84.
dug sel 53.
gdugs dkar 61.
bdud gžom sogs 33.
mda ka 31.
rdo rje mc'u 100.
rdo rje sder mo 122.
rdo rje mi p'am 97.
rdor gcod (rdo rje gcod pa)
    11—27.
sdud pa 8.

*nag po c'en po 118.
nam mk'ai sñiñ po 37.
nor rgyun 80.
*nor bu bzañ po 64.
nor lha 49.
rnam rgyal 59.
rnam ojoms 119.
*rnam par snañ mdzad 44.
*snañ ba brgyad 134.

dpañ skoñ 1.
dpal c'en mo 96.
*dpal rdo rje ojigs byed 132.
spyan ras gzigs 85.
*sprin c'en 36.
p'an p'yin drug sogs 89.
p'yir zlog rnam rgyal 92.
*p'yir zlog pa 93.
p'yir bzlog 95.
*p'yogs bcui mun sel ba 38.
bar du gcod pa 63.
*bar du gcod pa sel ba 73.
bu mañ po 129.
byams pa dam bcas 76, 77.
byams pai mts'an 74.

---

1) Die mit * bezeichneten, in den Handschriften nicht vorhandenen Randtitel sind von mir auf Grund der Aussen- und Innentitel hinzugefügt.

byam(s) smon 121.
byams žus 29.
dbugs c‘eno 101.
dbyig ldan 52.
o bum c‘uñ 6.
o bru spel 115.

*man ñag gi rgyud 3.
mi rgod 72.
mi t‘ub pa 87.
mes poi žal luñ 4.
*mig nad rab tu ži ba 67.
mig rnam par spyod 66.
*smon lam gyi rgyal po 105.

*gtsug tor o bar ba 62.
gtsug nor 56.
brtsegs 39.

ts‘e sñiñ 127.
ts‘e mdo 82.
ts‘ogs bdag 81.

*gžañ o brum rab tu ži ba 69.

*zla bai o od 45.
gza yab 125.
gza yum 79.

bzañ spyod 120.

o od zer 50.

yañ dag šes 124.
yañs pa 71.
yi ge drug 57.
yum mts‘an 9.

ri k‘rod ma 94.
rin c‘en zla ba 34.
*rims nad rab tu ži ba 70.

lag na rdo rje 98.
lam rim o k‘rid 136.

*šākyai t‘ub pa sñiñ po 43.
šer sñiñ 41.
*šes rab p‘ar p‘yin pa 58.
*šes rab p‘ar p‘yin pa sdud pa 7
šloka 88.

sa sñiñ mts‘an brgya 75.
sai mdo 109.
*sañs rgyas rin c‘en gtsug tor
  can 46.
*gser o od dam pa 47, 48.

lha mo rol 123.

Indische Übersetzer.

Die Zahlen verweisen auf die Nummer der Handschrift.

Gayadhara 128.
Jinamitra 29, 32, 35, 65, 86,
  97, 104, 129.
Dānaśīla 29, 35, 52, 65, 86, 97,
  104, 129.
Devacandra 34.
Padmākaravarman 109.
Puṇyasambhava 83.

Prajñāvarman 87, 101.
Vāgīśvara 85 (ñag di dbañ p‘yug).
Vidyākarasiṁha 34, 38.
Viśuddhasiṁha 34.
Śīlendrabodhi 32.
Sarvajñādeva 107.
Surendrabodhi 30, 71.

Tibetische Übersetzer.

Klui rgyal mts‘an (Nāgadhvaja) 38.
dKon cog c‘os op‘el (Ratnadharmavardhana) 136.
s Kyo o od o byuñ 103.
Groñ smad pa o p‘rin las rgya mts‘o, Grammatiker 136.
dGe dpal (Kalyāṇaśrī) 34.

Ṅag dbaṅ blo bzaṅ rgya mts'o, 5. Dalai Lama 136.
Ñi ma rgyal mts'an dpal bzaṅ (Sūryadhvajaśrībhadra) 28.
Tāranātha 82, 119.
Devacandra 34.
Don yod rdo rjei žal sṅa nas (Amoghavajra) 50, 103.
Dharmapālabhadra[1]) 7, 8, 112.
dPal brtsegs (Śrīkūṭa) 107.
Ba ts'ab ñi ma grags 83.
rTsaṅs de Bendrarakṣita 38.
Ža lu lo c'en, s. Dharmapālabhadra.
Ye šes sde (Jñānasena) 29, 30, 32, 35, 52, 65, 71, 86, 87, 97,
    101, 104, 129.
Rin c'en grags pa (Ratnakīrti) 50.
Rin c'en bzaṅ po (Ratnabhadra) 109, 112.
Śākya ye šes (Śākyaprajña) 128.
Šoṅ blo gros brtan pa 112.
Seṅge bzaṅ po (Simhabhadra) 7, 8.
'Atīśa 6.
'Ānanda śrīi žal sṅa nas 28.
'Amoghavajra, s. Don yod rdo rje.

Klöster, in denen Übersetzungen stattfanden.

T'ar pa gliṅ 29.
Byams sbran in Maṅ yul 128.
o Bras spuṅs 136.

Druckorte.

sKu rab rnam rgyal rtse 112.
dGa ldan p'un ts'ogs gliṅ 82, 112.

Asula snaṅ, nepalesischer Übersetzer 6.
dga ba = nandikā 136.
rgyud bži 3, 4.
Kanjur und Tanjur 136.
Kun dga pa, Ānanda, Bildnis 12.
Lha t'o t'o ri sṅan šal 1.
man ñag, Teil des *rgyud bži*. 3.
Medizinische Schriften 3, 4.
Nepalesische Verse 6.
Pralamba, Örtlichkeit in Indien 136.
Rab o byor, Subhūti, Sthavira, Bildnis 12.
Ses rab p'ar p'yin ma, Bildnis 12.
Sum pa, Volk und Sprache, 2.
T'ub pa dbaṅ po, Śākyamuni, Bildnis 12.
Yum bu bla mk'ar, Königspalast, 1.
Zaṅ žuṅ, Sprache von, 2.

---

1) S. Sitzungsberichte d. Bayer. Ak. 1898, p. 524—526.

028

两则密勒日巴的传说（圣徒故事）

Archiv

für

Religionswissenschaft

in Verbindung mit

Professor D. W. Bousset in Göttingen, Hofrat Crusius in Heidelberg, Professor Dr. H. Gunkel in Berlin, Professor Dr. E. Hardy in Würzburg, Professor Dr. A. Hillebrandt in Breslau, Professor Morris Jastrow in Philadelphia, Dr. J. Karlowicz in Warschau, Professor Dr. E. Mogk in Leipzig, Professor Dr. R. Pietschmann, Direktor der Universitätsbibliothek in Greifswald, Professor Dr. W. Roscher, Gymnasialrektor in Wurzen bei Leipzig, Geh. Kirchenrat Professor D. B. Stade in Giessen, Professor Dr. E. Stengel in Greifswald, Professor Dr. A. Wiedemann in Bonn, Professor Dr. H. Zimmern in Leipzig

und anderen Fachgelehrten

herausgegeben

von

Professor Dr. **Ths. Achelis**

in Bremen.

Vierter Band.

Tübingen und Leipzig

Verlag von J. C. B. Mohr (Paul Siebeck)

1901.

I. Abhandlungen.

Zwei Legenden des Milaraspa.

Von

Berthold Laufer.

Mit der Ermordung des den Buddhismus ausrottenden
Königs gLang dar ma (um 915 A. D.) erreicht die erste
Periode der tibetischen Geschichte ihren Abschluss. Dieser
bedeutet das Ende des Königtums überhaupt: Versuche zur
Bildung einer neuen Centralgewalt werden nicht mehr gemacht.
Es folgt eine Zeit trostloser politischer Verwirrung: das Land
ist in eine Reihe kleiner und schwacher Fürstentümer ge-
spalten. Im 11. Jahrhundert findet die Wiederherstellung des
Buddhismus statt, vor allem mit dem Namen des indischen
Gelehrten Atiça verknüpft, der 1042 von Indien nach Tibet
kommt. Tiefer als je zuvor schlägt nun der Buddhismus im
Lande Wurzel. Sektenbildungen treten auf, Gründungen von
Orden und Klöstern sind in stetem Wachsen begriffen, und
deren Leiter beginnen ihren Einfluss auf das politische Gebiet
hinüberzuspielen und allmählich durch Aufsaugung des Grund-
besitzes der Macht über die Gemüter die Herrschaft über die
Leiber hinzuzufügen. In diese Periode der Keimung des später
voll entwickelten hierarchischen Systems fällt das Auftreten

des Milaraspa, eine der merkwürdigsten Erscheinungen des tibetischen Mönchslebens. Ausser einigen kurzen Notizen ist bisher wenig über seine Persönlichkeit bekannt geworden, obwohl eine umfangreiche Biographie von ihm existiert, die seinem Schüler Ras chung zugeschrieben wird. Einige Daten über sein Leben findet man bei WADDELL, The Buddhism of Tibet, p. 65, mitgeteilt. Er soll von 1038 bis 1122 gelebt haben. Er war ein Schüler des Marpa und gehörte mit diesem zur bKa rgyud pa Sekte, als deren Stifter Nāro gilt. Das sich mit Milaraspa befassende grosse Legendenbuch führt den Titel: *rje btsun Mi la ras pai rnam t'ar rgyas par p'ye ba mgur ,bum,* d. h. ,Die das Leben des ehrwürdigen Milaraspa auseinandersetzenden hunderttausend Gesänge' und umfasst in dem mir zur Verfügung stehenden Holzdruck 263 Folioblätter. Ueber die Autorschaft ist in dem Werke selbst nichts bemerkt[1]; dieselbe dem Milaraspa selbst zu vindizieren erscheint mir zweifelhaft. JÄSCHKE[2] giebt über dieses Buch folgendes Urteil ab: „Verglichen mit der langweiligen Eintönigkeit und häufigen Geistlosigkeit der dem Buddha selbst in den Mund gelegten Erzählungen des Dsanglun zeichnen sich die des Milaraspa durch Mannigfaltigkeit der Situationen und des Inhalts überhaupt, sowie durch anschauliche Lebendigkeit der Schilderung äusserst vorteilhaft aus, und gewähren ausserdem, da sie mitten aus dem Leben des Volkes hervorgewachsen sind, einen Einblick in das Denken und Treiben desselben, wie man ihn nicht leicht anderswoher erlangen kann." Die beiden ersten hier in Text und Uebersetzung vollständig mitgeteilten Legenden bestätigen vollauf dieses Urteil.

---

[1] Ebenso ROCKHILL in Proceedings of the American Oriental Society 1884 p. 208.

[2] Zeitschrift der Deutschen Morgenländischen Gesellschaft, XXIII S. 544.

1. Milaraspa, der Holzsammler.

namo guru.

rnal ₀byor gyi dbang pʿyug rje btsun Mi la ras pa de ñid | ₀Cʿong lung kʿyung gi rdzong na ₀od gsal pʿyag rgya cʿen poi ngang la bzhugs pai dus | nam zhig gi tsʿe | ₀tsʿo bai sta gon la bzhengs pas | pʿye tsʿa cʿu rdor dag lta ci mos te | go kʿa na šing yang mi ₀dug | tʿab kʿa na cʿu dang me yang mi ₀dug pas | ngai ₀di ha cang rang yang blos blang grags ₀dug pas | šing zhig ₀tʿu ru ₀gro dgos dgongs nas byon te | šing tʿu ba gang tsam rñed pai tsʿe | rlung cʿen po zhig blo bur du langs nas | ras bzung tsʿe šing kʿyer | šing bzung tsʿe ras kʿyer ba las | tʿugs dgongs la |

ngas sngar ri kʿrod du de tsam zhig bsdad rung da dung bdag ₀dzin blos ma tʿongs par ₀dug | bdag ₀dzin blos ma tʿongs pai cʿos dang sgrub pas

Verehrung dem Meister!

Als der Gebieter der Yogin, der ehrwürdige Milaraspa, in der Art und Weise der Mahāmudrā prophetischer Erleuchtung in ₀Chong lung Khyung gi rdzong verweilte, geschah es einst, dass, als er sich erhob, um Vorbereitungen für seine Mahlzeit zu treffen, gar nichts vorhanden war, weder Mehl noch Salz noch Wasser noch Gewürz, ja nicht einmal Holz in der Vorratskammer noch Wasser und Feuer auf dem Herde. Da dachte er: „Mein Thun bedeutet doch wohl eine zu grosse Vernachlässigung meiner Person; ich muss doch Holz sammeln gehen." Er machte sich auf, und nachdem er soviel Holz gesammelt, als er gefunden hatte, erhob sich plötzlich ein grosser Sturm, der, wenn er sein Baumwollentuch fasste, das Holz, und wenn er das Holz fasste, sein Baumwollentuch zu entreissen drohte. Endlich kam ihm der Gedanke: „Obwohl ich früher in dieser Bergwildnis schon so lange verweilt habe, habe ich dennoch die Selbstsucht noch nicht auf-

1*

ci byed sñam | ras dga na ras kʻyer | šing dga nas šing kʻyer ysung gñis ka blo lang te bzhugs pa |

„tsʻo ba ngan pai stobs kyis | bser mai rnam pas yug gcig lʻugs dran med du lʻim pa las bzhengs pai tsʻe | rlung zhi ba dang ras de šing sdong gcig gi rtse la ling nge „dug pa dang | lʻugs skyo bai ñams šar bai ngang la | pʻa bong lug ro tsam zhig gi kʻar mñam par bzhag pai mtʻar |

šar pʻyogs Gro bo lung gi pʻyogs su sprin dkar po zhig ling byung bas | sprin pʻa kii „og na Gro bo lung gi dgon pa yod | de na ngai bla ma sgra bsgyur Mar pa lo tstsʻa de bzhugs te sñam pa dang | bla ma yab yum mcʻed grogs rdo rje spun gyi „kʻor dang bcas pai dbus na | rgyud kyi bšad pa dbang dang gdams ngag rnams | gnad tsʻul dran nas | da lta zhugs na | ci la

gegeben. Was soll ich thun, um die Gewohnheit des Haftens an der Selbstsucht zu überwinden?" „Wünschst du das Tuch," so rief er, „nimm das Tuch weg; wünschst du das Holz, so nimm das Holz!" So gab er beides dahin und setzte sich nieder.

Unter dem Einfluss der schlechten Ernährung und infolge des kalten Windes sank er eine Zeit lang in Bewusstlosigkeit. Als er sich wieder erhob, hatte sich der Sturm gelegt, und sein Tuch hing flatternd an der Spitze eines Baumzweiges. Da regten sich in seiner Seele Gedanken des Ueberdrusses; er liess sich eine Zeit lang auf einen Felsblock nieder, bis er das Gleichgewicht des Gemütes wieder gewonnen.

Im Osten zog in der Richtung von Grobolung her am Himmel schwebend eine weisse Wolke herauf. „Unter dieser Wolke", dachte er, „liegt das Kloster Grobolung: dort hat mein Lama, der sprachgewandte Marpa, der Gelehrte, gelebt." Er erinnerte sich an die im Kreise des Lama, der ihm Vater und Mutter war, seiner geistlichen Brüder und Vajra-geschwister empfangenen kraftvollen Erklärungen der

t'ug rung mjal du ₀gro ba yin
le dgongs pa dang | sngar gyi
skyo šas dei k'ar | bla ma
dran drags t'ugs skyo ba la
ts'ad las ₀das pa ₂hig gi ngang
nas spyan c'ab mang po bsil
₂hing | t'ugs skyo bai gdung
dbyangs bla ma dran drug
mgur ₀di gsungs so.

Tantra und an die Lehren, in
der Fingerstellung der Meditation. Er dachte: „Was mir
auch begegnen möge, wenn er
jetzt dort weilte, ich will ihm
meine Hochachtung bezeigen.“
Da überwand er trotz seiner
früheren Missstimmung durch
die lebhafte Erinnerung an den
Lama das Uebermass von Ver-
druss: ein Thränenstrom stürzte
ihm aus den Augen, und in der
Erinnerung an den Lama stimmte
er folgendes Lied an, einen
Sang wehmutsvoller Sehnsucht.

p'a dran pas gdung sel
Mar pai ₂habs | sprang gdung
dbyangs t'al lo.

In der Erinnerung an den
Vater, zu den Füssen des den
Sehnsuchtschmerz stillenden
Marpa, erhob er einen flehenden
Sehnsuchtsang.

Mar pa rje |
brag dmar ₀C'ong lung šar
p'yogs na
c'u ₀dzin sprin dkar lang ma
ling
sprin dkar po lding bai ₀og
₂hig na
rgyab ri glang c'en ₀gying
bai mdun ₂hig na

Marpa, Gebieter!
Im Osten des Rotenfels
₀Chong lung
Wogt eine Wasser bergende
weisse Wolke.
Unter der schwebenden weissen Wolke,
Vor dem Berge hinter mir,
der sich wie ein Elephant
gähnend streckt,

5 mdun ri seng c'en ₀gyings
₀drai steng ₂hig na

5 Auf dem Berge vor mir, der
sich wie ein grosser Löwe
gähnend streckt,

gnas c'en Gro bo lung gi dgon pa na	Liegt der heilige Ort, das Kloster von Grobolung.
rdo c'en ˌa mo li kai k'ri stengs na	In diesem steht ein Thron aus dem grossen Steine Amolika;
stan k'ri sñan sa lei lpags stengs na	Auf dem Polstersitz liegt ein glänzendes Argalifell;
de na su bzhugs su mi bzhugs	Wer sass darauf, wer sass nicht darauf?
10 de na sgra bsgyur Mar pa bzhugs	10 Darauf sass der sprachenbewanderte Marpa!
dus da lta bzhugs na dga ba la	Wenn er jetzt dort weilte, wie freute ich mich!
ngha' mos gus c'ung yang mjal sñing ˌdod	Wie schwach auch meine Andachtsglut, mein Herz begehrte zu ihm zu eilen!
gdung sems c'ung rung mjal sñing ˌdod	Wie klein auch meine Sehnsucht, mein Herz begehrte zu ihm zu eilen!
nga bsam zhing mts'an ldan bla ma dran	In meinen Gedanken erinnere ich mich des herrlichen Lama,
15 bsgom zhing Mar pa lo tsts'a dran	15 In meinen Betrachtungen erinnere ich mich an Marpa, den Gelehrten.
yum ma las lhag pai bdag med ma	Er ist der die Mütter übertreffenden unverehelichten Frau gleich.
dus da lta bzhugs na dga ba la	Wenn er jetzt dort weilte, wie freute ich mich!
sa bskor t'ag ring yang mjal sñing ˌdod	Wenn der Umweg noch so weit, mein Herz begehrte zu ihm zu eilen!

lam bgrod par dka yang mjal sñing „dod	Wenn der Weg auch mühsam zu wandeln wäre, mein Herz begehrte doch zu ihm zu eilen!
20 *bsams šing mts'an ldan bla ma dran*	20 In meinen Gedanken erinnere ich mich des herrlichen Lama,
bsgoms šing Mar pa lo tsts'a dran	In meinen Betrachtungen erinnere ich mich an Marpa, den Gelehrten.
rgyud zab mo dgyes pa rdo rje de	Er ist der das Herz tief erfreuende Edelstein.
dus da lta bzhugs na dga ba la	Wenn er jetzt dort weilte, wie freute ich mich!
nga šes rab c'ung yang „dzin sñing „dod	Wenn auch mein Wissen arm, mein Herz begehrte es zu fassen!
25 *blo gros c'ung rung skyor sñing „dod*	25 Wenn auch mein Verstand gering, mein Herz begehrte des Lehrers Worte zu wiederholen!
bsams šing mts'an ldan bla ma dran	In meinen Gedanken erinnere ich mich des herrlichen Lama,
bsgoms šing Mar pa lo tsts'a dran	In meinen Betrachtungen erinnere ich mich an Marpa, den Gelehrten.
sñan brgyud brda yi dbang bzhi de	Wenn er die vier Kräfte der Worte mündlicher Unterweisung
da lta bskur na dga ba la	Jetzt überlieferte, wie freute ich mich!

30 *nga „bul ba c'ung c'ung rung zhu sñing „dod*

    dbang yon med rung zhu sñing „dod

    k'rid zab mo Nāro c'os drug de

    da lta gsung na dga ba la

    nga sñing rus c'ung rung zhu sñing „dod

35 *nga sgom sran c'ung rung bsgom sñing „dod*

    bsam zhing ts'an ldan bla ma dran

    bsgoms šing Mar pa lo tsts'a dran

    mc'ed grogs dbus gtsang dang „dus de

    da lta bzhugs na dga ba la

40 *nga ñams rtogs ngan rung bsdur sñing „dod*

    go ba zhan rung bsdur sñing „dod

30 Obwohl meine Gaben nur klein sind, begehrte mein Herz zu bitten;

Obwohl ich nicht über Opferspenden verfüge, begehrte mein Herz zu bitten.

Wenn er die tiefe Belehrung, Nāro's sechs Vorschriften

Jetzt verkündete, wie freute ich mich!

Wiewohl meine Ausdauer nur schwach ist, begehrte mein Herz zu bitten;

35 Wiewohl der Faden meiner Beschauung nur klein ist, begehrte mein Herz Beschauung.

In meinen Gedanken erinnere ich mich des herrlichen Lama,

In meinen Betrachtungen erinnere ich mich an Marpa, den Gelehrten.

Wenn er, sich in den reinen Kreis der geistlichen Brüder gesellend,

Jetzt dort weilte, wie freute ich mich!

40 Wie schlecht auch meine Erkenntnis, mein Herz begehrte doch sie zu sammeln;

Wie schwach auch mein Verständnis, mein Herz begehrte doch sie zu sammeln.

*bsam zhing mts'an ldan bla
ma dran*

*sgom zhing Mar pa lo tsts'a
dran*

*sprang mos gus kyi ngang
nas ‚bral med kyang*

45 *bla ma sñing nas dran pa
yis*

*‚dod pas gdungs nas bzod
blags med*

*dbugs stod du ‚ts'angs nas
skad ma t'on*

*bui gdung ba sel cig bka
drin can*

*ces gsungs pas | sprin de dar
sna lngai yug brkyang ba lta
bu zhig tu gyur pai rtse la |
rje Mar pa de ñid sngar bas
kyang gzi brjid c'e ba zhig
seng ge dkar mo rgyan du
mas brgyan pa gcig la c'ibs
nas byon te | t'u c'en k'yod
da lan gdung ba drag pos
nga ‚bod pa de ci byung | bla
ma dang yi dam dkon mc'og
la yid c'ad dam | rtog pa
rkyen ngan gyi yul p'yir
‚brengs sam | sgrub k'ang du*

In meinen Gedanken erinnere
ich mich des herrlichen
Lama,

In meinen Betrachtungen erinnere ich mich an Marpa,
den Gelehrten.

Obwohl mein Wesen von in-
brünstiger Andacht un-
zertrennlich,

45 Ist durch die Erinnerung an
des Lama Herz,

Durch das Verlangen und die
Sehnsuchtqual meine Ge-
duld erschöpft.

Der Atem stockt in meiner
Brust, dass mir die Stimme
versagt.

Stille deines Sohnes Sehnsuchtqual, du Gnaden-
reicher!

Als er so gesprochen, erschien
auf der Spitze jener Wolke,
welche die Gestalt eines fünf-
farbigen, ausgebreiteten Seiden-
gewandes angenommen hatte,
der erhabene Marpa selbst, in
herrlichem Glanze wie nie zuvor,
auf einer reich geschmückten
weissen Löwin reitend, und
sprach: „Du mein älterer Bruder,
sprich, welcher heftige Schmerz
ist dir widerfahren, dass du mich
rufst? Verzweifelst du an dem
Lama, den Schutzgottheiten und

*c'os brgyad kyi bar c'ad zhugs
sam | re dogs kyi ,gong pos
blo rtser*

*yid sun nam | yang yar bla
ma dkon mc'og la zhabs tog |
mar rigs drug gi sems can
la sbyin gtong | bar du k'yod
rang la sdig sgrib dag pa dang
yon tan skye bai ,t'un rkyen
bzang po byung ba yin nam |
ci yin rung k'yod dang nga
,bral t'abs med kyis | sgrub pas
bstan pa dang sems can gyi
don gyis šig gsung bai ñams
zhig šar bas | dges drags bzod
blags med pai ngang nas |
ñams šar gyi gsung lan du
mgur ,di bzhes so.*

*p'a bla mai zhal t'ong gsung
t'os pa*

50 *sprang poi sñing rlung ñams
su šar*

*bla mai rnam t'ar dran pa
yis*

*rtogs pa mos gus gting nas
skyes*

dem höchsten Gut? Spürst du
den Dingen mit unheilvollem
Skrupel nach? Oder haben sich
in deiner Betklause Hindernisse
durch die Lehren der Welt ein-
gestellt? Bist du vom Dämon
der Furcht und Hoffnung gequält und missmutig gemacht?
So sei denn wieder oben der
Lama und des höchsten Gutes
Diener! Bring' unten den Wesen
der sechs Klassen Gaben dar!
Dann werden dir selbst im Zwi-
schenraum die Sündenflecken
rein und gute Gelegenheiten zur
Erzeugung von Vorzügen zu teil
werden. Oder was es auch sein
mag, ich werde mich nicht von
dir trennen. Indem du durch
beschauliches Bannen lehrst,
stifte den Wesen Nutzen!" Diese
Worte flössten ihm Mut ein,
und in grosser Freude sang er
zum Dank für die Ermutigung
folgendes Lied:

Da ich des väterlichen Lama
    Antlitz geschaut, seine
    Worte gehört habe,
50 Hat sich des Bettlers Herzens-
    kummer aufgerichtet.
In der Erinnerung an des
    Lama Lebensgeschichte
Ist beschauliche Andacht in
    der Seele Tiefen erwacht.

t'ugs rje byin brlabs dngos
su zhugs

c'os min snang ba t'ams cad
.gags

55 bla mai dran pai gdung
dbyangs .di

rje btsun sñan la gzan lags
kyang

sprang la snang ba .di las
med

da dung bgyid do t'ugs rjes
skyongs

sñing rus sdug sran gyi sgrub
pa .di

60 p'a bla ma mñes pai zhabs
tog yin

gcig spur ri k'rod .grim pa
.di

mk'a .gro mñes pai zhabs
tog yin

rang rtsis med pai dam c'os
.di

sangs rgyas bstan pai zhabs
tog yin

65 ts'e dang bsgrub pa sñoms
po .di

mgon med sems can gyi
sbyin gtong yin

na dga ši skyid kyi sñing
rus .di

Des Erbarmers Segen hat sich
verkörpert,

Unfromme Bilder sind alle
entrückt.

55 Dieser Sehnsuchtsang der Erinnerung an den Lama

Erklingt zwar in des Ehr-
würdigen Ohren,

Doch dem Bettler bleibt nichts
als die blosse Vorstellung
davon.

Noch einmal will ich ver-
suchen: Erbarmer, schütze
mich!

Der ausharrende, gegen Ungemach abgehärtete Asket
hier

60 Ist des väterlichen Lama freudiger Diener,

Ist der allein in der Bergwildnis umherwandelnden
Ḍâka freudiger Diener,

Ist der dünkellosen heiligen
Religion,

Der Buddha-Lehre, Diener,

65 Gleichgültigkeit gegen Leben
und Streben

Bedeutet Gabenspenden für
die schutzlosen Wesen;

Freudiges Ausharren im
Wechsel von Krankheit,
Alter und Tod

las sdig sgrib sbyong bai c'ags śing yin	Erzeugt Reinigung von der Sündenbefleckung durch die Werke;
sdig zas spangs pai dka t'ub ͵di	Sündige Speisen meidende Askese
70 *ñams rtogs skye bai ͵t'un rkyen yin*	70 Fördert die Entstehung guter Erkenntnis.
p'a bla mai drin lan sgrub pas ͵jal	Des väterlichen Lama Wohlthaten vergelte ich durch beschauliche Bannungen;
bu t'ugs rjes skyongs cig gu ru rje	Deinen Sohn schütze erbarmungsvoll, Lehrer und Herr!
sprang ri k'rod bzin bar byin gyis rlobs.	Segne des Bettlers Wahl der Bergeinsamkeit!

ces gsung | t'ugs spro sing nge bai ngang nas ras gos de bzhes | śing mts'on cig k'yer nas sgrub k'ang du byon pas | spyil po na lcags kyi ͵a tsa ra mig p'or k'og tsam bgrad pa lnga ͵dug pa las | gcig rje btsun gyi gzim mal na bsdad nas c'os ͵c'ad | gñis kyis ñan | gcig gis llo yyo | gcig gis p'yag dpe rnams yyeng zhing ͵dug pas | t'og mar zil bun pa zhig byung | de rjes gzhi bdag ma dga pai c'o ͵p'rul yin pa ͵dug | nga gnas gang du sdod rung glor ma btang rgyu ni med | bstod pa re ma byas pa med pas | gnas ͵di la ͵ang bstod pa zhig byed dgos pa yin yong

So sprach er. Von grosser Freude erfüllt, ergriff er sein Baumwollgewand und kehrte mit einer Handvoll Holz in seine Betklause zurück. Da waren in der Hütte fünf eiserne Kobolde (Atsara) mit Augen wie Gefässe hervorquillend. Einer sass auf des Ehrwürdigen Lagerstätte und trug aus einem geistlichen Buche vor, zwei lauschten seinem Wort; der vierte bereitete Essen; der fünfte war damit beschäftigt, die übrigen Bücher zu durchstöbern. Anfangs überlief ihn doch ein kleiner Schauder. Dann kam ihm der Gedanke: „Das ist das Zauberspiel, an dem die Ortsgöttin ihre Freude hat; an wel-

dgongs te | gnas la bstod pai
mgur ˏdi gsungs so ||

chem Orte ich auch weilen mag,
ein Streuopfer brauche ich nicht
darzubringen. Doch da ich die
edle Herrin noch nicht gepriesen
habe, muss ich wohl ein Lob-
lied auf diesen Ort verfassen."
Damit stimmte er folgenden
Lobgesang auf den Ort an:

ˏe ma ri kʻrod dben pai gnas

O du Einsiedelei in der Bergeinsamkeit,

75 *mcʻog rgyal bas byang cʻub*
brñes pai sa

75 Stätte, wo die herrlichen Jina
die Bodhi erlangen,

grub tʻob rnams kyis bshugs
pai šul

Gefilde, wo die heiligen
Männer weilen,

ming rang gcig bur ˏdug pai
yul

Ort, wo ich der einzige Mensch
jetzt bin!

brag dmar ˏCʻong lung kʻyung
gi rdzong

Rotfels ˏChong lung, Adler-
horst,

steng na lho sprin kʻor ma
kʻor

Ueber dir ballen sich des
Südens Wolken,

80 *ˏog na gtsang cʻab rgya ma*
gyu

80 Unten schlängeln Flüsse sich
in schnellem Lauf,

bar na rgod po lang ma ling

In der Luft schwebt der Geier
kreisend;

rtsi šing sna tsʻogs ban ma
bun

Die artreichen Waldbäume
säuseln,

ljon šing gar stabs šigs se
šigs

Prachtbäume wiegen sich nach
Tänzerart;

bung ba glu len kʻor ro ro

Bienen summen ihr Liedchen
khorroro,

85 *me tog dri ngad cʻil li li*

85 Blumen strömen Duft aus
chillili;

bya rnams skad sñan kyur
ru ru

Vögel zwitschern wohllautend
kyurruru;

*de ˛drai brag dmar ˛C'ong
lung na*

*bya dang byiu šog rtsal
sbyong*

*spra dang spreu yang ˙rtsal
sbyong*

90 *ri dags sna ts'ogs baṅ rtsal
sbyong*

*nga Mi la ras pa ñams rtsal
sbyong*

*ñams rtsal byang c'ub sems
gñis sbyong*

*nga dang dben gnas gźhi
bdag mt'un*

*k'yed ˛dir ts'ogs ˛byung po
mi ma yin*

95 *byams sñing rjei bdud rtsi
˛di ˛t'ung la*

*rang rang so soi gnas su
dengs*

*ces gsungs pas | ˛a tsa ra de
rnams rje btsun la mi dga bai
rnam pas p'an ts'un sdang
mig gi ˛gros šig šig pai ngaṅ
nas | ˛a tsa ra gñis bsnan te
bdun du song bai la las rngams
gtsigs gdams | la la so k'rig
k'rig byed cing mc'e ba gtsigs |
la la rgod cing skad drag po
˛don | kun dril nas brdeg
stangs dang sdigs mo byed
du byung bas | mi ma yin gyi
bar c'ad brtsom par dgongs*

Auf diesem Rotfels ˛Chong
lung

Ueben Vögel und Vöglein des
Fittichs Behendigkeit,

Ueben Affen und Aefflein sich
im Wettsprung,

90 Ueben Hirsch und Reh sich
im Wettlauf;

Ich, Milaraspa, übe geistige
Geschicklichkeit,

Geistige Geschicklichkeit und
innere Heiligkeit übe ich:

Ich bin mit der Ortsgottheit
der Einsiedelei in fried-
licher Eintracht.

Gespenstige Unholde, die ihr
hier versammelt seid,

95 Trinkt den Saft der Liebe
und des Erbarmens

Und weicht, jeder an seinen
Ort, von hinnen!

So sang er. Die Kobolde aber,
aus Aerger über den Ehrwürdigen, rollten zornfunkelnd die
Augen hin und her. Die Kobolde
vermehrten sich um zwei, und
als sie so sieben geworden,
fletschten einige wutschnaubend
die Zähne, andere knirschten und
wetzten die Hauer, andere lachten
und brüllten in grauenvollen
Tönen. Dann stürzten alle,
zu einem Knäuel geballt, in
Kampfbereitschaft mit drohen-

*nas | kʻro boi lta stangs kyi
ngang nas drag sngags bzlas
rung ma song | sñing rje cʻen
po tʻugs la ͻkʻrungs nas cʻos
bšad rung ͻgror ma ͻdod pai
tsʻe | tʻugs la nga la lho brag
Mar pas cʻos tʻams cad rang
sems su ngo sprad cing | rang
sems ͻod gsal stong par kʻo
tʻag cʻod pa la | gdon bgegs
pʻyi rol du bzung nas | sod
na dga ba ͻdi byed don mi
ͻdug dgongs | mi ͻjigs pai
gdeng mngon du gyur nas |
lta ba gdeng dang ldan pai
mgur ͻdi gsungs so.*

den Gebenden auf ihn zu. Da
sann er, den Unholden Unheil
zu bereiten. Mit magischem
Zornesblick rezitierte er schreckliche Beschwörungen, aber sie
gingen dennoch nicht. Voll von
grossem in seinem Herzen erwachenden Erbarmen trug er
ihnen geistliche Lehren vor,
dennoch bezeigten sie keine Lust
zu weichen. Da dachte er:
„Mir hat zwar Marpa aus Lho
brag alle Lehren aus seinem
Innersten erklärt; doch da mein
Innerstes der Erleuchtung bar
ist, werde ich von Verzweiflung
ergriffen. An die Wirklichkeit
der Kobolde und Dämonen muss
ich wohl glauben; sie könnten
mich daher töten, und ihnen
diese Freude zu bereiten wäre
nutzlos." Indem sich seine un-
erschütterliche Zuversicht so
offenbarte, stimmte er folgendes
beschauliche und zuversichtliche
Lied an:

*pʻa bdud bzhii dmag dpung
las rgyal ba*

Vor dem Vater, dem Sieger
über das Heer der vier
Māra,

*sgra bsgyur Mar pai zhabs
la ͻdud*

Dem sprachgewandten Marpa
zu Füssen verneige ich
ᵛmich!

*ming ngag nga rang mi
zer te*

Meinen Namen nenne ich
nicht.

100 *nga ni ngar seng dkar moi*   100 Ich bin der starken weissen
     bu   Löwin Sohn:

¸a mai mngal nas rtsal gsum   Im Mutterleibe besass ich die
     rdzogs   drei Fertigkeiten vollen-
                                 det;

p῾rug gui lo la ts῾ang du ñal   In den Kinderjahren lag ich
                                 im Neste,

t῾ong poi lo la ts῾ang sgo   In den Jahren der Jugend-
     bsrungs   blüte hütete ich des Nestes
                                 Thür,

dar mai lo la gangs stod   In den Jahren der Mannheit
     ¸grims   wanderte ich über die
                                 Gletscher.

105 *nga gangs bu yug ¸ts῾ubs*   105 Wenn auch Schneesturm die
     rung ya mi nga   Gletscher umwirbelt, bebe
                                 ich nicht;

brag γyangs c῾e rung bag   Ist der Felsenabgrund auch
     mi ts῾a   gross, ich fürchte nicht.

ming ngag nga rang mi zer te   Meinen Namen nenne ich
                                 nicht.

nga ni bya rgyal k῾yung gi   Ich bin des Königs der Vögel,
     bu   des Adlers Sohn:

sgo ngai nang nas gšog sgro   Im Inneren des Eies ent-
     rgyas   wickelten sich die Fittiche;

110 *p῾rug gui lo la ts῾ang du*   110 In den Kinderjahren lag ich
     ñal   im Neste,

t῾ong poi lo la ts῾ang sgo   In den Jahren der Jugend-
     bsrungs   blüte hütete ich des Nestes
                                 Thür,

k῾yung c῾en dar mai lo la   In den Mannheitsjahren des
     nam ¸p῾angs cad   grossen Aars kreuzte ich
                                 des Himmels Höhe.

nga nam mk῾a c῾e rung ya   War der Himmel auch gross,
     mi nga   ich bebte nicht;

sa lung srol dog rung bag mi ts'a	Wenn die Thäler der Erde auch dunkelten, ich fürchtete nicht.
115 *ming ngag nga rang mi ẕer te*	115 Meinen Namen nenne ich nicht.
nga ni ña c'en yor moi bu	Ich bin der Sohn der Woge, des grossen Fisches:
‚*a mai mngal nas gser mig „k'yil*	Im Mutterleibe rollte ich meine goldenen Augen;
p'rug gui lo la ts'ang du ñal	In den Kindheitsjahren lag ich im Neste,
t'ong pai lo la k'yu sna drangs	In den Jahren der Jugendblüte zog ich an der Spitze der wimmelnden Schar,
120 *ña c'en dar mai lo la mts'o mt'a bskor*	120 In den Mannheitsjahren des grossen Fisches umkreiste ich die Grenzen des Meeres.
nga mts'o rba rlabs drag rung ya mi nga	Waren auch schrecklich die Meereswogen, ich bebte nicht;
dol lcags kyu mang rung bag mi ts'a	Waren der Fischer Eisenhaken auch zahlreich, ich fürchtete nicht.
ming ngag nga rang mi ẕer te	Meinen Namen nenne ich nicht.
nga ni bka brgyud bla mai sras	Ich bin der Sohn des Lama der mündlichen Ueberlieferung:
125 ‚*a mai mngal nas dad pa skyes*	125 Im Mutterschosse erwuchs mir der Glaube,
p'ru gui lo la c'os sgor ẕhugs	In den Kindheitsjahren trat ich in das Thor der Religion ein,

Archiv für Religionswissenschaft. IV. Band, 1. Heft. 2

- 93 -

t'ong pai lo la slob gñer byas	In den Jahren der Jugend-blüte widmete ich mich dem Studium,
sgom c'en dar mai lo la ri k'rod ₀grims	In den Mannheitsjahren des grossen Mystikers durch-wanderte ich die Berg-wildnis.
nga ₀dre gdug rtsub c'e rung ya mi nga	Ist auch der Geister Bosheit gross, ich bebe nicht;
130 *gdon c'o ₀p'rul mang rung bag mi ts'a*	130 Wie oft auch der Dämonen Gaukelspiel wiederkehrt, ich fürchte nicht.
seng gangs la ₀gying ba spar mi ₀k'yag	Der auf dem Gletscher sich reckende Löwe friert nicht an den Pfoten:
sengge gangs la spar ₀k'yag na	Wenn der Löwe auf dem Gletscher an den Pfoten fröre,
rtsal gsum rdzogs pa don re c'ung	Wäre der Nutzen der drei vollendeten Fertigkeiten wahrlich gering!
k'yung mk'a la ₀p'ur ba llung mi srid	Der am Himmel fliegende Aar kann nicht herab-stürzen:
135 *k'yung c'en mk'a nas llung srid na*	135 Wenn der grosse Aar vom Himmel herabstürzen könnte,
gšog sgro rgyas pa don re c'ung	Wäre der Nutzen der ent-wickelten Flugkraft wahr-lich gering!
ña c'ab la ₀p'yo ba ₀ts'ubs mi srid	Der im Wasser schwimmende Fisch kann nicht ertrinken:
ña c'en c'u yis ₀ts'ub srid na	Wenn der grosse Fisch im Wasser ertrinken könnte,

c͑u nang skye pa don re
c͑ung

140 lcags kyi p͑a p͑ong rdos mi
šigs

lcags kyi p͑a p͑ong rdos
šigs na

zhun t͑ar byas pa don re
c͑ung

nga Mi la ras pa ͒dres mi
͒jigs

Mi la ras pa ͒dres ͒jigs na

145 gnas lugs rtogs pa don re
c͑ung

k͑yed ͒dir byon ͒dre gdon
bgegs kyi ts͑ogs

da lan yongs pa ngo mts͑ar
c͑e

mi rings blong la rgyun du
bzhugs

bka mc͑id gsung gleng zhib
tu bgyid

150 rings rung do nub cis kyang
sdod

͒o skol sgo gsum rtsal ͒gran
zhing

dkar nag c͑os kyi c͑e k͑yad
lta

Wäre der Nutzen der Geburt
im Wasser wahrlich gering!

140 Ein eiserner Felsblock wird
durch Steine nicht zerstört:

Wenn ein eiserner Felsblock
durch Steine zerstört würde,

Wäre der Nutzen des Schmel-
zens wahrlich gering!

Ich, Milaraspa, fürchte keine
Geister:

Wenn͑ Milaraspa Geister
fürchtete,

145 Wäre der Nutzen von der
Kenntnis des Wesens der
Dinge wahrlich gering!

Ihr, die ihr hier erschienen
seid, Geister und Kobolde,
Scharen der Widersacher,

Vor allem meinen Dank für
eueren Besuch!

Wenn ihr keine Eile habt,
weilet nur immer hier!

Ergeht euch nur in ausführlichen Reden und Dis-
kursen!

150 Wenn ihr's auch eilig habt,
bleibt unter allen Um-
ständen heute Abend hier!

Im Wettstreit mit der Ge-
wandtheit von Leib, Wort
und Gedanke

Wollen wir einmal den Wert
der weissen und schwarzen
Lehre messen!

2*

k'yed bar c'ad ma ts'ugs mi
ldog yong

bar c'ad ma t'sugs log gyur
na

155 *da lan yongs pa k'a re*
skyengs.

zhes gsungs nas t'ugs dam gyi
nga rgyal bzhengs te nang du
t'al gyis byon pas | ͺa tsa ra
rnams ͺjigs skrag dngangs
nas mig rig rig byed cing |
lus ͺdar bai stobs kyis p'ug
pai nang kun kyang ͺdar yam
me ba zhig byas te

rings stabs su gcig la kun
t'ims pai mt'ar | gcig po yang
rlung ts'ub k'yil pa zhig byas
nas med par song ba dang |
rje btsun gyi t'ugs dgongs la |
bgegs kyi rgyal po Bhi na ya
kas glags ts'ol du ͺdug ͺsnga
mai rlung de yang k'oi c'o
ͺp'rul yin par ͺdug ste | bla
mai t'ugs rjes glags ma rñed
dgongs pa dang ͺde rjes t'ugs
dam la bogs bsam gyis mi
k'yab pa byung ba yin no |

Ihr wollt nicht zurückkehren,
ohne mir einen Streich zu
spielen!

Doch wenn ihr, ohne mir
einen Streich zu spielen,
zurückkehren werdet,

155 Müsst ihr euch schämen, noch
einmal wiederzukommen!

Mit diesen Worten erhob er
sich im Stolz seiner Beschauung
und ging ohne Zaudern hinein.
Da wurden die Kobolde von
Furcht und Schrecken ergriffen
und blickten mit scheuen Augen
um sich. Sie zitterten so am
ganzen Leibe, dass die ganze
Höhle im Innern zitterte und
bebte.

Schleunigst schwanden die Sieben zu Einem zusammen, und
auch dieser Eine fuhr schliesslich in einem kreisenden Wirbel-
wind von dannen. Der Ehr-
würdige dachte: „Der Dämonen-
könig Vināyaka trachtet nach
einer Gelegenheit, mir zu schaden;
der Sturm vorher wird wohl
auch sein Zauberwerk gewesen
sein. Aber dank dem Erbarmen
meines Lama wird er weiter
keine Gelegenheit mehr finden."
Darauf wurde ihm eine unermessliche Förderung in der Meditation zu teil.

bgegs kyi rgyal po Bhi na ya
kas rkyen byas pa | bla ma
dran drag gi bskor ram | brag
dmar mC'ong tung gi skor
ram | Mi la šing ťun gyi bskor
zhes kyang bya ste | don gcig
la mts'an gsum ldan gyi bskor
ćo dgeo ||

Dieser Abschnitt heisst: Des Dämonenkönigs Vināyaka Spuk und die glühende Sehnsucht nach dem Lama, oder der Rotenfels ‚Chong lung, oder Mila, der Holzsammler, auch kurz der Abschnitt mit drei Namen.

2. Milaraspa auf dem La phyi.

rnal ‚byor gyi dbang p'yug
rje btsun Mi la ras pa de ñid |
‚c'ong lung brag la bzhugs pa
las | bla mai bka sgrub p'yir
ťod la brgyud de | la p'yi
gangs kyi ra ba la bsgom du
byon pas la p'yii gnas sgo
gña nang rtsar mar p'yag p'ebs
pai ts'e | rtsar ma pa rnams
ć'ang sa ć'en po zhig byed
cing yod pai gral gtam la | da
lta Mi la ras pa zhes bya ba |
‚ts'o ba dka ťub rang la brten
nas | gnas mi med kyi ri k'rod
rang la bzhugs pai ć'os rnams
dag rang zhig yod par ‚dug
ces | rje btsun gyi sñan pa
brjod cing yong pai ts'e | rje
btsun k'ong ts'oi sgo drung du
p'ebs pas | bud med gzhon nu
ma gzhin dang rgyan bzang
pos spras pa zhig p'yir ťon
byung ba de legs se ‚bu | ma
yin te mo na re | rnal ‚byor

Nachdem der Gebieter der Yogin, der ehrwürdige Milaraspa, auf dem Felsen ‚Chong lung verweilt hatte, überschritt er, um das Gebot seines Lama zu erfüllen, die Schwelle seines Hauses und machte sich auf den Weg, um sich auf dem gleichsam eingezäunten Schneeberg La phyi der Meditation hinzugeben. Als er im Dorfe rTsar ma in gÑa nang am Aufgang zum La phyi anlangte, da waren gerade die Männer von rTsar ma zu einem grossen Gelage in ihrer Bierschenke versammelt und ergingen sich in Gesprächen: „Da ist doch jetzt ein gewisser Milaraspa, der sich nur heilenden Bussübungen hingiebt; ganz allein haust er dort nach den Vorschriften just in der menschenleeren Bergwildnis.“ So unterhielten sie sich

pa k'yod gang nas yin zer ba
la | rje btsun gyis nga nges
med kyi ri la bsdod pai sgom
c'en Mi la ras pa zer ba zhig
yin | yon bdag mo k'yod la
bza btung zas kyi „brel pa
zhig slong bas ster dgos gsungs
pas | mo na re | „ts'o ba drangs
pas c'og | k'yed Mi la ras pa
zer ba de yin nam zer bas |
rje btsun gyis | de la brdzun
zer don med gsungs pas | mo
dga nas slab stob tu nang du
p'yin te gral ba rnams la
snga gong t'ag ring na „dug
zer bai c'os pa dga mo de |
da lta rang rei sgo drung na˙
p'ebs gdao byas pas | kun p'yir
t'on te la la p'yag „bul | la
las lo rgyus zhib tu dris nas |
rje btsun yin par nges pa
dad nang du spyan drangs
bsñen bkur p'un sum ts'ogs
pa p'ul te | kun dad dad mos
mos su „dug pa la | gral mgo
na yon bdag p'yug po gzhon
nu ma zhig „dug pa de | gsen
rdor mo yin te | k'os rje btsun
la zhal ta zhib rgyas zhus
pai t'ar | da bla ma gang du
„byon zer ba la | nga la p'yi
la bsgom du „gro gsungs pas |
k'o na re | de bas nged kyi
„dre lung skyog mor bzhugs

in Lobsprüchen über den Ehrwürdigen. In diesem Augenblick trat der Ehrwürdige an
ihre Thür. Eine junge, wohl-
gebildete und schön geputzte
Frau, die einer hübschen, sich
öffnenden Rose gleich war, trat
gerade heraus und sagte: „Heiliger, woher kommst du?" Der
Ehrwürdige erwiderte: „Ich bin
der hier irgendwo im Gebirge
hausende grosse Mystiker Milaraspa. Mildthätige, du solltest
mir ein wenig Speise und Trank
reichen, da ich dich darum bitte."
Die Frau sprach: „Du magst
Nahrung vorgesetzt erhalten.
Bist du denn wirklich jener
Milaraspa oder sagst du es
bloss?" Der Ehrwürdige antwortete: „Welchen Vorteil hätte
es, zu lügen?" In freudig auf-
geregter Hast eilte die Frau
hinein und sagte den Biergästen:
„Jener heilige Mönch, von dem
ihr vorhin gesagt habt, dass er
weit entfernt wohne, ist hier an-
gekommen und steht jetzt vor
unserer Thür." Alle stürzten
hinaus. Einige brachten ihm
sogleich Verehrung dar, andere
forschten ihn zuerst umständlich
über seine Geschichte aus, und
als sie die Ueberzeugung ge-

nas sa gzhi byin gyis rlob dgos | zhabs tog kyang ,o mi rgyal ba ,bul zer bas | de nas ston pa zhig ,dug pa na re | yon mc'od gñis t'un byung | la p'yi ka ,dre lung skyog moi ming yin | da bla ma der bzhugs pa byung na bdag kyang ci lcogs kyi zhabs tog ,bul zhing c'os zhu zer te |

wonnen, dass es wirklich der Ehrwürdige sei, führten sie ihn hinein und erwiesen ihm ausgezeichnete Ehrenbezeugungen. An der Spitze des Tisches befand sich die Wirtin, eine wohlhabende junge Frau mit Namen gŠen rdor mo, die den Ehrwürdigen mit aller Sorgfalt bediente und ihn schliesslich fragte: „Wohin begiebt sich jetzt der Lama?" „Ich gehe zu beschaulicher Uebung auf den La phyi", sagte er. Sie sprach: „Du solltest doch erst ein wenig in deinem ,Dre lung skyog mo bleiben und den Erdboden dort segnen; wir werden unermüdlich zu deinen Diensten sein." Darauf sagte ein dort gerade anwesender Lehrer: „Gabenspender und Gabenempfänger mögen sich einigen! Der La phyi ist die Bezeichnung von ,Dre lung skyog mo. Wenn nun der Lama dort verweilen wird, werde ich ihn nach Kräften mit allem Nötigen versehen und dafür religiöse Belehrung erbitten."

ston pa Šākya gu na yin no | de sjes yon bdag de na re | des na ,a la la nged kyi ,brog sa dga mo rang zhig yod pa la | ,drei gnod pa mngon sum du

Der Sprecher war der Lehrer Çākyaguna. Darauf sagte die Wirtin: „Das ist wahrhaftig vortrefflich! Dir erscheint die Einsamkeit sogar als Genuss, wäh-

yod pa mang drags ₉jigs nas
₉brog ts'ugs su ye ma ₉dod
na myur ₉p'ebs par zhu zer
ts'ogs pa kun gyis p'yag p'ul
bas | rje btsun gyi zhal nas |
nga myur du ₉gro ste k'yed
kyi ₉brog gi p'yir ₉gro ma
yin | nga la bla mai bka zhig
yod pas de sgrub tu ₉gro ba
yin gsungs pas | k'ong rnams
na re nged la skyes c'og | nga
₉ts'o c'as bzang po dang bcas
te zhabs tog pa gtong zer bas |
rje btsun gyi zhal nas | nga
ri k'rod du grogs dang ₉ts'o
ba bzang po rang brten pai
mi zhig min pas | t'og mar
nga rang gcig pur ₉gro | k'yed
rnams zhabs tog byed pa ngo
mts'ar c'e bas rting sor lta
gsung | rje btsun rang la p'yi
gangs la byon pas |

la rtser p'ebs pa nas mi
ma yin gyi c'o ₉p'rul drag po
byung | lai gzug p'ebs ₉p'ral
nam mk'a ts'ubs | drag poi

rend wir andere uns vor den
sichtbaren Streichen der Kobolde schrecklich fürchten und
gar keine Lust haben, uns in
der Bergeinsamkeit anzusiedeln,
vielmehr lieber schnell hinab-
steigen." Die ganze Versammlung bezeigte ihm Verehrung.
Da sprach der Ehrwürdige: „Ich
muss bald gehen, aber der Einsamkeit wegen, wie du sagst,
gehe ich nicht; ich habe nämlich ein Gebot von meinem
Lama erhalten, und um dieses
zu erfüllen, muss ich gehen."
Sie erwiderten: „An Geschenken
soll es dir nicht fehlen. Wir
werden dir mit guten Nahrungsmitteln zu Diensten sein." Der
Ehrwürdige sprach: „Ich habe
in der Bergwildnis keinen
Freund, keine gute Nahrung
noch einen zuverlässigen Menschen, doch fürs erste will ich
selbst allein gehen; für euere
Dienste danke ich euch, später
werden wir das Weitere sehen."
Da machte sich der Ehrwürdige
selbst zum Schneeberge La p'yi
auf.

Als er die Passhöhe über-
stieg, entstand ein schreck-
licher Gespensterspuk. In dem
Augenblick, als er den Berg-

*„brug lding zhing | glog „k'yug
pa dang | lung pa p'an ts'un
gyi ri nur nas | p'u c'u „k'yil
te rba rlabs drag po „k'rug
pai mts'o c'en po zhig tu gyur
pa la | rje btsun gyis lta stangs
mdzad p'yag mk'ar bsnan pas
ts'o zhabs nas zags te med
par song ba la rmu rdzing du
grags |*

*de nas mar cung zad byon
pas | mi ma yin rnams kyis
ri p'an ts'un bsñil bai bud la |
p'a p'ong mang po rba rlabs
„k'rug cing byung pai ts'e mk'a
„gros lung pa p'an ts'un gyi
bar du | ri sbrul t'ur du rgyugs
pa „dra ba zhig gi lam p'ul
baš | rbab zhi bai lam de la
mk'a „gro sgang lam du grags |*

*de nas mi ma yin stobs
c'ung ba rnams rang zhin la
subs | c'e ba rnams glags mi
rñed rung da dung glags ts'ol
ba la mk'a „gro sgang lam
rdzogs „ts'ams su | rje btsun*

gipfel erreichte, umwölkte sich der Himmel. Furchtbarer Donner krachte, Blitze zuckten rings umher, Berge zu beiden Seiten der Thäler zerbröckelten, und Giessbäche stürzten wirbelnd zusammen und bildeten einen grossen See mit furchtbar tosendem Strudel. Der Ehrwürdige entsandte seinen magischen Blick und schlug mit seinem Stabe auf. Da floss der See auf dem Boden ab und verschwand. Er ist bekannt als der „Teufelsteich" (rmu rdzing).

Als er darauf ein wenig weiter nach unten kam, schleuderten die Geister zu beiden Seiten des Berges in einer Staubwolke viele Felsblöcke wie tosende Wogen hinab. Da erschienen zu beiden Seiten des Thales die Ḍâka und machten gleich herabschiessenden Bergschlangen den Weg frei. So hörte denn das Trümmergeröll auf. Dieser Weg ist bekannt als der „Pfad des Felsvorsprungs der Ḍâka".

Darauf liessen die Geister von schwacher Kraft von selbst ab; die stärkeren indessen suchten, obwohl sie keine Gelegenheit fanden, noch einmal nach einer Gelegenheit, ihm beizukommen.

gyis log „dren ꞩil gnon gyi
lta ꞩtangs ꞩhig mdꞩad pas |
c̓o „p̓rul kun ꞩhi nas bꞩhugs
ꞩa der rdo la ꞩhabs rjes
„byung |

de nas cung ꞩad ꞩhig byon
pas | nam mk̓a dangs nas
t̓ugs ꞩpro bar gyur te ꞩgang
ꞩhig tu bꞩhugs pas | sems can
rnams la byams pai ting nge
„dzin „k̓rungs pas t̓ugs dam
la bogs ꞩin tu c̓e bar byung
ꞩte bꞩhugs ꞩa de la byams
ꞩgang du grags |

de nas c̓u bꞩang du byon
nas c̓u bo rgyun gyi rnal
„byor gyi ngang la bꞩhugs pai
tꞩ̓e | me p̓o ꞩtag gi lo ꞩton
ꞩla ra bai tꞩ̓es bcui nub |

bal poi Bha roi rnam pai
gdon c̓en po ꞩhig gis gtso

Da wurden sie gerade an der Stelle, wo der Pfad des Felsvorsprungs der Ḍâka endete, durch den die Anfechtungen besiegenden Zauberblick des Ehrwürdigen zurückgeschlagen. Nun war der ganze Spuk zu Ende. An jener Stelle blieb in einem Steine seine Fussspur zurück.

Darauf schritt er eine Strecke weiter. Der Himmel heiterte sich auf, und in froher Stimmung liess er sich auf einem Felsvorsprung nieder. Dort versenkte er sich in eine Betrachtung der Liebe zu allen Geschöpfen, wodurch ihm reicher Gewinn an innerer Förderung zu teil ward. Davon heisst dieser Platz der „Felsvorsprung der Liebe".

Darauf kam er zum Flusse Chu bzang (gutes Wasser). Als er sich in die Betrachtung des strömenden Wassers versenkte, ging der 10. Tag des ersten Herbstmonats des Feuer-Tiger-Jahres (10. September 1085)[1] zur Rüste.

Da erschien, von dem grossen nepalesischen Dämonenfürsten

---

[1] Die Jahreszahl 1083 giebt die chronologische Tafel Reu mig an als die Zeit, zu der Milaraspa Askese übte, um die Heiligkeit zu erlangen. Siehe Journal of the Asiatic Society of Bengal 1889, p. 43.

byas pai mi ma yin gyi dmag |
c'u bzang lung pai gnam sa
gang ba ˌongs nas | rje btsun
la ri bsñil ba dang | t'og la
sogs te mts'on c'ai c'ar drag
po ˌbebs pa dang | mts'an nas
bos te bzung zhig | gsod cig
la sogs te mi sñan pai sgra
sgrogs cing | mi sdug pai gzugs
du ma ston du byung bas | mi
ma yin gyis glags ts'ol du
ˌdug dgongs nas rgyu ˌbras
bden pai c'os ˌdi mgur du
gsungs so ||

Bharo angeführt, ein ganzes Heer von Geistern und erfüllte Fluss, Thal, Himmel und Erde. Sie schleuderten auf den Ehrwürdigen einen schrecklichen Regen von Felstrümmern, Donnerkeilen und anderen Waffen herab, riefen ihn beim Namen und stiessen ein Geschrei wie: fange! töte! und andere grässliche Worte aus. Auch liessen sie viele hässliche Gestalten erscheinen. Im Gedanken an die Nachstellungen der Geister trug er ihnen die wahrhaftige Lehre von der Frucht der Thaten in folgendem Gesange vor.

bla ma rnams la p'yag ˌts'al lo
rje bka drin can la skyabs
  su mc'i
yong ˌk'rul snang mig gi
  rnam šes la
gdon gnod sbyin p'o moi c'o
  ˌp'rul byung

ˌa tsa ma yi dags sñing re
  rje
mi nga la k'yed dag mi
  gnod de
sngon las ngan bsags pai
  ˌbras bu des

Verehrung sei den Lama!
Dir, huldreicher Herr, befehle ich mich!
Indem das Auge die Illusionen gewahrt,
Entstehen die Gaukelspiele der männlichen und weiblichen Dämonen und Yaksha.

Ach, ihr armseliges Hungervolk der Hölle,
Mir könnt ihr keinen Schaden zufügen!
Die angesammelte Frucht euerer früheren schlechten Thaten

160 *da lta rnam smin gyi lus la spyod*

yid mk'a la rgyu bai gzugs can de

bsam ngan ñon mongs kyi kun slong dang

sbyor ba lus ngag gi gdug rtsub kyis

gsod gcod brdeg ,ts'og byed do zer

165 *rtog med rnal ,byor Ras pa ,di*

blo yang med pa lta bai gdeng

dpa bo seng gei spyod ,gros kyis

lus ni lha skui rdzong la zhen

ngag ni sngags kyi rdzong la zhen

170 *sems ni ,od gsal gyi rdzong la zhen*

ts'ogs drug rang gi nga bos stod

k'o bo de lta bu yi rnal ,byor la

k'yed yi dags kyi t'o ,ts'am mi sto ste

spyir las dge sdig gi ,bras bu bden pas na

160 Ist jetzt durch die gereifte Vergeltung in euerem Körper wirksam.

In der Form der in der Luft wandelnden Seelen

Thun sie durch die Erregung der Sünden schlechter Gedanken

Und durch das scharfe Gift von Leib und Wort

Nichts anderes als töten, schneiden, schlagen und stechen, sagt man.

165 Einfachen Herzens wandle der Einsiedler Ras pa,

Einfältigen Sinnes, voll Zuversicht auf seine Beschauung,

Dem Wandel des tapferen Löwen gleich!

Der Leib sehnt sich nach dem Hort des Götterbildes,

Die Rede sehnt sich nach dem Hort des Zaubers,

170 Das Herz sehnt sich nach dem Hort der Erleuchtung.

Die sechs Arten der Sinneswahrnehmungen preise ich.

Mich, der ein solcher Yogin ist,

Kann euer, der Preta, Höhnen nicht verletzen.

Da die Belohnung der Tugenden und Sünden wahr ist,

175 *rgyu „t'un rnam smin bsags*
bsags nas

ngan song du „gro ba yi re
mug

„a tsa ma ñon mongs yi dags
kun

don gnas lugs ma rtogs
sñing re rje

dngos ngan Mi la ras pa
„di

180 *ngag rig byed glu yi c'os*
bšad pa

nang bcud kyis bsdus pai
sems can rnams

gñen p'a mar ma gyur gang
yang med

de drin can yin pas sdug
par p'ongs

k'yod de bas gnod sems slar
log la

185 *las rgyu „bras bsams na ci*
ma legs

c'os dge bcu spyad na ci ma
legs

ts'ig legs par gzung la rtog
dpyod t'ong

don go bar gyis la p'yir la
ñon

175 Sammeln die Wesen Vergeltung durch Missgeschick

Und verzweifeln in der Hölle.

Ach, ihr Sündenvolk der
Hölle,

Armselige, die ihr Sinn und
Wesen der Dinge nicht
wisst!

Ich, der abgemagerte Milaraspa hier,

180 Habe die Religion mit einem
Liede belehrender Rede
erklärt,

Das den durch das Lebens-
prinzip geeinigten Wesen

Gleichsam ein elterlicher Beistand geworden

Und sie kraft seiner Wohlthätigkeit vor Unglück bewahrt.

Wendet euch nun ab von
der Bosheit!

185 Wenn ihr über die Vergeltung
der Werke nachdächtet,
wäre vortrefflich;

Wenn ihr die zehn Tugenden
der Religion übtet, wäre
vortrefflich.

Versteht mein Wort gut und
prüft es mit Ueberlegung!

Höret noch einmal, erfasst
seinen Sinn!

*zhes gsungs pas | „dre dmag
rnams na re | k'yod kyi k'a
sbyang gis | nged mgo mi „k'or |
c'o „p'rul zhi nas k'yod bde
bar blang mi yong zer | dmag
mang du song c'o „p'rul c'er
song ba las | rje btsun gyi
l'ugs la dgongs pa ltar | „dre
dmag rnams la da bla mai
bka drin gyis gnas lugs rtogs
pai rnal „byor pa | bar c'ad
bdud kyi c'o „p'rul sems kyi
rgyan yin pas | da dung c'er
l'ong byang c'ub mc'og la sybor
bar „dug gis ysung nas | rgyan
du c'e pai bdun gyi mgur „di
gsungs so ‖*

So sang er. Doch die Dämonenheere riefen: „Mit deiner Zungenfertigkeit kannst du uns nicht den Kopf verdrehen; wir sind nicht gekommen, um unsere Magie einzustellen und dich hübsch in Ruhe zu lassen." Ihre Heere wurden zahlreich, und der Spuk mehrte sich. Der Ehrwürdige überdachte die Sache und rief den Dämonenheeren zu: „Der dank der Gnade seines Lama das Wesen der Dinge durchschauende Yogin empfindet den Teufelsspuk der Widersacher als einen Segen für seine Seele. Meinetwegen treibt es noch schlimmer! Ich verharre hier mit der höchsten Bodhi vereinigt." Darauf trug er den Sang von den sieben grossen Zierden vor:

*rje sgra bsgyur Mar pai
zhabs la „dud*

Zu des Herrn, des sprachgewandten Marpa, Füssen verneige ich mich.

190 *nga gnas lugs rtogs pai rnal
„byor pa*

rgyan du c'e bai glu cig len

190 Ich, der das Wesen der Dinge durchschauende Yogin,

Singe ein Lied von den grossen Zierden.

*k'yed „dir „ts'ogs gnod sbyin
p'o mo rnams*

Ihr hier versammelten männlichen und weiblichen Schadenstifter (Yaksha),

*yid ma yengs rna bai dbang
po blod*

Leiht mir aufmerksam euer Ohr.

dbus ri bo mcʿog rab mcʿod rten la	In einem Caitya des in der Mitte der Welt gelegenen herrlichsten Berges
195 lhor raidūryai ₒod gsal ba	195 Strahlt im Süden der Lichtglanz des Lasurs:
ₒdzam gling nam kʿai rgyan du cʿe	Dem Himmel von Jambudvîpa ist er eine grosse Zierde.
ri gña šing ₒdzin gyi gong rol na	Auf der Oberfläche des Berges Yugaṁdhara
ñi zla zuny cig ₒod gsal ba	Strahlt der Lichtglanz des Sonne-Mond-Paares:
gling bzhi yongs kyi rgyan du cʿe	Allen vier Kontinenten ist es eine grosse Zierde.
200 klu byang cʿub sems kyi rdzu ₒpʿrul gyis	200 Durch den Zauber der Nâgabodhi
nam mkʿai mtʿongs nas cʿar bab pa	Fällt vom Himmelsgewölbe der Regen:
dog mo sa yi rgyan du cʿe	Der engen Erde ist er eine grosse Zierde.
pʿyi rgya mtsʿo cʿu yi rlangs pa las	Aus dem Wasserdampf, der aus dem Ocean aufsteigt,
bar snang yongs la lho sprin byung	Bilden sich im ganzen Luftraum die Südwolken:
205 lho sprin bar snang gi rgyan du cʿe	205 Die Südwolken sind eine grosse Zierde des Luftraums.
ₒbyung ba drod gšer gyi rten ₒbrel gyis	Aus der Verbindung der Wärme und Feuchtigkeit der Elemente
dbyar spang ri logs la mja tsʿon bkra	Entsteht im Sommer gegenüber den Almen der buntfarbige Regenbogen:
ₒja mtsʿon spang rii rgyan du cʿe	Der Regenbogen ist eine grosse Zierde der Almen.

nub ma dros mts'o las c'ab babs pas	Durch den im Westen aus dem Manasarovara - See hervorströmenden Fluss
210 *lho ,dzam bu gling gi rtsi šing rgyas*	210 Gedeihen die Obstbäume von Jambudvîpa im Süden:
skye ,gro yongs kyi rgyan du c'e	Allen Wesen sind sie eine grosse Zierde.
nga rnal ,byor ri k'rod ,dzin pa la	Mir, dem die Bergeinsamkeit wählenden Klausner,
sems stong ñid bsgoms pai nus mt'u yis	Sind durch die Kraft der Meditation über die Leere
,dre gnod byin p'o moi c'o ,p'rul byung	Die Gaukelspiele männlicher und weiblicher Kobolde und Yaksha entstanden:
215 *c'o ,p'rul rnal ,byor gyi rgyan du c'e*	215 Die Gaukelspiele sind eine grosse Zierde des Klausners.
k'yed legs par ñon dang mi ma yin	So hört mir richtig zu, ihr Geister!
mi nga ,o k'yed kyis šes sam ci	Ihr wisst doch wohl, wer ich bin?
nga ngo k'yod kyis ma šes na	Wenn ihr mich noch nicht kennt,
nga rnal ,byor Mi la ras pa yin	So vernehmt, ich bin der Klausner Milaraspa,
220 *byams sems kyi me tog gling nas ,k'rungs*	220 Geboren aus dem Blumenschoss der Liebe!
glu šnan pai dbyangs kyis brda sbyar nas	Mit des Liedes Wohllaut weiss ich zu deuten,
ngag bden pai ts'ig gis c'os bšad de	Mit den Worten wahrhaftiger Rede die Lehre zu erklären
p'an pai sems kyis gros cig ,debs	Und hülfreichen Sinnes Rat zu erteilen.

k'yod byang c'ub mc'og tu
sems bskyed nas
225 *,gro ba gzhan don ma byung*
yang

mi dge bcu po spangs nas
ni
rang zhi bde t'ar ba ci mi
t'ub
mi nga zer ñan na don c'en
,grub

,p'ral dam c'os byas na yun
du bde

ces gsungs pas | mi ma yin
de rnams p'al c'er rje btsun
la dad cing gus par gyur nas |
c'o ,p'rul zhi bar byas te | rnal
,byor ba ngo mts'ar c'e | yin
lugs kyi brda ma bkrol zhing |
rtags ma mt'ong na nged rnams
kyis kyang ma rtogs par ,dug |
da k'yod la nged kyi bar c'ad
mi ts'ugs ,dug gis | sngar las
,bras kyi c'os gsungs pa bka
drin c'e na ,ang nged ngan pai
bag c'ays mt'ug pa dang blo
gros c'ung bas yid la ma zin |
nga ts'ig ñung la don c'e ba |
go sla la ,k'yer bde bai c'os
cig zhu zer ba la | rje btsun
gyis yin pa bdun gyi mgur
,di gsungs so ||

Richtet eueren Sinn auf die
höchste Erkenntnis,
225 Und wenn auch ein Nutzen
für andere Wesen nicht
daraus erwächst,
Wie solltet ihr nicht, wenn
ihr die zehn Sünden meidet,
Seelenfrieden und Befreiung
erlangen?
Wenn ihr auf meine Worte
hört, werdet ihr grossen
Segen erzielen;
Wenn ihr sogleich die heilige
Lehre übt, werdet ihr in
Zukunft glücklich sein!

So sang er. Da waren die
Geister der Mehrzahl nach schon
soweit gebracht, dass sie dem
Ehrwürdigen gläubige Verehrung
bezeigten und ihr Gaukelspiel
einstellten. „Einsiedler, sei bedankt", sprachen sie, „doch
wenn wir ohne Erklärung der
Natur der Dinge deine Argumente nicht erfassen, müssen
wir in unserer Unwissenheit verharren. Da wir dir ja nun gar
keinen bösen Streich gespielt
haben, so bitten wir, wenn auch
der Segen der verkündeten Lehre
von der Frucht der früheren
Thaten gross war, um einige
neue kurzgefasste, nützliche,
leichtverständliche und freund-

lich anziehende Lehren; denn
unser Hang zur Schlechtigkeit
ist stark, und unser Verstand
ist klein und von schwacher
Fassungskraft." Auf ihre Bitte
hin trug der Ehrwürdige folgendes Lied von den sieben rechten
Dingen vor:

230 *sgra bsgyur Mar pai zhabs*
　　la dud

230 Dem sprachgewandten Marpa
　　zu Füssen verneige ich
　　mich!

byang sems byongs par byin
　　gyis rlobs

Gewähre deinen Segen, dass
　　die Bodhi vollendet werde!

don bden pai ts'ig dang ma
　　brel na

Wenn nicht mit Worten wahr-
　　haftigen Sinnes verbunden,

glu gre ggur sñan yang pi
　　wang yin

Sind Lied und Stimme, wenn
　　auch noch so wohltönend,
　　nur einer Zither gleich.

c'os t'un dpe yis ma mts'on
　　pai

Wenn nicht durch Gleich-
　　nisse, im Einklang mit der
　　Religion, erläutert,

235 *ts'iy sdeb sbyor mk'as k'yang*
　　sgra brñan yin

235 Sind Worte, wenn auch mit
　　Geschick gesetzt, nur ein
　　leeres Echo.

c'os lag len rgyud la mi
　　gel na

Wenn die praktische Aus-
　　übung der Religion nicht
　　von Herzen kommt,

ngas šes so zer yang mgo
　　bskor yin

Ist es, auch wenn man sagt,
　　dass man sie kennt, eine
　　blosse Täuschung.

don sñan brgyud kyi gdams
　　pa mi sgom na

Wenn man über die Vor-
　　schriften der Unterweisung
　　in der Wahrheit nicht nach-
　　denkt,

gnas ri k'rod bzung yang rang sdug yin	Ist es, auch wenn man die Bergeinsamkeit wählt, eigenes Missgeschick.
240 *p'an nges pai dam c'os mi byed na*	240 Wenn man die heilige Religion des wahren Heils nicht übt,
„p'rul so rnam rem yang sdug sngal yin	Ist die Magie, auch wenn man darin sehr stark ist, nur ein Unglück.
las rgyu „bras zhib tu mi rtsi na	Wenn man die Vergeltung der Werke nicht genau berechnet,
ts'ig k'a lta c'e yang smon pa yin	Sind Worte und Ratschläge, wenn auch noch so gross, nur ein frommer Wunsch.
ts'ig don lag tu mi len na	Wenn man den Sinn der Worte nicht bethätigt,
245 *„p'ral bšad lo byas kyang k'ram pa yin*	245 Ist man, auch wenn man im Augenblick viele Worte macht, nur ein Lügner.
las mi dge sangs na ngang gis „grib	Reinigt euch von den Lastern, und sie werden sich allmählich vermindern;
las dge ba bsgrub na šugs la „grub	Erstrebt die Tugend, und ihr habt innere Kraft gewonnen.
gag gciy la gril zhing ñams su long	Richtet euere Gedanken auf einen Punkt und nehmt es euch zu Herzen!
ts'ig mang poi bšad pas p'an pa c'ung	Aus wortreichen Erklärungen entsteht nur geringer Nutzen;
250 *don de bzhin ñams su len „ts'al lo*	250 In diesem Sinne nehmt es euch, bitte, zu Herzen!

3*

ces gsungs pas | c'os zhu ba
de rnams rje btsun la dad
cing gus par gyur te | p'yay
dang bskor ba lan du ma byas
nas | kun rang gnas su song
rung | gtso bo de Bha ro ̦gai
k'or dang bcas pa slar yang
sngar ltar c'o ̦p'rul ston du
byung ba las | rje btsun gyis |
yang rgyu ̦bras kyi c'os ̦di
mgur du gsungs so.

So sang er. Da zollten sie, die Belehrung nachgesucht, dem Ehrwürdigen gläubige Verehrung, verneigten sich und umkreisten ihn viele Male, dann entfernten sich alle. Nun erschien aber erst jener Fürst Bharo selbst mit einigen Begleitern und liess noch einmal wie vorher einen Spuk losgehen. Da trug der Ehrwürdige wiederum die Lehre von der Frucht der Thaten in einem Gesange vor:

drin can Mar pai zhabs la
* ̦dud*

Dem huldreichen Marpa zu Füssen verneige ich mich!

k'yod da dung ñon dang mi
* ma yin*

Ihr Geister, höret mich noch einmal!

lus mk'a la rgyu ba t'ogs
* pa med*

Euer in der Luft wandelnder Leib ist ungehindert;

bsam ngan bag c'ags rgyud
* la ̦t'as*

Böse Gedanken und schlimme Leidenschaften haben in euerem Herzen Wurzel geschlagen;

255 *ñon mongs kyi mc'e bas*
* gzhan la brngams*

255 Mit den Fangzähnen der Sünde stürzt ihr euch auf andere;

gzhan sdug gi sdig pas rang
* ñid mnar*

Unter der Sünde, anderen Uebles zugefügt zu haben, müsst ihr selbst leiden:

rgyu ̦bras kyi bden pa stor
* sa med*

Die Wahrheit von der Vergeltung der Werke geht nicht verloren,

rnam smin nus pas gtong mi srid	Die Macht der vergeltenden Gerechtigkeit kann euch nicht fahren lassen,
k'yod rang gis rang ñid mnar sdang byed	Ihr selbst bereitet euch die Qualen der Verdammnis!
260 *a tsa ma yi dags kyi k'rul pa la*	260 Ach, Aermste! Wenn ihr über den Wahnsinn der Höllengeschöpfe,
yi re mug las ngan gyi nus pa la	Ueber die Verzweiflung, die durch schlechte Werke bewirkt wird,
di dra rang bsams na sñam de rlung ldang	Wenn ihr über dergleichen selber nachdenkt, erhebt sich der Wind solcher Gedanken.
k'yod sngon yang las ngan bsags bsags pas	Ihr habt schon früher Schlechtigkeit auf Schlechtigkeit aufgehäuft,
da dung las ngan sog sñing dod	Und noch immer begehrt euer Herz, Böses zu sammeln:
265 *gsod gcod kyi sdig pas xin pa yi*	265 Gepackt von der Sünde des Schlachtens,
xas su ša dang k'rag la dga	Liebt ihr zur Nahrung Fleisch und Blut;
las su gro bai srog gcod pa	Ihr raubt, als übtet ihr ein Gewerbe, das Leben der Wesen:
rigs drug gi nang nas yi drags lus	Darum werdet ihr unter den sechs Klassen der Wesen in der Gestalt der Preta,
sdig la spyad pas ngan gror ltung	In Sünden wandelnd, zur Hölle fahren.

270 *a re p'angs c'os la snang*   270 O lasst ab! Wendet eueren
    ba bsgyur   Sinn der Lehre zu!
    re dogs med bde ba myur   Wenn ihr ohne Hoffnung
    du grub   und Furcht seid, ist die
                                  Glückseligkeit schnell er-
                                  reicht.

    ces byas pas ¦ k'ong rnams   Darauf sprachen jene: „Durch
  na re ¦ k'yod nged ts'o la c'os   deine uns gegebenen Erklä-
  bšad mk'as mk'as su dug pas   rungen der Lehre sind wir nun
  go dug ¦ go ba ltar k'yod rang   gar kenntnisreich und gut-
  gis ñams su blangs pai gdeng   begreifend. Solchem Verständ-
  ci yod zer du byung ba la ¦   nis entsprechend sage uns jetzt,
  rje btsun gyis gdeng ts'ang   wie du dir selbst diese Zuver-
  ldan pai mgur di gsungs so.   sicht erworben hast." Da trug
                                  der Ehrwürdige folgenden zu-
                                  versichtsvollen Sang vor:

    mts'an ldan Mar pa zhabs   Zu den Füssen des vortreff-
    la dud   lichen Marpa verneige ich
                                  mich!

    nga don dam rtogs pai rnal   Da ich, der das Wesen der
    byor pa   Dinge durchschauende
                                    Klausner,

    gzhi skye med nyang la   Darauf vertraue, dass es
    gdeng bcas nas   keinen Ursprung und kein
                                    Entstehen giebt,

275 *lam gag med kyi rtsal k'a*   275 Bin ich durch die Stufen der
    rims kyis rdzogs   den Weg frei machenden
                                    Fertigkeiten vollendet.

    t'abs sñing rje c'en pos brda   Die rechten Mittel mit
    sbyar nas   grossem Erbarmen er-
                                    läuternd,

    don c'os ñid kyi glod nas   Singe ich, von den Lehren
    dbyangs cig len   der Wahrheit selbst an-
                                    gespornt, ein Lied.

k'yod las ngan sgrib γyogs
mt'ug po yis

nges don gyi gnas lugs mi
go sle

280 *drang don gyi c'os cig da*
dung ₀c'ad

sngon t'ams cad mk'yen pa
sangs rgyas kyis

bka dri ma med pai mdo
rgyud las

las rgyu ₀bras kyi c'os la
nan tan gsungs

de ₀gro ba kun gyi gñen gcig
po

285 *don mi slu nges pai bden*
ts'ig yin

rje byams pai gsung la k'yod
kyang ñon

nga ñams rgyud la sbyangs
pai rnal ₀byor ₀di

p'yi ₀k'rul snang gi bgegs la
p'ar bltas pas

sems skye med kyi c'o ₀p'rul
yin par go

290 *nang rig pai sems la ts'ur*
bltas pas

rtsa bral gyi sems ñid ye
nas stong

de bla ma brgyud pai byin
brlabs dang

Da ihr, mit Sünden böser
Thaten dicht bedeckt,

Der mystischen Erkenntnis
Wesenheit nicht begreift,

280 Soll euch die Wahrheit der
Schrift noch einmal erklärt
werden.

Einstens hat der allwissende
Buddha

In den Sûtra und Tantra des
fleckenlosen Wortes

Die Lehre von der Vergeltung
der Werke eifrig verkündet:

Diese ist allen Wesen ein
grosser Helfer,

285 Ein Wort von untrüglichem
Wahrheitssinn.

So lauscht auch ihr der Verkündigung des Erbarmers!

Wenn ich, der Einsiedler, mit
gereinigtem Geiste,

Mich nach den Dämonen der
äusseren Illusionen um-
schaue,

So begreift der Geist, dass
die Gaukelspiele keine Entstehung haben.

290 Wenn ich in das Innere
meiner Seele schaue,

So ist der ursprungslose Geist
vollständig leer.

Da ich den Segen genoss,
jenen Lama zum Lehrer
zu haben,

rang gcig pur bsgoms pai yon tan gyis	Da ich den Vorzug habe, ganz allein der Beschauung obzuliegen,
rje Nâ ro c'en poi brgyud pas rtogs	Da ich den grossen Gebieter Nâro zum Lehrer hatte, begreife ich
295 *don ma nor rgyal bai dgongs pa bsgoms*	295 Und übe in der Wahrheit untrügliche, siegreiche Beschauungen.
t'abs zab mo rgyud sdei dgongs pa rnams	Da die Meditationen der Tantra über die tiefen Mittel
bla ma rje yis gnad bkrol te	Der Lama, der Gebieter, in der Hauptsache erklärt hat,
lam bskyed rdzogs brtan par sgoms pai mt'us	So kenne ich kraft der durch die Erlangung des rechten Pfades vollendeten starken Meditation
nang rtsa gnas rten ,brel šes pai p'yir	Die Verbindung des inneren Fasernetzes
300 *p'yi k'rul snang gi bgegs la nga mi ,jigs*	300 Und fürchte daher nicht die Dämonen der äusseren Illusionen.
deng bram ze c'en poi rgyud pa la	Heute sind die Heiligen, die grosse Brahmanen zu Lehrern haben,
dpal nam mk'a lta bui rnal ,byor mang	Zahlreich wie der prächtige Himmel.
yid gñug mai don la sbyangs sbyangs nas	Wenn man beständig den Sinn eines einfachen Gemütes übt,
k'rul snang gi dran pa dbyibs su yal	Schwinden die Illusionen aus dem Bewusstsein:

305 gnod bya dang gnod byed
ngas ma mt'ong

c'os sde snod kyi gsung rab
zhal p'ye yang

don di las med par blo t'ag
c'od

ces gsungs pas | Bha ro k'or
bcas rang rang gi t'od rnams
p'ud nas p'yag dang bskor ba
mang du byas nas

zla ba gcig gi ts'o ba bul
zer ja yal ba bzhin song ba
las nang par nam langs ñi
ma šar ba dang | dang gi Bha
ro de rnams | bha ri ma rgyan
bzang po btags pai k'or du
mas bskor nas byung ste | rje
btsun la rgun c'ang la sogs
pai c'ang sna mang pos rin
po c'ei snod du ma bkang |
bras c'an dang ša la sogs te
zas sna mang pos k'ar gzhong
bzang po rnams bkang nas
dren cing | da p'yin c'ad bkai
bangs bgyid cing | ci gsung gi
bka bsgrub par bgyio | zer
p'yag dang bskor ba mang du
byas te mi snang bar gyur pa
de la ni rgyal po t'ang grem
zhes kyang bya ste | lha c'en
po ts'ogs kyi bdag po yin no

305 Weder Harm noch Harm-
stifter sehe ich!

Die heiligen Schriften des
Piṭaka haben sich euch
geöffnet:

Dass es keinen anderen als
diesen Segen giebt, steht
fest.

So sang er. Bharo und sein
Gefolge legten ihre Turbane ab,
verneigten sich und umkreisten
ihn oftmals.

Mit dem Versprechen, ihm
einen Monat lang Nahrungsmittel zu liefern, entfernten sie
sich, wie ein Regenbogen verschwindet. Am dritten Tage,
bei Sonnenaufgang, erschien
Bharo, von zahlreichem, mit
Bharimaschmuck schön ge-
schmücktem Gefolge umgeben
und brachte dem Ehrwürdigen
Wein, Bier und viele andere
Arten gegohrener Getränke, in
kostbare Gefässe gefüllt, Reis-
brei, Fleisch und viele andere
mannigfaltige Speisen in schönen
Bronzeschüsseln. Mit dem Versprechen, von jetzt ab seinem
Gebote zu gehorchen und seinen
Worten Folge zu leisten, verneigten sie sich, umkreisten ihn
viele Male und verschwanden.
Der König derselben ist der

*des rje btsun la yang t'ugs
dam bogs c'e ba dang | sku
k'ams bde bai ngang nas zla
ba gcig gi bar du bkres snang
ye med par gyur to |*

*de nas rje btsun la p'yi c'u
bzang gi gnas t'ugs k'ongs su
c'ud nas | la p'yi gnas ,t'il gzigs
su byon pai lam |*

*ram bui t'ang c'en po zhig
gi dbus na brag skyibs dang
bcas pai p'a p'ong c'en po
zhig ,dug pa der re šig bzhugs
pai ts'e | mk'a ,gro mang pos
p'yag btsal ,dod yon sna ts'ogs
kyi mc'od pa p'ul bskor ba
byas pas | rdo la mk'a ,groi
zhabs rjes gñis byung ba dang |
,ja yal ba bzhin song ngo |*

*de nas cung zad byon pa
dang | mi ma yin gyi c'o ,p'rul
,byon lam t'ams cad du mo
mts'an c'en po du ma bstan
byung bas | rje btsun gyis*

sogenannte Thang ,grem, d. i.
der grosse Gott Gaṇapati (Tshogs
kyi bdag po).

Dadurch wurde die innere
Förderung des Ehrwürdigen so
gross und seine Gesundheit so
vortrefflich, dass ihm einen Monat lang nicht einmal der Gedanke an Hunger kam.

Als nun der Ehrwürdige die
Gegend des La phyi und des
Chu bzang genugsam kannte,
machte er sich auf den Weg,
um gNas ,thil am La phyi zu
besuchen.

Als er sich eine Weile auf
einem grossen Felsblock nieder-
liess, der sich unter einem über-
hängenden Felsen inmitten der
grossen Ebene von Ram bu befand, brachten viele Ḍâka unter
Verneigungen Spenden mannigfacher begehrenswerter Gaben
dar und umwandelten ihn ehr-
furchtsvoll, so dass in einem
Steine zwei Fussspuren der
Ḍâka zurückblieben, worauf sie
wie ein verblassender Regenbogen verschwanden.

Deinde cum aliquantum iti-
neris progressus esset, in illa,
qua larvae lemuresque trans-
formationes magicas ostenderant,
via, muliebria multa et magna

kyang lta stangs kyi ngang nas |
gsang rdor las su rung bar
mdzad nas byon pas | mts'an
ma dgu ,das pai ,ts'ams su |
gnas kyi bcud ,dus pai rdo ba
zhig la gsang gnas brdor
zhing lta stangs mdzad pas
c'o ,p'rul kun zhi bai sa la |
la dgu lung dgur grags |

de nas gnas ,t'il du p'ebs
su c'a bai ts'e | yang sngar
gyi Bha ro des bsu ba byas
mc'od pa p'ul c'os k'ri brtsigs
nas c'os zhus pa la | rje btsun
gyis kyang rgyu ,bras kyi c'os
mang po ysungs pai mt'ar | c'os
k'rii mdun na p'a p'ong c'en
po zhig ,dug pa de la t'im
song ngo |

de nas rje btsun gnas ,t'il
du byon pas | t'ugs šin tu dges
te zla ba gcig tsam bzhugs
nas | gña nam rtsar mar p'ebs
te | yon bdag rnams la ,dre
lung skyog mo rang du ,dug
ste | ngas ,dre rnams t'ul nas
da sgrub gnas su song yod |
nga yang gdod kyi p'yin te

ubique videbantur. Milaraspa ille venerabilis vi magica sua videndo peni sedato iter perrexit. Eodem loco quo vestigia novem evanescebant, cum membrum, ut sucus eius conflueret, in saxo perfricasset, vultu magico oculos in locum convertit, quoad praestigiis finem afferret, unde locus ille appellatur La dgu lung dgu, id est ,Furcae novem, Valles novem'.

Als er seinen Weg fortsetzte, um nach gNas ,thil zu gelangen, ging ihm wieder der vorher genannte Bharo entgegen, brachte ihm Ehrengaben dar und errichtete in dem Verlangen nach einem religiösen Vortrag eine Kanzel. Wiederum predigte der Ehrwürdige über viele Dogmen der Vergeltungslehre, bis er endlich in einem vor der Kanzel befindlichen grossen Felsblock verschwand.

Darauf begab sich der Ehrwürdige nach gNas ,thil, wo er in froher Stimmung einen ganzen Monat lang zubrachte. Dann kehrte er nach rTsar ma in gNa nam zurück, und als er in ,Dre lung skyog mo selbst war, sagte er zu den Gabenspendern: „Ich habe die Dämonen bekehrt

sgom pa yin gsungs pas | k'ong rnams mc'oy tu dad par gyur to ||

   la p'yi la byon pa c'u bzang gi bskor ro.

und will jetzt in meine Betklause zurückkehren; zuerst will ich gehen und mich der Beschauung hingeben." Da wurden sie alle des Glaubens voll.

Dies ist das Kapitel von der Reise nach dem La phyi oder von dem Flusse Chu bzang.

029

黑龙江诸部落的装饰艺术
附：书评二则

WHOLE SERIES, VOL. VII. ANTHROPOLOGY, VOL. VI.

MEMOIRS

OF THE

American Museum of Natural History.

VOLUME VII.

PUBLICATIONS OF

THE JESUP NORTH PACIFIC EXPEDITION.

I.—The Decorative Art of the Amur Tribes.

BY BERTHOLD LAUFER.

January, 1902.

The Knickerbocker Press, New York

MEMOIRS

OF THE

AMERICAN MUSEUM OF NATURAL HISTORY.

The Jesup North Pacific Expedition.

I.—THE DECORATIVE ART OF THE AMUR TRIBES.

BY BERTHOLD LAUFER.

PLATES I–XXXIII.

THE material published and described in the following pages was obtained under the auspices of the Jesup North Pacific Expedition during my two years' researches among the various tribes of Saghalin Island and the Amur region.

There is not much literature as yet bearing on the decorative art of these tribes. Schrenck, in his fundamental work " Reisen und Forschungen im Amur-Lande " (Vol. III, pp. 399–401), makes a few remarks on the subject, emphasizing the peculiarity of the Gilyak ornaments, which are totally different from those of all other Siberian peoples. He sees in them an evident Chinese influence. No explanations of the ornamental figures are unfolded in his book. Further, H. Schurtz, in his paper " Zur Ornamentik der Aino " (Internationales Archiv für Ethnographie, Vol. IX, pp. 233–251), has considered to some extent the ornamentation of the Amur tribes so far as known to him. It would carry us too far to enter into a minute discussion of the leading problems there treated, the alleged solution and disentanglement of which fall to the ground when compared with the results of investigations in the field. I must confess, I adhere to the principle that ornaments should not be regarded as enigmas which can be easily puzzled out by the homely fireside. Neither are ornaments of primitive tribes like inscriptions, that may be deciphered : they are rather productions of their art, which can receive proper explanation only from the lips of their creators. They are comparable to modern symphonic compositions, that are incomprehensible without the printed synopsis in the hands of the auditors. The writing of such guides can only be accomplished by consulting the native artist as to his own fancy concerning the significance of the ornaments evolved from it. The human and bear heads which Schurtz claims to have ferreted out exist not in the

[1]

minds of the natives.   Artistic representations of the bear in wood-carving are limited to the Gilyak, for use at their bear-festival.   This animal, however, is never reproduced in drawings or paintings, either in natural or conventional form, according to the verbal testimony of both the Gilyak and Gold.   Neither have I myself discovered even a trace of the bear-heads suggested by Schurtz. Of the existence of his eye-ornaments, apparently a mere outcome of his enthusiasm, my authorities were also entirely ignorant.

I am under obligations to Professor A. Bastian for permitting me to take advantage of those collections in the Königliche Museum für Völkerkunde in Berlin which relate to the Amur region, and which were made by Captain A. Jacobsen.   I have also to thank the authorities of the Königliche Kunstgewerbe Museum in Berlin for placing at my disposal fourteen specimens of Chinese and Japanese weavings.   These objects from the two Berlin museums have been drawn by Mr. W. von den Steinen ; drawings for the other illustrations were prepared by Mr. Rudolf Weber.   A list of plates is given at the end.

HISTORICAL ASPECT. — The history of the decorative art of the Amur tribes is shrouded in mystery, since no written records give any account of it ; nevertheless we may be able to make some historical observations regarding its development.   A comparison of the artistic material found in my collections with that obtained by Schrenck nearly half a century ago, and illustrated in his work " Reisen und Forschungen im Amur-Lande," affords instructive evidence that the forms of this sphere of art have remained unaltered up to the present time, notwithstanding all political turbulence and change that have affected the Amur region in the mean time.   Although Russian influence is nowadays all-powerful, yet it has not been able to suppress or eradicate native art, nor to replace it by something better, for the apparent simple reason that the Russian settlers had indeed nothing better to offer.   Whereas Russian " culture " tended to shatter the entire life of the natives, its effect is the more striking and remarkable in view of the fact that the native art has been retained pure and intact.   From this we may be justified in inferring that their artistic conceptions have taken deep root in the hearts of the people, and have acquired a high value in their intellectual world. The tenacity with which the style of art survives should be counted as evidence of its national character, at least of an ancient naturalization on the soil in which it was planted.   On the other hand, we observe at first that the forms and conceptions of this ornamentation are imbued, for the most part, with a Chinese spirit ; but considering the historical feature just mentioned, and, moreover, the fact that the present aspect of the wide propagation and the skilful execution of this art all over the Amur region can be the result only of long-enduring tradition, it can hardly be designated *en masse* as a Chinese importation.   Its basis rests undeniably in China.   In the course of time the Amur tribes appropriated Chinese forms to themselves, and very likely further developed them independently.   The introduction of Chinese devices must surely date as far back as the earliest connection of the Chinese with the Amur region and with Tungusian tribes.   This art was perhaps first introduced as a mere fashion, which overruled taste, then gradually infused itself into the minds of the people, who in this way absorbed and assimilated a part of the Chinese art, as the nations of Europe imbibed classic art in the period of the Renaissance.   It was due no less also to a congeniality of the minds of the two peoples.   At present it is hardly possible to define exactly the historical relation between Chinese and East Siberian art, especially since the art of China, and particularly its ornamentation, has as yet been so little explored.

We read in the annals of Chinese history that the great body of Tungusians knew nothing further than the use of wooden tallies with certain rude conventional marks, which served as bonds in case of contracts ; and that then A-paou-ke, the first emperor of the Liao Dynasty, employed a great number of Chinese ; and they instructed him, by an adaptation of the official Chinese writing, with certain additions and contractions, how to construct several thousand characters, by which the engraved contract-tallies were replaced, these new forms referring to the beginning of the tenth century.   Although the Khitans thus early took the lead, their example was not followed by their neighbors, at least not for many years ; for up to the twelfth century we still find the Niüchi chiefs issuing their orders by the old device of an arrow with a notch in it, while matters of urgency were distinguished by three notches.   On their establishment as the Kin Dynasty, however, they for the first time gained a knowledge of written characters.[1]

Since writing forms a most important part of art, according to Chinese views, we may conclude that the introduction of ideograms among Tungusian tribes became at the same time the incentive for adopting also ornamental and decorative forms.   So, too, we may be sure that the ornamentation of these Tungusian tribes can have been but very poor before ; and from this point of view it is still more likely that they felt themselves under the necessity of adopting Chinese ornaments.   From remote times the forms and figures of Chinese ornature may have been handed down among the Amur tribes for many centuries ; and thus it may even be the case that traditions regarding the meaning of certain patterns are fuller, and have been better preserved in the minds of these naïve unlettered tribes than in the fast-fading memories of a writing nation. If the patterns of the Amur tribes were derived from China, it is most astounding that exactly corresponding devices have never before been discovered in that country, nor adequate explanations obtained for related ones.   It is true that we know very little about Chinese ornaments ; nevertheless, from the fact that the inhabitants of the Amur country have now given us the first clew to patterns of apparent Chinese origin, we seem to be justified in concluding that they are founded on a better-preserved oral tradition there.   Further, we may infer that examples similar to those in our ornaments are necessarily still to be found in the large province of Sino-Japanese art.   Those Chinese and Japanese designs which

---

[1] Wylie, Chinese Researches, Vol. II., p. 254.

I shall here compare with our Siberian devices cannot prove, of course, the direct historical connection between the practices of both arts : they are merely material chosen to demonstrate some characteristic congruous features, which may bear witness, if not to the exact degree of relationship, yet to a general one.

What is necessary, first of all, to sift out, is the ornamental art of the Manchu, and those Chinese peoples in the northern part of the Celestial Kingdom bordering on Siberia. The source from which the Amur peoples have drawn may be discovered there. After all, it is clearly too intricate a problem, thus far, to distinguish accurately between what of their art the Amur tribes owe to their masters, and what to themselves. A great many features should be attributed to direct Chinese transmission. On the whole, therefore, the standpoint to be taken, in a consideration of the decorative art of these tribes, must for the present be one that looks upon it as an independent branch of East Asiatic art, which sprang from the Sino-Japanese cultural centre. The exact historical position of this domain of art in the grand framework of this culture has yet to be ascertained.

The dependence of the art of the Amur tribes on the Chinese arises, in the next place, from the fact that Chinese models are immediately copied by the Gold. The explanation of such reproductions can be drawn only from the realm of Chinese conceptions. Many Chinese designs are simply based on a play upon words ; that is, abstract ideas are symbolized by an object the name of which is homonymous with that of the former, although written with different characters. On Plate 1 are combined three designs, apparently Goldian copies of Chinese originals. Fig. 1 represents somewhat more than half of a sleeping-mat covered with a silk embroidery. Around the central circle, surrounded by a key pattern, are grouped four bats and four butterflies, alternating with each other. The meander is repeated in semicircles in the four corners. The bat is called in Chinese *fu ;* there is another word *fu* with the meaning " good luck." The butterfly is designated *tieh ;* this same complex of sounds means also "aged." The abstract idea of this pattern is therefore that it may convey to the possessor old age and good luck. Fig. 2 shows the design on the top of a Goldian teatable. The centre is occupied by the dragon in the exact style of that seen so frequently on Chinese porcelain boxes and other objects. On either side it is beset by two bats. To the right and left of these is a vine bearing three blossoms. Fig. 3 represents a square kerchief of bluish-green silk lined with red cloth and edged with a black border. In the middle we see a conventionalized form of the Chinese character *shou* ("long life"). Around it are four butterflies hovering over plum-blossoms. They are embroidered in the most variegated colors. Plum-blossom is called *mei,* and is looked upon as the symbol of beauty, as *mei* also signifies " beautiful." This pattern presents, accordingly, an allusion to long life, old age, and beauty.[1]

Among other Chinese forms which we meet with in East Siberian art we

---

[1] See W. Grube, Zur Pekinger Volkskunde (Veröffentl. aus dem Mus. f. Völkerkunde, Berlin, Vol. VII, p. 138).

Decorative Art of the Amur Tribes.

find the svastika and the triskeles. Furthermore, the animals which appear in the designs of the Amur natives are just like those which play an important part in Chinese art and mythology. It is indeed most remarkable that animals, such as the bear,[1] the sable, the otter, and many others which predominate in the household economy, and are favorite subjects in the traditions as well as in daily conversation, do not appear in art, whereas the ornaments are filled with Chinese mythological monsters which are but imperfectly understood. In the progress of this paper we shall see, further, that the cock, the fish, the dragon, and other creatures are also loans. As with the Chinese, the representations of animals are not connected with concrete ideas : they have merely an emblematic meaning, and they symbolize abstract conceptions. The art of the Amur peoples is lacking, therefore, in realistic character, and merges into the formative. Objects of nature are not reproduced ; but foreign samples handed down from generation to generation, and at last assimilated, are continually being copied. Many women retain in their memories a great variety of patterns, and cut them out of paper with a speed and dexterity that are worthy of admiration.

SOME GENERAL CHARACTERISTICS OF ART AND ARTISTS. — Generally activity in the province of art is limited to the decoration of surfaces. The sense for plastic representations is lacking. These occur rarely, and are to be found only under exceptional circumstances. Animal carvings are met with on the richer sepulchral monuments of the Gilyak. Dishes and spoons, for use at the bear-festival, are adorned with carved bears. For a boy's toy, the bear is also crudely carved out of wood, and perforated above at the back to allow of a string passing through it, on which the figure is moved up and down. Other animals also — as, for instance, dogs, frogs, lizards, carp, salmon — are cut out of wood by the Gilyak as well as by the Gold, for use as playthings. To the prow of a boat is sometimes attached, especially among the Gold, a wooden duck, generally of rude workmanship. The wooden *burchans* — images of deities — which are manufactured according to the direction of the shaman, for the purpose of curing disease, — a new effigy on each occasion, — can by no means claim a place among works of art, since they embody only the particular attributes required in the special case in question, and, for the rest, remain a *rudis indigestaque moles*. Most striking is the lack of ability to draw human faces or forms ; the more so, since, on prehistoric monuments of the Amur region, petroglyphs have been found which doubtless represent human heads. Where such occasionally occur,— as, for instance, in certain paintings on Goldian paper charms (so-called *boachi*),— they reveal an appalling crudeness. In fact, human faces are never met with in

---

[1] One of the principal faults of Schurtz's studies, cited on pp. 1 and 2, lies in the fact that the single forms of ornaments have been extricated from the larger groups in which they occur, and the connection they originally had has thus been dissolved. Ornamental forms have ever-varying significations, according to the combinations in which they are used. Fig. 1, on p. 235 of Schurtz's paper, borrowed from Schrenck, and interpreted by him as a bear's head, is the ingredient of a composition covering the back of a Goldian or Gilyak fish-skin garment. The whole figure should be inverted, and then we see obviously the cock with fish in its beak, and perched on an ornamental figure intended to represent a tree.

decorative art; and even where we should imagine they might be, they are earnestly disclaimed by competent native judges.   Nevertheless, Schrenck states substantially, "Crude and primitive representations of the human face by means of a pair of circles with a point in the middle, a vertical line between them, and a horizontal below them, as eyes, nose, and mouth respectively, occur not seldom on utensils of the Gilyak, and owe their origin, I believe, to the idea that by placing them on an object the influence of evil spirits may be avoided, and the use of the implement attended with success."[1]   I have not succeeded in discovering the slightest vestige of proof of such a statement.

The materials used by the Amur tribes for expressing their ornaments are wood, birch-bark, fish-skin (especially salmon and sturgeon skin), elk and reindeer skin, cotton, and silk.   All decorations are executed by means of a long, sharp, pointed knife.   As regards special points of technique, they will be found at the proper place.

All needlework is done by women, and clever embroiderers especially enjoy a high reputation among their countrymen.   To be skilful in such work is regarded as a great merit, and increases exceedingly the value and esteem of a girl in the eyes of her father, who, a careful calculator, includes the amount brought in from this talent in the purchase-price due from his son-in-law. Men, on the other hand, aspire to possess a woman experienced in this line of art, and take great pride in her work; while wives are proud of dressing up their husbands with all the costly and gaudy art expedients available, and vie with their fellow-artists in their zeal to produce the most striking effects.

GEOGRAPHICAL ASPECT. — Schrenck, in his book previously mentioned, says (p. 401), that, besides among the Gilyak, the same style of ornamentation as is met with on clothing and other objects is also to be found among all other peoples of the lower Amur region, from the Gilyak upstream, along the main river as well as along its tributaries, as far as the Sungari River.   "In spite of the fact that these tribes are of Tungusian origin," continues that author, "still they have nothing in common with the Russian-Siberian Tungus regarding the ornaments used by them, but follow the Chinese and Gilyak.   In this connection one is struck by the fact that the sense for ornamentation, and its display in the Amur country, do not decrease, but increase, with distance from the most influential cultural people, the Chinese, and culminate among the Gilyak, who live farthest away from them."   The reason for this is sought by Schrenck, not in the natural dispositions of the peoples in question, but in political conditions.   The Gilyak remained independent of the rule of the Chinese, and thus attained to greater opulence than the Tungusian Amur tribes subject to the Chinese.   Secure in their property, they were necessarily better able to enjoy it, and to feel an incentive to adorn and embellish their clothing and implements.

This statement and its accompanying hypothesis are decidedly erroneous.

---

[1] Reisen und Forschungen im Amur-Lande, Vol. III, p. 402.

Schrenck's investigations were unduly devoted to the Gilyak, and in his predilection for these he likes to hold them up as superior to all other tribes of the Amur country. As I visited first the east coast, and afterward the interior and the western part, of Saghalin Island, later the entire lower Amur region from the mouth of the river up to Khabarovsk, I had an opportunity to study and judge of the activity of the people in branches of art also from a geographical point of view. On my journeyings my observations led, first of all, to the deduction of a prevailing law: namely, that the nearer the people live to a centre of Chinese culture, the higher the development of their art; the farther they recede from it, the less their sense of the beautiful. The art of the Gilyak of Saghalin is very poor and undeveloped; they possess a limited number of ornaments, and are unable to produce complicated compositions like those found on the mainland, as they themselves assured me. The farther east one goes the more destitute, and the farther west the more gorgeous, is the display of art, which reaches its climax in and around Khabarovsk. Indeed, the most artistic embroideries of our collection all came from this metropolis, where the Gold dwell in the immediate neighborhood of the Chinese, and have frequent intercourse with them. It is evidently owing to this influence solely that the Gold have attained to such extraordinary skill in the art of silk-embroidery, the knowledge of which, in its highest degree of perfection, is restricted to those inhabiting that area.

This geographical observation confirms anew the establishment of the historical truth regarding the affiliation of the arts of both groups. As the Gold are generally the most talented representative of the Amur tribes, so they are also those who possess the best understanding of decorative art and the largest number of individual artists. From the correspondence of the Gold and Gilyak patterns, it may be concluded that the Gilyak have derived the greater part of their motives from the Gold. Perhaps only the band-ornaments belonged originally to the former. This tallies with other cultural phenomena, for in all probability the Gilyak have adopted a considerable portion of their material culture, as well as a large mass of traditions and religious conceptions and institutions, from the intellectually superior and more versatile Gold. The decorative art of the Amur tribes is accordingly to be regarded, on the whole, as that of the Gold, who occupy the most prominent place in it.

This manner of geographical dissemination explains the uniformity of character of this art; so that diversities, if any exist, lie much less in a varying distribution of the patterns over geographical provinces than in the different grades of execution dependent on the tendency of artists in one community to concentrate their individual minds on particular lines of work, in which, in the course of time, their unequally allotted talents have received special training. The Gold, as a rule, are well versed in all branches of art, and excel all other tribes in proficiency in embroidering; the Gilyak may be superior to others in wood-carving; and the Tungusian tribes of the Amgun and Ussuri Rivers are unsurpassed in cutting ornaments for decorating birch-bark baskets. At all events, if we consider the geographical distribution of decorative design in these regions, the art industry carried on by the Gold in Khabarovsk and its environs remains the central circle from which the practices of the other tribes radiate, and lose in light and warmth toward the periphery.

Although the elaboration of ornaments is still actively going on, and in no more danger of dying out than the Gold and Gilyak themselves, yet the people, whose interests are more and more absorbed by recent demands of Russian intercourse, seem to overlook the relics of the past; at least at times they fail to understand their own singularity, for I came across but few individuals who were able to "read" their ornaments. To the great mass of the people they are indeed a mystery. Perhaps, however, they have never paid much attention to decoration, which may always have been confined to the initiated. If the common people are questioned as to the significance of a particular ornament, their usual answer is, accompanied by a shrug of the shoulders, that it is only for decorative purposes. Very few expert artists are able to give approximately satisfactory information, and even what they do give is fragmentary, and probably a mere skeleton of what must have been known about the subject in previous times; so that out of these shreds it is hard to piece together the perfect original fabric. The following account is an objective, although somewhat disconnected, record of the ideas which the native artists of to-day know how to develop on their productions. I think the clews obtained from this source of interpretation should form the impassable boundary to our knowledge in this domain of research, beyond which limit we should not attempt to go; for we should neither pretend nor strive to know more about things than the people who have made them. Gaps may in many cases be filled in, perhaps, by comparisons of single pieces one with another. An explanation for a definite form cannot be transferred unhesitatingly to another homologous one, except on the condition that the latter appears in the same connection of lines and structures as the former, — the same rule as holds regarding comparisons of traditions of various tribes. Just as in a fragmentary manuscript many a missing link may be guessed at, inferred, or even restored, so may it also happen in ornamentation; nevertheless one ought never to be off one's guard, but should adopt the expectant method until new sources are opened from allied provinces, be our present knowledge never so meagre and even in shreds.

Our investigation starts with an analysis of the simple component forms of this ornamentation, i. e., the band and the spiral, and will then deal with the usual forms of animals.

BANDS.—The fillet or band ornament occurs primarily on handles of spoons. Such decorated spoons are now things of the past. At the present day they are used only by the Gilyak, on the occasion of the bear-festival, having been superseded in every-day life by spoons of Russian make. The specimens represented

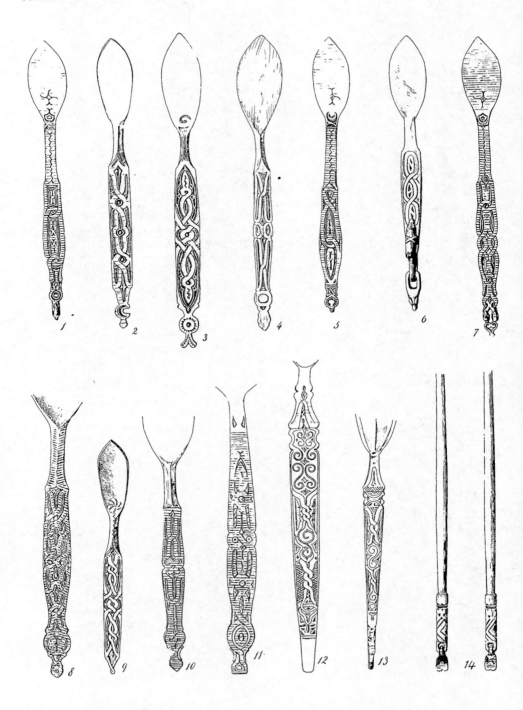

Decorative Art of the Amur Tribes.

in Figs. 1–11, Plate 11, are old pieces from the most remote villages of the Liman, and have long been in use. These spoons are made by special artists, for partic- ular use a short time previous to the bear-festival, and are characterized by their elegant and graceful shapes and by their elaborate ornamentation. The ends of the handles are carved into forms which, in most cases, have special reference to the bear-festival. Some present sculptured bear-figures, others figures of sun or moon. Before describing the bands, I will discuss these carvings.

The handles of the spoons illustrated in Figs. 5 and 9 of this plate were originally surmounted by bear-figures, which, unfortunately, had already been broken off when the specimens were obtained. Fig. 1 shows a bear in the act of walking, on top of which another bear originally stood, but it is now missing. The handle of Fig. 7 is surmounted by an open-work carving, the main portion of which consists of two bear-cubs side by side.

In Fig. 6, Plate 11, is represented a very realistic scene bearing upon certain events of the feast itself. Near the end of the handle may be seen the image of a standing bear bound around its body with two ropes, which cross each other over the back. This has reference to the first of the ceremonies connected with the festival, when the bear is taken from its cage, tied with ropes, and led to the scene of festivities. The extreme end of the handle consists of a movable link carved out of the same piece of wood as the perforation through which it passes. This link terminates in the figure of a bear-head, which is intended to represent the head of the bear that is shot with bow and arrows at the close of the feast, and exhibited in the house of the host.

The bowls of the specimens, Figs. 1, 5, 7, Plate 11, which are adorned with carved figures of bears, are further decorated with svastika-like figures, the central part having the form of a rhombus or lozenge. In the latter two there is a St. Andrew's cross within the lozenge. Each of the two vertical arms of the svastika branches off into two curved tips, while the extremities of the two hori- zontal arms bend upward (as in Fig. 7), or one curves upward and the other downward (as in Fig. 1), or both point downward (Fig. 5). At the base of the bowl is a primitive representation of the sun, which implies a symbolic meaning connected with that of the svastika and the bear-carvings. On Fig. 9 there is a variation of the svastika, perhaps developed by the insertion of a triskeles in such a way that its arms alternate with those of the svastika.

On the bowl of Fig. 3, Plate 11, we observe the figure of a crescent hooked at one end, while the handle is surmounted by a carved ring, the two incised concentric circles on which represent the sun. The outer circle is set with a row of small triangular figures symbolizing rays. A more primitive representation of the sun is to be seen on specimen Fig. 4, and a crescent surmounts the handle of Fig. 2.

On Fig. 6, Plate 11, we meet with the simplest form of the fillet-ornament, which here runs around in two windings, forming three loops. If we take into consideration the earliest stages in the development of this special ornament, it will at once be understood, that, owing to its form, it was readily employed on spoon-handles; for it is easily adapted to the space available, since it admits of lengthening or shortening to suit the decorative field, and, besides, contributes in a high degree to the gracefulness and elegance of the spoon; furthermore, if we regard the special purpose of these spoons, we may perceive a certain connection between this pattern and the representations of the bear. In the case of the spoon represented in Fig. 6 the fillet-ornament may be considered as the continuation of the ropes with which the carved standing bear is bound, and this may be the underlying reason for the employment of this ornament on spoons specially designed for use at the banquet of the bear-festival. Not alone from this example, but from other instances as well, may it be seen that a deeper connection exists between the fillet and the object, or the purpose of the object, on which it appears. At all events, the bear-figures in combination with the fillet-ornament should not be regarded as merely accidental. From this point of view, spoons decorated in similar style, but without bear-carvings, should be ascribed to a secondary stage of development.

In almost all spoons there is a narrow curved portion between the bowl and the handle proper. Seen from the side, this narrow section, in most cases, forms, with the edge of the bowl, nearly a half-circle. In Figs. 1, 5, 7, 9, Plate II, this part of the handle adjoining the bowl is decorated with a simple zigzag line, which appears to be a single thread drawn out from the main ribbon symbolizing the band-ornament. Among the spoons in the collection, this serpentine line occurs on those specimens only which have carvings of bear-figures in combination with the svastika. The majority, however, are undecorated on this part, while a few bear an incised figure composed of lines parallel with the curved line of the edge, as in Fig. 3. On the last-named specimen a short zigzag appears at the upper, and another at the lower, end of the handle proper, inside of the fillet. A few spoons, as those in Figs. 1, 5, 7, and 8, have rib-like designs on this part.

The Gold have no bear-festival, and naturally, therefore, possess no spoons decorated with symbols like those above described. Neither do we find any serpentine lines on their spoons.

The bands on these spoons are all formed of the raised portions lying between two parallel incised lines, the latter being of a negative character only; that is to say, the incised parts serve merely to outline the ornament, and in some cases also to fill in otherwise vacant spaces. In themselves they are not ornamental.

On Fig. 1, Plate II, we observe a band twisted at two points. Inside of this band are designs identical with those of the arm of the svastika on the bowls. On the next specimen (Fig. 2) two bands intertwine, forming two circular knots. The portion between the knots curves out at the centre on either side, admitting a third knot, indicated by an incised circle. The negative parts at both ends are filled up with short parallel horizontal lines, and in the centre with single vertical lines. The ornament in Fig. 3 consists of three bands. The middle one forms a circular knot at both ends and a large rhombus in the centre. Two short side-

bands are so intertwined as to form a circle within the rhombus.   The ornamentation on the handle of Fig. 4 is made up of a pair of cords or lacings, one simply
thrown over the other a short distance from either end, which form at the centre
two contiguous ellipses.   In Fig. 5 the same principle of the band-ornament is
employed as in Fig. 1.   In each of the two ellipses formed by the band-ornament
are two peculiar designs which happen to look very much like old Assyrian cunei-
forms.   On Fig. 7 are three bands artistically twisted.   One band runs along both
sides of the handle, bending at the centre into two contiguous curves ; the second
band forms a rhombus in the middle ; and the third intertwines with the first, and
then continues in the direction of the bowl.   The middle part of the handle of
Fig. 8 is occupied by two bands interlaced with each other and closed at both
ends.   They are joined by a short double band at each end for the purpose of
rounding off the ornament.   The spoon in Fig. 9 is ornamented in a style similar
to that in Fig. 3 ; the small rhombuses in the interstices also occur.   The handle
of Fig. 10 represents two double bands arranged in a manner similar to those on
Fig. 4.   They are interrupted in the centre, however, by a different figure.   This
ornament, which is also to be seen at the extremity of the handle of Fig. 8,
occurs frequently in later examples in connection with the spiral.

The decoration represented on the ladle Fig. 11, Plate 11, is likewise composed of the band-ornament, but it differs from the designs hitherto explained in
that the band is indicated at two places only — once in the middle and again at the
end — by short connecting lines, and that the negative parts, between which one
has to look sharply to discover the band, are more prominent (cf. Figs. 4–6,
Plate IV).

Figs. 12 and 13, Plate 11, illustrate spoons of Goldian origin.   The former
represents the handle of a large fish-ladle ; the latter, that of a spoon for eating.
In the fillet on the Goldian spoons the pure and rigid forms of the Gilyak are not
adhered to, and much less space is required for it, as it alternates with spiral-
ornaments.   Fig. 14 shows a pair of chopsticks, — a mere imitation, of course, of
Chinese-Japanese work, — which are interesting here because they show an incised
crescent at their ends, and terminate in movable pieces, as in Fig. 6.   The decoration on the handle consists of short parallel converging lines which meet in
acute angles.   Chopsticks are used but seldom, and only by such of the wealthy
and noble as lay great stress on etiquette and are fond of imitating foreign customs.   The most common method of eating is to use one's fingers, and finally to
lick the plate with the tongue.

SPIRALS. — We shall now enter into an examination of the kinds of spirals to
be found in this sphere of artistry, and discuss a series of objects on which they
occur.

Fig. 1, Plate III, represents an eye-protector, which is tied with a string
around the forehead, and shades the eyes from the snow in sledge-driving.   It is
especially worn during the transition period between winter and spring, when the
snow begins to melt.   It is made of cloth, and has a simple spiral-ornament

stitched into it.   There are two outer and two inner spirals corresponding to each other symmetrically, the latter two coinciding at the centre.

In Fig. 2 of this plate we see one of the two symmetrical halves of a design painted on the upper edge of a pair of leggings.   The ornamentation is on a piece of fish-skin, which is sewed to the material of the leggings.   The trapezoidal section across the top, the narrow stripe under it, and the lower border-line, as well as five of the large dots inside, are red; all the rest, deep black.   The ornament starts with a spiral winding round to the left, the centre of which is indicated by a small thickened circle.   To this spiral is attached, on the right-hand side, a figure the foundation of which appears as a simple wave-line from which proceed three scroll-like branchlets.   The upper ones run in the same direction as the main spiral.   The branch nearest the main spiral sends out a smaller offshoot in the form of a triskeles.

The wooden Gilyak box of cylindrical form, shown in Fig. 3, Plate III, is decorated with an ornament that offers a typical example of a compound spiral. From one and the same centre proceed two spiral bands, one within the other, and both running in the same direction.   The line forming the spiral is made up of three incised lines, close together and parallel to each other, which throw out in relief the two intervening spaces.   The spaces between these groups of lines form bands, which continue from one spiral into the next, producing alternately two knots and one knot, that serve to connect two adjoining spirals.   In the upper and lower edge of the spirals are twisted knots; so that one may look upon this pattern also as a very artistic interlacement of bands, which sometimes results in knots, and sometimes in spirals.   All together, there are four such spirals covering the convex surface of the cylinder.   On the inside of the bottom of this box is found a peculiar variation of the svastika, in that the design has two additional arms on the sides.

Fig. 4, Plate III, shows the cover of a decorated tobacco-box.   It is ornamented with three spirals, the central one smaller than the others, which are treated at the same time as band-ornaments.   Each spiral figure is composed of two spiral lines of the same direction, one within the other.   All three figures are solid spirals.   In the spirals at the top and bottom the regular circuit of the windings is interrupted by two circular inlaid bands which cross the spiral lines at two places; that is to say, they run under them: in this way in one semicircle are combined eight parallel bands.   The band of the central spiral is structurally connected with those on either side of it.   To the left of this central spiral are two parallel, frequently interlaced bands, and to the right of it is a band plaited into three knots, and tied to a boat extending along the side of this pattern.   This forms a conspicuous example of the essential principle of the band-ornament in connection with a realistic motive.   Around the rim of this cover, which is not visible in the drawing, runs a continuous chain-band whose form corresponds to that on Fig. 6, Plate II.

In Fig. 5, Plate III, is represented a Goldian knife which was obtained in

Decorative Art of the Amur Tribes.

the village of Sakhacha-olen. This knife is used, especially by the women, in lieu of scissors, which they do not possess. The end of the handle is sloped off with a slight curve. The carving, which covers only one side of the handle, is very roughly and inartistically worked out : it consists of two groups of spirals. Above, nearest the blade, are two, below three, simple spirals combined into one figure, which are bordered on either side by semicircles parallel to them.

A Goldian fish-scraper made of elk-bone, and decorated with a combination of incised spirals, is seen in Fig. 6, Plate III. These spirals are composed of double lines between which are short cross-lines. The ornament is symmetrically distributed over both of the roof-like sides of the bone. If one looks at the object horizontally, the scheme of the ornament appears as a wave-line from which proceed spirals with one winding, that here and there have lateral offshoots.

Fig. 7, Plate III, shows a board of modern Goldian work, for cutting fish on. The end of the board is shaped into a fish-tail. The board proper is divided into three fields, — a square at each end, and a rectangle between them. The latter is unornamented, and serves to cut the fish on. Thus there are three decorated fields, — the fish-tail and the two squares. The incised lines stand out from the black background. The foundation of the ornament is the double spiral, which occurs six times, and is surrounded by equidistant curves which run out into little spirals on the upper end of the board. Here appear also some leaf-ornaments, — combinations of three and four lobed leaves, the latter occurring twice between two double spirals. In the square adjoining the fish-tail both the spirals are pointed toward that side, and consequently they correspond to the acute angle formed by the combination of the border-curves. On the other side, however, the spirals preserve their usual forms ; whereas the border-lines do not meet, but are connected with each other by a short straight line to make room for a trifoliate leaf.

In Figs. 8–10 of this plate are represented metal objects of Yakut origin which are attached to the ceremonial garment of the shaman. They illustrate the use of the spiral farther in the interior of Siberia. Fig. 10 shows tendrils twining into spiral-like forms.

BAND AND SPIRAL ORNAMENTS.—In Fig. 1, Plate IV, we see a reel, the two horizontal arms of which are decorated on both sides alike. On the upper arm, in the middle, is a small rectangle bearing one link of a chain-ornament. On either side of it is a band-ornament consisting of a thrice-intertwined band. On the raised rectangle of the under arm is a knot similar to the one above, and the short pieces of bands on both sides show merely the single negative parts, whereas the connecting lines for indicating the direction of the band are missing. It is evident that also in this case a definite relation exists between the use of the band-ornament and the purpose of the object, which serves for winding up the ropes in netting.

The interlacement-band also occurs in the art of the Gold, although much less frequently than in the decorative art of the Gilyak. We meet with a band of this kind on an awl made of elk-bone (Fig. 2, Plate IV).   Just below the point of the instrument we observe two short bands plaited into a knot in the middle.   From this point another band starts, and fits into the sides of one of the vertical acute angles.   This band is plaited in the form of two lozenges, and ends with half of a third lozenge.   The bands added on both sides of the lozenges would seem to indicate the continuation of the latter indefinitely on both sides.   Inside of these rhomboids is a vertical row of three round dots, and on the bands themselves a series of smaller dots placed close together, and having the appearance of a dotted line.   This kind of decoration should be considered in a symbolical sense, since it suggests the use to which the instrument is put.

Another symbolical device is met with in Fig. 3 of the same plate, which represents a girdle-ornament made of antler, and shows a simple double-knotted band.   As this object serves to fasten the girdle, a reference to this purpose is obviously implied in the ornament.

Band-ornaments are especially employed on the ends of large dishes cut out of one piece of wood (Figs. 4–6, Plate IV), and used for fish and rice at large social gatherings.   Both ends of each piece show the same ornamentation, the bands projecting in relief above the incisions.   In Fig. 4 is represented a chain-band composed of three links and forming two knots.   Of the two bands on the right and left sides, little more than half is visible, but it should be imagined that they continue in the same way as the middle one.   The central band is linked to the two lateral ones, and is itself crossed in the middle.

In Fig. 5, Plate IV, we see a band running up and down, alternating with a horizontal one.   The former is twisted into two circular knots; the horizontal band is so treated that an ellipse is produced both above and below, the bands coinciding at the centre, the whole presenting a sort of flattened hourglass-shaped figure.   To fill up the centre of the ornament a circular band, over which is a semicircle, is made use of.   Fig. 6 shows two lateral bands, one crossed over the other, and a pair of horizontal bands twisted into a knot, which is indicated by a circle and two connecting strokes tangent to it.   These two bands are coiled at their ends, forming four circles.

Next on this plate we see a dish (Fig. 7) which shows a different ornamentation on each end, due to their difference in form.   On the trapezoidal-shaped piece are visible two interlaced bands which form a rhombus in the centre. In the triangle on the other side is a double ornament,— a simple band-ornament, and attached to it another band-ornament the negative parts of which are made up of two central combined facing spirals, a simple scroll on either side of them, and an engrailed line along the edge of the dish.

Fig. 8, Plate IV, represents a small square box with separate cover.   The ornament on the side of the box consists of two double spirals treated as bands, and surrounded by a band following the windings of the spirals.   The cover (Fig. 8 a) shows a combination of two pairs of simple facing band-spirals.   The sides of the box seen in Fig. 9, which is shaped like a horse's hoof, are covered

Decorative Art of the Amur Tribes.

Decorative Art of the Amur Tribes.

with a continuous series of compound double spirals.   Whereas the cover (Fig. 9 a) shows the band-ornament in a very impressive way, the form of the compound spiral in the central rhombus reminds one strongly of the Japanese *futatsutomoye* (see Figs. 2–9, Plate xvi).

The dish shown in Fig. 1, Plate v, is decorated with a different pattern on each end.   The arrangement of the band-ornament here differs widely from the other representations of it: on the end to the left is a continuous band formed by two parallel outlines, one of which is placed near to and parallel with the edge of the dish proper, the other being combined with the terminations of the two central facing spirals, resulting in a very strange figure, which has properly no ornamental or symbolical significance.   In a similar way a still stranger figure is produced in the longer tapering end on the other side.   The negative parts here require so much space that at first sight one might consider them as expressing the ornament proper.   If, however, we bisect the four-armed figure longitudinally, we shall recognize that the arms with the adjoining outlines of the figure are simply portions of spirals, and that merely their combination, and their adaptation to the space available, have given rise to this peculiar kind of figure.   Very striking is the lack of symmetry displayed in the two halves of this device.   To an X-shaped figure on the one side corresponds a hammer-shaped figure on the other side, just as the condition of size varies.   A combination of two such figures at the apex of this triangle is represented in an anchor-shaped figure, to which, however, no positive ornamental meaning is attributed, but it simply designates the course of the band.   Fig. 2 of this plate is the cover of a tobacco-box of ellipsoidal shape.   The central and lateral portions stand out a little beyond the two half-elliptical sections, which show the same ornamentation in correspond-ing symmetrical arrangement, — two double spirals treated as bands.   In the middle raised part is a pair of intertwined bands which coil at the ends into band-spirals.

Figs. 3–5, Plate v, illustrate drills, — three old rare pieces from the village of Chomi on the Liman.   While the upper and lower parts are ornamentally carved, the middle portion is covered with a band consisting of incised parallel oblique lines, to symbolize, as it were, the turning motion of the instrument.   The patterns show again a combination of the band-ornament with the spiral.   That the spiral however, is not to be considered as the fundamental element of the ornament, is seen from the neighboring auxiliary figures, which run parallel to the winding of the spiral, and have no other purpose than to indicate the direction of the band. Thus we see in the lower part of Fig. 3, in the centre, facing spirals surmounted by a pointed arch, the two curves of which run parallel to the windings of the spirals above ; and under these facing spirals is a strangely shaped figure which has arisen through the four curves of which it consists being drawn parallel to the adjoining spirals in making room for the band.   The circle symbolizes its terminal knot.   On the upper part of this drill occurs again a similar combination of spiral and of interlacement-bands.   The ornamentations on Figs. 4 and 5 are only variations of the same principle, the lower part of Fig. 5 showing negative incisions similar to those in Fig. 5, Plate IV.

DECORATIONS ON BOATS. — The bows of wooden rowboats are sometimes adorned, both inside and outside, with paintings. For this purpose, stencils are cut out of strong birch-bark, applied to the parts to be decorated, and brushed over with black paint. In all cases the negative cut-out parts, which naturally appear as the positive portions on the object to be decorated, should be regarded as the ornament proper. They have therefore been blackened in the drawings. The outline of the stencil is sometimes adapted in a certain degree to the form of the pattern. The most frequent motive employed in this case is the double-spiral fillet, although the compound spiral is also used.

FIG. 1 (⁴⁄₁₀). Birch-bark Stencil. Tribe, Gold. Length, 43 cm.

The ornaments shown in Figs. 1 and 2 are constructed on one and the same principle. The two double spirals forming the main part of the pattern are surrounded by a simple band that runs parallel to the outer curves of the spirals. This band merges above, on either side, into

FIG. 2 (⁴⁄₁₀). Birch-bark Stencil. Tribe, Gold. Length, 44 cm.

a central head-shaped ornament, the upper part of which consists of a pair of short symmetrical spirals, while the under part results from the combination of two conventionalized fishes.[1]   The lower edge of the pattern is in the form

[1] See p. 29.

of an engrailed line. In Fig. 3 the band surrounding the two facing spirals is decorated with six small scrolls, which branch off from it at a tangent.

In Fig. 4, four pairs of facing spirals are placed around a star-like rosette (so-called "star-cross"). The design corresponds to or recalls the anchored or forked cross of mediæval heraldry with convoluted flukes. In Fig. 5 the spirals join a somewhat square-shaped figure, in the angles of which are found flower-buds with four circles in front of them, and in the centre a rhombus with curved sides.

Figs. 6 and 7 represent the beginnings of two decorations placed longitudinally in the interior of the prow of a boat. They may be ex-

FIG. 3 (⁷⁰⁄₈₆). Birch-bark Stencil. Tribe, Gold. Greatest width, 31 cm.

tended at either end at will. Fig. 6 consists of a star-cross and a double design whose elements are formed according to a principle similar to that employed in Fig. 2, except that the spirals have only one winding, and face each other. Whereas Fig. 6 consists of a series of two different designs, Fig. 7 is composed of only one figure. As this ornament, like the preceding one, is executed with

FIG. 4.

FIG. 5.

FIGS. 4 (⁷⁰⁄₁₇₈₀), 5 (⁷⁰⁄₁₇₈₈). Birch-bark Stencils. Tribe, Gold. Diameter, 20 cm., 18 cm., respectively.

perfect symmetry, there is but one motive, the scroll, which terminates in a horn-like offshoot with an adjoining semicircle.

OTHER BIRCH-BARK PATTERNS. — Figs. 8 and 9 are patterns cut out of birch-bark, used for embroidering ear-lappets. Both consist largely of spirals. In Fig. 8 is seen a lower row of four spirals surmounted by a triangular field filled with

fanciful figures that are characterized as derivations from the conventional form of
the fish-ornament in so far as they do not appear at the outset to be mere space-
fillers.   This figure runs out into a face-shaped head-piece which at first sight
one might take to
be a convention-
alized human
face : in this case
the eyes would
be denoted by
spirals, the mouth
by the figure con-
nected with these,
and the four
tooth-like forma-
tions would repre-
sent tusks not

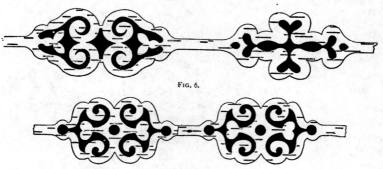

FIG. 6.

FIG. 7.

FIGS. 6, 7 (₈₈₈ a, b). Birch-bark Stencils. Tribe, Gold. Length, 28 cm., 25 cm., respectively.

unlike those identified by Hein [1] on the demon-shields of the Dayak. Nevertheless,
in this as well as in the following figure (9), the Gilyak in the village of Chai on
the northeast coast of Saghalin Island, from whom these patterns were obtained and
information concerning them sought, decidedly
denied that these figures have any relation to the
human figure ; and it seems also that the form of
the outline of this pattern is solely due to an

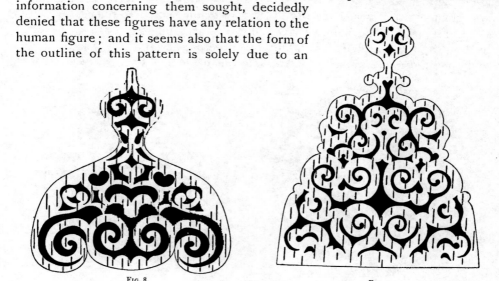

FIG. 8.

FIG. 9.

FIGS. 8, 9 (₈₈₈ a, b). Birch-bark Patterns for Embroidery. Tribe, Gilyak. Height, 16 cm., 20 cm., respectively.

adaptation and assimilation to the space occupied by the object itself. Fig. 9
consists of a structure of spiral ornaments, whose width gradually lessens as it
proceeds upward from the broad base, until it ends in a narrow neck surmounted

---

[1] A. R. Hein, Die bildenden Künste bei den Dayaks auf Borneo (Wien, 1890), pp. 41–85.

by a figure apparently treated like a face, but in semblance only, — a circle above, two crescents on the sides, their convex sides turned towards each other, and below them a lozenge standing on its point. In the trapezoidal under figure we observe three rows of spiral ornaments placed one above another. In this design are found numerous conventionalized fishes.

CIRCLE-ORNAMENTS. — In a few ornaments the circle is also used as a fundamental form. Figs. 10 and 11 are both Gilyak birch-bark patterns used for an embroidery that was plaited into the hair of little children in ancient times, but is now out of fashion. Both of these consist of combinations of circles and spirals. In Fig. 10 the motives are arranged in a series of horizontal rows, only two of

FIG. 10 (70/880 d). Birch-bark Pattern for Embroidery. Tribe, Gilyak. Height, 17 cm.

FIG. 11 (70/1287). Birch-bark Embroidery-Pattern. Tribe, Gilyak. Length, 26 cm.

which contain spirals. In Fig. 11 five rows may be distinguished, the two upper ones being perfectly symmetrical.

In Fig. 6, Plate xv, we observe a new motive of the circle-ornament, obtained by describing circles so that they intersect each other.

THE COCK. — The animal which plays a predominant part in the ornamental art of all the Amur peoples, and is more frequently reproduced than all other animals together, is the cock. This circumstance is the more conspicuous, since the cock is not a native of the Amur country, but was introduced from China, and recently, of course, by the Russians. Nowadays there are some Gold who raise poultry in their houses. The Gilyak on the northeastern coast of Saghalin, excepting a few who had chanced to see a Russian village, never saw a cock, but they know and explain it by their ornaments. They call it *päkx*, a word apparently derived from the Goldian and Olcha word *pökko*, that may be traced back to *fakira gasha* of the Manchu language. Another Goldian term, *chokó*, appears likewise in Manchu, and is perhaps allied to the Mongol *takiya*.

Since the cock is a new-comer in that region, it is not surprising that it plays no part in the mythology of the natives, as it does with the Chinese. In their opinion, the cock is a symbol of the sun, because it announces the rising of the sun. Besides the earthly cocks, there is a heavenly cock, which, perched on a tree, sings at sunrise. This tree is the willow, which also symbolizes the sun. The cock is sometimes called in Chinese " he who enlightens the night ; " and the sun, " the golden cock." Besides, it belongs to the class of animals that protect man from the evil influences of demons. Live white cocks are sometimes used in funeral rites.

Regarding the representation of the cock in Chinese art, only a few general facts may be stated, as this branch of research is little explored, and investigations of ornaments have unfortunately been almost neglected. Japanese art is based wholly on Chinese, and the ground on which it stands is somewhat better known. The ordinary domestic fowls are frequently depicted by Japanese artists, the cock being the favorite among them. It is painted on hanging scrolls, and modelled in wood, bronze, porcelain, and other materials. Most frequent and admired is the painted design of a cock standing on a drum (*taikō*) ; and in this case the sides (or one side) of the drum are decorated with a triskeles (*tomoye* or *mitsutomoye*). This is the well-known circular diagram divided into three segments (see Figs. 2–9, Plate XVI).

Single Cocks. — On Plate VI we have four examples of the cock drawn true to nature. Fig. 1 shows the typical form of cock cut out of paper, and used as a pattern for embroidery. Head and eye are circular, the beak semicircular. On the head is a bipartite crest shaped like a fish-tail. To the back is attached a quadrifid wing, and a tripartite tail almost convoluted. The feet are missing. On the body is a conventionalized fish, the upper border-line of which runs parallel to the outline of the cock's body ; the under border-line, shaped partly like a brace, partly like an invected line, being composed of three portions, indicating head, body, and tail. Fig. 2 is an embroidery-pattern representing a similar type of cock, but with some remarkable differences. This cock holds a fish in its beak. The motive is, of course, far from being realistic. It does not convey the idea that the cock devours the fish : its meaning is purely emblematical. The wing-feathers are indicated by four teeth, projecting from a line generated from the beak, which line continues into a scroll parallel to the outline of the body, and representing a fish-tail. The tail-feathers are highly developed, showing six parallel flukes. The body is cut into a double spiral. The space between the beak and neck of the cock is so formed as to represent a bird's beak.

Fig. 3, Plate VI, is a weaving-pattern, whether of Chinese or Japanese provenience is uncertain. It consists of circular fields in which are designed realistic cocks, whose somewhat stiff forms are attributable to the technique of weaving. Comb, beak, eyes, feet, feathers, plumage, and, in all, nine wing and tail feathers, are indicated. There are slight but delicate differences in

Decorative Art of the Amur Tribes.

the forms of the bodies, the attitudes, and the manner of stepping, of the single birds.

Fig. 4, Plate VI, is an ornamentation on the surface of a birch-bark hat, the rim of which is covered with an uninterrupted sequence of double spirals sending off little branchlets. In the main field three naturalistic cocks standing upright are observed, beak and eye being indicated, a circle being placed over their heads, and, what is most singular, two long stretched-out feet with spurs are to be seen. The pinions are represented by a semicircle with a recurved hook on one side; the tail, by a spiral with short appendage of a form similar to that on the rim, above which are three tail-feathers. On the body we see a picture of a fish consisting of two parts, — the head and the tapering body. It is worthy of note, in what graceful forms the outlines of the bodies of the two animals are adapted and assimilated to each other.

Fig. 1, Plate VII, represents the side of a Goldian birch-bark basket. The decorative field is enclosed by a triple border consisting of a meander, an invected line, and a row of braces which are apparently derived from the form of the cock-spur, and which I have therefore styled "spur-ornaments." The rectangle is divided into two parts which are separated by three figures, — a carp realistically drawn (a), a large conventionalized fish with ·long prominent fin (b), and a small conventionalized fish (c). In the field on the right two cocks are visible, their heads turned downward, and in their beaks trichotomous fishes conventionalized in the same form as in c. The four tail-feathers are turned upward; and the bent feet, stretched out to both sides, are remarkably long. The feet of the bird on the right terminate in a mucronated process, in a style assimilated to that of the tail-feathers, whereas on the other side they run parallel to each other. The space between these two birds is filled with drawings of fishes, — between their heads the rather natural-looking fish d; between their tail-feathers the tail of a fish (e), apparently lacking a head; and between their bodies the rosette f, the four leaves of which show the same form as the fish-body in d. On the other side of this rectangular field we see two cocks, one above the other, the upper of which is looking toward the left, and the under one toward the right. The style of drawing of these two birds tallies with that of the other two, except that the eyes are not indicated by dots, as in those, and that they hold in their beaks, not one fish, but each two fishes. Both of these fishes are scalloped on the upper edge; but in the lower fish, that terminates in a scroll, the scallops are more sharply cut. In the body of the under fish the design of a bipartite fish is represented, and runs parallel to the outlines of the fish. In the upper cock the pinions are symbolized by a spiral, which, however, is disconnected from its body; whereas in the lower cock, in lieu of spirals, are two comma-shaped figures (g, h) which seem to be derived from the fish-body. To the extreme left, beyond h, is a very curious form of a conventionalized fish, made up of a circle and a curved serpentine stripe. The feet of the cocks are fashioned in the same way as those of the neighboring birds on the other side, except that here

an oval-shaped figure is inserted, connecting the body with the stripes indicating the feet.

Combined Cocks, Type A. — Figs. 2 and 3, Plate VII, represent two Gilyak
bear-spears made of iron.   The greater part of the design is inlaid with silver, and
the portions shown in hachure are inlaid copper and brass.   On the blade of Fig. 2
are two single cocks symmetrically arranged, each with a circle in front of the
beak, the body and tail shaped like a fish.   This circular object was explained to me
by natives as a grain of wheat that the bird is about to swallow ; but this explanation seems to have arisen after the true and original meaning had been forgotten.   It is rather more probable that the circle which is generally between two
cocks facing each other, or in front of a single one, represents the sun, which,
according to Chinese mythology, belongs to the cock.   In fact, the sun is represented on mythological pictures of the Gold as a simple circle, or as two concentric circles, with two diameters at right angles to each other.   This particular
type of single cock appears doubled in Fig. 3 in such a way that the two roosters
face each other, and hold one circle in common between their beaks ; this is the
attitude called "combatant" in heraldry, and this frequently occurring typical
device we shall designate for brevity "Type A."   On this blade we see, all
together, two symmetrical pairs of such combatant cocks, easily distinguished as
birds, particularly by their crests ; feet and wings being omitted, as in the preceding case, and only the tail-feathers denoted.   Besides these approximately naturalistic cocks, which are explained and recognized as such also by the natives,
there are other purely geometrical designs on these blades, which seem to have a
certain connection with the cock-ornament, although we cannot prove that they
are derived and developed directly from it.   The native interpreters deny that
they mean cocks, or have anything to do with them.   In Fig. 2 we see, next to
the single cock, a combined figure placed around a circle, and terminating in
a helical line ; if there were an indication of a comb, the components of this
figure might be regarded as cocks.   The following figures consist of a pair of
combined triskeles connected by an oval, two arms of the triskeles being spirals.
There is also a geometrical repetition of the combatant cocks on Fig. 3, the heads
ending in simple scrolls, and the tails in convoluted forms.   The animal at the
upper end of this blade is explained to be a fox devouring a carp, and that on the
raised medial line at the lower end is said to be a lizard.   There is also a lizard
on the corresponding part of the other spear, and, a little farther below, a
flat fish.

Fig. 4, Plate VII, shows an embroidery made of reindeer-hair, probably of Yakut origin.   We observe here the type of the two combatant footless cocks, whose
beaks, heads, eyes, long-extended bodies, and four tail-feathers each, are distinctly
marked.   In the beak of each is an oval object, the two uniting into one figure.
Under these cocks we observe a symmetrical geometric figure composed of
spirals and curves, which, however, is nothing more than an ornamental sketch
showing, as it were, the reflected images of the cocks above.   Corresponding

Decorative Art of the Amur Tribes.

parts in the real image and its counterpart are designated by the same letters. The head of the cock *a*, for instance, is expressed by the spiral *a' ;* the oval *b* corresponds to the portion *b'*, somewhat more extended, and connected with the body *c'*. The tail-feathers *d* and *e* are reproduced below in a scroll *d'*, with a semicircle *e'* attached to it.

Fig. 5, Plate VII, represents an ornament on the upper part of the leg of a pair of boots from the Orochon on Ussuri River. These boots are made of elk-skin. The decorated section consists of two fields, the upper ornaments being painted on fish-skin in red, blue, yellow, and black ; those below being cut out of fish-skin dyed black, and attached with red, yellow, and blue thread to a piece of cloth, which is sewed to the elk-skin. There are two combatant cocks standing upright in the lower design, and, what is most remarkable, they even have spurs in the form of a brace, which is rarely found on other patterns. On the paint-ings the same picture is reproduced, showing the cocks also with spurs and a spiral and some strange figures in the body, the latter of which may perhaps be traced back in part to the design of a spur.

Fig. 6, Plate VII, is an embroidery-pattern cut out of paper, and is used on the upper of a woman's shoe. It shows the cocks, Type A, in a nearly heraldic attitude, the heads treated merely ornamentally, the wing-feather as a scroll, and the tail as a fish-tail. Fig. 7 is a paper pattern for embroidering gloves, the larger portion on the right being used on the back, and the other for the thumb, the motive being exactly the same as in the foregoing figure with slight modifi-cations in form. These show how the same pattern is assimilated to an altered space, additions and omissions being made according to the variation in the space to be filled. In this way on the larger design the forms of the body, tail, and wings have been correspondingly enlarged. On the smaller piece the comb has been omitted on account of lack of space, and the two-lobed wattle of the larger cocks has shrunk into one small knob. In both groups, fishes are attached to the wing-feathers : on the right side of the pattern a little fish is clinging to the outer line of the scroll, whereas in the smaller cocks it lies inside of the scroll, and forms its starting-point. In both figures the cocks lean toward a wave-line, having on the under part curved prongs agreeing in form with the cock's tail-feathers. In Fig. 6, where the same motive occurs, we see a close connection between this part and the cock itself, so that they form a real unit. In Fig. 7, however, the cock itself has a highly developed tail-feather immediately adjoining its body, so that we meet with two tail-feathers, one above the other, on these designs. The question arises, Does the under tail-feather suggest the existence of another, strongly conventionalized cock, or is it merely an ornamental addition ?

Figs. 1 and 2 on Plate VIII represent embroideries designed for trimming the pocket of a shirt. Here two combatant cocks are grouped around a central ver-tical axis. In the one figure, head, eye, body, wing and tail feathers, are clearly to be distinguished. The feet are missing, and on the body of each bird is a con-ventionalized fish, the head and tail of which are discernible as separate parts.

These two cocks are resting on a simple geometrical figure, which is perhaps to be regarded as a strong conventionalization of another cock. The pattern Fig. 2 becomes intelligible by comparing it with the preceding one. It represents a stage of conventionalization much further advanced than is seen in the first one. The body is merely indicated by a spiral, neck and head simply by the continuation of the scroll-line bent upward and slightly curved to the side, the tail being in the form of an ornamental double fish-tail. The bifurcated arms projecting on either side above the two cocks are meant for fishes, which are essentially characterized by the form of the tail.

The question as to how the motives hitherto discussed, especially the combatant cocks, were derived from Chinese art, whether entirely or partially, cannot as yet be satisfactorily answered. Notwithstanding this fact, some material may be adduced from which to draw nearer to the solution of this problem.

Some Chinese Prototypes. — Fig. 3, Plate VIII, a Chinese weaving-pattern of the nineteenth century, is inserted here simply to illustrate the idea of the combatant birds as employed in Chinese art. Here two pairs of such birds are grouped around a floral device, so that the style of the head and the body of the bird depends largely on that of the foiling, and thus shows certain deviations from the forms seen in our Siberian patterns.

Figs. 4 and 5 of the same plate are likewise weaving-designs, from the Königliche Kunstgewerbe Museum in Berlin. Although the origin of these fabrics is given in the catalogue as "Orient, 17th–18th centuries," yet without doubt they are of Chinese creation, at least as regards the pattern. In the centre of Fig. 4 we see a conventionalized tree (*a*), at the top of which are two bird-heads (*b*) beak to beak. Two heads of the same style are visible at the foot of the tree, but at a short distance from each other. A perfect representation of the cock appears in *c*: the comb (designated by three lines), the pinions, the tail-feathers, the feet (indicated in a way similar to that of the comb), are all shown. Under the throat is a rectangle, which seems to correspond to the circular object that the bird usually holds in its beak in the Siberian designs. That all forms are square here which are round there, is due solely to the technique of weaving. The cock with outspread wings, its head stretched forward, is represented in the figures marked *d*; whereas *e* reproduces the bird in a walking attitude, with head erect. Both these cocks (*d* and *e*) are placed sideways, so that their heads nearly touch each other. In the corresponding figures, *f*, one may recognize the cock in a standing or squatting position, the two being combined in the picture *g* into one escutcheon-like unit, the heads looking in opposite directions. The smaller design above (*h*) shows an advanced stage of conventionalization of the same conception; and in *i*, still farther above, only the head and crest of the cock are distinguishable. The combination of these three figures, *g*, *h*, and *i*, together with the lateral types *b* and the additional ramifications (*j*) on either side, seem to indicate that the artist may have intended to suggest, in the figure as a whole, a tree. On this pattern, then, we meet with five different types of cock. The style of design shown in

Decorative Art of the Amur Tribes.

Fig. 5 is allied to the preceding. In the centre are two birds perching to the right and left of a tree. Very likely a cock is intended in this case, the crest on the head being visible, the wattle under the throat, the feet, the outspread wings, and the tail-feathers. Above the tree and the two large birds we observe three small birds, one of which seems to be perched on the top of the tree, which recalls perfectly the perching cocks on Siberian fish-skin garments (see Plates xxix and xxx). The form of the tail, and, still more, the object held in the beak, admit no doubt as to its significance. Also the various patterns employed on either side of this picture remind one, in their exterior form, of some compositions occurring on the fish-skin garments, although an exact identification is impossible. The parts projecting from the hatched portions, and particularly those branching off from the vertical line below, appear to represent bird-heads. Still more difficult to explain is the row of six figures across the bottom. In the upper part of these may be recognized cocks, which in the three figures on the right-hand side hold their heads bent to the right, and on the other side to the left. In the rooster on the left the eye is missing, which is not the case in the birds on the right side. Slight deviations from symmetry may also be noticed in the figures above. It is possible, and very probable too, that other pictures of cocks or other birds may be contained in this composition.

Combined Cocks, Type B. — There are also combinations of two cocks, their backs contiguous, and necks bent in opposite directions, which for convenience we shall call " Type B." This occurs on Fig. 1, Plate ix, a pattern cut out of birch-bark, that serves as a foundation for an embroidered pocket. The bodies of the birds consist here of mere compound spirals. The heads are not represented, from lack of space, but the oval objects belonging to them are visible. The curved offshoots at the bottom of the spirals seem to indicate feet.

The ornaments represented in Fig. 2, Plate ix, are cut out of birch-bark and sewed to a birch-bark hat. They are put on in three rows around the hat, each row containing four double cocks executed in an ornamental style. In the outermost row on the border the tail-feathers are easily discerned. The body is indicated by a spiral, to which a circle is joined. The two heads are placed together so as to form a rhomboidal figure. These eight cocks are dyed blue. On the edge between the tail-feathers are four single pieces dyed black. These are ornamental survivals of the cock's spurs. The cocks in the middle row have their heads distinctly marked, and two circles on each side of the neck. Their bodies have nearly the shape of the triskeles. These are colored red, but the heads are not dyed at all. The circles are blackened. The cocks of this row are ornamentally connected with those in the outer circle at their heads, and with those of the inner row at their tails. This central row shows the most conventionalized forms of the cock. If we imagine a line drawn through the two points where the tail-feathers of the outer row come in contact and where those of the middle row meet, we shall be able to distinguish the two united cocks of the third row. Here the two heads have coalesced into an ellipsoid which has a circle on either side, and the bodies are adapted to the top of the conical hat. Heads and circles are colored blue, and the other parts are blackened.

In Fig. 3, Plate IX, is reproduced a painting on the upper part of a pair of fish-skin leggings. Across the extreme upper edge is a border of black; and of the two ornamented fields below, the upper is red, the under one blue. The latter is edged with a narrow red band. The lines of the pattern are painted with black China ink. In the illustration, only a portion of the leggings is represented. The ornament, however, is continued to both sides, and terminates at some distance from the seam. The design is painted on a special piece of fish-skin, which is sewed to the material of the leggings proper. In the under section we see on the extreme sides a band-spiral terminating in a fish-tail. This one is continued toward the other side so as to form a double spiral, one of which is disconnected and represents a conventionalized fish of semicircular form. Under the central pair of spirals are three pairs of corresponding fishes, one below another, the undermost of which is connected with the spirals by a hook resembling a bird-head. The upper of these three pairs of fishes finds its counterpart in the figure placed above the central spirals by way of rounding off the design. That this device also has arisen from the combination of two conventionalized fishes, is perceived by a glance at the corresponding design on the upper section, in which the head of the fish is distinctly marked off from its body. On this border we see spirals connected with fish-tails, and in the centre compound but disconnected spirals. Above, on the spirals, are visible tiny offshoots, looking, as it were, like survivals of bird-heads.

An embroidery-pattern made of birch-bark, for use on an ear-lappet, is shown in Fig. 4, Plate IX. Above and at the base of this pattern are two distinct groups of roosters. In the upper group the heads (*a*) are turned upward and away from each other; the body is represented by a closed spiral, the interior of which assumes the conventional form of a fish. In the under group the head of the cock (*b*) is cut out of birch-bark, body and feet being likewise indicated, and the pinions by an exaggeratedly large spiral. Here also occurs the type B. What significance the additional arms terminating in the circles *c*, in the centre of the figure, may have, is hard to decide: they are either ornamental fillings, or perhaps reminiscences of the circular figures connected with the cock. The same may be said of the circular forms *d*.

Fig. 5, Plate IX, represents a boot, the leg of which is made of cloth, the foot part of seal-skin, and the upper, the ornaments on which are cut out in relief, of sturgeon-skin. The lower part of this ornament is made up of two pairs of facing spirals, which are connected with each other by a heart-shaped figure. On this heart are drawn two ovals combined as in an 8, that is usually placed between the beaks of two combatant cocks. For this reason it may very probably be correct to suggest that the two curves bent toward the border-lines to the right and left represent birds' necks, and that we have here the same cock type as occurs in Fig. 1 of this plate. The upper double spirals continue

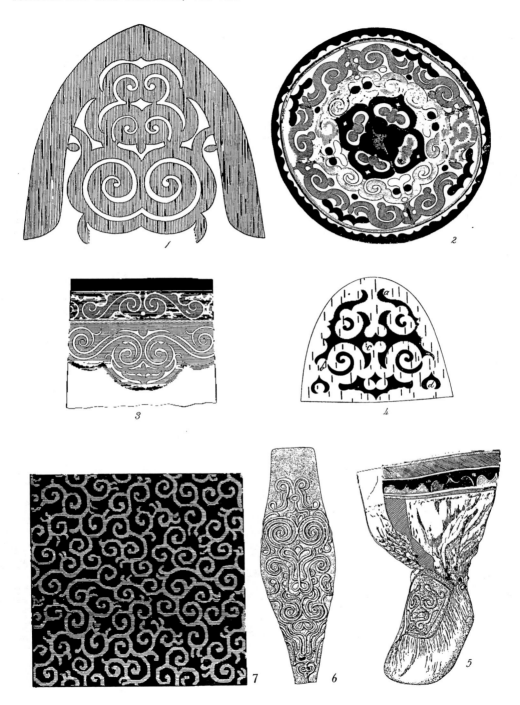

Decorative Art of the Amur Tribes.

into figures resembling the capital letter R. In this appendage the usual form of the conventionalized fish may easily be recognized.

The knife-case seen in Fig. 6, Plate IX, is manufactured of sturgeon-skin. The lines of the decorative design, which stand out in relief, are produced by cutting the cuticle away with a knife. Considering the brittleness of the material and the difficulty of its execution, the regularity of the forms and the graceful sweep of the lines are most admirable. The ornamentation consists of a clever combination of compound spirals and conventionalized fishes, so arranged that the whole forms a coherent structure. Above are to be noted two conventionalized fishes with round heads at both ends. Then follow two spirals of four windings each, terminating in a double loop. Below them is a palmetto-like figure, from which spirals arise on both sides, and under these are single conventionalized fishes. Then follow two spirals joined to the figure above, which seem to be conceived of as cock-bodies, since the curved band evolving from them corresponds to the form of the cock's neck.

A tapestry of rather old Japanese workmanship is represented in Fig. 7, Plate IX. The motive of the pattern is a double spiral, of a somewhat angular form, due to the technique of weaving. Here are seen spirals with two offshoots, and again others in pairs, one of which has four branchlets and the other one branchlet. These embellished spirals appear in the same manner on our Siberian patterns, where they have undoubtedly been proved to be closely connected with the cock-ornaments. This Japanese pattern may also tend to symbolize conventionalized birds, in which heads, body-spirals, and feet may be discerned.

Fig. 1, Plate X, shows part of the decoration on the side of a birch-bark basket. The leading motive is the realistic representation of cocks of Type B, the tail-feathers being turned toward each other, and the heads on opposite sides, but so turned that the birds look at each other. Under each of their beaks is a circle. The wing-feathers are characterized by three, the tail-feathers by four, lobes. Fishes are drawn on their bodies, but of two different forms, which agree in the two outer and two inner cocks respectively. In the former the tail tapers to a point, while in the latter it is bifurcated. The two inner cocks have, besides, a circular figure just below the neck, which is missing in those placed outside. The cocks are encircled by double lines, which are gracefully adapted to the form of the body, and are adorned above with flame-shaped lobes approaching in form those representing wing and tail feathers. In the centre of the design are two strongly conventionalized cocks, whose heads lie on the extreme sides, turned in the direction of the cocks above mentioned. Body and wing-feathers are marked by two equal spirals. This entire decoration is bordered above and below by a chessboard-like pattern. The vivid power of motion which pervades this whole composition shows a wonderful degree of artistic ingenuity.

In the next illustration (Fig. 2, Plate X) is seen a painting across the top of the right-hand portion of a pair of fish-skin leggings. The ornament itself is blue ; the two under border-lines, blue and red.   In the centre we see two roosters with distinctly marked heads and eyes.   The body is symbolized by a band-spiral which starts from a circle in the centre.   The neck is a continuation of the border-line running parallel to the winding of this spiral.   To the beak of each rooster is attached a spiral wound to the right, the termination of which is made into a bird's beak with a circle just like the central spiral.   Above it is a conven-tionalized fish with distinct tail ; and below it a scroll rolled into a beak at the end, with a circle at its tip.   Particularly worthy of note is the abrupt manner in which the well-executed head of a cock, with the usual oval object in front of it, is placed under the scroll to the right, — another evidence of the fact that any spiral can be made to symbolize a cock's body.

Fig. 3, Plate x, represents an embroidery on the upper part of a pair of leg-gings made of Chinese silk.   The background consists of black silk cloth (hatched in the drawing), the spaces between the embroidered lines being covered with white silk.   This design is mainly filled up with two opposite swans fashioned like cocks, the long arched necks running parallel to their oval-shaped bodies. The heads, treated in the form of scrolls, are turned away from each other, and are provided with beaks stretched outward.   Each body is divided into halves for the purpose of ornamentation.   In the outer half are designed neck and head of a cock, a circle in front of the latter ; the inner halves are formed of simply-wound spirals.   Between the two halves is inserted a conventionalized fish, which continues above into another, tripartite fish, standing upright.   It is an example of the same kind of opposite fishes rampant as in Figs. 1 and 9 (pp. 16, 18), and frequently elsewhere.   The ornament is bordered above by a spur-line embroidered in green.   On the upper edge are represented triskeles, singly and combined.

Fig. 4, Plate x, represents part of an ornament embroidered on a girdle of black velvet, out of which the patterns are cut.   The edge of the velvet is seamed with chain-stitching.   The threads selected are all of dull colors.   All negative parts are filled up with light yellow cloth (hatched in the drawing) ; the positive ornament, therefore, is formed by the cut velvet parts.   It is a device composed of a succession of two different figures, one of which has a circular form, the other an ellipsoidal.   Complete symmetry is carried out inside of these figures on either side, as well as above and below.   The velvet, being a somewhat stiff and un-handy material, is not favorable to the formation of lines, and thus a degree of conventionalization is attained that makes the development of the ornament in the oval drawing hardly recognizable.   What we observe are a fish-tail, an oval body, and a figure distributed symmetrically on both sides, in which, as a rule, the beaks of the opposite cocks are united.   If we take for granted that the cock's head is in this part, its tail must needs be recognized in the fish-tail, which occurs not rarely.   If, however, which is also possible, we see the cock's head in the smaller, under branch of the fish-tail, and attribute to the larger branch perhaps the hint of a wing-feather, then we have here also the type of the opposite cocks, as in Figs. 1, 2, etc.   It is therefore possible to recognize in this case the head of

Decorative Art of the Amur Tribes.

the cock at both ends of the body, a kind of Janus-cock. A still more striking example of the same case is met with in the two following specimens.

These are the decorated sides of birch-bark boxes. In Fig. 5, Plate x, the whole trapezoidal piece is divided into three fields ; the narrow strip below, along the edge, being filled up with a spur-ornament. The trapezoid above it contains two peculiarly conventionalized cocks of Type B. The two outer cocks have their heads turned downward, and on their bodies a realistic fish, in which eye, gill, and the conventional picture of a fish are drawn. The tail of the fish is identical with the cock's head. Each of the two inner cocks shows two spirals, the larger of which seems to denote the tail-feathers, and the smaller one the pinions, completed by two parallel projections above it. The upper of these is shaped into the conventional form of a fish by the addition of a semicircular figure indicating the fish's head. The two cocks in either half of this symmetrical ornament are combined with each other by a wave-line, the concavities of which the cocks occupy. The same motive is met with in the ornaments above this trapezoid, only that here conventionalization has advanced much further. The heads may easily be found by a comparison with the cocks in the centre below, the bodies being spirals, and the tail-feathers joining above in purely geometrical figures. On the ground of this stage of conventionalization, also, the two tapering figures in the right and left corners are to be explained. These corner figures proceed below, and each forms here a fish rather true to nature, although placed in a kind of scroll. Head, eye, the curved body, a ventral fin, and the bipartite tail are discernible.

The pattern Fig. 6, Plate x, is doubly symmetrical above and below, as well as on the right and the left. The whole ornament can be traced back to a figure which represents a wave-line, in the hollows of which two types of cock are placed. One of these types has a remarkably long process at the end of the head, a conventionalized fish between the head and this part, its body being indicated by a spiral. The other type strongly approaches that in Fig. 5, in the interior of the central trapezoid, with the fish-body attached to the back part of the head, as there. The wing-feather has here become a spiral, as have also the body and the tail.

THE FISH. — That the fish plays a very important part in the decorative art of the Amur tribes has already become evident from various examples in which it occurred in connection with the cock, sometimes drawn on its body, mostly in strongly conventionalized form. We shall now enter upon a special examination of the subject, and demonstrate by some designs how this conventionalization has developed from the realistic picture of the fish.

In Figs. 1 and 1a, Plate xi, the long and short sides of a birch-bark basket of the Gold are represented, in which portions of the pattern are cut out of bark and sewed on the bark forming the basket. In Fig. 1 the arrangement of the pattern is very gracefully executed by a finely drawn wave-line, the course of which is interrupted in the centre by two combatant cocks designed almost true

to nature.   Their beaks cohere.   The tops of their heads are combined by
a figure composed of two fish-bodies.   Besides, in the body of the two cocks conventionalized fishes are incised, and from under their throats depend two fishes
with heads downward, the eyes of which are clearly distinguished.   The pinions
are symbolized by a simple scroll.   These latter parts described as such have at
the same time another function : they form the body of a conventionalized cock,
being above the realistic one and cut out of bark.   The long outstretched heads
of this pair of cocks almost touch each other ; to the head is attached the upper
outline of the body in a slight curve, which joins an upward-extending hook indicating the tail.   Over these combined figures of cocks are two odd independent
bird-heads, obviously with crooked beaks.   To the right and left of this central
group we see a repetition of the same picture.   Here we are immediately confronted by two naturalistic fishes that were explained as carp.   In these the head,
the eyes, the ventral fins, and the tail are expressed.   Incised on these carp is
the image of a conventionalized fish, whose tail is turned toward the eye of the
realistic fish.   The form of the body of the latter is assimilated to the wave-line
above it, which runs off below into a scroll.   The knob forming the starting-point
of this scroll represents at the same time the round object held in the cock's
beak ; and this cock's head is really represented in the succeeding circuit of the
curve.   These cocks' heads, in their turn, rest on the spirals below, symbolizing
their bodies ; and these spirals are executed in such a way that in their interior a
fish-body with plainly distinguished head, body, and tapering tail, is clinging closely
to them.   Furthermore, the spirals above, surrounding the realistic carp, are at
the same time symbolical representatives of a cock's body, except that here, unlike the case below, the cocks' heads are put in a realistic design, but in a manner
similar to that in the upper part of the central figure, in which conventionalized
heads are shown ornamentally connected with the adjoining diametrical line
of the body and tail.   The short side of the basket (Fig. 1a) is bisected by
wave-lines of a form allied to that in Fig. 1.   The separation of the fields is
effected by a cluster of three downward-extending fish-heads.   On either side of
these are two standing realistic roosters with hooked beaks, triskeles-shaped
pinions, fishtail-formed tail-feathers, and feet.   On both the extreme ends are
drawn two carp true to nature, in an attitude as if about to dive.   They are components at the same time of a spiral ; the form of their triskeles-shaped tails is
nearly identical with the pinions of the two central roosters.   In the body of both
carps is seen again a conventionalized bipartite fish.   The whole of this fish-
spiral symbolizes at the same time the body of a cock, as is clearly pointed out by
the incised bird-heads visible above it.

       Figs. 2 and 2a, Plate XI, represent the ornaments on the rim of the cover
of a lacquered tobacco-box, the cover of which is seen in Fig. 18.   They are
painted red and light green, and bordered by black lines.   On the front side of
the rim (Fig. 2) we observe two small and five large equal triangles.   In the small
triangle to the left there is a fancifully combined figure, showing in its centre the

Decorative Art of the Amur Tribes.

picture of a fish, from which radiate three cock-heads with circles before their beaks. The two on the left side are shaped like triskeles, the lower one ending in a fish-tail. The bird branching off from the fish on the right side holds its head turned downward to the left, wing and feet being slightly symbolized by a short crooked foil, and terminating in a disproportionately extended fish-tail. The following isosceles is occupied by two conventional, almost heraldic forms of animals explained by natives as musk-deer (see p. 41), whose bodies are usually treated like that of the cock, their ears standing upright, their faces turned away from each other, two legs being indicated above the body, and the other two below it; the tail a bushy tuft consisting of five curved broaches. The animal is shaped intentionally in the form of a spiral; its mouth being the starting-point, and one of the fore-legs the terminus. Its body is in the form of a fish, and, besides, in its interior is a bipartite fish adapted to the outline of the body. The head of this fish is directed downward, its eye distinctly marked by a circle and its gill by a crescent, which is somewhat bigger in the animal on the right side. The tendency to introduce fishes into these decorations goes so far as to affect even the clearness of the fundamental design. In this case the ends of the foremost lobe of the tail and of the adjoining hind-leg are so connected with each other as to leave space for the design of a conventional form of a fish on the hind part of the deer. To see to advantage the picture contained in the third triangle, one should invert the illustration. Then it is possible to observe the two longstretched heads of the roosters holding a green-colored circle between their beaks, the pinions being duly indicated by a tricorned branch, the bodies being formed by graceful wave-lines to which cling two carp full of life and vigor, characterized in the usual way, and having, besides, a spinal fin. To the tail of each is joined an S-shaped fish, to the head of which is attached another fish with a triskeles-like tail. Only one half of the following central triangle is shown in the figure: consequently it contains but one of the combatant cocks. In front of its beak is visible a conventionalized tripartite fish with a bifurcated tail. From below the head of the bird branches out to the left a conventionalized fish consisting of head and body. From the same point proceeds a curve representing the bird's breast, and continuing to the left into the pinions, sending three prongs to the left and one to the right. The last two offshoots form jointly a fish-tail. Oddly enough, the foot is symbolized by a three-lobed leaf joined on either side by a branchlet which seems to signify part of a spur. The body, horizontally placed, follows the outlines of the fish drawn into it. The tail attached to this part is formed in a style widely different from the usual cock-tail, being simply an imitation, or rather adaptation, of the musk-deer tail in the preceding triangle.

Fig. 2 a shows a portion of the ornamentation a continuation of which constitutes the back of the same rim. The division of the ornamented surface is executed here by an elegant sweep of a wave-line. In the first concavity the under part of this wave-line forms at the same time the outline of a fish that clings closely to it (*a*).   The head of the fish is surmounted by a slight curve terminating in two parallel branchlets to the right (*b*) and a beak-like figure to the left (*c*). It is hard to decide whether this part is intended for an independent cock, or merely for the tail-feathers of that larger cock whose head joins the body of the spiral-formed fish (*d*) and sends off a long beak (*e*) in the form of a semicircular wave.   On its head is a crest, the shorter component of which is treated as a cock's beak (*f*), with an oval in front of it ; the other makes a wide curve terminating in a fish-tail (*g*), one branch of which contains a dot for an eye, thus indicating a bird's beak, which is corroborated by placing an ellipsoid in front of it. This whole offshoot has the shape of a triskeles.   Returning to the large cock filling the middle ground of this area, we see the outlines of its body rendered true to nature, and a scroll on its hind part (*d*), to which two feathers are attached, indicating the tail, or, if the figure above the fish-head is correctly to be interpreted as a cock's tail, the pinions.   In the smaller intervening part the two cocks rampant are easily discernible, their feet united, the long falciform beaks directed upward and the tails downward, the latter being connected by a pair of small ellipsoids.   In accordance with this, the remainder of this ornament is self-explanatory : the cock rampant is to be seen single in one of the following fields.

Fig. 3, Plate XI, is an ornament cut out of paper, which was to serve as an embroidery-pattern of a bag for a strike-a-light.   The exquisite gracefulness of lines and the fine taste here displayed deserve special mention.   The two artistic fishes in the extreme lobes are explained as crucians, with ciliated mouth ; the whiskers are formed like a cock's comb, the under arm having the shape of a conventionalized fish.   The middle of the centre is filled up with two large cocks rampant, facing each other, heads and necks recurved, their beaks joining in a long curve to which two conventionalized fishes are attached.   The outline of the body of this cock, generally speaking, has the form of a spiral, and is a well-designed fish at the same time.   Above the head of this fish is a spiral with a closely adjoining strongly conventional form of a fish, this whole figure being already familiar to us as symbolizing the cock's wing-feathers.   Below the eye of the fish we note a conspicuous crescent accompanied by an ellipse, apparently derived from the cock's beak.   Close by it is a well-developed fish-tail, which, if the drawing be inverted, signifies the tail of this latter cock ; in this case the spiral containing the fish should be considered as its body.

Fig. 4, Plate XI, is an ornament cut out of red paper and pasted on a triangular cartoon.   This object, hung to the wall, was used as a holder for newspapers by a rich Gold living in Khabarovsk, who, as a gentleman, was proud of having Russian papers, although he could neither read nor write.   The ornament spread out along the border consists of a succession of spiral triskeles. The main field is occupied by two spirals with two fishes adapted to the outline of the first winding.   Attached to the ends of these spirals is a triskeles with two long arms and one short one.   The outer arm is continued by a double triskeles below ; the longer arm of the latter, as well as the adjoining one of the preceding, has the form of the cock's beak and neck. These two arms form a rounding whose space is filled up by a kind of trident. In the upper part of this decoration two triskeles, with one part shaped like a fish-tail and the other like a cock's head, are placed together in a way similar to that below.

Fig. 5, Plate xi, is an embroidery-pattern the ornament on the under part of which consists of four spirals. To the outside of each of the inner spirals clings a conventionalized fish, the form of which has become somewhat stiff owing to the kind of work. To the outer spirals below is added a triskeles of the well-known fish-tail form, and at the extreme ends a conventionalized fish holding the head downward. In the upper part of this ornament we note double combinations of triskeles in which one arm is rolled in like a scroll, and to which ovals or circles are joined, as they appear elsewhere in front of the cocks' beaks, so that here the idea of conventionalized cocks may be the underlying conception.

Fig. 6, Plate xi, shows an embroidered collar. The ornament consists of a double row. The element of the inner row is formed by a wave-line, which joins in the centre in a pair of spirals. In the concavities on either side are two distinctly designed cocks with fishes for their bodies. In this case, the plumage is marked on the body, not, as elsewhere, outside of it. The outer row of ornaments is composed of single pieces reproducing, as it were, diagrams of the cock holding the fish in its mouth, the heads of the cocks being recurved so that they are turned toward each other ; the fishes have the shape of the triskeles.

Fig. 7, Plate xi, represents the embroidery on a wristlet, the design exemplifying a very curious amalgamation of the cock with the fish. The beak of the cock with the circular object under it is clearly visible, as well as the form of its body. All the remaining parts, however, are shaped like a fish,— two dorsal fins, at the same time the bird's pinions ; the bipartite fish-tail, at the same time the cock's tail-feathers. The close association of both animals, and their ornamental harmony, have advanced to such a degree, that one might speak in this case of a cock-fish or a fish-cock, according to the predominance of the one or the other element. The other spirals occurring here, and the S-shaped cocks with two circles between their beaks, require no further explanation.

Fig. 1, Plate xii, represents the ornamentation on the inside of a birch-bark basket. It consists of three rows, one above another, the uppermost and undermost containing the same design. The middle field shows at both ends two naturalistic cocks with long, bent beaks, and bodies in the form of fishes ; the head of the latter is indicated by a scroll. This fish-body is embedded in the curve of the wave-line, between the upper end of which and the cock's head is the design of a crescent-like conventionalized fish, indicating at the same time pinions. To recognize the pictures in the two central hollows, the plate must be inverted. Then we see the cock, Type B. It is worthy of note that the conventionalized bipartite fishes are attached to the neck of the cock in the rounding between the neck and the spiral body. In the edge ornaments we observe again a series of fish-spirals. The fishes form here the starting-point of the spirals ;

their heads, eyes, gills, bodies, and tapering tails are plainly marked by black lines. These somewhat angular spirals are rounded off above by a projection in which a conventionalized fish is drawn.

Fig. 2, Plate XII, is an embroidery for trimming the front and upper edges of a garment. This is a very good example for illustrating the amalgamation of the fish and spiral. It is a continuous fish-spiral pattern. The basis of this ornament is a combination of a pair of double spirals contiguous to each other in the corner between the upper and front edges. In these spirals we see head

FIG. 12 (⅛⅛). Section of Paper Pattern. Tribe, Gold. ⅔ nat. size.

body, and tail of the fishes thoroughly characterized. In the pair of spirals next below this corner the spirals retain their rigid forms, and the conventionalized fishes combined with each other are placed around them independently. Their forms have the usual fish-outlines, as shown by a comparison with the preceding fishes. The next spiral below differs from the corresponding first one in that the head is not separated by a special line from the trunk, but only indicated in its form.

The principle of the ornament on the embroidery-pattern in Fig. 3, Plate XII, is based on the combination of facing spirals, treated partly as cocks, partly as fishes. In the pair at the upper end we see the type B inverted; beak and head shaped like a fish-tail, a bifurcation forming the pinions, and leaving between them and the head a space outlining the conventional form of a fish. The ovals, semicircular curves, and the compound braces placed around the spirals are well-known appearances to be traced back to the cock-ornament. The next pair of spirals shows the pronounced figure of a fish, its two sides running parallel to each other, its head not being especially marked off, as in that seen in Fig. 2 of this plate. The left side ends in a fish-tail, to which is closely joined a triskeles or a triskeles-shaped fish, suggesting perhaps that the short hook of the first-mentioned tail should be regarded as at the same time the beak of a cock holding a fish.

FIG. 13 ($\frac{74}{648}$). Section of Paper Pattern. Tribe, Gold. About ⅛ nat. size.

Fig. 12 offers a very interesting pattern as showing the predominating, all-governing influence of the fish-ornament, for the sake of which all other forms are remodelled. In *a* the cock is clearly to be recognized as a bird; its beak, however, assumes the form of a fish-tail. On the left, to its trisulcate wing-feather a conventionalized bipartite fish is attached. Its body is, of course, conceived of as a fish, and ends above in a fish-tail. In *b* we see a figure to be defined neither as cock nor as fish, but to be designated only as cock-fish. The body has the outer form of a cock's body and a fish incised on it. To the right, in the interior of the spiral, a fish-head is appended, whereas to its narrow neck two conventionalized fishes are annexed. Between *a* and *b* we see a spiral to which a fish-body is clinging, with a head at each end. *c* is a variation of the type seen in *a*. *d* is a cock without tail-feathers. *e* is allied to *b*, but adorned with leaf motives. Also *f* is a cock-formation composed of fishes. In *g*, *h*, and *i*, fishes and cocks are, as it were, like arabesques, amalgamated into one composite whole, the separate parts of which are hard to single out, since one always glides into the other. *j* and *k* show the principle of the cocks of Type B. In *j* the backs of the cocks, strange to say, are replaced by two conventionalized bipartite fishes rampant. The acme of all these phenomena of cock and fish is reached, however, in *g*. Here we observe in the middle a large realistic carp, on the tail of which is drawn again a fish. At three points this carp is ornamented with bird-beaks and the usual ovals ; so that this design, when viewed from three different sides, has the appearance of as many cocks.

FIG. 14 ($\frac{70}{556}$). Section of Paper Pattern. Tribe, Gold. About $\frac{1}{2}$ nat. size.

Figs. 13–17 are likewise large compositions based on the principle of the combination of conventionalized, mostly bipartite, fishes with spiral and cock ornaments. In these conglomerations are shown the endless variation of which this ornament is capable, and the great effectiveness of the forms in the composition of larger structures. In Fig. 16, moreover, a lavish use is made of leaf and floral ornaments. In special beauty of forms the large realistic cock on the left side of Fig. 17 excels.

THE DRAGON. — The Chinese dragon (*lung;* Gold, *mudur*) holds a prominent place in the mythology of the Gold, and is believed by both these peoples to

Decorative Art of the Amur Tribes.

produce rain and thunder. This monster is very popular throughout eastern Asia, and is a favorite subject in ornamentation.[1]

Fig. 4, Plate XII, is a decoration cut out of paper, in which the picture of the dragon is repeated four times. It is laid out in the form of a double spiral, the starting-point of which is, on the one side its head, on the other side its tail. To the single spirals are now added offshoots representing the legs of the monster, so that one might distinguish the form of the dragon as well as the combination of two triskeles. The curved lines outlining the dragon's body run

parallel to each other, and are covered with a row of small triangles indicating scales. The upper part of the head has almost a helmet-like shape, its mouth being strongly promi-nent, and its tongue quivering. On the face of the dragon (a) there is a very remarkable design : in the three objects, eye, semicircle, body, is reproduced the image of the con-ventionalized fish as it usually appears on the naturalistic fish - body or the cock. The con-ventionalized fish oc-

FIG. 15 ($\frac{7}{9}\frac{0}{4\frac{0}{7}}$). Section of Paper Pattern. Tribe, Gold. About $\frac{1}{4}$ nat. size.

curs once again under the head, where it is formed by the outline of the latter and an added triskeles, representing in this case the whiskers of that mythical creature. The horns on its head are so shaped as to remind one of the cock's tail-feathers. The one four-broached portion of the horns is also identical with the design of the dragon's tail on other pictures. On its neck are three claws, representing its wings or flag-feathers. It is a striking fact that the four fields at the ends of the dragon-tails are filled with cocks (b), each holding a fish in its beak, and having their bodies formed like fishes, the symbolical design of the conventionalized fish being cut out of them. This is placed also around the spiral wing-feathers in the characteristic manner. The tail belongs rather to the dragon than to the bird, for it consists, which is unusual for a cock's tail, of two bisected parts, each made up of two offshoots. The picture marked c may be interpreted as showing certain stages in the development of the cock-fish ornament.

[1] The dragon of purely Chinese type has already been referred to on p. 4.

Fig. 5, Plate XII, is another paper pattern representing dragons, likewise in the form of double spirals, but in much simpler form than in the preceding case. The head (*a*) is extended, and has only one horn of the shape of the familiar conventionalized bipartite fish. The mouth projects and is wide open. An oval figure is put under its lower jaw, significant of a bright pearl.[1] The serpentine body is covered, not with scales, but with an ornamental spur-line. The tail (*b*) is a bushy tuft with four branches, one of which corresponds to the form of a conventionalized fish. Feet and claws are not indicated.

Fig. 6, Plate XII, shows a painting on the upper edge of a pair of leggings, colored in red and black. In this case two dragons are placed side by side, the

FIG. 16 (₆₁₉⁷₉). Section of Paper Pattern. Tribe, Gold. About ¼ nat. size.

faces turned away from each other. These are strongly conventionalized, indicating in reality only the open jaws, the scales marked as in the foregoing example. The tops of the heads are connected by a brace. Both above and below the body is a foot with claw. The tail consists of a circle and adjoining triskeles-shaped fish-tail.

Fig. 2, Plate XIII, represents half of the decoration on the cover of a lacquered tobacco-box, the edge of which is adorned with a conventionalized design of eight dragons in the form of spirals. These are paired so that their faces, which consist of bifid ovals, are turned away from each other. The horns are fashioned after the cock's tail-feathers. The heads of the dragons on the long

[1] See Grünwedel in Sitzungsberichte der preussischen Akademie der Wissenschaften in Berlin, 1901, p. 215.

side are connected by a brace. Six dentiform projections stand out from the back. The body terminates in a fish-tail, the shorter arm of which signifies a foot with four sharp claws, the other being supplied with a bushy tail resem-

FIG. 17 (₈₄₄). Section of Paper Pattern. Tribe, Gold. About ⅓ nat. size.

bling cock-feathers. In the centre of the middle field is a rosette, the elements of which are made up of conventional designs of fishes. Above follow two cock-fishes, each in a spiral. Connected with the central rosette by a narrow band is an elliptical figure with an ornamental ring inside, between which and the periphery of the ellipse are delineated triskeles and conventionalized cocks. Beyond this figure are two opposite dragon-heads.

Fig. 18 is also a deco-
ration on the cover of a
lacquered tobacco-box, but
here the main field is taken
up by eight large finely
drawn dragons. This type
approaches in its form
very nearly that of the
cock-ornament. On each
side of the cover the two
dragons above and below
are placed in the form of a

FIG. 18 (₈₇₈). Cover of a Tobacco-box. Tribe, Gold. Extreme length, 54 cm.

double spiral, tails contiguous, faces turned so that they look at each other. At the end of both the upper and the lower jaw is a triskeles, apparently signifying cocks, as the upper ones have ovals connected with them.   Likewise the horns on one side are characterized by a triskeles in the form of a fish-tail.   The two central dragons are connected by two combatant carp.   The dragon-tails are wholly fashioned after the form of the cock-tails.   These may be designated either as dragons or as cocks with dragon-bodies, so that one may speak of cock-dragons as well as of dragon-cocks.

On Fig. 1, Plate XIII, a dragon in front view is sprawled over the cover of a Goldian tobacco-box, the greater part of which it occupies, in fanciful connection with a chain-band pattern.   The head looks similar to that of an elephant. In the mouth are designed a pointed tongue and two pairs of front teeth.   In general the ornamental treatment recedes as much as possible into the background, that the animal character may receive more emphasis.   Ingenuity is given free scope, in this case, by the introduction of the perfect representations of the four feet stretched out to both sides, each with four claws.[1]   The bobtail also is not a decorative part, but consists simply of seven natural-looking furcations.   Between the two hindmost claws on each foot is inserted an oblong object which they seem to hold.   The band-ornament along the edge of the cover is so placed around the monster as to suggest that the animal might be bound with ropes.   The bands start from the ends of the upper and under jaw, and are twisted into three loops above and three below, which show two, and in one case four, prominent tips.   Several S-shaped figures, which also presumably represent portions of the band, are inserted between the single knots.

This monster, conveying the impression rather of an enormous python, is very likely the embodiment of the rain-dragon soaring in the clouds, but hampered by its fetters in pouring out its blessings on the thirsty land.   In this connection mention should be made of the Chinese and Japanese "cloud-and-rain patterns," simple illustrations of which are given in Fig. 3, Plate XIII, and in Fig. 1, Plate XIV.   The former is a cloud pattern composed of spirals with cocks resting on them, and of clusters formed by a central spiral with six scrolls around it.   The latter is the device on a Japanese weaving belonging to the period between the eighteenth and the nineteenth centuries.   It is constructed of a combination of semicircles and spirals, and adds another to the already wide range of objects for which the spiral is a symbol, as, in this case, for cloud-formation.

Fig. 4, Plate XIII, an old Chinese weaving-pattern, gives a somewhat more graphic account of Chinese notions of atmospheric phenomena.   The upper and lower edges are taken up with two variations of the meander, while the intervening part is occupied with an evidently emblematic effigy.   This special representation is designated in Chinese art as a "cloud-and-thunder picture."   It reproduces a dragon, which, as Hirth[2] sets forth, in its aerial abode starts the thunder a-rolling with its hind-paw upraised and stretched backward.   The

---

[1] Five-clawed feet are only accorded to the Imperial dragon.
[2] See Hirth, Verhandlungen der Berliner Anthropologischen Gesellschaft, 1889, p. 493.

thunder is symbolically characterized by a triskeles, one arm of which is enclosed in a semicircle.   Below it the lightning is represented by a trident.   The meandrian patterns are also a symbolical equivalent for thunder, so that this whole representation might be called the best illustration of the deductions of Hirth in the paper quoted before.

THE MUSK-DEER. — Foremost among the animals which play an important part in the productions of this art, after the cock and the dragon, is the musk-deer ; at least the creature portrayed in the following examples is explained by the natives as such.   It is rather naturalistic on some of the larger zoöphoric compositions ; but, under the pressure of the leading gallinaceous motive, it undergoes such conventional transformations, especially in its double character, that the difference between the construction of its forms and those of the cock is hardly perceptible.

In Fig. 2, Plate XIV, not only has the deer retained the form of head of the cock, but it has also been invested with its beak grasping the fish.   The head is adorned with antlers which are made up of two triskeles joined by a heavy dot.   On the body, and parallel with its outline, are cut out two conventionalized fishes side by side.   The two hind-legs are formed in the same way as the tail, consisting of

FIG. 19.   Side of Birch-bark Basket.   (From Museum für Völker kunde, Berlin, Cat. No. IA 1207.)
Tribe, Gold.   Length, 21 cm.

two slightly undulating curves.   The two animals are rampant, their fore-legs united in a straight bar.

Fig. 3, Plate XIV, represents a paper pattern for embroidering a pair of ear-lappets.   The two figures (*a*) on both sides are combatant musk-deer of more conventionalized form than the preceding ones ; only their heads, with ears upright and mouths open, have a somewhat natural appearance.   Their bodies are shaped like the fish whose form is cut out of them.   Two large dots serve to express the feet.   The tails consist of one falcation and a combination of two triskeles with an oval knob.   The lines *b* are wave-lines ending below in a form reminding one of the cock's tail-feathers.   The ornamental figures *c* and *d* signify the last stage of development of the cocks of Type A, that is, of the combatant cocks, *d* showing two combatant fishes in lieu of cocks' bodies.   The oblong crenations (*e*) around the edge are apparently derived from the constituents of the cock's wing-feathers.

On the side of a birch-bark basket (Fig. 19), are delineated two combatant musk-deer in crouching attitude, and invested with cocks' crests. The feet unite below in a trefoil. On their bodies are fishes, gracefully outlined. Each deer runs out into a fish-body, the forked tail of which is visible, and to which a collateral fin is attached. The style of execution of these fishes is such, that the space between them and the body of the musk-deer remains the usual conventional form of the fish.

ANIMAL PIECES. — Of other animals which occur in the ornamentation of these tribes, aside from those hitherto noted, the following deserve mention : wild duck, wild goose, swan, eagle, swallow, elk, reindeer, roe, fox, dog, crucian, lizard, frog, snake, and insects. The following animal pieces demonstrate the supreme degree of zoöphily innate in the minds of these people, who display such a wonderful amount of creative power in these productions so full of freak and fancy.

Fig. 4, Plate XIV, shows a pattern cut out of paper, which is divided by winding curves into ornamental fields. The birds marked *a* were explained as wild ducks. In form they can hardly be distinguished from the cock. The bird seems to be conceived of by the artist as swimming. On its head is a horn-like piece, formed on the one side by a conventionalized fish, on the other by two parallel pikes. A conventionalized fish, consisting of two separate parts, is cut into the body as in the cock. The wing-feather is a scroll ; the tail, of the conventional fish-form. The figures *b* are two circle-ornaments to which are attached, above and below, birds' heads.

In the paper pattern, Fig. 5, Plate XIV, we see a very remarkable, graceful combination of various animals. In the centre, four musk-deer (*a*) are grouped around a lozenge-shaped figure. The head is formed in exactly the same style as that in Fig. 3 of this plate. The body looks very odd, because it is moulded like that of a fish, to the head of which cock-spurs are added to indicate the feet of the deer. In the heart-shaped fields above and below are two frogs (*c*) with four outstretched legs formed like fishes, and with two fishes indicated on their bodies. At the extreme ends are four crucians (*b*), covered with triangular scales. Between these and the musk-deer are placed four pairs of wild swans, each pair having one body in common, but distinct necks and heads, — one naturalistic head turned inward, whose gracefully arched neck rests on a wave-line, giving at the same time the outline for the bird's body ; the other head, turned outward, being ornamentally conventionalized.

Fig. 6, Plate XIV, represents a paper pattern showing a design for embroidering a shirt. In the centre is a circle, around which are grouped four tortoises (*a*), strongly conventionalized. Around it, on both sides, two bands forming four circles and two ellipses are symmetrically arranged. In every circle there is a roe (*Cervus capreolus* L.), *b ;* two snakes (muikí), *d ;* and a bird (*c*), called tewerkó, the species of which I have not yet been able to determine. Each ellipse contains a frog (*Rana temporaria* L.), *e ;* two spiders (atkomama), *f ;* and two gadflies (*shigaxtá*), *g.* Outside of these figures a number of animals are

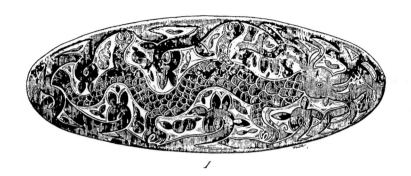

Decorative Art of the Amur Tribes.

represented standing along the edges of the pattern. There are four mosquitoes, *h;* four chimney-swallows (*Hirundo rustica* L.), *i;* four snakes, *d;* four stags (*Cervus elaphus* L.), *j;* and four fawns (*Cervus capreolus* L.), *k.*

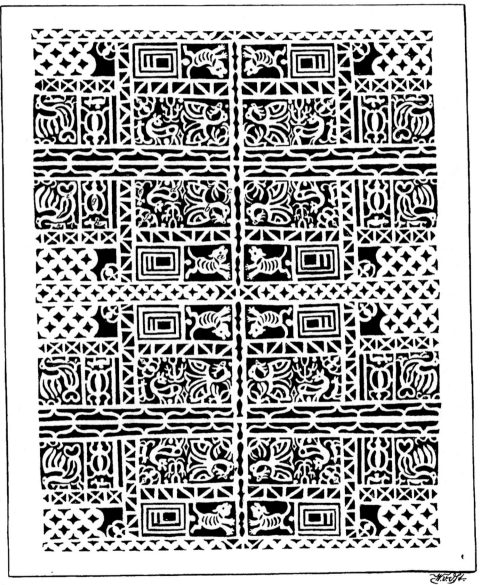

FIG. 20 ($\frac{70}{852}$). Paper Pattern. Tribe, Gold. $\frac{2}{3}$ nat. size.

In the paper pattern shown in Fig. 20 the same picture is represented eight times. In it the following animals are represented : a jumping tiger (*a*) with open jaws, the fore-legs with paws outstretched, with only one hind-leg, and tail upturned; an eagle grasping a fish in its beak (*b*), this conception being very likely derived from the cock holding the same little creature in its beak; a flying wild duck (*c*); a musk-deer with a design of a conventionalized bipartite fish on its body (*d*); a fox lying in ambush (*e*); frogs (*f* and *g*); a horse and its rider (*h*); an eagle flapping its wing and having a foot with three outstretched claws (*i*); what is said to be a glutton (*k*).

Fig. 21 represents nearly one-fourth part of a paper pattern that is divided again into two symmetrical parts. The ornaments are distributed over four large and twenty-six small quadrangular and ten triangular fields. In the large rectangle to the left are united four strongly conventionalized dragons, whose heads merge into the geometrical figure *a*, and whose tails are distinctly marked in the palmate figure *b*. The body itself is not drawn, but merely symbolized by spiral windings.

On the small squares surrounding the large figure just described, a number of animals are shown. The three small squares designated as *c* contain representations of spiders formed in a way similar to that of the dragon's tail. In one of them are drawn the outlines of three conventionalized fishes. *d* represents a raccoon-like dog (*Canis procyonoides* or *viverrinus*) with five cross-stripes; *e* is a young musk-deer; *f*, a frog; and *g* is a wild duck in the act of flying, the wing being marked by a trapezoid containing an inscribed smaller trapezoid; *h* is a roe with neck turned backward; *i*, a cock with outspread wing and erect tail-feathers; *j* is a wild reindeer; *k* is identical with *d;* *l* is a wolf; and *m* represents a wild goose.

Proceeding to the triangles at the bottom of the design, we find a doe looking backward (*n*); to the left of it a conventionalized deer with recurved cock-neck; *o* is a lizard.

The large squares on the right are divided by bands into a number of fields, which are also filled with animal figures. In the lower square we find to the left an elk (*p*), above which are three quadrupeds, one of them a stag; *q* is a double eagle with body in common and outspread wings; *r* is a wild duck; *s* is a panther; and *t*, a jumping tiger. In the second square are seen roes (*u*) with rebent necks; *v*, a duck; *w*, a swallow; *x*, a frog; *y*, a flying wild duck (cf. *g*); *z*, a galloping hound.

Fig. 22 is a pattern for a blanket, cut out of paper. There is a central piece with an upper and lower edge. The main ground is taken up with two dragons wound in the form of spirals, the heads (*a*) of which lie in the termini of these spirals. The space inside of the dragon-spirals is occupied by representations of animals, which correspond to each other on both sides. As a sort of decorative cæsura, a large frog (*b*) and a smaller adjoining one are inserted. The intervening spaces between the dragons above and those below are filled up with four tortoises (*c*). Close against the dragon's head a fox is leaning, followed by a wild duck (*d*) on the other side of the head. In *e* is represented a branch

1

2

3

4

5

6

Decorative Art of the Amur Tribes.

consisting of three parts, with leaves and blossoms. *f* is a musk-deer; *g*, a stag; *h*, the cock running out into a fish-body; *i*, a conventionalized bird; *j*, a roe. Outside of the dragons we observe a conventionalized tree with roots (three

FIG. 21 (⁷⁰⁄₉₅₇.)　Paper Pattern.　Tribe, Gold.　⅔ nat. size.

semicircles in succession), trunk, boughs, and foliage, by which is indicated symbolically the primeval forest, the so-called taigá, where the whole animal kingdom nests. *k* denotes a swan floating on the surface of the water, with a cross on its body[1] (the ellipse forming the body is repeated to furnish the connection with the dragon); above it is the design of a squirrel (*l*).   *m* is a duck with a fish in its bill; *n*, a lizard; *o*, an elk; *p*, a musk-deer; *q*, a cock; *r*, a duck perching on the side-branch of a tree, as on the fish-skin garments; *s*, a lizard; *t*, a carp; *u*, a swan with open, upturned beak; *v*, a roe.   The edge is cut by means of jagged lines into rectangles, and each of these again into four triangles.   In the first triangle at the extreme left is an eagle with outspread

FIG. 22 ($\frac{70}{811}$).   Section of Paper Pattern.   Tribe, Gold.   About ⅓ nat. size.

pinions, almost in the fashion of our escutcheon eagle.   In the opposite triangle there is a lizard, and below it a snake.   In the two central triangles are two eagles and two roosters standing opposite each other.

LEAF AND FLORAL ORNAMENTS. — Not only does the delineator manifest his artistic spirit as a skilful faunist, but, to a certain extent, the flora also occupies his attention.   Leaves and floral forms occur partly as independent ornaments in connection with other elements, partly in close combination with the cock and fish ornaments.   Especially single portions connected with the latter are treated

---

[1] See Globus, Vol. LXXIX, 1901, p. 70, and cf. Figs. 3 and 14.

Decorative Art of the Amur Tribes.

as leaves, chiefly the heads of conventionalized fishes, and the round object held in the beak of the cock.

Fig. 1, Plate xv, represents a carved Goldian wooden dish. In the middle, around a large circle, are grouped four smaller ones showing peculiar forms of svastika. Two opposite fields -inside of these have coarse cross-hatchings, two others fine ones. The same kind of hatching occurs in the central circle, which shows the heads of four realistic does holding in their mouths a young fawn. One of the large animals has seized it by the head, a second by the tail, a third by the fore-leg, and the fourth by a hind-leg. The deer are so drawn that their outlines form likewise a svastika. The rim of this dish is covered on the sides with clinging vines, leaves, and blossoms of various kinds and forms, and, on the ends, with flower-spikes. To the four corners at the extreme ends are attached four animal heads in open-work carving. It is hard to say what species of animal is meant.

In Fig. 2, Plate xv, is seen the cover of a wooden box. This composition is remarkable for the reason that the middle piece of the ornament is not shaped symmetrically, the only symmetry visible being in the arrangement of the ornaments across the upper and lower ends. Below we see a three-lobed leaf. Two leaves of the same kind, though not of the same rigid geometrical form, are found in the central part. The whole is intended, perhaps, to signify the bough of a tree, whence perhaps also arises the irregular arrangement of the single parts.

The next design shown on Plate xv (Fig. 3) is that of an embroidered tobacco-pouch, the edge of which is trimmed with sable. The stitches employed on the edge are a triple row composed of feather-stitch in the centre with chain-stitch either side of it. In the middle field chain-stitches are mostly used, the leaf parts being worked in satin-stitch. In the right and left upper corners of the central rectangle we observe two three-lobed leaves, under which are two cocks holding triskeles-shaped fishes. There is a red leaf near the fish and a light-green leaf on the cock's body, both seeming to represent the well-known round object. From these cocks branch off toward the middle two double spirals. The smaller, outer spiral has its starting-point in a large two-lobed leaf held in the beak of the cock; the other, inner spiral, from a petal with three lobes grouped, rosette-like, around a circle. Within this spiral is delineated a conventionalized fish, whose body is assimilated to the winding of the spiral, and whose tail tapers to a point. The heads of these fishes are worked in satin-stitch, in the same manner as the leaves, with dark red. On the lower edge of the rectangle are placed blossoms consisting of five petals in pyramidal arrangement.

Fig. 4, Plate xv, represents an embroidered border. In the lower part, on a black ground, we see leaf-forms in connection with triskeles, and in the centre two rosette-like blossoms at the starting-points of two spirals. In the upper portion, with red background, all fish-heads and circular forms are treated as leaves, their surface being filled in with satin-stitches, while the remaining parts are merely represented by lines of white chain-stitching. This process is therefore at the same time a device by which to make out easily the somewhat obscure cocks and the fishes.

The embroidered waistband, Fig. 5, Plate XV, consists of a rectangular central field and an ornamental border. In the central field a very interesting geometrical formation of the cock is met with, combined with single small curves and triskeles. In the illustration the outlines of this cock have been strengthened to show more clearly the type of this bird. The border is composed of a succession of continuous double spirals connected with two and three lobed flowers. At the same time the spirals symbolize cocks' bodies.

Fig. 1, Plate XVI, represents the upper front border of the half of a collar of a woman's embroidered dress. The ornament shows clearly the way in which leaves and blossoms appear in connection with spirals. One of the arms of the triskeles attached to the spirals is treated like a leaf.

Fig. 6, Plate XV, shows a paper pattern presenting a purely geometrical formation of flowers or blossoms, the single parts of which are circles, semicircles, and ovals.[1]

There is a certain power of attraction between cock and plant ornaments, leading sometimes to a perfect amalgamation, which may be illustrated in the following specimens.

In Figs. 12–14, Plate XVI, are reproduced paintings on three Goldian bows. Fig. 12 shows a combination of a tendril-like ornament with a cock-ornament. The outer side of the bow (Fig. 12 a) is divided into ten fields; but the five fields on the one side do not symmetrically correspond to the five on the other side, in which the same ornamental parts appear in different combinations. In this pattern the motives of the cock and of the fish ornament are so strangely mixed up with leaf and floral designs, and the two are so closely assimilated to each other, that it is sometimes hard to decide what is an ingredient of the cock and what of the plant ornament. In the centre of the field a (Fig. 12 a) there is an obvious representation of the cock, with head, body, and spur, holding a conventionalized fish in its beak, to which is attached, on the right, a petal. It would be difficult, however, to determine whether the first design on the left is meant to represent a leaf or a fish. In the field b we observe likewise a cirrose leaf, in the middle a conventionalized fish of the characteristic form, whereas all plant portions are adapted to this style of the fish, both here and in the central field c. In d we see another cock with a fish in its beak. Its tail-feathers, which in design are like a fish-tail, form at the same time the component of a petal. In a similar way, in field e is a cock with a fish, on a stalk proceeding from a five-lobed leaf. This ornament terminates at the other end in a trifoliate leaf.

On the inner side of this bow (Fig. 12 b, Plate XVI) the ornament on the left-hand side begins with a leaf-tendril, which is continued to the end by a long undulating line. It may be that in this wave-ornament the curve itself is con-

---

[1] Cf. what is said about the circle-ornament, p. 19.

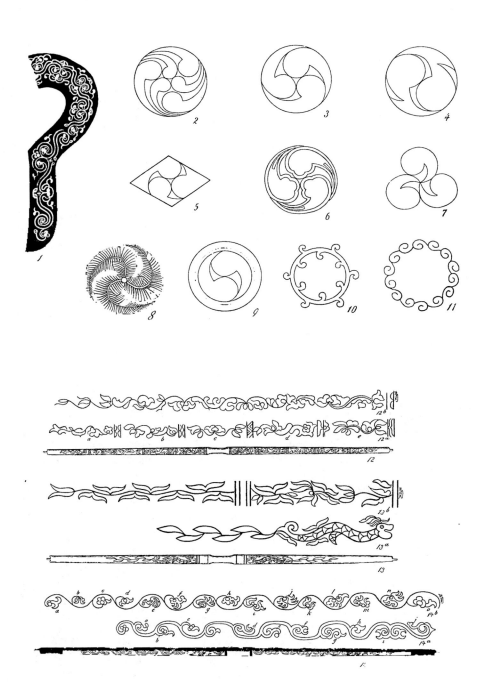

Decorative Art of the Amur Tribes.

Decorative Art of the Amur Tribes.

ceived of as something real, as it is entirely filled with conventionalized fishes, or, if one prefers, with leaves which have adopted their forms. These patterns coincide to such a degree, that, in the two designs terminating this ornament on the right side, it might appear doubtful whether they are to be looked upon as cocks, fishes, or leaves: they combine in their forms all three meanings.

The outer side of the specimen represented in Fig. 13, Plate xvi, is divided, for the purpose of decoration, into two fields, each of which contains the painting of a scaled dragon. The parts of its horns have the form of conventionalized fishes (Fig. 13 a). On the body, below, is visible a foot with scales like those on the body. To its body are attached, besides, three cock-spurs, which would here seem to indicate feet. The tail is coiled around into a spiral, as in the cock, and connected with it is a stem bearing six leaves. The inner side of the bow is divided into two unequal parts (Fig. 13 b). The element in the left field is a wave-line. Inside of each curve of the line, both above and below, is a cock-spur, which, in this connection, is meant to represent a leaf. This is one of the remarkable instances where the function of the ornament is different from what it would appear to be. In the field on the right side are represented similar figures, but with freer arrangement. Three oval-shaped leaves are added, and a flower-bud with two petals on either side of a central ovary. This ornament is completed on the right by the representation of three conventionalized fishes.

In Fig. 14, Plate xvi, is reproduced a bow, the outer side of which is separated into two equal parts in the same way as the others. The fundamental motive of this ornament is a wave-line from which extend either simple curved branchlets or ornaments in the form of triskeles. Connected with one of these triskeles we see a conventionalized fish (Fig. 14 a, a), whereas in the branchlets b and c we recognize heads of cocks, from the typical figure of the fish connected with it, and from the circle drawn in the fish. The head of the cock connected with the fish is easier to recognize in d; in e the figure of the circle appears at the point of the beak. In f a new combination is reached through the addition of a semicircle to the spiral, which forms with it nearly an X. g and h show the same type of cock with the fish, only in g the wing-feathers are indicated by the addition of a spiral. The fish in the beak has in both cases the same well-known form, the only part properly indicated being the bipartite tail. In i we see two triskeles united into one figure; and in j a new style of conventionalization of the cock-ornament. It is most remarkable that the artist has used new forms at each offshoot of this wave-line, and exercised his imagination to a great extent to obtain new and varied modifications of the same theme.

Fig. 14 b, Plate xvi, represents the decoration on the inner side of the same bow. In this case also there is no symmetrical treatment inside of the wave-line, but the maker has striven to vary as much as possible the motive in each con-cavity. This illustration is of great value for the study of the evolution of the cock-ornament, for it shows a great many stages in its development. At a we observe the beak of the cock holding a round object and at the same time the fish, the dimensions of whose tail are exaggerated in comparison with the other parts of the ornament.   In the field *b* may be seen, suspended from a tendril, two graceful little cocks rampant, under the beaks of which is a circle.   Their bodies approach closely the fish-ornament.   At first sight the figure *c* might seem to convey the impression of a leaf or flower ornament ; but the horn-like offshoot on the right side of this leaf cannot be explained in this case : it is obviously to be regarded as a cock holding a fish, as is especially shown on comparison with the following ornament ; the seeming flower-bud is a combination of fish-heads, and the circle drawn into it is that belonging to the cock.   In *d* are reproduced two combatant cocks, which, however, are distinguished from those in *b* by being joined together and placed around a circle.   The cock to the right has its tail turned upward, and that on the left side downward.   In the field *e* the cock may be recognized as drawn true to nature, with eye indicated, the circle at its beak.   Feet and spurs are designated by a long lobe.   The end of the tail terminates, strange to say, with the body of a fish.   *f* shows the cock, in spite of its conventionalization, clearly outlined : the circle in front of its beak, and in front of the circle the fish, consisting of three parts, — head, body, and curved tail.   Parallel to the fish-tail run the cock's feet, which are indicated by a long falcation, as in *d*.   The tail-feathers of the cock are conventionalized like the fish-tail.   In *g*, head, neck, four tail-feathers, and two concentric circles around the beak of the bird, are visible.   *h* represents a cock with fish, closely allied to that in *c*, the cock terminating in a fish-tail disproportionately large.   In *i* is shown one of the most remarkable and instructive designs within the scope of this entire ornamentation : there are two triskeles here, in one of which one arm is much shortened through adaptation to the available space ; that these pure triskeles, however, are interpreted as cocks, or at least were formerly so conceived, results from the fact that between them are two circles, as usually appear with combatant cocks.   *j* illustrates a type of combatant cocks with the circle between their beaks, but, for the rest, soaring with outspread wings, three feathers of which are indicated.   *k* presents the two cocks again in the form of pure triskeles in a way similar to that in *i ;* here, however, only one circle appears between them.   The field *l* offers a design analogous to *h*, except that in the former the fish-tail is turned upward, and to the cock to the right a prong indicating feet has been added.   The form *m* is allied to those in *i* and *k*, only that here three circles — two greater ones surrounding a lesser middle one — are represented.   The general style of form of the ornament in *n* is nearly identical with that in *j*, but with some slight modifications, while *o* is intermediate between the designs of *c* and *h*.

One would hardly imagine that the leaf-patterns thus far treated were originally invented by the East Siberian tribes.   The purely conventional forms in which they appear, as well as their connection with other ornamental parts, make their derivation from Sino-Japanese art very probable.   Primitive tribes generally pay little attention to the vegetable world ; and the Gilyak, and especially the Gold, reveal a surprising degree of ignorance concerning the plants in their immediate neighborhood, not to mention the large trees the wood of which is valuable to them as timber. As soon as I tried to gather information regarding the names of plants, I was directed in both tribes to consult the women, who indeed proved to have a much more detailed and deeper acquaintance with flowers and fruits than the men, apparently because they are accustomed to collect berries, roots, and certain herbs and leaves, as food for the household. This inefficient knowledge of the flora makes it difficult to realize that these peoples should have made an independent attempt to allot a space to plants in their ornamentation; and since the groundwork on which all its other parts rest is borrowed from their teachers, one would hardly err in supposing that this element also originated

Fig. 23. Japanese Weaving-Pattern. (From Kunstgewerbe Museum, Berlin, Cat. No. 81,650.)

from the same source. Although I am unable at this time to present exactly corresponding patterns from the realm of Chinese art, the weaving-patterns on Plates XVII, XVIII, and in Fig. 23, point out sufficiently well that leaf and floral ornaments occur in China and Japan in combination with spirals and triskeles, no less than on the Amur.

The Japanese weaving-pattern in Fig. 1, Plate XVII, is a composition of maple-leaves and chrysanthemums. The most remarkable feature here is the association of the conventionalized plants with the *mitsutomoye*. These *tomoye* seem to be devised in their outlines as serrated leaves. They are surrounded by a border showing forms of single and compound triskeles in exact accord with formations on our ornaments. A close connection, consequently, may exist between the triskeles and the *tomoye*. A selection of the latter, obtained from a native

Japanese book, is presented in Figs. 2–9, Plate XVI ; and in Figs. 10 and 11 of that plate are shown arabesque rings derived from the same source.

Fig. 2, Plate XVII, represents clusters of leaves as well as of triskeles, both arranged inside of circles. The foliage reproduced in Fig. 1, Plate XVIII, has developed shapes reminding one of the forms of our cock-ornaments. The two confronting creatures in its centre may be prototypes of our musk-deer. As to Fig. 2, Plate XVIII, as well as the ornamentally related Fig. 23, conventionalized cocks seem to be interwoven with vegetable ingredients ; the latter, particularly, illustrate stages of development almost identical with those represented on the Goldian bows on Plate XVI.

Up to this point in our investigations we have treated our subject from an analytical standpoint, defining the different elements as they occur in ornamentation. We shall now take into consideration its synthetical side, and show how the various motives are employed on different groups of ethnological objects.

BASKETS. — As to the technical methods employed in the designs on birch-bark baskets, the following occur : 1. The lines are incised in the bark material with the sharp point of a knife, and these incisions are sometimes partially dyed (Plate XIX) ; 2. Patterns are cut out of thick bark and sewed to the bark of the basket with a few short, hardly visible stitches ; 3. Only the uppermost layer of the bark is cut out, so that the ornament stands out in relief from the lower bark layer ; in this case the raised parts are usually blackened (Plates XX, XXI).

The ornamentation around the basket shown in Fig. 1, Plate XIX, is made up of two closely joined constituents. On the left side there is a pair of facing spirals, symbolical representatives of cocks' bodies, as suggested by the two down-stretched heads with pointed beaks, surmounted by two round figures. In the centre of the design on the right-hand side we note two lozenge-shaped figures placed one above the other, the upper one being connected on either side with a large triskeles, and forming with it the bipartite form of a conventionalized fish. These triskeles may stand as an abbreviation for the cock. They terminate below in a knob, the course of the spiral which might here be expected being interrupted, and a cock-spur inserted to fill the space. From the under lozenge a pair of facing spirals of one winding proceed downward. The edge above the main design is decorated with a continuous spur-line.

Fig. 2, Plate XIX, shows a design on the cover, and 2 a that on the side, of a box. The former is divided into four rectangular fields grouped around a lozenge. The two fields above and the two below contain two combatant cocks (white), the tips of whose beaks are connected by an ornamental figure the extremities of which are formed like cock-heads with pointed beaks. The body has an ellipsoidal form. A thickened knob and a somewhat larger projection apparently characterize the wing-feather. We see the tail in the shape of a fish-tail, one lobe being fashioned into a conventionalized bipartite fish, the other having the form of a bird's head and neck, under which another conventionalized fish is visible.

2

1

Decorative Art of the Amur Tribes.

Decorative Art of the Amur Tribes.

To the latter is attached a bird's beak with an oval under it, forming, with the adjoining corresponding figure in the field below, a geometrical, almost heart-shaped design.

The foundation of the ornament in Fig. 2 a, Plate XIX, based on simple symmetry only, is a double spiral; the body of the inverted cock, Type B, whose collateral curved branchlet distinctly marks the head, being formed by the inner spiral, the beak grasping a fish with down-stretched circular head, and tail upturned. The interior of the outer spiral may be described as a triskeles, or, better, as a fish-tail the two lobes of which are shaped like cock-beaks. The upper one holds a circle; the under one, a trichotomous fish, which it grasps between body and tail. There is a violation of the rules of symmetry here, in that the negative space between this fish and the cock's beak forms a bipartite fish on the right side only.

In Fig. 3, Plate XIX, is seen a front view, and in Fig. 3 a a back view, of a basket. The central figure on Fig. 3 was explained to me by a native as a human face; nevertheless I am distrustful of such an interpretation, which stands quite alone, and seems to be merely an invention of my informant. The ears and mouth would then be indicated by scrolls. On either side of this design are grouped several fishes in graceful arrangement. Above is a fish with broadened head. This head bears an incised conventionalized bipartite fish, which is above the large fish to the right of an incised fish-tail, and another in the lower right-hand corner. Under the large fish we observe a coiled fish with a roundish head.

In the upper part of the ornamentation on Fig. 3 a are two cocks rampant, having affixed to their beaks circles which coalesce with them. In the negative sections we see a cock's beak between this circle and the positive cock, and another beyond its neck and resting on the outline of its back. On the body, extended forward, is incised a conventionalized fish with tapering tail, which — a deviation from symmetry — cuts the whole body on the left-hand side only. The tail is formed of two parts, — a scroll, with a fish-tail cut out inside of it; and a long projection below, representing a bird's beak with attached head and large incised circular eye. From this head a spiral winds off downward, symbolizing, as it were, the body of this cock. The centre is taken up by a perforated lozenge-shaped figure, from which extend on both sides two conventionalized bipartite fishes. The two triskeles in the extreme corners at the base also represent fishes with scroll-like heads. The manner in which the negative portions are reflected from the positive images, in designs of this kind, is very remarkable.

Fig. 4, Plate XIX, which represents approximately a quarter of a birch-bark tray, shows the design incised on its bottom. It is reproduced here not so much because it offers especially characteristic features in this connection, but rather on account of its eminent beauty and the careful execution of work of similar technique. It belongs to the same category as the band and chain patterns already described.

The eye-like circles serve to mark certain termini and resting-places for the bands. Some of the negative portions have assumed the shape of fishes.

The ornament which occurs on the side of the basket shown in Fig. 1, Plate xx, is composed of three sections. The upper starts with a brace in the middle, forming on each side the upturned heads of two cocks with a circle in front of their beaks, except in the case of the cock on the extreme left, where it is missing. These facing birds are connected by two curves, producing a spur. The middle ornamental portion commences under the point of the brace above with two conventionalized bipartite fishes, whose long-extended bodies follow the outline of the upper brace-line, and finally terminate in a compound spiral. Three heavy dots, one between the heads, another on the body, the third over the coiled tail, denote the course of this ornamental fish. The third and lowest row in this design starts in the centre with two scrolls, appended to each of which is a fish-tail in triskeles form. Farther along, the outer winding of these spirals runs parallel to the fish-body above it, to form on the other side the outlines of a conventionalized bipartite fish. This is completed by a parabolic curve to which three leaves are attached; and this figure is so combined with the cock's head above, that it forms at the same time the body, tail, and wing-feathers of that bird.

In the centre of the decoration on another basket (Fig. 2, Plate xx) we observe a vertical axis to which are fastened two cocks (a) of Type B, standing erect, recognizable as such only by their attitude and feet. As for the rest, head and body bear the form of fishes. The same type, devised as fishes, is shown in e and f. The tail of e ends in two lobes, so arranged that its outlines form a conventionalized bipartite fish. The fishes at f are combined into a purely ornamental design. In b we see a different but simple style of fish. This figure forms, with the adjoining scroll, another fish. Between a and b is inserted a spiral, whose starting-point is adorned with two leaves. It passes over into another spiral (d.) This second spiral seems to symbolize the body of a cock, whose head lies in the base below, its beak holding the triskeles-shaped fish c.

The foundation of the ornament on the basket represented in Fig. 3, Plate xx, is based on the double spiral, whose ingredients b and g are doubled so as to form facing spirals. b is a compound spiral starting in a rounded fish-head, and is at the same time the symbolical expression for a cock's body, with head visible at a. The oval c, placed under the throat, is the same object which, in other cases, the bird seizes in its beak. The two opponent birds are connected by an arc consisting of two spurs, and sending down in the middle a cross-formed trefoil (d) which has its counterpart below, resting on a wave-line. The other spiral (g) is so shaped that it includes a fish, the head of which joins the two united triskeles e and f, the latter of which forms, with the adjoining curve, a conventionalized fish.

Fig. 4, Plate xx, shows a tall basket for holding spoons and chopsticks. It is usually suspended from the wall. The upper half has a cylindrical form; the lower, a quadrangular. There is a double ornament here. The upper one is cut

Decorative Art of the Amur Tribes.

out of a piece of blackened bark, which is sewed around the basket; the under design is incised into the bark. Between the double spirals *d* and *f* of the upper ornament is inserted the picture of a cock: its head (*a*) sending forth a long falcate beak; the usual oval (*b*) under its throat; its pinions symbolized by the scroll *c*, over which a crescent-like spur is placed; its spurred feet marked by a triskeles. Enclosed in the spiral *d* we see two conventionalized bipartite fishes, their heads contiguous. Under this spiral is the figure of a spur (*e*), which suggests that this spiral is considered as a cock's body. The scroll *f* symbolizes likewise a cock's body, as is indicated by the two parallel falcations with adjoining oval, in *g*, apparently signifying the bird's head and beak. The circle *h* above this spiral is the object usually found in connection with the cock, and above it is in reality a cock's head cut out of the bark, over which is placed the conventionalized bipartite fish with head turned downward; so that here a double cock is united in the same spiral. The design on the under portion is a triple structure. The central field is occupied by two cocks, heads pointed downward, an oval under each of their throats, the bodies indicated by scrolls, each encircling a conventionalized bipartite fish, the tails being simply prominent knobs. Above and under this bird are triskeles-shaped fish-tails, the outer arm of the upper one being shaped like a bird's beak, and the inner arm of the under one wound into a scroll.

Figs. 5, 5 a, Plate xx, and Figs. 1, 1 a, Plate xxi, represent the four sides of a basket. In Figs. 5 and 5 a the under and side edges are covered with key-ornaments; the upper edges with a chess-board decoration, which latter also appears in Fig. 1, Plate xxi. On these three designs the frequent use of the St. Andrew's cross is particularly noticeable. In Fig. 5 are two conventionalized cocks in the form of double spirals placed longitudinally, and combined ornamentally in a medial vertical axis. The heads are in the form of fish-tails, the beaks being characterized by prolongations of their under arms. The tail is a long tapering falcation stretched downward tangent to the circle filling the under half of the trapezoid. A similar type is met with in Fig. 1, Plate xxi.

The concavities in the upper part of Fig. 5, Plate xx, are taken up by two realistic carp, each with a crescent-like fin. It is rather singular that the drawing on these fishes should vary on the two sides. On the right, fish-head and eye are distinguished by two concentric circles. The conventionalized fish on its body shows a distinct head in circular form, and the body under it has the comma shape of the Japanese *magatama*. The head of the carp on the left-hand side is of ellipsoidal shape, its gill being specially indicated by a brace, one arm of which is prolonged into a semicircle from which depend two successive loops, — one large, the other small. Another remarkable departure from symmetry may be observed in Fig. 5 a, where are seen two conventionalized cocks, each holding two circles in its down-stretched beak. The right one shows the conventionalized bipartite fish under the tail, while in the left one the bipartition is replaced by the simple rounded fish.

The general framework of the ornament presented in Fig. 1, Plate xxi, is almost the same as that shown in Fig. 5, Plate xx, but particularly in the lower portion, where two facing scrolls are surrounded by two conventionalized fishes having a curved body in common. The cocks above have two circles in their mouths, as in Fig. 5 a, Plate xx. The field represented in Fig. 1a is treated merely in a geometrical way, two wave-lines filled in with triskeles extending along both sides.

On the basket, Fig. 2, Plate xxi, the ornamented portions are cut out of bark and appliquéed to the box. The ornaments are symmetrically arranged above and below, as is shown by the inserted auxiliary lines. The rectangle enclosed by them is the fundamental ingredient of the whole series; slightly varied, however, in the corresponding design beyond the vertical medial axis. Here occurs the interesting case of two cocks united in one figure. At the points *a* and *b* two combatant cocks meet, the right one (*a*) running out into a scroll to which the body of the cock *b* runs parallel, whereas on the other side the body of the cock corresponding to *a* only borders on the scroll which belongs to the body of the cock *b* on that side. To speak from a purely ornamental point of view, there is a lozenge in the centre (*c*) with two perforations, which sends forth four scrolls to the sides and a three-scalloped figure above and below.

Fig. 3 of the same plate represents a profusely and richly decorated basket, colored in red, black, and blue. The upper edge (*a*), divided into small sections, contains strongly conventionalized cocks of Type B. Those in the hatched parts have their necks, heads, and beaks lying at the extreme ends of a wave-line, their bodies being indicated by two united triskeles. In the other, larger fields the beaks are recurved; and between neck and spiral body is a circle, which seems to hint at a misplacement of the circle usual in front of the beak. In the central part there are several large fields (*b, c, d*) bounded by a wave-line. In field *c* there is a pair of facing spirals in the centre, framed by combined semicircles. Above this figure are two confronting cocks; under their two circles, a two-lobed leaf. There is a three-lobed leaf under the two spirals. On either side of these is a pair of fishes with heads contiguous. In field *d* prevails a tasteful composition of spirals, two upright fish-heads being inserted below. The lower edge (*e*) is composed of double spirals shaped into triskeles by tangential offshoots.

Fig. 3 a is the cover to the box represented in Fig. 3, the edge being adorned with the same decorative line as in Fig. 3, *e*. The central field shows in the middle the same spiral structure as in Fig. 3, *c*, around which six scrolls are grouped.

EMBROIDERY-PATTERNS. — Fig. 1, Plate xxii, is an embroidered border covered with a double row of ornaments. The upper row is based on a combination of two figures, — two conventionalized combatant fishes and two united cock-heads shaped like the letter X, large ovals being attached to the middle piece which joins them. In the under row there is a wave-line, the single components of which are fashioned like bird-heads with pointed beaks. A similar formation

Decorative Art of the Amur Tribes.

Decorative Art of the Amur Tribes.

is added to the facing spirals filling the hollows of the wave-line that open downward, while in those opening upward appears a doubling of the same figure as
that in the upper row; that is to say, from the middle vertical axis proceed two
pairs of beaks to both sides, the lower, smaller ones being adorned with ovals.

On the collar in Fig. 2, Plate XXII, occurs a design of similar style, except
that in the under part the two ovals are put side by side between the two outstretched heads.   The border (Fig. 2 a), made for the same robe as the collar, is
embroidered with a double pattern.   In that to the left the wave-line is formed
likewise of cock-heads, between which are confronting conventionalized fishes of
two different forms alternating with each other.

In the following embroideries the single parts constituting cock and fish
ornaments are more or less torn apart, displaced, and partly distorted; so that it
is hard to define in every case exactly what represents a cock or a fish, or where
the beginning or ending of these creatures is.   We see, for instance, in Figs. 3, 4,
4 a, and 5, Plate XXII, simple and compound triskeles in various styles and combinations, grouped together with spirals.   A comparison with the forms hitherto
described undoubtedly proves them to be derived from components of the cock.
The high degree of distortion gives so much individual freedom of choice as to
interpretation, considering the ambiguity of the significance of the single pieces,
that it would be a hopeless task not only for the Western student of these ornaments, but also for the cleverest native connoisseur, to draw any conclusion as to
the details of this ornament.   At this point a geometrical stage opens up, where
realistic explanation is hopeless, and beyond the pale of which no one can go.
That there is, however, an undeniably close continuity between these various
degrees of evolution is evidently shown by Figs. 6, 6 a, 6 b, all patterns belonging
to the same garment.   While, as regards Fig. 6, we can but feel like declaring
our *non possumus*, still we are able to decipher the two cocks with their downward-
bent beaks and oval bodies in Fig. 6 a, and even the two conventionalized fishes
placed together in a figure the geometrical character of which seems to be strongly
emphasized, at first sight, in the upper part of this ornament.   In Fig. 6 b it might
be possible to distinguish the cock-beaks, through the circles placed in front of
them, on the spiral to the left, as well as the cock filling the concavity of the following wave-line; but in this case it is next to impossible to state with certainty
which part is to be looked upon as head or tail, granting that these two possibili-
ties are admissible.

The preceding remarks apply also to Figs. 1–3, Plate XXIII.   Only the two
combatant cocks over the last pair of spirals in Fig. 2, and the two conventionalized fishes turned away from each other in Fig. 3, may be recognized as such
with any degree of certainty.

In the following designs a definite group of ornamentations is exhibited.
There is a double principle active in them,—that of displacement and that of
combination.

Fig. 4, Plate XXIII, shows a silk collar.   The design consists of two ele-

ments, — one being the figures cut out and buttonhole-stitched to the foundation ; the other, the designs embroidered on these pieces.   The former consists
of two pairs of cocks in disconnected parts.   The heads and beaks of the two are
formed by two triskeles united into one figure ; the bodies consist of two of the
cordate figures with appended fish-tails or scrolls.   The embroidery on these
body-pieces is composed of representations of contiguous cocks in two different
forms.   In both cases the animal is adapted to the cordate leaf on which it is
worked.   As to the one form, the neck is recurved in an arch.   A fish is substituted for the body ; a spiral with an adjacent parallel lobe, for the wing-feathers ;
and two huge, almost circular falcations, cleft in the middle, for the tail-feathers.
The other type has as body a spiral, the prolongated outer winding of which
forms the upstretched neck, whereas the plumage is indicated by an annexed
semicircle with an attached offshoot running downward and closing a two-foliated
leaf.   These two forms of cocks, so far as their relation to each other is concerned,
represent Type B.   The graceful cordate leaf-forms are reproduced in Figs. 4 a,
5, 5 a, 5 b, but more freedom is displayed in the use of foliage in the figures
inside of them.   In Fig. 4 a, even the fishes held in the cock's beak are embroidered
in the same style as leaves.   In the first two fields are two cocks curiously placed
one above the other, and connected with each other on the inner side by an arc.

Figs. 5, 5 a, and 5 b, Plate XXIII, show the foundations of an embroidery-
pattern, the ornaments being cut out of paper and pasted on the underlying cloth,
to be worked around.   In the first leaf on the left-hand side of Fig. 5 the combination of two cocks is clearly visible.   The one cock holds a realistic spiral-
formed fish in its beak, and has a fish-body whose head is indicated by a spiral
and the tail by a semicircular appendage.   The adjoining cock has seized in its
beak two circular objects adapted for embroidering as leaves, and has a strongly
marked tail of three long prongs.   In the following leaf the two cocks are united,
and hold between their beaks a large bipartite fish, while the three wing-feathers
of the lower cock have adopted the form of this same fish-body.   Also in Fig. 5 a
we meet with a field containing two superposed cocks.   In the two central leaf-
forms the upper birds are combatant, the lower ones opponent and inverted.
The upper cock has one leaf above, and another under, its neck, the origin of
which is to be explained by the fact that the upper leaf represents the leaf-like
treatment of the head, the under one that of the well-known circle.   For the
body of this cock is substituted a fish, and another realistic fish with recurved tail
is attached to the spiral above it.   At the place where the tail turns upward is a
leaf.   Two leaves supply the place of a scroll in the body of the lower cock.   On
the outer leaf the lower, inverted cock holds in its beak a bipartite fish with the
tail pointing upward.   Its spiral-formed body sends off to the side a branchlet in
the form of a bird's head with an oval under it, so that here again a cock seems
to be intended.   The superposed figure resembles one of the forms seen in
Fig. 4 of this plate.

Fig. 5 b, Plate XXIII, is constructed of three cordate leaves, so arranged

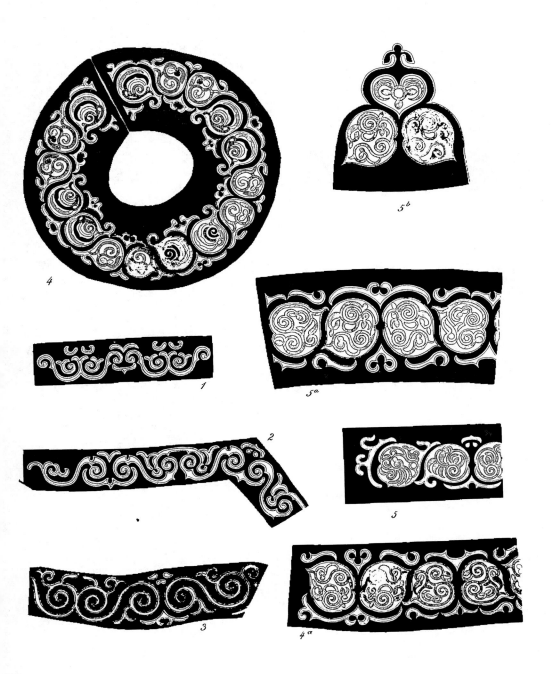

Decorative Art of the Amur Tribes.

Decorative Art of the Amur Tribes.

that the outlines of the three form another heart-shaped figure.   Under the indentation of the upper heart is a circle, three oblong leaves radiating from it, — one below, and two on the sides.   Its two crescent-shaped lobes are occupied by two confronting fishes coiled into spiral form with tapering body and tail. The two lower hearts agree, in the representations on them, with the two central ones in Fig. 5 a.

Fig. 1, Plate XXIV, is an embroidered pattern.   The upper part is taken up by two facing cocks rampant, head and body formed after the fish type.   As cocks they are recognizable merely by their two feet.   Between these feet are two bird-beaks combined into a geometrical figure in the centre of the lower, wider section of the design, bearing a triskeles, one arm of which is likewise beak-formed, and the other two convoluted into a scroll.   The oval into which the under arm runs out is at the same time the head of a conventionalized bipartite fish, which, as it would seem, is drawn on a cock's body whose tail is formed by the triskeles just referred to.   The beak is lowered as if about to touch the circle under the throat.   The branches intended to indicate crest and pinions are above the head.   The remarkable features in this representation are the manner in which the single portions making up the three cocks merge into one another, and the fact that many parts belong to the three types in common.

Fig. 2, Plate XXIV, is an embroidered pattern in triangular form.   In this pattern we observe on both sides three distinct single cock-beaks, — the uppermost bent upward, the middle one outward, the undermost still more curved and turned inward,— all three set with ovals or circles, probably survivals of head and eye. On the edge below are two separate long-stretched conventionalized tripartite fishes with spoon-formed tails.   The same types, standing upright, and connected below with each other, appear in the upper part of this ornament.   Also the long branches of the two facing spirals under them are composed of two pairs of cock-beaks which join at their points.   In verification of the fact that this principle of displacement occurs also in the area of Sino-Japanese art, a Japanese weaving-pattern is represented in Fig. 3, in which bird-heads having only a long neck are placed parallel to spirals and alternating with them, as in our ornaments. Also the long offshoots of the spirals agree with our conventionalized fishes, as well as the adjoining bipartite figures.

The next three embroideries (Figs. 4, 5, 6, Plate XXIV), each of double symmetry, are usually united into groups of four, and sewed to sleeping-mats.   In Fig. 4 there is a lozenge in the middle, around which cluster four compound spirals, between the inner and outer windings of which are spur-lines.   At the upper and lower extremities of this pattern are two smaller triskeles-spirals which proceed from the larger ones.   On either side of the large spirals two triskeles are placed, the two outer ones striving after the fish-form, the other two after the cock-form.   The square patterns seen in Figs. 5 and 6 are cut out of velvet and outlined in chain-stitch.   If we look at one of the quarters from one of its outer corners in the direction of its diagonal, we shall see that the fundamental element of the ornament in Fig. 5 consists of two superposed confronting cocks. The spirals which represent the tails are rolled outside in the upper pair, and inside in the lower ones. On either side of the upper cock is a conventionalized bipartite fish. Both of these and also the large cocks form ornamental figures with the adjacent corresponding cocks or fishes of the neighboring rectangles. Fig. 6 illustrates a structure related to that of the preceding one. Four heart-shaped figures (*b*) are clustered in the centre in a square. Above their points, in the direction of the diagonal of the square, in each of the four quarters, is a cross, its two side-arms terminating in spirals (*c*), and its rounded extremity (*a*) being adorned with a pair of fishtail-formed triskeles.

Fig. 1, Plate xxv, represents an embroidered quadrangular piece placed on its point, used on the cape of a winter hood. The ornament represents a spiral structure that decreases in size as it proceeds upward. In the centre (black) are two cocks rampant developed from the fish-form; to the right and left of these, two conventionalized birds, their necks and heads stretched upward. The confusion of the single ornamental parts here has been carried to such an extent that the circular object has been taken away from the beak and placed in front of the two falcations of the tail, which thus convey the impression that they are beaks. In corroboration of the idea of the wing-feathers, which are expressed by the upper of the two tail-flukes, appears by the side of it a parallel crescent. This distortion proves sufficiently well that the conception of the original meaning of the ornament has diminished in clearness. Almost all elements of this decoration, aside from the pure spirals, are either birds' necks with beaks, or spurs, or small ovals. The original types are on the verge of being dissolved into single disconnected and sometimes misunderstood parts: the principle of symmetrical and tasteful arrangement, however, is still observed.

An embroidered pattern for a pair of wristers is shown in Fig. 2 of this plate. The edges are decorated with single spur-lines above and below. The same style of line is also used to surround other figures. In the uppermost section of the ornamentation we find two facing combatant fishes, ending below in spirals, their heads surrounded by a figure formed of two spur-lines. Below them is a four-leaved rosette. Between two pairs of facing spirals are observed two conventionalized bipartite fishes in the act of swimming; farther below, two cocks rampant whose heads are connected by a semicircle. Over the spirals forming the tails of these birds are bipartite fishes.

We will now turn to some fantastic compositions occurring on embroidered material. On the triangular pattern of raised embroidery (Fig. 3, Plate xxv) are, in the upper part, two combatant musk-deer with two legs and scrolled tails. To the right and left of the compound facing spirals under them we see two long outstretched bird-heads, the upper line of which is formed by a brace. The outlines constituting the head continue downward into two parallel spirals. The ornament on the embroidered band (Fig. 4) is made up of two semicircles. In the two ends of each are visible two cock-heads side by side, with two circles

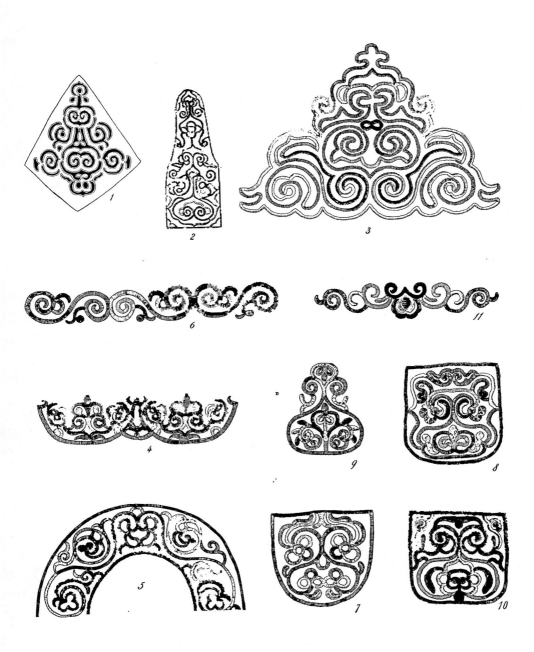

Decorative Art of the Amur Tribes.

in front of them forming together the figure 8. Their bodies are composed of a spiral with a semicircle resting on it. The two symmetrical figures are combined below by a spur, which they have in common.

On the collar (Fig. 5, Plate xxv) belonging to the same series of patterns, we observe, in the centre, two confronting upright fish-cocks. On either side of them are two others, whose bodies are produced by placing two fishes together in the form of a spiral. In the lower field the cock-beaks hold, instead of the usual fish, the figure of a plant-form, — a kind of trefoil.

The riband seen in Fig. 6, Plate xxv, belongs to the same garment as the collar above mentioned. Here are facing spirals. From the sides of those rolled upward project cock-heads with semicircles under the throats, resting on the outer windings of the spirals; from those rolled downward issue inverted cock-heads with ovals placed on the backs of their necks.

The ornaments on the following patterns (Figs. 7–10, Plate xxv,) are treated in arboreal style. The figures represented are used to trim shirt-pockets. Fig. 7 shows two cock-beaks turned downward, and encompassed on the sides by spur-lines, both holding a three-lobed rosette. At the lower extremities of the under arcs are two leaves, consisting of three circles each; and near the same ends of the arcs are two fishes moulded in the style of leaves. In Fig. 8 four different forms of conventionalized fishes lie close together (a, b, c, d), three of them (b, c, d) clustered around a circle. Farther below is a conventionalized tripartite fish (e); f is also an imitation of the fish-body, but is here developed into a palmetto-like floral pattern. A figure of similar character occurs in the lower part of Fig. 9. The latter was evidently intended for the trunk of a tree sending off spiral-formed boughs, the edges of which are adorned with three single leaves. Fig. 10 illustrates a plant-like design of allied style in the under part of the ornament, the lateral branches being indicated by long, narrow fish-forms (large ribbed leaves), and the centre filled with a small two-lobed leaf, below which issues another large one. In the upper part we have two conventionalized fishes attached to a pair of facing spirals. In the corners beyond the fishes are two triskeles-shaped cocks characterized as such by the conventional form of the fish in their respective beaks.

Fig. 11, Plate xxv, is added here because it shows a pattern pertaining to the same robe as the four preceding ones. There is a palmetto-like figure in the centre, from which branch off on both sides arabesques built up of triskeles.

Figs. 1 and 1a, Plate xxvi, show a woman's embroidered mitten made of reindeer-skin covered with cloth. The former represents the back, and the latter the palm. The spaces between the single lines are filled up with zigzag stitches. On the back of the mitten is a tree-like formation, in which two two-lobed leaves are attached to opposite sides of a stem, the two on the left being embroidered in green satin-stitch, those on the right in lilac. This tree is crowned with a heart-shaped figure enclosing a bifoliate red-colored leaf. The five leaves contained in the ellipse below are all light green. Tendrils adorned with triskeles grow round this tree. On Fig. 1a the embroidery is placed on the thumb of the mitten. The motive is here the same, — composed of a triskeles form, an S-shaped figure, the under part of which is cordate in shape and encloses two leaves in red.

The skin glove pictured in Fig. 2 of the same plate is covered with velvet bearing a chain-stitch embroidery in silk. The pattern is an artistic structure of fanciful combinations. On the top are two heraldic combatant cocks, whose heads are formed by an oval (a), from which the plumage goes off into three depending branches. On the marginal branches (b) is drawn, with the aid of a spur-line, a bipartite fish ; and a similar figure occurs also in c and d in connection with spirals. In e two leaves are enclosed again in a heart-form. f is the head of a cock placed sideways, and g its tail. In the pointed end of h are united two cock-heads holding in common the leaf i, while on the outside appears the exquisitely curved bird-neck j bearing the leaf k. In the interior of the undermost spiral is the body of a conventionalized bipartite fish embroidered as a leaf (l), the head of this fish being held by the beak m. This figure is surrounded by a line.

Fig. 3, Plate XXVI, represents an elk-skin garment, obtained from the Tungus on the Ussuri. A series of figures is spread over the surface of the back, the decorations being painted in blue, red, and yellow. Only the part over the hips is cut out of fish-skin and appliquéed to the garment. In a we see two opposite single cocks, built up essentially from purely geometrical ingredients. The head consists of two superposed semicircles, the lower of which runs out into a recurved arc. From that issues a branch in the opposite direction, to form with the scroll a triskeles, expressing the fish held in the cock's beak. The body is formed of three semicircles which unite at their ends, and enclose two crescent-like fishes. The feet are in the shape of an anchor-formed combination of two triskeles ; the outer arm of the outer triskeles in both cocks being shortened into a knob, the inner forming a semicircular claw. The cock on the right side has below it an additional figure that repeats a schematic outline of the foot. The tail is a very intricate formation, — below a spiral, which appears as a continuation of the under outline of the body. The upper outline is continued into a strongly conventionalized cock with a circle on its head and a fish-tail beyond. Between the tails of the principal cocks and those appended appear the R-formed figures enclosing the image of a bipartite fish. The cocks b and c stand sideways, and also consist of geometrical elements. The manner in which they are evolved is shown by a comparison of these figures with d. At the sides of d we observe anchor-formed appendages. These are carried out in b and c in such a way that one arm of the anchor forms the head and neck, the other the tail, of the cock. Whether the anchor-formed type d has been developed from b and c, or, better, whether d is a prius which served as a foundation for building up b and c, must still be regarded as an unsolved problem. In the same figures two additional groups have been produced, the one in the middle in combatant attitude with spiral body, the other at the top with recurved beaks. This latter, inverted

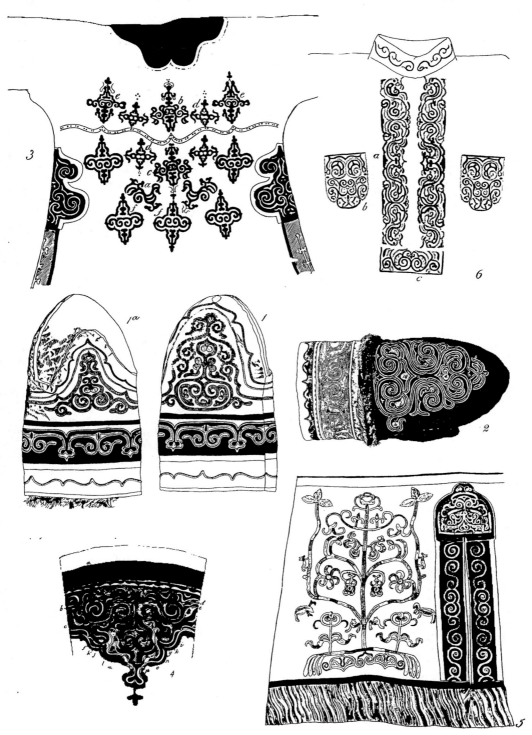

Decorative Art of the Amur Tribes.

form appears in normal position in the under portion of *e*, only that here a slight wave occurs in the comparatively long neck. The point of this figure (*e*) is crowned by a trefoil, under which, in the figure on the right-hand side, is a cock type closely allied to the under one, except that here the beaks are more extended in a downward direction without tapering. On the left-hand side is a case of asymmetry, since there, instead of thick lines like those in the figure on the right, occur simple lines of the same form, but inverted. There are five figures exactly alike (*f*). These form, above and below, a narrow tapering structure. In this picture the negative parts show the type of the confronting tripartite cocks in the fish style, which is so hard to distinguish from the conventionalized fishes themselves.

Fig. 4, Plate XXVI, represents a painting on the upper part of a pair of leggings made of elk-skin. The ornamental organization is executed here by two opposite double spirals near the upper edge, separated from each other by a longitudinal wedge, the fish-tail end of which joins a pair of facing spirals. Over the left double spiral is a long-stretched cock (*a*) in green, with a tail like that of a pheasant. The lozenge shape of the body and the engrailed line forming its edge are most remarkable. Strange and unique in its kind is the fact that this horizontally placed cock (*a*) occurs in combatant attitude with the vertically placed cock (*b*), whose body likewise is shaped like a lozenge ; the latter, however, does not run out into a tail, but into another inverted cock-head turned to the left, which, in its turn, is represented in combatant position with an inverted cock (*c*), whose body, also lozenge-shaped, runs off above into a fish-tail. We are again surprised in this design to note, on the corresponding side, an arrangement of types bearing the same relation to each other as *a* to *b*, but the two cocks *d* and *e* are placed on a horizontal plane. The cock *d* has likewise the tail of a pheasant. The line forming the back, however, is an uninterrupted curve, as the proper form of the cock's body is in general retained here in a much higher degree than in *a* and *b*. *d* and *e* have an engrailed line consisting of three arcs, marked more strongly in *e* than in *d*. Over *d* and *e* are two combatant cocks of more distinct forms than the two birds over *a*. The fish-like cock is represented in the designs *f* and *g*. Each of these holds two fishes in its beak,— a conventionalized one (*h, i*), and a rather realistic one (*j, k*) with the eyes marked. Worthy of note is the asymmetry between the two space-filling conventionalized fishes *l* and *m* on the one side, and the fishes *n* and *o* on the other. It is hardly necessary to call special attention to the cock-heads united in the figure *p*, nor to those on the adjoining leaf-forms below.

The embroidery in Fig. 5, Plate XXVI, is worked on the lower part of the back of a garment. The same design is found on both sides. In the middle is the trunk of a tree with an ornamental top, and sending off three main boughs to both sides. Two musk-deer with heads turned so that they face each other are embroidered at the place where the lowest pair of boughs branch off. The tips of these boughs are adorned with trifoliates ; on their sides are two roes which seem to be climbing up.   On the second boughs blossom two quinquefoliate flowers, the petals of which are grouped in the form of hooks around a circle. Between the two flowers are two large tortoises ; over these, cocks placed sideways, with a two-lobed leaf behind them.   To the right and left of the large tree-trunk are two smaller trees decorated at their tops with a trefoil surrounded by triskeles-formed branches.   From the trunks of these trees proceed to both sides cocks that appear, as it were, to be growing out from the tree.   The ground on which the trees stand is characterized by a line sending off downwards at both ends four offshoots corresponding to the cocks' plumage.   At a short distance from the tops of the small trees are two elks with antlers.   In the escutcheon-like piece in the upper right-hand corner of the embroidery are to be seen two conventional forms of musk-deer with faces turned away from each other.

Fig. 6, Plate xxvi, represents an embroidered shirt.   In the centre of the longitudinal border (*a*) are two combatant fish-cocks with a bifid crest on their heads and a spur-line combining these.   In the other concavities are pairs of strongly conventionalized opposite musk-deer, their necks recurved so that they face each other, with erect ears, spiral bodies, and two long curved legs.   This type is nearest to that described in the preceding figure.   The ornament on the two pockets (*b*) is composed of two portions.   The upper part contains two scrolls, oval in shape, the outer winding of which continues in the form of a conventionalized bipartite fish, the ends of which are connected by a spur-line. Lying within the scrolls are two inverted cocks, whose type is derived from that of the fish just mentioned, except that here a tripartition is employed.   Both from this fact and from the crest marked on the head, the gallinaceous character of this theme is indicated.   The under part is taken up by a group of two parallel tendrils, the lower of which encloses a quadrifoil ; and the upper, four spirals grouped around a lozenge-like rosette.   In the two under lateral tendrils, which issue from a branch, the conventionalized bipartite fish is used to connect the two.   The field *c* shows an interesting variety of the conventional dragon. There are two creatures represented in confronting attitude.   The heads are two simple scrolls.   The bodies are indicated by spirals wound three times.   In the outer windings a portion is marked off on which the scales are characterized by three short teeth.   The outer spirals, forming double spirals with the dragon bodies, are set with three claws to indicate the feet.   Over them the tail of the dragon is symbolized by three cock-feathers.   The two serpentine lines lying between the tails and the bodies are explained as snakes, a further ornamental expedient to characterize the animal nature of this creature.

Fig. 1, Plate xxvii, represents an embroidered shirt of the Gold. From the collar, down both sides of the front opening, is a border (*a*) composed of double spirals consisting of two parallel lines.   These double spirals are so inter-locked that the outer winding of the scroll at one end merges into the inner line of that at the other end, and *vice versa*, the outer line of both scrolls being adorned with a double triskeles.   In the fields *b* and *c* ornamental trees are

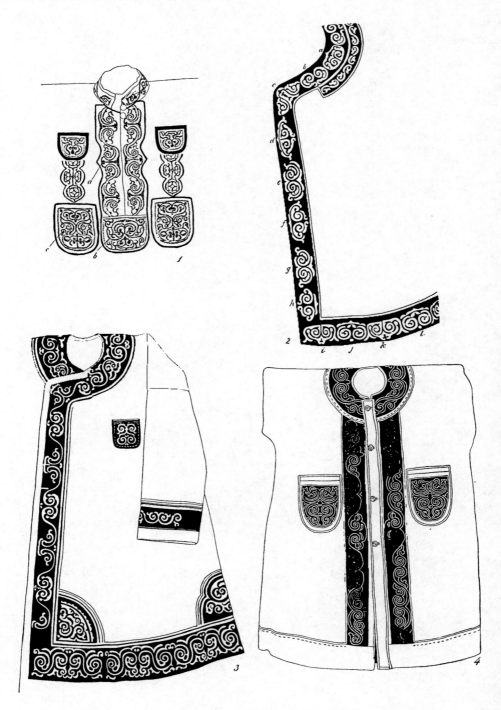

Decorative Art of the Amur Tribes.

designed.   Two opposite hearts divide the latter field.   In the upper heart are four semicircles used as supports for leaves ; the under heart shows a tree-trunk, the top of which is adorned with an oval leaf ; while semicircles, crescents, and ovals are represented as leaf-bearing boughs.   The lateral fields are occupied y fish-cocks with one head at the end of the fish-tail and two heads superposed over the fish-head.   In *b* two branches are carried out as bird-beaks, each holding two leaves.   There is a similar motive on the upper, smaller pockets, only there a cruciform leaf-cluster appears between the two deflected beaks.

On the border extending from the collar to the bottom of the woman's embroidered silk dress seen in Fig. 2, Plate xxvii, the artist has pictured single groups of facing spirals connected with conventionalized fishes, displaying the enormous variations of which these simple forms are capable.   Among the twelve consecutive groups *a–l*, there are only three corresponding pairs ; viz., *e* and *g*, *i* and *k*, *j* and *l*.

On the garment in Fig. 3, Plate xxvii, we see combinations of two fish-spirals.   One of them contains the same motive as *i* in Fig. 2 ; that is, spirals with two conventionalized bipartite fishes united into one figure and placed around them.   The other motive is a recurved spiral worked out as a fish-body, with tail in the form of a triskeles.   On the bottom edge, fish-bodies are gracefully twined close around the spirals, that terminate alternately, below, in cock's heads.

Fig. 4, Plate xxvii, shows a dress embroidered with white chain-stitching on a black background.   The ornaments on the two longitudinal borders might be designated as continuous cock-spirals, for, in spite of their scroll character, the original cock motive is still rather conspicuous.   Exactly in the middle we observe what are obviously cocks of Type B, their backs turned toward each other.   The beaks are strongly marked.   The ovals are under the throats, and the two bodies are connected above by an arc, inside of which two conventional-ized fishes are designed.   At the beginning no less than at the end of this pattern the beak with the roundish object in front of it is distinctly visible in the fishtail-shaped triskeles ; also in the other triskeles next to the central figure the head stands out distinct from the beak of the bird, and this motive occurs also on the collar.   On the upper part of the pockets sewed to both sides we note an odd figure not as yet met with.   Within two crescents we find two conventionalized fishes, and over their heads the head, eye, and beak of a cock ; while over each fish-tail rises the head of a musk-deer, its two ears erect.   Two combatant cocks and two deer, their heads turned so that they are looking at each other, are accordingly united in this one figure.

In Fig. 1, Plate xxviii, is represented a woman's silk-embroidered coat. On it are seen two perching cocks (*a*) standing opposite each other, and holding fishes in their mouths.   Under each of these single cocks is a pair of combatant cocks (*b*) showing a much more advanced stage of conventionalization.   In the cocks placed sideways (*c*) the pinions as well as the tail-feathers are expressed by conventionalized cocks. *d* is the terminating figure of a wave-line,—a spiral, with a fish-body attached below, and two cock-feathers above. The same motive is employed in the representation of the cock *e*. *f* shows the type of two inverted combatant cocks. The body is formed here by a fish, which continues into another fish placed around the spiral of the wing. The tail is indicated by three feathers, and the feet by a scroll with lateral offshoot. *g* corresponds almost to the type B, only that here the head of the fish in the bird's body is placed above, and its body below. Besides, the spaces between the cocks' bodies and the separating vertical axis form again conventionalized fishes. In *h*, *i*, and *j* are to be seen spirals adorned with cocks' wing and tail feathers. A remarkable design is *k*, where the two cock-feathers in the interior of the oval figures represent the missing spiral lines. Finally, in *l* leaf-ornaments have also been employed, partly in the form of two contiguous circles, partly in that of ellipses enclosing a heart-shaped figure.

SPECIMENS MADE OF FISH-SKIN. — We have several times met with chess-board patterns (see Fig. 1, Plate X; Figs. 5, 5a, Plate XX; Fig. 1, Plate XXI), notwithstanding the fact that the game of chess is not known to any of the tribes of the Amur region. Two other examples follow here.

Fig. 2, Plate XXVIII, shows the design on a tobacco-pouch made of roe-skin, the interior of which is covered with fish-skin. Here are quadrangular fields covered with chess-board patterns composed of pieces of white and black fish-skin, which alternate with other fields of plain roe-skin. The ornament cut out of fish-skin on the inner side of the lappet is subdivided into three parts. The upper part contains a pair of facing spirals, around which cling two conventionalized bipartite fishes, the eyes marked by small circles. Between their bodies is a trefoil. In each of the two lower symmetrical fields are two superposed spiral cocks, each of the under ones holding a trefoil in its beak.

Fig. 3 of this plate represents an apron which is a kind of fish-skin patchwork. There are three rows of squares containing alternately chess-board patterns and other decorations. In the former, light and dark strips are interlaced as in braid-work, the number of checks varying from seven to nine. The spaces between the squares and the separate rows are filled up with long stripes, alternately white and black, arranged in most cases diagonally. There are two different ornamental figures in the other squares. In the one are four pairs of facing spirals, grouped like a rosette around a figure consisting of two trefoils. This ornament is cut out of fish-skin and sewed on a piece of dark-red cloth; the other figure is sewed on black cloth. In this latter, four conventionalized cocks are grouped around a lozenge. The figures across the lower edge are likewise cut out of fish-skin, sewed on, and colored alternately light brown and bluish green. The fish-skin threads used here are red, green, blue, lilac, and violet.

Fig. 4, Plate XXVIII, is a Goldian hunter's cap made of roe-skin, lined with blue Chinese cotton. The crown is topped with two tassels and a sable-tail.

Decorative Art of the Amur Tribes.

2

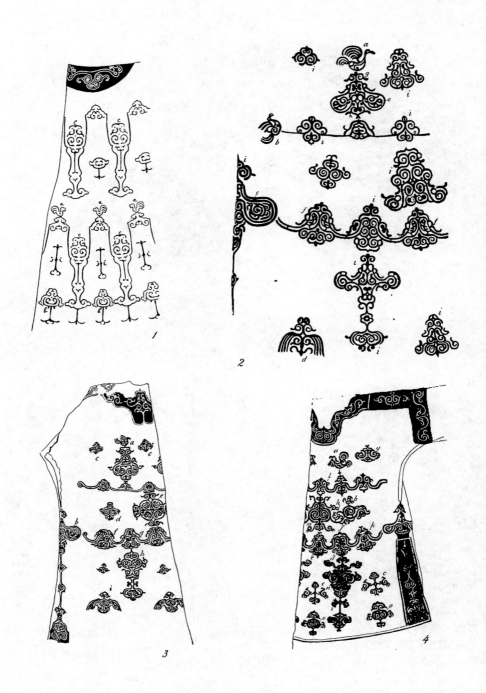

Decorative Art of the Amur Tribes.

The edge is covered with a strip of black and one of red cloth, between which are narrow stripes of yellow. These, as well as the ear-lappets, are trimmed with appliqué ornaments cut out of fish-skin. The elements of the ornament running around this cap are distorted cocks, arranged in pairs around trefoils in the lower row, and in double pairs attached to the upper and lower points of a quadrifoliate lozenge in the upper row. This case, together with the two preceding ones, proves that fish-skin is sometimes used merely as decorative material.

On Plates XXIX and XXX we have eight representations of decorated fish-skin garments, which are worn exclusively by women. The ornaments are cut out of pieces of fish-skin, and are generally colored blue; they are then sewed with fish-skin thread to a piece of fish-skin of a shape adapted to the size and form of the ornament. A great number of such single patterns are then symmetrically put together on the garment itself. A different method is employed only on the garment in Fig. 1, Plate XXIX. Here we have three layers of fish-skin, the undermost representing the skin of the garment proper; the uppermost showing the ornaments in their cut-out forms. Between these two layers is inserted a middle layer, which serves as a background to the ornament proper, throwing out distinctly the negative parts as well as the outline of the ornament. It extends a little beyond the edges of the uppermost layer, which is sewed to this one. The middle layer is dyed partly light red, partly blue, so that the edges of the negative parts of the ornaments appear in these colors, setting off the monotonous color of the underlying plain fish-skin. There are three neat naturalistic perching cocks (a) with trisulcate tails and open beaks. Very curious are the downward-stretched cock-heads in b, alternating in the intervening figures with triskeles corresponding to them. If the eye were marked in the latter, they could hardly be distinguished from these realistic heads. The constituents of the long-extended figures marked c may be analyzed in a similar way. There is an embroidery on the collar of this garment showing various two-lobed leaves and a trefoil surrounded by the outlines of a heart.

Nearly all forms of cock and fish ornaments are represented on the following specimens. We observe the cock with wings outstretched, in a of Fig. 2, Plate XXIX, probably perched on an ornamentally devised tree, and crowing, for its beak is open. Its body is shaped like a fish, the head of which, formed by a circle, lies in the back part; and another fish, enclosing a large dot, is marked off in this same body. The cock placed sideways (b) is similarly formed. It is likewise crowing; but tail-feathers and wing-feathers are represented by only three curved lines, whereas the former (a) shows four parallel curves for the tail, and even six for the wing. Inside of the fish-formed body b the head of the fish is marked by a scroll and a circle similar to that in a above it. The cock on the border to the left side (c) has undergone some further alterations, because the artist was obliged to adapt its shape to the double circular lines which enclose it. It shows a wattle under its throat, and has a fish-body. Its pinion is formed by a composition of two adjoining beaks, at the end of which is a two-lobed leaf. The form of the tail deviates from all other hitherto known forms, and is merely the product of a purely ornamental assimilation to the given space. The pattern *d* deals in a striking way with the subject of the two combatant cocks. The heads are distorted, and have shrunk into scrolls, including the circles attached to them. The wing-feathers placed under the heads are symbolized by spirals, each with a lateral process; i. e., the spiral-triskeles; but the symbolic expression of the tails claims an undue amount of space, quite out of proportion to that occupied by the parts just described. Four exaggeratedly long tail-feathers are indicated on either side, the space between them being filled with a pair of united triskeles, and under them a hook-formed figure. In the interior of the figure suggestive of an ornamental tree, below the cock *a*, we observe two realistic fishes (*e*), whose eyes and gills are characterized in the usual way; on their bodies is a conventionalized bipartite fish, the tail extending out into a compound triskeles, one of the arms of which is continued into a scroll. The figure which separates these fishes contains in its negative parts two upright bipartite fishes, which occur also on the sides of the design *f* in the familiar R-form. In the middle of *f* are two naturalistic fishes rampant, without any spirals on their bodies, but marked with two parallel ventral fins. The R-formed fishes are also to be found under the two facing triskeles marked *g*. In *h* two triskeles are conceived of as two combatant cocks, chiefly characterized by the two combined circles, one of them being held by each. The large figures marked *i* are compound, rather complicated, ornamental arrangements, which are built up of spirals, trigrams, leaves, and conventionalized fishes, and elements of the cock-ornaments.

The garment represented in Fig. 3, Plate xxix, in general resembles very much that in the preceding figure. There is a perching cock (*a*) with open beak and trichotomous wing and tail feathers. At the end of the fish-shaped body is a spiral, and on the under outline a small solid circle. The enclosed cock (*b*) with fish-body and oddly ornamental tail tallies exactly with the bird in *c*, Fig. 2. There are two inverted fish-cocks in *c*, a pair of fish-spirals in *d*. A very graceful group of four fishes is placed in the form of a spiral around a quadrifoil (*e*). Inside of *f* lie two fishes united in a horizontal position over a pair of facing spirals, the upper outline of the fishes forming a brace. In *g*, fishes of the same type are situated under the spirals; further, in the upper part are two opposite fishes rampant. Here as well as in *h* occur two-lobed leaves. There are two conventionalized lateral cocks in the design *h*. *i* should be compared with *d* in the preceding figure. The tail and wing feathers are ornamentally fashioned by the aid of spirals, triskeles, and leaves.

Whereas the ornamentation of the garments in Figs. 2 and 3, Plate xxix, is based on two horizontal rows, that of Fig. 4 of the same plate is composed of three rows. The cock *a* holds a distinctly marked fish in its beak; furthermore, a conventionalized fish is designed on its body, and another added to one division of its comb.   To the spiral forming the pinion is attached a cock's beak turned to the right, with a circle below it.   In the second row is the representation of another erect realistic cock (*b*) with bent beak and bipartite tail.   In the two figures *c* we note a conventionalized tree, on the two side-branches of which two cocks are perching.   Of very peculiar shape are the cock-fishes *d.*   Their heads and bodies have fish shapes ; they terminate, however, in a three-lobed cock-tail. Their heads are turned away from each other.   There are spiral fishes to be met with in *e, f,* and *g.*   *h* is a musk-deer with fish-body ; its hind-leg is a cock's beak with the oval.   In *i* are represented two cocks with their heads turned downward, which bear, strange to say, triskeles-formed fishes on their heads, and have, besides, fish-shaped bodies.   The fish-heads in which they end have two small erect prongs of the same form as the ears of the musk-deer.   In *k* and *l* the double wave-lines are made use of as supports for cock-heads, of naturalistic representation in *k*, of conventional form in *l.*   *m* and *n* show the use of the facing spirals, which are joined in *n* to fish-bodies above, and on the sides to conventionalized cocks.   In *o* the negative portions of the inner facing spirals are two conventionalized bipartite fishes ; on either side of them, and at the bottom of this figure, are two strongly conventionalized cocks holding circles in their beaks.

As regards the ornamentation on the garment in Fig. 1, Plate xxx, the rather naturalistic cock *a* is represented with four-pronged pinions and tail-feathers. A conventionalized bipartite fish is designed on its body.   Exactly the same forms are shown in cock *b*, placed sideways.   *c, d,* and *e* illustrate the combatant cocks fashioned as on an escutcheon.   They are most elaborate in *c.*   Here the head of the cock is designated by a heavy dot surmounted by a semicircle, the beak being characterized by a smaller semicircle.   Attached over its head is a triskeles-shaped fish.   The bodies of the two cocks are united into a heart-shaped figure, to which are joined on either side the strongly marked, long wing-feathers.   The tail added under the cordate figure is treated as an independent ornamental element, in which, properly speaking, a conventionalized cock with fish-shaped body, wing and tail feathers, is to be recognized.   *d* represents heads and bodies of cocks in the form of erect fishes.   Their gallinaceous character, however, is sufficiently preserved by the four-lobed tail and the spur below it, which latter they have in common.   In *e* the heads have vanished ; the heart-formed body, as in *c*, a spiral wing-feather, and a double-toothed tail-feather, are visible.   In *f* are two inverted conventionalized fishes placed around facing spirals, just as in *g :* in the latter case, however, the fish-heads are set with cock-beaks which run parallel to the winding of the spiral.   Of the different representations of the fish, the following are to be found here.   *h* contains two confronting erect fishes of naturalistic forms, with eyes, gills, a design on their backs, fins, and spiral-formed tails.   In *i* we see two fishes projecting from the sides in an almost straight horizontal direction, whose heads are set with two offshoots formed like cock-feathers.   In *j* the fishes arise likewise from the sides, but the heads are turned upward, and the bodies are coiled and have four fins.

Another fish proceeding from a wave-line is *k*.   Here the semicircular hook on the head, continuing the line of the gill, is placed toward the outside.   The two outlines making up the body do not unite to form the tail, but run parallel to each other.   Another group of fishes is connected with spirals, as, for instance, in *l*.   They are used as continuations of two facing spirals, together with which they are enclosed in a figure.   In *m* seems to occur a fish-cock, as the two hook-like offshoots from the head of the fish appear to show ; perhaps the same is the case in *n*.   In *o* the fish joining the spiral is not completely drawn, as its outer edge line remains parallel to the winding of the spiral.   The most conventional-ized design of all is *p*, the characteristics of which afford insufficient ground for explaining it positively as fish or cock.

The cock *a* in Fig. 2, Plate xxx, is composed in a striking way.   It holds a triskeles-formed fish in its beak ; its body is shaped like a fish.   Its pinion is represented as a cock-beak with a circle under the throat.   The tail consists of two sections.   The upper circular curve is combined with two conventionalized fishes (not visible in the illustration) ; the lower part is a cock-beak holding a fish, both so connected that they enclose a small circle.   In the same way is built up the cock *b*, which is placed sideways.   In *c* a conventionalized cock joins a spiral.   *d* and *e* are designs constructed from single ingredients of the cock and fish ornaments. *f* shows in the interior two confronting bipartite fishes in an upright position and two bipartite fishes proceeding from the scrolls on the outside.   In *g* two con-ventionalized combatant cocks unite in an ornamental device in the pointed upper structure.   The figure *h* below is identical with *e*.   In *i* we observe four coiled fishes grouped around a central lozenge.   Their bodies are scaled like that of the dragon.   Just as here, so in *j* we meet with two scaled, very realistically formed fishes in the concavities of the wave-line.   At its terminus sits a dragon (*l*) with open jaws and two three-clawed feet.   Its tail is a fish.   In form it is like two combined triskeles.

The back of the garment in Fig. 3, Plate xxx, is covered with a series of more compact figures.   The single ingredients of the same are generally to be traced back to cock-ornaments.   In *a* the heads of the two cocks side by side are turned away from each other, and a circular object is held in each of their beaks ; while in *b* the beaks are turned toward each other, and between them are two objects, just as in *c*, where the tails are better characterized.   *d* is composed of two pairs of superposed cocks, the lower of which are surrounded by border-lines.   In the figures *e*, *f*, and *g* the principle of the four spirals grouped around a lozenge comes into play.   In *f* two lateral cocks are added to the under spirals ; and underneath, two opponent cocks with conventionalized fishes in their beaks. Fig. *h* is identical with *d*.   An odder variation of the fish-cock is visible in *i*. There is here a spirally wound fish with a bipartite crest.   It terminates in a spiral tail, and has two cock feet placed as if in the act of walking ; *j* is a distorted cock in which the principle of misplacement is conspicuous.

The whole design of Fig. 4, Plate xxx, is built up of vertical and horizontal

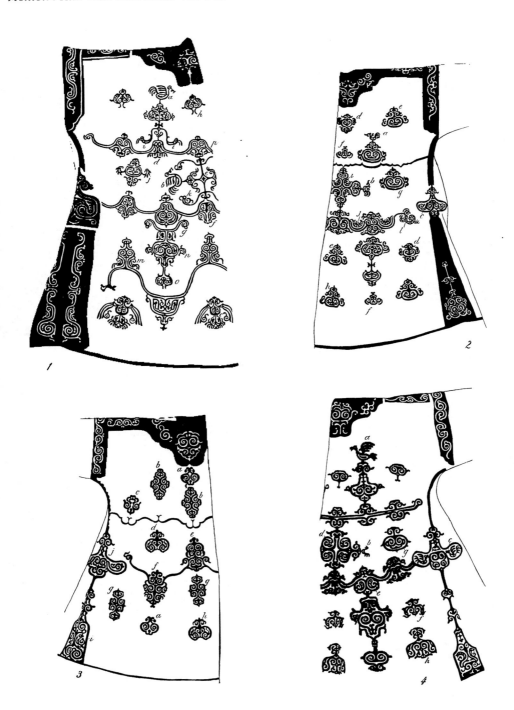

Decorative Art of the Amur Tribes.

rows.   On the top is perched a crowing cock (*a*), with open beak and a fish on its body.   Next to this in form come the cock placed sideways (*b*) and one flying (*c*); the latter surrounded by a line, its tail quadrilobate and its wing-feather trisulcate, whereas the reverse is the case in *a*.   Of fishes, we see four realistic ones coiled around four spirals (*d*) ; two appearing as continuations of spirals, and each enclosed in an oval (*e*) ; the cock-fishes with heads downward (*f*) ; and the fishes adjoining the spiral in *g*, consisting of one piece only.   As regards the design seen in *h*, it should be compared with *c* (Fig. 1) and *h* (Fig. 3) of this plate, and especially with *i* in Fig. 3, Plate XXIX.

AINU ORNAMENTATION. — We will now cast a brief glance at the ornaments of the Ainu.   This tribe still holds a rather exceptional position, due, on the one hand, to their isolation in the southern part of the Island of Saghalin, and, on the other hand, to their indolent, passive character.   Notwithstanding their resemblance to the neighboring Gilyak, many an invention and many an idea is met with which is wholly their own, and is not found in any other tribe.   Generally speaking, the subject of ornamentation among these people is a very intricate one, since three blended elements must be distinguished, — a special overwhelming Japanese influence ; loans from the neighboring Amur tribes ; and perhaps certain dregs of their artistic ideas, which are to be considered as almost wholly their own property.   There is no doubt that a great many figures and patterns might receive proper explanation by comparing them with the art of the Gilyak and the Gold.

Fig. 24 represents the coat of an Ainu chief from the east coast of Saghalin. It is of home make, and woven from nettle-fibres. The edges are adorned with dark blue, yellow, medium blue, and dark blue stripes of Japanese cotton, arranged

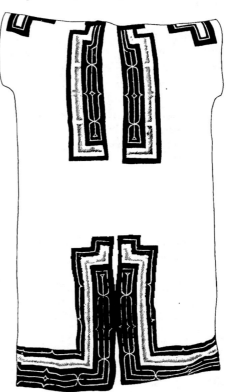

FIG. 24 ($\frac{7}{68}$).   Coat of Ainu Chief.

somewhat like a key pattern.   The dark blue stripes are broad along the inner side or slit of the garment, and narrow along the outer edge of the border.   The broad ones are covered with a design in embroidery, as are also the narrow ones in the under part of the coat.   The stitch used here is the so-called "couching-stitch."   The narrow band in the lower part shows what is called in our Goldian ornaments a continuous spur-line. This motive, only double, is employed likewise on the other borders, and so intertwined that a long-extended oval figure is produced. Through the middle of this figure is a red line forming a lozenge in the centre. The ornamentation across the bottom goes all the way around. This form of decoration is the typical style for all Ainu clothing. On the upper part of the back is a crest after Japanese fashion, showing a quadrifoil the leaves of which are cut out of bright red cloth, and edged with purple.

Fig. 1, Plate XXXI, shows a decorated attachment for a belt. Two such pieces are generally worn together, suspended from the side. A trapezoidal piece of whalebone is covered with dark red cloth, at both sides with a section of black cloth, and the upper and lower edges are set with blue glass beads fastened in clusters of three. The two ornaments sewed on with chain-stitching are applied in the same manner to both sides of the object. Both forms we have met with in the ornamentation of the Amur tribes. The same is applicable also to the decoration on Fig. 2, a bone implement for untying knots, which is adorned with a band-ornament showing, above and below, two knots especially marked by round incised hollows. Also here, as in the related objects of the Gilyak previously described, we see a connection between the ornament and the purpose of the object on which it occurs.

In Fig. 3 of the same plate is shown a knife-case inlaid with bone, obtained by Professor Bickmore from the Island of Yezo. On the middle longitudinal bone there are cross-hatched triangles. The other decorations are simple band-ornaments, the negative portions of which are indicated either by lozenges or by cross-hachures.

Fig. 4, Plate XXXI, represents a knife-case, from which the handle of the knife projects. It is likewise from Yezo, and was declared to be an old piece. The plant-ornament on the handle, the end of which is cut off slanting like a Gilyak knife, is incised, likewise the flower on the upper part of the case; but the group of leaves on the hatched part, like the leaves in the form of superposed semicircles, stand out in relief. These floral ornaments manifest, in both style and technique, an obvious Japanese influence.

Figs. 5–20, Plate XXXI, represent small wooden sticks (*ikuni* on Saghalin, *ikubashui* on Yezo) used in ceremonial drinking-bouts to lift the mustache and beard to prevent them from getting wet. The pieces represented in Figs. 5–9 were obtained by me on Saghalin. They are old family heirlooms, given away by their possessors only with reluctance. The following are explanations as to the carvings on them. The design on Fig. 5 is said to represent a human face wearing a pair of spectacles. Each of the two glasses is indicated by two concentric circles having a round hole in the middle. The connecting-piece between them is likewise cut through, forming a slit. The half-perforated oval projecting above the upper glass of the spectacles is supposed to be an eye, while below the spectacles two pairs of nostrils are represented in the form of pointed arches. The lowest larger hole, the point of which turns upward, indicates the outline of

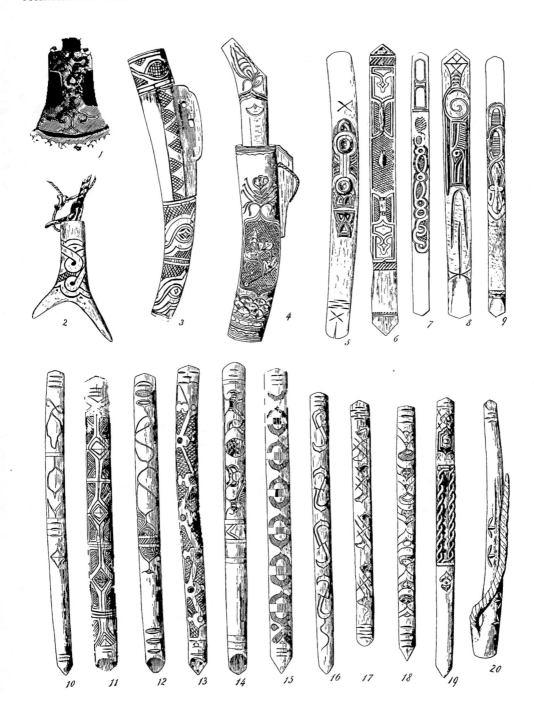

Decorative Art of the Amur Tribes.

the nose. On the specimen, Fig. 6, were three incised open-carved seals, one of which is unfortunately broken off : that in the middle is floating on the surface of the sea, which is symbolized by diagonally cross-hatched lines ; the other two animals are resting on shore, the beach being shown by parallel oblique lines on either side, enclosed in segments.   Fig. 7 was interpreted as the representation of a landscape.   All hatched parts signify mountains ; the hatchings themselves, grass and wood ; and the serpentine lines, valleys and roads.   From a merely ornamental point of view, an irregular interlacement of bands is here presented, the negative parts of which are taken up with parallel lines.   On the lower half of Fig. 8 is a netting-needle, above it the picture of a sturgeon.   Its head is represented in raised work in the form of a long rectangle.   The eye is in the middle, and the extended head with jaws is reproduced rather naturally in spite of the geometric treatment.   The body is symbolized by a spiral, adjoining which is the tail, — a rather realistic design with four rings and two crosses.   Fig. 9 portrays two sledges driving over the ice, one behind the other.   The back parts of the sledges rise over the stick in open-work carving.   In the centre there is a quadrifoil, the upper outlines of which are connected by means of a band with a sledge above it.   From the mere consideration of these five mustache-lifters it may be seen that the Ainu have a predilection for open-work sculpture, and use for representations the fauna of their surroundings and other objects familiar to them.   Moreover, it becomes clear that the forms are partially the same as with the Amur peoples, and that these very same forms are made to serve as the basis for a symbolical interpretation.   The explanations of these *ikuni* are of a purely personal character, being kept in the same family and handed down together with the ceremonial sticks ; so that under certain circumstances the same pattern might have different explanations in different families.   The pieces from Yezo (Figs. 10–20), in the collection of Mr. A. C. James, are given place here partly for comparison with those from Saghalin, partly as an incentive to further research regarding the peculiar ornamentation of these sticks, and in general of that of the Ainu.   There is certainly no specimen among those from Japan that shows so realistic and characteristic a mould as ours from Saghalin.   Particularly Figs. 11, 13, 15, 16, 17, and 18 show a strongly geometrical cast, owing to the continuous repetition of the same forms ; while others, like Figs. 10, 12, and 14, seem to tend towards realistic conventionalizations.   Chain-bands occur very frequently on drinking-sticks.   There are two such running side by side in open carving in Fig. 19.   On the stick, Fig. 20, from the lower left side proceeds a natural scion ; very interesting are four trefoils on it, the forms of which exactly agree with those on our Amur ornaments.

COLORING. — As already stated, a great number of the decorated specimens are very rich, and even extravagant, in the variety of their colors, which reaches its climax in the embroideries, since here the most beautiful dyed Chinese embroidery-silks are at the disposal of the artist. These silks abound in all imaginable tints and shades, whereas the selection of the pigments for painting purposes is, of course, not nearly so diverse. From the very fact that the silks as well as the coloring-matters are traded from the Chinese, we may infer that also in tasteful and artistic arrangement of colors this people has been the instructor of the Amur tribes. This statement is confirmed, moreover, by the fact derived from actual observation, that the more the natives are in contact with the Chinese, the nearer they dwell to a centre of Chinese culture, the more splendidly developed in beauty of color are their works ; while the farther one recedes from that centre, the poorer the color-sense seems to grow, and at last to vanish almost entirely. The choice of colors is not arbitrary, but subject to certain rules of taste, although no definite formulas can be deduced. It may be asserted that, throughout, to the symmetry of the pattern corresponds the symmetry of the colors. This symmetry, however, is not so strictly observed that symmetrical parts must be adorned in all cases with precisely identical colors: there should be different shades of the same ground-color, or even sometimes contrasting hues. There may also be the same set of colors in two symmetrical figures, but with a change in the arrangement. The brilliant colors occur mostly on shirt-embroideries with white background ; and the duller hues, in their various shades, on coats and other clothing. A better knowledge of Chinese art will no doubt throw more light also on this most attractive side of the Goldian works of art.

To illustrate the appearance and the effect of the colors, five paintings have been selected for reproduction here (Plates XXXII, XXXIII).

Fig. 1, Plate XXXII, represents the lateral continuation of the ornament shown in Fig. 4, Plate XXVI. I have already pointed out the peculiarity of the composition of this design ; here also is shown an entirely new motive not elsewhere observed : in the centre are two cocks (*a* and *b*), — *a* of red color ; *b*, one half red, the other half deep magenta, or the shade "American beauty," so called. Although the two cocks correspond to each other in their position, they are not constructed in symmetrical agreement, since the whole composition narrows off toward the side. The cock *a* is therefore in erect vertical, and *b* in reclining horizontal, attitude, the latter running out into a long body, with tail coiled into a spiral. The upper outline is here also an engrailed line, which causes an irregular lozenge to spring up in the middle of the body. The beaks of the two cocks *a* and *b* are curved downward and then recurved, and each holds a conventionalized bipartite yellow fish, grasping it at a point between head and body. Between the bodies of the fishes and cocks are two greenish cocks whose forms are assimilated to those of the adjoining cocks above. That below *a* is accordingly represented standing, and that belonging to *b* in a sitting position. In this way also the ends of the yellow fishes depend upon the forms of the cocks, and are influenced by them : the body of the fish *a* runs parallel to that of its cock, and terminates, like that one, in a broad plane ; whereas *b* tapers into a point at the tail, like the green cock belonging to it. The red cock *a* is joined by a yellow inverted cock, which sends off farther below a blue-colored fish. To the left of this one is a magenta

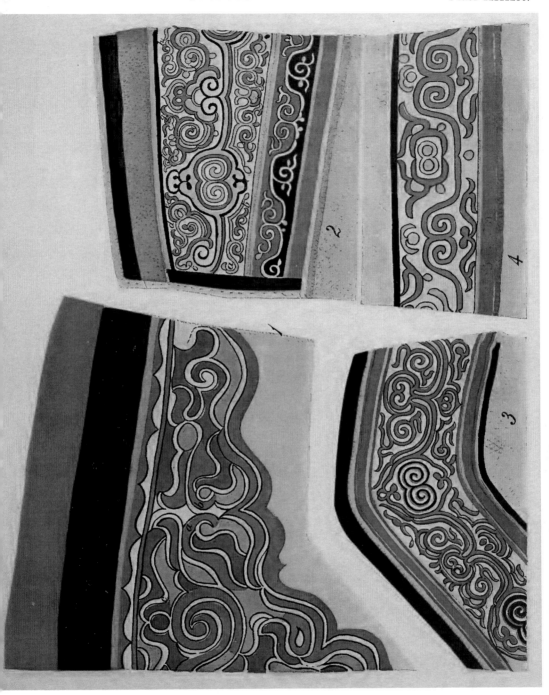

Decorative Art of the Amur Tribes.

cock-head, whose beak seizes a green fish. There are two fishes over the two cocks *a* and *b*. That above *a* has the head green, the body blue ; that above *b*, the head yellow, and the body blue. In this case, consequently, we can speak, at most, of a harmony, not of a symmetry, of colors. There is, for instance, the cock *c*, green as to neck and beak, while the rest of the body is yellow. On the other hand, the opposite cock is green, and only its upper tail-feather of a red hue.

Fig. 2, Plate XXXII, represents the painting on the upper edge of a boot made of elk-skin, the outside of which is tanned, while inside the hair is left on. The decoration is painted on a piece of salmon-skin which is sewed to the elk-skin. A comparison of this with the preceding specimen shows that fish-skin is a much better substance for painting, and gives the colors a brighter and more resplendent effect. An extraordinary feature of this ornament is, that parts of continuous geometrical arabesques, without stepping beyond the pale of their ornamental character, are shaped into fishes and cocks. Thus the spiral ornament at the lower edge starts with a bird-head (*a*) and terminates in a fish-head (*b*). Just so a merely ornamental line (*c*) is treated as cock-head and beak holding an inverted bipartite lavender-colored fish. Around the spirals are distributed a series of conventionalized cocks, all of which represent different variations of the same forms, that is, *d, e, f, g, h.* In *f* and *g* the feet of the cocks are fashioned as two-lobed leaves. To the pure spiral *i* corresponds the spiral *j*, the interior of which is formed like a cock. From the lavender-colored spiral line stand off conspicuously the red curved beak and the semicircular head-line. Here is demonstrated one of the reasons for the employment of contrasting colors to mark off distinctly one part of the body from another, and thus bring it into prominence. Another characteristic feature appears in the fact that the black tint serves to mark the wave-line terminating in scrolls, which helps to analyze the composition into its subdivisions : it affords, as it were, a frame for miniature pictures. To the yellow color is attributed merely a secondary significance : it serves as a filling for negative portions, mostly for narrow stripes. Of paramount importance is the color red, with which the essential parts are painted ; with it is interchanged, very happily and tastefully, a lavender color, which softens the glare of the red in a most agreeable way, lending a restful and harmonious effect to the whole composition.

Fig. 3, Plate XXXII, represents the upper front edge of a fish-skin garment which came originally, like the preceding specimen, from the Ussuri River. As regards the use of the colors in this ornamentation, first of all, it should be pointed out that a difference is made between spirals which serve exclusively for decorative purposes and those which claim, besides, a symbolical meaning. The former are painted with black China ink and surrounded with red lines, the latter with red color and black border-lines. For the representation of the cock, either red or blue, or both colors at the same time, are in use, while yellow is limited again to the filling of intervening spaces and stripes. In this way bipartite fishes, some of which occupy negative spaces, are better brought out.

Fig. 4, Plate xxxii, is a painting on the upper edge of a pair of leggings made of fish-skin. The picture is remarkable on account of its peculiar coloring, — a light red for the wave-line ending in scrolls, which effects the ornamental division; an exquisite magenta for the conventionalized cocks of the type B; and a light blue for the figures under the wave-line, which are composed of two united cocks, their heads being circles and running out into fish-tails; a greenish blue for circles, ovals, and united bird-beaks in the form of a crescent. The background is of a light buff hue; lemon-yellow is twice applied to the circular objects of the crescent-shaped cocks and for negative portions, twice for the heads of conventionalized fishes.

Plate xxxiii represents the back of a woman's dress of fish-skin. Part of the front edge of the same specimen was shown in Fig. 3 of the foregoing plate. The whole surface is covered with a magnificent painting. The decoration consists of three vertical rows, the two outer of which tally and are composed of three single figures each, while the middle series presents a coherent structure. The ornamental principle from which these have arisen is very simple: there is a pair of facing spirals in the middle, above and below them are two erect conventionalized bipartite fishes, and the whole is surrounded by a line corresponding to their forms. Whereas this figure remains constantly the same, the appendages on its sides, components of the cock-ornament, vary. Not to this formal change is due the special charm which this design offers, but rather to the harmonious variation of its colors, especially of red, blue, and black.

SOME GENERAL RESULTS. — If we cast a retrospective glance at the decorative art of the Amur tribes, we are struck most forcibly by the predominance of the cock and the fish, the manifold combinations in which these two motives appear, and the strange mingling of the two. These two inventions stamp the character of the whole ornamentation. If we ask for the reason, no other explanation can be found than that these particular animals have an extremely ornamental character because of the great permutations of their graceful motions, and thus lend themselves admirably to the spirit which strives after beauty of form. The reason, then, lies in their unquestionable availability for the ornamental. It is to their gracefulness and beauty of form that the cock and fish owe their popularity among artists, here as well as in the Chinese and Japanese pictorial arts. The part which the cock plays in the mythology and household economy of the Chinese is not so important as to justify so abundant a use of it in ornature. Since, besides, in the life of the Amur tribes it plays no part whatever, the mere artistic reason of its adaptability has decided its use. That such is exclusively the case is seen from all the various positions of fish and cock which are suggested solely by the tendency to create new and æsthetically effective forms. This strongly developed form-perception prevents the production of realistic representations, — which exist without doubt in embryo, and in early times existed perhaps to a much greater extent, —

Decorative Art of the Amur Tribes.

as shown in the designs of numerous animals, none of which have endured in their natural forms, but rather have deteriorated into a style of conventionality adapted to the cock and fish ornaments, as the musk-deer, the dragon, and so on. It would almost seem that other groups of animals gain favor and meet with approval, only so far as they are capable of conforming to the cock and fish pattern. In these last-mentioned figures we recognize at the same time stages, second in point of time, which probably arose after the development of the first-described ornaments.

If we now take into consideration the evolution of the cock and fish ornaments, we are impressed first by the fact that such differing and numerous stages of development are met with, frequently even in the same design; so that the development appears almost to be based on a juxtaposition in space rather than on a succession in time. In other words, the question arises, Are we correct in supposing a definite scale of gradation in the stages of development, from the cock and fish true to nature, down to the hardly recognizable conventional patterns? The whole series of forms does undeniably occur. These, however, should under no circumstances be regarded as of chronological sequence; for it is by no means true that the natural picture of the cock or fish is sunk in oblivion, and that the conventional form has exclusively taken its place. On the contrary, we see that the single phases of development are nothing more nor less than various forms of different kinds of adaptation to certain spaces or given geometrical forms, mostly spiral. This process of adaptation, constantly repeating itself in multitudinous ways, has created a large number of varieties, all co-existing side by side, like the varieties of a zoölogical species. One does not exclude the other, but each retains its separate existence, because art indulges in a wealth of forms, and requires an abundance of varieties for building up large ornamental compositions.

The strong inward impulse to create new forms is the primary underlying cause for the rise of the various degrees of conventionalization. Moreover, it is a further incentive to the simultaneous retention of all these manifold forms, a great number of which, without the influence of this law, would have perished. The form-character of this ornamentation had therefore a conservative effect, and is consequently responsible for its offspring. In spite of this form-character, however, conventionality is by no means a production of a purely rationalistic method of speculation. It should not be imagined that the creations of animal life continued to lose more and more of their original forms, and gradually shrunk into geometrical devices. On the contrary, the multifarious kinds of convention-alization have their final cause, last but not least, in a faithful observation of nature, especially in the ability to watch motions, so highly developed in the East Asiatic mind. The conception of a fish in the form of a spiral is based on a true observation of that animal in its natural state; it would never have been drawn in spiral form, never have clung to a spiral, without a foundation of fact. This very capacity of the fish for motion, together with the highly cultivated power of

the people to observe its motions, formed the reason for its adoption in ornamentation.   The same remark holds good for the cock.   Here we have, perhaps, not a
primitive form from which all others have genetically originated ; rather, a long
series of fundamental forms exists, based on the observation of the various
natural attitudes and motions of this ever-moving bird.   We have distinguished
a series of types ; we have found standing, reclining, perching, and perfectly erect
cocks, some with beaks turned downward, others with heads looking backward,
all types which exist side by side, without having developed one from another.
The conventionalizations proper have arisen only through the influence of the
fish-ornament on the cock-type.   This is the same process which was above
designated, in a more general style, as an assimilation to existing forms.   Thus
the cock, for instance, assumes a fish-body to get a spiral form more suitable for
the entire ornament ; or its tail is represented as a fish-tail, its pinion as a spiral.
Finally, forms are even found in which the whole cock is composed of geometrical constituents.   These have not been evolved from the form of the cock, but
they are the primary element, the material from which it is constructed.   This
ensues — and here we touch another important theoretical point regarding our
ornaments — from the diversity of function of the geometrical components.   The
spiral, for instance, may symbolically express all possible things.   It may serve
to indicate the cock's body, its pinion, its tail-feather.   It may even perform two
or more functions.   In Fig. 1, Plate XI, the large curve of a spiral is, first, a
geometrical element ; secondly, part of a wave-line serving to distinguish ornamental subdivisions ; thirdly, it forms the upper outline for the body of a fish
below, naturalistically drawn ; fourthly, it outlines the body of a cock, the other
parts of which are drawn above it.   It would be absurd to infer from this that
the spiral is the final result of the gradual conventionalization of such realistic
images : it is rather a given prius, — the origin of which is of no consequence
here,—which is employed for the symbolical expression of the most varied things,
since its forms are so convenient for this particular purpose.   Another example
is offered in the brace, signifying the cock-spur ; this symbol indicates also the
feet of the musk-deer (Fig. 5, Plate XIV), the feet of the dragon (Fig. 13a,
Plate XVI), and even the scales on the dragon's body (Fig. 5, Plate XII).   If a
conventionalized fish appears in place of the body of a cock or even of a musk-
deer, or if it even serves to indicate the horn of a dragon, no one, perhaps, will
conclude from this fact that the conventionalized fish has resulted from the cock
or deer body, but only that this particular form is used as the means to an end,
as an easy expedient for ornamental symbolism of the parts of the bodies of other
animals.

     From this proof proceeds another very important and far-reaching conclusion
as regards the triskeles.   This also is a given factum used as a foundation upon
which to build other ornaments.   The supposition that the triskeles has developed from the outlines of the cock does not prove true at all for the tribes of the
Amur.   In no case is the cock represented as a purely geometrical triskeles.   In

a few cases, as for instance on the spears and bows, where the cock has a triskeles-like appearance, it is always determined, first of all, by the added circle; secondly, the single arms of it are shaped in such a way that they admit of recognizing, in truth, the forms of the bird. We may clearly distinguish also, in these cases, beak, body, and feet. In the formation of the cocks and fishes the triskeles plays an active rôle for indicating single parts of the body, but not the whole creature, and so its functions are extensive. It sometimes interchanges with the spiral. It symbolizes the pinions of the cock, sometimes the fish-tail, sometimes the fish itself held in the cock's beak; in fact, even the horn of the musk-deer and the dragon's whiskers. As an independent element, having a definite meaning, however, the triskeles never occurs. It is rather a secondary expedient of multifarious significations, which, however, by no means appears as a resultant from the phases of the cock itself. This fact of the multiplicity of the functions of geometrical formations confirms again the thorough form-character of this decorative art, which sacrifices everything to the beauty of lines and forms.

The question may arise as to whether people like the Gold, who are able to produce such fine work, may justly be classified among primitive tribes. The Gold, at all events, are promising, and some time or other will undeniably advance to the rank of a civilized nation, like their ancestral relations the Niüchi and Manchu, but under more peaceable circumstances, relying on the cultivation of the soil, industry, and fine arts. There is no doubt but that they are chosen for their share in civilization, and that they will have a future, if only the Russian Government will continue to lend its assistance in improving the economic life-conditions of this intelligent tribe, which numbers so many good-natured and highly gifted individuals.

LIST OF PLATES.

[80]

PLATE VI. (See pp. 20, 21.)

Fig. 1. — Cock cut out of Paper. Tribe, Gold. Length, 14 cm. Cat. No. $\frac{70}{853}$ o.

Fig. 2. — Embroidered Cock. Tribe, Gold. $\frac{1}{2}$ nat. size. (Museum für Völkerkunde, Berlin. Cat. No. IA 1387 b.)

Fig. 3. — Chinese or Japanese (?) Weaving-Pattern. About $\frac{1}{2}$ nat. size. (Kunstgewerbe Museum, Berlin. Cat. No. 94,263.)

Fig. 4. — Birch-bark Hat. Tribe, Gold. (Museum für Völkerkunde, Berlin. Cat. No. IA 386.)

PLATE VII. (See pp. 21–23.)

Fig. 1. — Side of a Birch-bark Basket. Tribe, Gold. (Museum für Völkerkunde, Berlin. Cat. No. IA 1204.)

Figs. 2, 3. — Bear-Spears. Tribe, Gilyak. Length, 41.5 cm., 47 cm. Cat. Nos. $\frac{70}{1091}$, $\frac{70}{880}$.

Fig. 4. — Embroidery made of Reindeer-Hair. Tribe, Gilyak. (Museum für Völkerkunde, Berlin. Cat. No. IA 1824.)

Fig. 5. — Painting across Top of Elk-skin Boots. Tribe, Orochon (Ussuri). Depth, 14.5 cm. Cat. No. $\frac{70}{621}$.

Fig. 6. — Paper Pattern for embroidering Woman's Shoe. Tribe, Gold. Height, 11 cm. Cat. No. $\frac{70}{853}$ g.

Fig. 7. — Paper Pattern for embroidering Gloves. Tribe, Gold. Length, 33 cm. Cat. No. $\frac{70}{853}$ b.

PLATE VIII. (See pp. 23–25.)

Fig. 1. — Silk-embroidered Ornament for a Shirt. Tribe, Gold. Height, 17 cm. Cat. No. $\frac{70}{604}$ d.

Fig. 2. — Silk-embroidered Ornament for a Shirt. Tribe, Gold. Height, 16 cm. Cat. No. $\frac{70}{599}$ b.

Fig. 3. — Chinese Weaving-Pattern, 19th Century. $\frac{2}{5}$ nat. size. (Kunstgewerbe Museum, Berlin. Cat. No. 83,1758.)

Figs. 4, 5. — Oriental Weaving-Patterns, 17th–18th Centuries (?). $\frac{1}{2}$ nat. size. (Kunstgewerbe Museum, Berlin. Cat. Nos. 76,1090, 76,1085.)

PLATE IX. (See pp. 25–27.)

Fig. 1. — Birch-bark Pattern for embroidering a Pocket. Tribe, Gilyak. $\frac{1}{2}$ nat. size. (Museum für Völkerkunde, Berlin. Cat. No. IA 1781 b.)

Fig. 2. — Birch-bark Hat. Tribe, Gilyak. Diameter, 38 cm. Cat. No. $\frac{70}{883}$.

Fig. 3. — Painting on Boy's Fish-skin Leggings. Tribe, Gold. Height of painting, 14 cm. Cat. No. $\frac{70}{635}$ a.

Fig. 4. — Birch-bark Embroidery-Pattern for an Ear-Lappet. Tribe, Gilyak. Height, 12 cm. Cat. No. $\frac{70}{659}$ q.

Fig. 5. — Woman's Boot. Tribe, Gilyak. Length of foot, 27 cm. Cat. No. $\frac{70}{1195}$.

Fig. 6. — Knife-Case. Tribe, Gilyak. Length, 25 cm. Cat. No. $\frac{70}{1183}$.

Fig. 7. — Ancient Japanese Weaving-Pattern. $\frac{1}{2}$ nat. size. (Kunstgewerbe Museum, Berlin. Cat. No. 84,825.)

PLATE X. (See pp. 27–29.)

Fig. 1. — Side of Birch-bark Basket. Tribe, Gold. $\frac{1}{3}$ nat. size. (Museum für Völkerkunde, Berlin. Cat. No. IA 1217.)

Fig. 2. — Painting across Top of Fish-skin Leggings. Tribe, Gold. Width, 10 cm. Cat. No. $\frac{70}{634}$ a.

Fig. 3. — Embroidery on Silk Leggings. Tribe, Gold. Width, 23 cm. Cat. No. $\frac{70}{583}$.

Fig. 4. — Embroidery on a Velvet Girdle. Tribe, Gold. Width, 8 cm. Cat. No. $\frac{70}{585}$.

Fig. 5. — Side of a Birch-bark Box. Tribe, Gold. $\frac{1}{2}$ nat. size. (Museum für Völkerkunde, Berlin. Cat. No. IA 1724.)

Fig. 6. — Side of a Birch-bark Box. Tribe, Gold. $\frac{1}{4}$ nat. size. (Museum für Völkerkunde, Berlin. Cat. No. IA 1211.)

PLATE XI. (See pp. 29–33.)

Figs. 1, 1a. — Long and Short Sides of a Birch-bark Basket. Tribe, Gold. Length, 11.5 cm. Cat. No. $\frac{70}{669}$.

Figs. 2, 2a. — Design on the Rim of the Cover of a Tobacco-Box. Tribe, Gold. Height, 4.7 cm. Cat. No. $\frac{70}{775}$.

Fig. 3. — Paper Pattern for embroidering Bag for Strike-a-light. Tribe, Gold.

Fig. 4. — Paper Pattern for decorating Newspaper-Holder. Tribe, Gold. Length of base, 48 cm. Cat. No. $\frac{70}{654}$.

Fig. 5. — Embroidery-Pattern for a Shirt-Pocket. Tribe, Gold. Length of base, 22 cm. Cat. No. $\frac{70}{606}$ d.

Fig. 6. — Embroidered Collar. Tribe, Gold. Diameter, 38.5 cm. Cat. No. $\frac{70}{608}$ a.

Fig. 7. — Embroidery on a Wrister. Tribe, Gold. Length, 21 cm. Cat. No. $\frac{70}{615}$ a.

PLATE XII. (See pp. 33, 34, 37, 38.)

Fig. 1. — Design on the Inside of a Birch-bark Basket. Tribe, Gold. Height, 17 cm. (Museum für Völkerkunde, Berlin. Cat. No. IA 1571.)

Fig. 2. — Embroidered Border. Tribe, Gold. Width, 6 cm. Cat. No. $\frac{70}{607}$ n.

Fig. 3. — Embroidered Border. Tribe, Gold. Width, 7 cm. Cat. No. $\frac{70}{604}$ b.

Fig. 4. — Paper Pattern. Tribe, Gold. Width, 20 cm. Cat. No. $\frac{70}{653}$ a.

Fig. 5. — Paper Pattern. Tribe, Gold. Height, 12.5 cm. Cat. No. $\frac{70}{653}$ e.

Fig. 6. — Painting on Leggings. Tribe, Gold. Width, 10.5 cm. Cat. No. $\frac{70}{632}$ a.

PLATE XIII. (See pp. 38–41.)

Fig. 1. — Cover of a Tobacco-Box. Tribe, Gold. Length, 20 cm. (Museum für Völkerkunde, Berlin. Cat. No. IA 1786.)

Fig. 2. — Half of Cover of Tobacco-Box. Tribe, Gold. Extreme length, 50.5 cm. Cat. No. $\frac{70}{674}$.

Fig. 3. — Ancient Chinese Weaving-Pattern. $\frac{1}{2}$ nat. size. (Kunstgewerbe Museum, Berlin. Cat. No. 69,165.)

Fig. 4. — Ancient Chinese Weaving-Pattern. $\frac{1}{2}$ nat. size. (Kunstgewerbe Museum, Berlin. Cat. No. 81,690.)

PLATE XIV. (See pp. 41–43.)

Fig. 1. — Japanese Weaving-Pattern, 18th and 19th centuries. (Kunstgewerbe Museum, Berlin. Cat. No. 79,46.)

Fig. 2. — Paper Pattern. Tribe, Gold. Height, 6.5 cm. Cat. No. $\frac{70}{653}$ j.

Fig. 3. — Paper Pattern. Tribe, Gold. Height, 17 cm. Cat. No. $\frac{70}{653}$ i.

Fig. 4. — Paper Pattern. Tribe, Gold. Width, 36 cm. Cat. No. $\frac{70}{653}$ h.
Fig. 5. — Paper Pattern. Tribe, Gold. Width, 33 cm. Cat. No. $\frac{70}{653}$ d.
Fig. 6. — Paper Pattern. Tribe, Gold. Width, 33 cm. Cat. No. $\frac{70}{653}$ c.

PLATE XV. (See pp. 47, 48.)

Fig. 1. — Carved Dish. Tribe, Gold. Extreme length, 40.5 cm. (Museum für Völkerkunde, Berlin. Cat. No. IA 1231.)
Fig. 2. — Box-Cover. Tribe, Gold. Height, 15 cm. (Museum für Völkerkunde, Berlin. Cat. No. IA 1269.)
Fig. 3. — Embroidery on a Tobacco-Pouch. Tribe, Gold. Length, 23 cm. Cat. No. $\frac{70}{643}$.
Fig. 4. — Embroidered Border for Wristband. Tribe, Gold. Height, 10.3 cm. Cat. No. $\frac{70}{817}$ a.
Fig. 5. — Embroidered Girdle. Tribe, Gold. Length, 64 cm. Cat. No. $\frac{70}{584}$.
Fig. 6. — Paper Pattern. Tribe, Gold. Width, 19 cm. Cat. No $\frac{70}{653}$ n.

PLATE XVI. (See pp. 48–51.)

Fig. 1. — Embroidered Collar and Front of Woman's Coat. Tribe, Gold. Width of pattern, 6 cm. Cat. No. $\frac{70}{598}$.
Figs. 2–9. — Various Forms of Japanese Tomoye. (From a Japanese Book.) Fig. 8, bashō-mitsu-domoye.
Figs. 10, 11. — Arabesque Rings. (From a Japanese Book, the same as above.) Fig. 10, Karakusawa; Fig. 11, tsuruwarabi.
Figs. 12, 12a, 12b. — Bow: a, Design on the back; b, Design on the inside. Tribe, Gold. Bow, $\frac{1}{8}$ nat. size; designs, $\frac{2}{7}$ nat. size. (Museum für Völkerkunde, Berlin. Cat. No. IA 399.)
Figs. 13, 13a, 13b. — Bow: a, Design on the back; b, Design on the inside. Tribe, Gold. Bow, $\frac{1}{8}$ nat. size; designs, about $\frac{1}{4}$ nat. size. (Museum für Völkerkunde, Berlin. Cat. No. IA 1401.)
Figs. 14, 14a, 14b. — Bow: a, Design on the back; b, Design on the inside. Tribe, Gold. Bow, $\frac{1}{8}$ nat. size; designs, $\frac{2}{7}$ nat. size. (Museum für Völkerkunde, Berlin. Cat. No. IA 1400.)

PLATE XVII. (See pp. 51, 52.)

Fig. 1. — Ancient Japanese Weaving-Pattern. $\frac{1}{4}$ nat. size. (Kunstgewerbe Museum, Berlin. Cat. No. 75,922.)
Fig. 2 — Chinese Weaving-Pattern. $\frac{2}{5}$ nat. size. (Kunstgewerbe Museum, Berlin. Cat. No. 79,2062 a.)

PLATE XVIII. (See p. 51.)

Fig. 1. — Chinese Weaving-Pattern. $\frac{3}{7}$ nat. size. (Kunstgewerbe Museum, Berlin. Cat. No. 85,1741.)
Fig. 2. — Chinese Weaving-Pattern. $\frac{3}{7}$ nat. size. (Kunstgewerbe Museum, Berlin. Cat. No. 83,1742.)

PLATE XIX. (See pp. 52–54.)

Fig. 1. — Birch-bark Basket. Tungus (Amgun). Diameter, 24 cm. Cat. No. $\frac{70}{1234}$.
Figs. 2, 2a. — Side and Cover of Birch-bark Box. Tribe, Gold. Length, 17.5 cm. Cat. No. $\frac{70}{865}$.

Figs. 3, 3a.— Front and Back of Birch-bark Basket. Tribe, Gold. Depth, 20 cm. Cat. No. 70/870.

Fig. 4.— Design on a Birch-bark Tray. Tribe, Gilyak. Length of dish, 24.5 cm. Cat. No. 70/876.

PLATE XX.   (See pp. 54, 55).

Fig. 1.— Birch-bark Basket. Tribe, Gold. Length, 34 cm. (Museum für Völkerkunde, Berlin. Cat. No. IA 1207.)

Fig. 2.— Birch-bark Basket. Tribe, Gold. Height, 11 cm. Cat. No. 70/867 a.

Fig. 3.— Birch-bark Basket. Tribe, Gold. Height, 13 cm. Cat. No. 70/863.

Fig. 4.— Birch-bark Basket for holding Spoons and Chopsticks. Tribe, Gold. Height, 26 cm. Cat. No. 70/866.

Figs. 5, 5a.— Birch-bark Basket. Tribe, Gold. Width at base, 15 cm. (Museum für Völkerkunde, Berlin. Cat. No. IA 1572.)

PLATE XXI.   (See pp. 55, 56.)

Figs. 1, 1a.— Birch-bark Basket. Tribe, Gold. Length of base, 15.5 cm. and 14 cm. respectively. (Museum für Völkerkunde, Berlin. Cat. No. IA 1572.)

Fig. 2.— Birch-bark Basket. Tribe, Gold. Height, 21 cm. Cat. No. 70/868.

Figs. 3, 3a.— Birch-bark Basket and Cover. Tribe, Tungus (Ussuri). Height of basket, 19.4 cm.; diameter of cover, 34 cm. Cat. No. 70/948.

PLATE XXII.   (See pp. 56, 57.)   Embroideries.

Fig. 1.— Tribe, Gold. Width, 11 cm. Cat. No. 70/605 d.
Fig. 2.— Tribe, Gold. Width of embroidered portion, 11 cm. Cat. No. 70/605 a.
Fig. 2a.— Tribe, Gold. Width 11.5 cm. Cat. No. 70/605 b.
Fig. 3.— Tribe, Gold. Width, 19.5. cm. Cat. No. 70/612.
Fig. 4.— Tribe, Gold. Width, 4 cm. Cat. No. 70/600 c.
Fig. 4a.— Tribe, Gold. Width, 4.5 cm. Cat. No. 70/600 d.
Fig. 5.— Tribe, Gold. Width, 4.5 cm. Cat. No. 70/599 e.
Fig. 6.— Tribe, Gold. Length of side, 15 cm. Cat. No. 70/602 c.
Fig. 6a.— Tribe, Gold. Width, 13 cm. Cat. No. 70/595 d.
Fig. 6b.— Tribe, Gold. Width of embroidered portion, 11 cm. Cat. No. 70/595 a.

PLATE XXIII.   (See pp. 57–59.)   Embroidery-Patterns.

Fig. 1.— Tribe, Gold. Width, 5.5 cm. Cat. No. 70/602 d.
Fig. 2.— Tribe, Gold. Width, 5 cm. Cat. No. 70/604 c.
Fig. 3.— Tribe, Gold. Width, 7 cm. Cat. No. 70/602 a.
Fig. 4.— Tribe, Gold. Diameter, 34 cm. Cat. No. 70/603 a.
Fig. 4a.— Tribe, Gold. Width, 10 cm. Cat. No. 70/603 d.
Fig. 5.— Tribe, Gold. Width, 9 cm. Cat. No. 70/607 e.
Fig. 5a.— Tribe, Gold. Width, 11 cm. Cat. No. 70/607 j.
Fig. 5b.— Tribe, Gold. Height, 17 cm. Cat. No. 70/607 l.

PLATE XXIV.   (See pp. 59, 60.)

Fig. 1.— Embroidery. Tribe, Gold. Height, 19.7 cm. Cat. No. 70/606 e.
Fig. 2.— Embroidery. Tribe, Gold. Height, 16 cm. (Museum für Völkerkunde, Berlin. Cat. No. IA 1759.)

Fig. 3. — Japanese Weaving-Pattern. ½ nat. size. (Kunstgewerbe Museum, Berlin. Cat. No.
    81,784.)
Fig. 4. — Embroidery. Tribe, Gold. Height, 29 cm. Cat. No. $\frac{70}{610}$ a.
Fig. 5. — Embroidery. Tribe, Gold. Length, 21 cm. Cat. No. $\frac{70}{611}$ b.
Fig. 6. — Embroidery. Tribe, Gold. Length, 20.5 cm. Cat. No. $\frac{70}{611}$ d.

PLATE XXV. (See pp. 60, 61.) Embroideries.

Fig. 1. — Tribe, Gold. Height, 19.5 cm. Cat. No. $\frac{70}{640}$.
Fig. 2. — Tribe, Gold. Height, 18 cm. Cat. No. $\frac{70}{618}$ a.
Fig. 3. — Tribe, Gold. Length of base, 17 cm. Cat. No. $\frac{70}{601}$ h.
Fig. 4. — Tribe, Gold. Width, 5 cm. Cat. No. $\frac{70}{601}$ c.
Fig. 5. — Tribe, Gold. Diameter, 20.5 cm. Cat. No. $\frac{70}{601}$ a.
Fig. 6. — Tribe, Gold. Width, 3.6 cm. Cat. No. $\frac{70}{601}$ b.
Fig. 7. — Tribe, Gold. Width, 10 cm. Cat. No. $\frac{70}{608}$ b.
Fig. 8. — Tribe, Gold. Width, 9 cm. Cat. No. $\frac{70}{608}$ c.
Fig. 9. — Tribe, Gold. Height, 9 cm. Cat. No. $\frac{70}{601}$ j.
Fig. 10. — Tribe, Gold. Width, 9 cm. Cat. No. $\frac{70}{601}$ i.
Fig. 11. — Tribe, Gold. Width, 3.5 cm. Cat. No. $\frac{70}{601}$ e

PLATE XXVI. (See pp. 61–64.)

Figs. 1, 1a. — Embroidered Mitten. Tribe, Gold. Cat. No. $\frac{70}{626}$.
Fig. 2. — Embroidered Mitten. Tribe, Gold. Cat. No. $\frac{70}{619}$.
Fig. 3. — Painted Elk-skin Garment. Tribe, Tungus (Ussuri). Cat. No. $\frac{70}{581}$.
Fig. 4. — Painting on an Elk-skin Legging. Tribe, Orochon (Ussuri). Height, 33 cm. Cat.
    No. $\frac{70}{533}$.
Fig. 5. — Embroidery on Lower Part of Back of a Garment. Tribe, Gold. Height of tree
    design, 35 cm. Cat. No. $\frac{70}{580}$.
Fig. 6. — Embroidered Shirt. Tribe, Gold. Cat. No. $\frac{70}{587}$.

PLATE XXVII. (See pp. 64, 65.)

Fig. 1. — Embroidered Shirt. Tribe, Gold. Cat. No. $\frac{70}{589}$.
Fig. 2. — Embroidered Silk Dress. Tribe, Gold. Length, 92 cm. Cat. No. $\frac{70}{582}$.
Fig. 3. — Embroidered Coat. Tribe, Gold Length, 110 cm. Cat. No. $\frac{70}{595}$.
Fig. 4. — Embroidered Dress. Tribe, Gold. Length, 86 cm. Cat. No. $\frac{70}{596}$.

PLATE XXVIII. (See pp. 65–67.)

Fig. 1. — Woman's Silk-embroidered Coat. Tribe, Gold. Cat. No. $\frac{70}{581}$.
Fig. 2. — Tobacco-Pouch. Tribe, Tungus (Amgun River). Length, 28 cm. Cat. No. $\frac{70}{1088}$.
Fig. 3. — Design on Fish-skin Apron. Tribe, Tungus (Amgun River). Height, 48 cm. Cat.
    No. $\frac{70}{1184}$.
Fig. 4. — Hunter's Cap. Tribe, Gold. Cat. No. $\frac{70}{642}$.

PLATE XXIX. (See pp. 67–69.) Fish-skin Coats.

Fig. 1. — Tribe, Tungus (Ussuri River). Cat. No. $\frac{70}{581}$.
Fig. 2. — Tribe, Gold. Cat. No. $\frac{70}{623}$.
Fig. 3. — Tribe, Gold. Cat. No. $\frac{70}{624}$.
Fig. 4. — Tribe, Gold. Cat. No. $\frac{70}{626}$.

PLATE XXX. (See pp. 69–71.) Fish-skin Coats.

Fig. 1. — Cat. No. $\frac{70}{33}$.
Fig. 2. — Tribe, Gold. Cat. No. $\frac{70}{627}$.
Fig. 3. — Tribe, Gold. Cat. No. $\frac{70}{628}$.
Fig. 4. — Tribe, Gold. Cat. No. $\frac{70}{625}$.

PLATE XXXI. (See pp. 72, 73)

Fig. 1. — Girdle Attachment. Tribe, Ainu (Saghalin). Cat. No. $\frac{70}{974}$.
Fig. 2. — Bone Implement for Untying Knots. Tribe, Ainu (Saghalin). Length, 10 cm. Cat. No. $\frac{70}{973}$.
Fig. 3. — Carved Bone Knife-Case. Tribe, Ainu (Yezo). Cat. No. $\frac{1}{1020}$.
Fig. 4. — Dagger in Case. Tribe, Ainu (Yezo). Cat. No. $\frac{70}{193}$.
Figs. 5–9. — Ceremonial Drinking-Sticks. Tribe, Ainu (Saghalin). Cat. Nos. $\frac{70}{1268}$, $\frac{70}{1270}$, $\frac{70}{1288}$, $\frac{70}{1289}$, $\frac{70}{1287}$.
Figs. 10–20. — Ceremonial Drinking-Sticks. Tribe, Ainu (Yezo). Cat. Nos. $\frac{70}{28}$, $\frac{70}{26}$, $\frac{70}{32}$, $\frac{70}{28}$, $\frac{70}{28}$, $\frac{70}{33}$, $\frac{70}{30}$, $\frac{70}{31}$, $\frac{70}{27}$, $\frac{70}{24}$, $\frac{70}{23}$.

PLATE XXXII. (See pp. 74–76.)

Fig. 1. — Painting on an Elk-skin Legging. Tribe, Gold. Cat. No. $\frac{70}{638}$ a.
Fig. 2. — Embroidery on Elk-skin Boot. Tribe, Orochon (Ussuri). Cat. No. $\frac{70}{552}$ a.
Fig. 3. — Painting on Collar of Fish-skin Garment. Tribe, Orochon (Ussuri). Width, 9 cm. Cat. No. $\frac{70}{550}$.
Fig. 4. — Painting across Top of Fish-skin Leggings. Tribe, Gold. Cat. No. $\frac{70}{633}$ a.

PLATE XXXIII. (See p. 76.)

Painting on Back of a Fish-skin Garment. Tribe, Orochon (Ussuri). Full length, 106 cm. Cat. No. $\frac{70}{550}$.

American Museum of Natural History.

The publications of the American Museum of Natural History consist of the 'Bulletin,' in octavo, of which one volume, consisting of about 400 pages, and about 25 plates, with numerous text figures, is published annually; and the 'Memoirs,' in quarto, published in parts at irregular intervals. Also an 'Ethnographical Album,' issued in parts, and the 'American Museum Journal.'

MEMOIRS.

Each Part of the 'Memoirs' forms a separate and complete monograph, with numerous plates.

Vol. I·(not yet completed).

PART I.—Republication of Descriptions of Lower Carboniferous Crinoidea from the Hall Collection now in the American Museum of Natural History, with Illustrations of the Original Type Specimens not heretofore Figured. By R. P. Whitfield. Pp. 1–37, pll. i–iii, and 14 text cuts. September 15, 1893. Price, $2.00.

PART II.—Republication of Descriptions of Fossils from the Hall Collection in the American Museum of Natural History, from the report of Progress for 1861 of the Geological Survey of Wisconsin, by James Hall, with Illustrations from the Original Type Specimens not heretofore Figured. By R. P. Whitfield. Pp. 39–74, pll. iv–xii. August 10, 1895.. Price, $2.00.

PART III.—The Extinct Rhinoceroses. By Henry Fairfield Osborn. Part I. Pp. 75–164, pll. xiia–xx, and 49 text cuts. April 22, 1898. Price, $4.20.

PART IV.—A Complete Mosasaur Skeleton. By Henry Fairfield Osborn. Pp. 165–188, pll. xxi–xxiii, and 15 text figures. October 25, 1899.

PART V.—A Skeleton of Diplodocus. By Henry Fairfield Osborn. Pp. 189–214, pll. xxiv–xxviii, and 15 text figures. October 25, 1899. Price of Parts IV and V, issued under one cover, $2.00.

PART VI.— Monograph of the Sesiidæ of America, North of Mexico. By William Beutenmüller. Pp. 215–352, pll. xxix–xxxvi, and 24 text cuts March, 1901. Price, $5.00.

PART VII.—Fossil Mammals of the Tertiary of Northeastern Colorado. By W. D. Matthew. Pp. 353–446, pll. xxxvii–xxxix, and 34 text cuts. Price, $2.00.

Vol. II. Anthropology.

Jesup North Pacific Expedition.

PART I.—Facial Paintings of the Indians of Northern British Columbia. By Franz Boas. Pp. 1–24, pll. i–vi. June 16, 1898. Price, $2.00.

PART II.—The Mythology of the Bella Coola Indians. By Franz Boas. Pp. 25–127, pll. vii–xii. November, 1898. Price, $2.00.

PART III.—The Archæology of Lytton, British Columbia. By Harlan I. Smith. Pp. 129–161, pl. xiii, and 117 text figures. May, 1899. Price, $2.00.

PART IV.—The Thompson Indians of British Columbia. By James Teit. Edited by Franz Boas. Pp. 163–392, pll. xiv–xx, and 198 text figures. April, 1900. Price, $5.00.

(Continued on 3d page of cover.)

(*Continued from 4th page of cover.*)

Vol. II. Anthropology (*continued*).
Jesup North Pacific Expedition.

PART V.—Basketry Designs of the Salish Indians. By Livingston Farrand. Pp. 393–399, pll. xxi–xxiii, and 15 text figures. April, 1900. Price, 75 cts.

PART VI.—Archæology of the Thompson River Region. By Harlan I. Smith. Pp. 401–442, pll. xxiv–xxvi, and 51 text figures. June, 1900. Price, $2.00.

Vol. III. Anthropology (not yet completed).

PART I.—Symbolism of the Huichol Indians. By Carl Lumholtz. Pp. 1–228, pll. i–iv, and 291 text figures. May, 1900. Price, $5.00.

Vol. IV. Anthropology (not yet completed).
Jesup North Pacific Expedition.

PART I.—Traditions of the Chilcotin Indians. By Livingston Farrand. Pp. 1–54. June, 1900. Price, $1.50.

PART II.—Cairns of British Columbia and Washington. By Harlan I. Smith and Gerard Fowke. Pp. 55–75, pll. i–v. January, 1901. Price, $1.00.

ETHNOGRAPHICAL ALBUM.
Jesup North Pacific Expedition.

Ethnographical Album of the North Pacific Coasts of America and Asia. Part I, pp. 1–5, pll. 1–28. August, 1900. Sold by subscription, price, $6.00.

BULLETIN.

The matter in the 'Bulletin' consists of about twenty-four articles per volume, which relate about equally to Geology, Palæontology, Mammalogy, Ornithology, Entomology, and (in the recent volumes) Anthropology, except Vol. XI, which is restricted to a 'Catalogue of the Types and Figured Specimens in the Palæontological Collection of the Geological Department.'

Volume	I, 1881–86	Price, $5.00	Volume	IX, 1897	Price, $4.75
"	II, 1887–90	" 4.75	"	X, 1898	" 4.75
"	III, 1890–91	" 4.00	"	XI, Part I, 1898	" 1.25
"	IV, 1892	" 4.00	"	" " II, 1899	" 2.00
"	V, 1893	" 4.00	"	" " III, 1900	" 2.00
"	VI, 1894	" 4.00	"	" " IV, 1901	" 1.75
"	VII, 1895	" 4.00	"	XII, 1899	" 4.00
"	VIII, 1896	" 4.00	"	XIII, 1900	" 4.00

AMERICAN MUSEUM JOURNAL.

The 'Journal' is a popular record of the progress of the American Museum of Natural History, issued monthly, from October to May, inclusive,— eight numbers a year. Price, 75 cents a year, 10 cents a number.

For sale by G. P. PUTNAM'S SONS, New York and London ; J. B. BAILLIÈRE ET FILS, Paris ; R. FRIEDLÄNDER & SOHN, Berlin ; and at the Museum.

AMERICAN
ANTHROPOLOGIST

NEW SERIES

ORGAN OF THE ANTHROPOLOGICAL AND ETHNOLOGICAL SOCIETIES
OF AMERICA

EDITORIAL BOARD

VOLUME 4

NEW YORK
G. P. PUTNAM'S SONS
1902

series. The text and tables fill thirty pages of printed matter. Each of the 63 plates contains from three to many individuals, so that it is difficult to find where the author has neglected or omitted aught.

O. T. Mason.

The Decorative Art of the Amur Tribes. By Berthold Laufer. (Publications of the Jesup North Pacific Expedition: Memoirs of the American Museum of Natural History, vol. VII, number 1.) New York: 1902. 86 pp., 33 plates, 24 figures, 4°.

This monograph, like all the memoirs of the Jesup North Pacific Expedition, consists of the presentation of entirely new material. Sumptuously illustrated with 250 drawings, it deals with the decorative art, which is practically all the art, of the Gold, Gilyak, Orochon, and other tribes of the Amur region, including the Ainu. More articles of the Gold are described than all the other tribes together. Various arts are represented — carving in relief, ornamental painting, cutting of patterns in birch-bark and paper, and especially embroidering. A great variety of decorated objects are treated of, such as eye-protectors, mittens, spears, baskets, coats, and spoons.

Dr Laufer finds that there has been a strong Chinese influence on the art of the Amur country. Nothing, however, is actually known as to the history of the art-relations of the two regions. Dr Laufer's attitude on this matter is very conservative. He concludes that the art of the Amur tribes is old and deeply rooted, though its basis undeniably rests in China. He holds that the art is not an importation *en masse* from China, but must have had for its *conditio sine qua non* a congeniality in the minds of the two peoples; and that probably Chinese art was gradually absorbed and assimilated by the Amur tribes much as classic art was by the Europeans of the Renaissance.

The bulk of the book consists of a reproduction in illustrations of a large number of specimens of this art, and of an analysis in the text of the ornamental forms so shown. This analysis is carried out with great detail and much accuracy; it is so undeniably thorough as to make tedious reading to any one not specifically interested in problems of ornamentation. This care and thoroughness of analysis, however, give the book its value, for in the interpretation of decorative forms, superficial fancy has such an appalling opportunity that it is the great danger of study of this kind, and the condemning fault of much that has been published. Dr Laufer's analysis, in addition to being marked by caution and good sense, has the inestimable advantage of being founded on that of the natives.

The art of the Amur tribes is essentially ornamental. Its character is, to use the author's expression, formative, not realistic. Chinese art is largely emblematic or symbolic, but of this quality little has been adopted by the Gold. Chinese symbolic patterns are imitated without their symbolism being known. The decorative nature of the art is shown most decisively by the very extensive use of the cock, an animal that plays no part whatever in the mythology or daily life of any of the tribes of the region, and which by some of them has never even been seen; their knowledge of it, and their use of it in art, are due to Chinese influence.

The variety of degrees of conventionalization with which the cock and the fish, the two most important objects of representation in Amur art, are employed, and the way in which these decorative motives are used and abused and adapted to the purposes of decorative art expression, are shown very fully and convincingly in the course of the volume. The difference between the several conventionalizations of the same animal occurring on one object is sometimes very remarkable, and must be set down as one of the salient characteristics of the art. Such fully gradated series of cocks and fishes, as are illustrated, for instance, in figure 1 of plate xxx,—from those that are simply but quite effectively realistic, through others which the author's careful analysis and feeling for the spirit of the art make visible, to those forms, finally, where even his explanations end in a declaration of *non possumus*—are very unusual in primitive art. The cause for these various degrees of conventionalization the author does not find to be a gradual and progressive crystallization of originally realistic non-decorative designs. An influence of technique or material seems out of the question because the different conventionalizations are sometimes found together on the same object; and the author does not even consider this possibility. He wisely attributes this variety of degrees of conventionalization to the creative decorative spirit, or, as he calls it, the "inward impulse to create new [ornamental] forms."

His explanation of the Amur preference for the cock and the fish as decorative motives is at bottom the same: the cock and the fish are peculiarly available and adaptable ornamentally. "These particular animals have an extremely ornamental character because of the great permutations of their graceful motions, and thus lend 'themselves admirably to the spirit which strives after beauty of form." ("Form," here as elsewhere in the volume, is to be taken as equivalent to "decorative form.")

This explanation, however, does not seem to be sufficient. Another causal factor is required: the nature and spirit of Chinese and

Amur art.   The cock and the fish unquestionably were extremely available to this art; but it is too much to say absolutely that their availability is the cause of their prominence in it, for they were, *per se*, equally
available to European decorative art, which did not take them up and
use them to any extent.   An analogous case is the fleur-de-lys.   Finding its origin (partial if not ultimate) in heraldic symbolism, its wide
ornamental employment in European civilization today is due not so
much to any significance as to inherent ornamental availability; but to
allege this as the sole cause of its use would be obviously insufficient,
for were the character of European art other than it happens to be, the
fleur-de-lys would not have been so adaptable and available an ornament.   The triskeles and the rosette both have great ornamental possi-
bilities, yet the art of one culture uses only one of them and the art of
another only the other to any considerable extent.

A point of general bearing that is brought out in the conclusion, is
the essential and necessary connection between conventionalized ornamentation on the one hand, and the forms of nature on the other.   A
fish " would never have been drawn in spiral form, would never have
clung to a spiral, without a foundation of fact."   This clinging of the
most purely decorative (" formal ") arts to a certain amount of realism,
is an unexplained and perhaps unexplainable fact, but one that occurs
the world over and should never be forgotten in the study of ornamentation.

Two faults of omission can be found with the paper.   The internal
geographical relations of the art are nowhere made clear, and it is not
even stated, except incidentally and incompletely, whether, and to what
extent, the technique of the various tribes differs.   Another point on
which more clearness would have been desirable is as to the precise
degree to which the explanations of ornaments that are given belong
respectively to the author and to the natives.

Altogether Dr Laufer's volume is a valuable contribution both to
the ethnography of the Amur region and to the general study of
ornamentation.   A. L. KROEBER.

THE JOURNAL OF
AMERICAN FOLK-LORE
VOLUME XV

BOSTON AND NEW YORK

Published for The American Folk-Lore Society by

HOUGHTON, MIFFLIN AND COMPANY

LONDON: DAVID NUTT, 270, 271 STRAND

LEIPZIG: OTTO HARRASSOWITZ, QUERSTRASSE, 14

MDCCCCII

a man's life; and that it has given rise to much picturesque, if also to some flamboyant writing. In short, to write in praise of Indian summer is now a literary convention on three continents. So varied a history in little more than a century is certainly remarkable" (p. 36). As an interesting pendant to the Trumbull story, Mr. Matthews finds that the "alleged Indian legend" in explanation of the term "Indian Summer" dates only from 1839, while the term itself "had already been in existence among the whites for nearly half a century." As to the exact connotation of the word *Indian* in this term, the author says (p. 50): "We shall, therefore, be obliged to suspend judgment with respect to the origin of the name of the Indian-summer season until fresh evidence as to the early history of the term is produced." Mr. Matthews will welcome any further evidence on these doubtful points.

                                                           Alexander F. Chamberlain.

Memoirs of the American Museum of Natural History, vol. vii. Anthropology, vol. vi. Publications of the Jesup North Pacific Expedition, vol. i. THE DECORATIVE ART OF THE AMUR TRIBES. By BERTHOLD LAUFER. N. Y.: January, 1902. Pp. 86. Plates i.-xxiii. (Figs. 228).

The material discussed in this valuable and interesting monograph is the result of the author's two years' researches among the various tribes of Saghalin Island and the Amur region, under the auspices of the Jesup North Pacific Expedition. After considering the historical, general artistic, and geographical aspects of the subject, Dr. Laufer treats of band-ornaments (pages 8–11), spirals (11–13), band and spiral ornaments (13–16), decorations on boats (16–17), other birch-bark patterns (17–19), circle-ornaments (19), the cock (19–29), single and combined, the fish (29–36), the dragon (36–41), the musk-deer (41, 42), other animals (42–46), leaf and floral ornaments (46–52), basketry-designs (52–56), embroidery-patterns (56–60), specimens made of fish-skin (66–71), Ainu ornamentation (71–73), coloring (73–76), some general results (76–79). The character of the whole ornamentation of these tribes is stamped by "the predominance of the cock and the fish, the manifold combinations in which these two motives appear, and the strange mingling of the two." Here, as in China and Japan, the author believes, these creatures "have an extremely ornamental character because of the great permutations of their graceful motions, and thus lend themselves to the spirit which strives after beauty of form." The ability to watch motions is highly developed in the East Asiatic mind, and is a powerful adjunct of art. Many conventionalizations have arisen from the "influence of the fish-ornament or the cock-type." Dr. Laufer wisely says that the ornaments of primitive tribes are "productions of their art, which can receive proper explanation only from the lips of their creators." They are neither inscriptions to be deciphered, nor enigmas to be puzzled out by the homely fireside. The "bear-heads" of Giliak ornament, *e. g.*, exist only in the imagination of Schurtz, — his "eye-ornaments" are likewise "a mere outcome of his enthusiasm."

During the half-century since the time of Schrenck's investigations, "the forms of this sphere of art have remained unaltered up to the present time, notwithstanding all political turbulence and change that have affected the Amur region in the mean time." Moreover, in spite of the shattering of the whole life of these peoples by Russian " culture," it appears that " the native art has been retained pure and intact." The basis of the art of the Amur is to be found in China, whence, as a mere fashion, like classic art in Europe during the Renaissance, it gradually infused itself into the minds of the Tungusian peoples. But native development and transformation have their rôle also, and it must be concluded that the decorative art of the Amur tribes is " an independent branch of East Asiatic art, which sprang from the Sino-Japanese cultural centre." The swastika and the triskeles are due to Chinese influence. While animals prominent in the household economy and traditions of these tribes, and matter of every-day talk, do not appear in the art-designs, the animals which do occur in them " are just like those which play an important part in Chinese art and mythology." The art of the Amur is lacking in realistic character, and merges into the formative. The sense for plastic representations is largely absent. The lack of ability to draw human faces or forms is noteworthy, since " on prehistoric monuments of the Amur region, petroglyphs have been found, which, doubtless, represent human heads." The painted faces on Goldian paper-charms are very crude. The art-implement of the Amur tribes is a long, sharp, pointed knife. The materials used are wood, birch-bark, fish-skin (especially salmon and sturgeon), elk and reindeer skin, cotton, silk, etc. The needlework is done by women, and " clever embroiderers especially enjoy a high reputation among their countrymen." Dr. Laufer rejects Shrenck's view that the art-sense is most highly developed among the Giliak, who are farthest away from the Chinese, holding that the Gold (from whom the Giliak have borrowed the greater part of their *motifs*) are really *the* artistic people of this region, through whom the Chinese influence has permeated the others. Moreover, their close proximity to the Chinese and their long intercourse with them have enabled them to reach their great skill, especially in silk-embroidery. Only in wood-carving, perhaps, do the Giliaks excel, and the Tungusian tribes of the Amgun and the Ussuri " are unsurpassed in cutting ornaments for decorating birch-bark baskets." The elaboration of ornaments is still in active progress, and " in no more danger of dying out than the Gold and Giliak themselves," but the " reading " of the ornaments is becoming a lost art.

The band-decorated spoons, used only by the Giliak at the bear-festival (the Gold have no bear-festival), " have been superseded in every-day life by spoons of Russian make." In the art of the Gold, " the interlacement-band " is much less frequent than in the art of the Giliak. The cock, an animal not native to the Amur country, but introduced from China (and later by the Russians), is most conspicuous in the art-forms now under consideration, and " is more frequently reproduced than all other animals together." Dr. Laufer's development of the conventionalization of the cock and the fish is very interesting. The Chinese dragon " holds a prominent place in the mythology of the Gold, and is believed by both these people to produce rain and thunder." The symbolic treatment of thunder in art is curious. Even the musk-deer, "under the pressure of the leading gallinaceous motive, undergoes such conventional transformations, especially in its double character, that the difference between the construction of its forms and those of the cock is hardly perceptible." In the decorative art of the tribes of the Amur "leaves and floral forms occur partly as independent ornaments in connection with other elements, partly in close combination with the cock and fish ornaments." The purely conventional forms of leaf-patterns are probably of Sino-Japanese origin. In the art of the Ainu of Saghalin, the author detects three blended elements — "a special overwhelming Japanese influence; loans from the neighboring Amur tribes; and perhaps certain dregs of their artistic ideas, which are to be considered almost wholly their own property." With respect to color it may be said of the Amur tribes that "the more the natives are in contact with the Chinese, the nearer they dwell to a centre of Chinese culture, the more splendidly developed in beauty of color are their works; while the farther one recedes from that centre, the poorer the color-sense seems to grow, and at last to vanish almost entirely." The paper patterns seem to have a special development among these tribes, and "many women retain in their memories a great variety of patterns, and cut them with a speed and dexterity that are worthy of admiration." This monograph is of value to students of American Indian art, in that it suggests what would have happened had any well-defined Chinese influence been present upon the Northwest Pacific coast.

Apart from the special data in these pages, the following observation of the author on the broader human question involved is worth reproducing here : —

"The question may arise as to whether people, like the Gold, who are able to produce such fine work, may justly be classed among primitive tribes. The Gold, at all events, are promising, and some time or other will undeniably advance to the rank of a civilized nation, like their ancestral relations, the Niüchi and Manchu, but under more peaceable circumstances, relying on the cultivation of the soil, industry, and fine arts. There is no doubt but that they are chosen for their share in civilization, and that they will still have a future, if only the Russian government will continue to lend its assistance in improving the economic life-conditions of this intelligent tribe, which numbers so many good-natured and highly-gifted individuals " (p. 79).

Alexander F. Chamberlain.

DER MENSCHHEITSGEDANKE DURCH RAUM UND ZEIT. Ein Beitrag zur Anthropologie und Ethnologie in der "Lehre vom Menschen." A. BASTIAN. Berlin : 1901. F. Dümmlers Verlagsbuchhandlung. 2 Bde. Pp. 246, 257+35.

In characteristic fashion the *doyen* of German ethnologists treats of fate, deity, soul, right, feeling, force and matter, thought, being, the corporeal, metempsychosis, God, causality, the demiurge, songs of origin, the first

030

密勒日巴的生平和道歌

DENKSCHRIFTEN

DER

KAISERLICHEN

AKADEMIE DER WISSENSCHAFTEN.

PHILOSOPHISCH-HISTORISCHE CLASSE.

ACHTUNDVIERZIGSTER BAND.

MIT 1 TAFEL UND 14 ABBILDUNGEN IM TEXTE.

WIEN, 1902.

IN COMMISSION BEI CARL GEROLD'S SOHN

BUCHHÄNDLER DER KAIS. AKADEMIE DER WISSENSCHAFTEN.

II.

AUS DEN GESCHICHTEN UND LIEDERN
DES
MILARASPA.

BERTHOLD LAUFER.

Mit dem Namen des Milaraspa sind zwei umfangreiche Werke der tibetischen Literatur verknüpft, welche durch die Eigenartigkeit dieser Persönlichkeit ein besonderes Interesse verdienen. Das eine derselben führt den Titel: *rje btsun Mi la ras pai rnam thar rgyas par phye ba mgur ḥbum* ‚die des ehrwürdigen Milaraspa Lebensgeschichte ausführlich darlegenden Hunderttausend Gesänge‘ oder kurz nur *mgur ḥbum* ‚Die Hunderttausend Gesänge‘ genannt; ich besitze einen aus 263 fol. bestehenden, gut lesbaren Holzdruck desselben, ohne Angabe des Druckortes. Ausserdem befinden sich Exemplare im Besitze des Asiatischen Departemets in St. Petersburg (Nr. 520 des Katalogs),[1] der Congress-Bibliothek in Washington D. C. (322 fol., gedruckt in Peking) und der American Oriental Society (245 fol.);[2] wenn ich nicht irre, besitzt das Britische Museum das aus Jäschkes Nachlass stammende Exemplar, von dem eine lückenhafte und unvollständige, jetzt in der Bibliothek der Deutschen Morgenländischen Gesellschaft zu Halle befindliche Copie H. Wenzel angefertigt hat. Von diesen vier Exemplaren hatte ich bisher nur Gelegenheit, das der Library of Congress zu sehen, von dem ich aber leider nur wenige Blätter copieren konnte; dieselben stimmen mit denen meines Holzdruckes genau überein. Ausser dem angeführten Werke gibt es eine kürzere, ganz in Prosa geschriebene Biographie des Milaraspa, gewöhnlich nur als *Mi la ras pai rnam thar* citiert, die sich gleichfalls in Petersburg (Nr. 519 des As. Dep.) und Washington befindet, beides Pekinger Drucke von 139 fol. Als Verfasser dieser Schrift wird Milaraspa's Schüler Ras chung pa genannt.[3] Beide Bücher haben eine Uebersetzung ins Mongolische erfahren; Exemplare dieser Version sind im Asiatischen Departement unter Nr. 460 und 482, sowie in der Universitätsbibliothek von Kasan[4] vorhanden. Aus dem

---

[1] Vergl. Schiefner, Bericht über die neueste Büchersendung aus Peking, in Mélanges asiatiques I, 413.
[2] Proceeding of the American Oriental Society, Oct. 1884, p. VI.
[3] Waddell, The Buddhism of Tibet, p. 66. S. die 5. Erzählung.
[4] Kowalewski, Dictionnaire mongol-russe-français, p. X.

Denkschriften der phil.-hist. Classe. XLVIII. Bd. II. Abh. 1

Hor chos byung des *₀Jigs med nam mkha* erfahren wir, dass die Uebertragung beider Werke ins Mongolische durch Zheregethu Gushri Chos rje im Anfange des 17. Jahrhunderts stattgefunden habe.[1] Die Heranziehung der mongolischen Uebersetzungen könnte jedenfalls für das Verständnis des tibetischen Originales gute Dienste leisten.

Milaraspa oder kurz Mila genannt, sagt Jäschke[2] von ihm, ist der Name eines buddhistischen Asketen des elften Jahrhunderts, der zwischen seinen Meditationsperioden als Bettelmönch im südlichen Theile Mitteltibets umherwandernd, durch seine stets in gebundener Rede und Gesangsweise vorgetragenen Improvisationen Lehrbegierige unterrichtet, Weltlichgesinnte zum Glauben bringt, Ketzer niederdisputiert und bekehrt und mannigfache Wunder *(rdzu ₀p'rul)* verrichtet, und dessen nicht ohne Witz und Poesie geschriebene Legenden das beliebteste und verbreitetste Volksbuch in Tibet sind. Ob nun Milaraspa selbst der Verfasser des *mgur ₀bum* ist, steht einstweilen dahin. In dem Buche selbst lässt sich keine darauf oder überhaupt auf die Autorschaft desselben bezügliche Angabe nachweisen. Ebensowenig ist etwas über die Zeit der Abfassung gesagt. Nach der chronologischen Tafel *Ren mig*[3] hat Milaraspa von 1038 bis 1122 gelebt. Zunächst ist sicher, dass das vorliegende Werk keine posthume Erfindung vorstellen kann. Denn es führt die Ereignisse aus Milaraspa's Zeit mit solch lebendiger Frische und unmittelbarer Treue der Darstellung, mit einer Schärfe der Charakteristik des Trägers der Handlung und der dabei betheiligten Personen, mit solch feiner Wiedergabe der Orts- und Zeitfärbung vor, dass man nicht anders als dem Gedanken Raum geben kann, der Inhalt dieser Blätter müsse einer gleichzeitigen Aufzeichnung entstammen. Das gilt in erster Reihe von den zahlreichen Liedern, welche die erzählenden Prosapartien so stark überwuchern, dass das Werk mit Recht nach ihnen benannt worden ist. Sie rühren zweifellos von dem Dichter selbst her und sind sicher von ihm auch niedergeschrieben worden; sie sind zu sehr von der Subjectivität der Persönlichkeit erfüllt, als dass man mit gutem Grunde annehmen könnte, ein Schüler des Meisters habe dieselben, wie er sie aus seinem Munde gehört, aufgezeichnet und nachträglich bearbeitet. Gleichwohl kann nicht das ganze Werk in dem Zustande, wie es uns jetzt vorliegt, als Erzeugnis des Milaraspa angesprochen werden. Dagegen streiten verschiedene in den Prosatheilen vorkommende Stellen, welche deutlich auf ein späteres Zeitalter als das des Milaraspa anspielen. So werden wiederholt Stätten der Verehrung, die durch die Erinnerung an den Aufenthalt des verehrten Lama geweiht sind, mit ihren späteren Benennungen aufgeführt; solche Plätze konnten ja naturgemäss auch erst nach dem Tode des Meisters ihre rechte Bedeutung erlangen, als der Ruf seiner Thätigkeit und seiner Lieder sich weiter und weiter im Volke verbreitete. So heisst es in dem Prosaabschnitt nach dem ersten Liede der folgenden Proben, dass der Felsblock, auf dem Milaraspa einen hüpfenden Tanz aufführte und Spuren seiner Füsse und seiner Schneeschuhe hinterliess, in früherer Zeit ,der flache, weisse Felsblock', später dagegen ,Spuren der Schneeschuhe' genannt wurde. Diese Bemerkung kann wohl nur der Zusatz eines späteren Bearbeiters sein. Eine ähnliche Stelle begegnet auf fol. 6 b 3: Nach einem durch Dämonenspuk erregten Unwetter lässt sich Milaraspa auf einem Steine nieder, wo ebenfalls seine Fussspur zurückbleibt; dann geht er einige Schritte weiter, der Himmel heitert sich auf, in freudig gehobener Stimmung setzt er sich auf einen Hügelvorsprung, wo er sich einer Betrachtung der Liebe zu allen Wesen

[1] Huth, Geschichte des Buddhismus in der Mongolei, Bd. II, S. 248.
[2] Handwörterbuch der tibetischen Sprache, S. 419a.
[3] Journal of the Asiatic Society of Bengal 1889, 40, 45.

hingibt. Dieser Ort nun, so lautet ein folgender Zusatz, ist bekannt als *Byams syang* ‚der Hügelvorsprung der Liebe‘. In derselben Erzählung, fol. 9 b 5, erhält ein anderer, durch eine phantastische Vision Milaraspa's berühmt gewordener Ort den Beinamen *La dgu lung dgu* ‚die neun Pässe, die neun Thäler‘. Solche Namen werden im Volke erst in späterer Zeit, als die Geschichten des beliebten Lehrers populär wurden, zum Andenken an seine Wanderungen entstanden sein.

Die Beliebtheit des *mgur obum* wird einmal durch die Aussage von Jäschke[1] bezeugt, dass das bekannteste und am meisten gelesene Werk in Bezug auf Milaraspa eben diese Hunderttausend Gesänge sind, welche in verschiedenen Holzdruckausgaben in ganz Tibet verbreitet sind. sodann durch den Historiker *oJigs med nam mkha*, der an zwei Stellen seines Werkes sowohl das *rnam thar* als *mgur obum* zusammen mit philosophischen Abhandlungen als Vortrags- und Studienschriften in der Mongolei für die zweite Hälfte des 18. Jahrhunderts erwähnt.[2] An zwei anderen Stellen citiert derselbe Autor einmal drei und das andere Mal vier Verse aus Mila.[3]

Wenn wir auch vorläufig noch nicht imstande sind, das Datum der Abfassung des vorliegenden Werkes zu bestimmen, so gibt uns doch dieses selbst Hilfsmittel in die Hand, um die Zeit der Ereignisse festzusetzen, welche den Rahmen der vorliegenden Geschichten bilden. Auf fol. 6 b 3 findet sich nämlich die Angabe einer bestimmten Jahreszahl. Als Milaraspa zu dem Flüsschen *Chu bzang* (‚trefliches Wasser‘) gelangte und sich in die Betrachtung des fliessenden Wassers vertiefte, heisst es: *me pho stag gi lo ston zla ra bai tses bcui nub* ‚da ging der 10. Tag des ersten Herbstmonats des Feuer-Tiger-Jahres zur Rüste‘. Das Feuer-Tiger-Jahr ist das 60. Jahr des Cyklus, in diesem Falle des 1. Prabhava, also das Jahr 1085. Der erste Herbstmonat wäre nach der indischen Eintheilung der Jahreszeiten der September, nach tibetischer Anschauung der August. Dies ist die Zeit, zu der Milaraspa die Dämonen auf dem *La phyi* bekehrt. Darauf bringt er einen ganzen Monat in *gNas othil* zu, worauf er sich in das Dorf *gÑa nam* begibt (fol. 9 b 7). Wie lange er dort weilt, ist nicht mit Bestimmtheit gesagt, doch lässt sich die Zeit ungefähr aus der Angabe in unserer ersten Erzählung bemessen, dass unmittelbar nach seiner zweiten Besteigung des *La phyi* ein neun Tage und Nächte andauernder heftiger Schneefall, d. h. also der Winter eintritt. Dies Ereignis muss demnach im October stattgefunden haben, was durch zwei weitere Zeitangaben bestätigt wird. Im Pferd-Monat *(rtai zla ba)*, d. i. im 3. Monat des Jahres 1086, und zwar in der zweiten Hälfte desselben, bringen dem Meister seine Verehrer in dem Glauben, dass er während des strengen Winters auf dem Schneeberge umgekommen sei, ein Todtenopfer dar; im folgenden Monat *Saga* machen sie sich auf, ihn zu suchen, treffen ihn lebend an und kehren mit ihm nach *rTsar ma* zurück. In Lied 3, Vers 16, gibt Milaraspa die Zeit seines Winteraufenthaltes auf sechs Monate an, was also mit der Annahme, dass derselbe im October 1085 begonnen, übereinstimmt. In Lied 1, Vers 49, wird der Neujahrstag 1086 erwähnt. Der Zeitraum, den die Handlung unserer ersten Erzählung einnimmt, dehnt sich also von September 1085 bis April 1086 aus. Dem gegenüber ist die in Lied 1, Vers 8—10, gemachte Zeitangabe befremdend, dass Milaraspa am Jahresende des Tigerjahres, im Jahresanfang des Hasenjahres, am sechsten Tage des Monats *wa-rgyal*, der Weise des Kreislaufes überdrüssig, auf den *La phyi* ge-

---

[1] ZDMG, Bd. XXIII, 543.

[2] Huth, Geschichte des Buddhismus in der Mongolei, Bd. II, S. 352, 401.

[3] Huth, l. c., S. 245, 414.

1*

gangen sei. Unter dem Tigerjahr kann nur das Jahr 1085 *(me stag)* und unter dem Hasen-jahr nur das Jahr 1086 *(me-yos)* verstanden werden. Der Monat *wa-rgyal* wird in den Wörterbüchern nicht erklärt; *wa* ist aber Bezeichnung des 7. oder 8. Nakṣatra (= Skr. *açleṣa*) und *rgyal* Name des 6. Nakṣatra (= Skr. *puṣya*), und das Sternbild beider ist der Krebs. Darnach könnte es sich möglicher Weise um den Monat Juli handeln. Nach der voraus-gegangenen Darlegung ist es jedoch unverständlich, wie hier dieser Monat in Betracht kommen könnte, und ferner, in welcher Beziehung er zu den beiden vorhergehenden Jahreszahlen stehen sollte, aus denen man nur schliessen kann, dass es sich um den Winter von 1085 auf 1086 handelt. Man darf nicht übersehen, dass die Zeitangabe in jenen drei Versen eine stark poetische Färbung hat, wie aus den ungewöhnlichen Aus-drücken *yong stag* für *stag*, *yos bu* für *yos*, aus der Wiederholung von *lo* in V. 8 und 9 und dem sonst nicht belegten *wa-rgyal* hervorgeht; ebenso ist *ña* (V. 10) ganz ungewöhnlich statt *tshes*. Schon aus diesem Grunde kann diese Stelle nicht die gleiche Bedeutung bean-spruchen wie die übrigen in dem gewöhnlichen Stile der Prosa gegebenen Zeit-bestimmungen, deren Richtigkeit dadurch in keiner Weise angefochten wird. Die zweite, dritte und vierte Geschichte schliessen sich zeitlich eng an die erste an und spielen gleich-falls im Jahre 1086. Aus den Naturschilderungen des Liedes 19 in der dritten Erzählung geht hervor, dass die Zeit derselben der Spätfrühling oder Sommer sein muss. Beachtens-wert ist, dass die von Sum pa mkhan po verfasste chronologische Tafel *Reu-mig* ausser anderen Jahreszahlen aus Milaraspa's Leben die Vollziehung seiner Bussübungen zur Er-langung der Heiligkeit dem Jahre 1083 zuweist.[1]

Alle diese Zeitbestimmungen weisen deutlich auf die historische Grundlage des vor-liegenden Werkes hin. Einen weiteren Fingerzeig für dieselbe geben die Namen und Schil-derungen der Oertlichkeiten, die sich selbst bei unseren spärlichen Hilfsmitteln zum Theile noch im heutigen Tibet nachweisen lassen. Der Schauplatz von Milaraspa's Erzählungen ist das nordwestliche Tibet in den nördlichen Theilen des Himalaya zwischen der Grenze Nepals und dem Oberlaufe des Brahmaputra *(Ya ru gtsang po)*. Die Schönheiten der Ge-birgswelt und der Natur überhaupt sind wiederholt Gegenstand der Lieder. Die in den Geschichten genannten Dörfer liegen alle nicht weit von *Gung thang*, Milaraspa's Geburts-ort (vergl. Lied 24, Vers 10), der sich gleichzeitig rühmt, den Uebersetzer Rva lo zu seinen Söhnen zu zählen. Es ist eine anziehende Erscheinung, dass das Volk in diesem Districte die Erinnerung an Milaraspa noch bewahrt hat; denn in der tibetischen Geographie des Minchul Chutuktu (gest. 1839) wird bei der Erwähnung von *Gung thang* ausdrücklich der Grotte gedacht, in der Mila die Vollendung erlangte, und einer Insel *(Chu bar)*, wo er predigte.[2]

Ausser diesen äusseren Momenten der Zeit und des Ortes sprechen für den geschicht-lichen Hintergrund des Buches der historische Charakter der Persönlichkeit des Milaraspa selbst, sein Zusammentreffen mit anderen historisch beglaubigten Personen und die innere Wahrscheinlichkeit der in chronologischer Reihenfolge vorgetragenen Ereignisse, soweit sie nicht durch Wunder und andere rein legendenhafte Zuthaten ausgeschmückt sind. Ueber Milaraspa's Person ist uns freilich bis jetzt nicht viel bekannt geworden, und eine vollständige Kenntnis derselben wird sich erst aus einer Durchforschung des *mgur ḥbum* und *rnam thar* gewinnen lassen. Als ziemlich gesichert mag feststehen, dass er ein Schüler des Marpa war, mit dem er den Lehren der Secte der *bka rgyud pa* anhing. Wie der Name besagt,

---

[1] JASB 1889, 43.

[2] Wasiljev, Geographie Tibets (russisch), Pet. 1895, p. 12.

stützt sich diese Secte auf den Glauben an die mündliche Ueberlieferung des Buddhawortes, die neben der schriftlichen Tradition der heiligen Bücher durch eine ununterbrochene Reihe von Lehrern und Schülern fortgepflanzt sein soll. Als ihre Begründer werden die drei Grub chen Näropa, Telopa und Maitripa genannt.[1] Marpa's Lehrer Näropa oder Näro, ein Zeitgenosse des Padmasambhava und Atisha, erhielt seine Belehrung von Telopa,[2] der seinerseits eine unmittelbare Inspiration vom Buddha Vajradhara empfing[3] (vergl. Lied 12, Vers 30, 43). Marpa wird wiederholt als *bka rgyud bla ma* oder kurz *bla ma*, der Lama, bezeichnet; er wird am Anfang der meisten Lieder angerufen und stets mit dem Ausdrucke höchster Verehrung und innigster Liebe von Milaraspa genannt. Wenn sich auch Milaraspa's philosophische Anschauungen nicht eher werden systematisch darstellen lassen, als bis das ganze Buch bearbeitet ist, so lässt sich doch schon nach den hier mitgetheilten fünf Capiteln einigermassen sicher urtheilen, dass er weit davon entfernt ist, ein eigenes System zu begründen; er gibt nur die Lehren seiner Vorgänger in einer seinem stark ausgesprochenen persönlichen Geiste entsprechenden Form wieder, mit einer übermässig scharf hervorgekehrten Betonung der Meditation, wie sie in der von Asanga begründeten Yogācārya-Schule betrieben wurde. Viele Lieder beschäftigen sich eingehend mit den physiologischen und psychischen Grundlagen ekstatischer Beschauung und verlieren sich zuweilen in einen schwindelnden Abgrund bodenloser Mystik. Doch darf die grosse Wichtigkeit dieses Materiales zur Erkenntnis und Beurtheilung dieser merkwürdigen religiösen Richtung keineswegs unterschätzt werden. Aber gerade die interessantesten Züge an Milaraspa's Charakter sind die nichtbuddhistischen. So sehr er sich auch immer anstrengt, den starren asketischen Yogin hervorzukehren, gelangt doch der Tibeter und der Dichter in ihm wieder und wieder zum Durchbruch, und das bedingt eben einen der grössten Reize dieses fesselnden Werkes. Die verschrobensten Meditationstheorien vermögen nicht den Menschen, das Ich in ihm zu unterdrücken; er ist Yogin weit mehr in der Theorie als in der Praxis, er vollzieht keine fabelhaften Bussübungen wie seine indischen Vettern, er ist nicht zum stumpfsinnigen Brüter herabgesunken, der den Zustand seligsten Glückes in stumpfer Gedankenlosigkeit erreicht zu haben vermeint. Nein, Milaraspa thut Dinge und entwickelt Eigenschaften, die sich mit den strenggläubigen Anforderungen an den heiligen Beruf des buddhistischen Mönches und Yogin nimmermehr vereinigen lassen: er dichtet, er singt, er lacht, er scherzt, ist froh mit den Fröhlichen, ja, er tanzt vor Freude über die Freude des Volkes, trinkt tibetischen Gerstensaft *(chang)* und geniesst sogar Fleisch, er entwickelt Witz und Humor und zuweilen beissende Satire, d. h. kurz gesagt, er ist und bleibt Tibeter in seinem Herzen trotz allem Buddhismus. Hier bewahrheitet sich Herders[4] schönes Wort über Tibet: ‚Sonderbar ist der Unzusammenhang, in welchem die Sachen der Menschen sich nicht nur binden, sondern auch lange erhalten. Befolgte jeder Tibetaner die Gesetze der Lama, indem er ihren höchsten Tugenden nachstrebte, so wäre kein Tibet mehr. Das Geschlecht der Menschen, die einander nicht berühren, die ihr kaltes Land nicht bauen, die weder Handel noch Geschäfte treiben, hörte auf; verhungert und erfroren lägen sie da, indem sie sich ihren Himmel träumen. Aber zum Glück ist die Natur der Menschen stärker als jeder angenommene Wahn. . . . Glücklicher Weise hat die harte Mönchsreligion den Geist

---

[1] Pander-Grünwedel, Das Pantheon des Tschangtscha Hutuktu, S. 50.
[2] Ueber die verschiedenen Formen dieses Namens s. Schiefner's Uebersetzung des Tāranātha, S. 226, Nr. 5.
[3] Waddell, The Buddhism of Tibet, p. 64.
[4] Ideen zur Philosophie der Geschichte der Menschheit, 11. Buch, 3. Capitel.

der Nation so wenig als ihr Bedürfnis und Klima ändern mögen. Der hohe Bergbewohner kauft seine Büssungen ab und ist gesund und munter; er zieht und schlachtet Thiere, ob er gleich die Seelenwanderung glaubt, und erlustigt sich fünfzehn Tage mit der Hochzeit, obgleich seine Priester der Vollkommenheit ehelos leben. So hat sich allenthaben der Wahn der Menschen mit dem Bedürfnisse abgefunden: er dung solange, bis ein leidlicher Vergleich ward. Sollte jede Thorheit, die im angenommenen Glauben der Nationen herrscht, auch durchgängig geübt werden: welch ein Unglück! Nun aber werden die meisten geglaubt und nicht befolgt, und dies Mittelding todter Ueberzeugung heisst eben auf der Erde Glauben.' In Milaraspa spricht sich ein starkes Nationalgefühl aus, sowohl in der Anhänglichkeit an den Boden seiner Heimat wie in der Liebe zu seinem Volke, dessen Lehrer, Prediger und Ermahner er ist, dessen Leiden und Freuden er gerne theilt. Sein stark ausgeprägtes Selbstgefühl gelangt besonders in der Erzählung von der Begegnung mit dem Inder Dharmabodhi zum Ausdruck, in der gerade die wiederholte Betonung des Nationalitäten-unterschiedes einen eigenartigen Zug bildet.

Einen auffallenden Gegensatz bildet Milaraspa zu Padmasambhava, dessen Legendenbuch sich geradezu als die Ausgeburt einer wahnsinnigen Phantasie darstellt; es ist wie der tolle Tanz eines Schamanen; seine sogenannten Bekehrungen sind von rohestem Canni-balismus und widerlicher Unzucht begleitet.[1] Nichts von solchen Zügen findet sich in Milaraspa's Buche. Seine Bekehrungen erfolgen nie unter Anwendung von Gewalt, sondern einzig und allein durch die Macht seines Wortes und seiner Predigt in der Form des Liedes. Alle Erzählungen tragen einen milden und ruhigen Charakter, und selbst seine Spukmären von bösen Geistern entbehren nie eines schalkhaften Humors und einer anmuthigen Liebens-würdigkeit. In seinen beschaulichen Beobachtungen der Natur zeigt sich die Feinheit seiner Empfindungen, und zuweilen geräth in ihm, wie aus dem Schlusse des 19. Liedes hervorgeht, der Dichter mit dem Philosophen in Conflict. Es ist schwer, eine Definition dieser Erzäh-lungen aufzustellen. Mit den Jātaka und Avadāna haben sie nicht die mindeste Aehnlich-keit; auch liegt keine Uebertragung indischer Stoffe vor. Das *mgur ʼbum* ist vielmehr von echt tibetischem Leben erfüllt und wird gerade durch seine lebendigen Schilderungen des Volkslebens zu einem wichtigen Beitrag zur Culturgeschichte Tibets in der zweiten Hälfte des 11. Jahrhunderts. Einige Erzählungen könnte man als Heiligenlegenden im Sinne unseres Mittelalters bezeichnen, andere aber spiegeln wirkliche Erlebnisse Mila's wieder, wie z. B. unser erstes Capitel, andere historische Ereignisse, wie die fünfte Erzählung. Zu-weilen kann man sich kaum des Eindruckes erwehren, als wenn in einigen Gedanken abendländisch-christliche Einflüsse zutage träten.

Die Sprache des *mgur ʼbum* weist von der des Kanjur und Tanjur bedeutende Ab-weichungen auf. Schon Jäschke[2] bemerkt: ‚Dieses Legendenbuch gehört der späteren Periode der tibetischen Literatur an, deren Sprache sich in ihrem grammatischen Bau, wie in ihrem Wortschatze merklich von der älteren unterscheidet und der gegenwärtig in Mittel-tibet gesprochenen näher steht.' Eine eingehende Darstellung dieser Abweichungen muss einer späteren Untersuchung vorbehalten bleiben, ebenso die eigenthümliche Metrik der Lieder. Es kommen sieben-, acht-, neun- und zehnsilbige Verse vor, und zwar oft in bunter Abwechslung innerhalb desselben Liedes, ohne dass sich eine gesetzmässige Anwendung nachweisen liesse. Im ersten Liede finden sich 29 Siebensilbler, 42 Achtsilbler und

---

[1] Grünwedel, Ein Capitel des Ta-she-sung (Bastian-Festschrift), S. 5, 12 des Sonderabdruckes.
[2] ZDMG, Bd. XXIII, 543.

13 Neunsilbler. Von den vierzig Versen des zweiten Liedes sind ein Fünftel achtsilbig und vier Fünftel siebensilbig. Im dritten Liede, das aus 78 Versen besteht, gibt es siebensilbiger Verse 41, achtsilbiger 35, neunsilbiger und zehnsilbiger je einen. Neben dem gewöhnlichen trochäischen Rhythmus begegnet, wenn auch seltener, der daktylische. Vergl.

z. B. Lied 1, Vers 1:   *dpal dĕ riṅ bkra šĭs k'rĭ gduṅs lă*

Vers 3:   *ṅa rnăl ₒbyor Mĭ lă ras pă gñĭs*

Vers 12:   *ₒbrog lă p'yĭ gaṅs kyĭ ră bă der*

Vers 14:   *k'ŏṅ gnăm sa gñĭs pŏ gros byăs te*

Meine im Folgenden gegebenen Uebersetzungen sollen nur als ein vorläufiger Versuch angesehen werden; es befinden sich noch viele ungelöste Schwierigkeiten darin, die erst durch ein umfassendes Studium des gesammten Gegenstandes beseitigt werden können.

Die erste Erzählung behandelt den Angriff des Dämonenkönigs Bhinayaka auf Milaraspa, die zweite seinen Sommeraufenthalt auf dem La phyi und die Bekehrung des Dämonenfürsten Bharo, worauf er in das Dorf gÑa nam zurückkehrt. Hier setzt die erste im Folgenden mitgetheilte Geschichte ein.

I.

Milaraspa's Winteraufenthalt auf dem La phyi.

(fol. 10 a 2 — 16 b 5.)

namo guru.

rje btsun Mi la ras pa de ñid kyis | daṅ por la p'yi gaṅs gzigs su byon pas | lha ₒdre gdug pa can rnams btul ces pai sñan pa la brten nas | gña nam pa kun gyis p'yag mc'od kyi gnas su gyur ciṅ | k'yad par Jo mo °Ur mos c'os žus | dei ts'e ñam pa rgyags p'u ba sku na p'ra mo raṅ du yoṅ ba la | bdag gi bu ₒdi yaṅ c'er soṅ ba daṅ | rje btsun gyi p'yag p'yir ₒbul zer šin tu dad par gyur pai sa nas | rtsar ma pa rnams kyis spyan draṅs te | gŠen rdor mos žabs tog ₒbul bai ts'e | rje btsun groṅ gseb tu bžugs pa daṅ | ₒk'or bai bya ba rnams gzigs pas šin tu t'ugs ₒbyuṅ

Verehrung dem Meister!

Auf die Kunde, dass der ehrwürdige Milaraspa bei seinem erstmaligen Besuche des Schneeberges La phyi die gefährlichen Geister und Kobolde bekehrt habe, strömten alle Einwohner von gÑa nam herbei, um ihm ihre Ehrerbietung zu bezeigen. Besonders Jo mo ₒUr mo suchte religiöse Belehrung nach: ‚Gegenwärtig', sprach sie, ‚ist Ngam pa rgyags phu ba[1] noch klein und jung; doch wenn dieser mein Sohn erwachsen sein wird, werde ich ihn dem Ehrwürdigen zum Diener geben.' Infolge des in ihnen erwachten grossen Glaubens luden ihn die Bewohner von rTsar ma ein, und während gŠen rdor mo ihm ihre Dienste widmete, weilte der Ehrwürdige im Dorfe. Aus der Beobachtung der weltlichen Geschäfte gewann er eine ausserordentlich frohe Stimmung. Als er nun sagte, dass er auf den Schnee-

---

[1] Auf fol. 74 a—74 b 5 ist ihm eine kleine Erzählung unter dem Titel ‚Begegnung mit Ngam pa rgyags phu ba' gewidmet, in der berichtet wird, wie er sich als Diener und Schüler an Milaraspa anschliesst, als dieser, durch das Traumbild eines schönen Mädchens aufgefordert, die Reise vom La phyi nach dem Tise antritt.

*bai rnam pa mdzad | la p'yi gaṅs
la ₒgro gsuṅs pas | rtsar ma pa rnams
kyis rje btsun la sems can gyi don
las ni med | ṅed rnams kyi don laṅ
lo dgun ₒdir bžugs ₒk'rid re gsuṅ bar
žu | ₒdre t'ul rtiṅ nam byon ruṅ c'og
pas | naṅ par sos ṅed rnams kyaṅ
p'yags p'yi byas ₒgro zer ba daṅ |
k'yad par ston pa Çākyaguṇa daṅ |
gŚen rdor mos nam zla dgun du soṅ
ba daṅ | gaṅs la ₒo rgyal yoṅ lugs
sogs bŚol ₒdebs kyi žu ba ci p'ul ruṅ
ma gsan par | spyir ṅa Nā ro paṇ
c'en gyi bu brgyud gaṅs kyi ṅa ras
mi ₒjigs | Mar pai bka yaṅ ₒdu ₒdzi
daṅ yyeṅ ba spoṅs la | gnas mi med
kyi dben pa bsten gsuṅ ba yin | k'yad
par ṅa groṅ yul du gži p'ab sdod pa
ₒc'i bas ₒts'er gsuṅ ₒbyon par t'ag c'od
pa daṅ | rtsar ma pa rtsar ma pa
rnams kyis myur bar rje btsun gyi
ₒts'o c'as ₒbul ba daṅ | k'o raṅ ₒga
dgun c'as žur yoṅ bai c'ad mdo byas
te | ston pa Śākya gu na daṅ | gŚen
rDor mo sogs ban skyes drug gis gŚegs
skyems k'yer te la t'og bar ₒgro rtsis
la | la babs nas kyaṅ rmu rdziṅ bar
p'yin | de nas rje btsun gyis sku c'as
p'ye bre do ₒbras bre gaṅ | ša gzug
gcig | mar žogs gcig rnams yod pa
bsnams te bdud ₒdul p'ug mo c'er byon
nas bžugs | k'oṅ rnams kyis p'ar log
p'yin pas | la t'og tu gnas ₒk'riys |
bu yug ots'ubs nas lam p'ed pa yaṅ
dka mo raṅ byuṅ | pus goṅ man k'a
ts'ub daṅ ₒab ciṅ yoṅs pas yul du
mi ṅal ba tsam la sleb | dei nub mo
nas bzuṅ ste k'a ba ṅiṅ mts'an beva*

berg La phyi gehen wolle, baten die Einwohner von rTsar ma den Ehrwürdigen, indem Jeglicher sagte: ,Es gibt nichts anderes als den Nutzen der Wesen; damit uns Nutzen entstehe, bringe doch den Winter hier zu und leite uns!' Dann fügten sie hinzu: ,Auch wenn du zur Bekehrung der Dämonen später des Nachts auszichst, genügt es; im Frühjahr, am Morgen, wollen wir dich als deine Diener begleiten.' Besonders der Lehrer Çākyaguṇa und gŚen rdor mo wandten alle möglichen Bitten auf, um ihn durch den Hinweis auf die winterliche Jahreszeit und durch die Vorstellung der Lebensgefahr, der er sich auf dem Schneeberg aussetze, zurückzuhalten; dennoch schenkte er ihnen kein Gehör und entgegnete: ,Ich, der Sprössling des grossen Paṇḍita Nāro,[1] fürchte im allgemeinen die Luft der Schneeberge nicht; besonders muss ich nach dem Gebot des Mar pa[2] Weltgetümmel und Störung meiden und an einem menschenleeren einsamen Orte verbleiben. Ueberdies bereitet mir der dauernde Aufenthalt in einem Dorfe tödtlichen Ueberdruss.' Als er so zur Abreise fest entschlossen war, beeilten sich die Leute von rTsar ma dem Ehrwürdigen Lebensmittel zu besorgen, und einige von ihnen baten ihn, wenigstens im Winter von da wegzugehen, und versprachen zu kommen. Der Lehrer Çākyaguṇa und gŚen rdor mo und vier andere Leute, die den Abschiedstrunk mitnahmen, geleiteten ihn bis über den Pass und gelangten, den Pass hinabsteigend, zum Teufelsteich. Darauf übergaben sie die für den Ehrwürdigen mitgebrachten Vorräthe, zwei Mass Mehl, ein Mass Reis, ein Viertel Fleisch und einen Schnitt Butter.[3] Dann trat er in die grosse Dämonen bekehrende Grotte, wo er zu bleiben gedachte. Jene traten den Rückweg an. Als sie die Passhöhe überstiegen, umwölkte sich der Himmel, und es brach ein wirbelndes Schneegestöber aus. Mit grosser Schwierigkeit bahnten sie sich einen

---

[1] Ueber Nāro oder Nāropa s. Pander-Grünwedel, Das Pantheon des Tschangtscha Hutuktu, S. 50, Nr. 16; Grünwedel, Mythologie des Buddhismus in Tibet und der Mongolei, S. 40. Nāro war der Lehrer Marpa's, s. Waddell, The Buddhism of Tibet, p. 66, no.

[2] Marpa war der Lehrer des Milaraspa, s. die Einleitung.

[3] Mass, tib. *bre*, ein Hohlmass für trockene Dinge wie für Flüssigkeiten, etwa 1—2 Quart; nach Jäschke: *bre* = Skr. droṇa. Viertel, tib. *zug* oder *gzug*, die Hauptstücke beim Zerlegen eines Thieres; es wird sich hier wohl um Schaffleisch handeln. Schnitt Butter, tib. *mar žogs* (von *žog pa* ‚schneiden'), die Hälfte eines *mar ril* oder kugelförmigen Klumpens frischer Butter, etwa ein Pfund schwer, der nicht selten Reisenden als Ehren- oder Gastgeschenk gebracht wird (Jäschke).

brgyad babs pas | der sla ba drug tu
brin gña nam gyi ₒgrul ćad pas | slob
ₒbañs rnams kyis rje btsun groñs par
t'ag bcad te | t'ugs dgoñs rdzogs t'abs
kyi ts'ogs mc'od p'ul | p'yis sa gai
zla bar soñ ba dañ | yañ sñar gyi bu
slob rnams kyis gañs sta gris gsags
nas | rje btsun gyi gduñ gdan ₒdren
du p'yin te | gnas su sleb tu ća bai
ts'e | stegs žig tu ñal bso žiñ bsdad
pas | p'a p'oñ žig gi k'ar bsa ćen po
žig ₒdzegs nas bya sñeñ byas te | yun
riñ po žig bltas nas soñ skad | der k'oñ
rnams kyis | bsa ma gis rje btsun gyi
pur zos yod | da na bzai dum bu ₒdra
ba ₒam | dbu skra ₒdra ba yod dam |
gžan mi rñed ces gleñ žiñ | šin tu
ma dga nas ñu ruñ p'yin pas | bsa
babs pai šul na p'ar la mii rañ rjes
žig ₒdug de nas ₒp'red riñ žig la gzig
dañ stag tu rdzus nas soñ bai ₒp'red
lam de la | stag ₒk'red gzig ₒp'red du
grags so | der k'oñ rnams kyañ rtog pa
skyes te lha ₒdre yin nam sñam nas t'e
ts'om za žiñ ₒoñs pas | bdud ₒdul p'ug
mo ćer sleb tu ñe bai ts'e | rje btsun
gyis mgur bžes pa t'os pas | ₒo na
k'yirabas brgyags p'ul ba ₒam | gcan
zan gyis bsad pai t'eñ ₒdra rñed nas
ma groñs pa srid dum sñam nas p'yin
pas | bla mai žal nas | glen pa kun
da p'yi de tsa na ₒoñ gi ₒdug pa la
| da duñ mi sleb pa ci yin | zan dañ
ts'od ma yañ grañs ₒgro bar ₒdug | da
p'ug tu myur du šog gsuñs pas | der
k'oñ rnams dga ćes nas ñu bro ste |
kun rje btsun gyi p'yag žabs kun nas
rub rub ₒjus te ñus pas | rje btsun
gyi žal nas | da de skad ma byed par
za ma zo gsuñs pas | k'oñ rnams kyis
kyañ gdod kyis p'yag btsal sñun dris
nas brtags pas | sñar gyi ćas rnams

Weg, und bis über die Knie tief einsinkend, erreichten sie im Kampfe mit dem Schneesturm erst
bei Einbruch der Nacht ihr Dorf. Von jenem Abend
an fiel aber neun Tage und Nächte lang Schnee,
so dass während sechs Monaten das handelsthätige
gÑa nam von allem Verkehr abgeschnitten war. Die
Anhänger des Lehrers waren daher der festen Ueberzeugung, dass der Ehrwürdige verschieden sei, und
brachten zu seinem Gedächtnis in vollkommener
Weise Opfergaben dar. Als der folgende Saga-
Monat[1] kam, machten sich wiederum die früheren
Schüler auf den Weg, indem sie das Eis mit der
Axt spalteten, um des Ehrwürdigen Gebeine zu holen.
Während sie gingen, um zu dem Orte zu gelangen,
liessen sie sich, um sich zu erholen, auf einem Sitze
nieder. Da stieg ein grosser Schneeleopard einen
Felsblock hinauf und gähnte. Er hielt lange Umschau und liess seine Stimme ertönen. Da sprachen
jene: ‚Der Schneeleopard dort drüben wird wohl
den Leichnam des Ehrwürdigen verzehrt haben.
Ob nun etwas da ist, was seinem Haupthaar gleicht?
Etwas anderes werden wir wohl nicht finden.‘ Sehr
traurig und weinend schritten sie weiter. Auf dem
leeren Platze, von dem der Schneeleopard herabstieg,
zeigten sich weiter abwärts menschliche Fussspuren.
Darauf verwandelte sich das Thier in einen Leopard
und dann in einen Tiger; auf dem Querpfade, den
sie nun einschlugen, erschien es, in der Längsrichtung[2]
gesehen, als Tiger, quer gesehen als Leopard. Da
wurden jene nachdenklich und gingen in dem Gedanken, es möchte wohl ein Dämon sein, Zweifel
nährend weiter. Als sie sich der grossen, Dämonen
bekehrenden Grotte näherten, hörten sie den Ehrwürdigen Lieder singen. ‚Wie,‘ dachten sie, ‚sollten
ihm doch etwa Jäger Lebensmittel gebracht oder er
selbst etwa Aasreste aus dem Vorrath eines Raub-
thieres gefunden haben, dass er nicht verschieden ist?‘
Da sprach der Lama: ‚Es ereignet sich zuweilen,
dass die Thoren alle draussen stehen. Was ist denn,
dass ihr nicht eintretet? Brei und Gemüse werden
euch ja kalt! So kommt doch schnell in die Grotte

---

[1] = Skr. vaiçākha.

[2] Das Wort ₒk'red ist unbekannt; die Bedeutung ‚Längsrichtung‘ vermuthe ich nur daraus, dass es in Gegensatz zu ₒp'red
‚Querrichtung‘ gestellt ist.

Denkschriften der phil.-hist. Classe. XLVIII. Bd. II. Abh.

2

p'ye bre gaṅ las ma zad pai steṅ
nas ₒbras lu ša btab pai tsʼod ma
byas ₒdug pa la | ston pa Šākya gu
nas ṅed rnams kyi za mtsʼun mdzad
gda ba | bdag rnams ₒoṅ ba rje btsun
gyi tʼugs mkʼyen gyis gzigs pa lags sam
žus pas | rje btsun gyi žal nas | ṅas
pʼa pʼoṅ gi steṅ nas bltas pas | kʼyed
rnams ṅal bso žiṅ ₒdug pa mtʼoṅ gsuṅ
| yaṅ ston pa Šākya gu nas | bdag
cag rnams kyis ni pʼa pʼoṅ gi steṅ
na bsa gcig las rje btsun ma mtʼoṅ | de
dus rje btsun gaṅ na bžugs žus pas
| rje btsun gyi žal nas bsa de ka ṅa
yin mod | rluṅ sems la dbaṅ tʼob pai
rnal ₒbyor pa ₒbyuṅ ba bžii skye mcʼed
zil gyis non pas | lus ₒdod dgur bsgyur
bai rdzu ₒpʼrul ston nus pa yin | da
lan ṅas kyaṅ kʼyed rnams skal ldan
du ₒdug pas bkod pa lus kyi rdzu ₒpʼrul
bstan pa yin | ṅa de ltar byed pai
gtam mi la ma lab gsuṅ | yaṅ gŠen
rdor mos žus pa | rje btsun na niṅ
las kyaṅ sku mdaṅs šin tu bzaṅ bar
gda ba | lam sgo gṅis ni kʼa bas cʼod
de mii žabs tog ni med | lha ₒdres žabs
tog bgyis sam | ri dags kyi tʼeṅ ro
₀dra ṅed pa lags sam | ji ltar lags
žus pas | rje btsun gyi žal nas | ṅa
pʼal cʼer tiṅ ṅe ₀dzin lu ₀byams nas
zas za ma dgos | dus bzaṅ gi tsʼe mkʼa
₀gro rnams kyis tsʼogs kyi ₀kʼor loi
skal ba re yaṅ bskyal | res pʼye tʼur
mgo re tsam ₀gams | kʼyad par kʼa
saṅ rtai zla bai smad la | kʼyed bu
slob rnams kyis ṅai mtʼa bskor nas
bza btuṅ maṅ po draṅs pai ṅams
žig šar ba nas | žag maṅ poi bar du
za ma za sṅiṅ ye mi ₀dod pa žig byuṅ
ba | de dus kʼyed rnams kyis ci byas
pa yin gsuṅ | der kʼoṅ rnams kyis
dus rtsis pas | tʼugs dgoṅs rdzogs tʼabs
kyi mcʼod pa pʼul ba daṅ dus mtsʼuṅs

hinein!' Da hielten sie vor Freude kaum die Thränen
zurück und stürzten sämmtlich von allen Seiten herein
auf den Ehrwürdigen, um ihm Hände und Füsse zu
drücken, und weinten. Der Ehrwürdige aber sagte:
,Lasst das jetzt lieber sein und nehmt Speise zu
euch!' Jene machten zuerst eine Verneigung und
erkundigten sich nach seinem Befinden. Als sie sich
umschauten, fanden sie, dass von den früheren Vor-
räthen nicht mehr als ein Mass Mehl aufgebraucht
und dazu das in den Reis gelegte Fleisch nicht einmal
versucht war. Der Lehrer Çakyaguṇa sagte: ,Mit unseren Nahrungsmitteln richtest du ja ein Todtenmahl
her. Hat der Ehrwürdige vermöge seines Seherblickes
unsere Ankunft vorausgeschaut?' Der Ehrwürdige
erwiderte: ,Als ich oben von einem Felsblock her
Umschau hielt, sah ich euch, wie ihr euch ausruhtet.'
Wiederum fragte der Lehrer Çakyaguṇa: ,Wir haben
oben auf dem Felsblock nur einen Schneeleoparden,
den Ehrwürdigen aber nicht bemerkt; wo weilte denn
der Ehrwürdige zu jener Zeit?' Der Ehrwürdige entgegnete: ,Eben jener Schneeleopard war ich ja selbst:
Da nämlich die Yogin, welche Adepten der höchsten
Mystik[1] sind, die auf den vier Elementen beruhenden
Sinnesorgane durch ihren Glanz besiegen, so habe ich
die Fähigkeit erlangt, mich in einen beliebigen Körper verwandeln und so Trugbilder zeigen zu können.
Da ihr nun dessen würdig seid, habe ich euch diese
Verwandlung meines Leibes sehen lassen. Doch sagt
den Leuten nicht ein Wort davon, dass ich solches
thue.' Darauf fragte gŠen rdor mo: ,Das Aussehen
des Ehrwürdigen ist ja weit besser als im vorigen
Jahre. Haben etwa zu der Zeit, als die beiden Zugänge
des Weges durch den Schnee abgeschnitten waren
und kein Mensch dich bedienen konnte, Geister dir
Dienste geleistet? Oder hast du etwa dergleichen wie
Aasreste von Wild gefunden? Wie ging das wohl
zu?' ,Die meiste Zeit,' antwortete er, ,war ich ganz
in Beschauungen versunken, so dass ich keine Speise
zu essen brauchte. Zur Zeit der Festtage aber brachten
mir die Ḍāka von ihren Antheilen an den Opferspenden. Ich nehme immer eine Löffelspitze Mehl
auf einmal. Neulich überdies, in der zweiten Hälfte
des Pferdemonats (Caitra), wurdet ihr, meine Schüler,

---

[1] Ueber den Ausdruck rluṅ sems s. Jäschke, Dictionary 538a.

ₒdug ts'ul žib rgyas p'ul bas | rje btsun
gyi žal nas | ₒjig rten pa rnams kyis
dge rtsa bgyis pa bar dor p'an par ₒdug
ste | de bas kyaṅ da lta bar do gcod
pa stobs c'e gsuṅ | de nas k'oṅ rnams
kyis | rje btsun gña nam du gdan ₒdren
pai žu ba nan can p'ul bas | rje btsun
gyi žal nas | ṅa ₒdi ka raṅ ñams dga
žiṅ tiṅ ṅe ₒdzin ₒp'el bar ₒdug pas
mi ₒgro | k'yed raṅ rnams soṅ gsuṅ
bas | k'oṅ rnams kyis | da lan rje
btsun p'ar mi ₒbyon na | bdag rnams
la gña nam pa kun gyis k'yed kyis
rje btsun groṅs su bcug pa yin zer
gtam ṅan daṅ k'a c'ag c'en po raṅ
yoṅ | k'yad par Jo mo ₒUr mos | ñai
rje btsun k'yer la šog zer bai p'rin
ṅan yaṅ yaṅ bskur byuṅ bas | bla
ma mi gšegs na ñed rnams kyaṅ ₒdir
oc'i ba sgugs ciṅ bsdod žus pas | rje
btsun gyis k'oṅ rnams kyi nan ma
t'eg par ₒbyon par žal gyis bžes te |
rje btsun la dgun mk'a ₒgros | Mi la
ras pa k'yod la mi dgos ruṅ | ma ₒoṅs
pa na slob rgyud la dgos pas | gaṅs
gšog t'abs ₒdi ltar gyis zer | mk'a ₒgros
bslabs pa ltar byas pai dkyar de k'yer
| k'oṅ rnams byuṅ bai naṅ par ts'ur
byon pas | la t'og nas gŠen rdor mos
sṅon la p'yin rtsar ma pa sogs ña
ma rnams la | rje btsun sku ma noṅs
par ₒbyon žiṅ yoṅ bya bai gtam sñan
bsgrags | rje btsun dpon slob kyis kyaṅ
p'a p'oṅ dkar leb tu gro btab pai sar
| gtam sñan t'os pai ña ma p'o mo
rgan byis med pa kun | rje btsun gyi
gdoṅ bsu ste žal lta ru byuṅ ba kun
gyis | rje btsun gyi sku la ₒpyaṅs žiṅ
ñus smre sṅags kyi sgo nas sñun dri
žus | p'yag daṅ bskor ba byas | dei
ts'e rje btsun la p'yag ₒk'ar spai ber
ma lcag gcig yod pa ṅos ₒm̐ ts'ugs
mdzad | gaṅs gšog pai dkyar de žabs
la gsol te | p'a p'oṅ dkar leb kyi steṅ
nas | der ts'ogs pai ña ma rnams la

so lebendig in mir, dass ein solches Gefühl der
Sättigung durch reichliche Speise und Trank über
mich kam, dass ich viele Tage lang gar kein Bedürfnis fühlte, Nahrung zu mir zu nehmen. Was habt
ihr denn eigentlich zu jener Zeit gethan?' Jene rech-
neten die Zeit nach und fanden sehr genau heraus,
dass es zu ebenderselben Zeit geschah, da sie zu
seinem Gedächtnis in vollkommener Weise Opfergaben dargebracht hatten. Da sagte der Ehrwürdige:
,Ja wenn weltlich Gesinnte den Grund zur Tugend
legen, so gereicht das für den Zwischenzustand zum
Heil. Deshalb ist es jetzt auch von grosser Kraft, um
den Zwischenzustand zu verhindern.' Darauf drangen
jene mit Bitten in den Ehrwürdigen, einer Einladung
nach gÑa nam zu folgen. Der Ehrwürdige erwiderte aber: ,Ich bin von Freude darüber erfüllt,
doch da ich mit der Förderung meiner Meditation
beschäftigt bin, kann ich nicht gehen. Ihr selbst
aber mögt ziehen.' Da sagten jene: ,Wenn der
Ehrwürdige jetzt nicht mitkommt, werden uns alle
Einwohner von gÑa nam beschuldigen, dass wir
den Ehrwürdigen dem Tode preisgegeben haben;
so werden uns üble Nachreden und grosse Schmä-
hungen erwachsen.' Jo mo ₒUr mo sagte insbesondere: ,Da böse Nachrichten mit den Worten:
„Berichte doch von meinem Ehrwürdigen!" wieder
und wieder ausgestreut werden würden, so ist es
besser, dass auch wir, wenn der Lama nicht mitkommt, hier bleiben und den Tod erwarten.' Da
vermochte der Ehrwürdige ihrem Drängen nicht
länger zu widerstehen und versprach, mit ihnen zu
gehen. Im Winter hatten die Ḍāka zu dem Ehrwürdigen gesagt: ,Für dich, Milaraspa, ist es zwar
nicht nöthig, doch in Zukunft wird es für das
Geschlecht deiner Schüler nöthig sein: nimm daher
folgende Methode an, um einen Weg über das Eis
zu bahnen.' So trugen sie denn die nach Anweisung
der Ḍāka verfertigten Schneeschuhe, und als es mitt-
lerweile Morgen geworden war, machten sie sich
nach dieser Seite hin auf den Weg. Von der Passhöhe an eilte gŠen rdor mo voraus und verkündete den Leuten von rTsar ma und den übrigen
Zuhörern die frohe Botschaft: ,Der Ehrwürdige ist
heil und gesund und ist im Anzuge hierher!' Der
Ehrwürdige, Lehrer und Schüler nahmen auf einem

2*

sñun dri smed kyi lan du mgur odi
gsuñs so.

flachen, weissen Felsblock ihr Frühstück ein. Dahin eilten nun die Zuhörer, welche die frohe Botschaft vernommen, Männer und Frauen, Alt und Jung, alle ohne Unterschied, dem Ehrwürdigen entgegen. Alle Besucher hängten sich an den Ehrwürdigen, weinten und erkundigten sich unter Freudenthränen nach seinem Befinden. Sie verneigten sich vor ihm und umwandelten ihn. Der Ehrwürdige hatte ein Bambusrohr, das ihm als Stab diente; auf dieses stützte er beide Hände und das Kinn. Jene einen Weg über das Eis bahnenden Schneeschuhe hatte er an die Füsse gebunden und trug von dem flachen, weissen Felsblock herab den dort versammelten Zuhörern zur Antwort auf ihre Frage nach seinem Ergehen folgendes Lied vor:

1.

dpal de riñ bkra śis k'ri gdugs la

k'yed p'yag mjal gyi yon bdag p'o
mo dañ
ña rnal obyor Mi la ras pa gñis
oo skol ma śi p'rad pa blo ba dga

5 *ña mi rgan glu yi dkor mdzod yin*

bsñun dri smed kyi lan de glu yis ojal

sñan lhan ne gsan la t'ugs gtad ots'al

yoñ stay gi lo yi lo bžug³ la
yos bu yi lo yi lo mgo la
10 *wa rgyal zla bai ña drug la*
ña ok'or bai c'os la yid obyuñ nas
obrog la p'yi gañs kyi ra ba der

mi ñas kyañ dben pa sñog tu p'yin

Dass wir heute unter dem Thronbaldachin[1] des Glückes,
Ihr gabenspendende Männer und Frauen, die ihr mich ehrerbietig besucht,
Und ich, der Yogin Milaraspa,
Uns noch einmal vor dem Sterben getroffen haben, freut mein Herz!
Ich, alter Mann, bin eine Schatzkammer von Liedern:
So will ich mit diesem Liede die Antwort auf eure Frage nach meinem Ergehen zurückzahlen.
Ich bitte, die Ohrwärmer[2] abzunehmen und mir geneigtes Gehör zu schenken.
Am Jahresende des Tigerjahres,
Im Jahresanfang des Hasenjahres,
Am sechsten Tage des Wa-rgyal-Monates[4]
War ich des Kreislaufes Weise überdrüssig
Und begab mich in die Wildnis, das Gehege des Schneeberges La phyi,
Wo ich die ersehnte Einsamkeit fand.

---

[1] Tib. *k'ri gdugs*, nach Jäschke eine poetische Bezeichnung für die Sonne.
[2] Tib. *sñan lhan*; *lhan pa* ,flicken' = *glan pa*, *slan pa* ,flicken'; folglich ist auch *lhan* = *slan* in *rna lhan* oder respectvoll *sñan lhan*. S. über Ohrwärmer Rockhill, Notes on the ethnology of Tibet 695. Das Wort *ne* ist nicht belegt, kann aber dem Zusammenhang nach nur ,abnehmen, wegnehmen' bedeuten; vergl. bei Jäschke *nen pa* (westtibetisch) = *len pa*.
[3] *bžug* = *gžug*.
[4] Tib. *wa rgyal zla ba* lässt sich weder aus Jäschke's noch Desgodins' Angaben der Monatsnamen feststellen; vergl. die Einleitung, S. 4.

k'oṅ gnam sa gñis po gros byas te	Da hielten Himmel und Erde zusammen Rath
15 kyeṅ sir[1] rluṅ gi baṅ c'en btaṅ	Und entsandten als Eilboten den Wirbelwind. 15
ₒbyuṅ ba c'u rluṅ ɣyos ɣyos nas	Die Elemente des Wassers und Windes wurden ent-
	fesselt.
lho spriṅ dmug po mdun mar bsdus	Die dunklen Südwolken wurden zu einer Berathung
	versammelt.
ñi zla zuṅ gcig btson tu bzuṅ	Sonne und Mond, das Paar, wurden zu Gefangenen
	gemacht.
rgyu skar ñer brgyad star la brgyus	Die 28 Mondhäuser wurden festgebunden und ge-
	fesselt.
20 k'rims kyi gza brgyad lcags su bcug	Auf Befehl wurden die acht[2] Planeten in Eisen- 20
	ketten gelegt.
dgu ts'igs skya mo gtod la mnan	Die Milchstrasse wurde unsichtbar.
skar p'ran yoṅs la bud kyis btab	Die kleinen Sterne wurden ganz in Dunst gehüllt.
bud kyi mdaṅs ɣyogs t'a ma la	Als endlich alles vom Nebelglanz bedeckt war,
k'a ba ñin dgu mts'an dgu babs	Fiel Schnee neun Tage, neun Nächte lang,
25 c'a la ñin mts'an bco brgyad babs	Gleichmässig vertheilt auf achtzehn Tag-Nächte 25
	fiel er.
ₒbab c'e ste c'e ba bal ₒdab tsam	War grosser Schneefall, fielen die Flocken wie Woll-
	flausche,
ₒdab c'ags bya ltar ldiṅ žiṅ babs	Wie fliegende Vöglein herniederschwebend;
c'uṅ ste c'uṅ ba p'aṅ lo tsam	War kleiner Schneefall, kamen sie wie Spinnwirteln
	herunter,
buṅ ba lta bur ₒk'or žiṅ babs	Wie Bienen herumkreisend;
30 yaṅ c'uṅ sran ma yuṅs ₒbru tsam	Dann wieder fielen sie wie kleine Erbsen und Senf- 30
	körner,
k'u ₒp'aṅ bžin du ₒdril žiṅ babs	Gleich Spindeln, die sich rund drehen.
lar 'ka ba c'e c'uṅ ts'ad las ₒdas	Aber grosser und kleiner Schnee zusammen wächst
	zu einer unermesslichen Schicht:
mt'o gaṅs dkar gyi rtse mo dguṅ la	Schon berührt des hohen Schneeberges weisse Kuppe
reg	den Himmel;
dma rtsi šiṅ nags ts'al gñal žin mnan	Die niedrigen Bäume und Wälder liegen darnieder-
	gehalten am Boden.
35 ri nag po rnams la sku dkar gsol	Die schwarzen Berge kleiden sich in Weiss. 35
mts'o rba rlabs can la dar c'ags btab	Auf dem wogenden See bildet sich eine Eisdecke.
gtsaṅ c'ab sṅon mo sbubs su bcug	Der blaue Brahmaputra ist wie in einer Höhle ein-
	gesperrt.
sa mt'o dman med par t'aṅ du mñams	Der Boden, ob er sich hoch oder niedrig erhebt,
	gleicht sich zu einer Fläche aus.
ₒbab de ltar c'e bai raṅ bžin gyis	Und wie es denn bei so starkem Schneefall ge-
	schieht,

---

[1] Jäschke, Dictionary p. 7a, schreibt *kyiṅ sir* und hält dieses Wort für eine onomatopoetische Bildung, was ich indessen bezweifle; denn *sir* wird wohl mit *gsir ba* zusammenhängen, was ‚umdrehen‘ (zunächst von einer Spindel) und ‚schwingen‘ (einen Pfeil) bedeutet.

[2] Den gewöhnlichen sieben Planeten ist *Rāhu* als achter hinzugedacht.

40 *spyir mgo nag na mi la bzan btson gal*   Sind insgesammt die schwarzköpfigen[1] Menschen zur 40 Gefangenschaft gezwungen.

rkan bžii p'yugs la mu ge byun   Hungersnoth trifft das vierfüssige Vieh.

sgos ri dags dman mai ots'o ba bcad   Einzeln findet das arme Wild kein Futter mehr.

sten ₒdab c'ags bya la brgyags c'ad byun   Den beschwingten Vögeln oben ist die Nahrung ausgegangen.

ₒog bra ba byi ba gter du sbas   Unten verbergen sich Ziesel und Mäuse bei ihren Schätzen.

45 *gcan zan rnams la k'a c'ins byun*   Den Raubthieren sind die Rachen wie durch Fesseln 45 gesperrt.

de ₒdrai skyi ₒdan[2] skal ba la   In solchem allgemeinen Schicksal

na Mi la ras pai sgos skal du   War dies mein, Milaraspa's, besonderes Los.

sten nas ₒbab pai bu yug dan   Zwischen dem von oben herabfegenden Schneesturm,

dgun lo gsar sgan gi lhags pa dan   Dem kalten Windstoss des vollendeten Winterneujahrs

50 *na rnal ₒbyor Mi lai ras gos gsum*   Und meinem, des Yogin Milaraspa, Baumwollkleid, 50 zwischen diesen dreien

mt'o gans dkar gyi ltons su ₒk'rug pa šor   Entspann sich auf dem Gipfel des hohen, weissen Schneeberges ein Kampf.

k'a ba babs rgyala c'ab tu žu   Der herabfallende Schnee schmolz im Barte zu Wasser;

rlun ₒar ₒur c'e yan ran sar ži   Der Sturm legte sich trotz seines tosenden Brüllens von selbst;

ras gos me ltar ₒbar gyi gda   Das Baumwollkleid zerfiel wie von Feuer verzehrt.

55 *gyad dpe bžag gi ši srog de ru snol*   Am Ringer nahm ich mir ein Vorbild und kämpfte 55 sterbend um dieses Leben.

mts'on rtse rgyal gyi ral k'a de ru sprad   Mit siegreichen Waffenspitzen kreuzten wir die Klingen:

dpa la dor gyi ₒk'rug pa der rgyal bas   Des Feindes Stärke verachtend blieb ich Sieger in diesem Krieg.

spyir c'os pa yons la ts'ad cig gžag   Im allgemeinen ist in alle Geistliche ein Mass (von Kraft) gelegt,

sgos sgom c'en yons la ts'ad do bor   Und im besonderen sind allen grossen Mystikern zwei Masse bescheert.

60 *yan sgos gtum mo ras rkyan gi c'e ba ston*   Noch mehr im einzelnen erklärte mir die innere 60 Glut der Meditation den Vorzug eines einfachen Baumwolltuches.

nad ₒdu ba rnam bži sran la gžal   Die vier Ansammlungen der Krankheiten[3] wurden mir auf einer Wage zugewogen.

p'yis nan ₒk'rugs med par gtan k'rigs bgyis   Als aussen und innen der Aufruhr beigelegt war, wurde ein Vertrag geschlossen.

---

[1] Tib. *mgo nag*, häufiges Attribut der Menschen (besonders auch in mongolischen und türkischen Heldensagen). Vergl. Schlagintweit, Die Könige von Tibet, S. 833 no. 2; I. J. Schmidt, Die Thaten Bogda Geser Chan's, S. 7. Ebenso im tibetischen Gesarepos, s. Desgodins, in Annales de l'extrême Orient II, 133.

[2] Der hier zweimal vorkommende Ausdruck *skyi ₀dan* ist in den Wörterbüchern nicht erklärt.

[3] S. Mémoires de la Société Finno-Ougrienne XI, p. 84 v. *nad*.

rluṅ tsʻa graṅ gñis ka bcud la bor

pʻyis ci zer ñan par kʻa yis blaṅs

65 *skyi odaṅ[1] kʻa bai gdoṅ sri mnan*

pʻyis ɣyo ogul med par rtsigs la pʻab

sri[2] dmag dpuṅ bgyid yaṅ ma otsʻal bar

dus da res kyi okʻrug pa rnal obyor rgyal

ña mes kyi tsʻa bo stag slag can

70 *sñar wa glag gyon nas ogros ma myoṅ*

ña pʻa la bu skyes gyad kyi rigs

sdaṅ dgra la ru zur ocʻor ma myoṅ

ña gcan zan rgyal po seṅ gei rigs

gnas gaṅs kʻuṅs min par sdod ma myoṅ

75 *šoms kyaṅ bre mo bgyis pa lags*

ña mi rgan gyi kʻa la gtad yod na

pʻyis sgrub brgyud kyi bstan pa dar te mcʻi

grub tʻob oga re obyon te mcʻi

rnal obyor Mi la ras pa ña

80 *rgyal kʻams yoṅs la grags te mcʻi*

kʻyod bu slob rnams bsams šiṅ dad te mcʻi

gtam sñan žig pʻyi nas brjod te mcʻi

ña rnal obyor dbyar[4] kʻams bde bar byuṅ

kʻyed yon bdag rnams sku kʻams bde lags sam

žes gsuṅs pai tsʻe | ña ma rnams dga drags ñams our te bro okʻrabs pas | rje btsun yaṅ tʻugs ñams our ba ltar mdzad nas | žabs bro mdzad pas | pʻa

Der kalte wie der warme Wind, beides war mir gut bekommen.

Dann versprach mir der Feind, allen meinen Worten zu gehorchen.

Den Dämon, der das Gesicht des Schnees zeigte, 65 habe ich niedergeworfen.

Später, als Bewegung und Erschütterung aufhörte, liess ich mich in meinem Baue nieder.

Das Dämonenheer hatte die Lust zu handeln verloren:

So war damals der Yogin siegreich im Kampfe.

Als des Grossvaters Enkel besass ich einen Tigerpelz;

Vorher, als ich in ein Fuchsfell[3] gekleidet war, war 70 ich nie gegangen.

Als der dem Vater geborene Sohn gehöre ich zum Geschlecht der Ringer:

Dem übelgesinnten Feind bin ich nie entflohen (?).

Ich gehöre zum Geschlecht des Löwen, des Königs der Raubthiere:

Ich habe nie anders als mitten im Schnee der Berge gehaust;

Vorbereitungen sind dadurch unnütz gemacht (?). 75

Wenn ihr mir, dem Greise, Gehör schenkt,

Wird sich auch bei den künftigen Geschlechtern die Lehre ausbreiten.

Einige Siddha sind erstanden;

Ich, der Yogin Milaraspa

Bin in allen Ländern bekannt; 80

Ihr Schüler seid durch Nachdenken gläubig geworden.

Eine frohe Botschaft wird auch noch in Zukunft verkündet werden.

Ich, der Yogin, fühle mich wohl.

Ist auch euer Befinden gut, ihr Gabenspender?

Als er so gesprochen, führten die Zuhörer in brausendem Frohlocken einen stampfenden Tanz auf, und auch der Ehrwürdige stimmte in das Frohlocken ein und tanzte mit hüpfendem Fusse. Dadurch em-

---

[1] Siehe S. 14, Note 2.

[2] Der Holzdruck liest *srid*; das *d* ist wohl durch das folgende *d* von *dmag* veranlasst, Vergl. *sri* in V. 65.

[3] Tib. *wa glag*; *glag* steht hier wohl für *slag*. Vergl. oben S. 12, Note 2.

[4] Der Sinn von *dbyar* ist in diesem Zusammenhange unverständlich; vielleicht liegt ein Druckfehler vor.

p'oṅ ,dam rdzis pa bžin du nur nas
p'a p'oṅ gi steṅ t'ams cad žabs rjes
daṅ p'yag ok'ar gyi rjes kyis gaṅ
rked pa nur bai baṅ rim log ge ba
byuṅ ba de la | snar p'a p'oṅ dkar
leb zer ba la | p'yis dkyar rjes par
grags so | de nas ña ma rnams kyis
gña nam rtsar mar spyan draṅs nas
btaṅ rag gi bsñen bkur obul bai gral
du | legs se obum gyis da lan rje btsun
sku ma noṅs par p'ebs pa odi ci gaṅ
las kyaṅ dga ba žig byuṅ | sku mdaṅs
ni snar bas kyaṅ bzaṅ bar odug | t'ugs
dam bzaṅ po ni ye ok'ruṅs yin | žabs
tog mk'a ogro rnams kyis bgyis lags
sam zes žus pai lan du mgur odi
gsuṅs so.

pfing der Felsblock Eindrücke, wie wenn man in weiche Erde tritt, so dass der Felsblock oben von allen Fussspuren und den Spuren des Stabes voll war. Aus den Eindrücken des mittleren Theiles entstand ein umgekehrtes Bang rim;[1] früher wurde der Stein ,flacher weisser Felsblock' genannt, später war er als ,Spuren der Schneeschuhe' bekannt. Als ihn darauf die Hörer nach rTsar ma in gÑa nam einluden, sagte Legs se obum, die sich mit grossen Gaben[2] in der Reihe der Verehrer befand: ,Darüber, dass der Ehrwürdige diesmal heil und gesund angelangt ist, ist grosser Jubel entstanden. Dein Aussehen ist sogar vortrefflicher als früher. Ist es dir als Folge eines guten Gelübdes zutheil geworden oder sind es die Dienste, welche dir die Ḍāka geleistet haben?' Als Antwort auf ihre Frage trug er folgendes Lied vor:

2.

rje bla mai žabs la spyi bos odud

Mit dem Scheitel verneige ich mich dem Herrn, dem Lama zu Füssen!

byin brlabs kyi dṅos grub mk'a ogros byin

Die Siddhi des Segens haben mir die Ḍāka verliehen.

dam ts'ig gi bdud rtsi p'an pa c'e
dad pai mc'od pas dbaṅ po sos

Der Nektar des Gelübdes ist von grossem Nutzen, Durch gläubige Opferspenden wurden die Sinne wieder belebt;

5 *bu slob kyi ts'ogs sog bzaṅ nas byuṅ*

Die von den Schülern aufgehäuften Tugendverdienste entsprangen einem guten Herzen.   5

blta bai sems la stoṅ ñid šar

Im Sinne, welcher der Beschauung zugewandt ist, ist die Leerheit entstanden;

blta rgyui ṅo bo rdul tsam med

Das Wesen der Ursache der Beschauung ist nicht mehr als ein Atom:

blta bya lta byed stor nas t'al

Mit dem, der das zu Beschauende beschaut, ist es sonst vorbei;

btal bai rtogs ts'ul bzaṅ nas byuṅ

Doch die Art der Erkenntnis der Beschauung ist einem guten Herzen entsprungen.

10 *bsgom pa ood gsal c'u boi rgyun*

Die Erleuchtung durch die Meditation ist wie die Strömung eines Flusses.   10

sgom rgyui t'un ots'ams bzuṅ rgyu med

Man sollte nicht zum Zweck der Meditation Nachtwachen halten:

bsgom bya sgom byed stor nas t'al

Sonst ist es mit dem, der über das zu Meditirende meditirt, vorbei;

---

[1] Der treppenförmige Theil eines *mc'od rten* (stūpa, caitya).
[2] Tib. *btaṅ rag*, sonst nicht belegt; die gegebene Uebersetzung ist zweifelhaft.

sgom pai sñiṅ rus bzaṅ nas byuṅ

spyod pai bya byed ₒod gsal daṅ
15 *rten ₒbrel stoṅ par t'ag c'od pas*

spyad bya spyod byed stor nas t'al
spyod pai byed ts'ul bzaṅ nas byuṅ

p'yogs c'ai rtog pa dbyiṅs su yal

ṅo lkog c'os brgyad re dogs med

20 *bsruṅ bya bsruṅ byed stor nas t'al*

dam ts'ig bsruṅ ts'ul bzaṅ nas byuṅ

raṅ sems c'os skur t'ag c'od ciṅ

raṅ gžan don gñis grub pai p'yir

bsgrub bya sgrub byed stor nas t'al
25 *ₒbras bu grub ts'ul bzaṅ nas byuṅ*

mi rgan ṅa yi skyid glu ₒdi
dad can ṅa mai žus lan yin
sgrub pai sku ₒts'ams k'a bas bsdams
ₒts'o bai žabs tog mk'a gros bgyis
30 *yya gaṅs kyi ₒbab c'u btuṅ bai mc'og*

sus bsgrubs med par dpal las grub

las bya dgos med par so nam ts'ar

ₒp'ral sog ₒjog med par baṅ mdzod gaṅ

sems la ltas pas t'ams cad mt'oṅ
35 *dman sar bsdad pas rgyal sa zin*

mdun ma yaṅ rtse bla mai drin

bu slob yon bdag ₒk'or bcas rnams
dad pas žabs tog byed pa yi
drin lan c'os kyis ₒjal ba yin

Doch die Standhaftigkeit der Meditation ist einem
  guten Herzen entsprossen.
Das Vollziehen der Uebung beruht auf Erleuchtung;
Durch die Ueberzeugung von der Leerheit der 15
  Nidāna
Ist es mit dem, der das Auszuübende ausübt, vorbei:
Doch die Art des Vollziehens der Uebung ist einem
  guten Herzen entsprungen.
Die Bedenklichkeit, die aus der Eingenommenheit
  für die Welt entsteht, schwindet gänzlich;
Die acht Truglehren der Welt sind weder Gegen-
  stand der Hoffnung noch der Furcht:
Mit dem Bewahrer des zu Bewahrenden ist es sonst 20
  vorbei:
Doch die Art, wie das Gelübde bewahrt wurde, ist
  einem guten Herzen entsprungen.
Durch die Erkenntnis der Nichtigkeit der eigenen
  Seele
Ist es mit dem, der um des Strebens nach dem
  eigenen Heil und dem anderer
Das zu Erstrebende erstrebt, vorbei:
Doch die Art des Strebens nach der Frucht ist 25
  einem guten Herzen entsprungen.
Dies mein, des Greises, Freudenlied
Ist die Antwort auf die Frage der gläubigen Hörer.
Der Schnee hatte mich von der Welt abgeschnitten.
Lebensmittel setzten mir die Dāka dienend vor;
Das herabfliessende Wasser des Schieferbergs war 30
  das trefflichste Getränk.
Ohne dass mich jemand damit versorgte, ward es
  mir infolge meiner Vorzüge zuteil.
Ohne dass Arbeit erforderlich war, wurde die Wirt-
  schaft geführt.
Ohne die täglichen Bedürfnisse gesammelt und nieder-
  gelegt zu haben, war mein Vorratshaus ge-
  füllt.
In meine Seele schauend sah ich alles.
Auf niedriger Erde sitzend nahm ich einen Thron 35
  ein.
Der Gipfel befriedigender (?) Vollendung ist des
  Lamas Gnade.
Schüler sammt dem Gefolge der Gabenspender,
Den Dank für eure mir mit gläubigem Dienst
Erwiesene Güte habe ich euch durch religiöse Be-
  lehrung abgestattet.

3

40 *t'ugs dyes bar mdzod cig ⲟdir byon
    rnams

*žes gsuṅs pas | ston pa Śakya gu nas
p'yay p'ul te | da lan gyi k'a ba c'en
po de ⲟdra bas rje btsun sku ma noṅs
par byon pa daṅ | ñed slob ⲟbaṅs
rnams la ⲟaṅ ši ts'ad med par dpon
slob žal ⲟdzom pa ⲟdi šin tu dga ts'or
c'e ba žig byuṅ žiṅ | de riṅ gi skabs
su ñed rnams la drin lan du c'os kyi
skyes k'e gnaṅ gsuṅ ba bka drin c'e
bas | sku skyes la da lo dgun rje
btsun raṅ gi t'ugs ñams su ⲟk'ruṅs
pai c'os žig gnaṅ bar žu žus pas |
rje btsun gyis ston pa Śakya gu nai
žus lan ña ma rnams la byon skyes
su t'ugs ñams gnad kyi mdo drug ⲟdi
mgur du gsuṅs so.*

Freut euch in euerm Herzen, die ihr hierher ge- 40
kommen seid!

Als er so gesprochen hatte, sagte der Lehrer
Çäkyaguṇa, indem er sich verneigte: ‚Der Ehrwürdige ist dieses Mal heil und unversehrt von einem
solchen grossen Schnee hierher zurückgekehrt; auch
uns Schülern ist ausserordentlich grosse Freude darüber entstanden, dass wir vor dem Tode den Ehrwürdigen noch einmal getroffen haben. Für die Worte,
mit denen du uns am heutigen Tage als Gegengabe
den Nutzen eines religiösen Geschenkes gewährt
hast, sei vielmals bedankt! Gewähre uns nun bitte
als Geschenk eine religiöse Erbauung, wie sie im
Winter dieses Jahres in des Ehrwürdigen eigenem
Herzen erwuchs.‘ Da trug der Ehrwürdige als Antwort auf die Bitte des Lehrers Çäkyaguṇa zum Will-
kommengeschenk für die Hörer die sechs Kern-
punkte, wie sie in seinem Herzen entstanden waren,
in folgendem Liede vor.

3.

rje sum ldan bla mai žabs la ⲟdud

Dem Lama, der drei Gebieter hat, zu Füssen verneige ich mich!

*ṅas dben pa bsten pai sgom ñams
    rnams*

Ich, der sonst in der Einsamkeit lebt, weile mit den
Kräften meiner Meditation

*dguṅ do nub bkra šis gral gaṅ ⲟdir
k'yod ston pa Śakya gu nas dbu mdzad
    pai*

Am heutigen Abend in dieser Versammlung des Heils.
Du, Lehrer Çäkyaguṇa, das Haupt aller,

5 *gdan gaṅ ⲟdzom pai bu slob rnams*

Und ihr Schüler, deren Haus reich an allen Gütern 5
ist.

*smon lam dag pas ots'ams sbyar nas
yon bdag rDor mo bza ts'o daṅ*

Habt durch reines Gebet euern Geist vorbereitet.
Durch den Tischgenossen der Gabenspenderin rDor
mo

dam ts'ig c'os kyi ⲟbrel pa yi

Und die Wirksamkeit der Vorschriften seines Gelübdes

*k'yed bu slob rnams kyi žu p'ul ba
10 p'a byon skyes c'os žig gnaṅ ots'al skad*

Habt ihr, meine Schüler, ein Bittgesuch gestellt.
‚Wir bitten um Gewährung religiöser Belehrung als 10
väterliches Gastgeschenk,‘ lauteten eure Worte.

*don de la mc'id lan ⲟdi skad lo
ña ⲟk'or bai c'os la yid ⲟbyuṅ ste*

Zu diesem Zweck sei euch die Antwort mit folgender
Rede erteilt.
Des Kreislaufs Weise überdrüssig,

skyo nas la p'yi gaṅs la p'yin

Begab ich mich missmutig zum Schneeberg La
phyi.

dben gnas bdud odul p'ug pa der

15 rnal obyor Mi la ras pa ñas
zla drug sgom pai ñams šar ba
gnad kyi mdo drug glu ru len
p'yi yi yul drug dper bžag nas
naṅ gi skyon drug gtan la obebs
20 mi t'ar oc'iṅ bai sgrog drug la
t'abs kyis grol bai lam drug mt'oṅ

kloṅ drug gdeṅ du gyur pa las
ñams la bde ba rnam drug šar
glu de rjes bskyar nas ma blaṅs na

25 don de t'ugs su mi obyon pas
ts'ig brda yis ogrol na odi skad yin
yoṅs t'ogs brdugs yod na nam mk'a min

graṅs kyis laṅ na skar p'ran min

ɣyo ogul yod na ri bo min

30 p'ri bsnan yod na rgya mts'o min
zam pas sleb na c'u bo min
bzuṅ du yod na oja ts'on min
de ni p'yi yi dpe drug yin
yoṅs bza gtad yod na blta ba min

35 byiṅ rgod yod na sgom pa min
blaṅ dor yod na spyod pa min

rnam rtog yod na rnal obyor min
šar nub yod na ye šes min

skye oc'i yod na saṅs rgyas min
40 de ni naṅ gi skyon drug lags
yoṅs že sdaṅ c'e na dmyal bai sgrog
ser sna c'e na yi dags sgrog
gti mug c'e na byol soṅ sgrog

odod c'ags c'e na mi yi sgrog

Am einsamen Orte, in jener Dämonen bekehrenden
    Grotte
Erwuchs mir, dem Yogin Milaraspa,   15
Sechs Monate lang die Kraft der Meditation.
Nun singe ich das Lied von den sechs Kernsätzen.
Die sechs Gebiete der Sinne als Gleichnis nehmend,
Bringe ich die sechs inneren Mängel in Ordnung.
Für die sechs bindenden Fesseln der Nicht-Befreiung 20
Sehe ich die sechs Wege zur Befreiung durch die
    verschiedenen Mittel.
Aus den sechs zuversichtlichen Unermesslichkeiten
Entsteht das sechsfache geistige Wohlbefinden.
Wenn man dies Lied nicht auch später wiederholt
    singt,
Dringt sein Sinn nicht ins Herz;   25
Nun will ich ihn mit Worten erklären wie folgt.
Wenn sich ein Hindernis entgegenstellt, gibt es
    keinen Himmel;
Wenn sie gezählt werden können, gibt es keine
    kleinen Sterne;
Wenn Bewegung und Erschütterung ist, gibt es
    keine Berge;
Wenn Abnahme und Zunahme ist, gibt es kein Meer; 30
Wenn man über Brücken geht, gibt es keinen Fluss;
Wenn er ergriffen wird,[1] gibt es keinen Regenbogen;
Das sind die sechs Gleichnisse der Aussenwelt.
Wenn man mit reichen Vorräten lebt, ist keine
    Beschauung;
Wo Zerstreuung ist, gibt es keine Meditation;   35
Wo ein Schwanken zwischen Für und Wider[2] ist,
    gibt es keine Übung;
Wo Skepsis ist, gibt es keinen Yoga.
Wo Aufgang und Untergang ist, gibt es keine
    Weisheit.
Wo Geburt und Tod ist, gibt es keinen Buddha.
Dies sind die sechs inneren Mängel.   40
Wo grosser Hass herrscht, ist die Fessel der Hölle;
Wo grosser Geiz herrscht, ist die Fessel der Preta;
Wo grosse Unwissenheit herrscht, ist die Fessel der
    Tiere;
Wo grosse Leidenschaft herrscht, ist die Fessel der
    Menschen;

---

[1] Wohl von dem die Epilepsie verursachenden Planetendämonen, der herabsteigt, um von dem Wasser zu trinken, in welches das eine Ende des Regenbogens getaucht ist. S. Ramsay, Western Tibet, p. 130, v. rainbow.
[2] Tib. blaṅ dor, über diesen Ausdruck s. Sitzungsberichte der bayerischen Akademie 1898, S. 537.

3*

45 p'rag dog c'e na lha min sgrog
   na rgyal c'e na lha yi sgrog
   ₒdi mi t'ar ₒc'iṅ bai sgrog drug lags

   yoṅs dad pa c'e na t'ar pai lam
   mk'as btsun bsten na t'ar pai lam

50 dam ts'ig gtsaṅ na t'ar pai lam

   ri k'rod ₒgrim na t'ar pai lam

   gciy pur sdod na t'ar pai lam
   sgrub pa byed na t'ar pai lam
   ₒdi t'abs kyis ₒgrol bai lam drug lags

55 yoṅs lhan cig skyes pa gñug mai kloṅ

   p'yi naṅ med pa rig pai kloṅ

   gsal ₒgrib med pa ye ses kloṅ

   k'yab gdal c'en po c'os kyi kloṅ

   ₒp'o ₒgyur med pa t'ig lei kloṅ

60 rgyun c'ad med pa ñams kyi kloṅ

   de dag gdeṅ can kloṅ drug lags

   yoṅs lus la gtum mo ₒbar bas bde

   rluṅ ro rkyaṅ dbu tir c'ud pas bde

   stod byaṅ c'ub sems kyi rgyun ₒbab
   bde

65 smad daṅs mai t'ig les k'yab pas bde

   bar dkar dmar t'ug p'rad brtse bas
   bde

   lus zag med bde bas ts'im pas bde

   de rnal ₒbyor ñams kyi bde drug lags

Wo grosser Neid herrscht, ist die Fessel der Asura; 45

Wo grosser Stolz herrscht, ist die Fessel der Götter:

Dies sind die sechs bindenden Fesseln der Nicht-Befreiung.

-Grossen Glauben hegen ist ein Weg zur Befreiung;

Auf gelehrte Geistliche vertrauen ist ein Weg zur Befreiung;

Ein reines Gelübde haben, ist ein Weg zur Be- 50 freiung;

In der Bergwildnis umherwandern ist ein Weg zur Befreiung;

Allein leben ist ein Weg zur Befreiung;

Bannungen vollziehen ist ein Weg zur Befreiung.

Dies sind die sechs Wege zur Befreiung durch die verschiedenen Mittel.

Das Mitgeborenwerden ist die natürliche Uner- 55 messlichkeit;

Die Uebereinstimmung des Äusseren und Inneren ist die Unermesslichkeit des Wissens;

Die Uebereinstimmung von Licht und Schatten ist die Unermesslichkeit der Weisheit;

Das grosse Allumfassen ist die Unermesslichkeit der Religion;

Das Unwandelbare ist die Unermesslichkeit der Beschauung;

Das Ununterbrochene ist die Unermesslichkeit der 60 Seele;

Dies sind die sechs zuversichtlichen Unermesslichkeiten.

Wenn im Leibe die innere Glut entfacht ist, fühlt sich der Yogin wohl;

Wohl, wenn die Luft aus der rechten und linken Ader des Herzens in die mittlere eintritt;

Wohl im Oberleib durch das Herabfliessen der Bodhi;

Wohl im Unterleib durch die Verbreitung des Chylus- 65 samens;

Wohl in der Mitte durch die Liebe des Erbarmens beim Zusammentreffen des weissen Samens der rechten Ader und des rothen Blutes der linken Ader;

Der ganze Leib fühlt sich wohl durch die Befriedigung des glücklichen Gefühls der Sündlosigkeit;

Dies ist das sechsfache geistige Wohlbefinden des Yogin.

ńa gnad kyi mdo drug don gyi glu

70 zla drug bsgoms pai ńams dbyańs ᵒdi

k'yed bu slob ᵒts'ogs pai dga ston mdzod

lar mi dga ba ᵒts'ogs pa da res tsam

c'ań bdud rtsir ᵒtuń ba kun kyań mńes
mi rgan glu rgan len pa de

75 k'yed bu slob rnams kyi ńo ma c'ogs
da res kyi c'os skyes de la mdzod

t'ugs dges par bkra šis c'os la ᵒbuńs

las dag pai smon lam ᵒgrub par šog

ces gsuńs pas | gŠen rdor mos rje btsun
rin po c'e dus gsum sańs rgyas lta
bu dań mjal ruń | žabs tog byed pa
dań | p'yags p'yi la ᵒbreńs nas c'os
byed pa dag lta smos te | mos pa
tsam yań mi byed pa ᵒdi rnams byol
soń bas kyań glen pa ᵒduġ žus pas
| rje btsun gyi žal nas | ńa la dag
mos pa ma byas kyań ruń ste | mi
lus rin po c'e sańs rgyas kyi bstan
pa dar dus ᵒt'ob nas c'os mi byed pa
rnams | šin tu glen pa yin gsuń mgur
ᵒdi gsuńs so.

Dies ist mein Lied vom Sinn der sechs Kernsätze;
Dies ist der Sang der Seele, die sechs Monate medi- 70
tiert hat.
Ihr, meine Schüler, veranstaltet den Versammelten
ein Fest!
Jetzt, wo die Männer in froher Stimmung versammelt
sind,
Schlürft alle freudig den Nektar des Bieres!
Mit diesem alten Lied, das ich alter Mann gesungen,
Habe ich nun euren Wunsch, meine Schüler, erfüllt. 75
Es ist denen ein Schatz, welchen jetzt erst die
Lehre aufgegangen ist.
Werft euch freudigen Sinnes mit Macht auf die
Religion des Segens!
Möge das Gebet, das reiner That entspringt, in
Erfüllung gehen!
Als er so gesprochen, sagte gŠen rdor mo:
‚Es gibt Leute, die, obwohl sie mit dem ehrwürdigen
Rin po che,[1] vergleichbar dem Buddha der drei
Zeiten, zusammentreffen, thörichter als Tiere ihm
nicht die geringste Achtung bezeigen, geschweige
denn Dienste leisten und in seinem Gefolge als
Diener einen religiösen Wandel führen.‘ Der Ehr-
würdige erwiderte: ‚Mögen sie mir immerhin keine
Hochachtung bezeigen! Wer zur Zeit der Aus-
breitung von Buddha's Lehre in dem kostbaren
menschlichen Leibe[2] wiedergeboren wurde und keinen
religiösen Wandel führt, ist sehr thöricht.‘ Darauf
trug er folgendes Lied vor.

4.

sgra bsgyur Mar pai žabs la ᵒdud

k'yod gsan dań yon bdag dad pa cań
dam c'os t'ań mar gdal ba la

sdig pa lab bcol du byed pa šin tu
glen

Dem sprachgewandten Marpa zu Füssen verneige
ich mich.
Du, merke auf und auch ihr, gläubige Gabenspender!
Jetzt, da die heilige Religion wie in einer Ebene
ausgebreitet vor euch liegt,
Ist es sehr thöricht, durch irriges Reden Sünden
zu begehen.

---

[1] D. i. Edelstein, Titel eines jeden Lama höherer Classe (Jäschke).

[2] In der mongolischen Erzählung ‚von dem Knaben, der ohne Sattel auf dem schwarzen Ochsen ritt‘ (A. Popov, Mongolische Chrestomathie, Kasan, 1836, p. 20) heisst es: ene yirtintsü teghēre chamuk amitan ene amitu dsayaghan tu kümün ētse basa teghēre türlikü üghei bui; kümün-ü beye-yi olbasu erdeni-tü beye kemekteyü . . . ‚In dieser belebten Natur aller Geschöpfe auf dieser Welt wird kein höheres Wesen geboren als der Mensch; da ich nun den Leib eines Menschen erlangt habe, welcher der kostbare Leib genannt wird . . .‘

5 dka yaṅ t'ob pai dal ₒbyor lus po ₒdi

mi t'se stoṅ zad la skyₑl ba šin tu
glen

groṅ k'yer gyeṅ ra žig pai dur k'rod
du

dus rtag tu sₐlod pa šin tu glen

bza mi ts'oṅ ₒdus ₒgron po lta bu la

10 ṅan šags ₒt'ab mo byed pa šin tu glen

ts'ig šnan grags sgyu mai raṅ sgra la

že ṅan gal du ₒdzin pa šin tu glen

sdaṅ dgra me tog bžin du yal rgyu la

ₒt'ab mos raṅ srog skyel ba šin tu glen

15 gñen ts'an blo brid rdzun gyi zol k'aṅ
la

ši ts'e mya ṅan byed pa šin tu glen

nor rdzas zil pa lta bui ɣyar po la

ser snai mdud pas ₒc'iṅ ba šin tu glen

p'uṅ po mi gtsaṅ rdzas kyi rkyal ba
la

20 bzaṅ ₒdod byi dor byed pa šin tu
glen

gdams ṅag bdud rtsii zas mc'og ₒdi

zas nor don du ₒts'oṅ ba šin tu glen

glen pa maṅ po ₒts'ogs pa la

bcad na lha c'os raṅ mgo t'on

25 p'o druṅ na rnal ₒbyor ṅa bžin mdzod

ces gsuṅs pas | der ts'ogs pai ṅa ma
rnams kyis t'ugs rjes bzuṅ | bdag rnams
kyaṅ rje btsun lta bui bcaṅ po druṅ
po raṅ ma byuṅ ruṅ | glen pai zla
la ma lus tsam gcig e yod lta ba
lags pas | rje btsun yul p'yos ₒdir t'ugs
p'ab nas | ñed gson po rnams la c'os

Das Leben dieses Leibes, dessen innerer Friede 5
schwer zu erlangen ist,
Nutzlos zu vergeuden ist sehr thöricht.

Auf dem Begräbnisplatz einer Stadt mit verfallener
Erdmauer
Beständig seine Zeit hinzubringen ist sehr thöricht.
Unter Ehegatten, Marktleuten, Fremdlingen, unter
solchen
Durch schlechte Scherze Streit zu erregen ist sehr 10
thöricht.
Mit hochtönenden trügerischen Worten der eigenen
Stimme
Die Bosheit hochschätzen ist sehr thöricht.
Wenn der übelgesinnte Feind wie eine Blume ver-
schwindend abzieht,
Ist es sehr thöricht, durch einen Kampf sein Leben
aufs Spiel zu setzen.
Wenn die Verwandten in dem trügerischen Hause 15
der Tücke und Lüge
Verschieden sind, ist es sehr thöricht zu klagen.
Da der Reichtum nur ein geliehener Gegenstand,
einem Tautropfen gleich, ist,
Ist es sehr thöricht, vom Knoten des Geizes um-
schlungen zu sein.
Da der Leib nur ein mit unreinen Stoffen gefüllter
Sack ist,
Ist es sehr thöricht, sich mit Eitelkeit zu schmücken. 20

Der Nektar guter Lehren ist die köstlichste Speise:
Solche Speise gegen Geld zu verkaufen ist sehr
thöricht.
Ja, in grosser Zahl sind die Thoren angesammelt;
Wenn sie unschädlich gemacht sind, wird die Götter-
religion ihr Haupt erheben.
Wenn ein Mann klug ist, möge er handeln wie ich, 25
der Yogin!

Als er so gesprochen, wurden die dort versammelten Hörer von Mitleidsgefühl ergriffen. ‚Wir
sind zu der Ansicht gelangt,‘ sagten sie, ‚dass wir
alle Gefährten der Thoren sind, ausgenommen du
allein. Doch da ja ein so gescheiter und kluger
Mann wie der Ehrwürdige sich nicht von selbst
entwickelt, so bitten wir dich, weil du dich ja doch

gsuṅ žiṅ bla mc'od | gšin po rnams kyi yar ₒdren la rgyun du bžugs par žu žus pas | rje btsun gyi žal nas | ṅa la p'yi gaṅs la sgom dgos pa bla mai bka luṅ yod pas | t'og re tsam bsdod | gži bzuṅ nas k'yed yon bdag gi ṅo ₒdzin pai mi c'os ṅas mi šes | k'yed rnams kyaṅ ṅa la mig brñas k'yer ba las mi yoṅ gsuṅ mgur ₒdi gsuṅs so.

wohl in dieser Gegend anzusiedeln gedenkst, beständig hier zu wohnen, um uns Lebenden durch Verkündigung der Religion Geistlicher und Opferpriester zu sein und die Toten mit dir himmelwärts emporzuziehen.' Der Ehrwürdige erwiderte: ‚Ich muss auf dem Schneeberg La phyi der Beschauung obliegen: so will es das Gebot meines Lama; da oben will ich eine Zeit lang wohnen. Ich kenne nicht die Sitte der Leute, die gegen euch Gabenspender zu viel Rücksicht walten lassen, indem sie ihren Wohnsitz unter ihnen aufschlagen. Weil ihr mich am Ende mit verächtlichen Augen anschauen könntet, komme ich nicht.' Darauf trug er folgendes Lied vor.

5.

rje lho brag Mar pai žabs la ₒdud

k'yed ₒdir ts'ogs kyi yon bdag p'o mo rnams

ṅa rnal ₒbyor Mi la ras pa la

ₒgyur med kyi dad pa gtiṅ nas gyis

5 k'a že med pai gsol ba t'ob

gži gcig tu rgyun du bsdad gyur na

ₒp'ral mñam du ₒgrogs pas snaṅ ba sun

mig ₒdris c'es na gšis kyis brñas

ₒgrogs yun riṅs na re ₒk'aṅs maṅ

10 gšis ṅan ₒt'ab mos dam ts'ig dkrug

grogs ṅan gyis dge sbyor bzaṅ po spur

gcom dun[1] gyi k'a gzus las ṅan sog

yin min k'a rtogs dgra bo ₒk'or

p'yogs c'ai ɣyo sgyu sdig pa c'e

Dem Gebieter Marpa aus Lho brag zu Füssen verneige ich mich.
Ihr gabenspendende Männer und Frauen, die hier versammelt sind,
An mich, den Yogin Milaraspa,
Hegt von ganzem Herzen unwandelbaren Glauben!
Mögt ihr ungeheuchelte Bitten erlangen! 5
Wenn man beständig an einem und demselben Wohnsitz lebt,
Wird man bald durch das Aufgehen in der Alltäglichkeit der Erscheinungen müde.
Wenn sich Menschen sehr an einander gewöhnen, entsteht durch die Natur der Sache schliesslich Verachtung.
Wenn die Gemeinschaft lange dauert, ergeben sich viele Reibungen.
Streit, der aus ungünstigen Verhältnissen entsteht, 10 stört das Gelübde.
Schlechte Freunde verscheuchen die gute Andachtsübung.
Durch unüberlegte Worte des Stolzes werden böse Werke angehäuft.
Durch das Schwanken zwischen Recht und Unrecht schafft man sich Feinde.
Betrug zu eigenem Vorteil ist eine grosse Sünde.

---

[1] dun ist unerklärt; zu k'a gzu vergl. gzu lum.

15 *rań zas k'a lan bsam ńan ok'rug*

gšin zas kyi ro dom t'og t'abs med

k'yed mi nag gi bla mc'od šin tu lod

ogrogs kyin brñas na yi mug ldań

ts'ań mań gi dpon po oc'i kar sdug

20 *de bas ri k'rod kyi rnal obyor pa*
groń yul du las na gyod re c'e
ńa p'yogs med kyi ri k'rod ogrim
    du ogro
k'yed dad ldan rnams ts'ogs sog ńo
    mts'ar c'e
odir ts'ogs kyi yon bdag p'o mo rnams

25 *bla mc'od gyis zer bag c'ags bzań*

smon lam gyi obrel pas yań yań mjal

    žes gsuńs pas | ña ma rnams ńed
dag rna ba mi sun te | rje btsun
t'ugs sun dgoṅs pa yin pa odug | da
žu ba nan gyis p'ul ruń gsan dgoṅs
byuṅ dogs med | bdag cag rnams la
gzigs nas | la p'yi la yań yań myur
du obyon par žu | ces žus nas | obul
ba mań po p'ul ba rnams ma bžes
pa las | der ts'ogs pai bu slob kun
mc'og tu ńo mts'ar du gyur nas | dga
mgu yid rańs śiṅ | rje btsun la mi
p'yed pai dad pa t'ob par gyur to.

gańs mgur gyi bskor ro.

Wenn schon bei der Danksagung für die eigene 15
    Nahrung böse Gedanken in Aufruhr geraten,
Wie ist es dann möglich, die Sporteln des Leichen-
    schmauses zu empfangen!
Es wäre sehr leichtsinnig, wenn ich Geistlicher und
    Opferpriester von Euch Laien sein wollte.
Wenn aus dem Zusammenleben Verachtung entstände,
    würde mich Verzweiflung ergreifen.
Noch mehr als der Besitzer vieler Häuser sich in
    der Todesstunde unglücklich fühlt,
Empfindet der Yogin der Bergwildnis   20
Bei der Thätigkeit im Dorfe grosse Reue.
Ich gehe, um gleichgültig gegen die Welt in der
    Bergwildnis herumzuwandern.
Ihr Gläubige, seid für eure Gaben und Dienste be-
    dankt!
Ihr gabenspendende Männer und Frauen, die hier
    versammelt sind,
Eure Liebe, die sich in den Worten bekundet: ,Sei 25
    unser Geistlicher und Opferpriester!' ist schön.
Dank der Wirkung eures Gebets werde ich euch
    wiedersehen!

Als er so gesprochen, sagten die Zuhörer:
,Unsere Ohren sind noch gar nicht müde zu hören,
doch der Ehrwürdige wird wohl müde geworden
sein. Obwohl wir nun mit Bitten in dich gedrungen
sind, besorgen wir doch, dass dir nicht der Gedanke
kommen wird, denselben Gehör zu schenken. Daher
bitten wir dich, wieder bald auf den La phyi zurückzukehren.' Allein die zahlreichen Geschenke,
die sie ihm geben wollten, nahm er nicht an. Da
gerieten alle dort versammelten Schüler in das
grösste Staunen und erlangten, von höchster Freude
ergriffen, unerschütterlichen Glauben an den Ehrwürdigen.

Dieser Abschnitt enthält den Sang vom Schneeberg.

II.

Die Felsen-Rākshasī von Ling ba.

(fol. 16 b 5 — 23 a 7.)

Na mo gu ru.

*rje btsun Mi la ras pa de ñid
la | gña nam rtsar mai ña ma sogs
gña nam pa rnams kyis bžugs pai
gsol ba ci btab ruṅ ma gsan par | bla
mai bka sgrub p'yir skyid groṅ gi
ri bo dpal ₒbar la sgom du byon pa
las | liṅ bai brag p'ug la t'ugs soṅ
ste der re žig t'ugs dam skyoṅ žiṅ
bžugs pa las | nam žig nam srod byiṅ
soṅ bai ts'e | rje btsun bžugs pai γγon
p'yogs na | brag gi ser k'a žig yod
pa dei srubs la gšugs skad žig ts'er
ts'er ₒdug pa las bžeṅs te gzigs pas
ci yaṅ mi ₒdug | ña sgom c'en pa raṅ
snaṅ ₒk'rul bar ₒdug dgoṅs nas | yaṅ
gzims mal du bžugs pas brag gi ser
k'a na ts'ur | ₒod c'en po žig gi sna
la mi dmar po gla ba nag po la žon
pai sna bud med bzaṅ mo gcig gis
k'rid pa žig byuṅ nas | mi des rje
btsun la gru moi p'ul brdeg žig byas
nas rluṅ ts'ub žig tu yal soṅ | bud
med de k'yi mo dmar mo žig tu soṅ
nas žabs γγon pai mt'e boṅ bzuṅ ste
gtoṅ du ma ñan pas | brag srin moi
c'o ₒp'rul du mk'yen nas mgur ₒdi
gsuṅs so.*

Verehrung dem Meister!

Obwohl an den ehrwürdigen Milaraspa die Zuhörer von rTsar ma in gÑa nam und die anderen Leute von gÑa nam Bitten richteten, noch länger zu verweilen, hörte er nicht auf sie, sondern machte sich, um das Gebot seines Lama zu erfüllen, zum Zweck der Meditation auf den Weg nach dem Berge dPal ₒbar des Dorfes sKyid grong. Da fand er Gefallen an der Felsgrotte von Ling ba und verweilte dort einige Zeit lang, der Beschauung obliegend. Eines Abends, als die Dämmerung schon hereingebrochen war, liess sich zur Linken des Platzes, wo sich der Ehrwürdige befand, in der Ritze eines Felsspaltes zu wiederholten Malen ein Pfeifen vernehmen. Er erhob sich und schaute hin, doch es war nichts da. In dem Gedanken: ‚Ich grosser Meister der Meditation muss mich wohl in einer Täuschung befinden' legte er sich wieder auf sein Lager hin. Da erschien in der Felsspalte auf dieser Seite ein heller Lichtstrahl, an der Spitze desselben ein roter Mann, welcher auf einem schwarzen Moschustiere ritt, das von einer schönen Frau geführt wurde. Jener Mann versetzte dem Ehrwürdigen einen Stoss mit dem Ellenbogen und verschwand dann in einem Wirbelwinde. Jene Frau verwandelte sich in eine rote Hündin und packte den grossen Zeh seines linken Fusses, so dass er ihn nicht losmachen konnte. Da er wusste, dass dies das Zauberspiel einer Felsen-Rākshasī sei, trug er folgendes Lied vor.

6.

*drin can Mar pai žabs la ₒdud
mi ña la t'o ₒts'am glags lta žiṅ*

*mi sdug sprul pai gzugs ston pa
k'yod liṅ ba brag gi brag srin mo
5   las ñan ₒdre mo ma yin nam*

Dem huldreichen Marpa zu Füssen verneige ich mich!
Die du auf eine Gelegenheit lauertest, mir Hohn zuzufügen,
Und dich in hässlich verwandelter Gestalt zeigst,
Du Felsen-Rākshasī vom Ling ba-Felsen,
Bist du nicht eine übelthuende Dämonin?   5

Denkschriften der phil.-hist. Classe. XLVIII. Bd. II. Abh.

4

- 299 -

glu śñan pai ₒgyur k'uɡs ñas mi śes	Des Liedes Wohllaut zu gewinnen verstehe ich nicht,
ñag bden pai tśig la k'yod kyań sñan	Doch höre du auf die Worte wahrer Rede!
ya ki dguń śñon gyi dkyil žig na	In der Mitte des blauen Himmels dort oben
ñi zla zuń gcig yyań ćags pa	Erwächst das Glück des Sonne-Mondpaares;
10 *de ño mts'ar lha yi gžal yas k'ań*	Jenes wundervollen Götterpalastes 10
ₒod zer skye ₒgroi dpal du śar	Strahlen scheinen zur Wohlfahrt der Geschöpfe;
las su gliń bži bskor ba la	Wenn sie in ihrem Amte die vier Continente um-
	kreisen,
gza k'yab ₒjug dgra ru ma ldań cig	Möge Rahu nicht als Feind wider sie aufstehen!
lho nags mai gseb kyi ldiń k'ań du	Unter dem Laubdach im Dickicht des Südwaldes
15 *stag mo ri k'rai yyań ćags pa*	Erwächst das Glück der buntgestreiften Tigerin; 15
de gcan zan yoñs kyi gyad pa yin	Unter allen Raubtieren, ist sie die Ringerin;
dpa rtags su rań gi srog mi p'ańs	Zum Zeichen ihrer Stärke schont sie des eigenen
	Lebens nicht;
k'oń dog moi ₒp'rań la ₒgrim pai ts'e	Wenn sie auf den schmalen Felspfaden umher-
	wandelt,
sa mda dyra ru ma ldań cig	Möge die Falle nicht als Feind wider sie aufstehen!
20 *nub ma p'ań yyu mts'o mer ba la*	In dem länglichen Türkissee Manasarovara im 20
	Westen
lto dkar ña yi yyań ćags pa	Erwächst das Glück der weissbauchigen Fische;
de ₒbyuń ba ću yi yar mk'an yin	Sie sind die Tänzer im Element des Wassers;
ya mts'an du gser mig ₒk'yil pa yod	Wunderbar rollen sie die Goldaugen;
k'oń ₒdod yon zas p'yir ₒbreń bai ts'é	Wenn sie der Lust und der Nahrung wegen ein-
	ander folgen,
25 *lcags kyu dgra ru ma ldań cig*	Möge der Angelhaken nicht als Feind wider sie 25
	aufstehen!
byań gi brag dmar bsam yas la	Auf dem roten Felsen bSam yas im Norden
bya rgyal rgod poi gyań ćags pa	Erwächst das Glück des Geiers, des Königs der
	Vögel;
de ₒlab ćags bya yi drań sroń yin	Er ist der Einsiedler (Ṛishi) unter den beschwingten
	Vögeln;
ño mts'ar gžan gyi srog mi gcod	Wunderbarer Weise nimmt er andern nicht das
	Leben;
30 *k'oń ri gsum rtse la bzan ₒts'ol ts'e*	Wenn er auf den Gipfeln der drei Berge seine 30
	Nahrung sucht,
t'ag sñi la dgra ru ma ldań cig	Möge die aus Stricken gewundene Schlinge nicht
	als Feind wider ihn aufstehen!
brag rgod ts'ań can gyi liń ba la	Auf dem den Horst des Felsengeiers bergenden
	Ling ba
Mi la ras pai yyań ćags pa	Erwächst Milaraspa's Glück;
ₒdi rań gžan don gñis sgrub mk'an	Er trachtet nach seinem eigenen Heil und dem
yin	anderer;
35 *bden rtags su ts'e ₒdi blos btań ste*	Zum Zeichen der Wahrheit hat er auf dies weltliche 35
	Leben verzichtet
rgyu byań ćub mc'og tu sems bskyed ñas	Und zur Grundlage die höchste Bodhi gemacht;

dus c'e gcig lus gcig ₀di ñid kyi
rtse gcig sañs rgyas sgrub pa la
k'yod brag srin dgra ru ma ldañ cig

40 glu ño mts'ar dpe lña don dañ drug

ts'ig sñan ñags mod la ña gser t'ag
can
don mjal bar gda ₀am brag srin mo

las bsags kyin lci ba sdig pa yin

k'yod de la nan tan ma c'e bar
45 ñan sems gdug rtsub slar t'ul dañ
lar t'ams cad sems su ma ŝes na
rnam rtog gi ₀dre la zad pa med

sems ñid stoñ par ma rtogs na

₀dre log gis ldog par cañ gda ₀am

50 k'yod ma gnod ma gnod ₀dre mo ñan
mi ña la ma gnod slar log cig

ces gsuñs pas | žabs nas bzuñ ba de
ni ma btañ | gzugs med pai skad žig
gsi lan ₀di skad btab bo.

Zu derselben grossen Zeit, da dieser eine Leib
Unablässig nach der Buddhawürde strebt,
Mögest du, Felsen-Rākshasī, nicht als Feind wider
    ihn aufstehen!
So enthält dies Lied fünf wunderbare Gleichnisse und 40
    sechstens deren Sinn.
Ich, der goldene Fesseln trägt im Augenblick, wo
    das Gedicht entsteht,
Verstehe ich wohl seinen Sinn? Die Felsen-Rāksh-
    asī
Ist durch Ansammlung ihrer Werke mit schweren
    Sünden behaftet.
Da dein Eifer nicht gross ist,
Ist das wilde Gift deiner Bosheit wieder zu bändigen. 45
Doch wenn alle ihre Seele nicht kennen,
Gibt es eine unendliche Zahl von Dämonen der
    Wahrnehmung.
Wenn man nicht erkennt, dass die Seele selbst leer
    ist,
Wird sich dann wohl der Frevel der Dämonen ab-
    wenden?
Du Schadenstifterin, schadende böse Dämonin,   50
Ohne mir zu schaden, kehre zurück!

Als er so gesprochen, liess sie ihn, den sie
am Fusse gepackt hatte, dennoch nicht los und
sprach mit geisterhafter Stimme folgende Worte zur
Antwort.

7.

,e ma skal ldan rigs kyi bu
gcig pur rgyu bai sñiñ stobs can
ri k'rod ₀grim pai rnal ₀byor pa
dka bai spyod pai ño mts'ar can
5 mi k'yod kyis glu blañs rgyal poi bka

rgyal poi bka de gser bas lci
gser la ra ŝor ₀ts'añ du c'e

₀ts'añ de las slar ldog ma ŝes na

ñas sñon c'ad smras pa rdzun du
zad

O du Sprössling eines würdigen Geschlechts,
Der den Mut hat, allein zu wandeln,
In der Bergwildnis umherziehender Yogin,
Wunderbar in den Bussübungen!
Das von dir gesungene Lied ist wohl ein Gebot des 5
    Königs;
Des Königs Gebot ist schwerer als Gold.
Doch gegen Gold Messing auszutauschen ist ein
    grosses Vergehen;
Wenn man nicht versteht, sich von solchem Ver-
    gehen abzuwenden,
Erschöpfen sich meine vorhergehenden Worte in einer
    Lüge.

4*

10 *de yan ćad rje rgyal k'rims kyi glu*	Im ersten Teil will ich das Lied vom Königsrecht 10
	behandeln;
da dpe de las gras pai mćid cig ots'al	Nun bitte ich, eine aus jenen Gleichnissen aus-
	geschnittene Rede halten zu dürfen.
t'ugs ma yeńs dar cig gsan par žu	Geruhe einen Augenblick mit Aufmerksamkeit zu-
	zuhören!
ya ki dguń sňon gyi dkyil žig na	In der Mitte des blauen Himmels dort oben
o͜od gsal ñi zlai yyań ćags pa	Erwächst das Glück der leuchtenden Sonne und
	des Mondes;
15 *de ňo mts'ar lha yi gžal yas k'ań*	Jener wundervolle Götterpalast 15
gliń bžii mun pa sel lo skad	Vertreibt die Finsternis der vier Continente, sagt
	man:
k'oń las su gliń bži bskor bai ts'e	Wenn sie in ihrem Amte die vier Continente um-
	kreisen,
snań snań bya ra mi gtoń žiń	Nehmen sie sich nicht in Acht mit ihrer Helligkeit:
dkyil ok'or o͜od kyis ma slus na	Denn wenn Rahu nicht durch das Licht der Scheibe
	verführt würde,
20 *gza k'yab o͜juġ dgra ru ci la ldań*	Weshalb sollte er als Feind wider sie aufstehen? 20
šar gańs dkar šel gyi zur p'ud la	Im Haaraufsatz des Ost-Schneebergs ‚Weisser Kry-
	stall‘
ňar seń dkar moi yyań ćags pa	Erwächst das Glück des starken, weissen Löwen;
de byol soń bkod pai rgyal po ste	Er ist der die Tiere beherrschende König;
o͜bańs dud o͜gro k'rims la gnon no	Die unterthänigen Tiere hält er durch Gesetze
skad	nieder, sagt man.
25 *k'oń yya sňon zur la o͜bab pai ts'e*	Wenn er am Rande des blauen Schieferfelsens 25
	hinabsteigt,
k'ro gtum ňa rgyal mi će žiń	Ist sein Zornwüten und Stolz gross:
yyu ral sňon mo ma non na	Denn wenn er nicht mit seiner blauen Türkismähne
	den Schneesturm siegreich bestände,
bu yug dgra ru ci la ldań	Weshalb sollte er als Feind wider ihn aufstehen?
lho nags mai gseb kyi bliń k'ań na	Unter dem Laubdach im Dickicht des Südwaldes
30 *stag gu ri k'rai yyań ćags pa*	Erwächst das Glück des gestreiften jungen Tigers; 30
de gcan zan yoňs kyi gyad pa ste	Unter allen Raubtieren ist er der Ringer;
sder ćags zil gyis gnon no skad	Die Krallentiere besiegt er durch seinen Glanz,
	sagt man;
k'oń dog moi op'rań la o͜grim pai ts'e	Wenn er auf den schmalen Felspfaden umher-
	wandelt,
yań rtsal dregs pas mi šnems šiń	Ist er von Stolz auf seine hohe Geschicklichkeit
	aufgeblasen:
35 *ri mo o͜dzum gyis ma slus na*	Denn wenn die Falle nicht durch sein lächelndes 35
	Bildnis verführt würde,
sa mda dgra ru ci la ldań	Weshalb sollte sie als Feind wider ihn aufstehen?
nub ma p'ań yyu mts'o mer ba la	In dem länglichen Türkissee Manasarovara im Westen
lto dkar ňa yi yyań ćags pa	Erwächst das Glück der weissbauchigen Fische;
de o͜byuń ba ću yi gar mk'an te	Sie sind die Tänzer im Element des Wassers;

40 *lha draṅ sroṅ gzigs mor c'eo skad*

k'oṅ ₒdod yon zas p'yir ₒbreṅ bai ts'e

mi zas don du mi gṅer žiṅ

sgyu mai lus kyis ma slus na

lcags kyu dgra ru ci la ldaṅ

45 *byaṅ gi brag dmar bsam yas la*
bya rgyal rgod poi ɣyaṅ c'ags pa

de ₒdab c'ags bya yi draṅ sroṅ ste
ₒdab c'ags zil gyis gnon no skad

k'oṅ ri gsum gyi rtse la bzan ₒts'ol
ts'e

50 *ša k'rag bzan du mi ts'ol žiṅ*
gšog sgro ldem gyis ma slus na

t'ag sṅi dgra ru ci la ldaṅ
brag rgod ts'aṅ can gyi liṅ ba la

k'yod Mi la ras pai ɣyaṅ c'ags pa
55 *raṅ gžan don gṅis sgrub mk'an de*
rgyu byaṅ c'ub mc'og tu sems bskyed
nas
dus c'e gcig lus gcig ₒdi ñid kyis
rtse gcig saṅs rgyas bsgrubs nas kyaṅ
ₒgro drug gi lam sna ₒdren no skad

60 *k'yod rtse gcig bsam gtan sgom pai ts'e*
bag c'ags ₒt'ug pos rgyu byas te
raṅ sems ₒk'rul pai ñer len gyis

rnam rtog dgra ru ma laṅs na
ṅa brag srin dgra ru ci la ldaṅ

65 *l·ir bag c'ags bdud ₒdi sems las byuṅ*

sems kyi de ñid ma šes na

k'yod soṅ šig byas pas ṅa mi ₒgro
raṅ sems stoṅ par ma rtags na

Göttern und Einsiedlern sind sie ein grosses Schau- 40
spiel, sagt man;
Wenn sie der Lust und der Nahrung wegen ein-
ander folgen,
Trachten die Menschen nach dem Erwerb von
Nahrung:
Denn wenn der Angelhaken nicht durch ihren
Täuschungsleib verführt würde,
Weshalb sollte er als Feind wider sie aufstehen?
Auf dem roten Felsen bSam yas im Norden 45
Erwächst das Glück des Geiers, des Königs der
Vögel;
Er ist der Einsiedler unter den beschwingten Vögeln;
Alle Geflügelten besiegt er durch seinen Glanz, sagt
man;
Wenn er auf den Gipfeln der drei Berge seine
Nahrung sucht,
Trachtet er nach Fleisch und Blut zur Nahrung: 50
Denn wenn die aus Stricken gewundene Schlinge
durch sein Flügelschlagen nicht verführt würde,
Weshalb sollte sie als Feind wider ihn aufstehen?
Auf dem den Horst des Felsengeiers bergenden
Ling ba
Erwächst dein Glück, Milaraspa;
Nach deinem eigenen und dem Heil anderer trachtend, 55
Hast du die höchste Bodhi zur Grundlage gemacht;

Während zu derselben grossen Zeit dieser eine Leib
Unablässig nach der Buddhawürde strebt,
Bist du der Pfadführer der sechs Klassen der Wesen,
sagt man;
Während du im Dhyāna unablässig meditierst, 60
Wenn nicht infolge der starken Leidenschaft
Durch die ursprüngliche Ursache der Täuschung
der eigenen Seele
Die Skepsis als Feind in dir aufstände,
Weshalb sollte ich Felsen-Rākshasī als Feind wider
dich aufstehen?
Dieser Dämon der Leidenschaft aber entsteht aus 65
der Seele.
Auch wenn man nicht das Wesen der Seele kennt,
Weiche ich nicht, nachdem du gesagt hast: ,Gehe
fort!'
Auch wenn man nicht die Leere der eigenen Seele
erkennt.

odre ṅa bas mi ts'ad gźan yaṅ yod

70 raṅ sems raṅ gis ṅo śes na
mi mt'un rkyen rnams grogs su o c'ar
ṅa brag srin mo yaṅ obaṅs su me c'i
lar mi k'yod kyi yid la ok'u k'rig yod

da duṅ raṅ sems gtan la p'ob

 ces brag srin mos de skad žu ru
byuṅ bas | rje btsun gyi t'ugs la śin tu
o c'ad pa cig byuṅ ste | dei lan du
dran pai dpe brgyad kyi mgur odi
gsuṅs so.

Gibt es eine unendliche Zahl von anderen Dämonen ausser mir.

Auch wenn man selbst die eigene Seele kennt, 70
Erscheinen ungünstige Umstände als Freunde.
Auch ich Felsen-Räkshasī bin andren unterthan.
In deinem Gemüt jedoch herrschen Bosheit und Begierde;
Bring daher deine Seele noch besser in Ordnung!

Als die Felsen-Räkshasī diese Rede vorgetragen hatte, entstand darob im Herzen des Ehrwürdigen grosse Befriedigung, und er trug zur Antwort folgendes Lied von den acht Gleichnissen des Bewusstseins vor.

8.

k'yod bden no bden no odre mo ṅan
ts'ig de bas bden pa yod re skan
ṅa sñon c'ad rgyal k'ams ogrim ogrim nas
glu de las sñan pa t'os ma myoṅ

5 mk'as pa brgya yi žal bsdur yaṅ

don de las lhag pa yod mi srid

odre k'yod kyi k'a nas legs bśad byuṅ

legs bśad gser gyi t'ur ma de
mi k'o boi rnam śes sñiṅ la rgyab
10 ṅaṅ dṅos por odzin pai sñiṅ rluṅ bsal

ma rig ok'rul pai mun nag saṅs

blo padma dkar po k'a yaṅ p'ye

raṅ rig gsal bai sgron me sbar
dran pai ye śes t'ur gyis sad
15 dran pa sad dam ma sad pa
dguṅ sñon gyi dkyil du yar ltas pas

c'os ñid stoṅ pa t'ur gyis dran

Du hast Recht, hast Recht, böse Dämonin!
Wahrere Worte als diese gibt es überhaupt nicht.
Auf meinen einstigen Kreuz- und Querzügen durch die Welt
Habe ich ein wohltönenderes Lied als dieses nie gehört.
Wenn auch hundert Gelehrte zum Vergleich herangezogen würden, 5
Etwas Vorzüglicheres als den Sinn desselben gibt es nicht.
Aus deinem Munde, Dämonin, ist eine treffliche Erklärung gekommen;
Die goldene Stange deiner trefflichen Erklärung
Schlägt mich mitten in meine menschliche Seele.
Der Herzenskummer, der aus dem Glauben an die 10
Wirklichkeit der Existenz entspringt, ist entfernt:
Die schwarze Finsternis der auf Unwissenheit beruhenden Täuschung ist verscheucht;
Die weisse Lotusblume des Verstandes öffnet ihren Kelch;
Die Fackel des klaren Selbstwissens ist angezündet;
Die Weisheit des Bewusstseins erwacht deutlich.
Ist das Bewusstsein wirklich erwacht oder nicht? 15
Wenn ich hinauf zur Mitte des blauen Himmels schaue,
Kommt die Leere der Existenz deutlich zum Bewusstsein;

ńa dńos poi ćos la ńam ńa med	Ich fürchte nicht die Lehre von der Wirklichkeit.
ńi zla gńis la p'ar ltas pas	Wenn ich den Blick auf Sonne und Mond richte,
20 sems ńid ₒod gsal tur gyis dran	Kommt die geistige Erleuchtung deutlich zum Be- 20
	wusstsein;
ńa byiṅ rgod gńis la ńam ńa med	Ich fürchte nicht Ermattung und Erschlaffung.
ri boi rtse la p'ar ltas pas	Wenn ich den Blick auf den Gipfel der Berge
	richte,
ₒgyur med tiṅ ₒdzin tur gyis dran	Kommt die unwandelbare Beschauung deutlich zum
	Bewusstsein;
ńa ₒp'o ₒgyur rtog pas ńam ńa med	Ich fürchte nicht immer wechselnde Grübelei.
25 ću boi gźuṅ la mar ltas pas	Wenn ich nach unten in die Mitte des Flusses schaue, 25
rgyun ćad med pa tur gyis dran	Kommt das Ununterbrochene deutlich zum Bewusst-
	sein;
blo bur rkyen gyis ńam ńa med	Ich fürchte nicht plötzliche Geschehnisse.
ₒja ts'on ri mo mt'oṅ tsa na	Wenn ich das Bild des Regenbogens sehe,
snaṅ stoṅ zuṅ ₒjug tur gyis dran	Kommt die Leere der Erscheinungen im zuṅ ₒjug[1]
	deutlich zum Bewusstsein;
30 ńa rtag ćad gńis la ńam ńa med	Ich fürchte nicht das Dauernde und das Vergängliche. 30
gzugs brńan ću zla mt'oṅ tsa na	Wenn ich das Spiegelbild des Mondes im Wasser sehe,
ₒdzin med raṅ gsal tur gyis dran	Kommt die von Interessen freie Selbsterleuchtung
	deutlich zum Bewusstsein;
ńa gzuṅ ₒdzin rtog pas ńam ńa med	Ich fürchte nicht die an Interessen geknüpften
	Erwägungen.
raṅ rig sems la ts'ur ltas pas	Wenn ich in meine eigene Seele schaue,
35 bum naṅ mar me tur gyis dran	Kommt die Lampe im Innern des Gefässes[2] deutlich 35
	zum Bewusstsein.
ńa gti mug rmoṅs pas ńam ńa med	Ich fürchte nicht Thorheit und Dummheit.
ₒdre k'yod kyi k'a nas de t'os pas	Da ich jene Worte aus deinem Munde, Dämonin,
	vernommen,
ńo bo raṅ rig tur gyis dran	Kommt meine Natur und meine Seele deutlich zum
	Bewusstsein;
ńa bar ćod bgegs la ńam ńa med	Ich fürchte nicht den Widersacher.
40 k'yed de tsam legs bśad śes śes nas	Da du so viele treffliche Erklärungen weisst 40
sems kyi de ńid go go nas	Und das Wesen der Seele so gut verstehst,
da lta lus ńan ₒdre mor skyes	Bist du jetzt als Dämonin in einem schlechten
	Leibe wiedergeboren.
las ńan gnod ciṅ ₒts'e bar byed	Durch böse Thaten stiftest du Schaden und Harm.
de rgyu ₒbras k'yad du bsad pas lan	Das ist der Lohn für die Verachtung der Wieder-
	vergeltung.
45 da yaṅ ₒk'or bai ńes dmigs bsoms	Nun halte dir die Strafe vor, die im Durchwandeln 45
	des Kreislaufs besteht!

---

[1] Kunstausdruck der praktischen Mystik, das Hineinzwängen des Geistes in die Hauptarterie (dbu ma), um bei der Meditation der Zerstreuung vorzubeugen (Jäschke). Vergl. Lied Nr. 3, 62—68.

[2] Tib. bum (pa) scheint hier im Sinne von snod gebraucht zu sein.

las mi dge bcu bo gtan nas spoṅs	Meide die zehn Todsünden ganz und gar!
ṅa seṅ ge lta bui rnal obyor yin	Ich bin der einem Löwen gleiche Yogin.
ñam ṅa bag tsʿa med pa yin	Ohne Angst und Furcht bin ich.
mi ṅa yis ku re byas pa la	Ich hatte nur einen Scherz gemacht:
kʿyod bden no ma sñam odre mo ṅan	Ich glaube nicht, dass du Recht hast, böse Dämonin!
odre kʿyod kyis do nub tʿo otsʿams pas	Da du, Dämonin, mich heute Abend verhöhnst,
sṅon gnod sbyin ša za spun lṅa daṅ	Habe ich zuvor gemäss der Stärke der fünf Yaksha- und Piçāca-Geschwister
rgyal po byams pai stobs bžin du	Und des liebreichen Königs
smon lam obrel bas otsʿams sbyar nas	Durch die Kraft des Gebets einen Entschluss gefasst;
byaṅ cʿub sems daṅ ldan gyur te	Wenn du den Sinn der Heiligkeit erlangt hast,
dus pʿyi ma ṅa yi gdul byar šog	Möge später mein Bekehrungswerk erfolgen!
ces gsuṅs pas | brag srin mo šin tu dad par gyur nas | žabs nas bzuṅ ba btaṅ ste | yaṅ lus med pai skad sñan pa žig gis | bar snaṅ na tsʿur la žu ba odi ltar pʿul lo.	Als er so gesprochen, wurde die Felsen-Rākshasī sehr gläubig und liess ihn, den sie am Fusse gepackt hielt, los. Wiederum mit geisterhafter, aber wohlklingender Stimme richtete sie folgende Bitte an ihn, der sich auf dieser Seite in der Luft befand.

9.

kye ma skal ldan rnal obyor pa	O würdiger Yogin,
bsod nams bsags pas dam cʿos spyod	Tugendverdienste ansammelnd übst du die heilige Religion.
ṅo mtsʿar gcig pur ri la bžugs	Es ist wunderbar, dass du allein auf den Bergen wohnst.
tʿugs rjei spyan rgyaṅ ogro la gzigs	Soweit das Auge des Erbarmens reicht, sorgst du für die Wesen.
ṅa padma tʿod pʿreṅ brgyud pa odzin	Ich besitze den Lehrer mit dem Lotuskranz als Kopfschmuck;
dam cʿos tsʿig gi pʿreṅ ba ñan	Ich höre den Kranz der Worte der heiligen Religion.
tsʿig tʿos pa yod kyaṅ žen pa cʿe	Obwohl ich solche Worte gehört, ist meine Sehnsucht noch gross.
ṅa rnal obyor yoṅs kyi tsʿogs kʿaṅ ogrims	Ich will von einem Versammlungshaus der Yogin zum andren ziehen.
las opʿro can rnams dge la okʿod	Wer durch frühere Werke glücklich ist, verweilt in der Tugend;
skal ba can rnams don daṅ spyod	Die Würdigen pflegen auch die Wahrheit.
sems bzaṅ snaṅ ba dkar lags kyaṅ	Obwohl die Begriffe meines guten Sinnes rein sind,
lus ṅan ma sos ltogs tsʿor cʿe	Empfindet mein schlechter Leib, wenn er nicht genährt wird, grossen Hunger;
las su odzam gliṅ groṅ kʿyer ñul	In meiner Thätigkeit durchwandre ich die Städte von Jambudvipa
zas su ša daṅ kʿrag la dga	Und nehme gern Fleisch und Blut zur Nahrung.

1 5 mi tsam po yoṅs kyi sems la ₀jug

dman ₀p'yor mo kun la sñiṅ rluṅ sloṅ
p'o ₀p'yor po' kun la mts'al ris btab

mig gis kun la ltad mo ltas
sems kyi rgyal k'ams že la mnan
20 lus kyis kun la γγeṅ ₀deγs byas
gnas ni liṅ bai brag la gnas
de ts'o ṅa yi spyod ₀gros yin
k'yed kyi druṅ du žu ba lags
lan gcig glu ru blaṅs pa yin
25 ₀o skol mjal bai dga spro yin
da lta mts'ar bai žu don yin
rnal ₀byor k'yed kyi žabs tog yin
ṅa raṅ dad pai gsal ts'ig yin
glu draṅ gtam p'ul bas mñes par šog

ces zer bas | rje btsun gyi t'ugs la | mi
ma yin gyi rigs ₀di la nan tan gyi
dri ba byas | dam la btag dγos dgoṅs
nas moi žus lan du mgur ₀di gsuṅs
so.

Ich werde in die Seele des ersten besten Mannes 15
fahren, der mir begegnet;
Eine schöne Frau erregt allen Kummer;
Ein schöner Mann ist allen wie ein Bild in roter
Farbe gemalt;
Mit den Augen bietet er allen ein Schauspiel dar;
Im Reich der Seele unterdrückt er die Neigungen;
Mit dem Leibe stützt er allen die Unbeständigkeit. 20
Auf dem Felsen von Ling ba hause ich.
Das ist mein Wandel.
An dich habe ich eine Bitte:
Noch einmal sei ein Lied gesungen.
Möge aus unserer Begegnung Freude entstehen! 25
Jetzt habe ich eine schöne Bitte gethan.
Zu deinen Diensten bin ich, Yogin.
Ich selbst bin des Glaubens klares Wort.
Gib ein aufrichtiges Lied zum besten, an dem wir
uns freuen mögen!

Als sie so gesprochen, richtete der Ehrwürdige
eifrige Fragen an sie über diese Klasse von Geistern,
und indem er dachte, dass er sie bannen[1] müsse,
trug er zur Antwort auf die Bitte des Weibes folgendes Lied vor.

10.

₀o na ñon cig dman mo k'yod
slob dpon bzaṅ ste slob ma ṅan
lha c'os t'os bsam byas pa rnams

don ma go ts'ig gi p'reṅ ba bzuṅ

5 k'as legs legs smas kyaṅ lag len med

k'ram gtam k'a yi stoṅ bšad des
raṅ rgyud dri ma ₀dag mi srid

k'yod sṅon gyi bag c'ags ṅan pa daṅ
₀p'ral du las ṅan bsags pai p'yir
10 sdom pa dam ts'ig k'as len ñams

Wohlan, merke auf, o Frau!
Der Lehrer ist gut, doch der Schüler ist schlecht.
Die, welche nachgedacht haben über das, was sie
von der Götterlehre gelernt,
Haben den Kranz der Worte verstanden, deren Sinn
unbegreiflich ist.
Obwohl ihr Mund treffliche Fragen stellt, fehlt es 5
ihnen an Übung.
Indem ihr lügnerischer Mund das Leere erklärt,
Können die Flecken ihrer eigenen Seele nicht ge-
reinigt werden.
Du hast wegen deiner früheren bösen Leidenschaft
Und gegenwärtig angesammelten bösen Werke
Verpflichtung, Gelübde und Versprechen verletzt 10

---

[1] Tib. *dam la ₀dogs pa* Dämonen bannen, aber nicht durch Zaubermacht, sondern durch gütliche Überredung zu dem Versprechen, keinen Schaden mehr zu thun (Jäschke).
Denkschriften der phil.-hist. Classe. XLVIII. Bd. II. Abh. 5

de yi stobs kyis dman por skyes

lus sdug bsṅal ša zai groṅ du rgyug

ṅag mi bden ṅo lkog rdzun tsʿig maṅ
sems mi dge ˳gro bai srog la gnod
15 lus ṅan pai gzugs su skyes pa de

las rgyu ˳bras kʿyad gsod cʿes pas lan

da ˳kʿor bai ñes dmigs bsam šes na

las mi dge byas tsʿad bšags pa gyis

dge ba bsgrub par kʿa yis loṅ
20 ṅa seṅ ge lta bur ˳jigs pa med
glaṅ cʿen lta bur ñam ṅa na med
smyon pa lta bur gza gtad med

ṅa yis kʿyod la bden tsʿig brjod
kʿyod kyis da duṅ draṅ gtam smros
25 mi ṅa la bar cʿad mtʿo ˳tsʿams pa

˳dre kʿyod kyi gtan gyi so nam yin
cʿos smon lam ˳brel bai stobs bžag nas

ma ˳oṅs dus na rjes su ˳dzin
de rtog dpyod mdzod cig dman mo
˳pʿrul

     žes gsuṅs pas | brag srin mos lus sṅa
ma ltar bstan nas draṅ gtam gyi žu
don ˳di glu ru pʿul lo.

Und bist kraft dessen als niederes Wesen wieder-	

Und bist kraft dessen als niederes Wesen wieder-
     geboren,
In unseliger Körperform in der Sphäre der Piçāca
     wandelnd.
Unwahre Rede, Trug, viele Lügen,
Sündiger Sinn schaden dem Leben der Wesen.
Diese Wiedergeburt in der Gestalt eines schlechten 15
     Leibes
Ist der Lohn für die Verachtung der Lehre von der
     Vergeltung der Werke.
Wenn du dir nun die Strafe vor Augen führst, die
     im Durchwandeln des Kreislaufs besteht,
Sühne deine Sünden nach dem Masse, wie du sie
     begangen hast,
Und versprich, nach der Tugend zu streben.
Ich bin wie ein Löwe, aber nicht furchtbar; 20
Ich bin wie ein Elephant, doch habe ich keine Furcht;
Ich bin wie rasend, doch nicht der Wirkung der
     Planeten anheimgefallen.
Ich spreche wahre Worte zu dir.
Lass du noch einmal aufrichtige Rede hören!
Denn mir Hindernisse zu bereiten und mich zu ver- 25
     höhnen
War dein beständiges Geschäft, du Dämonin.
Erfasse nun die Kraft der Wirkung der Religion
     und des Gebets
Und bewahre sie in Zukunft im Gedächtnis!
Erwäge und prüfe das, du Verwandlung eines Weibes!

     Als er so gesprochen, zeigte sich die Felsen-
Rākshasī wie zuvor und richtete mit aufrichtiger
Rede ihre Bitte an ihn in folgendem Liede.

11.

rje dus gsum saṅs rgyas kun gyi gtso
rdo rje ˳cʿaṅ cʿen draṅ sroṅ lus
ṅo mtsʿar bstan pai bdag po mdzad
ya mtsʿan byaṅ cʿub sems bskyed bzaṅ

5 ñed dman mo miṅ sriṅ bsod nams cʿe

kʿyed kyi gsuṅ tʿos go ba skyes

Herr, Gebieter aller Buddha der drei Zeiten,
Der den Leib des Einsiedlers des Mahāvajrapāṇi hat,
In den Besitz der wunderbaren Lehre bist du gelangt.
Dass du in mir den Sinn erstaunlicher Bodhi geweckt
     hast, ist vortrefflich.
Für mich, das Weib, und meine Geschwister ist es ein 5
     grosses Glück.
Verständnis der Reden, die ich von dir vernommen,
     ist mir erwachsen.

ṅas daṅ por slob dpon gñan gyi bka

dam c'os lha c'os t'os bsam byas

bar du las ₒp'ro ṅan la žugs

10 *ñon moṅs gdug rtsub bzod sran med*
des na lus ṅan dman mor skyes

ₒgro ba sems can t'ams cad la
p'an yaṅ btags te gnod yaṅ bskyal
dus na niṅ soṅ bai sñon rol na
15 *liṅ bai brag la sgom c'en k'yod*

byon nas gcig pur sgrub pa mdzad

ṅed kyaṅ res dga res ma dga
dga bai p'yir na do nub mjal
ma dgai p'yir na žabs la ₒjus
20 *de gnod pai p'yir na rjes la bšags*

dus da nas dman moi gdug rtsub spoṅ
c'os sñiṅ nas byed la gtoṅ grogs byed
ma ₒoṅs dus na bdag cag rnams
bde c'en ljon pai bsil grib kyis
25 *dug lhas ñin mts'an gduṅs pa yi*
las ṅan ₒdre mo skyab tu gsol
k'yod kyis bka la brten nas su
ₒdi nas bzuṅ ste byaṅ c'ub bar
gdug pai bsam pa kun ži ste
30 *rnal ₒbyor skyob pai bsruṅs ma byed*
sgrub pa po yi gtoṅ grogs byed
sgom c'en kun gyi ₒbaṅs mo byed

c'os mdzad kun gyi k'a ₒdzin byed
dam ts'ig can gyi stoṅ grogs byed

35 *bstan pa bsruṅ žiṅ žabs tog byed*

ces smon lam bzaṅ po brjod ciṅ | sgrub
pa byed pa t'ams cad kyi mgon byed
| gnod pa mi skyel bar k'as blaṅs dam
bcas ṡin tu dad par gyur pa daṅ |
rje btsun gyis brag srin mo rjes su ₒdzin
pai mgur ₒdi gsuṅs so.

Zum ersten Male habe ich durch des Lehrers strenges Wort

Über das, was ich von der heiligen Religion, der Religion der Götter gehört, nachgedacht,

Doch inzwischen bin ich ins Unheil der Folgen meiner Werke geraten;

Das wilde Gift der Sünden ist unerträglich. 10

Deshalb bin ich als Weib in schlechtem Leibe wiedergeboren.

Allen Geschöpfen

Erwies ich bald Nutzen, fügte bald Schaden zu.

Im verflossenen Jahre

Bist du, grosser Meister der Beschauung auf dem 15 Felsen von Ling ba,

Hierhergekommen und vollzogst Bannungen ganz allein.

Bald freute ich mich, bald freute ich mich nicht.

Da ich mich freute, besuchte ich dich heute Abend;

Da ich zürnte, ergriff ich dich am Fusse.

Da ich dir so schadete, habe ich darauf meine Sünden 20 bekannt.

Von jetzt ab will ich das wilde Gift des Weibes meiden,

Will von Herzen die Religion üben und ihr Freund sein.

In Zukunft werden wir

Im kühlen Schatten des Baums der Seligkeit ruhen.

Nun Tag und Nacht von den fünf Giften gequält, 25

Bitten wir übelthuende Dämoninnen um Hilfe.

Wenn wir auf dein Wort vertrauen

Und damit beginnend bis zur Bodhi fortschreiten,

Ist der Hang zur Wildheit gänzlich beruhigt.

Der Yogin ist unser helfender Schützer. 30

Wir sind Freundinnen des Bannenden;

Wir sind unterthänige Dienerinnen aller grossen Meister der Meditation;

Wir sind der Beistand aller Pfleger der Religion;

Wir sind die Helferinnen der durch ihr Gelübde Geweihten.

Die Lehre bewachend leisten wir Dienste. 35

Als sie so ein treffliches Gebet gesprochen, wurde sie sehr gläubig durch das Versprechen und Gelübde, Schützerin aller Bannenden zu sein und ihnen keinen Schaden mehr zuzufügen. Da trug der Ehrwürdige folgendes Lied vor, damit es die Felsen-Rākshasī im Gedächtnis bewahre.

5*

12.

ṅa ok'or ba spaṅs pai ban dʰe yin
bla ma dam pai sras po yin
gdams ṅag rin c'en gter mdzod yin
dam c'os sñiṅ nas sgrub mk'an yin

5 *c'os ñid rtogs pai rnal obyor yin*
sems can rnams kyi ma rgan yin
sñiṅ stobs can gyi skyes bu yin
šākya t'ub pai srol odzin yin
byaṅ c'ub sems kyi bdag ñid yin
10 *byams pa sṅon nas bsgoms pa yin*

sñiṅ rjes gdug pa odul ba yin
liṅ ba brag la gnas pa yin
yeṅs pa med par sgom mk'an yin
k'yod dka bar sems sam dman mo op'rul

15 *dga bar ma ts'or k'yed cag ṅan*
odre k'yod pas c'e ba ṅar odzin yin
odre k'yod pas maṅ ba odu šes yin
odre k'yod pas ṅan pa bsam ṅan yin
odre k'yod pas rgod pa rnam rtog yin
20 *odre k'yod pas laṅ šor bag c'ags yin*
odre odre ru bzuṅ na gnod pa yin

odre stoṅ par šes na sod pa yin

odre c'os ñid du go na grol ba yin

odre p'a mar šes na zin pa yin

25 *odre sems su šes na rgyan du oc'ar*

de yin lugs šes pas t'ams cad grol
odre dman pos žu ts'ig brda ru bstan
mi k'o bos gdul byar dam la btags
k'yed c'ag dam bcas ts'ig bžin bsgrubs
30 *bdag cag rdo rje odzin pa yi*
dam ts'ig gñan poi gsuṅ ma bcag
t'ugs rje drag po sun ma obyin
lus ṅag bsam pai bar ma gcod

gal te dam la odas gyur na

Ich bin der den Kreislauf meidende Bandhe.
Ich bin des heiligen Lama Sohn.
Ich bin die kostbare Schatzkammer der guten Lehren.
Ich bin der die heilige Religion mit ganzem Herzen
    Übende.
Ich bin der das Sein erkennende Yogin.   5
Ich bin der Wesen alte Mutter.
Ich bin der mutvolle heilige Mann.
Ich bin der Nachfolger der Gewohnheit Çākyamuni's.
Ich bin die verkörperte Bodhi.
Ich bin der von Anfang an über die Liebe Medi- 10
    tierende.
Mit Erbarmen vernichte ich das Gift.
Auf dem Ling ba-Felsen hause ich.
Ohne Ablenkung liege ich der Meditation ob.
Bist du von freudigen Gedanken erfüllt, du Ver-
    wandlung eines Weibes?
Wenn du keine Freude empfindest, bist du schlecht. 15
Grösser als ihr Dämonen ist die Eigensucht.
Zahlreicher als ihr Dämonen sind die Wahrnehmungen.
Schlechter als ihr Dämonen sind schlechte Gedanken.
Wilder als ihr Dämonen ist die Skepsis.
Hartnäckiger als ihr Dämonen ist die Leidenschaft. 20
Wenn man die Dämonen für Dämonen hält, ist es
    ein Schaden;
Wenn man weiss, dass die Dämonen leer sind, ist
    es erfreulich;
Wenn man das Wesen der Dämonen begreift, ist es
    Befreiung.
Wenn man die Dämonen gleichsam als Eltern auf-
    fasst, ist es Besessenheit.
Wenn man den Geist der Dämonen kennt, erweist 25
    es sich als Segen.
Wenn man ihre Umstände kennt, sind alle befreit.
Du, niedere Dämonin, hast ein Bittgesuch dargelegt;
Ich habe dich durch die Bekehrung gebannt.
Du erfülle nun dein Versprechen!
Unseres Vajradhara   30
Wort des strengen Gelübdes brich nicht!
Mache mein grosses Erbarmen nicht zu Schanden!
Begehe keine Frevel des Leibes, des Worts und des
    Gedankens!
Wenn du dein Gelübde brichst,

5 rdo rje dmyal bar ₒgro bar ṅes
de ṅo gal će bas laṅ gsum bskyar

de go bar gyis la lag len skyoṅs
ₒo skol smon lam bzaṅ poś p'rad

ma ₒoṅs tśe na bde ćen žiṅ
10 rab ₒbyam bsam gyis mi k'yab par
byaṅ ćub sems daṅ k'yod ldan nas
gdul bya ₒk'or gyi daṅ por skye

rdo rje sems dpai bud med bya

žes gsuṅs te | dam la btags pas |
brag srin mos rje btsun la p'yag daṅ
bskor ba maṅ du byas | ci gsuṅ sgrub
par k'as blaṅs nas ₒja yal ba bžin
soṅ bai mt'ar | nam laṅs ñi ma šar
ba daṅ | brag srin mo miṅ srin ₒk'or
daṅ bcas pa | mi skyes pa daṅ bud
med rgyan bzaṅ po btags pa | mdzes
šiṅ yid du ₒoṅ bai rnam pas | tś'ogs
daṅ mć'od pa maṅ po k'yer nas byuṅ
ste | rje btsun la mć'od pa p'ul nas |
brag srin mos bdag las ṅan gyis ₒdre
moi lus blaṅs | bag ćags ṅan pas k'a
bsgyur | bsam pa ṅan pas ṅa ci mt'o
ₒtś'ams pa bzod par gsol | da p'yin
ćad rnal ₒbyor pa k'yed gsuṅ bai bka
ṅan ciṅ ₒbaṅs su mć'i bas | da rje
btsun k'yed raṅ gi t'ugs la ₒk'ruṅs pai
ṅes don gyi ćos cig gnaṅ bar žu žus
nas | žu don du glu ₒdi p'ul lo.

Wirst du sicher in die Vajra-Hölle fahren. 35
Da dies von grosser Wichtigkeit ist, wiederhole es
dreimal.
Um das zu begreifen, wende dich der Übung zu.
Wir haben uns mit trefflichem Gebete zusammen-
gefunden.
In Zukunft wird dir Seligkeit zu teil:
Unendlich und unfassbar 40
Von dem Sinn der Heiligkeit erfüllt,
Wirst du als erste aus dem Gefolge der Bekehrung
wiedergeboren
Und zum Weib des Vajrasattva gemacht werden.

Als er sie durch solche Worte gebannt hatte,
bezeigte die Felsen-Rākshasī dem Ehrwürdigen Verehrung und umkreiste ihn oftmals. Mit dem Versprechen, alle seine Worte zu erfüllen, entfernte sie
sich endlich wie ein verschwindender Regenbogen. Als
bei Tagesanbruch die Sonne aufging, erschien die
Felsen-Rākshasī mit dem Gefolge ihrer Geschwister,
schön geschmückter Männer und Frauen mit lieb-
lichem und anmutigem Wesen, und brachten eine
Menge von Dingen und viele Opfergaben. Sie überreichten dem Ehrwürdigen die Opfergaben, und die
Felsen-Rākshasī sprach: ‚Ich habe durch meine
schlechten Werke den Leib einer Dämonin empfangen; böse Leidenschaft beherrscht mich. Geruhe, mir
zu verzeihen, was ich dir mit böser Absicht geschadet
habe. Von jetzt ab will ich auf das von dir, Yogin,
verkündete Wort hören und dir unterthänig dienen.
Nun, Ehrwürdiger, teile uns bitte religiöse Belehrung
mit über die mystische Erkenntnis, die in deinem
eigenen Geiste entsteht.‘ So richtete sie in folgendem
Liede ein Bittgesuch an ihn.

13.

‚e ma ya rabs ₒdzaṅs kyi bu
bsod nams bsags pai skal ba can

brgyud pa bzaṅ poi byin brlabs can
sgrub pa mdzad pai sñiṅ rus can

5 gcig pur bžugs pai sñiṅ stobs can

O Sohn aus vornehmem Adelsgeschlecht,
Würdiger, der verdienstliche Handlungen angesammelt
hat,
Der den Segen eines trefflichen Lehrers geniesst,
Der Standhaftigkeit besitzt in der Vollziehung von
Bannungen,
Der den Mut hat, allein zu wohnen! 5

zab mo don gyi sgrub la brtson	Nach der Erreichung der tiefen Wahrheit[1] strebend,
ᴏdre bar c'od yoṅ bar mi srid pas	Können dir die Hindernisse der Dämonen nichts anhaben.
naṅ rtsa rluṅ rtags kyi rten ᴏbrel las	Durch Ursache und Wirkung der Anzeichen der im Innern befindlichen Aderluft
sgyu ma gar gyis don ston pas	Lehrst du auf Grund des Tanzspiels der Illusion die Wahrheit.
10 *ṅed cag k'yed kyi sems daṅ sdoṅs*	Wir vereinigen uns mit deiner Seele: 10
smon lam bzaṅ pos sṅon nas ᴏbrel	Obwohl wir durch die uranfängliche Wirkung guten Gebets
sṅon c'ad grub t'ob maṅ mjal yaṅ	Schon früher mit vielen Siddha zusammengetroffen sind,
bka drin byin brlabs k'yed la t'ob	Haben wir erst durch dich den Segen der Gnade erlangt.
da dman mo bdag gi žu ᴏbul ba	Nun richtet ein Weib eine Bitte an dich.
15 *c'os draṅ don t'eg dman mgo bskor ᴏdis*	Durch die hinsichtlich der Wahrheitserkenntnis aus 15 den heiligen Schriften veranlasste Täuschung des Hīnayāna
las ñon moṅs t'ul ba dka bar mc'is	Ist es schwer, die sündigen Werke zu zähmen.
k'a bšad ston pai p'yor ᴏgron de	?[2]
rkyen sdug bsṅal byuṅ na ᴏbros par ṅes	?
c'os kyis bred pai slob dpon ṅa	Ich, ein Lehrer, der sich vor der Religion fürchtet,
20 *raṅ don ma grub k'ro bar ᴏdug*	Bin erzürnt, dass mein eigener Nutzen noch nicht erreicht ist. 20
rje dus gsum saṅs rgyas sprul pai sku	Gebieter, Nirmāṇakāya der Buddha der drei Zeiten,
c'os ñid ṅes pai don rtogs pa	Der du das Sein und die mystische Erkenntnis begreifst,
t'ugs la ᴏk'ruṅs pa zab moi don	Ein Sang der Unterweisung, der die in deinem Geist entstehende tiefe Wahrheit
gnad la dril bai man ṅag dbyaṅs	In den Kernpunkten zusammenfasst,
25 *ṅes don mt'ar t'ug skyal sa de*	Sei unser Geleite bis ans Endziel der mystischen 25 Erkenntnis!
bdag cag miṅ sriṅ ᴏk'or bcas la	Mir samt dem Gefolge meiner Geschwister
rdo rjei gsaṅ ts'ig dam pai don	Geruhe den heiligen Sinn des Vajra-Geheimwortes,
ye šes c'en po ᴏod gsal ba	Die grosse erleuchtende Weisheit,
mc'og gi ᴏod gsal gnaṅ du gsol	Die höchste Erleuchtung zu gewähren.
30 *ṅes don zab mo gsaṅ bai rgya*	Da wir das Zeichen des tiefen Geheimnisses der 30 mystischen Erkenntnis
t'os pas ṅan soṅ lhuṅ mi srid	Vernommen, können wir nicht mehr zur Hölle fahren.
bsgoms pas ᴏk'or bar k'yam mi srid	Da wir nachgedacht haben, können wir nicht mehr im Kreislauf herumwandern.
sba sri med par gnaṅ bar žu	Teile bitte deine Gaben verschwenderisch aus!

---

[1] Tib. *zab mo don*, synonym mit *ṅes don* gebraucht.

[2] Da der Ausdruck *p'yor ᴏgron* unbekannt und die Verbindung *k'a bšad* nicht sicher erklärt werden kann, so ist eine Deutung dieses Verses ausgeschlossen, und deshalb lässt sich auch der folgende V. 18 nicht verstehen, der in wörtlicher Uebersetzung besagt: „Wenn unglückliche Ereignisse entstehen, ist es sicher, dass man entflieht.‘

žes žus pas | rje btsun gyis | da
duṅ k'yed c'os ṅes don sgom mi nus
| gal te -nus na ṅa la srog sñiṅ daṅ
dam bca drag po ₒbul dgos gsuṅs pas
| ₒgyur ba med pai srog sñiṅ p'ul nas
da p'yin c'ad rje btsun ci gsuṅ gi bka
sgrub ciṅ | c'os mdzad t'ams cad kyi
gtoṅ grogs byed par dam bca p'ul bas
| rje btsun gyis kyaṅ moi žus lan du
ṅes pa don gyi c'os t'im pa ñer bdun
ₒdi mgur du gsuṅs so.

Als sie diese Bitte gethan hatte, sagte der Ehrwürdige: ‚Du bist nicht imstande, in der mystischen
Erkenntnis der Lehre zu meditieren; wenn du aber
dazu imstande bist, so musst du mir dein Leben zum
Pfande einsetzen und ein starkes Versprechen geben.
Nachdem sie unabänderlich ihr Leben zum Pfande
eingesetzt und das Versprechen gegeben, in Zukunft
alle Gebote des Ehrwürdigen zu erfüllen und allen
Pflegern der Religion in Freundschaft beizustehen,
trug er zur Antwort auf die Bitte des Weibes die
Lehre von der mystischen Erkenntnis in folgendem
Liede von den 27 verschwindenden Dingen vor.

14.

rje sbas pai saṅs rgyas mi gzugs can

mts'an brjod par dka bai lo tsa ba
p'a bka drin can gyi žabs la ₒdud
ṅa rig byed glu mk'an ma yin te
5 *ₒdre k'yed kyis glu loṅ glu loṅ zer*
da res gnas lugs kyi dbyaṅs cig len
yoṅs ₒbrug daṅ glog daṅ lho sprin gsum
byuṅ yaṅ nam mk'a raṅ las byuṅ
t'im yaṅ namk'a raṅ la t'im

10 *ₒja daṅ na ₒun k'ug sna gsum*
byuṅ yaṅ bar snaṅ raṅ las byuṅ
t'im yaṅ bar snaṅ raṅ la t'im

rtsi bcud daṅ lo tog ₒbras bu gsum
byuṅ yaṅ sa gži raṅ las byuṅ
15 *t'im yaṅ sa gži raṅ la t'im*
nags ts'al daṅ me tog lo ₒdab gsum
byuṅ yaṅ ri bo raṅ las byuṅ
t'im yaṅ ri bo raṅ la t'im

c'u kluṅ daṅ c'u lbur c'u rlabs gsum
20 *byuṅ yaṅ rgya mts'o raṅ las byuṅ*
t'im yaṅ rgya mts'o raṅ la t'im
bag c'ags dan žen c'ags ₒdzin c'ags gsum
byuṅ yaṅ kun gži raṅ las byuṅ
t'im yaṅ kun gži raṅ la t'im

Dem Gebieter, der Buddha's verborgene Menschen-
gestalt hat,
Dem Übersetzer schwer auszusprechender Namen,
Dem huldreichen Vater zu Füssen verneige ich mich.
Ich bin kein berufsmässiger Sänger;
Ihr Dämonen sagt: singe, singe ein Lied! 5
Deshalb singe ich jetzt ein Lied vom Wesen der Dinge.
Donner, Blitz und Südwolke,
Wenn sie entstehen, entstehen aus dem Himmel selbst,
Wenn sie verschwinden, verschwinden am Himmel
selbst.
Regenbogen, Nebel und Dunst, 10
Wenn sie entstehen, entstehen aus der Luft selbst,
Wenn sie verschwinden, verschwinden in der Luft
selbst.
Fruchtsaft, Ernte und Frucht,
Wenn sie entstehen, entstehen aus dem Boden selbst,
Wenn sie verschwinden, verschwinden im Boden selbst. 15
Wald, Blumen und Laubwerk,
Wenn sie entstehen, entstehen aus den Bergen selbst,
Wenn sie verschwinden, verschwinden in den Bergen
selbst.
Flüsse, Wasserschaum, Wogen,
Wenn sie entstehen, entstehen aus dem Meere selbst, 20
Wenn sie verschwinden, verschwinden im Meere selbst.
Leidenschaft, Begierde und Habsucht,
Wenn sie entstehen, entstehen aus der Seele selbst,
Wenn sie verschwinden, verschwinden in der Seele
selbst.

25 *raṅ rig daṅ raṅ gsal raṅ grol gsum*
byuṅ yaṅ sems ñid raṅ las byuṅ
t'im yaṅ sems ñid raṅ la t'im

skye med daṅ ₀gag med brjod med
    gsum
byuṅ yaṅ c'os ñid raṅ las byuṅ
30 *t'im yaṅ c'os ñid raṅ la t'im*
₀*drer snaṅ daṅ ₀drer ₀dzin ₀drer rtogs*
    gsum
byuṅ yaṅ rnal ₀byor raṅ las byuṅ
t'im yaṅ rnal ₀byor raṅ la t'im
lar bgegs rnams sems kyi c'o ₀p'rul te

35 *raṅ snaṅ stoṅ par ma rtogs pai*

₀*dre raṅ rgyud du ₀dzin na rnal ₀byor*
    ₀*k'rul*
lar ₀k'rul pai rtsa ba sems las byuṅ

sems kyi ṅo bo rtogs pa las
₀*od gsal ₀gro ₀oṅ med par mt'oṅ*

40 *p'yi yul gyi snaṅ ba ₀k'rul bai sems*

snaṅ bai mts'an ñid brtags pa las
snaṅ stoṅ gñis su med par rtogs

lar sgom pa ñid kyaṅ rtog pa la
mi sgom pa yaṅ rtog pa ste
45 *sgom daṅ mi sgom gñis su med*

gñis ₀dzin lta ba ₀k'rul bai gži

mt'ar t'ug don la blta ba med

₀*di kun sems kyi mts'an ñid de*
nam mk'ai mts'an ñid dper bžag nas
50 *don c'os ñid gtan la ₀bebs pa yin*
da k'yod lta ba po ₀baṅs kyi don la
    ltos
bsgom pa yeṅs med kyi ñaṅ du žog

Selbstwissen, Selbsterleuchtung, Selbstbefreiung,   25

Wenn sie entstehen, entstehen aus dem Geiste selbst,
Wenn sie verschwinden, verschwinden im Geiste
    selbst.
Nicht-Wiedergeboren werden, das Unbehinderte, das
    Unaussprechliche,
Wenn sie entstehen, entstehen aus dem Sein selbst.
Wenn sie verschwinden, verschwinden im Sein selbst.   30
Was als Dämon erscheint, was als Dämon gilt, was
    als Dämon erkannt wird,
Wenn es entsteht, entsteht aus dem Yogin selbst,
Wenn es verschwindet, verschwindet im Yogin selbst.
Da nun die Dämonen ein Täuschungsspiel der Seele
    sind,
So ist ein Yogin, wenn er, ohne die Leere der eigenen   35
    Gedanken zu erkennen,
Die Dämonen in seiner eigenen Seele begreift, im
    Irrtum befangen.
Die Wurzel des Irrtums aber ist aus der Seele ent-
    standen;
Aus der Erkenntnis der Natur der Seele
Ersicht man, dass die Erleuchtung weder geht noch
    kommt.[1]
Wenn die in den Erscheinungen der Aussenwelt sich   40
    täuschende Seele
Die Theorie der Erscheinungen erkannt hat,
So erkennt sie, dass zwischen den Erscheinungen
    und der Leere kein Unterschied besteht.
Wenn man in der Erkenntnis der Beschauung
Die Nicht-Beschauung erkennt,
Ist zwischen Beschauung und Nicht-Beschauung kein   45
    Unterschied.
Zwei Dinge als verschieden betrachten ist die Ursache
    des Irrtums.
Dann sind die Gedanken nicht auf den Nutzen der
    Erreichung des Endzieles gerichtet.
Wenn man die Natur der Seele
Mit der Natur des Äthers vergleicht,
Ist das Wesen der Wahrheit in Ordnung gebracht.   50
Sorge du nun als Dienerin der Beschauenden für
    ihren Nutzen!
Schliesse dich der Sache der unablenkbaren Medi-
    tation an!

---

[1] D. h. beständig an ihrer Stelle bleibt.

spyod pa šugs ‿byuṅ ‿gag med skyoṅs

‿bras bu re dogs t'a sñad spoṅs

5 ‿dre k'yod kyi c'os skal de la mdzod
ṅa stod dbyaṅs glu la ‿byams mi k'om

k'yod dri rtog ma maṅ k'a rog bsdod

‿dre glu loṅ zer bas blaṅs pa yin

da lta k'o boi smyon ts'ig yin
0 ‿dre k'yod kyis ñams su len nus na
zas su bde c'en zas la zo
skom du zag med bdud rtsi ‿t'uṅ
las su rnal ‿byor gtoṅ grogs mdzod

ces gsuṅs pas | brag srin mo ‿k'or
bcas mc'og tu dad par gyur nas | p'yag
daṅ bskor ba maṅ du byas | bka drin
c'e žes bjyod nas ‿ja yal ba bžin soṅ
ṅo | de pjin c'ad rje btsun gyi bka
bžin gnas der sgom c'en su bžugs ruṅ
mi gnod ciṅ | c'os daṅ ‿t'un pai gtoṅ
grogs byed do.
liṅ ba brag srin moi bskor ro.

Schütze die unbehinderte Kraftentfaltung ihrer Aus-
übung!
Meide das Gerede von Hoffnung und Furcht der
Vergeltung!
Gib den Dämonen Anteil an deiner Religion! 55
Ich kann keine Gelehrsamkeit in Lobliedern ent-
falten.
Ohne viele Fragen und Erwägungen sitzt ihr schwei-
gend da.
Nur weil die Dämonen sagten: ,Singe ein Lied!'
habe ich gesungen.
Jetzt habe ich ein thörichtes Wort gesprochen.
Wenn ihr, Dämonen, es zu beherzigen vermögt, 60
Verzehrt als Nahrung Seligkeit
Und trinkt als Trank kummerfreien Nektar!
Seid in euerm Thun Freunde des Yogin!

Als er so gesprochen, wurde die Felsen-Räkshasī samt ihrem Gefolge sehr gläubig, bezeigte ihm
Verehrung und umkreiste ihn oftmals. ,Vielen Dank!'
sagte sie und entfernte sich wie ein verschwindender Regenbogen. Von da an that sie nach dem Gebot
des Ehrwürdigen, so lange er an jenem Orte in
grosser Beschauung verweilte, keinen Schaden mehr
und war eine Helferin im Einklang mit der Religion.
Dies ist der Abschnitt von der Felsen-Räkshasī
von Ling ba.

III.

Milaraspa in Rag ma.

(fol. 23 b 1 — 26 b 7.)

na mo gu ru.
rje btsun Mi la ras pa de ñid |
liṅ ba brag nas ri bo dpal ‿bar la
sgom du ‿byon dgoṅs nas byon pai
ts'e | rag mai yon bdag rnams la ri
bo dpal ‿bar du ‿byon pai lo rgyus
gsuṅs pas | k'oṅ rnams na re | ri bo
dpal ‿bar bas | ri bo dpal ‿bar gyi ‿gag
na gnas dgon rdzoṅ skyid po raṅ žig
yod | der bžugs na dga | ri bo dpal ‿bar
gyi lam rgyus ñed ts'o la ‿aṅ med |

Verehrung dem Meister!
Als der ehrwürdige Milaraspa von Ling ba brag
auf den Berg dPal ‿bar zum Zweck der Meditation
zu gehen gedachte, entstand, während er sich auf
den Weg machte, unter den Gabenspendern von Rag
ma über seine Wanderung nach dem Berge dPal ‿bar
ein Gerede. Sie sprachen: ,Was den Berg dPal ‿bar
betrifft, so bietet dGon rdzong skyid po selbst den
einzigen Durchgang zum Berge dPal ‿bar. Dort verweilt er auch gerne, aber einen Führer auf den Berg
dPal ‿bar haben wir nicht. Wenn er dort auf dGon

Denkschriften der phil.-hist. Classe. XLVIII. Bd. II. Abh.

6

- 315 -

dgo rdzoṅ der bžugs na ṅed kyis lam
mk'an btaṅ gis zer bas | t'ugs dgoṅs
la t'og mar gnas der bsdad | de nas
ri bo dpal ₒbar la ltar ₒgro dgos te |
k'oṅ ts'oi lam mk'an mi dgos dgoṅs nas
| kyed kyi lam mk'an mi dgos ṅa raṅ
gis rṅed gsuṅs pas | lam mk'an med
pai rtsad c'od pa med | kyed la lam
mk'an yod pa ,e yin zer ba la | yod
gsuṅs pas | su ji skad bya ba yod zer
bai lan du mgur ₒdi gsuṅs so.

rdzong verweilen will, wollen wir ihm einen Führer senden.' In dieser Erwägung blieb er anfangs an diesem Orte. Darauf kam ihm der Gedanke, dass er auf den Berg dPal ₒbar gehen müsse, aber eines Führers von jenen nicht bedürfe, und sagte: ,Ich brauche euern Führer nicht, ich werde selbst den Weg finden.' ,Wir forschen nicht nach, da es keinen Führer gibt. So hast du denn wirklich einen Führer?' fragten sie. ,In der That,' sagte er. ,Wer ist es, und wie ist sein Name?' Darauf gab er Antwort in folgendem Liede.

15.

mts'an ldan bla ma dam pa de
mun pa sel bai lam mk'an yin
graṅ dro med pai ras rkyaṅ ₒdi

žen pa spoṅ bai lam mk'an yin
5 *gdams ṅag bsre ₒp'o[2] bskor gsum ₒdi*

bar do sel bai lam mk'an yin

rluṅ sems las su ruṅ ba ₒdi
rgyal k'ams bskor bai lam mk'an yin
p'uṅ po gzan du kyur[3] ba ₒdi
10 *bdag ₒdzin ₒdul bai lam mk'an yin*
dben pai gnas su sgom pa ₒdi
byaṅ c'ub bsgrub pai lam mk'an yin
lam mk'an drug gis sna draṅs nas
byaṅ c'ub rdzoṅ la ₒdug kyaṅ ₒts'al

ces gsuṅs nas rag ma p'ui gnas
su byon te | de man c'ad gnas de la
byaṅ c'ub rdzoṅ du grags so | de nas
rje btsun gyis gnas der c'u bo rgyun
gyi tiṅe ₒdzin la bžugs pa las | dus
nam žig gi ts'e | nam guṅ la bab pa
daṅ | dmag gi duṅ sgra daṅ bso sgra
maṅ po byuṅ bas | yul pa rnams la

Jener ausgezeichnete heilige Lama[1]
Ist ein Führer, um die Finsternis zu vertreiben.
Dies einfache Baumwolltuch, das nicht kalt noch
    warm ist,
Ist ein Führer, um auf Wünsche zu verzichten.
Dies dreimalige Umwandeln der Berggipfel, mit guten Lehren verbunden,
Ist ein Führer, um den Zwischenzustand zu vernichten.
Diese vollkommene Errungenschaft höchster Ekstase
Ist ein Führer, um die Länder zu durchwandern,
Dieser geringschätzig behandelte Leib
Ist ein Führer, um die Selbstsucht zu bezwingen.
Dies Meditieren in der Einsamkeit
Ist ein Führer, um die Bodhi zu erreichen.
Da ich von sechs Führern geleitet werde,
Begehre ich, auf Byang chub rdzong zu verweilen.

Als er so gesprochen, begab er sich an das obere Ende von Rag ma. Der Ort unterhalb davon heist Byang chub rdzong. Darauf verweilte dort der Ehrwürdige, in die Betrachtung des fliessenden Wassers vertieft. Einstmals geschah es, als die Mitternacht heranrückte, dass Töne von Kriegstrompeten und Losungsworte in grosser Zahl hörbar wurden. Da dachte er, ob vielleicht ein Feind über die Ein-

---

[1] D. i. Marpa, Milaraspa's Lehrer.
[2] *₀p'o* sehe ich als eine Variante zu *₀po* ,Berggipfel' an.
[3] *kyur = skyur*.

dgra byuṅ ba yin nam dgoṅs nas |
sñiṅ rje drag po ok'ruṅs pai tiñe odzin
gyi ṅaṅ la bẑugs pa las | je ñe je ñe
soṅ ste | ood dmar po c'en po ẑig t'al
byuṅ ba las | ci yin nam sñam gzigs
pas | t'aṅ t'ams cad mes ok'rigs śiṅ sa
daṅ bar snaṅ t'ams cad ojigs śiṅ ya
ṅa bai dmag ts'ogs kyis gaṅ odug pa
rnams kyis me sbor ba daṅ | c'u ok'rug
pa daṅ | ri sñil ba daṅ | sa ɣyo ba
daṅ | mts'on bsnun pa la sogs pa c'o
op'rul sna ts'ogs ston ciṅ | k'yad par
bsgrub p'ug bśig pa la sogs pa daṅ
| mi sñan pa sna ts'ogs brjod du byuṅ
ba las | rje btsun gyi t'ugs la | mi ma
yin gyi bar c'ad daṅ t'o ots'am yin par
dgoṅs na | ,a tsa ma srid pa t'og med[1]
dus nas las ṅan bsags pas | rigs drug
gi gnas su ok'yams | dei ṅaṅ nas kyaṅ
mk'a la rgyu bai yi dags su skyes |
bsam ṅan sbyor rtsub kyi gẑan la gnod
sems daṅ ldan pa odi rnams kyis | skye
ogro maṅ poi srog la ots'e ẑiṅ gnod pa
byed bar odug pas | ṅa yaṅ t'ar med
kyi dmyal bar skyes nas bzod med kyi
sdug bsṅal myoṅ dgos pa odi rnams
sñiṅ re rje | dgoṅs nas mgur odi gsuṅs
so.

wohner hereingebrochen sei, und als er sich darüber
einer sein tiefstes Mitleid regenden Betrachtung hin-
gab, kam es immer näher und näher. Ein grosses
rotes Licht leuchtete vorbeifahrend auf, und während
er sich noch besann, was es sein möchte, schaute
er hin: siehe, da war die ganze Fläche von Feuer
bedeckt, unter dem Erde und Luft ganz verschwanden. Schrecken erregende Heerscharen waren da,
die das Feuer anzündeten, Wasser in Bewegung
setzten, Berge hinabschleuderten, Erdbeben, stechende
Waffen und andern mannigfachen Spuk erscheinen
liessen. Insbesondere zerstörten sie seine Grotte und
stiessen misstönende Worte aus. Da stieg dem Ehrwürdigen der Gedanke auf: ‚Das sind die Nachstellungen und Verhöhnungen der Geister.‘ Und weiter
dachte er: ‚Ach! Seit undenklicher Zeit haben sie
schlechte Werke aufgehäuft und wandeln am Orte
der sechs Klassen der Wesen. Unter ihnen gibt
es welche, die unter den in der Luft herumziehen-
den Preta wiedergeboren werden. Böse Gedanken
sinnend, voll roher Schadenfreude, stellen sie dem
Leben vieler Wesen nach und gehen nur darauf aus,
Harm zu stiften. Ach, ich bedaure sie, die ohne Befreiung in der Hölle wiedergeboren werden und un-
erträgliche Qualen erdulden müssen.‘ In diesem Gedanken trug er folgendes Lied vor.

16.

t'ugs byams pai namk'a ỿaṅs pa la
sñiṅ rjei c'u odzin rab bsdus te
p'rin las kyi c'ar rgyun p'ab nas kyaṅ
gdul byai la tog smin mdzad pai
5 sgra bsgyur Mar pai ẑabs la odud

ogro ba namk'a daṅ mñam pai sems
can rnams
rnam mk'yen saṅs rgyas kyi go op'aṅ
t'ob par byin gyis brlobs

Der an dem weiten Himmel seiner Liebe
Die Wolken des Erbarmens versammelnd
Den Regenguss der Thaten herabströmen
Und die Ernte der Bekehrung reifen liess,
Dem sprachgewandten Marpa zu Füssen verneige 5
  ich mich!
Auf dass ich dessen, welcher die dem Himmel gleich
  zahllosen Geschöpfe
Kennt, des Buddha Würde erlange, möge er mich
  segnen!

---

[1] *srid pa t'og med* ‚von den existierenden Dingen, der Welt nicht behindert oder sie durchdringend‘, d. i. wohl in Verbindung mit *dus* ‚undenkliche Zeit, graue Vorzeit‘. Der Ausdruck ist sonst nicht belegt.

6*

kyed ₒdir ₒts'ogs gnod sbyin mi ma yin	Ihr hier versammelten Yaksha und Gespenster,
yid¹ mk'a la rgyu ba mk'a ₒgroi ts'ogs	Ihr in der Luft wandelnden Scharen der Dāka,
10 *zas yid la btags pai yid btags rnams*	Ihr Preta, deren Sinn nur auf Speise gerichtet ist, 1
las mi dge spyad pai rnam smin gyis	Durch die gereifte Vergeltung euerer sündigen Werke
dus da lta yid btags lus su skyes	Seid ihr gegenwärtig im Leibe der Preta wiedergeboren,
ts'e ₒdir yaṅ gźan la gnod pai mt'us	Und weil ihr zu dieser Zeit anderen Schaden zufügt,
p'yi ma dmyal bai gnas su skye	Werdet ihr später im Ort der Hölle wiedergeboren werden.
15 *de rgyu ₒbras zur tsam ston pai glu*	Dieses die Vergeltung nur in kurzem Umriss be- 1 handelnde Lied
ṅa dgos don gtan la p'ab tsa na	Habe ich, soweit für das Verständnis erforderlich, verfasst.
ṅa ni bka brgyud bta mai bu	Ich, der Sohn des Lama, welcher der Überlieferung anhängt,
gźi dad pa skyes nas c'os la źugs	Bin in den geistlichen Stand getreten, als die Grundlage, der Glaube, erzeugt war;
las rgyu ₒbras śes nas dka ba spyad	Habe Bussübungen vollzogen, als ich die Werke und ihre Frucht erkannte.
20 *lam brtson ₒgrus skyed nas sgoms pai mt'us*	Nun da ich den rechten Pfad durch meinen Eifer 2 gewonnen, sehe ich kraft der Meditation
ₒbras bu sems kyi gnas lugs mt'oṅ	Das innere Wesen des Sinnes der Vergeltung.
ṅas snaṅ ba t'ams cad sgyu mar śes	Ich weiss, dass alle Erscheinungen eine Täuschung sind.
bdag tu ₒdzin pai nad las grol	Von der Krankheit der Eigensucht bin ich befreit,
bzuṅ ₒdzin ₒk'or bai ₒc'iṅ t'ag bcad	Die Fesseln des das Ergriffene festhaltenden Saṁsāra sind durchschnitten.
25 *ₒgyur med c'os sku rgyal sa zin*	Den Thron des unveränderlichen Dharmakāya habe 2 ich in Besitz genommen.
blo yaṅ bral bai rnal ₒbyor la	An dem vom Denken losgelösten Yogin
k'yod gnod pai sems kyis bar gcod pa	Scheitert eure Schadenfreude.
lus ṅal ba tsam źig byuṅ bar zad	Sein Leib ist zwar ermattet und erschöpft,
sems slar yaṅ k'oṅ k'ro skye bai rgyu	Doch das bewirkt, dass in seinem Sinn aufs neue der Zorn erwacht.
30 *mi ṅa yi śes rgyud kun gźi la*	Da mein Wesen die treibende Ursache ist, 3
ₒdre k'yed pas ts'aṅs pai ₒjig rten nas	Wenn auch zahlreicher als ihr Dämonen, von der Welt des Brahma
dmyal k'ams bco brgyad yan c'ad kyi	Bis zu den achtzehn Höllenreichen
ₒgro ba rigs drug dgrar laṅs kyaṅ	Die sechs Klassen der Wesen als Feinde wider mich aufstünden,
ₒjigs so sñam pa yoṅ re skan	Könnte mir der blosse Gedanke an Furcht nimmermehr kommen.
35 *da ₒdir ₒts'ogs gnod sbyin mi ma yin*	Die ihr nun hier versammelt seid, Yaksha und Ge- 35 spenster,

---

¹ Das Wort *yid* in Verbindung mit *mk'a* ist unverständlich; es ist vielleicht nur durch die beiden im folgenden Verse enthaltenen *yid* veranlasst.

k'yed mt'u stobs rdzu ₀p'rul dpuṅ bskyed
la
mi ṅa la gnod pa ma ₀k'yol bar
k'yed ₀di nas gžan du soṅ gyur na
sṅan c'ad byas pa don med yin
40 slar yaṅ k'yed raṅ ṅo re ts'a
rem cig rem cig ₀dre ts'ogs rnams

ces gsuṅs nas | c'os ñid kyi dad
du mñam par bžag pas | ₀dre ts'ogs
rnams slar dad par gyur te | p'yag
daṅ bskor ba maṅ du byas | žabs spyi
bor blaṅs te | k'yed brtan pa t'ob pai
rnal ₀byor par gda ba | ñed kyis ṅo
ma šes | sṅan c'ad mt'o ₀ts'ams pa bzod
par gsol | da p'yin c'ad ci gsuṅ gi bka
sgrub pas c'os ₀brel re yaṅ gnaṅ bar
žu zer ba la | rje btsun gyis ₀o na |
sdig pa ci yaṅ mi bya žiṅ | dge ba
p'un sum ts'ogs par bya | ces gsuṅs
pas | k'oṅ rnams kyi yin lugs brjod |
srog sñiṅ pul nas | bkai ₀baṅs bgyid
par k'as blaṅs te | raṅ raṅ so soi gnas su
soṅ ba de rnams ni | maṅ yul gyi gsoi
lha mo daṅ | ri bo dpal ₀bar gyi gži
bdag yin no | rje btsun gyis kyaṅ | ri
bo dpal ₀bar gyi gži bdag ₀dir byuṅ
rtiṅ | ri bo dpal ₀bar la c'ed du sgom
du ₀gro mi dgos par ₀dug dgoṅs ste |
gnas de kar žag ṡas bžugs pa las | t'ugs
dam šin tu ₀p'el bar byuṅ nas mgur ₀di
bžes so.

Ihr habt durch Zauberkraft und Gaukelei ein Heer
    hervorgebracht;
Wenn ihr, ohne mir Schaden zuzufügen,
Von hier anderswohin abziehen werdet,
So sind all eure früheren Bemühungen unnütz,
Immer wieder müsst ihr Scham empfinden.   40
Auf, frisch ans Werk, frisch ans Werk, ihr Kobold-
    scharen!

Als er so gesprochen, brachte er sie durch den
Glauben an die Lehre zur Ruhe. Die Koboldscharen
wurden gläubig, bezeigten ihm Verehrung, umwandelten ihn oftmals und setzten seinen Fuss auf ihren
Scheitel. ‚Dass du ein Yogin bist, der die Festigkeit
erlangt hat, wussten wir nicht. Wir bitten daher,
uns die früheren Verhöhnungen zu verzeihen. In
Zukunft wollen wir alle deine Befehle vollziehen;
gewähre uns bitte doch ein wenig religiöse Belehrung,‘ sagten sie. Der Ehrwürdige sprach: ‚Wohlan,
begeht keine Sünden, übt herrliche Tugenden!‘ Als
er so gesagt hatte, erzählten sie ihm ihre Umstände.
Dann weihten sie ihm Leben und Herz und versprachen ihm, Diener seines Worts zu sein, worauf
sie an ihren Ort zurückkehrten. Es waren die näh-
rende Göttin von Mang yul[1] und die Ortsgottheit des
Berges dPal ₀bar. Der Ehrwürdige dachte: ‚Nachdem
die Ortsgottheit des Berges dPal ₀bar hier erschienen
ist, brauche ich nicht absichtlich zur Meditation auf
den Berg dPal ₀bar zu gehen.‘ So verweilte er denn
einige Tage an jenem Orte, und als seine Meditation
sehr stark wuchs, stimmte er folgendes Lied an.

17.

byaṅ c'ub rdzoṅ gi dben gnas na
byaṅ c'ub bsgrub pai Mi la ṅa
byaṅ c'ub sems la dbaṅ bsgyur žiṅ
byaṅ c'ub sems kyi rnal ₀byor skyoṅ

5 byaṅ c'ub c'en po myur t'ob nas

In der Einsamkeit der ‚Feste der Bodhi‘
Weile ich, Mila, der die Bodhi erreicht hat.
Den auf die Bodhi gerichteten Sinn besitzend,
Bin ich der Meditation des auf die Bodhi gerichte-
    ten Sinnes ergeben.
Da ich die grosse Bodhi schnell erlangt habe,   5

---

[1] Eine an Nepal grenzende Provinz Tibets, in welcher das Fort und Dorf ₂Kyid grong gelegen ist (Jäschke). Vergl. Wasiljev,
Tibetische Geographie des Minchul Chutuktu (russisch), Pet. 1895, p. 11, wo sich die Schreibung dMang yul findet.

mar gyur ₒgro ba ₒdi dag kun
byaṅ c'ub mc'og la sbyor bar šog

    ces gsuṅ žiṅ t'ugs dam la brtson par
gyur to ‖ de nas žag ₒga nas | yon bdag
gcig gis šiṅ k'ur gcig daṅ | p'ye k'al p'yed
tsam p'ul nas | na bza srab po raṅ gda
bas bser du yoṅ | lho ri naṅ nas rag
ma ₒdi braṅ žiṅ | dei naṅ nas kyaṅ
brag t'og ₒdi nan tar graṅ ba žig lays
pas | bžes na bdag gis t'ul pa cig ₒbul
| bla ma k'yed ji skad bya ba lags zer
ba la | rje btsun gyis | yon bdag k'yod
raṅ gi miṅ ci yin gsuṅs pas | lha ₒbar
bya ba lags zer ba la | rje btsun gyis
miṅ legs te | k'yod kyis p'ye daṅ t'ul pa
la sogs pa rnams la ltos mi dgos pa
žig yin na ₒaṅ | p'ye dbul ba ňo mts'ar
c'e t'ul ba mi dgos | ṅa ₒdi ltar yin gsuṅs
nas | lha ₒbar la ṅgur ₒdi gsuṅs so.

Mögen alle diese wie eine Mutter geliebten Wesen
Mit der höchsten Bodhi vereinigt werden!

So sprach er und beeiferte sich der Meditation.
Darauf kam nach einigen Tagen ein Gabenspender,
der ihm eine Ladung Holz und einen halben Scheffel
Mehl brachte. Er sagte: ‚Da dein Gewand so dünn
ist, wirst du frieren. Zwischen den Südbergen ist
dieses Rag ma gelegen, und unter diesen ist gerade
dieses Felsdach ausserordentlich kalt. Willst du ihn
nehmen, so gebe ich dir einen Pelzmantel. Lama,
was sagst du dazu?‘ Der Ehrwürdige fragte: ‚Gabenspender, wie ist dein Name?‘ ‚Lha ₒbar heisse
ich,‘ erwiderte er. Der Ehrwürdige sagte: ‚Der Name
ist gut. Obwohl du nicht für Mehl, einen Mantel
und anderes zu sorgen brauchst, so sage ich dir
doch für die Mehlspende meinen Dank, den Mantel
aber brauche ich nicht. Ich bin nun einmal ein
solcher.‘ Dann trug er dem Lha ₒbar folgendes
Lied vor.

18.

ňa rigs drug ₒk'rul pai groṅ k'yer du

rnam šes ₒk'rul pai k'yeu c'uṅ ₒk'yams

  las kyi ₒk'rul snaṅ sna ts'ogs myoṅ
  res ₒga ltogs pai ₒk'rul snaṅ byuṅ
5 *ro sňoms ldom bu zas su zos*
  res ni dka t'ub rde ₒc'a byas
  res ₒga stoṅ ňid zas su zos
  res ₒga t'abs c'ag sdug sran skyed
  res ₒga skom pai ₒk'rul snaṅ byuṅ
10 *ɣyai bsil c'ab sňon mo ₒt'uṅ*
  ras ₒga raṅ byuṅ dri c'u bsten
  res ni sňiṅ rjei c'u rgyun ₒt'uṅs
  res ₒga mk'a ₒgroi dam rdzas ₒt'uṅs
  res ₒga ₒk'yags pai snaṅ ba byuṅ
15 *ras gos rkyaṅ re gos su gyon*

Ich wanderte umher in der Sphäre der Täuschungen
    der sechs Klassen der Wesen
Als kleiner Knabe, der sich in den Wahrnehmungen
    täuscht.
Manche Illusionen des Thuns habe ich erfahren:
Bald entstand in mir die Illusion des Hungers,
Milde Gaben und Almosen bildeten meine Nahrung. 5
Bald habe ich zur Bussübung Steinchen gemacht;[1]
Bald habe ich die Leerheit zur Speise genommen,
Bald erlangte ich Notbehelfe[2] und Abhärtung.
Bald entstand in mir die Illusion des Durstes.
Das blaue Wasser des kühlen Schiefers trank ich, 10
Bald genoss ich selbst-entstandenes duftiges Wasser,
Bald trank ich von dem Strome des Erbarmens,
Bald trank ich von den heiligen Spenden der Ḍāka.
Bald entstand in mir die Illusion des Frierens.
In ein einfaches Baumwollgewand kleidete ich 15
    mich.

---

[1] Die Übersetzung des Verses ist nicht ganz sicher.
[2] Tib. *t'abs c'ag* Notbehelf, ärmliches Auskunftsmittel, Surrogat (Jäschke).

res ₒga gtum moi bde drod sbar

Bald entfachte ich die glückliche Wärme der inneren Glut.

res ni t'abs c'ag sdug sran skyed

Bald erlangte ich Notbehelfe und Abhärtung.

res ₒga grogs kyi ₒk'rul snaṅ šar

Bald erhub sich die Illusion der Freundschaft in mir.

rig pa ye šes grogs su bsten

Auf Wissen und Weisheit vertraute ich als meinen Freunden

20 *dkar po dge bcui las la spyad*

Und übte die zehn reinen Tugenden. 20

yaṅ dag blta bai ñams len byas

Vollkommene Beschauung habe ich mir angeeignet,

raṅ rig sems kyi rtsa gdar bcad

Den Sinn des Selbstwissens habe ich genau geprüft.

ṅa rnal ₒbyor mi yi seṅ ge yin

Ich, der Yogin, bin der Löwe der Menschen,

lta ba bzaṅ poi yyu ral rgyas

Dessen Türkisglanzmähne trefflicher Beschauung ausgebreitet ist,

25 *bsgom pa bzaṅ poi mc'e sder can*

Der die Fangzähne und Krallen trefflicher Medi- 25 tation hat;

ñams len gaṅs kyi ltoṅs su byas

Merkverse mache ich auf dem Gipfel der Berge

yon tan ₒbras bu t'ob tu re

Und hoffe, die Frucht der Tugend zu erlangen.

ṅa rnal ₒbyor mi yi rgya stag yin

Ich, der Yogin, bin der Königstiger der Menschen,

byaṅ c'ub sems kyi rtsal gsum rdzogs

Der die drei vollendeten Fertigkeiten des Bodhi-Sinnes besitzt,

30 *t'abs šes dbyer med gra ₒdzum can*

Der in der unzertrennlichen Verbindung von Stoff 30 und Geist (gra?), lächelnd,

ₒod gsal sman ljoṅs nag pa la bsdad

In dunklen, an Arzneien der Erleuchtung reichen Thälern gelebt hat.

gžan don ₒbras bu ₒbyuṅ du re

Ich hoffe, dass die Frucht der Uneigennützigkeit (parārtha) erzielt wird.

ṅa rnal ₒbyor mi yi rgod po yin

Ich, der Yogin, bin der Geier der Menschen,

bskyed rim gsal bai ₒdab gšog rgyas

Der die Flügel der lichten Meditationsstufe des Utsakrama ausgebreitet hält,

35 *rdzogs rim brtan pai ldem sgro can*

Der mit den Schwingen der starken Meditationsstufe 35 des Sampannakrama schlägt,

zuṅ ₒjug c'os ñid mk'a la ldiṅ

Der am Himmel der Lehre vom zung ₒjug schwebt,

yaṅ dag don gyi brag la ñal

Der auf dem Felsen der mystischen Erkenntnis ruht,

ₒbras bu don gñis ₒgrub tu re

Der hofft, dass die Frucht in den beiden Arten der Erkenntnis erlangt wird.

ṅa rnal ₒbyor mi yi dam pa ste

Ich, der Yogin, bin der Heilige der Menschen,

40 *ṅa ni Mi la ras pa yin*

Ich bin Milaraspa. 40

ṅa ni snaṅ ba gdoṅ ₒded mk'an

Ohne Rücksicht auf andere verfolge ich meinen Weg.

ṅa ni ₒdun ma gaṅ byuṅ mk'an

Ich bin der Ratschaffer in allen Fällen.

ṅa ni ṅes med rnal ₒbyor pa

Ich bin der heimatlose Yogin.

ṅa ni gaṅ byuṅ gtaṅ med mk'an

Ich lasse nicht fahren, was mir zu teil geworden.

45 *ṅa ni zas med ldom bu pa*

Ich bin der Bettler ohne Nahrung, 45

ṅa ni gos med gcer bu pa

Ich bin der Nackte ohne Kleidung,

ṅa ni nor med sloṅ mo pa

Ich bin der Besitzlose, der von Almosen lebt.

ña ni p'yi ts'is[1] bsam med mk'an
ña ni ₒdir sdod ₒdir gnas med
50 *ña ni spyod pa byuṅ rgyal mk'an*

ña ni sod pa ši skyid mk'an
ña ni caṅ med dgos med mk'an
dgos pai yo byad bsgrub dgos na

mi k'yod la ñon moṅs dka ts'egs yod
55 *yon bdag ₒo rgyal log pa bžud*

rnal ₒbyor gaṅ byuṅ ₒdun ma byed
t'ugs·bsam pa bzaṅ žiṅ dge ba yis

sbyin gtoṅ byed pa t'ugs la btags

ts'e ₒdir ts'e riṅ nad med ciṅ

60 *dal ₒbyor bde skyid loṅs spyad nas*
p'yi ma dag pai žiṅ k'ams su
mjal nas c'os la spyod dar šog

de nas gžan don ₒgrub par šog

  ces gsuṅs pas | k'o šin tu dad par
gyur te | k'yed grub t'ob Mi la yin par
gda bas | de kas c'og par gda ste | ñed
mi nag pa ts'ogs rdzogs p'yir du ₒdir
bžugs rin ₒts'o c'as bdag gis sgrub pas
| cis kyaṅ bžes pa žu žes žus nas |
byaṅ c'ub rdzoṅ du bžugs rin gyi ₒts'o
brgyags p'yug po lha ₒbar gyis p'ul
lo ‖ de nas rje btsun la t'ugs dam
šin tu bogs c'e bar byuṅ bas | t'ugs
mñes bžin pai ñaṅ la rag mai ña
ma ₒga mjal du byuṅ ba rnams na
re | gnas la t'ugs ₒgro ba daṅ t'ugs
dam dmar po byuṅ ñam žu ba la
| rje btsun gyis gnas la yid mgu |
dge sbyor yaṅ ₒp'el bar byuṅ gsuṅs pas
| k'oṅ rnams na re | de ka yoṅ žus
| gnas dga mo ₒdi la bstod pa žig
daṅ | k'yed raṅ gi t'ugs dam mdzad

Ich bin der nicht an Berechnung Denkende.
Ich bin nicht hier wohnend, hier weilend.
Ich bin der König derer, denen die Übung der Meditation obliegt.
Ich bin der erfreuliches Glück Besitzende.
Ich bin, der nichts hat und nichts braucht.
Wenn ich die notwendigen Lebensbedürfnisse selbst erwerben müsste,
Wäre das eine schwere Sünde von dir.
Die Mühen der Gabenspender werden sich abwenden und dahingehen,
Denn der Yogin weiss für alle Fälle Rat zu schaffen.
Durch die guten und tugendhaften Gedanken deines Herzens
Geschenke darbringend hast du dich meiner liebevoll angenommen.
Mögest du in dieser Existenz langes Leben, Freisein von Krankheit,
Das Glück inneren Friedens geniessen!
Möchten wir uns in Zukunft in dem reinen Gefilde
Wieder begegnen! Möge religiöser Wandel sich ausbreiten
Und darauf das Heil anderer erzielt werden!

Als er so gesprochen, wurde jener sehr gläubig und sagte: ‚Da du Mila bist, der die Heiligkeit erlangt hat, und das eben mich befriedigt, so bitte ich, der nur ein einfacher Laie ist, um die Ansammlung der Verdienste zu vollenden, dich während deines Aufenthaltes hier mit wertvollen Lebensmitteln versehen zu dürfen und alles anzunehmen.‘ So gab ihm, während er in Byang chub rdzong verweilte, Lha ₒbar wertvolle reichliche Lebensmittel. Als darauf dem Ehrwürdigen grosse Förderung der Meditation zu teil wurde, wodurch er in frohe Stimmung geriet, erschienen einige seiner Anhänger von Rag ma und sagten: ‚Findest du Wohlgefallen an diesem Orte, und ist die Beschauung gewinnreich gewesen?‘ Der Ehrwürdige erwiderte: ‚An dem Orte habe ich meine Freude, auch in der Kunst der Meditation habe ich gewonnen.‘ ‚Vortrefflich!‘ sagten sie, ‚geruhe uns ein Preislied auf diesen lieblichen Ort und die Art, wie du deine Beschauung vollziehst, zum besten zu geben.‘

---

[1] *ts'is*, nach Jäschke wahrscheinlich secundäre Form von *rtsis*; *p'yi* in dieser Verbindung ist nicht recht verständlich.

ts'ul žig gnan bar žu žu bai lan du mgur ₒdi gsuns so.	In Erwiderung ihrer Bitte trug er folgendes Lied vor.

19.[1]

byan c'ub rdzon gi dben gnas ₒdi
p'u na lha btsan gans dkar mt'o

ₒda na yon bdag dad ldan man
rgyab ri dar dkar yol bas bcad

5 mdun na dgos ₒdod nags ts'al spuns
span gšons ne bsin c'e la yans
dri ldan yid ₒon padma la
rkan drug ldan pai dar dir can
rdzin bu lten ka c'u nogs la
10 c'u bya mgrin pa skyogs nas lta
ljon šin rgyas pai yal ga la
mdzes pai bya ts'ogs skad snan
sgyur
dri gžon ser bus btab pa la
rkan t'un yal gas gar stabs byed

15 mt'o žin gsal bai ljon šin rtser
spra sprei yan rtsal sna ts'ogs byed

sno ₒjam yans pai bsin ma la
rkan bži dud ₒgro bzas la bkram
de dag skyon byed p'yugs rdzi rnams
20 glu dan glin bui skad snan sgyur
ₒjig rten sred pai k'ol po rnams
zan zin las byed sa gži k'ebs
de la lta bai rnal ₒbyor na
kun gsal rin c'en brag stens na
25 snan ba mi rtag dpe ru ₒdren

Das ist die Einsiedelei von Byang chub rdzong.
Oben ragt der weisse hohe Gletscherberg mächtiger
  Geister.
Unten stehen viele gläubige Gabenspender.
Der Berg hinter mir ist mit einem weissen Seiden-
  vorhang bedeckt.
Vor mir dehnen sich wunschstillende Wälder aus. 5
Da sind grosse und weite Rasengründe und Matten.
Auf den duftenden lieblichen Blumen
Schweben summend die sechsfüssigen Insekten.
Am Strand der Teiche und Weiher
Späht der Wasservogel, den Hals drehend. 10
Im weitverästelten Gezweig der Bäume
Singt lieblich die schöne Vogelschar.

Vom Düfte tragenden Winde bewegt
Wiegen sich die Zweige der Bäume[2] tanzend hin
  und her.
Im Wipfel der hohen, weit sichtbaren Bäume 15
Zeigen Affen und Äfflein ihre mannigfachen Geschick-
  lichkeiten.
Auf dem grünen, weichen, weiten Wiesenteppich
Breitet sich weidend vierfüssiges Vieh hin.
Die das Vieh hütenden Hirten
Singen und entlocken der Flöte liebliche Töne. 20
Die Knechte weltlicher Habsucht
Stapeln auf dem Boden ihre Waren auf.
Wenn ich, der Yogin, darauf hinabschaue,
Auf meinem weithin sichtbaren herrlichen Felsen,
Betrachte ich die vergänglichen Erscheinungen als 25
  ein Gleichnis.

---

[1] Dieses Lied wurde als Probe aus Milaraspa in Text und Übersetzung mit ausführlichen grammatischen und lexikalischen
Noten von Jäschke in ZDMG, Bd. XXIII, p. 543—558 mitgeteilt. Unabhängig von dieser Arbeit ist, wie es scheint,
Rockhill's Übertragung desselben Liedes in Proc. AOS 1884, p. CCX, in welcher V. 1—3 und 13—14 fehlen. Meine
Übersetzung weicht in einigen Punkten von Jäschke's Auffassung ab. Rockhill's Text, der nach seiner eigenen Aussage
sehr uncorrect ist, muss von den Jäschke und mir vorliegenden Originalen starke Abweichungen bieten. Ganz unverständlich
ist mir Rockhill's Übersetzung von V. 21—25.
[2] Jäschke's Erklärung von rkan t'un (l. c., p. 553): ‚eig. Kurzfuss, poetische Benennung für Baum‘ kann ich nicht billigen,
da das Wort offenbar eine Übersetzung von Skr. pādapa ist (t'un = d'un trinken).

₀dod yon mig yor c'u ru sgom	Die sinnlichen Genüsse sehe ich als ein Spiegelbild im Wasser an.[1]
ts'e ₀di rmi lam sgyu mar lta	Dieses Leben halte ich für die Täuschung eines Traumes.
ma rtogs pa la sñiñ rje sgom	Gegen die Unständigen hege ich Mitleid.
namk'a stoñ pa zas su za	Den leeren Raum nehme ich zur Speise,
30 yeñs pa med pai bsam gtan sgom	Ungestörter Contemplation weihe ich mich.
sna ts'ogs ñams la ci yañ ₀c'ar	Wie alle Bilder, die in unserem Geiste aufsteigen,
,e ma k'ams gsum ₀k'or bai c'os	Ach, nach dem Gesetz des Kreislaufs der drei Welten
med bžin snañ ba ño mts'ar c'e	Nicht vorhanden sind, so auch nicht die herrlichen Erscheinungen der Welt.[2]

ces gsuñs pas ǀ k'oñ rnams dad bžin log soñ ño ǀ rag mai skor sña mao.	Als er so gesprochen, kehrten jene gläubig zurück. Dies ist der erste Abschnitt von Rag ma.

IV.[3]

Milaraspa auf dem rKyang phan namkha rdzong.

(fol. 27 a 1 — 28 b 5.)

na mo gu ru.	Verehrung dem Meister!
rje btsun Mi la ras pa de ñid ǀ	Als der ehrwürdige Milaraspa von Rag ma nach
rag ma nas rkyañ p'an namk'a rdzoñ	dem rKyang phan namkha rdzong gekommen war,

---

[1] Jäschke fasst *mig yor* als ‚Luftspiegelung, Fata Morgana‘ und übersetzt: die Lustgenüsse betrachte ich als durch die Mirage vorgespiegeltes Wasser. Doch da über die Bekanntschaft der Tibeter mit dieser Naturerscheinung nichts sicheres feststeht, scheint mir die obige ebenso gut mögliche Auffassung die einfachere zu sein.

[2] Jäschke's Übersetzung der V. 31—33 kann ich nicht zustimmen:

> Mannigfach Gedanken steigen auf;
> der drei Weltgebiete Kreiseslauf
> wird zum Nichts vor mir! O Wunder gross!

Die dazu gegebene Erklärung: ‚Das Erscheinen der Lehre vom Kreislauf der drei Welten als nicht existierend ist mir ein Wunder gross! d. h. dass ich mir die Lehre vom Kreislauf, oder nach der anderen Erklärung: den Kreislauf, die Welt selbst, als nicht existierend denken kann, ist mir ein (dankenswertes) Wunder‘ ist keineswegs plausibel. Denn was sollte dem Buddhisten an der Erkenntnis des Saṃsāra wunderbares sein? Viel näher scheint mir der Wahrheit Rockhill's Wiedergabe zu kommen:

> All the different images which may appear —
> Forsooth, 'tis but the universal law of things —
> They all, whate'er we see, are of a truth unreal.

Beide Übersetzer haben die Vergleichspartikel *bžin* in V. 33 übersehen und die Interjektion des Bedauerns ,e ma in V. 32 nicht scharf genug erfasst. Der Kreislauf, und damit die Nichtexistenz (*med*) der Dinge, bezieht sich sowohl auf das Ich (V. 31) wie auf das Nicht-Ich (*snañ ba*); wie unser Denken nicht wirklich ist, so ist auch die uns umgebende, sichtbare Natur irreal, und das ist vom Standpunkt des Dichters bei ihrer wunderbaren Schönheit, die er in den vorhergehenden Versen geschildert, lebhaft zu beklagen (,e ma). Mit dieser grammatisch wie sachlich annehmbaren Erklärung, die den Gedankengang des Gedichtes folgerichtig abschliesst, fällt der von Jäschke gesuchte Widerspruch des V. 31 mit dem vorhergehenden Verse von selbst.

[3] Dieser kurze Abschnitt wurde als Probe aus Milaraspa von Rockhill in Proc. AOS. 1884, p. VI—VIII (oder p. CCVIII—CCX der ganzen Serie) übersetzt; für diese Arbeit gilt dasselbe wie oben zu III Nr. 10 Bemerkte. Den Anfang in Prosa und Lied Nr. 20 hat Sandberg, The Nineteenth Century 1899, p. 618, ferner Nr. 21 und 22 ibid. p. 627 übertragen. Meine Übersetzung ist von den beiden genannten völlig unabhängig und vor der Bekanntschaft mit denselben entstanden. Von

du byon te | bžugs pai dus nam žiɥ
gi ts'e | spreu ri boṅ la žon pa žig
šo mai p'ub gyon | sog mai mda gžu
t'ogs nas glags bltar byuṅ ba la | rje
btsun bžad mo žig šor bas | k'o na re
| k'yod ₒjigs su re nas ₒoṅs pa yin te
| mi ₒjigs na ₒgro zer ba la | rje btsun
gyis ṅa snaṅ ba sems su t'ag c'od ciṅ
| sems ñid c'os skur ṅo ₒp'rod pas |
₀dre k'yod kyi c'o ₒp'rul ci bstan yaṅ
| rnal ₒbyor ṅa yi gad moi gnas ‖ ces
gsuṅs pas | k'os žabs tog bsgrub par
k'as blaṅs te ₒja yal ba bžin soṅ ba
de | gro t'aṅ rgyal po yin no | de
nas gro t'aṅ gi yon bdag | rje btsun gyi
žal ltar byuṅ ba rnams na re | gnas
₀di la yon tan ci gda žu bai lan du
mgur ₒdi gsuṅs so ‖

geschah es eines Nachts, als er dort verweilte, dass
ein auf einem Hasen reitender Affe vor ihm erschien,
der einen Pilz als Schild trug und Bogen und Pfeil
aus Stroh hielt, als wenn er auf eine Gelegenheit
lauerte. Der Ehrwürdige stiess ein Gelächter aus.
Jener sprach: ‚In der Hoffnung, dich zu vernichten,
bin ich hierher gekommen; doch wenn du nicht zu
vernichten bist, gehe ich weg.‘ Der Ehrwürdige erwiderte: ‚Da ich die sichtbare Welt für imaginär
halte und meine eigene Wesenheit in der Nichtexistenz wahrnehme,[1] sind deine, eines Kobolds Trug-
spiele, welche auch immer du zeigen mögest, ein
Gegenstand des Gelächters für mich, den Yogin!‘
Da verpflichtete sich jener ihm Dienste zu leisten
und verschwand wie ein verblassender Regenbogen.
Es war der Herrscher von Gro thang. Darauf kamen
die Gabenspender von Gro thang, um den Ehrwürdigen zu besuchen, und sagten: ‚Welche Vorzüge
haften an diesem Orte?‘ Als Antwort auf ihre Frage
trug er folgendes Lied vor.

20.

bla ma rje la gsol ba ₒdebs

gnas ₒdii yon tan šes ma šes
gnas ₒdii yon tan ma šes na
dben gnas rkyaṅ p'an namk'a rdzoṅ

5 *namk'a rdzoṅ gi p'o braṅ na*
steṅ na lho sprin dmug po ₒt'ibs
₀og ṅa gtsaṅ c'ab šnon po ₒbab
rgyab na brag dmar namk'ai dbyiṅs
mdun na spaṅ po me tog bkra

Zu meinem Lama, dem Gebieter, sende ich mein
    Flehen!
Kennt ihr, kennt ihr nicht die Vorzüge dieses Ortes?
Wenn ihr die Vorzüge dieses Ortes nicht kennt,
So wisst, die Einsiedelei ist rKyang phan namkha
    rdzong.
In dem Palast dieser Himmelsfestung   5
Sammeln sich oben purpurfarbene Südwolken,[2]
Unten strömt der blaue Brahmaputra[3] dahin,
Hinter mir rote Felsen gleich dem Himmelsraum,[4]
Vor mir Wiesen mit buntfarbigen Blumen;

Abweichungen habe ich nur die wichtigsten in den Noten angemerkt; auf Rockhill's Übertragung im einzelnen einzugehen
verbietet schon der Umstand, dass ihm ein stark verdorbener Text vorgelegen hat. Sandberg's Arbeit ist ein populärwissen-
schaftlicher Aufsatz mit fragmentarischen Proben; keine Geschichte ist vollständig mitgeteilt. Schon deshalb liegt mir jeder
Gedanke an eine Polemik gegen diesen Artikel völlig fern, der ein ganz anderes Ziel als meine Abhandlung verfolgt.

[1] Rockhill's Übersetzung ‚My mind has embraced the body of the truth (dharmakāya!)‘ muss als verfehlt bezeichnet werden.
Sandberg bietet: ‚understanding the imagination itself to be as impalpable as the body of Buddha in Nirvana‘.

[2] Sandberg hat irrtümlich: above it, to the south, lie clouds etc.: *lho sprin* stellt bei Milaraspa stets ein Compositum vor.

[3] Rockhill: the crystal stream; Sandberg: waters transparent and green. Mit *gtsaṅ c'ab* ist aber der Brahmaputra
gemeint.

[4] Rockhill: behind it the red rocks and heaven's expanse, und übereinstimmend Sandberg: behind are red rocks and the
expanse of the heavens. Ich kann diese Auffassung nicht teilen, sondern construiere diesen Vers als Parallelvers
zu dem folgenden: Wie die Wiesen solche sind, die bunte Blumen haben, so haben die Felsen die Ausdehnung des
Himmels, was ganz im Geiste des Dichters gedacht ist.

7*

10 zur na gcan zan ṅar skad ₒdon
logs la bya rgyal rgod po ldiṅ

mk‘a la sbraṅ c‘ar zim bu ₒbab

rgyun du buṅ bas glu dbyaṅs len
šva rkyaṅ ma bu rtse bro brduṅ

15 spra daṅ spreu yaṅ rtsal sbyoṅ
lco ga ma bu ₒgyur skad maṅ

lha bya goṅ mo glu dbyaṅs len
rdza c‘ab sil mas sñan pa brjod

dus kyi skad rigs ñams kyi grogs

20 gnas ₒdii yon tan bsam mi k‘yab
ñams dga glu ru blaṅs pa yin

gdams ṅag k‘a ru bton pa yin
ₒdir ₒts‘ogs yon bdag p‘o mo rnams

miṅ p‘yir ₒbreṅs la ṅa bžin mdzod

25 las sdig pa spoṅs la dge ba bsgrubs

 ces gsuṅs pas | k‘oṅ rnams kyi naṅ
na sṅags pa žig ₒdug pa de na re |
rje btsun lags | ṅed rnams la žal mjal
bai dga ston nam | byon skyes su | blta
sgom spyod pai ñams len go sla la ₒk‘yer
bde ba žig gnaṅ bar žu žu bži lan
du mgur ₒdi gsuṅs so.

Am Rande stösst das Raubtier sein Gebrüll aus, 10
An den Seiten schwebt der Geier, der König der
 Vögel;[1]
In der Luft fallen Insekten wie ein feiner Staubregen
 herab,
Und ohne Unterlass summen die Bienen ihr Liedchen.
Hirsche und Wildesel, Mutter und Junges, spielen
 und springen wie im Tanz;
Affen und Äfflein üben ihre Geschicklichkeit. 15
Lerchen, Mutter und Junges, trillern im Wechsel-
 sang.
Der Göttervogel, das Schneehuhn, singt sein Lied.
Über seinem Thonschieferbett murmelt der Bach mit
 melodischem Plätschern.
Das sind die jeweiligen Stimmen [der Natur], die
 Freunde des Herzens![2]
Die Vorzüge dieses Ortes sind unermesslich: 20
Deshalb singe ich ein Lied der Herzensfreude auf
 ihn.
Belehrung fliesst aus meinem Munde:
Ihr Gabenspender, Männer und Frauen, die ihr hier
 versammelt seid,
Folgt mir um meines Namens willen und thut wie
 ich!
Meidet sündige Werke und strebt nach der Tugend! 25

 So sang er. Unter den Anwesenden befand
sich ein Kenner der Mantra, welcher sagte: ‚Ehrwürdiger! Als ein Fest oder Gastgeschenk für unsere
Pilgerfahrt hierher gewähre uns bitte einige zu
Herzen gehende, leicht verständliche und gut zu bewahrende Belehrung über die Ausübung der Meditation!‘ Als Antwort auf diese Bitte trug er folgendes
Lied vor.

---

[1] Dieser Vers fehlt bei Sandberg.
[2] Rockhill übersetzt diesen Vers: ‚The voice of time and unworthy friends‘, mit der Bemerkung, dass der Text uncorrect
scheine, doch dass er nicht sehe, wie derselbe zu verbessern sei, und fährt dann fort: ‚Trouble not the dream of this
place's sweetness‘. Sandberg gibt die Übertragung: ‚Their voice is the voice of Time — of friends whose friendship has
degenerated‘. Der Sinn, den beide Übersetzer mit ihren Auslegungen verknüpfen wollen, ist mir unverständlich. Das
‚Zeit‘ erscheint im Tibetischen nicht als ein philosophisches Abstraktum; dus kyi hat vielmehr nach Jäschke's ausdrücklicher
Angabe bei Milaraspa den Sinn ‚zeitweilig, jeweilig, happening sometimes‘, eine Bedeutung, die in den Sinn der obigen
Stelle sehr gut passt. Der Dichter meint nämlich offenbar die in den vorhergehenden Versen aufgezählten, zeitweilig oder
von Zeit zu Zeit ertönenden Stimmen der verschiedenen Tiere, die zugleich seinem Herzen (ñams) ein lange befreundeter
Klang sind. Wäre ñams im Sinne von hurt, injured, imperfect zu fassen, wie dies Rockhill und Sandberg thun, deren
Übersetzungen unworthy und degenerated auch dann noch nicht gerechtfertigt erscheinen, so würde der Text auch ñams
pai und nicht ñams kyi bieten. Zum Überfluss citiert auch Jäschke, Dictionary p. 185 b, ñams kyi grogs ‚companions of
the soul, viz. the murmuring springs and rivulets in the solitude of alpine regions‘.

21.

bla mai byin brlabs sems la žugs	Möge der Segen meines Lama bei mir einziehen!
stoṅ ñid rtogs par byin gyis rlobs	Möge er mich segnen, dass ich die Leere erkenne!
yon bdag dad pai p'yag lan du	Zum Dank für die Verehrung der gläubigen Gaben-
	spender
lha yi dam mñes pai glu cig len	Will ich ein Lied singen, an dem sich Götter und
	Schutzgottheiten erfreuen.

5 snaṅ daṅ stoṅ daṅ dbyer med gsum — Die Erscheinungen, die Leerheit, das Unzertrenn- 5
liche,

odi gsum blta bai mdor bsdus yin — Diese drei bilden den Kern der Beschauung.
gsal daṅ mi rtog ma yeṅs gsum — Das Verständliche, das Unbegreifliche, die Nicht-
erregbarkeit des Geistes,

odi gsum bsgom pai mdor bsdus yin — Diese drei bilden den Kern der Meditation.
c'ags med žen med mt'ar bskyol gsum — Leidenschaftslosigkeit, Wunschlosigkeit und Standhaftigkeit,

10 odi gsum spyod pai mdor bsdus yin — Diese drei bilden den Kern der Lebensführung. 10
re med dogs med ok'rul med gsum — Hoffnungslosigkeit, Furchtlosigkeit, Irrtumslosigkeit,
odi gsum obras bui mdor bsdus yin — Diese drei bilden den Kern der Vergeltung.
ṅo med lkog med zol med gsum — Reinheit im öffentlichen wie im privaten Leben,
Freisein von Betrug,

odi gsum dam ts'ig mdor bsdus yin — Diese drei bilden die Grundlage des Gelübdes.

*žes gsuṅs pas | k'oṅ rnams dad
bžin du log soṅ ba las | yaṅ žag oga
nas ña ma maṅ po mjal du byuṅ ba
las | sṅar gyi de rnams kyis rje btsun
sku k'ams daṅ ok'yer so bde lags sam
| žes bsñun dris pai lan du mgur odi
gsuṅs so.*

Als er so gesprochen, kehrten jene gläubigen Sinnes zurück. Nach einigen Tagen erschienen wiederum zahlreiche Zuhörer, um ihm ihre Verehrung zu bezeigen. Jene, die vorher da gewesen waren, fragten: ‚Ist das Befinden und der Zustand des Ehr- würdigen gut?‘ Auf ihre Frage nach seiner Gesund- heit erwiderte er in folgendem Liede.

22.

bla mai dam pai žabs la odud	Zu den Füssen des heiligen Lama verneige ich mich.
gnas mi med dben pai nags odabs na	Im unbewohnten einsamen Walde
Mi la ras pai sgom lugs bde	Ist Milaraspa's Beschauungsart segensreich.
odzin c'ags med pai ogro odug bde	Glücklich ist, wer da wandelt frei von der Leiden-
	schaft zu besitzen,

5 na ts'a med pai sgyu lus bde — Glücklich, wessen Leib frei ist von brennendem 5
Schmerz.

ñal ba med pai odug lugs bde — Glücklich, wessen Wesen der Trägheit bar ist.
rtog pa med pai tiṅ odzin bde — Glücklich, wer einfachen Herzens Beschauung übt.
graṅ ba med pai gtum mo bde — Glücklich, wer, ohne kalt zu sein, die innere Glut
besitzt.

žum pa med pai brtul žugs bde
10 *rtsol ba med pai so nam bde*
ɣyeṅ ba med pai dben gnas bde
de tšo lus kyi oḱyer so yin
t'abs šes gñis kyi t'eg pa bde

skyed rdzogs zuṅ ojug ñams len bde

15 *rluṅ ogro ooṅ med pai dran pa bde*

lab grogs med pai smra bcad bde

de tšo ṅag gi oḱyer so yin
ṅos bzuṅ med pai blta ba bde

rgyun c'ad med pai sgom pa bde

20 *ñam ṅa med pai spyod pa bde*
re dogs med pai obras bu bde

de tšo sems kyi oḱyer so yin
ogyur med rtog med ood gsal bde

bde c'en rnam dag dbyiṅs su bde

25 *ogag med oc'ar sgoi kloṅ du bde*

šin tu bde bai dbyaṅs c'uṅ odi
ñams myoṅ glu ru blaṅs pa yin
blta spyod zuṅ du sbrel ba yin

slaṅ nas byaṅ c'ub bsgrub pa rnams
30 *ñams su len na de ltar mdzod*

ces gsuṅs pas | ña ma rnams na
re | bla mai sku gsuṅ t'ugs kyi oḱyer
soi bde lugs de rnams šin tu ṅo mtsar
c'e bar gda | de rnams gaṅ las byuṅ
žus pas | de rnams sems rtogs pa
las byuṅ ba yin gsuṅs pas | oo na ṅed
rnams la yaṅ bde ba de lta bu raṅ
ma byuṅ ruṅ | dei c'a odra tsam yoṅ
du re bas | sems rtogs t'abs kyi sgom
lugs odi ltar gyis gsuṅ bai c'os go sla

Glücklich, wer ohne Furcht Bussübungen vollzieht.
Glücklich der Landwirt, der nach nichts trachtet. 10
Glücklich, wer die ungestörte Einsamkeit wählt.
Das alles sind Vorzüge des Leibes.
Glücklich, wer das Fahrzeug der Materie und des Geistes hat,
Wer die Meditationsstufen des skyed rdzogs und zung ojug erreicht hat,
Glücklich, wer sich bewusst ist, dass zwischen Aus- 15 strömen und Einatmen der Luft kein Unterschied besteht.
Glücklich, wer, ohne Freunde, zu reden nicht gebunden ist.
Das alles sind Vorzüge des Wortes.
Glücklich, wessen Anschauungen frei von Selbstsucht sind,
Glücklich, wer sich beständig ununterbrochener Betrachtung weiht.
Glücklich, wessen Wandel frei von Furcht ist. 20
Glücklich, wer die Belohnung der Hoffnungs- und Furchtlosigkeit gewonnen.
Das alles sind Vorzüge des Sinnes.
Glücklich, wer unwandelbar, einfachen Herzens und erleuchtet ist.
Glücklich, wer in der reinen Sphäre des höchsten Segens weilt.
Glücklich, wer in der Tiefe schrankenloser Gedan- 25 ken weilt.
Diesen sehr segensreichen kleinen Sang,
Dies Lied der Wonne habe ich gesungen.
Beschauung und Übung sind eng mit einander verbunden.
Die ihr durch eure Bitten nach der Bodhi strebt,
Nehmt es euch zu Herzen und handelt so! 30

Als er so gesprochen, sagten die Zuhörer: ‚Wir sind von grossem Staunen befangen über den glückseligen Zustand, den der Lama in Leib, Wort und Sinn geniesst.‘ Als sie darauf fragten, woher derselbe entstanden sei, antwortete er, dass das alles von der Erkenntnis seiner selbst herrühre. Da sagten sie: ‚Wenn nun uns auch eine solche Glückseligkeit selbst nicht zu teil wird, so könnten wir doch wenigstens hoffen, dass uns ein Teil davon zufallen wird. Geruhe daher, uns eine leicht verständliche

la ₀kʻyer bde ba žig gnaṅ bar žu žus
pai lan du | rje btsun gyis sem don
bcu gñis ma ₀di mgur du gsuṅs so.

und gut zu bewahrende Lehre vorzutragen, in der
du die Art und Weise der Meditation vermittelst der
Selbsterkenntnis auseinandersetzest.' Zur Erwiderung
auf ihre Bitte trug der Ehrwürdige ein Lied von den
zwölf geistigen Gütern vor.

23.

bla ma dam pai žabs la ₀dud
yon bdag sems ñid rtogs ₀dod rnams

ñams su len na ₀di ltar mdzod

dad daṅ mkʻas daṅ btsun daṅ gsum
5 *₀di gsum sems kyi srog šiṅ lags*
btsugs na brtan la ₀jug na bde

srog šiṅ mdzad na de la mdzod
cʻags med žen med rmoṅs med gsum

₀di gsum sems kyi go cʻa lags
10 *gyon na yaṅ la mtsʻon kʻar sra*
go cʻa mdzad na de la mdzod
sgom daṅ brtson ₀grus sdug sran gsum
₀di gsum sems kyi rta pʻo lags
rgyugs na ₀gyogs la ₀bros na tʻar

15 *rta pʻo mdzad na de la mdzod*
raṅ rig raṅ gsal raṅ bde gsum
₀di gsum sems kyi ₀bras bu lags
btab na smin la zos na bcud

₀bras bu mdzad na de la mdzod
20 *sems don rnam pa bcu gñis ₀di*
rnal ₀byor blo la šar nas blaṅs
kʻyed yon bdag dad pai pʻyag lan
mdzod

ces gsuṅs pas | koṅ rnams dad par
gyur nas | pʻyis kyaṅ žabs tog pʻun sum
tsʻogs pa bsgrubs par gyur to | rje
btsun yaṅ yol mo gaṅs ra la tʻugs
cʻas so | rkyaṅ pʻan namkʻa rdzoṅ gi
bskor ro.

Dem heiligen Lama zu Füssen verneige ich mich.
Ihr Gabenspender, die ihr der Seele Erkenntnis be-
gehrt,
Wenn ihr sie euch zu Herzen nehmt, so handelt
darnach!
Glaube, Klugheit und Güte,
Diese drei sind der Lebensbaum[1] der Seele. 5
Wenn ihr diesen pflanzt und sorgfältig hütet, seid
ihr glücklich;
Macht den Lebensbaum und handelt darnach!
Leidenschaftslosigkeit, Wunschlosigkeit, Freisein von
Thorheit,
Diese drei sind der Panzer der Seele.
Wenn ihr ihn anlegt, seid ihr stichfest. 10
Macht diesen Panzer und handelt darnach!
Beschauung, Eifer und Abhärtung,
Diese drei sind der Hengst der Seele.
Wenn er dahineilt und ihr geschwind auf ihm
flieht, seid ihr befreit.
Schafft euch diesen Hengst und handelt darnach! 15
Selbstwissen, Selbsterleuchtung, Selbstglückseligkeit,
Diese drei sind die Frucht der Seele.
Wenn ihr sie angepflanzt habt und sie reif verzehrt,
ist sie Nahrung.
Erlangt diese Frucht und handelt darnach!
Diese zwölf geistigen Güter 20
Entstehen im Geiste des Yogin, der sie empfängt.
Ihr Gabenspender, verschafft sie euch zum Dank
für eure gläubige Verehrung.

Als er so gesprochen, wurden jene gläubig
und erwiesen ihm später ausgezeichnete Dienste.
Dann entschloss sich der Ehrwürdige, auf den Yol
mo gangs ra zu gehen. Dies ist der Abschnitt von
rKyang phan namkha rdzong.

---

[1] Das Bild ist von der eigentlichen Bedeutung von *srog šiṅ* hergenommen ‚Stützstange in der Mitte eines Stūpa'.

V.

Dharmabodhi.

(fol. 142 b 5 — 145 a 7.)

na mo gu ru.

rje btsun Mi la ras pa de ñid |
Ras c'uṅ pa la sogs pai bu slob rnams
daṅ bcas te gña nam groṅ p'ug na |
sñiṅ po don gyi c'os ₒk'or bskor žiṅ
bžugs pai ts'e | la stod na gu ru Ts'ems
c'en | diṅ ri na dam pa Saṅs rgyas
| bal po na Ši la bha ro | rgya gar
na Dharmabodhi | gña nam na Mi
la ras pa ste | grub t'ob lṅa dus
mts'uṅs pa las | Ši la bha ros Dharma-
bodhi spyan draṅs nas bal po rdzoṅ
na c'os ₒk'or bskor žiṅ bžugs pai ts'e
| bal bod kyi mi maṅ po Dharma-
bodhii žal blta ru ₒgro žiṅ ₒdug pa
las | rje btsun gyi bu slob rnams kyaṅ
mjal bar ₒdod nas | Ras c'uṅ pas rje
btsun la Dharmabodhi daṅ mjal na
legs pai rgyu mts'an maṅ po žus pas |
rje btsun gyis lan du mgur ₒdi gsuṅs so.

. Verehrung dem Meister!

Als der ehrwürdige Milaraspa im Verein mit Ras chung pa und den übrigen Schülern in der Grotte des Dorfes von gÑa nam verweilte und die Gebetsmühle, deren Nutzen den Kernpunkt (der Religion) bildet, drehte, gab es gleichzeitig fünf Siddha, nämlich den Guru Tshems chen oben auf dem Passe, den heiligen Sangs rgyas auf dem Ding ri, Shilabharo in Nepal, Dharmabodhi in Indien und Milaraspa in gÑa nam. Shilabharo lud Dharmabodhi ein, und als sie die Gebetsmühle drehend in Bal po rdzong[1] verweilten, kamen viele Männer aus Nepal und Tibet dorthin, um Dharmabodhi zu besuchen. Auch die Schüler des Ehrwürdigen hatten den Wunsch, dem Dharmabodhi ihre Aufwartung zu machen. Ras chung pa[2] setzte daher dem Ehrwürdigen mit vielen Gründen auseinander, weshalb es sich empfehlen würde, Dharmabodhi zu treffen. Der Ehrwürdige trug zur Antwort folgendes Lied vor.

24.

rje bla mai byin brlabs grub t'ob
    maṅ
ṅo mts'ar saṅs rgyas bstan pa dar
bde skyid ₒgro bai dpal du šar
rje grub t'ob mjal mi maṅ bde

5 *skal ldan ₒga re yoṅ bai rtags*

diṅ rii dam pa Saṅs rgyas daṅ
la stod gu ru Ts'ems c'en daṅ
bal po Ši la bha ro daṅ

Dank dem Segen des Herrn, des Lama, gibt es
    viele Siddha;
Buddha's wundervolle Lehre breitet sich aus,
Seligkeit erscheint im Wohlbefinden der Wesen,[3]
Durch die Begegnung mit den Siddha sind viele
    Menschen glücklich,
Und einige Würdige tragen die Vorzeichen an sich, 5
    dass sie es noch werden.
Der heilige Sangs rgyas vom Ding ri,
Der Guru Tshems chen auf der Passhöhe,
Shilabharo aus Nepal,

---

[1] D. h. nepalische Festung.
[2] D. h. der mit dem kleinen Baumwolltuch; *Mi la ras pa,* d. i. Mila mit dem Baumwolltuch. An anderer Stelle werden seine Schüler *c'a med pai ras pa rnams* ,die unvergleichlichen mit einem Baumwolltuch Bekleideten' genannt.
[3] Der Unterschied zwischen *bde skyid* und *dpal* ist so zu fassen, dass ersteres inneres und letzteres äusseres Glück bezeichnet.

rgya gar Dharmabodhi daṅ
10 guṅ t'aṅ Mi la ras pa rnams
sems rnam ri ga tsam re kun la yod
bsgom raṅ mgo t'on pa su la ₒaṅ yod

sems raṅ gsal sems ṅo sus kyaṅ śes

sprul bsgyur rdzu ₒp'rul kun gyis nus
15 stoṅ ñid sñiṅ rje kun gyis ₒbyoṅs

ₒp'ral ṅo mts'ar ltad mo k'oṅ ts'os ston

ñams byuṅ rgyal glu laṅ raṅ mk'as

sñiṅ rus žen log ṅa raṅ c'e

gžan rnams k'yad par ma mc'is pas
20 ṅa ni mjal du mi ₒgro bas
bu k'yed rnams mjal du cis kyaṅ soṅ

gžan skyon yod pa ma yin te
ṅa na so rgas pa mi t'on ₒdug
dpal ₒu rgyan yul du mjal bar smon

25 t'e ts'om ma byed blo gdeṅ yod

ces gsuṅs pas | skyon med na ka
mi rnams kyis | Mi la ras pa c'ags
sdaṅ gis mi ₒbyon par ₒdug zer skur
pa žu bar ₒdug pas | cis kyaṅ byon
pa ₒt'ad lags žus pai lan du mgur ₒdi
gsuṅs so.

Der Inder Dharmabodhi,
Milaraspa aus Gung thang, 10
Eines Jeglichen Geist ist so gross wie ein Berg.
Doch da ich durch meine Beschauung allein auf
mich selbst gestellt bin, wem gehöre ich an?
Wer kennt die Geistesklarheit meines eigenen
Geistes?
Zu Verwandlungen und Zaubereien sind alle fähig,
Durch die Leere werden alle zum Mitleid hin- 15
gerissen,
Jene sind schnell bereit, wunderbare Schauspiele zu
zeigen,
Doch ich selbst bin geschickt, siegreiche Lieder
zu singen, die meinem Herzen entsprungen sind;
Ich selbst bin gross in Standhaftigkeit und Re-
signation.
Die andern sind nicht so vortrefflich!
Ich gehe nicht, ihn zu besuchen. 20
Ihr, meine Söhne, mögt ihn auf jeden Fall be-
suchen.
Doch es ist nicht eines andern Schuld,
Nur meines Alters wegen ziehe ich nicht von hier.
In das herrliche Land von Udyāna möchte ich
pilgern.
Zweifel hege ich nicht, Zuversicht ist da! 25

Als er so gesprochen, sagten die Männer,
welche die Berufung auf das Alter nicht gelten
liessen: ‚Milaraspa will aus Eifersucht nicht gehen'.
So schmähten sie ihn. Als sie nun meinten, dass
es passend sei, jedenfalls hinzugehen, trug er zur
Antwort folgendes Lied vor.

25.

rje grub t'ob rnams la gsol ba ₒdebs
ñes ltuṅ dag par byin gyis rlobs
mi k'a byur gyi k'a lab la

t'e ts'om zan raṅ ñid ₒk'rul
5 sgom sgrub sñiṅ nas byed dus su

ₒgro ₒdug maṅ ba bar c'ad yin
bla ma rje la mjal dus su

An die gebietenden Siddha richte ich mein Gebet!
Mögen durch ihren Segen die Sünden rein werden!
Wenn man unseliges Gerede nach dem Geschwätz
der Leute führt
Und Zweifel nährt, täuscht man allein sich selbst.
Wenn zu der Zeit, wo man mit ganzer Seele der 5
Beschauung obliegt,
Viele Wesen zugegen sind, so ist das ein Hindernis.
Wenn zur Zeit des Besuchs bei dem Lama, dem
Gebieter,

bcos ma maṅ na lha grogs ₒk'rug

Viel künstliches Ceremoniell ist, werden die Götterfreunde gestört.

gsaṅ sṅags zab moi t'abs lam la

Wenn man an dem Wege der rechten Mittel zu den tiefen Geheimzaubersprüchen

10 *yid gñis byas na ₒgrub mi nus*

Zweifelt, ist man zur Vollendung nicht befähigt.   10

grub t'ob byin brlabs c'e lags te

Obwohl der Segen der Siddha gross ist,

₀k'or dpuṅ maṅ ba ts'ig pa za

Herrscht Unwille, wenn die Schar des Gefolges zahlreich ist.

bu Ras c'uṅ grogs mc'ed cis kyaṅ soṅ

Du, mein Sohn Ras chung, brüderlicher Freund, geh jedenfalls hin!

žes gsuṅs pas Ras c'uṅ pas | mi maṅ pos sdig sog pa ₒdug pas cis kyaṅ ₒbyon par žu | ṅed rnams la yaṅ p'an par ₒdug žus pas | ₒo na Dharmabodhi la p'yag len du ₒgro gsuṅs pas | Ras c'uṅ pa sogs grva pa rnams mgu nas | rje btsun ₒbyon na | rgya gar ba gser la rtsi ba yin pas | gser gcig bsgrubs nas byon pa grag žus pai lan du mgur ₀di gsuṅs so.

Als er so gesprochen, sagte Ras chung pa: ‚Viele Männer, die doch Sünden ansammeln, gehen sicher zu ihm hin, so wird es auch segensreich für uns sein; deshalb wollen wir zur Übung der Religion zu Dharmabodhi gehen.' Ras chung pa und die übrigen Schüler sagten freudig: ‚Wenn der Ehrwürdige hingeht, mag er auf indisches Gold rechnen; wenn er Gold gewonnen hat, wird ihm die Reise Ruhm bringen.' Als Antwort darauf trug er folgendes Lied vor.

26.

rje grub t'ob rnams la gsol ba ₀debs sbraṅ ₀dod pa zad par byin gyis rlobs

An die gebietenden Siddha richte ich mein Gebet! Mögen durch ihren Segen die Wünsche des Bettlers erschöpft sein!

byas ts'ad c'os su ₀gyur bar šog

Möchte das rechte Mass des Thuns zum Gesetz werden!

byas ts'ad c'os su ma gyur na

Wenn das rechte Mass des Thuns nicht zum Gesetz wird,

5 *byaṅ sems bsgoms pai don go c'uṅ*

Wäre das Verständnis vom Wesen der Beschauung   5 des reinen Sinnes gering.

sgom tiṅ ₀dzin skyes pa grogs mi ₀dod

Die Mitwirkung von Freunden zur Erzeugung der Beschauung und Betrachtung begehre ich nicht;

sgom raṅ grol šar bya grogs ₀dod na

Wenn ich Freunde begehrte zur Mitwirkung bei der Entstehung der Selbstbefreiung durch die Beschauung,

rgyun du bsgoms pa don go c'uṅ

Wäre mein Verständnis vom Wesen der ununterbrochenen Beschauung wahrlich gering.

Mi la ras pa nor mi sgrub

Milaraspa will keinen Reichtum gewinnen;

10 *Mi la ras pa nor sgrub na*

Wenn Milaraspa Reichtum gewänne,   10

bya ba btaṅ ba don go c'uṅ

Wäre sein Verständnis vom Wesen des Weltverzichts wahrlich gering.

Dharmabodhi gser mi ₀dod

Dharmabodhi's Gold begehre ich nicht;

Dharmabodhi gser ₒdod na
grub pa t'ob pa don go c'uṅ

15 *Ras c'uṅ rdor grags k'ye ma ₒdod*

rdo rje grags pa k'e ₒdod na

bla ma bsten pa don go c'uṅ

žes gsuṅs nas | k'yed raṅ rnams
sṅon la soṅ cig | ṅas p'yi nas ₒoṅ gis
gsuṅ | bu slob rnams sṅon la brdzaṅs
pas | rje btsun ₒbyon nam mi ₒbyon
nam bsam pai blo ts'om log ge bai
ṅaṅ la bal po rdzoṅ du sleb tu c'a
ba la | rje btsun gyis | sku šel gyi
mc'od rten du sprul nas | skar mda
ₒp'aṅs pa bžin du t'ogs brdug med
par namk'a la byon nas | bu slob
rnams kyi naṅ du byon pa | Dharma-
bodhis mt'oṅ bas ṅo mts'ar bar gyur
ciṅ | bu slob rnams t'e ts'om za ba rje
btsun namk'a nas byon pas | dga mgu
yi raṅs par gyur te | dpon slob rnams
ₒgrig pa Dharmabodhii druṅ du mi
maṅ dus² p'ebs pa las | rgya gar gyi
Dharmabodhi k'ri las babs te | bod
kyi Mi la ras pa la p'yag p'ul bas
mi kun Dharmabodhi bas kyaṅ Mi
la ras pa bzaṅ bar yid c'es šiṅ | gñis
ka la ts'ogs pa kun gyis saṅs rgyas
kyi ₒdu šes daṅ mi ₒbral bar gyur to
| de nas grub t'ob gñis k'ri gcig la
bžugs nas | p'an ts'un mñes pai gsuṅ
gleṅ mdzad pa na | Darmabodhis
rje btsun la | k'yed gcig pur bžugs
pa raṅ la t'ugs dges par yoṅ ₒdug
pa ṅo mts'ar c'e gsuṅs pas | rje btsun
gyis dei lan du mgur ₒdi gsuṅs so.

Wenn ich Dharmabodhi's Gold begehrte,
Wäre mein Verständnis vom Wesen der Heiligkeit
    wahrlich gering.
Ras chung, nach dem Gewinn köstlichen Ruhmes 15
    verlange ich nicht;
Wenn ich nach dem Gewinn köstlichen Ruhmes
    verlangte,
Wäre mein Verständnis vom Wesen des Vertrauens
    auf den Lama wahrlich gering.

Als er so gesprochen, fügte er hinzu: ‚Geht ihr selbst zuvor; ich werde später kommen.' So entsandte er seine Schüler zuerst. Indem sie unrichtige Zweifel hegten, ob der Ehrwürdige wohl kommen würde oder nicht, gelangten sie endlich nach Bal po rdzong. Der Ehrwürdige verwandelte seinen Leib in ein krystallenes Caitya und ging wie eine Sternschnuppe[1] unbehindert am Himmel hin. Als Dharmabodhi, der unter seine Schüler getreten war, ihn erblickte, wurde er von Staunen ergriffen, seine Schüler nährten Zweifel. Doch als der Ehrwürdige vom Himmel herabstieg, freuten sie sich ausserordentlich. Darauf traten der Lehrer und die Schüler vereint vor Dharmabodhi hin. Der Inder Dharmabodhi stieg von seinem Sessel herab und verneigte sich vor dem Tibeter Milaraspa. Alle glaubten, dass Milaraspa besser sei als Dharmabodhi. Gegenüber diesen beiden konnten sich alle Versammelten von der Vorstellung Buddha's nicht losreissen. Darauf liessen sich beide Siddha auf einen und denselben Sessel nieder und führten einander erfreuende Gespräche. Dharmabodhi sagte zu dem Ehrwürdigen: ‚Du, der allein dahinlebt und sich selbst Befriedigung gewährt, sei für dein Kommen bedankt!' Der Ehrwürdige trug zur Antwort folgendes Lied vor.

---

[1] Tib. *skar mda ₒp'aṅs pa* bedeutet wörtlich: abgeschossener Pfeil eines Sternes.

[2] *mi maṅ dus?*

27.

bla ma sprul pai sku la gsol ba ₒdebs	Zum Nirmāṇakāya des Lama sende ich mein Flehen!
bka brgyud grub t'ob rnams kyis byin gyis rlobs	Mögen die der Überlieferung anhängenden Siddha gesegnet sein!
rgya gar Dharmabodhis gtso mdzad pai ₒdir ts'ogs skal ldan bal bod k'rom pa la	Der Inder Dharmabodhi ist der erste hier Unter der hier versammelten Menge würdiger Nepaler und Tibeter;
5 bod kyi rnal ₒbyor Mi la ras pa ñas	Ich, der tibetische Yogin Milaraspa 5
ñams myoñ ye šes glu ru len	Singe ein Lied der Wonne und Weisheit;
ñams myoñ glu ru mi len du	Ohne ein Lied der Wonne und Weisheit
grub t'ob sprul skui ño ma c'ogs	Liesse ich es an der schuldigen Rücksicht gegen den Nirmāṇakāya der Siddha fehlen.
rtsa yon po lña po rluñ gis bsrañ	Die fünf krummen Adern werden durch den Wind gerade gemacht,
10 rluñ log pa lña po mal du bsad	Die fünf widrigen Winde werden an ihrer Stelle vernichtet, 10
k'ams sñigs ma lña po me la bsregs	Die fünf unreinen Niederschläge des Körpers werden im Feuer verbrannt.
sems ñon moñs dug lñai sdoñ po sgyel	Der Stamm der fünf Sündengifte der Seele wird zu Boden gerissen:
dgra las rluñ rnam rtog dbu mar bsad	Der Zweifel, der aus diesem Feinde kommende Wind, wird in der mittleren Ader vernichtet.
dgra gnod byed t'ul bai dpa bo la	Der den schädigenden Feind bezwingende Held
15 grogs ñan pai ₒk'ri šal[1] bya rgyu med	Sollte gegen schlechte Freunde nicht ergeben sein. 15

Darauf trägt Milaraspa auf Dharmabodhi's Bitte noch zwei kurze Lieder von 18 und 11 Versen vor, in demselben Stil wie das vorhergehende und gleichfalls Gegenstände der Mystik behandelnd. Da das Verständnis derselben durch die Dunkelheit des Themas an sich wie durch unsere Unkenntnis mancher technischer und anderer Ausdrücke sehr erschwert ist, so verzichte ich hier auf deren Wiedergabe. Dann führt der Text folgendermassen fort:

ces gsuñs pas | rnal ₒbyor pai lta sgom spyod pa ño mt'sar c'e gsuñ bas | rje btsun gyis Dharmabodhi la k'yed kyi ñams len gyi gnad zab mo rnams kyañ gsuñ bar žu gsuñs pas | Dharmabodhis mgur ₒdi gsuñs so.	Als er so gesprochen, sagte Dharmabodhi für das Lied von der Übung der Meditation des Yogin seinen Dank. Der Ehrwürdige sprach zu Dharmabodhi: ‚Geruhe, mir die tiefen Hauptpunkte deiner Merkverse vorzutragen.' Da trug Dharmabodhi folgendes Lied vor.

28.

ña lta ba bzañ poi bsgrub brgyud la	An den, der die Ausführung der trefflichen Beschauung lehrt,
ₒdir ₒts'ogs skal ldan gsol ba ₒdebs	Richte ich mein Gebet in dieser würdigen Versammlung.

---

[1] šal bya, ₒk'ri šal bya, Bedeutung unbekannt.

laś kyi ₒbrel ba bzaṅ po yis
myur du mjal bar byin gyis rlobs
5 *rnam rtog bag c'ags ma žig pai*
sems la bltas pa ci la p'an
bdag ₒdzin skyid ₒdod ma spaṅs pai

yun du bsgoms pas ci la p'an
ₒgro bai don la mi brtson pai

10 *ṅa rgyal spyod pas ci la p'an*
bla mai gsuṅ la mi ñan pai
dga ₒdus ok'or gyis ci la p'an
ṅo lkog ɣyo sgyu k'rel gyi gži

ₒbras bu gžan don ma byas na

15 *bla med byaṅ c'ub ₒgrub mi ₒgyur*
lar yin lugs bśad pas gnod la ₒp'og

ₒk'rug pa byas na p'uṅ gži c'e

k'a rog bsdad pa gdams ṅag zab
k'yed bod kyi rnal ₒbyor glu dbyaṅs
sñan
20 *ṅa glu dbyaṅs sñan po mi ₒdug ste*
spro ba skyes nas blaṅs pa yin
dpal bde ba rgyas pai žiṅ k'ams su

dbyaṅs len par spro bas myur du mjal

ces gsuṅs śiṅ | gžan yaṅ mñes pai
gsuṅ gleṅ maṅ po mdzad de | rgya
gar gyi Dharmabodhi p'ar byon | bod
kyi Mi la ras pa dpon slob rnams yar
byon pa las | gña nam kyi ña ma rnams
kyis rje btsun dpon slob la byon skyems
žus žal mjal ts'ul rnams dris pas | dei
lan du rje btsun gyis mgur ₒdi gsuṅs so |

dguṅ ñi zla mjal bas gliṅ bži gsal

ma bu p'rad pas gduṅ pa ži

Möge er durch die gute Wirkung der Werke
Uns segnen, dass wir uns bald wieder treffen!
5 Wenn der Hang zur Skepsis nicht zerstört wird,
Wozu nützt dann die Beschauung des Gemüts?
Wenn das Verlangen nach dem Glück der Eigensucht nicht gemieden wird,
Was nützt dann langdauernde Meditation?
Wenn man sich nicht um das Heil der Wesen bemüht,
Was nützt dann stolzer Wandel? 10
Wenn man auf des Lama Wort nicht hört,
Was nützt dann das festlich versammelte Gefolge?
Das ist die Ursache von Trug, Täuschung und Scham.
Wenn man als Frucht nicht die Uneigennützigkeit erwirbt,
15 Wird die unvergleichliche Bodhi nicht erreicht.
Doch durch die Darlegung deiner eigenen Umstände trifft dich Schaden;
Wenn man Krieg beginnt, ist grosse Ursache zum Verfall da.
Schweigend dasitzen ist eine tiefe Lehre.
Wohlklingend ist dein Gesang, tibetischer Yogin!

20 Mein Gesang aber ist nicht wohlklingend;
Im Drang der Freude nur habe ich gesungen.
Der du in den an Pracht und Glück reichen Gefilden
Lieder zu singen froh bist, besuche mich bald wieder!

Als er so gesprochen, führten sie noch weiter viele erfreuende Gespräche. Dann ging der Inder Dharmabodhi weiter, der Tibeter Milaraspa. der Lehrer samt seinen Schülern, zog aufwärts. Die Zuhörer von gÑa nam boten dem ehrwürdigen Meister den Willkommenstrunk und fragten ihn nach den Umständen seines Besuches. Zur Antwort trug der Ehrwürdige folgendes Lied vor.

29.

Durch die Begegnung von Sonne und Mond am
    Himmel werden die vier Continente erleuchtet,
Durch das Zusammentreffen von Mutter und Sohn
    wird die Familie beruhigt,

Denkschriften der phil.-hist. Classe. XLVIII. Bd. II. Abh. 9

- 335 -

drod gšer mjal bas rtsi t'og smin	Durch die Begegnung von Wärme und Feuchtigkeit reift das saftige Obst,
grub t'ob mjal bas rgyal k'ams bde	Durch die Begegnung der Siddha wird das Weltreich beglückt.
5 *bal po rdzoṅ gi nags gseb tu*	In den Wald von Bal po rdzong 5
Dharmabodhi p'yag p'ebs pas	War Dharmabodhi angelangt;
Mi la ras pas mjal du p'yin	Milaraspa ging hin, ihn zu besuchen;
k'oṅ Dharmabodhi sku bžeṅs nas	Er, Dharmabodhi, erhob sich
Mi la ṅa la p'yag byas pas	Und verneigte sich vor mir, Mila.
10 *k'rom pai mi rnams t'e ts'om zos*	Die versammelten Männer hegten Zweifel. 10
k'oṅ sgyu ma lus kyi p'yag rgya yi	Er mit der Mudrā des Täuschungsleibes
zuṅ ₒjug t'al sbyar pus mo btsugs	Faltete in der Meditationsstufe des zung ₒjug die Hände und kniete nieder
sñun dri med c'os kyi dbyiṅs su byas	Und wirkte im Dharmadhātu, der frei von Leiden und Flecken ist.
lan ni p'yag rgya c'en po btab	Noch einmal vollzog er die Mahāmudrā,
15 *gñis med dag pai dgon pa ru*	Dann führten wir aufrichtig[1] in der reinen Einsamkeit 15
ₒod gsas rjed med gsuṅ gleṅ byas	Klare unvergessliche Gespräche.
smon lam bzaṅ poi ₒbrel pa žig	Das ist die Wirkung trefflichen Gebets!
sṅon rgyal bai dus na yod par ṅes	Sicherlich hat er schon früher zur Zeit des Jina gelebt.
deṅ grogs mc'ed mjal p'rad legs pa de	Jene schöne Begegnung der brüderlichen Freunde
20 *rgyal k'ams yoṅs la grags pa yin*	Gereicht heute dem ganzen Weltreich zum Ruhme. 20

*ces gsuṅs pas	ṅa ma rnams dga mgu yi raṅs nas	ṅo mts'ar bar gyur ciṅ	rje btsun la Dharmabodhis p'yag byas pai stobs kyis sñan pa daṅ sku bsod yaṅ rgyas par gyur to.*	Als er so gesprochen, wurden die Zuhörer von hoher Freude und Verwunderung ergriffen. Infolge der ihm von Dharmabodhi bezeigten Verehrung mehrte sich der Ruf und das Wohlbefinden des Ehrwürdigen.
Dharmabodhi daṅ mjal bai bskor ro.	Der Abschnitt vom Besuch bei Dharmabodhi.

---

[1] Tib. *gñis med* scheint Skr. advayant zu entsprechen oder nachgebildet zu sein.

031

莱辛《中国青铜器》书评

通 報

T'oung pao

ARCHIVES

POUR SERVIR À

L'ÉTUDE DE L'HISTOIRE, DES LANGUES, DE LA GÉOGRAPHIE ET
DE L'ETHNOGRAPHIE DE L'ASIE ORIENTALE

(CHINE, JAPON, CORÉE, INDO-CHINE, ASIE
CENTRALE et MALAISIE).

RÉDIGÉES PAR MM.

GUSTAVE SCHLEGEL

Professeur de Chinois à l'Université de Leide

ET

HENRI CORDIER

Professeur à l'Ecole spéciale des Langues orientales vivantes et à l'Ecole libre des
Sciences politiques à Paris.

Série II. Vol. IV.

LIBRAIRIE ET IMPRIMERIE
CI-DEVANT
E. J. BRILL.
LEIDE — 1903.

français complétement différent mais aussi tout à fait indépendant de celui des autres manuscrits jusqu'ici connus; il ne reproduit pas un certain nombre de phrases du texte latin, bien que d'ordinaire il le suive presque mot à mot. M. Omont pose au sujet de ce nouveau manuscrit la question suivante: «Ne serait-il pas un représentant du texte primitif, écrit par Nicolas Falcon sous la dictée de Hayton, puis traduit en latin, tandis que la rédaction française, considérée jusqu'ici comme l'original, ne serait au contraire qu'une traduction postérieure et dont la forme plus élégante aurait assuré le succès aux dépens de la rédaction première?»

H. C.

Chinesische Bronzegefässe. Text von Julius Lessing. Vorbilder-Hefte aus dem Königl. Kunstgewerbe-Museum, Heft 29. Berlin 1902.

———

Vergebens haben wir beim Durchblättern dieses Heftes nach dem Texte gesucht, den wir doch erwarten sollten, um uns als Führer durch diese Abbildungen zu dienen. Stattdessen finden wir nur ein vorgedrucktes Verzeichnis, das im besten Falle als eine »Liste der Tafeln" zu bezeichnen ist, mit höchst dürftigen, mageren und teilweise ganz schiefen Angaben.

Seinen »Text" beginnt der Herausgeber folgendermassen:

»Die auf Tafel 1—14 abgebildeten 37 Bronzegefässe sind sämtlich ältere chinesische Arbeiten und stammen fast alle aus einer Sammlung, welche der Kaiserl. Gesandte, Herr v. Brandt, für die Zwecke des Kgl. Kunstgewerbe-Museums angelegt hatte. Mit dem sichern Geschmack, welcher diesen vorzüglichen Kenner ostasiatischer Kunst auszeichnet, sind für unsere über 100 Stücke enthaltende Sammlung nur Gefässe ausgesucht, welchen reine Schönheit der Form eigen ist. Die Sammlung steht daher in starkem Gegensatz zu sonstigen Sammlungen chinesischer Stücke, welche durch reichen Auftrag wunderlicher Zierformen zu glänzen suchen. Die Stücke tragen nur zum kleinen Teile Stempel, welche eine Datierung ermöglichen, aber es ist bekannt, dass ältere geschätzte Stücke in China stetig und einschliesslich der alten Stempel nachgeahmt worden sind. Die Unterscheidung ist für uns überaus schwer, aber für alle Stücke der Sammlung steht wenigstens so viel fest, dass es nicht Stücke sind, welche für den Export gearbeitet sind, sondern ausgewählt gute ältere Arbeiten aus chinesischem Gebrauch. Unter den hundert Stücken der Sammlung ist keines dem andern völlig gleich, jeder Typus ist auf das mannigfaltigste variirt".

Soweit Herr Lessing. Weiter kann man die Naivität kaum treiben, besonders nicht in der Furcht vor dem bösen Export, der in Bronzen niemals stattgefunden hat und wohl auch nie stattfinden wird. Als wenn die Chinesen die alten Typen ihrer Bronzen nur für die Herren Europäer nachmachen würden und einzig und allein auf deren Kauflaune angewiesen wären! Die alten echten Stücke kommen eben deshalb nie in fremde Sammlungen, weil die chinesischen Liebhaber solch horrende Preise dafür bezahlen, an die ein Europäer, und

zumal ein Museum, niemals denken würde. Den Siegeln bringt also Herausgeber kein besonderes Vertrauen entgegen, aber doch macht es drei Stücke namhaft als mit der Regierungsdevise des Ming-Kaisers Hsüan-teh gekennzeichnet. Wer aber weiss, dass die Bronzen dieser Periode wegen ihrer Anziehungskraft auf das Publikum von jeher und bis auf den heutigen Tag mit dieser Marke in Massen fabrikmässig hergestellt werden, um die allgemeine Nachfrage zu befriedigen, und dass schwerlich *ein* wirklich echter Hsüan-teh auf dem Markte zu haben ist, — die echten sind eben längst in den festen Händen reicher chinesischer Familien, — der wird jenem Siegel kaum eine historische Bedeutung zumessen. Diese Zeitbestimmung ist aber wenigstens annehmbar, da sie sich auf einen, wenn auch vielleicht nur scheinbaren, Anhaltspunkt stützt. Was sollen nun jedoch die übrigen chronologischen Definitionen besagen, wie da sind: »Ältere chinesische Arbeit", »Arbeit aus der Ming-Dynastie, XVI.

19

Jahrhundert", »China, XVI. Jahrhundert", »Chinesische Arbeit neuerer Zeit", »China, XVIII. Jahrhundert"?? Die Begründung für diese vagen und ganz wertlosen Angaben hat Herausgeber für sich behalten. Dreissigmal erklärt er Stücke einfach schlecht und recht als »ältere chinesische Arbeit". Gleichwohl lassen sich in diesen Stücken nur die letzten Ausläufer der gesamten chinesischen Bronzeentwicklung aus einer retrospektiven Epoche der Nachahmung erkennen, die vom Standpunkt des Chronologen gar nicht anders als durchaus moderne Arbeiten zu bezeichnen sind. Man vergegenwärtige sich doch, dass die Blütezeit der Bronzeindustrie jetzt fast drei Jahrtausende zurückreicht und in die Glanzzeit der vorchristlichen Dynastieen Shang und Chou fällt.

Noch allgemeiner und unsicherer als in der Chronologie ist Herausgeber in der Bestimmung der Bedeutung der einzelnen Stücke, obwohl er für diesen Zweck aus Paléologué bekanntem Buche L'Art Chinois sich manchen Rat hätte holen können. Er operiert ausschliesslich mit den drei Stichwörtern »Vase, Kanne, Kessel", auf die auch jeder Unbefangene auf den ersten Blick von selbst verfallen würde. Das erinnert mitunter an die antiquierte Etiquettierungsmethode einiger Museen, in welchen z.B. buddhistische Skulpturen mit der Bezeichnung »ein Götze", »ein andrer Götze" u.s.w. abgefertigt wurden. Der auf Tafel II*a* so bezeichnete »phantastische Vogel" ist ein stilisierter Hahn, eine ganz gewöhnliche Erscheinung in der chinesischen Kunst, und der andere »phantastische Vogel" *b* auf derselben Tafel stellt eine Ente vor. In Ermangelung von Thatsachen scheint sich Herausgeber mit Vorliebe auf die Phantasie zu legen. So sagt er, ganz von seinem Standpunkt als Ästhetiker: »Der enge Hals der meisten Blumenvasen erklärt sich daraus, dass sie nicht für grosse Sträusse, sondern für einzelne schlanke Blüstenzweige, Pfauenfedern oder dergleichen (was?) bestimmt sind". Thatsache ist, dass eben der Blumenstrauss in unserem

Sinne in China unbekannt ist und nur einzelne Blütenzweige in Vasen gesteckt werden, aber keine Pfauenfedern oder dergleichen, die in eigenen Bambusschachteln aufbewahrt werden. Im Text und am Fusse der Tafeln sind die Blumenvasen nirgends als solche gekennzeichnet. Der als »Kessel" beschriebene Dreifuss auf Tafel IV*c* dient als Räuchergefäss, ebenso *c* auf Tafel X. Die Vase *b* auf letzterer Tafel mit dem flaschenförmigen Halse wird von den Chinesen mit dem treffenden Namen »Hammerstiel", 鎚把 *ch'oei pa*, charakterisiert. Die vier Gefässe auf Tafel VI*b*, VII*b*, XIII*b*, XIV*b* sind Nachahmungen alter Sacralgefässe. Über die Ornamentik hätte sich nach den Originalen viel mehr feststellen lassen als hier geboten wird. So sind auf Tafel III*a* in dem ovalen Felde Pflaumenblütenzweige und Elstern (喜鵲 *hsi ch'üeh*) [1]) und auf dem Deckel ein junger Löwe mit Leichtigkeit erkennbar.

Leider ist in der Anordnung der Bronzen auf den Tafeln jede Systematik ausser Acht gelassen. Es ist ein buntes, principienloses Allerlei, das nur einen winzigen Bruchteil chinesischer Bronzekultur widerspiegelt, und zwar keineswegs weder in ihren typischen Stücken noch in ihren besten Leistungen, denn alle abgebildeten Proben stammen aus der jüngsten Verfallzeit. Auch die Anschauungen, die der Herausgeber mit dem Werte dieser Sammlung verbindet, können wir leider nicht teilen. Der Publikation mag ein aesthetischer Wert nicht abzusprechen sein, archäologischen Zwecken dient sie nicht. Der Hauptfehler des Herausgebers ist darauf zurückzuführen, dass er es unterlassen hat, bei seiner Arbeit einen Sinologen zu Rate zu ziehen, der die einheimische archäologische Litteratur kennt.

Dr. B. LAUFER,

z. Z. Hankow.

---

1) Über die Lesung dieses Ornaments vergl. Grube, Beiträge zur Pekinger Volkskunde, S. 140.

032

中国的宗教宽容度

GLOBUS

Illustrierte

Zeitschrift für Länder- und Völkerkunde

Vereinigt mit den Zeitschriften „Das Ausland“ und „Aus allen Weltteilen“

———

Begründet 1862 von Karl Andree

Herausgegeben von

H. Singer

Sechsundachtzigster Band

Braunschweig

Druck und Verlag von Friedrich Vieweg und Sohn

1904

Religiöse Toleranz in China.

Von Dr. B. Laufer.

Vor wenigen Wochen ist der zweite Band eines umfangreichen Werkes fertig geworden, das die Frage der Glaubensfreiheit in China behandelt und durch das aktuelle Interesse des Gegenstandes auch die Teilnahme weiterer Kreise beanspruchen dürfte. Verfasser ist der bekannte Sinologe J. J. M. de Groot, und sein Buch führt den Titel: Sectarianism and Religious Persecution in China, a Page in the History of Religions, zwei Bände, Amsterdam, Joh. Müller, 1903 und 1904.

Diese voluminöse Arbeit ist eine starke Erweiterung desselben Themas, das de Groot unter dem Titel: Is there religious liberty in China? bereits in den Mitteilungen des Seminars für orientalische Sprachen, Bd. V, S. 103 bis 151, Berlin 1902, angeschlagen hatte. Das Buch ist „allen in China arbeitenden Missionaren jeden christlichen Bekenntnisses" gewidmet und kennzeichnet schon dadurch seine engere Tendenz. Der Gedankengang des Verf. ist kurz folgender: Die gegen die Missionare in China aus Anlaß der traurigen Ereignisse von 1900 vielfältig gerichteten Vorwürfe und Beschuldigungen sind gänzlich unbegründet und nur als Erzeugnis einer leichtsinnigen Journalistik zu betrachten, die Missionare sind im Gegenteil von den besten und reinsten Absichten geleitete Menschen, die auf unsere vollste Sympathie und Achtung Anspruch haben. Die Schuld liegt einzig und allein an der chinesischen Regierung, von der Verf. durch eine große Zahl kaiserlicher Edikte und anderer chinesischer Dokumente nachweist, daß sie die intoleranteste und verfolgungssüchtigste aller irdischen Regierungen sei; die bisherige Annahme chinesischer Toleranz ist eine Schimäre, die aus unserem Gedankenkreise verbannt werden muß. Verf. hofft Sinologen und Diplomaten von ihrem verhängnisvollen Vorurteil zu bekehren und gelangt zu der Schlußfolgerung, daß das Christentum in China ohne den Schutz der fremden Mächte nicht bestehen und blühen könne, daß ohne diesen Schutz Vernichtung sein Los sei, daß schon eine schwache Haltung der Gesandtschaften und Konsulate, ein Ausdruck, ein Beweis ihrer Gleichgültigkeit für die Mission überall und in jedem Augenblicke für fanatische Präfekten und Unterpräfekten ein Signal zur Belästigung der Christen und blutiger Verfolgung abgeben können.

An dieser Argumentation läßt sich Verschiedenes aussetzen. Die in unserer Prämisse eingeschlossene Ansicht, daß die Mission in China gut und notwendig sei, begründet de Groot in keiner Weise. Und doch gibt es, abgesehen von philosophischen Köpfen und Leuten, die nur mit ihrem einfachen gesunden Menschenverstand denken, auch viele sehr christlich denkende Männer, die mit triftigen Gründen über diesen Punkt anderer Anschauung sind. Die Richtigkeit der Beweisführung zugegeben, ist es sehr fraglich, ob die Schlußfolgerung allgemeine Anerkennung finden wird. Es ist sogar zweifelhaft, ob die Missionare in China selbst sie im ganzen Umfange teilen werden.

Es gibt gegenwärtig genug einsichtige Missionare dort, die von dem Ruf nach Kanonen zum Schutz des Christentums nichts wissen wollen und die sich in der Unabhängigkeit von ihren Regierungen weit größere Erfolge versprechen. Und mit Recht verlangt Prof. Bälz in seinem geistvollen Vortrage „Die Ostasiaten" (Stuttgart 1901), S. 46, daß sich der Missionar entnationalisiere, damit nicht andere leiden müssen, weil er gelitten hat, und bemerkt von den alten Heidenbekehrern: sie haben

nicht an Konsuln appelliert, sie haben nicht nach Kriegsschiffen gerufen, aber sie haben die Welt erobert.

Es wäre auch gar nicht erforderlich gewesen, auf 595 großen Oktavseiten mit dem gesamten Arsenal sinologischer Gelehrsamkeit den Nachweis zu erbringen, daß die Missionare des Schutzes bedürfen; denn in der Tat ist von seiten der Mächte alles geschehen, um ausreichende Sicherheit für Leben und Eigentum der Missionare zu erwirken, und es ist mit de Groot nicht einzusehen, was eigentlich noch mehr getan werden könnte. Man kann sogar dreist behaupten, daß es kaum eine Berufsklasse in der ganzen Welt gibt, die sich ausgedehnterer Schutzrechte und daraus folgender Privilegien und persönlicher Vorteile erfreute als gerade die christlichen Missionare in China. Der Arbeiter an der Maschine oder im Kohlenschacht, der Forschungsreisende, der Soldat im Felde und Vertreter anderer Berufe setzen ihr Leben mehr aufs Spiel als der in bequem eingerichteten Häusern sehr friedlich und gut lebende Missionar in China. Es hätte der hochgelehrten Arbeit von de Groot nur zum Vorteil gereicht, wenn sie sich dem Fahrwasser politischer Tendenzen ganz fern gehalten hätte, da er für die Missionsfrage nur geringes Verständnis zeigt. Das alles wäre aber sehr unbedeutend, wenn ihn nicht sein Standpunkt zu einer solchen erstaunlichen und betrübenden Einseitigkeit in der Behandlung seines Themas verführt hätte, daß man nicht anders kann, als de Groots Buch als fast intoleranter zu bezeichnen als alle von ihm angeführten intoleranten Edikte und Handlungen der chinesischen Regierung zusammengenommen. Seine Begriffe von religiöser Freiheit und Duldsamkeit definiert er nicht, faßt sie aber, wie aus seinen Ausführungen hervorgeht, in einem absoluten Sinne, nicht in ihrem historisch-relativen Werte, so daß ihm jeder Maßstab zur Beurteilung der von ihm zitierten Beispiele fehlt. Alle Bestimmungen und Gesetze, die sich mit Klöstern, religiösen Vereinen, Sektierern, Schwärmern und Geheimgesellschaften befassen, fallen nach de Groot unter den Begriff der religiösen Einschränkung und Verfolgung, als wenn der chinesische Staat nicht wie alle anderen auch das Recht hätte, dem Triebe der Selbsterhaltung zu gehorchen! In allen Staaten hat es zu allen Zeiten ein herrschendes Glaubenssystem gegeben, das, mit der Form der Regierung aufs engste verknüpft, sich fremde eindringende Religionen unterordnete. In China ruht die Grundlage des Staates auf der Konfuzianischen Ethik, in welcher seine Leiter die Größe und Macht der Nation erkannten, weshalb sie sich für berechtigt hielten und vom historischen Standpunkt auch unzweifelhaft berechtigt waren, die Grundsätze des Staates zu vertreten und zu verteidigen und sich gegen Angriffe auf das herrschende System zur Wehr zu setzen. Wenn die chinesische Regierung ein wachsames Auge auf Politik treibende religiöse Geheimbünde hatte und gegebenenfalls gegen diese vorging, so kann ihr das niemand verübeln; eine solche unumschränkte Freiheit, wie sie de Groot vorzuschweben scheint, gibt es nicht und hat es nie gegeben. Legt man aber an China den einzig möglichen Maßstab, nämlich den der Geschichte, so zeigt sich China in weit günstigerem Lichte als die christlichen Länder während des Mittelalters und die mohammedanischen Völker. China hat keine Hexen verbrannt, keine Inquisition gehabt und keine uralten Kulturen wie die von Mexiko und Peru vernichtet. Jeder Chinese hat das Recht des

Übertritts zu einer von ihm beliebten Religionsform, während im modernen Rußland jeder Abfall vom orthodoxen Glauben und jeder Versuch der Verleitung zu einem solchen mit Verbannung nach Sachalin bestraft wird.

Der zweite Fehler, den de Groot begeht, besteht darin, daß er allen Handlungen der chinesischen Regierung in Sachen fremder Religionen das einzige ganz unpsychologische Motiv eines blinden, grausamen Verfolgungswahns unterschiebt und alle Äußerungen von Toleranz mit dem Schlagwort Verstellung und Heuchelei abtut. Seine Untersuchung ist die eines starren Dogmatikers, der um jeden Preis sein Dogma will triumphieren sehen, nicht die des nach Ursachen und Wirkungen forschenden, gerecht abwägenden Geschichtschreibers. Bewegungen gegen fremde Religionen waren in China niemals von reinem Religionshaß diktiert worden, sondern hatten, wie fast überall, ihren Grund in politischen und wirtschaftlichen Fragen. Das kolossale Anwachsen der buddhistischen Klerisei und die Vereinigung von Volksvermögen in der toten Hand der Kirche bildete für China unzweifelhaft eine große Gefahr, der die Kaiser mit Recht von Zeit zu Zeit zu steuern suchten. de Groot aber sieht auch in der Verfolgung und Bestrafung übler und staatsgefährlicher Vertreter der Religion einen Angriff auf diese selbst, in jenen alles Gute und in den Akten der von dem Recht der Notwehr Gebrauch machenden Regierung alles Häßliche und Schwarze. In China haben alle Religionen der Welt eine Zuflucht gefunden und geblüht, und nur wenn sie sich in Angelegenheiten der Politik einmischten oder dem sozialen und Wirtschaftsleben des Volkes Gefahr drohten, hat sich das Geschick gegen sie gewandt. Man denke nur an die Geschichte des Niederganges der Jesuiten, deren großer Einfluß am kaiserlichen Hofe die Eifersucht der Dominikaner und Franziskaner erregte, bis nach langen Streitigkeiten der Orden untereinander der Papst die Partei der Dominikaner nahm, der Kaiser sich dagegen für die Jesuiten aussprach. Als sodann eine päpstliche Bulle erschien, welche entgegen der kaiserlichen Entscheidung die jesuitische Auffassung verdammte, verlor Kaiser Khanghsi die Geduld und erklärte, in seinem Lande sei er der Herr und lasse sich vom Papste nicht dreinreden. Die Orden befeindeten sich immer gehässiger, und der Kaiser, angeekelt durch ihre Zänkereien und wegen der Einmischung des Papstes besorgt, beschränkte den Einfluß der Missionare mehr und mehr, bis sein Nachfolger die weitere Verkündigung der christlichen Lehre völlig verbot. War es nun chinesische oder nicht vielmehr christliche Intoleranz, welche in diesem Fall den Untergang des Christentums in China herbeiführte? Auch die Unterdrückung des Taipingaufstandes ist aber nach de Groot eines der glänzenden Beispiele chinesischer Unduldsamkeit und religiöser Verfolgung, wobei er gänzlich vergißt, daß die Taipings den Sturz der bestehenden Mandschudynastie planten, einen Staat im Staate gründeten und dementsprechend als Revolutionäre behandelt wurden, was ihnen in jedem andern Gemeinwesen aller Wahrscheinlichkeit nach auch passiert wäre.

Der dritte und schwerste Fehler in de Groots Buche ist aber seine Parteiischkeit: er reiht Edikte der Intoleranz eines an das andere und schweigt die Toleranzedikte einfach tot. Schon der Laie muß sich die Frage vorlegen: wenn alle Religionen der Welt in China Aufnahme gefunden haben, wenn schon in alten Zeiten Buddhisten, Parsen, Manichäer, Mazdäer, Nestorianer, Juden, Mohammedaner ihren Kultus ungestört dort ge[...] haben, woher denn der Erfolg dieser Religionsges[...] schaften, woher die noch heute nach vielen Million[...] zählenden Anhänger des Buddhismus und Islams, we[...] China der intoleranteste aller Staaten ist? Der Ve[...] bedenkt eben nicht, daß papierene Erlasse und Ergü[...] der Kaiser, die er aus verstaubten toten Büchern a[...] gegraben hat, und die lebendige Gesinnung u[...] Stimmung des Volkes zwei durchaus verschiedene Din[...] sind. Das chinesische Volk muß eben seit frühester Z[...] in ganz hervorragendem Maße auf religiösem Gebi[...] tolerant gewesen sein, wie der praktische Erfolg [...] nach China eindringenden Religionen schlagend bewe[...] Und wer hätte nicht von den vier großen Gönnern [...] Christentums auf dem chinesischen Kaiserthron gehö[...] T'aitsung von der T'angdynastie, Kubilai Chan von [...] Dynastie der Yüan und Shunchih und Khanghsi aus d[...] gegenwärtigen Herrscherhause? Wo bleibt der A[...] druck religiöser Duldung, der in Chinas größtem epig[...] phischen Denkmal, der nestorianischen Inschrift v[...] Hsi-ngan-fu, niedergelegt ist? In der ältesten erhalten[...] Inschrift der chinesischen Juden von K'ai-fong-fu, dati[...] 1489, wird die hohe Toleranz der Mingdynastie in A[...] drücken größter Bewunderung und Dankbarkeit gerüh[...] Worauf wir aber am meisten Gewicht legen, ist die Ta[...] sache, daß die Toleranz der gegenwärtigen Dynastie v[...] niemand geringerem proklamiert worden ist als von d[...] modernen Jesuiten in China. Dieselben haben im Jah[...] 1883 in ihrer Druckerei zu Sikkawei bei Schanghai e[...] zweibändiges Werk von 292 Oktavseiten Umfang he[...] gestellt, in welchem alle kaiserlichen Erlasse zu ihr[...] Gunsten in chinesischer Sprache authentisch abgedruc[...] sind. Im Bücherkatalog von Sikkawei ist dasselbe un[...] Nr. 44 verzeichnet und folgendermaßen beschrieb[...] „Vera Religio publica auctoritate laudata a P. Pet[...] Hoang: Documenta publica, acta officialia et edicta I[...] peratorum (ab anno 1635 ad 1826) quibus demonstrat[...] Sanctam nostram Religionem in magna aestimatio[...] fuisse apud Gubernium Sinarum." Im folgenden Jah[...] 1884 erschien die Fortsetzung dazu unter dem Tit[...] „Collectio praecipuorum edictorum in favorem nostr[...] Religionis a Mandarinis, praesertim ab anno 1846 [...] 1883 publicatorum auctore P. Petro Hoang", 340 Seite[...] Ferner haben die Jesuiten unter dem Titel: „Edicta I[...] peratorum Sinarum in gratiam Religionis Catholica[...] (Katalog Nr. 248) die hervorragendsten Toleranzedik[...] der chinesischen Kaiser auf vier zum Aufhängen b[...] stimmten Rollen herausgegeben, die sich gleichfalls i[...] Besitz des Ref. befinden. Ohne von dieser Tolera[...] überzeugt zu sein, ist es gewiß nicht denkbar, daß d[...] Jesuiten solche Veröffentlichungen in die Welt geset[...] hätten. Wir hoffen, daß de Groot sich anstelle [...] zur Abfassung einer Geschichte der religiösen Duldsa[...] keit in China bedienen wird, um der Sache Gerechtigke[...] widerfahren zu lassen, die in seiner Darstellung in [...] schroffer Einseitigkeit und Verzerrung erscheint. So w[...] sein Buch ist, kann es nur mehr Unheil als Nutze[...] stiften, denn seine Tendenz ist ausgesprochen die, An[...] mosität gegen China zu erzeugen und eine bish[...] günstige Meinung von ihm in das Gegenteil umzustin[...] men. Dieses Bestreben ist durchaus nicht zeitgemä[...] Wir wollen Frieden mit China haben, wir wollen Chi[...] und sein Volk besser verstehen lernen. Am Ende h[...] das sündige Europa Gründe genug, um dem unglüc[...] lichen Lande zuzurufen: „Und vergib uns unse[...] Schuld!"

033

佛教朝圣图

mäen bekannt, in deren Gesellschaft ich auch — und mit Erfolg — die seltene Spur aufnahm. Von November 1903 bis März 1904 kamen mir in dem östlichen Teil des oben beschriebenen Verbreitungsbezirkes vier Okapifährten zu Gesicht; das entspricht also, sagen wir, vier Paaren. Drei Decken, alle von verhältnismäßig jungen Tieren, sammelte ich während dieser Zeit. Man kann also, in Anbetracht der unglaublich schwierigen Zugänglichkeit der Reviere und der schweren Auffindbarkeit von Fährten und Wild, nicht gerade sagen, daß die noch überlebenden Okapis sehr selten sind. Das sagen auch die Schwarzen. Einer unserer Schwarzen hielt sich mehrere Jahre im Ituri-Semlikiurwald in einem Dorfe auf und behauptet in glaubwürdiger Weise, oft von dem Fleische der Okapia gegessen zu haben. Das gestreifte Fell der Keulen und Läufe ist bei Pygmäen und anderen Waldstämmen äußerst beliebt als Leibgürtel. Ich besitze einige solche, mit Schnallen versehen, deren primitive Herstellung einem paläolithischen Menschen alle Ehre machen würde.

Die erste Erwähnung, die dem Okapi zuteil wird, findet sich in Junkers „Reisen", Bd. III, S. 299; diese Erwähnung erweitert, — wenn kein Irrtum unterläuft — den oben umrissenen Verbreitungsrayon ganz bedeutend. Junker sah im Jahre 1878/79 im Uëlle-oder Nepokogebiet — von woher jede weitere Bestätigung des Vorkommens der Okapia fehlt — einen Teil der Decke mit den charakteristischen Streifen; Kopf und die besonders bezeichnenden Luser fehlten. Die betreffende Stelle im Werke Junkers nachzulesen hat mich ganz besonders interessiert; man nannte das Tier „Makapé", und Junker hielt es für ein Moschustier. Jetzt sieht man, salvo errore, das Helladotherium als seinen ihm am nächsten stehenden Verwandten an.

Nun kennen aber die Pygmäen noch ein anderes Antilopenwild, das sie „Soli" nennen. Ein im Walde von uns aufgelesener alter Schädel gehört ihm vielleicht an; das wird spätere Untersuchung zu zeigen haben. Es ist eine Antilope mit ganz kurzen, festen Hornzapfen. Das Cranium gleicht durchaus dem der Okapia, besonders

durch die langgestreckte Ausdehnung des Occiput un dessen mediane Schwellung. Die Pygmäen geben m an, das Tier sei ebenfalls gestreift, aber viel größe dunkler, und vorn heller rot. Ich vermute, daß man mit zwei Varietäten zu tun hat, da ich nicht nur a meinen drei Fellen, sondern auch an den in Brüssel un London ausgestellten Exemplaren Verschiedenheiten d Grundfarbe und der Hörnchen beobachtete. Die ein besitzt Hornzapfen, die andere keine und nur verhärte Integumenthörnchen.

Die Auffindung der ersten Haut und des Schädels d Okapia und damit die Entdeckung dieses Genus durc Sir Harry Johnston, 1899, und die Studien Slaters sin Ihnen in Europa besser bekannt als mir hier am Edwardse

Ich besitze nun noch das Magenpräparat und Schleim häute der Mundhöhle und Lippenränder in Alkoho Wüßte ich nur, wie solcher Transport zu bewerkstellige wäre!

Als Wild hält sich die Okapia nicht nur etwa a Sumpfstellen, Bachbetten und Unterholz, sondern s verzieht auch über steile, laubbedeckte und von Unte holz teilweise entblößte Halden und waldige Felslehne hinauf. Ich fand, daß ihr Gesicht entschieden schlechte war als dasjenige der Graslandantilopen. Das ist b den meisten Tieren des so äußerst dichten äquatoriale Urwaldes so (mit Ausnahme der Affen). Elefanten un alle Arten Schweine lassen einen in unglaubliche Näl herankommen. Entsprechend der fast stets herrschende Windstille spielt auch die Nase gewiß keine sehr gro Rolle, außer beim Vermeiden frischer und eventue Nachteil bringender Fährten und bei der Nahrungssuch Dagegen ist das Gehör bei weitem der vorherrschend Sinn, und wenn auch schon auf kürzeste Distanz al Gerüche der Fährten und der Gegenwart des Mensche durch die scharfen Bodenausdünstungen des Moderwalde verwittert sein mögen, so verrät doch das allerleises Geräusch jede Annäherung von etwas Lebendigem i Urwalde, und dann bricht auch die volle Flucht lo durch krachendes Gezweig und auf Nimmerwiedersehe

Ein buddhistisches Pilgerbild.

In Nr. 11, Vol. XII (November 1897), p. 24, 25 der Zeitschrift „The Hansei Zasshi" hat Takakusu ein Bildnis des berühmten chinesischen Buddhisten und großen Reisenden Hsüan Tsang (602 bis 664 n. Chr.) nach einer gut beglaubigten Kopie eines unbekannten Malers mitgeteilt, deren Original bis in das 13. Jahrhundert zurückgehen soll. Eine weit bessere Reproduktion eines ähnlichen Bildes findet sich im Katalog der Kollektion Hayashi. Das Musée Guimet in Paris besitzt eine lackierte Bronzestatuette, in der eine Darstellung des Hsüan Tsang vermutet wird (s. de Milloué, Petit guide illustré au Musée Guimet, 1897, p. 132).

Die nebenstehende Abbildung gibt ein auf Papier gemaltes japanisches Kakemono wieder, das sich im Museum für Völkerkunde in Köln als Nr. 218 unter den von Herrn W. Joest hinterlassenen Sammlungen befindet. Die Malerei ist 49,5 cm lang und 18,5 cm breit und wurde laut der am unteren Rande angebrachten chinesischen Beischrift 1825 in Nagasaki verfertigt, wahrscheinlich nach einer älteren Vorlage; weder der Name des Malers noch der des Kopisten werden darin genannt, ebenso wenig erscheint darin der Name des Hsüan Tsang. Daß aber er in dem Pilger im Reiseanzuge rechts im Vordergrunde des Bildes beabsichtigt ist, scheint mir nach einem Vergleich mit den anderen bekannten Darstellungen

zweifellos zu sein. Von diesen unterscheidet sich unser Vorlage dadurch, daß hier Hsüan Tsang nicht als selbs ständiges Porträt, sondern in einer großen Gruppe m Göttern und Menschen vereint erscheint. Die Leben schicksale des großen Reisenden sind seit Juliens Über setzung seines Lebens und seiner Tagebücher so bekan geworden, daß sie an dieser Stelle nicht wiederholt z werden brauchen; auch ist es überflüssig, auf die eminent Bedeutung seiner Schriften für die Geographie und Archä logie des alten Indien und Turkestan hinzuweisen, die se den letzten Jahren immer mehr in den Vordergrund ge treten ist.

Die Intention des Künstlers, der unser Bild geschaffe scheint die gewesen zu sein, den Abschied des Meister von seinen Freunden im Kloster Nālanda am Gange wo er fünf Jahre zugebracht hatte, vor dem Antritt seine Rückreise nach China (wahrscheinlich im Jahre 641) fest zuhalten, im Anschluß an die Stelle in seiner Biographi wo es heißt: „Er verabschiedete sich zuerst bei de Mönchen von Nālanda, trug die Bücher und Statuen weg die er gesammelt hatte, und schloß seine Vorträge; a 19. Tage danach nahm er Abschied von dem König un wollte zurückkehren" (s. Julien, Histoire de la vie d Hiouen-Thsang, Paris 1853, p. 251).

In den beiden hinter dem Pilger stehenden Gestalte

vermute ich seine beiden Gönner, die indischen Könige Harsha Çilāditya und Kumāra, und in der sich ihnen anschließenden Frau, die die Hände zum Gebet faltet, ist wohl des ersteren Königs Schwester zu erkennen, die mit seltener Intelligenz begabt war und mit Hsüan Tsang wegen seiner vortrefflichen Auseinandersetzung der Mahāyānalehre und seiner Bekämpfung des Hīnayāna sympathisierte (s. Julien, l. c., p. 241). Auf der linken Seite steht ein Mönch mit gefalteten Händen vor der Brust, offenbar als Vertreter der Mönche des Klosters von Nālanda gedacht; in dieser Haltung nehmen auch jetzt noch buddhistische Mönche in China Abschied voneinander und von Laien. Das übrige sind den Tempelraum füllende Statuen, Buddha Tathāgata oben in der Mitte tronend, umgeben von Mañjuçrī auf dem Löwen und Samantabhadra auf dem weißen Elefanten, dann auf beiden Seiten verteilt die 16 Geisterfürsten (chin. shen wang), Schutzgötter der buddhistischen Lehre.

Was nun Hsüan Tsang selbst betrifft, so kann natürlich keine Rede von einem individuellen Porträt sein; das Gesicht und auch die Tracht sind vielmehr stark japanisiert. Er trägt ein weißes Unterkleid, dessen Ränder auf der Brust und am Ende der Ärmel zum Vorschein kommen, darüber das gelbbraune Mönchsgewand, über das er einen hellgrünen Mantel mit dunkelgrünen Streifen gezogen hat. Auf beiden Schultern ist dieser mit einer roten Borte besetzt. In der linken Hand hält er ein von einem roten Band umschlungenes Buch mit blauem Umschlag in dem bekannten Langformat der chinesisch-japanischen buddhistischen Bücher. Die Rechte faßt den braunen Pilgerstab (khakkara), der am oberen und unteren Ende eine Zickzacklinie beschreibt. Über dem Kopfe befindet sich ein schirmartiges Sonnendach. Auf dem Rücken trägt er ein Büchergestell, vielleicht ein goldlackierter Kasten, der die von ihm später ins Chinesische übersetzten Schätze an Sanskrit-Manuskripten verwahrt. Die Zahl dieser seiner Übersetzungen beläuft sich auf 76 verschiedene Texte in 1335 Bänden, und doch ließ er bei seinem Tode noch mehr als die

Hälfte der von ihm mitgebrachten Bücher unübersetzt zurück.

Es ist wiederholt die Hoffnung ausgesprochen worden, daß sich in dem Tempel Tzʻǔ ngen ssǔ, wo sich Hsüan Tsang seit 649 aufgehalten und die Pagode Ta yen tʻa erbaute, Reliquien aus seinem Leben und seiner Tätigkeit finden ließen. Eine Skizze der Geschichte dieser Pagode gibt P. Havret, La stèle chrétienne de Si-ngan-fou, II, p. 127, Note. Als ich im Hochsommer 1903 in

Buddhistisches Pilgerbild.
(Kölner Museum.)

47*

Hsi an fu weilte, stattete ich dem 4 km südlich von der Stadt gelegenen Tempel einen Besuch ab in der Absicht, den Spuren der Vergangenheit unseres Helden nachzugehen. Aber alle meine Erwartungen wurden bitter enttäuscht: nichts war mehr dort vorhanden, was an ihn erinnerte, nicht einmal ein Bild oder eine seiner Schriften, keine Inschrift, die seinen Namen vor Vergessenheit bewahrte. Die Mönche des Tempels wußten nichts von ihm und lauschten nicht wenig erstaunt, als ich ihnen von ihrem großen Vorgänger erzählte, seinen weiten Reisen und der hohen Wertschätzung, die er unter unseren Gelehrten genießt. Der große Pan des Buddhismus ist tot in China, tot vielleicht für immer, und ein neues Geschlecht wandelt gleichgültig über die Trümmer der Geschichte hinweg, ohne Verständnis für die Größe und Herrlichkeit der Vergangenheit.   B. Laufer.

Der Ursprung der Religion und Kunst.

Vorläufige Mitteilung von K. Th. Preuß.

(Schluß.)

Wir sahen (Kap. II), daß der altmexikanische Schmetterling durch sein Urinieren den Regen gibt. Wer hat aber je den Schmetterling urinieren sehen? Und doch ist er in diesem Akt in dem Codex Vaticanus Nr. 3773 (S. 63) abgebildet. (s. vorher Abb. 9.)

So bedeutet denn der Imeotanz das Gedeihen der Vegetation und der Tierwelt, und zwar direkt ihre Erneuung, denn von den Bakairi am Paranatinga und Rio Novo wird angegeben [133], daß ihre Tänze zur Zeit der Ernte abgehalten werden, wo — wie ich vorhin (Kap. III) ausgeführt habe — die Erneuung bei vielen Völkern als notwendig empfunden wird. Es ist auch nicht die Vermehrung der einzelnen Tiergattung gemeint, die vielleicht zugleich als Nahrung dient. Das sehen wir aus den Intichiuma-Vermehrungszeremonien des Arunta-Stammes in Zentralaustralien [134]. Jede Totemgruppe übt zur Zeit, wo die Erneuung der Pflanzen- und Tierwelt bevorsteht, eine Zauberzeremonie zur Vermehrung des betreffenden Totemtieres oder der Totempflanze aus und meint, sich dadurch reichliche Nahrung überhaupt zu verschaffen. Das Totem ist mit dem Menschen selbst identisch, und so sind die Vermehrungszeremonien des Totems den phallischen Riten der Watschandi gleichzusetzen, die wir schon kennen (Kap. III).

Dieses ist die eine zu Tiertänzen führende Idee. Sie kann ebenso die Darstellung von größeren Tieren im Tanze hervorrufen, da ihnen ebenfalls derartige Kräfte zur Herbeiführung von Regen, Sonnenschein und von Witterungszuständen aller Art innewohnen. So ist der Hirsch, dessen Sprünge bei den Tarahumara den Regen bringen sollen, im Altmexikanischen die Flamme, d. h. er gibt Feuer und Sonnenwärme, ebenso wie der sich aus dem Hirsch entwickelnde anthropomorphe Gott Camaxtli [135]. Aber bei diesen Tieren kommt noch ein anderer Zweck der Tänze in Betracht, nämlich der Erfolg auf der Jagd.

Wenn die Känguruhs in der Zeit der Dürre spärlich werden, so begeben sich die Westaustralier zum Känguruhtarlow — einem Steinhaufen —, „der manchmal 50 bis 70 km entfernt liegt, und vollziehen dort gewisse Zeremonien, indem sie in Nachahmung der Sprünge des Känguruhs immer um den tarlow herumhopsen, nach Känguruhart aus hölzernen Trögen trinken, die man auf den Boden stellt, den tarlow mit Speeren, Steinen und whackaberries (Kampfkeulen) schlagen usw." Sie haben auch tarlows für Zeremonien zum Herbeiziehen von Zwergtrappen, Habichten, Leguanen und Kakadus. „Beim Emu-tarlow wird der Gang und Lauf des Emu nachgeahmt, und es ist erstaunlich, wie genau die Schwarzen jede Bewegung dieses Vogels darstellen können. Bei dieser Gelegenheit wird viel Schmuck aus Emufedern getragen [136]."

Der angegebene Zweck ist, daß die Nahrung reichlicher wird, als sie ist. Wir haben hier aber nicht an einen Zauber zur natürlichen Vermehrung der Känguruhs zu denken, sondern der Nachdruck ist einfach auf das Vorhandensein der Beute gelegt, gleichgültig, wie das zustande kommt, oder besser noch auf das Antreffen der Tiere. Denn bei diesen Zeremonien wird ein großes Gewicht darauf gelegt, daß alles zur Jagd und Tötung der Känguruhs Notwendige reichlich zur Stelle ist. Ebenso sind beim Fisch-tarlow-Ritus Fischnetze und eine Giftpflanze, „kurraru", mit der man den Fisch betäubt, indem man sie ins Wasser legt, überall zur Schau gestellt usw.

Das wird noch deutlicher, wenn wir uns z. B. die Büffeltänze der Prärieindianer ansehen. Es ist bekannt, daß diese den zauberischen Zweck verfolgen, die Büffelherden herbeizuziehen, um so eine erfolgreiche Jagd zu haben. In der Tat begegnet man auch in den nordamerikanischen Mythen häufig dem Gedanken, daß ein Indianer durch eine „Medizin", ein Schwitzbad oder eine sonstige Zeremonie in den Stand gesetzt ist, das Wild zu finden. Vorhanden ist es schon, es ist nur nicht, wo man es sucht [137]. Die als Büffel verkleideten Tänzer werden aber zuweilen am Schluß des Tanzes von den anderen Indianern scheinbar mit stumpfen Pfeilen erschossen. Es wird also eine Büffeljagd dargestellt, worauf es heißt, man habe nun Fleisch in Hülle und Fülle [138]. Das Ganze endet demnach wiederum, sobald Wild herangelockt ist, mit einer Art Analogiezauber.

Bevor ich die zugrunde liegende logische Verbindung der Handlung und des erzielten Erfolges darlege, sei noch das merkwürdigste Beispiel erwähnt. Bekanntlich hat K. v. d. Steinen [139] mit großem Scharfsinn aus der Technik, Anordnung und Bemalung der Xingumasken erkannt, daß der scheinbar als Gesicht der Maske hervortretende Teil ursprünglich ein in einen Strohbehang eingesetztes Netz ist, das dann mit Lehm ausgeschmiert und mit den Maschen des Netzes bemalt wurde. In diese Maschen zeichnete man je einen Mereschufisch. Dazu brachte man später auch menschliche Gesichtsteile darauf an. In der Tat ist auch beim Eremotanz der Nahuqua das Verhüllen des Kopfes der Tänzer durch ein Fischnetz beobachtet worden [140]. Soll nun das Ganze nicht eine sinnlose Maskerade sein

---

[133] A. a. O., S. 297.
[134] Spencer und Gillen, The Native tribes of Central Australia. London 1899, S. 167 ff.
[135] Vgl. den Beweis in meiner Arbeit „Der Ursprung der Menschenopfer". Globus, Bd. 86, S. 116.

[136] Clement, E., Ethnographical Notes on the Western Australian Aborigines. Intern. Archiv f. Ethnogr. XVI, S. 6 f.
[137] Z. B. Washington Matthews, The Mountain Chant. Ann. Rep. Bureau of Ethnology, S. 389.
[138] Prinz von Wied, Reise in das innere Nordamerika, II, S. 180, 181. Catlin, Illustrations of the Manners, Customs and Condition of the North American Indians, London 1851, I, S. 128, 157.
[139] Unter den Naturvölkern Zentralbrasiliens, S. 319 f.
[140] A. a. O., S. 99, 321 und Tafel VII.

034

中国犹太人的历史

GLOBUS

Illustrierte

Zeitschrift für Länder- und Völkerkunde

Vereinigt mit den Zeitschriften „Das Ausland" und „Aus allen Weltteilen"

Begründet 1862 von Karl Andree

Herausgegeben von

H. Singer

Siebenundachtzigster Band

>>>✳<<<

Braunschweig

Druck und Verlag von Friedrich Vieweg und Sohn

1905

GLOBUS.

ILLUSTRIERTE ZEITSCHRIFT FÜR LÄNDER- UND VÖLKERKUNDE.

VEREINIGT MIT DEN ZEITSCHRIFTEN: „DAS AUSLAND" UND „AUS ALLEN WELTTEILEN".

HERAUSGEGEBEN VON H. SINGER UNTER BESONDERER MITWIRKUNG VON Prof. Dr. RICHARD ANDREE.

VERLAG von FRIEDR. VIEWEG & SOHN.

| Bd. LXXXVII. Nr. 14. | BRAUNSCHWEIG. | 13. April 1905. |

Zur Geschichte der chinesischen Juden.

Von Dr. Berthold Laufer.

Die Geschicke der chinesischen Juden haben seit den ersten Tagen ihrer Entdeckung durch die Jesuiten im Anfang des 17. Jahrhunderts die Teilnahme der gesamten gebildeten Welt wachgerufen. Das Problem, das ihre vielhundertjährige Anwesenheit in der Hauptstadt der Provinz Honan aufgab, ist wiederholt von berufener und unberufener Seite in Angriff genommen und im allgemeinen mit der Theorie beantwortet worden, daß die Juden zur Zeit der Han-Dynastie im ersten nachchristlichen Jahrhundert über Persien und Zentralasien nach China eingewandert seien. Diese Anschauung haben namhafte Sinologen vertreten, wie T. de Lacouperie, Cordier und noch jüngst Tobar, dem wir die beste Herausgabe und Übersetzung der drei großen Inschriften von K'ai fong fu verdanken [1]). An dieser Meinung ist die asiatische Landroute reine Hypothese, die auch nicht durch den Schein eines Arguments gedeckt wird, während die zeitliche Voraussetzung sich im wesentlichen auf die mündliche Tradition der Juden selbst stützt, die zum ersten Male in dem Briefe des Pater Gozani vom Jahre 1704 gemeldet wird und die weiterhin durch eine Bemerkung in der zweiten, 1512 datierten Inschrift bestätigt werden könnte, daß sich die jüdische Lehre vom Anfang der Han-Zeit ab in China verbreitete, eine Angabe, die sich in der ältesten Inschrift nicht findet. Diese zweite Inschrift trägt aber einen durchaus dogmatischen Charakter, indem sie die in der ersten Inschrift dargelegten religiösen Grundsätze nach speziellen Gesichtspunkten erweitert, und wiederholt im übrigen nur die geschichtlichen Daten der ersten Inschrift vom Jahre 1489. Die dritte und jüngste Inschrift, vollends vom Jahre 1663, verlegt die Anfänge des Judentums in China in die grauen Tage der Chou-Dynastie, das ist 1122 bis 249 v. Chr. Je mehr sich also die Dokumente verjüngen, in desto entlegenere Zeiten versteigt sich die übereifrige Tradition. Diesen vagen Angaben einer der Chronologie sich nur schwach bewußten Überlieferung ist nicht der geringste historische Wert beizumessen, ebensowenig als wenn sich die erste Inschrift der chronologischen Ungeheuerlichkeiten leistet, daß Abraham im 146. Jahre der Dynastie Chou, das ist etwa 977 v. Chr., und Moses im 613. Jahre derselben Herrscherfamilie, also 510 v. Chr., gelebt hätte. Es mag sich bei diesen allgemeinen Zeitbestimmungen „Han" und „Chou" um gern gehegte Anschauungen im Schoße der Volksüberlieferung

handeln, die daran interessiert ist, mit ihrer Phantasie das höchste Altertum zu erreichen, aber es muß denselben jedwede historische Wirklichkeit abgesprochen werden.

Die Frage ist nun, was können wir aus einer kritischen Untersuchung der geschichtlichen Angaben unserer Inschriften für die Frage der Einwanderung der Juden nach China lernen? Mit voller Deutlichkeit ist in den Inschriften nur die eine Tatsache ausgesprochen, worauf bisher niemand hingewiesen hat, daß Indien als das Ausgangsland für diese Einwanderung zu betrachten ist. Dafür sprechen die folgenden vier Stellen:

1. Der historische Teil der ersten Inschrift von 1489 (Tobar p. 43) wird mit den Worten eröffnet: „Die ununterbrochene Tradition unserer Religion geschah von selbst: aus dem Lande T'ien chu, das ist Indien, ist sie gekommen (oder sind wir gekommen). Wir empfingen die Bestimmung und kamen hierher (nach China), 70 Familien stark. Wir brachten Baumwollzeuge der Westländer als Tribut dar an den Hof der Sung. Der Kaiser sprach: „Ihr kommt nach China, ehret die Vorfahren und bewahret eure Bräuche!" So ließen wir uns in Pien liang, das ist K'ai fong fu, nieder." Diese Einwanderung einer kleinen jüdischen Kolonie aus Indien in die damalige Residenz der Sung-Dynastie (960 bis 1278 n. Chr.) ist das erste entgegentretende historische Faktum, das mit großer Wahrscheinlichkeit in die erste Hälfte des 12. Jahrhunderts anzusetzen sein dürfte, da mit dem Bau der Synagoge in K'ai fong fu im Jahre 1163 begonnen wurde (Tobar p. 44, 58, 72). Ferner weist das Geschenk der Baumwollzeuge mit Bestimmtheit auf Indien hin, denn es kann sich bei der Baumwolle der Westländer nur um die in China in hoher Wertschätzung stehende indische Ware handeln, und da die tributbringenden Völkerschaften der chinesischen Annalen stets die Handelserzeugnisse ihrer Länder an den Kaiserhof bringen, so ist es sehr wahrscheinlich, daß es der Baumwollhandel war, in dessen Interesse die Juden nach China kamen.

2. Im Anfang des historischen Teiles der dritten Inschrift vom Jahre 1663 (Tobar p. 72) heißt es wiederum: „Die Religion der Juden nahm ihren Anfang in Indien (T'ien chu)."

3. Im dogmatischen Abschnitt der zweiten Inschrift vom Jahre 1512 wird der Ursprung des ersten Vorfahren, Adam, auf Indien (T'ien chu hsi yü) zurückgeführt. Tobar, p. 57, übersetzt diesen Passus: Adam leitete seinen Ursprung aus dem Lande Hsi yü in T'ien

[1]) Inscriptions juives de K'ai fong fu par le P. Jérôme Tobar, S. I, Variétés sinologiques No. 17, Schanghai 1900.

chu ab, eine ganz unhaltbare Übersetzung. Ebensowenig richtig ist seine Anmerkung, daß der Ausdruck auch „die westlich von Indien gelegenen Länder" bedeuten könne. Er bedeutet in der Tat nichts mehr als einfach „Indien", da Hsi yü nur ein Synonym für T'ien chu ist, und kommt in dieser Zusammensetzung schon in der historischen Literatur der Han-Zeit vor (z. B. Hou Han shu, Kap. 72)[2]).

4. In der vierten der in der Synagoge angebrachten Vertikalinschriften (Tobar p. 23) werden viele Philosophen erwähnt, welche das Prinzip der Existenz des Himmels, der Erde und des Menschen zu erkennen suchten, und die Heimat dieser Philosophen wird in Indien (Hsi chu) lokalisiert.

Zu diesen vier inschriftlich bezeugten Daten treten noch zwei andere Beweisstücke hinzu. Pater Gaubil[3]) berichtet, daß die Juden von K'ai fong fu nach dem Brande ihrer Synagoge in der Periode Wan li (1573 bis 1619) zum Ersatz für die verbrannten Bücher ein Exemplar der heiligen Schrift von den Juden aus Indien erhalten hätten, und damit mag der Umstand zusammenhängen, daß auffallende Ähnlichkeiten zwischen den Pergamenten des Pentateuchs von K'ai fong fu und dem nun Buchanan gefundenen Codex malabaricus der schwarzen Juden von Indien bestehen (s. Tobar p. 94). Sehr wesentlich fällt nun die Tradition der Juden selbst ins Gewicht, der zufolge sie sich selbst als „die Religion von Indien" bezeichnen. Dieser merkwürdige Umstand wird bereits von den beiden vom Bischof Smith nach K'ai fong fu gesandten christlichen Chinesen ausführlich erwähnt. Ihr Bericht lautet wörtlich wie folgt: „Als am Abend unser Freund erschien, um uns zu besuchen, fragten wir ihn: „Wie nennt ihr eure Religion?" Er sagte: „Früher trugen wir den Namen T'ien chu chiao oder indische Religion (oder indische Sekte); doch jetzt haben denselben die Priester in T'iao chin chiao verändert, d. h. die Religion derer, welche die Sehne ausreißen, weil allem, was wir essen, ob Hammel, Rind oder Geflügel, die Sehne herausgenommen werden muß." Da ehedem die Juden in K'ai fong fu in einen Tumult mit den Chinesen gerieten, hat der Priester eben deshalb den Namen der Religion in den erwähnten verändert." „Einige Personen", so fährt der Bericht der Delegierten fort, „mögen die Laute T'ien chu chiao (Lehre von Indien) mit T'ien chu chiao (das ist Lehre des Himmelsherrn, Katholizismus) verwechseln. Als wir daher die Laute hörten, baten wir ihn, die Charaktere niederzuschreiben, worauf er schrieb T'ien chu (Indien) chiao; da verstanden wir, daß er die Religion von Indien und nicht die römisch-katholische Lehre meinte[4])." Von gegenwärtig in Schanghai lebenden chinesischen Juden konnte ich feststellen, daß die Bezeichnung „indische Religion" auch jetzt noch als die einzig offizielle fortlebt. Der Name T'iao chin chiao, der offenbar ein von den Chinesen beigelegter Spitzname ist, findet sich in der Literatur zum ersten Male in Gozanis Brief von 1704, so daß der von den Delegierten erwähnte Namenswechsel sicher schon im 17. Jahrhundert stattgefunden haben muß.

Angesichts dieser sechs Fakten, von welchen vier auf die Inschriften selbst zurückgehen, ist kein Grund

zu einem Zweifel an der Herkunft der chinesischen Juden aus Indien vorhanden. Nur das von der alten Hypothese der zentralasiatischen Einwanderung verschuldete Vorurteil ließ diese richtige Tatsache übersehen und interpretierte in die chinesischen Bezeichnungen für Indien Begriffe hinein, die niemals darin lagen und nie darin hätten gesucht werden sollen. Selbst Wylie war der Ansicht, daß in diesem Falle unter T'ien chu Syrien zu verstehen sei[5]). Syrien aber heißt stets Ta ts'in, und T'ien chu bedeutet stets Indien[6]). Dem P. Tobar sind diese indischen Anspielungen gänzlich entgangen; im achten Abschnitt seines Buches „De l'entrée des Juifs en Chine" erwähnt er derselben mit keiner Silbe. Am nächsten meiner Ansicht kam bisher Prof. Pelliot[7]) in Hanoi, der es wahrscheinlicher fand, daß die Juden wie die Buddhisten und Mohammedaner den Landweg sowohl als den Seeweg benutzt hätten, ohne indessen die auf Indien bezüglichen Stellen zur Erklärung heranzuziehen.

Die indische Herkunft der chinesischen Juden steht nun ferner im besten Einklang mit dem gesamten Gange der Geschichte der Juden in Asien und in China insbesondere. Die indischen Juden sind aus Persien eingewandert, und persischer Einfluß ist, wie erwiesen, in der Sprache wie in den heiligen Schriften der Juden von K'ai fong fu vorhanden[8]). Leider liegt die Chronologie der Geschichte der jüdischen Kolonien in Indien sehr im argen, und bis jetzt sind nur wenige Daten erschlossen. Das älteste vorliegende Dokument ist der auf drei Kupfertafeln eingeschriebene Erlaß des letzten Vizekönigs von Malabar, der dem Joseph Rabbân Land, Gerechtsame und Königswürde verleiht. Nach der neuesten Untersuchung von Prof. Gustav Oppert wäre das wahrscheinlichste Datum dieser Urkunde das Jahr 379 n. Chr.[9]). Von anderen uns interessierenden Daten ist das Jahr 1511 zu erwähnen, in welchem sich die ersten spanischen Juden in der jüdisch-indischen Kolonie Kotschin niederließen[10]), das Jahr 1523, in welchem Cranganore von den Portugiesen genommen und befestigt wurde, und 1524, in welchem die Mohammedaner die Juden bei Cranganore angriffen, ihre Häuser und Synagogen zerstörten und eine große Anzahl töteten und vertrieben. Solche Ereignisse mögen auch den Anstoß zu Auswanderungen in östlicher Richtung gegeben haben, aber im großen und ganzen müssen wir uns die Tatsache vor Augen halten, daß es die ausgedehnten und großartigen Handelsfahrten der Araber nach Ost-

---

[2]) Siehe auch Hirth, Die Länder des Islam nach chinesischen Quellen. Leiden 1894, S. 45, Note 2.

[3]) Lettres édifiantes, vol. XXXI, p. 358.

[4]) The Jews at K'ae-fung-foo: being a Narrative of a Mission of Inquiry to the Jewish Synagogue at K'ae-fung-foo, on behalf of the London Society for promoting Christianity among the Jews. With an Introduction by the Right Revd. George Smith. Shanghai 1851, p. 28/29. Derselbe Passus abgedruckt in: Chinese Repository, vol. XX, p. 449.

[5]) „Syria appears to have been included by the Chinese formerly under the designation T'ien chu, and is no doubt so intended here, although the term is generally translated „India". A. Wylie, Israelites, p. 6 in Chinese Researches.

[6]) Vgl. besonders G. Hirth, China and the Roman Orient, p. 46, wo sich die Verbindung Ta Ts'in T'ien chu „Syrien und Indien" im Sung shu findet.

[7]) Bulletin de l'Ecole française d'Extrême-Orient, vol. I, No. 3, p. 264, 1901.

[8]) Die wesentlichsten Punkte der Übereinstimmung sind die 53 Abteilungen des Pentateuchs und die Zählungen von 27 Buchstaben des Alphabets bei chinesischen und persischen Juden (s. Tobar, p. 28, Note 5 und p. 29, Note 1). Ferner sind in K'ai fong fu gefundene hebräische Handschriften von Noten begleitet, in denen manche persische Wörter mit hebräischen Buchstaben geschrieben sind. Die erste Inschrift von 1489 bezeichnet den Rabbi mit dem persischen Worte ustád, in chinesischer Transkription wu-se-ta (s. Tobar, p. 44, Note 1). Es braucht nicht besonders hervorgehoben zu werden, daß die Geschichte Christi und der Talmud den chinesischen Juden unbekannt waren.

[9]) G. Oppert, Über die jüdischen Kolonien in Indien. In Semitic Studies in Memory of Rev. Dr. Alexander Kohut, Berlin 1897, p. 396—419.

[10]) Es wäre also nicht ausgeschlossen, daß auch spanische Juden, etwa im Verein mit den Portugiesen, nach Makao und Kanton oder anderen Häfen Südchinas gekommen wären.

asien waren, welche die persisch-indischen Juden nach dem Osten mit fortrissen, ebenso wie das Vordringen des Islams im westlichen Asien, in Afrika, Spanien und Sizilien für die Juden mit einer neuen Epoche freieren Schaffens und geistigen Fortschritts verbunden war. Denn bereits im Jahre 878 finden wir die Anwesenheit von Juden zusammen mit Mohammedanern in Südchina durch den Bericht eines arabischen Autors bezeugt [11]). Marco Polo, der von 1271 bis 1295 in China reiste, tut im fünften Kapitel des zweiten Buches seiner Reisebeschreibung der Juden daselbst Erwähnung. S. Yule's Marco Polo, 3. ed. by Cordier, vol. I, p. 346, 348; in vol. II, p. 375 werden auch die Juden in Coilum an der Malabarküste erwähnt. In die Zeit seines Aufenthaltes in China, und zwar in das Jahr 1279, fällt das in der ersten und zweiten Inschrift berichtete Ereignis, die Fortsetzung des 1163 begonnenen Baues der Synagoge von K'ai fong fu (s. Tobar, p. 45, 58). Aus dem 14. Jahrhundert liegen uns die folgenden Daten vor:

1326. Andreas von Perugia, Bischof von Zaitun (chinesisch Ch'üan chou fu, Küstenstadt nördlich von Amoy), klagt darüber, daß unter den Mohammedanern und Juden keine Bekehrungen erzielt würden.

1329. In den chinesischen Annalen der mongolischen Dynastie der Yüan (Kap. 33) werden die Juden bei Gelegenheit der Wiedereinführung des Gesetzes betreffend Einziehung der Abgaben von Andersgläubigen erwähnt [erste Erwähnung der Juden in der chinesischen Literatur unter dem Namen Dju-hu(t)] [12]).

1340. Unter diesem Jahre wird in den Annalen der Yüan-Dynastie (Yüan shih) Mohammedanern und Juden die Leviratsehe untersagt [13]).

1342. Der Franziskanermönch Jean de Marignoli disputiert gegen die Juden in Khanbâliq (Peking).

1346. Der arabische Reisende und Schriftsteller Ibn Baṭûṭa weist auf die Existenz von Juden in China, besonders in Hangchow (Khansa) hin und betont den großen Reichtum ihrer hervorragenden Männer.

1354. Zufolge dem 43. Kapitel der Annalen der Yüan-Dynastie werden wegen mehrerer Aufstände in China reiche Mohammedaner und Juden nach der Hauptstadt Peking berufen und aufgefordert, sich dem Heeresdienst anzuschließen [12]).

Für die zweite Hälfte des 15. Jahrhunderts ist uns aus der ersten und wichtigsten Inschrift von 1489 das Vorhandensein einer jüdischen Kolonie in Ningpo bezeugt, die sich im Besitz der heiligen Schriften befindet und Verkehr mit der Gemeinde in der Provinz Honan unterhält [14]). Mit Recht bemerkt Pelliot [15]) im Zusammenhang mit dieser Stelle, daß Ningpo, wo es noch eine Straße der Perser gibt, zu allen Zeiten und besonders unter den T'ang und den Sung (7. bis 13. Jahr-

hundert) einer der großen Seehäfen des Ostens war, wo, wie in dem ganzen Gebiet der Yangtsemündung, Abenteurer und Händler den großen persischen Dschunken entstiegen, Leute aller Rassen und aller Kulte, Manichäer und Mazdäer, Mohammedaner und Nestorianer, und daß es seltsam genug sein müßte, wenn die Juden allein sich außerhalb dieses mächtigen Stromes gehalten hätten. Aus den wenigen abgerissenen Tatsachen aber, welche uns die Geschichte aufbewahrt hat, erkennen wir deutlich zwei wesentliche Umstände: 1. daß die persisch-indischen Juden in der Gefolgschaft der seefahrenden Araber und Perser ihren Weg nach China fanden, und 2. daß sie langsam und allmählich, gleichsam etappenweise, ihre Route aus dem südlichen in das nördliche China verfolgten. Wir finden sie in Khanfu, Zaiton, Hangchow, Ningpo, vielleicht auch Nanking [16]) und sehen sie dann in nördlicher Richtung nach K'ai fong fu und Peking vorrücken, um in die Metropolen des Reiches zu gelangen, wo sie die besten materiellen Bedingungen für ihre Existenz zu erringen hofften. Hier zeigt sich in der Tat ein natürlicher und gesetzmäßiger Weg ihrer geographischen Verbreitung während des Mittelalters in China, in historischer Folgerichtigkeit, die sich bei der Annahme eines asiatischen Überlandweges weder erkennen noch erschließen läßt. Da der historische Zusammenhang zwischen den Etappen Ningpo und K'ai fong fu dank unserer ältesten Inschrift mit vollkommener Sicherheit feststeht, und da zudem die jüdische Kolonie daselbst inschriftlich als aus Indien stammend bezeugt ist, so weist K'ai fong fu eben auf den Süden Chinas zurück, nicht aber auf den Westen des Landes oder Innerasien. Während die indische Herkunft der chinesischen Juden alle mit ihrer Geschichte verknüpften Tatsachen in befriedigender Weise erklärt, bleibt bei der Theorie der asiatischen Landwanderung der Umstand rätselhaft, daß sich die Juden gerade in der Stadt K'ai fong fu niederließen, die erst seit 960 unter dem Namen Pien liang Residenz der Sung-Dynastie wurde. Wären die Juden bereits unter der Han-Dynastie nach China gedrungen, so sollte man mit Bestimmtheit erwarten, Spuren ihrer damaligen Existenz in den Hauptstädten jener Zeit, Ch'ang an und Lo yang, zu finden, was aber ebensowenig der Fall ist wie im übrigen zentralen und westlichen China.

Der islamische Einfluß auf das chinesische Judentum war außerordentlich groß. Die jetzt nicht mehr vorhandene Synagoge von K'ai fong fu war in ihrer Architektur und inneren Einrichtung eine Moschee, ihre Kultusgeräte eine Nachahmung der entsprechenden des Islams. Die gesamte chinesische Terminologie, wie sie uns in den jüdischen Inschriften entgegentritt, ist von der mohammedanischen abhängig und mit dieser so gut wie identisch. Das Judentum ist nicht älter, wie man früher annahm, sondern jünger als der Islam in China.

Interessanter ist auch die Tatsache, daß es in der Gegenwart in Hongkong und Schanghai eine ansehnliche Kolonie sogenannter orientalischer Juden gibt, welche das Arabische als ihre Muttersprache sprechen und sämtlich entweder auf dem Wege über Indien oder aus Indien selbst nach China eingewandert sind. Wir sehen also bis in die Jetztzeit dieselben historischen Faktoren wirksam, die seit dem 9. Jahrhundert eingesetzt haben.

---

[11]) Siehe Chinese Repository, vol. I, p. 8. Schefer, Relations des Musulmans avec les Chinois in Centenaire de l'école des langues orientales vivantes, p. 5, Paris 1895. Chavannes in Journal Asiatique 1897, p. 79.

[12]) Siehe Journal of the North China Branch of the Royal Asiatic Society, Shanghai, New Series, vol. X, p. 38, 1876.

[13]) Siehe China Review, vol. XXV, p. 92.

[14]) Siehe Tobar, l. c. p. 49 und Havret, La stèle chrétienne de Hsi an fu, vol. II, p. 289. Meine Auffassung dieser Stelle wie mancher anderen weicht von der Übersetzung des P. Tobar ab. Da indes nicht der Ort zu philologischen Auseinandersetzungen ist, so erspare ich diese Erörterung auf eine andere Untersuchung dieses Themas an anderem Ort.

[15]) L. c. p. 263.

[16]) Nach dem Bericht von Semedo, s. Chinese Repository, vol. XIX, p. 309.

31*

035

莱茵河流域罗马时代的中国文物

Redner sich ausdrückte, ein Delta hineinbaue in das Haff. Manche geologische Erscheinungen, über deren Ursache man sich den Kopf zerbreche, seien auch auf die Tätigkeit der — Wasserbaubeamten zurückzuführen. — Fossile Dünenformen im norddeutschen Flachlande behandelte hierauf Dr. Solger-Berlin unter Vorführung von Karten und Lichtbildern. Diese Dünen finden sich nicht an der Küste, sondern mehr im Innern, am Rande der breiten ost-westlichen diluvialen Flußtäler, z. B. in Brandenburg und zwischen Warthe und Netze bei Birnbaum, und rühren aus der Steppenperiode her, die dem Zurückweichen des Eises folgte. Sie fallen mit der Steilseite, die bogenförmigen Grundriß zeigt, nach Osten ab. Dieser Grundriß rühre aus der Steppenperiode her, das Profil sei das Ergebnis der heutigen Westwinde. Neben diesen Bogendünen kommen auch fossile Strichdünen vor, deren Bildung aber mit der der ersteren eng zusammenhängt. — Den Beschluß dieser Sitzung machte die Vorführung von Karten des Danziger Geographen Wied (16. Jahrhundert) durch Dr. Michow-Hamburg, der auch eine Schrift über Wied dem Geographentage gewidmet hatte.

Die fünfte und Schlußsitzung am 15. Juni vormittags leitete Professor Oberhummer-Wien. Beratungsgegenstand war die Landeskunde Westpreußens und der Nachbargebiete. Nach Annahme der bereits erwähnten Anträge, zu denen noch ein Dank an die geologische Sektion des Naturforscher- und Ärztetages dafür kam, daß sie die auf den Geographieunterricht gerichteten Bestrebungen des Geographentages zu den ihrigen gemacht habe, und nach Wahl des nächsten Tagungsortes hielt Dr. Seligo-Danzig an Stelle des verhinderten Dr. Lakowitz-Danzig einen Vortrag über die Temperaturverhältnisse der westpreußischen Seen. Eingehende Untersuchungen über die Temperatur hat Seligo im Klostersee bei Carthaus angestellt, während der Vortragende einige Beobachtungen aus dem Barnowitzer und dem Hintersee bei Stuhm beisteuern konnte. Die zahlreichen Beobachtungen aus dem Klostersee, einem kleinen, durch einen Querriegel in einen 10 und in einen 24 m tiefen Teil getrennten Gewässer, zeigen, daß im Sommer die kälteste Schicht, im Winter die wärmste Schicht über dem Boden lagert, und daß der Ausgleich Ende März und Ende September eintritt. Die Temperaturschwankungen sind erheblich, und der Wechsel der Jahreszeiten macht sich noch in 20 m Tiefe bemerkbar. In den Stuhmer Seen kommen Sprungschichten vor. Aus der Diskussion waren die Bemerkungen von Professor Halbfaß-Neuhaldensleben beachtenswert, der auf einige Aufgaben der westpreußischen Seenforschung hinwies. Man könne aus regelmäßigen Temperaturmessungen durch die ganze Tiefe eines Sees Schlüsse auf die Wärmemengen ziehen, die er im Laufe eines Jahres aufnimmt und abgibt. In einzelnen Ländern, wie Schweden und Italien, werden solche Messungen bereits ausgeführt. Wenn womöglich die Seen der ganzen Erde nach solchem Grundsatz beobachtet

würden, so ergäben sich daraus für die Klimatologie vielleicht bedeutendere Resultate als aus den Messungen der Lufttemperaturen. In diesen Seetemperaturen spiegelten sich z. B. harte Winter deutlich wider. Von besonderem Wert seien dabei natürlich die Resultate aus Seen mit großem Volumen, also aus umfangreichen und tiefen Gewässern, und er lenke daher auch die Aufmerksamkeit auf die westpreußischen Seen, unter denen es ja geeignete Objekte gäbe.

Professor Schubert-Eberswalde besprach hierauf das Thema „Wald und Niederschlag in Westpreußen, Posen und Schlesien". Von besonderem Interesse war seine näher begründete Ansicht, daß der Einfluß des Waldes auf die Regenbildung und Regenmenge viel zu sehr überschätzt werde. Die Präzision der Beobachtungen — auch der des Regens mit dem Regenmesser — lasse noch viel zu wünschen übrig; was das in Rede stehende Gebiet angehe, so könne man aber sagen, daß, wenn man dort ein Zehntel des Waldbestandes abholze oder ein Areal vom zehnten Teil desselben aufforste, die Niederschlagsdifferenz nur etwa 2 Proz. betragen würde. Demgemäß sei auch der Unterschied der Niederschläge auf bewaldeten und nicht bewaldeten Flächen nicht sehr erheblich, er schwanke hier zwischen 2 und 10 Proz. Der Wald liefere wenig Wasserdampf im Gegensatz zur See, die in Westpreußen ja großen Einfluß ausübe. — Den letzten Vortrag hielt Professor Kumm-Danzig über die Pflanzengeographie von Westpreußen. Er erörterte die verschiedenen Faktoren, die auf die heutige Flora der Provinz eingewirkt haben, so nacheinander die Eiszeit, die See, der diluviale Riesenstrom der Weichsel und auch der Mensch. Zum Schluß wurde die geographische Verbreitung der typischen Baumarten Westpreußens — Kiefer, Fichte, Buche, Erle u. a. — besprochen.

Mit Dankesworten schloß der Vorsitzende mittags den Geographentag. Auf dem Programm standen noch mehrere Ausflüge, so eine Weichselfahrt bis zur Grenze, Touren nach der Kassubei, nach dem Mündungsgebiet der Weichsel, nach dem Oberländischen Kanal, nach Cadinen und Kahlberg. Den stärksten Zuspruch fand mit 90 Teilnehmern die 2½ tägige Weichselreise. Die Herren aus dem Westen und Süden Deutschlands dürften auf diesen Ausflügen mehr Interessantes gefunden haben, als sie vielleicht erwartet hatten.

Die reichhaltige Ausstellung im Franziskanerkloster war im allgemeinen eine provinzielle, doch war auch Ostpreußen vertreten. Man sah dort zahlreiche alte und neue Karten, Photographien, Zeichnungen, die sonst nicht allgemein zugänglich sind und von Behörden, Vereinen, Städten und Privatpersonen zur Verfügung gestellt waren. Außerdem waren alte und neue Kriegsschiffsmodelle vorhanden, z. B. solche aus der Hansazeit Danzigs, und der Leiter der Ausstellung, Hafenbaudirektor Gromsch, der früher in Tsingtau war, hatte Karten und Photographien aus Kiautschou zur Ansicht ausgelegt.

H. Singer.

Chinesische Altertümer in der römischen Epoche der Rheinlande.

Aus Anlaß der gegenwärtig in der technischen Versuchsanstalt bei der Kgl. Porzellanmanufaktur in Charlottenburg angestellten Untersuchungen der Terra sigillata möchte ich mir gestatten, die Aufmerksamkeit auf einige jetzt ziemlich verschollene Funde angeblicher chinesischer Tongefäße aus der Römerzeit unserer rheinischen Erde zu lenken. Von dem ersten derartigen Funde wird uns

gemeldet in dem noch jetzt in archäologischen Kreisen geschätzten Prachtwerke von Ph. Houben und F. Fiedler, Die Denkmäler von Castra Vetera und Colonia Traiana, Xanten 1839, S. 48 mit Tafel XVI. Es handelt sich um ein Römergrab, das sich auf einem Felde in der Nähe der Porta zwischen anderen Gräbern befand und von Houben am 27. November 1829 geöffnet wurde. Dasselbe enthielt eine schöne gelbe Urne mit Gebeinen und einer Kupfermünze des Kaisers Vespasian aus seinem achten Konsulat im Jahre 77 n. Chr., ferner drei Schalen von Terra sigillata und eine zerbrochene Lampe. Außerdem fanden sich in demselben Grabe noch zwei Teller,

a
Abb. 1. a Vase von chinesischer Form, gefunden 1846 bei Harzheim. b In dieser Vase gefundene Bronzefigur.

Nr. 5 und 6 auf der zitierten Tafel, von feiner roter Erde, aber nicht Terra sigillata, deren Arbeit und Verzierung von der an römischen Gefäßen ganz verschieden ist, und zwei Kännchen mit Deckeln in Form eines Vogels, Nr. 7 und 8. „Diese vier Gefäße", sagt Houben wörtlich, „haben große Ähnlichkeit mit rotem chinesischen Porzellan (soll natürlich heißen: Ton!). Wenn nicht das glaubwürdigste Zeugnis vorläge, daß sie in

diesem Römergrabe, so wie es hier vorliegt, gefunde wurden, so würde man sie für unecht halten müssen. I ist nicht unwahrscheinlich, daß diese gewiß sehr kos baren Gefäße, die ich für chinesische Arbeit halte, durc Handel zu den Römern am Rhein kamen, wie sie zu u noch gelangen. Wir können sie murrhinische (Va murrhina) nennen, wenn es unter den Archäologen sche ausgemacht wäre, daß jene fremden Gefäße chinesische Ursprungs sind." J. Freudenberg, ein Referent d Houbenschen Werkes in den Bonner Jahrbücher Heft 3, 1843, S. 166 bis 182, bemerkt dazu auf S. 17 „Merkwürdig sind die Tafel XVI abgebildeten vier einem römischen Grabe gefundenen chinesischen Gefä Ein zweites Exemplar von Nr. 8 findet sich nach Klem S. 93 in Dresden. Ein Seitenstück bietet nur die b Mainz gefundene chinesische Specksteinfigur." Ohne d in Rede stehenden Gefäße selbst gesehen und geprüft z haben, kann man naturgemäß kein endgültiges Urte fällen. In der Annahme, daß sich die Houbensche Sam lung noch in Xanten befinde, wandte ich mich dah dorthin und erhielt am 23. August vorigen Jahres vo Vorsitzenden des niederrheinischen Altertumsvereins d selbst die Auskunft, daß sich der Verbleib der Hoube schen „chinesischen" Gefäße dort nicht mehr feststell lasse, da dieselben mit der ganzen Sammlung am 4. Ju 1860 durch J. P. Heberle in Köln verkauft worden seie Daraufhin wandte ich mich an diese Firma, die mir a 27. August mitteilte, daß die Sammlung zum Teil na London verkauft, zum Teil in Köln versteigert word sei, daß sie aber heute über den Verbleib einzeln Gegenstände keinen Aufschluß mehr geben könne. Betre der oben erwähnten Specksteinfigur schrieb ich an d Germanische Museum in Mainz. Herr Museumsdirekt L. Lindenschmit erwiderte unter dem 21. August: „ unseren Sammlungen ist die fragliche chinesische Spec steinfigur nicht aufbewahrt. Möglicherweise befand s sich ehemals im Privatbesitz in Mainz. Wir besitz überhaupt keine Gegenstände, die auf chinesischen L sprung zurückgeführt werden könnten." Teller, wie s bei Houben beschrieben und abgebildet sind, sollen si nach einer Notiz in den Bonner Jahrbüchern, Heft 7 1882, S. 170 im Museum zu Wiesbaden befinden. A eine darauf bezügliche Anfrage hin teilte mir jedo Herr Prof. E. Ritterling, Direktor des Wiesbadener Alt tumsmuseums, mit, daß sich solche Teller im dortig Museum nicht befänden. Um den letzten derartig Fund vorwegzunehmen, weil sich der Typus desselb der Houbenschen Kanne anreiht, verweise ich auf d Artikel von Konstantin Koenen, Neuß, Ein Römergr bei Norf und in einem solchen gefundenes chi sisches Gießgefäß aus der Mitte des ersten Jahrhunde unserer Zeitrechnung, Bonner Jahrbücher, Heft 73, 188 S. 169 bis 171. Es handelt sich hier um ein Gefä Gestalt eines phantastisch gebildeten Vogels mit u gewendetem Kopfe, das 1873 zwischen römischen Gefä neben einem Grabe in Norf gefunden wurde, das d ersten Jahrhundert n. Chr. angehört. Auf dem Rück desselben, bemerkt der Verfasser, befindet sich ei Öffnung, die zum Eingießen von Flüssigkeiten bestim und durch ein kleines Deckelchen verschlossen ist. I Brust des Vogels zeigt ein Röhrchen, das zum Ausgieß des Gefäßinhalts Verwendung gefunden haben ma während der Schweif des Vogels die Anhabe bild Die Masse der Verfertigung besteht aus jener roten, h klingenden Tonmasse, wie wir sie noch heute an d bekannten chinesischen Ware benutzt finden; sie ist n etwas dunkler in der Farbe. Mit der Masse der V fertigung stimmt auch der Stil und höchst eigentümlic Charakter überein, und zwar so, daß man das Gießge für ein modernes chinesisches Erzeugnis halten würde, wenn nicht die Umstände der Auffindung es in die Mitte des ersten Jahrhunderts setzen würden; denn abgesehen von diesem Funde sind auch anderwärts im Rheinlande Gefäße desselben Stiles in römischen Gräbern dieser Zeit gefunden worden. Nun kommt er auf die Houbenschen Funde zu sprechen und schließt, daß um die Mitte des ersten Jahrhunderts unserer Zeitrechnung in irgend einer Weise Gefäße von jenem durch die Natur von allen Ländern abgesperrten merkwürdigen Volke in das Rheingebiet gelangt seien, falls nicht der Nachweis geliefert werden könne, daß so Gefäße chinesischen Stils damals sonstwo angefertigt wurden. Zur historischen Erklärung bemerkt Koenen: „Zur Zeit der batavischen Freiheitskriege fand bekanntlich ein Wechsel der rheinischen Legionen statt. Es kann daher recht wohl möglich sein, solche Gefäße aus Asien rekrutierten Mannschaften zuzuschreiben. Wahrscheinlicher jedoch erscheint mir

chinesischen Handels siehe den Vortrag von H. Nissen, Der Verkehr zwischen China und dem römischen Reiche, Bonner Jahrbücher, Heft 95, 1894, S. 1 bis 28, und die übrige Literatur bei Schiller und Voigt, Die römischen Staats-, Kriegs- und Privataltertümer, München 1893, S. 438, Note 10.

Die Möglichkeit eines Imports chinesischer Tongefäße ins römische Reich muß vom historischen Standpunkte zugegeben werden; daraus folgt natürlich nicht, daß dem auch tatsächlich so war.

Schließlich bleibt ein ganz vereinzelt dastehender Gefäßtypus zu besprechen. Derselbe wird beschrieben und abgebildet in dem Aufsatze: Übersicht über die neuesten antiquarischen Erwerbungen der Frau Sibylla Mertens-Schaaffhausen (d. h. geborene Schaaffhausen), mitgeteilt von der Besitzerin, Bonner Jahrbücher, Heft 13, 1849, S. 136 bis 142, mit Tafeln III und IV. Unsere Abbildung 1 a und b ist nach einer Photographie jener

Abb. 2a bis d. Formen alter chinesischer Sakralgefäße („Ku").

die Möglichkeit, daß diese Gefäße auf dem Wege des Handels in derselben Weise wie schon nachweislich in einer Zeit, in der die Dampfrosse noch unbekannt waren, nach hier gebracht worden sind. Wenn wir die Waffen der Chinesen um die Zeit, der unsere Gießgefäße angehören, bis an das Kaspische Meer vordringen sehen, wenn wir ferner wissen, daß dadurch China zuerst mit fremden Kulturgewächsen versehen wurde, dann sollte man doch, wie nach der vergleichenden Gefäßkunde, so auch an der Hand der Geschichte jenen Export chinesischer Ware für höchstwahrscheinlich halten." Man vergleiche dazu auch die Bemerkung von Schaaffhausen, Bonner Jahrbücher, Heft 50 u. 51, 1871, S. 291: „Daß vornehme Römer aus allen Teilen des römischen Reiches an den Rhein kamen, hier wohnten und starben, darüber geben uns die zahlreichen römischen Grabsteine Aufschluß. Die Funde kostbarer Gold- und Silbergeräte wie die reichen Villen beweisen, daß nicht nur römische Legionen hier ihr rauhes Lagerleben führten, sondern daß einzelne Römer hier bleibenden Aufenthalt nahmen und sich mit allem Luxus römischer, griechischer und asiatischer Kultur umgaben." Zur Geschichte des römisch-

Tafel III hergestellt. Hören wir zunächst, was die Verfasserin darüber zu sagen hat: „Voran unter den Gegenständen, welche die Gunst des Zufalls mir zuführte, steht ein Gefäß von gebranntem Ton, gefunden im Dezember 1846 zu Harzheim bei Mainz, in einem Weinberge, nebst drei römischen Ziegeln, bezeichnet Leg. XXII. mit dem Delphin, und mehreren römischen Bronze- und Silbermünzen. Wenn letzteres kein Irrtum ist, was kaum anzunehmen, da die Aussage des Finders als schriftliches, durch den Bürgermeister von Mainz beglaubigtes Dokument mir vorliegt, so haben wir allen Grund, dieses Gefäß jedenfalls vor die Zeit der 30 Tyrannen zu setzen, indem nach Kaiser Probus die römischen Silbermünzen so selten wurden, daß sie nur vereinzelt vorkommen; selbst sehr bedeutende Münzfunde mit Geprägen jener Epoche entbehren der Silberdenare ganz.

„In dem mit Erde gefüllten Gefäße lag eine kleine Bronzefigur echt indischen Ursprungs (Abb. 1, Nr. b), eine männliche, mit anliegendem Waffenkleide, breitem Leibgurt, Schwert und Dolch gerüstete Gestalt, über dem Kopfe einen Elefantenrüssel, deren Beine in Elefantenfüße ausgehen, und an der sich eine Schlange heraufwindet. Die Inschrift an dem Postament, auf welchem die Figur steht, erkannten Lassen und Bopp als Pehlewischrift und lasen den Namen des indischen Kriegsgottes »Skandadeva«.

„Das Gefäß selbst (Abb. 1, Nr. a) ist gebrannt aus demselben roten Ton, den wir in den chinesischen Gefäßen älterer Fabrik erkennen, und gleicht seiner Form nach jenen schlanken, hohen Blumenvasen, die so häufig in buntbemaltem, reich vergoldetem chinesischen und indischen Porzellan uns begegnen; es ist leider am oberen Ende stark fragmentiert. Die ganze Oberfläche desselben ist mit kurzen Strichen geritzt, und die flachen Reliefs, welche es verzieren, sind besonders geformt und aufgesetzt. Viele derselben sind abgefallen, ein Beweis, daß sie vor dem Aufsetzen schon einmal gebrannt waren, wie auch die Vasen; doch lassen die Darstellungen sich aus dem scharfen eingeritzten Kontur erkennen. Vermutlich wurde nach dem Aufsetzen der Reliefs das Ganze noch ein oder mehrere Male gebrannt; die Härte des Stoffes und die Textur des Bruches weisen auf öfteres Brennen hin. Alle Darstellungen, die uns hier begegnen, wiederholen sich auf beiden Seiten des Gefäßes. Diese Reliefs zeigen uns eine ebenso sonderbare als rätselhafte Zusammenhäufung von Symbolen der verschiedenartigsten nationalen Mythen, durch welche das Gefäß selbst ein der gelehrten Forschung und Deutung gewiß willkommener Gegenstand wird. Die auferstehende gekauerte, himmelan schwebende bhudaistische (sic!) Gestalt weist nach Indien, die Tauben (der Semiramis?) nach Assyrien, Mithra und Diademe nach Persien, der dagonartige Drache mit Menschenantlitz nach Phönizien, die mit heiligen Binden umwundenen Spieße (Thyrsusstäbe) nach Ionien hin; die kuhähnliche Maske (Jo?) vielleicht nach Ägypten und Äthiopien. Der ganze Orient, von den Küsten des Ionischen Meeres bis nach Hinterindien, ist hier in seinen religiösen Emblemen repräsentiert, und das Ei nebst dem ihm gesellten Symbol, welches vielleicht als Lingam zu deuten wäre, vervollständigen die synkretistischen Darstellungen mythischer Begriffe, denen sich noch andere in dem oberen fragmentierten Teile dieses gewiß einzigen Gefäßes anreihten, welche wir leider nicht mehr klar zu erkennen vermögen. Vielleicht waren es Medusenmasken. Vermutlich kam diese Vase durch den Handel nach Europa und wurde nebst dem darin gefundenen Bronzeidol an den Rhein gebracht durch einen jener Römer, welche nach dem Geschmacke damaliger Zeiten die Altertümer und Kunsterzeugnisse ferner Länder mit ebensolchem Eifer sich aneigneten, wie die Sammler unserer Tage tun. Die Deutung jener Symbole überlasse ich geeigneteren Kräften; ich vermochte nur deren Bezeichnung zu geben, in der Hoffnung, daß uns recht bald eine Erklärung derselben erfreuen wird."

Bevor ich die vorstehenden Bemerkungen erörtere, führe ich der bibliographischen Vollständigkeit an S. Reinach, La représentation du galop dans l'art ancien et moderne, Paris 1901, p. 104, der bei Gelegenheit einer Besprechung chinesisch-europäischer Beziehungen die oben erwähnten Funde kurz streift und die Litteratur zitiert, aber offenbar ohne sie gelesen zu haben; denn er erklärt diese Gefäße für Porzellan und spricht von Pseudo-Entdeckungen, die nur leichtgläubige Liebhaber täuschen könnten. Aber es ist bei unseren Funden nur die Rede von rotem Ton, wenn auch Houben infolge Mißverständnisses den Ausdruck rotes Porzellan gebraucht; die Funde stehen daher durchaus nicht, wie Reinach annimmt, auf derselben Stufe mit den längst als Fälschungen erkannten, angeblich in ägyptischen Gräbern gefundenen chinesischen Porzellanfläschchen. Reinach erwähnt auch

unseren Stücken verwandte aus Frankreich in Argenton und zitiert dafür Congrès archéol. de France, Châteauroux 1874, p. 626. Dieses Werk ist mir leider nicht zugänglich.

Was die oben beschriebene Vase der Frau Mertens-Schaaffhausen betrifft, so habe ich versucht, dem Verbleib derselben in Bonn nachzuforschen, bisher ohne Erfolg; eine an das Bonner Provinzial-Museum gerichtete Anfrage ist unbeantwortet geblieben. Was sich nach der Abbildung und Beschreibung sagen läßt, vorausgesetzt daß die Vase keine Mystifikation ist, ist vor allem, daß die Form derselben echt chinesisch ist. Sie gehört zu dem Typus jener uralten, schon in der Schang-Dynastie auftretenden Sakralgefäße, die unter dem Namen „Ku" von den chinesischen Archäologen beschrieben werden. Bekannt sind solche Vasen aus Bronze, Nephrit und Porzellan. Abb. 2 a bis e zeigt die gebräuchlichsten Formen nach einheimischen Abbildungen unter Weglassung der Ornamente. Die beinahe lilienblattförmigen Zacken, wie sie d und e in Übereinstimmung mit Abb. 1 aufweisen, begegnen gerade in Verbindung mit diesem Gefäßtypus. Die Zoneneinteilung auf der rheinischen Vase ist ebenfalls chinesisch. Das American Museum of Natural History besitzt ein Ku von Bronze, in welchem die Anordnung der Ornamente in drei Zonen, und zwar 1. Zone mit Zacken, deren Spitzen nach oben gerichtet sind, 2. Mittelzone, 3. Zone mit Zacken, deren Spitzen nach unten gerichtet sind, in genau derselben Weise unserem rheinischen Funde entspricht. Die an den Seiten angebrachten Tigerköpfe sind wiederum eine chinesische Erfindung, wie sie ganz besonders häufig an den Bronze- und Tonvasen der Han-Dynastie sind, deren Zeit mit des römisch-chinesischen Handels zusammenfällt. Die Keramik dieser Epoche ist erst in den letzten Jahren durch Ausgrabungen spekulativer chinesischer Händler in der Provinz Schensi ans Licht gelangt. Ich habe in Hsi an fu eine Sammlung dieser Töpferei zusammengebracht, die sich jetzt in dem oben genannten Museum befindet und von mir demnächst eingehend behandelt werden wird. Tatsächlich herrscht in den keramischen Erzeugnissen der Han-Zeit der rote Ton vor, und zwar sowohl unglasiert als auch in grünen und braunen Glasuren. Ebenso sind Reliefornamente für dieselben durchaus charakteristisch. Ob das vorliegende Gefäß nun tatsächlich in China selbst fabriziert worden ist, bleibt einstweilen eine offene Frage und im Hinblick auf die fremdartigen nicht-chinesischen Figuren und Verzierungen sogar recht zweifelhaft. Da die Reliefs aufgelegt sind, könnte man zu der Erklärung die Zuflucht nehmen, daß dieselben teilweise spätere Zutaten sein mögen, wie denn z. B. in Sibirien chinesische Bronzespiegel gefunden worden sind, in die Eingeborene nachträglich Ornamente eingraviert haben. Oder es wäre auch, wie man etwa aus der indischen Bronzefigur schließen könnte, die Möglichkeit der Herstellung in Indien oder im vorderen Orient nach chinesischen Mustern vorhanden.

Bei den Houbenschen Gefäßen ist der chinesische Ursprung aus den Formen nicht so eklatant zu erschließen. Ein Urteil über ihre Ornamentik würde ich mir auf Grund der kleinen, ungenügenden Zeichnungen überhaupt nicht erlauben. Ich möchte meine Ansicht vorläufig etwa so formulieren, daß diese Gefäße an sich nichts Unchinesisches in ihrem Gesamtcharakter aufweisen, daß sie sehr wohl chinesisch sein können, aber es trotzdem nicht unbedingt sein müssen. Oktagonale Teller, wie die bei Houben abgebildeten, und auch sexagonale werden aus Ton und Porzellan in China fabriziert und dürften, wie fast alle gegenwärtigen Typen der Keramik, in ein hohes Altertum zurückgehen. Ebenso sind Gefäße in Form von Tieren und insbesondere von Vögeln zu allen Zeiten in Ton, Bronze und Nephrit in China gearbeitet worden, doch ist mir ein der Houbenschen und Koenenschen Vogelkanne genau entsprechender Typus bisher nicht bekannt geworden.

Die Lösung der Frage, ob wir es in diesen Gefäßen wirklich mit chinesischen Erzeugnissen oder mit der Keramik eines anderen Gebietes zu tun haben, hängt von einer gründlichen, an den Objekten selbst vorgenommenen Untersuchung ab, ohne welche ein endgültiges Urteil unmöglich ist. Besonders wichtig wäre es, durch Analyse festzustellen, ob und wie der Ton derselben von der Terra sigillata verschieden ist, wie Houben anmerkt, und ob derselbe etwa identisch ist mit dem Ton der Gefäße der Han-Dynastie. Es werden daher alle, welche über den Verbleib der Houbenschen Gefäße, der Kanne von Norf und der Vase der Frau Mertens-Schaaffhausen Auskunft geben können, im Interesse der Sache gebeten, einschlägige Mitteilungen an den Unterzeichneten zu richten.

Dr. B. Laufer,
American Museum of Natural History, New York C.

Zur Anthropologie der Mongolen.

Eine unter diesem Titel veröffentlichte Studie von Dr. F. Birkner (Archiv für Rassen- und Gesellschaftsbiologie 1904, Heft 6, S. 809) faßt das Rassenbild der Mongolen im wesentlichen wie folgt zusammen: Daß die Ostasiaten (Japaner, Chinesen und Koreaner) in körperlicher Hinsicht zusammengehörig sind, zeigt die einfachste Beobachtung, wenn die Unterschiede der Kleidung und besonders der Haartracht hinwegfallen. Chinesisch gekleidete Japaner sind von wirklichen Chinesen kaum zu unterscheiden und umgekehrt. Man darf daher annehmen, daß in den ostasiatischen Reichen im wesentlichen die gleichen Rassenelemente verbreitet sind. Doch sind innerhalb dieses „einheitlichen" Rassetypus gewisse Variationen vorauszusetzen, schon wegen der großen Ausdehnung des im Laufe der Zeiten von den Ostasiaten erworbenen Wohnsitzes. Außer älteren, zum Teil noch erhaltenen wenig bekannten Stämmen (Miautse, Mantse, Lolo im Südwesten von China, Ainu auf Jesso) sind im wesentlichen zwei Gruppen der Ostasiaten zu unterscheiden: eine malaio-mongolische und eine mandschu-koreanische. Letztere, der feinere Typus von Bälz, umfaßt die heute in China herrschenden Mandschu, viele Nordchinesen, die Mehrzahl der Koreaner und einen nicht sehr großen Teil der Bevölkerung Japans. Am reinsten und verbreitetsten ist er in der Gegend des Sugari, sowie an der mandschurisch-koreanischen Grenze, wo nach Bälz sein Ursprungszentrum zu suchen ist. Da er auch im Südwesten der japanischen Hauptinsel ziemlich reichlich vertreten ist, ist es wahrscheinlich, „daß er durch die kalte Polarströmung dahin kam". Der zweite, malaio-mongolische Typus, der grobe japanische Typus von Bälz, ist nicht leicht in die ihn zusammensetzenden Elemente aufzulösen. Seine Domäne ist China, während er in Korea nur im Südwesten reichlicher auftritt. Daß er auch nach Japan kam, ist nicht zu verwundern bei der großen Seetüchtigkeit der Südchinesen und Malaien, zumal die warme Äquatorialströmung, die über Formosa und die Liukiuinseln nach Japan geht, dies begünstigte. Das summarische Bild des groben (malaio-mongolen) Typus ist: gelbe Haut, straffes dunkles Haar von dickem, rundem Querschnitt, spärlicher Bartwuchs und dito Körperbehaarung, geringe Körpergröße, langer Rumpf, kurze Beine, meso- bis brachykephale Kopfform, abwärts spitz-ovales Gesicht, das durch Vorspringen der Wangenbeingegend flach und breit erscheint, an der Wurzel der Flügeln breite Nase mit relativ wenig vorspringendem Rücken und dito Spitze, vortretende Augen mit „Mongolenfalte" und mit schmaler, schief nach außen-oben gerichteter Lid-spalte. — Der nordchinesische, mandschu-koreanische Typus zeigt jenes mongolische Exterieur in abgeschwächter Form; es ist ein stattlicher, relativ aristokratischer Menschenschlag, mit 70 Proz. über 165 cm Körpergröße und absolut und verhältnismäßig größerem Kopf, schmälerem, langem Antlitz, weniger vorstehender Wange, abgerundetem Untergesicht, Mongolenauge, besser entwickelter, oft starker und hoher Nase. In Hautfarbe, Behaarung usw. bestehen sonst keine Unterschiede, auch sind blaue Kinderflecke regelmäßig. Offenbar haben wir es bei den Nordchinesen mit wirklichen Mongolen zu tun bzw. mit echten Vertretern der gelben Rasse, die nirgends über den Rahmen der auch für Volksstämme anderer Rassen geltenden Variationen hinaustreten. Das Verhältnis anderer Stämme, so der Malaien, zu den eigentlichen Mongolen ist noch näher festzustellen und dabei auf die anthropo-geographischen Einflüsse Rücksicht zu nehmen. Die bisherigen Untersuchungen genügen noch nicht, um das ganze Variationsbild der mongolischen Rasse auch nur annähernd zu ermessen. Es liegt aber schon einiges Material dazu, besonders aus Russisch-Asien vor, auf das der Verf., da er zunächst nur die Ostasiaten im engeren Sinne behandelt, noch nicht einging.

Mit der Bezeichnung der Mongolen als „gelbe" Rasse scheint es trotz aller Kontroversen, die darüber bestehen, doch seine Richtigkeit zu haben, wenigstens soweit Ostasien in Betracht kommt. Der japanische Anatom Adachi hielt diese Farbenbezeichnung anfangs für unzutreffend, zumal die gelbe Rasse selbst in der braunen Grundfarbe ihrer eigenen Haut keinen gelben Ton herausfindet. Als er aber einige Zeit in Europa gelebt und sich ihm dadurch offenbar die weiße Hautfarbe der Europäer fest eingeprägt hatte, erkannte er den Ton der Haut seiner Landsleute und sah sie so gelb, als ob sie an Ikterus leiden würden. Diese Beobachtung machen alle Japaner in Europa. Es kommt hier augenscheinlich viel auf Übung und Vergleichung an. Die nach Europa fahrenden Japaner finden, nach einer Beobachtung Adachis, auf der Reise in Hongkong, wo die Dampfschiffe kurze Zeit verweilen, viele Leute, mit denen sie immer Japanisch zu sprechen anfangen. Wenn sie nach Monaten oder Jahren zurückkommen, finden sie nicht mehr so viele Japaner in Hongkong. Der Grund liegt nach Adachi darin, daß jene vermeintlichen Japaner in Wirklichkeit meistens Portugiesen waren, und die heimkehrenden Japaner durch den steten Anblick der weißen Hautfarbe der Europäer besser für das Erkennen des gelben Tones der japanischen Haut geschult waren.

R. W.

Bücherschau.

Das Königreich Württemberg. Eine Beschreibung nach Kreisen, Oberämtern und Gemeinden. Herausgegeben vom Kgl. Statistischen Landesamt. 2. Band: Schwarzwaldkreis. VI und 683 Seiten. Mit Abbildungen und 1 Karte. Stuttgart, W. Kohlhammer, 1905.

Der erste Band des wichtigen Werkes ist bereits an diesem Orte besprochen worden. Der zweite ist natürlich ganz im gleichen Geiste gearbeitet, denn das Prinzip und die Autoren sind die gleichen geblieben, indem nur Dr. Schneider durch Dr. Mehring ersetzt und für die Trachtenbilder der Kunstmaler Lauxmann gewonnen wurde. Wie man weiß, hat man es nicht mit einem eigentlich geographischen Werke, wohl aber mit einem solchen zu tun, welches für die Geographie die wertvollsten Beiträge enthält und demjenigen, der sich dereinst einmal zur Abfassung einer schwäbischen Landeskunde im modernen Sinne berufen fühlt, große Dienste leisten kann. Die topographischen und ortsgeschichtlichen Abschnitte, höchst wichtig in ihrer Art, berühren uns hier nicht, um so mehr aber die Einleitungen, welche der Sonderbeschreibung eines jeden der politischen Bezirke vorangehen. Dieselben beginnen mit einer umfassenden Literaturübersicht, in der alle nur das fragliche Gebiet bezüglichen Schriften und Aufsätze, offenbar mit großer Ausführlichkeit, zusammengestellt sind, und daran reiht sich eine von sachkundiger Hand herrührende Charakteristik der physikalisch-geographischen Verhältnisse. Die geognostische Struktur des Landes wird durchgehends gründlich behandelt, wobei allerdings die für einen Württemberger sich von selbst verstehende Voraussetzung gemacht wird, daß man mit der Quenstedtschen Bezeichnung der jurassischen Ablagerungen ordentlich vertraut sei. Alle morphologischen Besonderheiten, Seen, Quellen, Höhlen- und Karbildungen, auffällige Bodengestalten, finden entsprechende Erwähnung. Auch die wichtigen Pflanzen fehlen nicht, und es wird der Einfluß, welchen die Vegetation auf die Gestaltung des Landschaftsbildes ausübt, allenthalben gewürdigt. Als auf ein typisches

036

玄奘朝圣图

erzinsung des Anlagekapitals je nach Kaufpreis und üte des Landes von 1½ bis 3 M. pro Zentner. Der urchschnittspreis für Weizen fiel während der letzten ahre niemals unter 4 M. pro Zentner und stand mehr-ch auf 5 und 5½ M., so daß der Landwirt in Wash-gton sehr wohl mit seinem Lose zufrieden sein kann. den Tälern der zahlreichen Flüsse, sowie überall dort, o künstliche Bewässerung möglich ist, treffen wir auf uttergräser und Kleeäcker, die eine ebenso rationell etriebene Viehzucht ermöglichen.

Diese hat sich in den letzten Jahren allerdings als eniger nutzbringend erwiesen, indem durch den Fleisch-rust die Preise für Schlachtvieh sehr gedrückt werden. nmerhin hat man von vornherein Wert darauf gelegt, ur gute Rassen zu züchten, und wir treffen auf den erschiedenen Farmen prachtvolles Milch- und Schlacht-ieh an. Für die Schafzucht scheinen augenblicklich ie Verhältnisse noch am günstigsten zu liegen, und an hat gleichfalls mit Angoraziegen gute Erfahrungen emacht. Da der Kongreß aber endlich entschlossen egen das Trust-Unwesen in den Vereinigten Staaten orgehen will, so darf man auch der Rindviehzucht ge-rost bessere Tage versprechen. Die Pferdezucht ist icht sehr bedeutend, doch waren die erzielten Resultate, oweit man sich überhaupt mit dem Heranziehen von assepferden befaßte, sehr zufriedenstellend.

Als drittes Arbeitsfeld des Landwirtes bleiben dann och die Obstkulturen zu erwähnen, die sich besonders m Yakima- sowie im Wenatchee- und Snake River-Tale zu großer Bedeutung entwickelt haben. Besonders ie beiden erstgenannten Täler sind sehr regenarm und roduzieren nur unter künstlicher Bewässerung. Wo iese aber eingeführt wurde, da stellte der Boden, was Fruchtbarkeit anbelangt, jeden anderen Teil des Staates

in den Schatten. Das Land kostet durchschnittlich 240 M. pro Hektar. Alle Obstsorten der gemäßigten Zone gedeihen hier in vorzüglicher Güte, besonders Pfirsiche und Äpfel, und zur besseren Ausnutzung sind die Obstgärten außerdem noch mit Futtergras und Klee bepflanzt, deren Heu sich vorteilhaft in den Bergbau-distrikten verkauft.

Man kann nach 4 Jahren pro Hektar auf eine Ernte von 350 sogenannten Kisten Äpfel rechnen, die an Ort und Stelle zum Preise von 4 M. pro Kiste glatten Ab-satz finden. Für besonders gute Sorten wird entsprechend mehr bezahlt. Eine Kiste Äpfel wiegt annähernd 20 kg.

In den letzten Jahren ist der Hopfenbau sehr in Aufnahme gekommen, und es macht fast den Eindruck, als ob Washington bestimmt sei, ein erstklassiges Bier-land zu werden; denn die hier gezogene Gerste soll sich ganz hervorragend für Brauzwecke eignen und wird in großen Mengen nach Europa und besonders nach Belgien exportiert.

Wenn wir nun noch erwähnen, daß das Klima des Staates demjenigen von Oberitalien ungefähr gleich-kommt, so darf man ihn getrost als das Land bezeichnen, das „von Milch und Honig fließt". Zieht man dann ferner die geographische Lage in Betracht, die Washing-ton mit seinem Riesenhafen Puget Sound zum natür-lichen Ausgangs- und Einfuhrtor von Amerika nach dem asiatischen Osten mit seinen Millionen Menschen macht, die in absehbarer Zeit bei ihrer ungeheuren Vermehrung auf amerikanisches Getreide und amerikanisches Fleisch angewiesen sein werden, so wird man zugeben, daß sich hier alle Faktoren vereinigen, die nach menschlichem Ermessen eine wirtschaftliche Entwickelung hervorrufen müssen, die bisher in der Geschichte der Welt nichts Ähnliches aufzuweisen hat.

Zum Bildnis des Pilgers Hsüan Tsang.

In Bd. 86, S. 386, dieser Zeitschrift habe ich eine japani-che Malerei besprochen, die den buddhistischen Reisenden Hsüan Tsang darstellt. Ich möchte darauf hinweisen, daß das ort erwähnte von Takakusu publizierte Porträt in vorzüg-ichem Farbenholzschnitt in Nr. 96 der japanischen Zeit-chrift Kokka reproduziert ist und hier gleichfalls einem un-bekannten chinesischen Maler des 13. Jahrhunderts zugeschrie-ben wird. Im zweiten Bande des Kataloges von T. Hayashi, Objets d'art et peintures de la Chine et du Japon, Paris 1903, p. 273, findet sich eine Heliogravüre, die ein Bild des Hsüan Tsang darstellt und in allen Punkten mit dem Bilde der Kokka übereinstimmt. Beide Wiederholungen gehen offenbar auf dasselbe Original zurück. Sonderbarerweise aber schreibt Hayashi dieses Bild, das 1900 in Paris ausgestellt war, dem japanischen Maler Kondara no Kawanari zu, der 853 starb. Ob er für diese Datierung besondere Gründe hat, weiß ich nicht, da er keine dafür anführt; jedenfalls ist es nichts Neues und Überraschendes, wenn von Japan aus ein Kunst-werk, das vermutlich dem 13. Jahrhundert angehören und eher jünger als älter sein mag, zum Besten der kauflustigen fremden Barbaren ein halbes Jahrtausend älter gemacht wird.

In der Nephritsammlung von Heber R. Bishop, die sich jetzt im Metropolitan Museum of Art, New York, befindet, ist eine kleine chinesische Nephritfigur (Nr. 440), die der Periode K'ang hsi (1662 bis 1722) zugeschrieben wird. Sie stellt einen buddhistischen Mönch dar, in sitzender Stellung, mit gekreuzten Beinen und gefalteten Händen, die auf den aufwärts gerichteten Fußsohlen ruhen; seine Lippen sind ge-öffnet, als wenn er die Lehre erklären wollte. In dem ge-druckten Verzeichnis dieser Sammlung (s. Metropolitan Mu-seum, Handbook Nr. 10, p. 43) wird diese Statue für den Priester Hsüan Tsang erklärt; auf welcher Grundlage diese Identifikation beruht, vermag ich jedoch nicht zu sagen.

Merkwürdig ist die in chinesischen Berichten über Tibet mitgeteilte Nachricht, daß sich im Tempel Jo-K'ang in Lhasa ein Freskogemälde befinde, das den Meister Hsüan Tsang der T'ang-Dynastie und drei seiner die heiligen Schriften suchen-

den Schüler darstellt (s. Rockhill in Journal of the Royal Asiatic Society 1891, p. 263, 282). L. A. Waddell, der jenen Tempel besucht und in seinem Buche „Lhasa and its My-steries", London 1905, beschrieben hat, bemerkt dort (p. 366), daß die drei Wände der inneren Veranda mit verfallenen Fresken bedeckt seien, deren hervorragendste acht Fuß hoch sei und den mongolischen Fürsten Gushi Khan vorstelle; dies scheine ihm das Bild zu sein, das die chinesischen Beamten in Lhasa irrtümlich für Hsüan Tsang gehalten hätten. Diese Auffassung ist mir durchaus nicht einleuchtend. Die Dar-stellung eines mongolischen Kaisers und eines chinesischen Mönches sind durch eine so weite Kluft getrennt, daß es keinem gebildeten Chinesen einfallen kann, sie miteinander zu verwechseln, ebensowenig als bei uns ein Schulknabe das Porträt eines deutschen Kaisers für ein Papstbild halten würde. Zudem habe ich durch langjährige Vergleichung chinesischer Berichte mit den Befunden der Wirklichkeit gelernt, zu den chinesischen Quellen ein mindestens ebenso großes Vertrauen zu haben als zu unseren eigenen, so daß ich die Richtigkeit von Waddells Vermutung von vornherein als ausgeschlossen betrachte. Im besonderen ist das Wei Tsang t'u chih, aus dem die obige Nachricht Rockhills geschöpft ist, nicht nur das beste chinesische Buch über Tibet, sondern überhaupt eines der vorzüglichsten und zuverlässig-sten Quellenwerke über diesen Gegenstand, dessen Treue ich im Laufe eines Jahrzehntes zu erproben genugsam Gelegen-heit hatte. Ma Shao-yün und Mei Hsi-sheng, dessen Ver-fasser, die 1791 schrieben, erwähnen das Bild des Hsüan Tsang im Tempel Jo-K'ang an zwei verschiedenen Stellen ihrer Schilderung, und sie sind nicht die einzigen Autoren, die dies berichten. Beim Durchsuchen der chinesischen Lite-ratur über Tibet finde ich dieselbe Nachricht bei Sheng Shêng-tsu, Verfasser einer undatierten Beschreibung von Tibet, die in der großen geographischen Enzyklopädie Hsiao fang hu chai yü ts'ung ch'ao abgedruckt ist, und in dem jüngsten und vollständigsten Werke über Tibet, Wei Ts'ang t'ung chi, verfaßt von einem Mandju namens Chien-hsi Ts'un-shê und veröffentlicht 1896 in acht Bänden. Die bezügliche Stelle findet sich in Buch 6, fol. 4. Dieser Autor entwirft eine selbständige detaillierte Schilderung des Jo-K'ang mit allen dort vorhandenen Inschriften und Kunstwerken und erwähnt ausdrücklich das Porträt von Hsüan Tsang und seinen Schülern. Es könnte ja sein, daß die Malerei inzwischen verschwunden ist, da Waddell angibt, daß die Fresken in verfallenem Zustande seien. Es scheint mir aber unzweifelhaft zu sein, daß das Bild einst vorhanden war. Die letzte und oberste

Instanz für die Entwickelung dieser Frage wären die tibe[tische] Klosterchronik von Jo-K'ang oder andere historisc[he] Quellen, die älter als das Jahr 1791 sind. Und ich bin übe[r]zeugt, daß einst aus solchen die Glaubwürdigkeit unser[er] chinesischen Berichterstatter wird bestätigt werden.

Berthold Laufer.

Bücherschau.

Die Beteiligung Deutschlands an der Internationalen Meeresforschung. 1. und 2. Jahresbericht, erstattet von dem Vorsitzenden der wissenschaftlichen Kommission Dr. W. Herwig, Wirkl. Geh. Oberregierungsrat. X u. 112 S. Mit zahlreichen Karten, Plänen und Bildern. Berlin, Otto Salle, 1905.

Die beiden in diesem stattlichen Hefte vereinigten Berichte beziehen sich auf die Jahre 1902 und 1903. Im Jahre 1899 fanden in Stockholm und 1900 in Christiania Beratungen über die Organisation einer zielbewußten Erforschung der germanischen Meere statt, zu denen der bekannte schwedische Ozeanograph Pettersson den ersten Anstoß gegeben hatte. Der Zweck, dem die Konferenzen dienen sollten, war in erster Linie ein biologisch-nationalökonomischer, indem es sich um „die Vorbereitung einer rationellen Bewirtschaftung des Meeres auf wissenschaftlicher Grundlage" handelte. Daß hierbei auch geographische Fragen gar häufig berührt werden müßten, verstand sich von selbst. Und lediglich unter diesem Gesichtspunkte kann natürlich an diesem Orte das an sich sehr reichhaltige Material besprochen werden.

Die vier Regierungen des Deutschen Reiches und der drei skandinavischen Staaten setzten einen „Zentralausschuß" mit dem Wohnorte Kopenhagen ein; Geheimrat Herwig wurde dessen Präsident und Prof. Pettersson Vizepräsident. Der Generalsekretär Dr. Hoek domiziliert in Kopenhagen, während das „Zentrallaboratorium" in Christiania der Leitung F. Nansens unterstellt ward. Die deutsche Abteilung erhielt für ihre Arbeiten den „Reichsforschungsdampfer Poseidon" zur Verfügung gestellt, dessen Vignette das Titelblatt des vorliegenden Bandes ziert. Der Kommission gehören als deutsche Mitglieder noch vier Professoren an, nämlich die Zoologen Brandt (Kiel), Heincke (Helgoland), Henking (Hannover) und der Geograph Krümmel (Kiel). In dieser letzten Stadt besteht schon seit längerer Zeit ein mit zwei Sektionen für Hydrographie und Biologie versehenes Laboratorium der k. preuß. Kommission zur Erforschung der deutschen Meere; dieses, sowie die Biologische Anstalt der Insel Helgoland und das Laboratorium des Deutschen Seefischereivereines in Hannover wurden den Bestrebungen der neuen Zentralinstanz dienstbar gemacht. Daß auch noch verschiedene jüngere Gelehrte für Spezialuntersuchungen herangezogen werden mußten, geht aus dem weitaussehenden Programm, das die Kommission aufstellte, als ganz natürlich hervor.

Einen namhaften Teil des Ganzen nehmen die auf die Lebensweise und Verteilung der Fische und der ihnen zur Nahrung dienenden Lebewesen bezüglichen Ermittelungen in Anspruch; obwohl auch unter dieser Fülle von Tatsachen gar manche sich befindet, der tiergeographische Bedeutung zukommt, dürfen wir uns unser Referat doch nicht auf die Wiedergabe von Einzelheiten einlassen. Unmittelbar gehen uns zunächst an diesem Orte an drei Berichte von Prof. Krümmel. Der erste von ihnen (S. 19 bis 83) schildert das Wesen der viermal im Jahre unternommenen „Terminfahrten", die dabei zur Verwendung gelangten Apparate und Methoden und die Einrichtung der Arbeitsstelle. Als einstweilige Ergebnisse jener Fahrten erhalten wir Isohalinen-Diagramme, die ersehen lassen, wie in der Ost- und Nordsee der Salzgehalt von Osten gegen Westen rasch zunimmt. Die Fahrt, die in die Zeit vom 3. bis 12. August 1903 fiel, wird noch einer besonderen Besprechung unterzogen, um den Lesern klar zu machen, wie sich der Tagesbetrieb im Verlaufe einer solchen Seereise gestaltete. Ein weiterer Krümmelscher Bericht (S. 61 bis 67) gibt darüber Aufschluß, wie zu Anfang des Jahres 1903 salziges und relativ kaltes Wasser weit östlich noch als aus der Beltsee stammend nachgewiesen werden konnte. Es drang sogar bis in die Danziger Bucht vor, aber in anderen Jahren läßt sich dieser Zustrom nicht wahrnehmen. Sehr interessant sind auch die Angaben, die Dr. Raben über die Chemie des Meerwassers macht. Es erhellt unter anderem aus ihnen, daß Stickstoffverbindungen in den nördlichen Meeren zwar auch nur eine ganz geringe quantitative Vertretung haben, immerhin aber in stärkerem Ausmaße auftreten, als dies nach Natterer, der die „Pola"-Fahrt mitgemacht hat, im Mittelländischen und Roten Meere der Fall

ist. Endlich verdienen noch die durch gute Abbildunge[n] erläuterten Beschreibungen der modernen Dredschprozedur[en] die Aufmerksamkeit eines jeden, der sich mit der Tiefse[e]fauna zu beschäftigen hat. Die Studien, die Dr. Reibisc[h] und Dr. Süßbach über die am Meeresgrunde lebenden O[r]ganismen begonnen und teilweise schon ziemlich weit gefüh[rt] haben, versprechen noch viele merkwürdige Aufschlüsse.

München.   S. Günther.

Neumanns Orts- und Verkehrslexikon des Deutsch[en] Reiches. Vierte, neubearbeitete und vermehrte Auflag[e] herausgeg. von Dr. M. Broesike und Direktor W. Ke[il] 2 Bände. Mit 2 Karten und 40 Stadtplänen. Leipzi[g] Bibliographisches Institut, 1905. 18,50 M.

Das geschätzte Nachlagebuch, das einer besonderen Em[p]fehlung nicht bedarf, liegt hiermit in einer neuen Aufla[ge] vor, die auf den jüngsten Zählungen und Informationen b[e]ruht. In nicht weniger als 75000 Artikeln wird über al[le] topographischen Objekte innerhalb des Deutschen Reiche[s,] darunter über alle Siedelungen von irgend welcher Bede[u]tung, knapp, aber doch erschöpfend das Wissenswerte üb[er] Lage, Verwaltungs- und Gerichtsverhältnisse, kirchliche, g[e]werbliche und landwirtschaftliche Verhältnisse, Geschichte us[w.] mitgeteilt. Bei den Orten, die eine Eisenbahn nicht habe[n,] ist die nächste Station angegeben und die Entfernung z[u] nannte. Die Übersichtskarte und die Verkehrskarte, sow[ie] die Stadtpläne sind alle neu bearbeitet und zeichnen sic[h] durch Klarheit und praktische Einrichtung aus.

Georg Wislicenus, Auf weiter Fahrt. Selbsterlebnisse z[ur] See und zu Lande. Deutsche Marine- und Kolonialbibli[o]thek. IV. Bd. XVIII und 318 Seiten, mit 19 Abbildunge[n.] Leipzig, Wilhelm Weicher, 1905. 3,60 M.

Der Band enthält neben den mehr feuilletonistische[n] Beiträgen auch solche mit einem geographischen Kern; [es] seien davon genannt: v. Liebert, Uhehe als deutsches Sied[e]lungsland; v. Schkopp, Aus Marokkos Vergangenheit un[d] Gegenwart; v. Morgen, Kamerun; Goedel, Japanisches; Finsc[h,] Kaiser-Wilhelms-Land; Heims, Die chinesische Mauer; Le[h]ner, Aus meinen Kameruner Briefen; Fritschi, Kreuz un[d] quer durch Peking; Kuhn, Ein Ritt ins Sandfeld von Sü[d]westafrika. Der Band bietet einem größeren Leserkreise vi[el] Interessantes.

Dr. Adolf Harpf, Morgen- und Abendland. Vergleichen[de] Kultur- und Rassenstudien. XV und 348 Seiten, Stuttga[rt] Strecker & Schröder, 1905. 5 M.

Eine Reise den Nil hinauf bis Omdurman gibt dem Ve[r]fasser Gelegenheit, nicht nur von seinen Beobachtungen un[d] Erlebnissen zu erzählen, sondern auch allerlei Fragen z[u] erörtern, über die die meisten Menschen sich aus Bequemlich[-]keit hinwegzusetzen pflegen, die aber dem Völker- und Kultu[r]psychologen als äußerst wichtig erscheinen müssen. Aben[d-] und Morgenland trennt eine so tiefe Kluft, daß eins da[s] andere in der Regel nicht versteht, trotz jahrtausendelang[er] Wechselbeziehungen. Wir wundern uns, daß uns die Orientale[n] nicht begreifen, wir nehmen ihnen das sogar sehr übel, w[eil] wir die alleinseligmachende, die beste Kultur und Moral i[n] Erbpacht zu haben glauben. Eins aber schickt sich nich[t] für alle. Das „kulturwidrige" Morgenland verfügt jedenfal[ls] über seine eigene, für seinen Bedarf zweckdienliche Kultu[r,] und seine Moralbegriffe können sich neben die unserige[n] wohl erst recht ganz gut sehen lassen. Es ist Überhebun[g,] daran zu zweifeln. Dieser Gedanke durchzieht die Harp[f]schen Betrachtungen, es sind ihm eine Reihe sehr lese[ns]werter Kapitel gewidmet. Auch die historische Seite de[r] Kultur, die kulturgeschichtliche Entwickelung und die Kultu[r]ziele sowie Rassefragen werden anregend behandelt.

E. Neuweiler, Die prähistorischen Pflanzenreste Mi[t]teleuropas mit besonderer Berücksichtigung der schwe[i]zerischen Funde. Zürich, Albert Raustein, 1905. 2,40 M[.]

Seit 1866 O. Heer seine Flora der Pfahlbauten herau[s]gab, ist von botanischer Seite keine zusammenfassende Arbe[it]

037

唐代的耶稣受难像

Hafen von Duluth und Superior City in unmittelbarer Nähe des Wassers liegen, und in das feuchte Element fördert, wo sie bis zur endgültigen Bestimmung schwimmen. Da aber die Küstengewässer des Oberen Sees für mindestens 4 Monate fest zufrieren, so muß für diese Zeit die Sägerei ruhen.

Das Zersägen geschieht in den Sägemühlen (Sawmills), die so viel als möglich mit selbsttätigen Maschinen versehen sind und daher während der Saison eine gewaltige Leistungsfähigkeit entfalten. Der Vorgang selbst ist lehrreich und höchst spannend zugleich. Eine solche Sawmill besteht aus einem ausgedehnten Holzplateau, das an einer Seite an das Wasser grenzt. Hier schwimmen, dicht aneinander gedrängt, die Baumstämme. Auf einem Holzgerüst vor der Mühle stehen mehrere Männer, welche die Logs mit ihrem Ende zu einer schräg in die Höhe führenden Gleitbahn leiten, wo sie in das Bereich der Dampfkraft kommen und, von dieser gezogen, aufwärts zur Plattform der Sägerei marschieren. Hier angelangt, werden sie nach der Dicke und Güte sortiert und den verschiedenen Sägeanstalten zugewiesen. Das geht so schnell, daß man den Vorgang eben mit den Augen verfolgen kann. Dabei sieht es ungemein drollig aus, wenn auf einmal, veranlaßt durch einen Arbeiter, aus der Tiefe ein plumper eiserner Klotz auftaucht und dem gerade heraufgehenden Stamme, der vielleicht 1 m im Durchschnitt hält, in seine richtige Lage klopft oder ihm ein paar tüchtige Rippenstöße versetzt, um ihn auf eine andere Gleitbahn zu bringen. Von den Sägen, deren jede ständig hin und her gehen und deren jede etwa von drei Mann bedient wird, werden die Stämme zuerst zu vierseitigen Balken zurecht geschnitten, wobei das Wenden sehr rasch und exakt vor sich geht, dann in Bohlen und Bretter von verschiedener Dicke zerlegt. Diese verlassen die Sägestelle zunächst in gerader Richtung vorwärts, fallen dabei auf ihre Breitseite und werden sortiert. Jede Gattung verfolgt ihren eigenen Weg. Die tadellosen brechen im rechten Winkel um und laufen aus der Plattform heraus auf dort bereit stehende Wagen. Die gewöhnlichen Sorten werden weggeschafft und zum Trocknen im Freien aufgestapelt, wo sie bis zu ihrer Verwendung liegen bleiben. Die besseren Sorten dagegen werden erst dann aufgeschichtet, wenn sie in einem besonderen Warmhause mittels künstlicher Hitze getrocknet worden sind. Geschähe dies nicht, so würden sie beim Lagern an den Außenseiten schwarz werden und an Verkaufswert erheblich einbüßen. Die Rindenteile, die zuerst von den Stämmen abgesägt werden, sowie die aus irgend einem Grunde aussortierten Bohlen und Bretter gehen geraden Weges ein Stück weiter, bis sie zu Stellen gelangen, wo sie zu Latten, Leisten und Schindeln zerschnitten werden. Was dazu nicht verwendet werden kann, sowie aller sonstiger Abfall marschiert langsam auf einer langsam aufsteigenden Bahn zu einem massiven, mit einer durchlochten Metallhaube versehenen Turme, wo alles verbrannt wird. Tag und Nacht glüht hier ein gewaltiges Feuer, das enorme Massen Holz zu Asche verwandelt, weil man in den Sägewerken damit nichts anfangen kann. Diese Massen aufzuspeichern, würde zu viel Raum und Arbeitslohn erfordern. Von den Abfällen der Sägemühlen werden nur die Sägespäne benutzt, und zwar zum Heizen der Dampfkessel. Früher verbrannte man auch die Abfälle nicht, sondern breitete sie auf dem tiefliegenden Lande aus oder warf sie in das seichte Wasser und die sumpfigen Stellen, um diese aufzuhöhen und dadurch brauchbaren Lagerboden zu gewinnen. Tatsächlich gibt es im Mündungsgebiete des St. Louis River ausgedehnte Strecken solchen Holzbodens.

Bei der enormen Leistungsfähigkeit der Sägereien, die jährlich Millionen von Stämmen zerschneiden, sind entsprechende Lagerflächen erforderlich, von denen aus die Fabrikate entweder unmittelbar an Verbraucher und Händler verkauft oder zur Ausfuhr gebracht werden. Letzteres geschieht in eigens für diesen Zweck gebauten Schiffen, bei denen ähnlich wie bei den sogenannten Bockschiffen auf unseren deutschen Flüssen der Laderaum in einem einzigen Zusammenhange durch die ganzen Fahrzeuge geht, während die Maschine, die Kabinen usw. an den Enden derselben angebracht sind. Die Gesamtproduktion der Sägereien im Distrikte von Duluth belief sich im Jahre 1902 auf 960,76 Millionen laufende Fuß; in Brettern, gedacht zu 20 Fuß Länge, gibt dies rund 48 Millionen Stück. Einzelne Sägereien verarbeiten täglich bis zu 5000 Stück Stämme. (Schluß folgt.)

Ein angebliches chinesisches Christusbild aus der T'ang-Zeit.

Mit 3 Abbildungen.

In seinem kürzlich veröffentlichten Buche „An Introduction to the History of Chinese Pictorial Art" bemerkt Herbert A. Giles, Professor des Chinesischen an der Universität Cambridge (England), am Ende des Vorworts, daß in seinen Illustrationen ein noch unbekanntes Bild von Christus enthalten sei, das seine Inspiration wahrscheinlich von nestorianischen Priestern empfangen habe und vom 7. Jahrhundert an in Holzschnittreproduktionen überliefert worden sei. Dieses sensationelle Bild findet sich auf S. 37 des Buches und wird auf S. 40 folgendermaßen erklärt: „Der sehr merkwürdige Holzschnitt, betitelt »Drei in Eins«, besteht aus einem Bilde von Christus und einem nestorianischen Priester, der zu seinen Füßen kniet und eine Hand zum Segen emporhält, während ein anderer Priester hinter ihm steht. Das nestorianische Christentum verschwand bald aus China und hinterließ die berühmte Steintafel in Hsi an fu als Zeugen dafür zurück, daß es den fernen Osten erreicht hatte — eine Ehre, die in Zukunft von diesem anspruchsvollen Bilde geteilt werden muß, das einen weiteren Beitrag zu den frühen Porträts von Christus bildet. Drei chinesische Schriftzeichen zur Linken bedeuten »Darf nicht gerieben werden« = heilig und wurden wahrscheinlich auf Veranlassung der nestorianischen Priester eingeschaltet."

Die Quelle, aus der dieses „Christusbild" stammt, gibt Herr Giles nicht an, obwohl der Gegenstand wichtig genug wäre, daß die Quelle einer eingehenden Erörterung unterzogen wird, die naturgemäß jeder Leser erwarten würde. Und erwartet Herr Giles von seinen Lesern, daß sie seine durch nichts gestützte, durch nichts bewiesene, einfach aus der Luft gegriffene Erklärung dieses Bildes ohne weiteres hinnehmen sollen?

Auf den ersten Blick erkannte ich, daß das Bild einem Holzschnitt des wohlbekannten chinesischen Buches Fang shih mo pu entnommen ist, und daß es nichts anderes darstellt als die Dreiheit Confucius, Laotse und Buddha, ein in der chinesischen und japanischen Kunst häufiges Motiv, das jeder noch so ungebildete Chinese oder Japaner unzweifelhaft erkennen würde. Das Fang

37

shih mo pu, das ist das Tuschen-Buch des Herrn Fang, wurde 1588 herausgegeben[1]). Der Verfasser war ein berühmter Fabrikant von Tuschstücken, auf denen vermittelst gravierter Holzblöcke künstlerische Darstellungen, wie Figuren, Landschaften, Blumenstücke oder Ornamente aufgepreßt wurden. In diesem Werke publizierte er eine Sammlung der auf seinen Tuschen vorkommenden Muster zum Besten seiner kauflustigen Abnehmer. Das Buch ist nicht sehr häufig, doch habe ich in Peking mit

Giles' chinesisch-englischem Wörterbuche, einen Bele[g] gibt, bezieht sich nur auf die Tusche und bedeutet, da[ß] man dieses Tuschstück nicht reiben, das heißt die Tusch[e] nicht zum Schreiben benutzen soll, wegen der verehrung[s]würdigen Personen auf der Darstellung; ein solches Tusch[e]stück sollte lediglich als Kunstwerk aufbewahrt werde[n]. Wer in aller Welt ist so mit Blindheit geschlagen, daß e[r] in der Figur links auf Abb. 1, die Giles für Christus häl[t], nicht den typischen indischen Buddha erkennen würde[?]

Abb. 1. **Chinesischer Holzschnitt aus dem Fang shih mo pu, darstellend Buddha mit Laotse und Confucius.**
Nach der Interpretation von Giles: Christus und zwei nestorianische Priester.

einer großen Anzahl anderer illustrierten Seltenheiten ein Exemplar desselben erworben. Hier finden wir denn auf S. 2 des 3. Buches die genaue Vorlage zu dem Gilesschen „Christus", die ich in Abb. 1 nach einer Photographie reproduziere, um jedermann in die Lage zu versetzen, sich ein eigenes Urteil in der Sache zu bilden. Abb. 2 zeigt die Inschrift, die sich auf der Rückseite des Tuschstückes befand und zu lesen ist: han san wei, das heißt „(das Bild) umfaßt drei Personen, die eine Einheit bilden", was nur so zu verstehen ist, daß das Bild die Begründer und Vertreter der drei Hauptreligionen Chinas darstellt, mit der Idee, die enge einheitliche Verbindung der drei Glaubensformen in der Vorstellung und Praxis des Volkes symbolisch zum Ausdruck zu bringen. Der Ausdruck „pu k'o mo", den Giles durch „heilig" erklärt, wofür es in der ganzen chinesischen Literatur, nicht einmal in

a.   Abb. 2.   b.
Inschrift auf der Rückseite des Tuschstücks, auf dessen Vorderseite Abb. 1 eingraviert war.
Nat. Gr. Reproduziert nach dem Fang shih mo pu. a ist die Inschrift in altem, ornamentem Stil, b die entsprechende moderne Schreibweise derselben Charaktere, vom Verf. hinzugefügt.

Wer sähe nicht die Tonsür des indische[n] Mönchs auf dem Scheitel (und hat es je[,] solange die Welt steht, einen Christu[s] mit Tonsur gegeben?), wer sähe nich[t] die spiralischen Stirn- und Bartlocken[,] wer nicht das indische Mönchsgewand[,] alles für den Buddha charakteristisch[e] Dinge? Der „vor Christus kniend[e] nestorianische Priester" des Herrn Gile[s] ist Laotse, der überhaupt nicht kniet[,] sondern aufrecht dasteht, und der „an[-] dere Priester" hinter ihm ist Confucius[.]

Abb. 3 gibt eine Variante desselbe[n] Sujets wieder. Sie ist nach der Photo[-]graphie eines Holzschnitts im Ch'ëng[-] shih mo yüan hergestellt, das gleich[-] falls eine Sammlung von Gravierunge[n] auf Tuschstücken enthält. Das Werk[e] muß kurz nach 1605 erschienen sein[,] wie sich aus den Datierungen der ver[-]schiedenen Vorreden ergibt, die zwische[n] 1594 und 1605 liegen. Es ist typogra[-]phisch das hervorragendste Werk de[r] chinesischen Buchdrucker- und Holz[-]schneidekunst. Unsere Abbildung be[-]findet sich dort in Buch IIIb, S. 11[.]

[1]) S. A. Wylie, Notes on Chinese Literature. Zweite Ausgabe, S. 146.

e Stellung der drei Religionsvertreter ist auf diesem ilde geändert. Buddha nimmt die Mitte ein, und seine onsur ist hier noch deutlicher sichtbar als auf dem origen Bilde. Rechts von ihm (links in der Abbildung) eht Confucius, und links von Buddha Laotse. Im vierten efte der japanischen Zeitschrift „Kokka" findet man einen orzüglichen Farbenholzschnitt, dasselbe Motiv darstellend, ach einer Malerei von Masanobu Kanô (1453 bis 1490).

Die Vorlage, die dem „Christus" von Giles zugrunde egt, ist nicht älter als die Mitte des 16. Jahrhunderts. Vie nun Giles dazu kommt, das Bild an das Ende des . Jahrhunderts hinaufzuschieben und noch vermutungs-

Ich habe es für meine Pflicht erachtet, diese Tatsache klarzustellen, um zu verhindern, daß sich diese angebliche große Entdeckung, die ein Londoner Verleger auf besonders gedruckten Reklamekarten in alle Welt ausposaunt hat, in die Handbücher der Kunstgeschichte einschleiche und etwa ein „Gemeingut der Wissenschaft" werde. In seinem Vorwort bemerkt Professor Giles, daß Professor Hirth von der Columbia-Universität vor einigen Jahren ein Buch in deutscher Sprache plante, das seinen Band überflüssig gemacht hätte. Es gewährt mir großes Vergnügen, an dieser Stelle anzukündigen, daß Professor Hirths große Arbeit über die chinesische

Abb. 3. **Chinesischer Holzschnitt aus dem Ch'êng shih mo yüan, Buddha, umgeben von Confucius und Laotse, darstellend.**
$\frac{1}{7}$ größer als das Original.

weise mit dem Namen eines Malers Yen der T'ang-Zeit zu verknüpfen, ist mir und wohl auch jedem anderen ein vollkommenes Geheimnis.

Malerei schon längst abgeschlossen und gegenwärtig im Druck ist.

New York. B. Laufer.

Neues über den Urmenschen von Krapina.
Von Dr. Ludwig Wilser.

Der Entdecker dieses unsere Kenntnisse von den ersten Anfängen und der frühesten Verbreitung des Menschengeschlechts wesentlich bereichernden Lebewesens, Prof. Gorjanovic-Kramberger in Agram, hat seinen verschiedenen schriftlichen (Mitteilungen der Anthropolog. Gesellsch. in Wien, Bd. 31, 2, Bd. 32, 3/4, Bd. 34, 4/5) und mündlichen Berichten (Naturforscherversammlung in Kassel 1903, Wanderversammlung der Wiener Anthropologischen Gesellschaft in Agram 1904 und Vereinigung der deutschen und österreichischen Anthropologen in

Salzburg 1905) im neuesten Doppelheft der Wiener Mitteilungen (Bd. 35, 4/5) als vierten Teil einen dritten Nachtrag mit 3 Tafeln und 13 Abbildungen im Text folgen lassen. Da derselbe allerlei bemerkenswerte, die ursprüngliche Auffassung des Verfassers in mancher Hinsicht berichtigende Einzelheiten enthält, möchte ich mir, mit Bezugnahme auf meine früheren Bemerkungen (Bd. 82, S. 147 und Bd. 86, S. 399) erlauben, die Leser des Globus davon in Kenntnis zu setzen.

Im August 1903 wurden die Untersuchungen der

37*

038

中国的鼻环

通 報

T'oung pao

ARCHIVES

POUR SERVIR À

L'ÉTUDE DE L'HISTOIRE, DES LANGUES DE LA GÉOGRAPHIE ET DE L'ETHNOGRAPHIE DE L'ASIE ORIENTALE

(CHINE, JAPON, CORÉE, INDO-CHINE, ASIE CENTRALE et MALAISIE).

RÉDIGÉES PAR MM.

HENRI CORDIER

Professeur à l'Ecole spéciale des Langues orientales vivantes et à l'Ecole libre des Sciences politiques à Paris

ET

ÉDOUARD CHAVANNES

Membre de l'Institut, Professeur au Collège de France.

Série II. Vol. VI.

LIBRAIRIE ET IMPRIMERIE
CI-DEVANT
E. J. BRILL.
LEIDE — 1905.

ANNEAUX NASAUX EN CHINE

PAR

B. LAUFER.

———✳———

Dans deux notes insérées au troisième volume du *T'oung Pao*,
p. 209 et 319, le Prof. Schlegel, d'après une source japonaise, a
traité de l'usage de porter des anneaux nasaux parmi les Ainou
d'une des îles Kouriliennes, et d'après les études de L. v. Schrenck,
aussi parmi des tribus tongouses de la région de l'Amour. Dans la
première de ces notes, le Prof. Schlegel soutient que cette coutume
est «totalement inconnue dans le domaine japonais et chinois».
Or, cette assertion ainsi généralisée est inexacte. J'ai eu l'occasion
d'observer cet usage en vigueur aux environs de Shanghai. Me
rendant au Collège de Nanyang situé à l'ouest de la ville, en
avril de 1904, j'ai remarqué que beaucoup d'ouvriers des villages
que j'avais à traverser portaient des boucles d'oreille et un anneau
nasal. Ces anneaux sont ouverts et faits d'un fil d'argent très fin
tordu et de 1,5—1,7 cm. de diamètre. Le fil s'atténue graduellement
vers les deux bouts qui sont enfoncés dans la cloison du nez laquelle pour ce but a été percée avec une épingle importée. Ces
anneaux sont fabriqués par tous les orfèvres de Shanghai pour les
laboureurs de la partie septentrionale de la province de Kiangsou,
à laquelle, d'après mes recherches, cette coutume est restreinte en
Chine. Elle n'existe dans aucune autre partie du pays (à l'exception

peut-être de Canton, v. infra) et est inconnue aux Chinois du nord que j'ai interrogés à ce sujet.

Il est remarquable que dans cette région de la Chine les anneaux nasaux ne sont portés que par les hommes, tandis que sur l'Amour ce sont à présent les femmes qui les portent exclusivement.

Les Chinois près de Shanghai portant de tels anneaux me donnèrent les informations suivantes: Ils sont portés dans le but de rendre un homme robuste et bien portant. La même vertu est attribuée aux boucles d'oreille. Les uns et les autres protègent contre une mort prématurée. Quiconque les porte ne peut plus les ôter sans exposer sa santé ou même sa vie; et le corps est enseveli avec ces ornements. L'idée, assurent-ils, a été inspirée par l'anneau nasal du bœuf ou du buffle, afin de transmettre à l'homme la force et la vigueur de cet animal. J'ai fouillé les encyclopédies chinoises, mais je n'ai rien trouvé sur ce sujet. D'après un mot du Dictionnaire français-cantonnais du Père L. Aubazac, Hongkong 1902, qui, p. 14, cite «anneau pour le nez» 鼻圈 *pi hun*, on pourrait supposer que ces anneaux sont communs à ou près de Canton. La même expression se trouve dans «An English-Chinese Vocabulary of the Shanghai Dialect», Shanghai 1901, p. 404, avec la romanisation *bih-choen*.

Le fait que la coutume tongouse frappa les Chinois du nord, est prouvé par la citation suivante prise du livre 皇清職貢圖 qui représente et décrit les peuples de l'Amour au commencement de son troisième ch'üan. Dans la description des 恰喀拉 Shih-k'o-la, tribu tongouse de la Mandchourie dispersée autour de 渾春 Hun-ch'un (v. Playfair, Cities no. 2513) il est dit: 男女俱於 鼻傍穿環綴寸許 «Tous les hommes et toutes les femmes percent la cloison du nez et passent un anneau de plus d'un pouce» [1]).

1) Schrenck, Reisen und Forschungen etc., vol. III, 2ième partie, p. 417, dit que cet ornement est limité au sexe féminin, mais il observe que Maack l'a vu aussi sur un

La langue des Goldes et d'autres dialectes tongouses possèdent un mot spécial pour cet anneau, à savoir sandaxá, que M. W. Grube, dans son Goldisch-Deutsches Wörterbuch, p. 89, compare au mot mandchou songgiha 'cheville nasale des chameaux' et au mot niutchi shuang-kih 'nez' (comparez aussi Sakharov, Dictionnaire mandchou-russe, p. 624a). Margaritov, dans son traité Оъ Орочахъ Умператорской Гаваыи, St. Pét. 1888, p. 11, constate l'existence de l'anneau nasal chez les femmes des Orotches de Port Impérial et donne le dessin d'un exemplaire dans la planche V, k. Cet anneau est tordu en forme de spirale, avec un crochet qui entre dans le nez. Aux Ghilyak et aux Ainou de Saghalin et de Yezo, l'anneau nasal est inconnu.

---

individu mâle. Moi aussi, je ne l'ai trouvé que chez les femmes des Goldes, mais on pourrait supposer de la relation chinoise ci-dessus ·qu'au 18ième siècle l'usage s'était étendu aux hommes aussi.

039

《西藏的寺庙》导言

Das
Kloster Kumbum
in Tibet

Ein Beitrag zu seiner Geschichte

Von

Wilhelm Filchner
Leutnant im k. b. 1. Infanterie-Regiment König
kommandiert nach Berlin,
korrespondierendem Mitglied der K. K. Geographischen Gesellschaft in Wien

Mit 39 Tafeln, 3 Karten und Abbildungen im Text

Berlin 1906
Ernst Siegfried Mittler und Sohn
Königliche Hofbuchhandlung
Kochstraße 68—71

In dankbarem Gedenken an die Vertretung des Antrags, mich zur Bearbeitung der wissenschaftlichen Ergebnisse meiner Expedition China—Tibet nach Berlin zu kommandieren, erlaube ich mir,

dem k. bayerischen Militärbevollmächtigten

Herrn Oberst Freiherrn von Gebsattel

den ersten Band meiner Veröffentlichungen

verehrungsvoll zu widmen

Leutnant Wilhelm Filchner

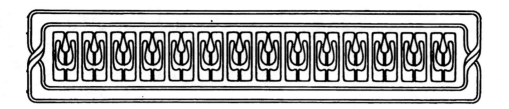

Zur Einführung.

— ⸺

Der liebenswürdigen Aufforderung des Herrn Leutnant Wilhelm Filchner nachzukommen und seinen Beitrag zur Geschichte des Klosters Kumbum mit einigen Geleitsworten auf seine Wanderung in die Welt auszusenden, gewährt mir großes Vergnügen. Wird doch hier zum ersten Male eines der fesselndsten Kapitel aus der Geschichte des Lamaismus im Rahmen einer Monographie behandelt, die sich auf Selbsterlebtem und Selbsterlauschtem aufbaut und mit dem Feuer der Jugend und der Bravour und Schneidigkeit des deutschen Offiziers den letzten Schleier fortreißt, der bisher über den Geheimnissen des Klosterlebens von Kumbum gelagert! Filchners Buch ist nicht nur die gründlichste und vollständigste Beschreibung dieses Gebietes, die wir jetzt haben, und die stets ihren wissenschaftlichen Wert behaupten wird, sondern überhaupt die umfassendste Schilderung eines lamaischen Gemeinwesens mit seinem vielseitigen Leben und Treiben, mit seinen Tempeln, Denkmälern und Sagen, die in unserer Literatur vorhanden ist. Die Topographie des Klosters wird uns mit sorgfältiger Gründlichkeit vor Augen geführt und ein Bild des Lebens seiner Bevölkerung in all ihren bunten Farben treffend und mit guter Beobachtungsgabe, vereint mit gesundem Humor, entrollt. Die Untersuchungen über den berühmten heiligen Baum von Kumbum machen dem Wissens- und Forschensdrang des Verfassers alle Ehre und müssen als abschließend betrachtet werden.

Ich möchte mir an dieser Stelle den Hinweis erlauben, daß dieser Baum des Tsongkapa mehr als eine auf Erwerbszwecke abgesehene, lediglich künstliche Machenschaft der Lamas ist, worin erst eine sekundäre oder tertiäre Entwicklung der demselben zugrunde liegenden Idee gesucht werden muß, vielmehr das Fortleben des uralten Gedankens der Baumverehrung darstellt, die wir in ganz Ostasien ausgeprägt finden, in Indien sowohl als in China. Das Gefühl der Hochachtung und Bewunderung vor alten, majestätischen Bäumen wirkt auch in

uns fort, und der Mensch in den Anfängen des religiösen Empfindens be-
gegnete dem stattlichen Baume mit heiliger Scheu und erblickte in ihm den
Sitz einer Gottheit. In historischen Zeiten wurden besonders schöne und ver-
ehrte Bäume dem Andenken großer Persönlichkeiten geweiht, zu ihrem Leben
in Beziehung gesetzt und mit geschichtlichen Erinnerungen und sagenhaften
Überlieferungen umwoben. Dies ist das sekundäre Moment. Weit später tritt
dann erst das letzte Stadium der Entwicklung ein — die zunehmende soziale
Bedeutung dieser heiligen Bäume, wie dies bei Kultstätten überhaupt der
Fall ist, ihre große Wirkung auf eine sich mehr und mehr verbreiternde Volks-
masse, die gläubig zu ihnen hinpilgert, die ihnen dann von Priestern zugeschriebenen Wunderwirkungen, die daran anknüpfende Reklame und Kapita-
lisierungsfähigkeit des Unternehmens. Eine solche Entwicklung ist eine
allgemein menschliche Erscheinung und überall unter ähnlichen gegebenen Verhältnissen möglich und tatsächlich eingetreten; sie ist durchaus nicht spezifisch
tibetisch oder lamaisch. Die Wand meines Arbeitszimmers schmückt ein einfaches gepreßtes Efeublatt, auf dem in goldfarbigen Lettern die Verse „Alt-
Heidelberg, du feine!" aufgedruckt sind. Es ist ein käufliches Andenken an
Heidelberg und stammt angeblich von dem Efeu des Schlosses. Was anders
ist es, wenn die Lamas von Kumbum mit Bildern oder Sprüchen bedruckte
Blätter vom Baume des Tsongkapa an die frommen Pilger verkaufen? Was
anders, wenn an unseren großen öffentlichen Denkmälern Photographien,
Albums oder Ansichtskarten mit Darstellungen derselben feilgeboten werden
und stets auf das schaulustige Publikum ihre ungeschwächte Anziehungskraft
ausüben? Haben wir nicht sogar die Naturerscheinungen zum Teil mit Be-
schlag belegt und ihren Genuß zinspflichtig gemacht? Wer den Rheinfall von
Schaffhausen von der vorteilhaftesten Seite aus beschauen will, wird sich wohl
oder übel dazu verstehen müssen, das dort ein für allemal geforderte Eintritts-
geld zu erlegen. Ich bin weit davon entfernt, den Schein eines Tadels auf
diese gastliche Stätte zu werfen, sondern führe sie nur als ein Beispiel dafür
an, daß in unserem Kulturleben sowohl, wie in dem fremder Nationen, verwandte und analoge Erscheinungen herrschen, deren angebliche Schattenseiten
wir nur zu gern hervorkehren, ohne uns der Gleichförmigkeit der menschlichen
Natur und der gleichartigen Entwicklung psychischer und geschichtlicher Vorgänge bewußt zu bleiben. Wir haben kaum Grund, uns über die Lamas zu
Gericht zu setzen wegen ihres geschäftlichen Vertriebs der heiligen Blätter,
nicht etwa, weil wir nicht besser wären, sondern weil wir in diesem Vorgang einen jenseits von Gut und Böse liegenden, von persönlichem Willen
unabhängigen Endprozeß einer langen religiös-sozial-ökonomischen Entwicklung erkennen, die sich überall abgespielt hat und abspielen kann.

Die richtige religionsgeschichtliche Erklärung für den heiligen Baum von Kumbum ist bereits von Albert Grünwedel*) gegeben worden. Tonpasten mit Darstellungen des Gautama Buddha aus Pagan sind mit solchen, die zu Buddhagayā in Indien gefunden worden sind, nahezu gleich. Gautama ist die Mittelfigur in diesen Darstellungen; auf dem Hintergrund ist ein Tempel, der Tempel von Gayā, sichtbar, hinter welchem der Bodhibaum durch ein paar Zweige angedeutet ist. Diese Pasten haben die Form eines Feigenblattes und sind so den als Andenken oder Reliquien aufbewahrten Originalblättern nachgebildet. Die rituelle Weiterbildung dieser Idee, bemerkt Grünwedel, finden wir nun in Tibet in dem heiligen Baume vom Kloster Kumbum vor, welcher das Bild des Tsongkapa oder wenigstens tibetische Inschriften auf seinen Blättern trägt. Solche Tonpasten mit Buddhabildern in Form eines Blattes kommen auch in China vor, und ich habe solche im Boden gefunden in der Nähe von Wu t'a sze, des Fünf-Pagoden-Tempels, nordwestlich von Peking.

Eine andere hochbedeutende Frage, die durch das Buch Filchners aufs neue angeregt wird, ist die nach dem Ursprung des Lamaismus. Hat sich der Lamaismus mit den eigentümlichen Formen seines Kultus und seiner streng geregelten Hierarchie selbständig aus dem indischen Buddhismus entwickelt, oder hat sich diese Entwicklung unter dem Einfluß fremder religiöser Anschauungen und in Anlehnung an einen anderen Kultus oder Kulte vollzogen? Dieses Problem steht an Bedeutung und Tragweite nicht hinter dem der Abhängigkeit oder Selbständigkeit der amerikanischen Kulturen zurück, aber noch keine wissenschaftliche Untersuchung hat bisher zu dem Versuche einer Lösung den Weg geebnet. Wir sind noch nicht über recht vage Vermutungen und bloße Anschauungen hinausgekommen, die mehr im Gefühl als in der Logik der Tatsachen wurzeln. Auf allgemeine Ähnlichkeiten in den Kulthandlungen der Lamas und ihren geistlichen Rangstufen mit denen der katholischen Kirche ist schon seit Jahrhunderten hingewiesen worden, und die ersten katholischen Missionare in Tibet konnten sich die lamaische Religion nicht besser erklären als durch die Annahme, daß sie die bewußte Schöpfung einer satanischen Travestie sei. Solche Übereinstimmungen sind in der Tat nicht wegzuleugnen; nimmt man sie aber unter die Lupe einer scharfen Analyse, so erkennt man in vielen Fällen, daß es sich um scheinbar oder wirklich gleichartige Endergebnisse handelt, die aus durchaus heterogenen Entwicklungsquellen stammen, aus verschiedenen Assoziationen von Ideen resultieren. Andere gemeinsame Erscheinungen, wie z. B. das Falten der Hände zum Gebet, finden ihre Erklärung

---

*) Buddhistische Studien, Veröffentlichungen aus dem Königlichen Museum für Völkerkunde, Band V, Berlin 1897, Seite 126 bis 127.

in der Tatsache, daß sie bereits früh vom Christentum aus dem Buddhismus
geschöpft worden sind. Bei der Vergleichung der lamaischen und katholischen
Hierarchie springt die psychologische Verwandtschaft am meisten in die Augen;
aber gerade hier muß man sich sehr vor übereilten Schlüssen hüten. Ein
prinzipieller Unterschied zwischen den beiden Systemen, den ich besonders her-
vorheben zu müssen glaube, da er den bisherigen Beobachtern entgangen zu
sein scheint, ist der, daß die katholische Hierarchie die Organisation eines weltlichen Staatswesens mit einer weltlichen Bureaukratie darstellt, in der das
Mönchtum eine unter- und eingeordnete Stellung einnimmt (gleichsam von
der Art in ein Imperium eingesprengter Fürstentümer), während der Lamaismus weiter nichts als eine Einteilung des Mönchtums in Rangstufen ist und
jeglichen Instituts von Weltgeistlichen entbehrt. Die Parallele des Dalai
Lama mit der päpstlichen Würde ist rein äußerlicher Natur, während beide
Einrichtungen aus weit verschiedenen psychischen, religiösen, politischen und
historischen Ursachen hervorgegangen sind. Das Amt des Dalai Lama hat sich
folgerichtig aus der tibetischen Geschichte heraus als eine politische Schöpfung
der mongolischen Kaiser gestaltet und wurde aus Gründen der Politik von
den chinesischen Herrschern aufrecht erhalten. Spuren eines fremden Einflusses lassen sich nicht darin erkennen. Mit den bloßen Vergleichspunkten ist
natürlich überhaupt nichts getan und nicht weiter zu kommen, und um eine so
schwerwiegende Frage, ob das Christentum bestimmend auf die Geschicke der
lamaischen Kirche eingewirkt habe, gewissenhaft zu entscheiden, müssen wir
einen streng exakten, auf zuverlässige Quellen gegründeten historischen
Beweis verlangen und den Standpunkt einnehmen, daß, im Falle dieser Beweis nicht geliefert werden kann, die Annahme eines solchen Einflusses abzu-
lehnen ist. Solche authentischen Belegstücke müssen vor allem in der tibetischen
Literatur gesucht werden: wenn es z. B. gelänge, in den Lebensbeschreibungen
derjenigen Persönlichkeiten, denen ein wesentlicher Anteil an der Ausbildung
der lamaischen Hierarchie zugeschrieben werden muß, Nachrichten über die
Beeinflussung ihrer Anschauungen durch christliche, z. B. nestorianische Lehrer
zu ermitteln, oder wenn in der religiösen Literatur jener Zeit ein weitgehender
Grad der Abhängigkeit von christlichen Lehren in objektiv überzeugender Weise
nachgewiesen werden könnte, so müßten wir uns vor der Macht der Tatsachen
beugen. Die Hauptquellen, aus denen Rat zu erholen wäre, sind naturgemäß
die Biographie und die zahlreichen Schriften des Tsongkapa selbst, des eigentlichen Stifters des Lamaismus. Filchner teilt uns nun die bereits von Huc
kurz skizzierte, noch jetzt im Volke lebendige Tradition über das Leben dieses
eigentümlichen Mannes mit, in der erzählt wird, daß er den Unterricht eines
fremden Lehrers aus dem Westen mit langer Nase empfangen habe. Huc

dachte dabei an einen katholischen Missionar, während andere die Nestorianer*)
vorschieben, die nach Marco Polo um jene Zeit in Sining angesiedelt waren.
Ich will hier weder die eine noch die andere Anschauung bekämpfen, noch auch
den Wert jener immerhin interessanten Überlieferung erörtern, sondern nur
so viel bemerken, daß eine lange Nase wohl alle Zeit ein sehr unsicheres
historisches Kriterium bleiben wird. Die Hauptsache ist vorläufig, daß diese
Tradition nicht durch die tibetische Geschichte bestätigt wird. Wir kennen den
Lebensgang des Tsongkapa ziemlich gut aus der eingehenden Bearbeitung
seiner Biographie in dem großen Geschichtswerke des o Jigs-med nam-mkha,
das die treffliche Übersetzung von Georg Huth**) zugänglich gemacht hat. Aus
dieser ersehen wir, daß Tsongkapa seit seinem siebenten Lebensjahre von einem
tibetischen Lama erzogen wurde, und daß er, erst 17 Jahre alt (im Jahre 1372),
zum Zwecke theologischer Studien nach Lhasa ging, wo es damals keine Spur
von Christentum gab, und daß alle von ihm studierten Werke indisch-buddhistische waren. Von irgendwelchem christlichen oder fremden Einfluß auf Tsongkapa ist in seiner beglaubigten Biographie, soweit wir sie bis jetzt kennen —
das Studium des weit umfangreicheren Originalquellenwerkes mag ja neue
Aufschlüsse geben — durchaus keine Rede. Die bloße Tatsache der Anwesenheit von Nestorianern an den Grenzen Tibets genügt auch keineswegs zur
Herstellung des Abhängigkeitsbeweises, wenn nicht direkte nestorianische Beziehungen zu Tibet und zum Lamaismus historisch erhärtet werden können.

Ethnographen werden mit Interesse die Schilderung des Hutfestes in
Kumbum lesen. Auch dies ist natürlich keine erst von den Lamas eingerichtete
Sitte, sondern der Rest uralter Bräuche, die wir bei manchen indochinesischen
Völkern finden. Auf der einen Seite treffen wir bei den Indochinesen das
Recht der freien Gattenwahl, wie noch heutzutage bei allen Aboriginerstämmen
des südlichen und westlichen China, aus der die hohe Stellung der Frau erwuchs, die bereits Herodot an den mit den Tibetern identifizierten Issedonen
hervorhebt, auf der anderen Seite die Hingabe von Frauen an Fremde, eine
Sitte, von der uns Marco Polo so drastische Schilderungen hinterlassen, und
die bei einzelnen Stämmen verschiedene Formen angenommen hat. Auch die
von denselben gegebenen Deutungen dieser Bräuche sind verschieden, und es
scheinen sexuelle und religiöse Motive darin so durcheinander gewirrt zu sein,
daß der objektive Ursprung schwer zu enträtseln ist. Das beste Gegenstück zu
dem Hutfest in Tibet bietet die Feier des Frühlingsanfangs bei den Hua Miao
in Kuei-tschau, China, einem der zahlreichen Stämme der Miao-tse, die nach

---

*) So besonders Bonin in Journal Asiatique 1900, Seite 592.
**) Geschichte des Buddhismus in der Mongolei. II. Teil. Straßburg 1896.

meiner Ansicht eng mit der Shan-Familie verwandt sind und daher im weiteren Sinne auch mit den Tibetern. Ein chinesischer Bericht erzählt, daß zu diesem Fest Frauen und Männer gepudert und geschminkt, in neuen Gewändern, an einem wenig besuchten Orte zusammenkommen; die Männer blasen Rohrflöten, die Frauen schwingen Glocken und führen Rundtänze mit Gesang und allerlei Scherz und Lustbarkeit auf. Bei Sonnenuntergang verfertigen sie Hütten aus Fichtenzweigen, in denen die Paare ihrer Wahl die Nacht verbringen, und bei Tagesanbruch geht jedes seinen Weg.

New York, 18. November 1905.

Berthold Laufer.

040

乔治·胡斯博士讣告

T'OUNG PAO

通報

OU

ARCHIVES

CONCERNANT L'HISTOIRE, LES LANGUES,
LA GÉOGRAPHIE ET L'ETHNOGRAPHIE
DE
L'ASIE ORIENTALE

Revue dirigée par

Henri CORDIER
Professeur à l'Ecole spéciale des Langues orientales vivantes
ET
Edouard CHAVANNES
Membre de l'Institut, Professeur au Collège de France.

SÉRIE II. VOL. VII.

LIBRAIRIE ET IMPRIMERIE
CI-DEVANT
E. J. BRILL
LEIDE — 1906.

NÉCROLOGIE.

━━◦✺⊙✺◦━━

Dr. Georg HUTH. †

Am 1. Juni dieses Jahres verschied plötzlich, und leider zu früh für seine Wissenschaft, Dr. Georg Huth, Privatdocent an der Universität Berlin für tibetische und mongolische Sprache und Geschichte des Buddhismus. Kurz vor seinem Ende war ihm der Auftrag zu teil geworden, einen systematischen Katalog sämtlicher in der Kgl. Bibliothek zu Berlin vorhandenen tibetischen Handschriften und Drucke abzufassen, eine Arbeit, die er seit bereits zehn Jahren geplant und erhofft hatte. Eine längere Krankheit warf ihn zu Anfang des Jahres darnieder, von der ihn der Arzt als geheilt betrachtete und ihm den Ausgang gestattete. Ein seltsam tragisches Verhängnis raffte ihn dann am ersten Tage seiner Thätigkeit in der Kgl. Bibliothek dahin.

Georg Huth wurde am 25. Februar 1867 in Krotoschin, Provinz Posen, als Sohn des Rektors A. Huth, späteren Leiters der Naun'schen Waisenhaus-stiftung in Berlin, geboren. Unter der Leitung Albrecht Weber's, an den er sich eng anschloss, und für den er stets die höchste Verehrung bewies, wid-mete er sich dem Studium des Sanskrit und der indischen Litteratur, und aus diesem Gebiet entstammte auch seine Doktordissertation [1]). Schon als Student entfaltete er eine eifrige literarische Thätigkeit, mit besonderer Vorliebe auf dem Felde vergleichender Märchenkunde [2]). Unter Georg v. d. Gabelentz war er auch den ostasiatischen Sprachen näher getreten, aber im Tibetischen, das er sich später als specielles Feld erkor, war er im wesentlichen Autodidakt. In diesem Fache wie im Mongolischen habilitirte er sich im Jahre 1892 als Privatdocent an der Berliner Universität; seine Antrittsvorlesung war eine

---

1) *Die Zeit des Kâlidâsa.* Mit einem Anhang: *Zur Chronologie der Werke des Kâlidâsa.* Berlin, Ferd. Dümmler, 1890. 68 p. und 2 Tafeln.

2) *Zur vergleichenden Litteraturkunde des Orients.* Separatabdruck aus den *Mitteilungen des Academisch Orientalistischen Vereins,* No. 2. Berlin, 1889. 16 p.

Die Reisen der drei Söhne des Königs von Serendippo. Ein Beitrag zur vergleichenden Märchenkunde. Einzel-Abdruck aus „Zeitschrift für vergleichende Litteraturgeschichte und Renaissancelitteratur". Berlin, 1891. 71 p.

geistreiche Studie über mongolische Volkslieder mit wohlgelungenen Über-
setzungsproben [1]). Seine Arbeiten waren fürs erste auf die Erforschung der
tibetischen Litteratur [2]) und centralasiatischen Geschichte gerichtet, mit einer
ausgesprochenen Neigung für die Inschriftenkunde, der er einen ungewöhnlichen
Scharfsinn und seltene Ausdauer entgegenbrachte. Schon sein erster epigraphischer Versuch über die damals noch unentzifferten Yenissei-Inschriften [3]), der
kaum noch in dem Gedächtnis der jetzt lebenden Generation liegen dürfte,
weist, obwohl in seinem Endergebnis verfehlt, diese Vorzüge auf. Diese Arbeit
war auch für die weitere Richtung seiner Studien von Bedeutung, denn sie
lenkte die Aufmerksamkeit der interessirten Kreise der Petersburger Akademie
auf ihn, was eine Reihe weiterer inschriftlicher Forschungen zur Folge hatte [4]).
In diese Zeit fallen auch seine Untersuchungen über den Tanjur, die gerade für
die älteste Epoche der tibetischen Litteratur von besonderer Bedeutung sind [5]).
Sein Hauptwerk war die Herausgabe und Bearbeitung einer tibetischen Geschichte
des Buddhismus in der Mongolei, die ein vorzügliches Quellenwerk über diesen

---

1) Dieser Vortrag wurde damals in der *Frankfurter Zeitung* abgedruckt; da ich ihn
leider nicht besitze, bin ich nicht imstande, ein genaues Citat zu geben.

2) *The Chandoratnâkara of Ratnâkaraçânti.* Sanskrit Text with a Tibetan Translation.
Edited with critical and illustrative Notes. Berlin, Ferd. Dümmler, 1890. 34 p. Vergl.
Recension von FRANKE in *Göttinger Gelehrte Anzeigen,* 1892, Nr. 12, pp. 478—497. —
Die tibetische Version der Naiḥsargikaprâyaçcittikadharmâs. Buddhistische Sühnregeln
aus dem Pratimokshasûtram. Mit kritischen Anmerkungen herausgegeben, übersetzt und mit
der Pâli- und einer chinesischen Fassung, sowie mit dem Suttavibhaṅga verglichen. Strassburg, Karl I. Trübner, 1891. 51 p. —
Das buddhistische Sûtra der „Acht Erscheinungen". Tibetischer Text mit Übersetzung
von Julius Weber. Herausgegeben von Georg Huth. *Zeitschrift der Deutschen Morgenländischen Gesellschaft,* Vol. XLV, pp. 577—591.
Eine tibetische Quelle zur Kenntnis der Geographie Indiens. In Gurupujâkaumudî,
Festgabe zum 50-jährigen Doctorjubiläum Albrecht Weber dargebracht. Leipzig, 1896,
pp. 89—92.

3) *Die Inschrift von Karakorum.* Eine Untersuchung über ihre Sprache und die
Methode ihrer Entzifferung. Autographirt. 25 p. Berlin, Ferd. Dümmler, 1892.

4) *Die Inschriften von Tsaghan Baišiṅ.* Tibetisch-Mongolischer Text. Mit einer Übersetzung, sowie sprachlichen und historischen Erläuterungen herausgegeben. Leipzig, 1894.
Gedruckt auf Kosten der Deutschen Morgenländischen Gesellschaft. 63 p. und 1 Tafel. —
Sur les inscriptions en langue tibétaine et mongole de Tsaghan Baišiṅ, et sur le rapport de ces monuments avec „l'Histoire du Bouddhisme en Mongolie", composée en tibétain
par ₀Jigs-med nam-mkʻa, in *Verhandlungen des X. Internationalen Orientalisten-Congresses
zu Genf,* 1894. —
Note préliminaire sur l'inscription de Kiu-yong Koan. Quatrième partie: *Les inscriptions
mongoles: Journal Asiatique,* 1895 (mars-avril), pp. 351—360.

5) *Verzeichnis der im tibetischen Tanjur,* Abteilung mDo (Sûtra), Band 117—124,

Gegenstand bildet [1]). Die Übersetzung ist mit denkbar grösster philologischer Genauigkeit und dabei doch in einem klar fasslichen und lesbaren Stil abgefasst und kann als Muster für die Methode gelten, wie aus dem Tibetischen übersetzt werden muss. Es war der Plan des Verfassers, Noten erklärenden und historisch-kritischen Inhalts sowie Indices der Eigennamen und Termini in einem besonderen dritten Band zu veröffentlichen, zu dessen Herausgabe er leider nie gekommen ist, und was einen höchst bedauernswerten Verlust darstellt.

Inzwischen wurde nämlich seine Thatkraft auf ein anderes Forschungsgebiet gelenkt, das der Niüchi-Inschriften. Angeregt durch das Werk von W. GRUBE über die Sprache und Schrift der Jučen (Leipzig, 1896), versuchte sich Huth an der Entzifferung der zuerst von Devéria behandelten Stele von Yent°ai (in *Revue de l'Extrême-Orient*, Vol. I, pp. 173—185) [2]). Die Lesung und Übersetzung der Überschrift derselben ist ihm glücklich gelungen. Diese Arbeit führte ihm die längst gefühlte Notwendigkeit einer eingehenden Erforschung der tungusischen Sprachen vor Augen, und so unternahm er im Jahre 1896 im Auftrage der Petersburger Akademie seine erste Forschungsreise nach Ostsibirien und betrieb linguistische Studien unter den Tungusenstämmen am Yenissei [3]). Nach seiner Rückkehr wurde er als Hülfsarbeiter an das Museum

enthaltenen Werke. Sitzungsberichte der *Kgl. Preussischen Akademie der Wissenschaften zu Berlin*, 1895, pp. 267—286. Auch als Separatabdruck. —

Nachträgliche Ergebnisse bezüglich der chronologischen Ansetzung der Werke im tibetischen Tanjur, mDo, Band 117—124. *Zeitschrift der Deutschen Morgenländischen Gesellschaft*, Vol. 49, 1895, pp. 279—284.

1) *Geschichte des Buddhismus in der Mongolei*. Mit einer Einleitung: Politische Geschichte der Mongolen. Aus dem Tibetischen des ₒJigs-med nam-mk°a herausgegeben, übersetzt und erläutert. Erster Teil: Vorrede. Text. Kritische Anmerkungen. Strassburg, Karl J. Trübner, 1893. Zweiter Teil: Nachträge zum ersten Teil. Übersetzung. Ibid., 1896. —

Hor c°os byuñ: *Geschichte des Buddhismus in der Mongolei*, in tibetischer Sprache. *Transactions of the Ninth International Congress of Orientalists*, London 1893, Vol. II, pp. 636—641.

2) *Zur Entzifferung der Niüči-Inschrift von Yen-t'ai*. Bulletin de l'Académie Impériale des Sciences de St. Pétersbourg, 1896. T. V, No. 5, pp. 375—378.

3) Ein kurzer Bericht über diese Reise „Über die Tungusen Ostsibiriens" erschien in den *Jahresberichten des Frankfurter Vereins für Geographie und Statistik*, 1899, pp. 59—61. —

Ferner „Meine Reise zu den Tungusen am Yenissei", *Vossische Zeitung*, 1898, Nr. 249, 251, 253, 255, 257, 259, 261, 263. — Bericht über einen Vortrag von Huth über seine Reise in Beilage zur *Allgemeinen Zeitung*, Vol. 50, 1899, p. 7. Vergl. *Jahresberichte des Württemberger Vereins für Handelsgeographie*, Vols. XVII—XIX, p. 271. — *Die tungusische Volkslitteratur und ihre ethnologische Ausbeute*. Separat-Abdruck aus dem „Bulletin de l'Académie Impériale des Sciences de St.-Pétersbourg", Vᵉ Série. T. XV, No. 3 (Octobre 1901), pp. 293—316. — Bericht über einen Vortrag von Huth über die Volkspoesie der Tungusen. *Zeitschrift des Vereins für Volkskunde*, Vol. X, 1900, p. 243,

für Völkerkunde berufen. 1901 begann er mit der Entzifferung der türkischen Mahaban-Inschriften [1]). 1902—3 begleitete er Albert Grünwedel auf seiner Expedition nach Turfan in Ost-Turkistan, wo er nach der Rückkehr der anderen Teilnehmer bis 1904 verblieb und sich hauptsächlich in Osh, Chokand und Samarkand aufhielt. Er machte sich dort die türkische Sprache zu eigen und brachte eine grosse Sammlung türkischer Volksüberlieferungen, Märchen und Lieder, zusammen [2]).

Huth's Arbeiten sind alle durch grosse Objektivität und scrupulöseste Gewissenhaftigkeit ausgezeichnet. Er war in erster Linie ein *matter-of-fact* Gelehrter, dem nichts ferner lag als voreiliges Theoretisiren, und er hat darum die Wissenschaft mit neuen Thatsachen, nicht mit Ideen, bereichert. Und auch Thatsachen hat er nur nach reiflicher Überlegung und allseitiger Durchdringung des Gegenstandes, fast möchte man sagen, in etwas zu ängstlicher Weise, von sich gegeben, aber darum dürfen sie als gesicherte Ergebnisse und Grundlage für weitere Forschung gelten. Gewiss hätte er der Wissenschaft noch weit grössere Dienste leisten können, wäre ihm eine friedliche materielle Existenz beschieden gewesen. Zeitlebens hatte er mit den grossen und kleinen Sorgen des Daseins zu ringen, und meist ohne Einkommen, musste er, oft kümmerlich genug, von den Erträgen schriftstellerischer Arbeiten (er war ein treuer Mit-

---

1) *Erste Probe der Entzifferung der Mahaban-Inschriften.* Autographirt, 4 p. Berlin, 5. März 1901. —

Die Entzifferung der Mahaban-Inschriften. Vorläufige Mitteilung. *Sitzungsberichte der Königl. Preussischen Akademie der Wissenschaften zu Berlin*, 1901, pp. 218—220. Auch als Separatabdruck. —

Neun Mahaban-Inschriften. Veröffentlichungen aus dem *Königlichen Museum für Völkerkunde.* Supplementheft. Berlin, W. Spemann, 1901. 19 p., 4°. Mit Photographieen der Inschriften. —

Zur Frage der Mahaban-Inschriften. Orientalistische Litteratur-Zeitung, Vol. VIII, 1905, pp. 530—535; Vol. IX, 1906, pp. 3—20 (mit einer Photographie).

Im letzten Jahre seines Lebens hat sich Huth fast ausschliesslich der weiteren Entzifferung der Mahaban-Inschriften gewidmet. Eine Abhandlung betreffend Entzifferung von Siegeln und Stempeln, die bei seinem Tode fast druckfertig vorlag, ist Herrn Prof. Sachau behufs Veröffentlichung übergeben worden. Ausserdem sollen noch 45 Entzifferungen in seinem Nachlass vorliegen.

2) Vergl. seinen Bericht „Die Turfan-Expedition", *National-Zeitung*, 1904, Nr. 479 (11. August). —

Die neuesten archäologischen Entdeckungen in Ost-Turkistân. Verhandlungen der Berliner anthropologischen Gesellschaft, Vol. XXXIII, 1901, pp. (150)—(157). Vergl. *Correspondenz-Blatt der Deutschen Gesellschaft für Anthropologie*, Vol. XXXI, p. 48; *Mitteilungen der Geographischen Gesellschaft in Hamburg*, Vol. XVII, pp. 249—251; *Deutsche Geographische Blätter*, Vol. XXIV, p. 45; *Jahresberichte des Württemberger Vereins für Handelsgeographie*, Vols. XVII—XIX, p. 321.

arbeiter der *Vossischen Zeitung*) oder öffentlicher Vorträge leben, um immer wieder zu seiner stillen Gelehrtenthätigkeit zurückzukehren. Er war von einer unglaublichen, fast buddhistischen Bedürfnislosigkeit und Bescheidenheit, und wie und wovon er eigentlich lebte, war seinen Freunden meist ein Rätsel. Er war ein selbstloser Idealist von dem jetzt immer seltener werdenden Typus des deutschen Gelehrten, dem die Welt gleichgültig, und dem die Wissenschaft und Arbeit reines Ideal ist. In seiner akademischen Lehrthätigkeit gab er ausser sprachlichen Kursen ein geistvolles Kolleg über die Geschichte des Buddhismus und die Kulturgeschichte der Mongolen, die stets einen teilnahmsvollen Zuhörerkreis fanden. Als Lehrer entfaltete er grosses Geschick in der Interpretation tibetischer Texte, er war stets wohlwollend und hilfsbereit gegen seine Studenten, die ihm gewiss ein warmes Andenken bewahren werden.

B. LAUFER.

Samuel I. Joseph SCHERESCHEWSKY [1]), D.D. 施 *Chē.*

Né en mai 1831, à Tauroggen, en Lithuanie, d'une famille juive, S. s'était rendu en Amérique, où devenu protestant, il fut envoyé comme missionnaire en Chine par le Bureau des Missions étrangères de l'Eglise épiscopale protestante des Etats-Unis; il arriva à Chang-hai le 22 déc. 1859 et en 1862, il se rendit à Pe-king. Nommé évêque en 1877, il fut consacré le 31 octobre à Grace Church, New York. Il avait demandé, mais n'avait pas obtenu des Chinois, l'autorisation de résider parmi ses anciens frères, les juifs de K'ai-foung. S., fort bon sinologue, est surtout connu par la grande version de la Bible qu'il a donnée en chinois de l'hébreu; quoique paralysé des mains il préparait une traduction chinoise de la version grecque du Testament lorsqu'il est mort à Tsukiji, au Japon, le 15 oct. 1906.

H. C.

Lucien FOURNEREAU.

Nous avons le regret d'annoncer la mort de M. *Michel Louis Lucien* FOURNEREAU, architecte, Inspecteur de l'Enseignement du Dessin et des Musées, décédé le 19 déc. 1906, dans sa 61ᵐᵉ année, en son domicile à Paris, rue Beautreillis 22. M. F. était bien connu par ses explorations en Indo-Chine dont les résultats ont paru dans les ouvrages suivants: *Les Ruines d'Angkor, Étude artistique et historique sur les monuments khmers du Cambodge siamois*, en collaboration avec M. Jacques PORCHER, cent planches et une carte (1890); *Les Ruines khmères Cambodge et Siam Documents complémentaires d'Architecture, de Sculpture et de Céramique*, cent dix planches (1890); *Le Siam ancien, Archéologie, Epigraphie, Géographie*, dont la première partie seule parue forme le tome XXVII des *Annales du Musée Guimet*.

H. C.

---

1) Il est plus exact d'écrire *Scherschewski.*

041

主编《鲍亚士纪念文集》并序

BOAS ANNIVERSARY VOLUME

ANTHROPOLOGICAL PAPERS

WRITTEN IN HONOR OF

FRANZ BOAS

PROFESSOR OF ANTHROPOLOGY IN COLUMBIA UNIVERSITY

PRESENTED TO HIM ON THE TWENTY-FIFTH ANNIVERSARY OF HIS DOCTORATE

NINTH OF AUGUST

Nineteen Hundred and Six

NEW YORK
G. E. STECHERT & CO.
1906

Was hat der Mensch dem Menschen
Grösseres zu geben als Wahrheit?

SCHILLER

PRESS OF
THE NEW ERA PRINTING COMPANY
LANCASTER, PA.

COMMITTEE OF ARRANGEMENT.

PRESIDENT NICHOLAS MURRAY BUTLER, *Chairman.*

HON. ANDREW D. WHITE.	DR. A. JACOBI.
MR. JACOB H. SCHIFF.	MR. CARL SCHURZ.†
MR. MORRIS K. JESUP.	PROF. W J McGEE.
MR. EDWARD D. ADAMS.	PROF. EDUARD SELER.

HARLAN I. SMITH, *Treasurer.*

BERTHOLD LAUFER, *Secretary and Editor.*

iii

SUBSCRIBERS.

MR. EDWARD D. ADAMS.

DR. FELIX ADLER.

DR. I. ADLER.

MR. EMIL L. BOAS.

MR. CHARLES P. BOWDITCH.

MR. ARTHUR V. BRIESEN.

MR. ANDREW CARNEGIE.

REV. JOHN W. CHAPMAN.

MR. WILLIAM DEMUTH.

LIEUT. G. T. EMMONS.

PROF. AMOS W. FARNHAM.

DR. MAURICE FISHBERG.

MR. G. B. GRINNELL.

MR. STANSBURY HAGAR.

MRS. JOHN HAYS HAMMOND.

MRS. PHOEBE A. HEARST.

MRS. ESTHER HERRMAN.

MR. GEORGE G. HEYE.

MR. B. TALBOT B. HYDE.

DR. A. JACOBI.

MR. W. S. KAHNWEILER.

DR. FRED. KAMMERER.

MR. ANTONIO KNAUTH.

MR. HERMAN C. KUDLICH.

DR. GEORGE F. KUNZ.

DR. G. LANGMANN.

PROF. MORRIS LOEB.

HON. SETH LOW.

MR. JACOB MEYER.†

DR. WILLY MEYER.

MR. WM. BARCLAY PARSONS.

PROF. J. DYNELEY PRINCE.

PROF. JULIUS SACHS.

MR. JACOB H. SCHIFF.

MR. ISAAC N. SELIGMAN.

MR. JAMES SPEYER.

MESSRS. G. E. STECHERT & CO.

MR. FELIX M. WARBURG.

IV

CONTRIBUTORS.

Dr. O. Abraham . . . . Berlin.
Prof. Richard Andree . . . . Munich.
Miss H. A. Andrews . . . . New York.
Mr. A. F. Bandelier . . . . New York.
Prof. A. F. Chamberlain . . . . . Worcester, Mass.
Capt. George N. Comer . . . . . . East Haddam, Conn.
Dr. Jan Czekanowski . . . . Zürich-Warschau.
Dr. Roland B. Dixon . . . . Cambridge, Mass.
Dr. Henry H. Donaldson . . . . . Philadelphia, Penn.
Dr. George A. Dorsey . . . . Chicago, Ill.
Dr. Maurice Fishberg . . . . New York.
Dr. Pliny Earle Goddard . . . . . Berkeley, Cal.
Prof. Wilhelm Grube . . . . Berlin.
Mr. Stansbury Hagar . . . . New York.
Mr. C. V. Hartman . . . . Pittsburg, Penn.
Dr. Franz Heger . . . . Vienna.
Mr. George G. Heye . . . . New York.
Prof. Friedrich Hirth . . . . New York.
Prof. W. H. Holmes . . . . Washington, D.C.
Dr. E. M. v. Hornbostel . . . . . Berlin.
Mr. George Hunt . . . . Fort Rupert, B.C.
Dr. Aleš Hrdlička . . . . Washington, D.C.
Mr. Waldemar Jochelson . . . . . Zürich.
Dr. William Jones . . . . Chicago, Ill.
Prof. J. Kollmann . . . . Basel.
Dr. Friedrich S. Krauss . . . . . Vienna.
Dr. A. L. Kroeber . . . . San Francisco, Cal.
Dr. Berthold Laufer . . . . New York.
Prof. Rudolf Lehmann . . . . . . Berlin.
Dr. Carl Lumholtz . . . . New York.
Mr. Charles W. Mead . . . . . . New York.
Capt. James S. Mutch . . . . . . Peterhead, Scotland.
Mr. W. W. Newell . . . . Cambridge, Mass.
Mrs. Zelia Nuttall . . . . Mexico.
Mr. George H. Pepper . . . . . . New York.

v

Prof. Johannes Ranke . . . . Munich.
Dr. Ernst Richard . . . . New York.
Prof. Karl Sapper . . . . Tübingen.
Dr. J. D. E. Schmeltz . . . . . . Leiden.
Prof. Eduard Seler . . . . Berlin.
Mr. Harlan I. Smith . . . . . . New York.
Dr. Leo Sternberg . . . . St. Petersburg.
Dr. John R. Swanton . . . . . . Washington, D.C.
Mr. James Teit . . . . Spences Bridge, B.C.
Dr. Alfred M. Tozzer . . . . . . Cambridge, Mass.
Dr. Clark Wissler . . . . New York.

PREFATORY.

ON Nov. 28, 1905, a circular was sent out to anthropologists of America and Europe, soliciting scientific contributions to be issued in honor of Professor Franz Boas on the twenty-fifth anniversary of his doctorate, Aug. 9, 1906. This call has met with a hearty, unanimous approval, to which the present volume bears witness.

The general sentiment of sympathy with this matter was well expressed by Mr. F. W. HODGE, editor of " The American Anthropologist," in a letter dated Dec. 15, 1905, as follows: " I am very glad of this proposed action, both because the great value of Dr. Boas's work, his energy in advancing the interests of anthropology, and his many personal qualities, are thus to be recognized, and because in this country we are rather prone, I fear, to forget that there is a sentimental side to scientific life."

Unfortunately Mr. Hodge was prevented from writing an article which he had planned for this occasion, owing to ill health.

There were many others who had expressed a desire to contribute, but who were thwarted in their good intents by expeditions or other circumstances. Miss ALICE C. FLETCHER of Washington was in the midst of an article on " Siouan Concepts of God," with which she was anxious to honor Professor Boas, when she was taken suddenly ill and had to abandon her plan, much to her regret.

Professor TH. FISCHER of Marburg, who undertook an exploring trip in February, 1906, wrote to the editor the following, under date Dec. 11, 1905: —

„Ich entnehme aus Ihrem freundlichen Schreiben mit dem grössten Vergnügen, dass der Plan besteht, das Doktorjubiläum von Franz Boas würdig zu begehen. Das interessirt mich im allerhöchsten Grade, da Franz Boas, wie Sie wohl auch wissen, ein Schüler von mir gewesen ist. Er. ist mit mir 1879 von Bonn nach Kiel übergesiedelt und hat dort mit einer meereskundlichen Dissertation promovirt. Dr. Boas' erste selbständige Arbeiten waren ja auch geographische, von denen er aber bald zur Ethnologie übergegangen ist, da auch ich ihn, der von der Physik zur Geographie gekommen war, auf die historisch-ethnographische Seite des Fachs hinweisen zu müssen glaubte. Schon seine Arbeit über den Cumberland-Sund zeugt davon. Mit den Eskimos haben seine epochemachenden Forschungen auf dem Gebiete der Ethnologie und Anthropologie überhaupt und Nord-Amerikas im besonderen begonnen.

„So gern ich nun dazu beitragen möchte, dass die Gabe, welche man dem trefflichen Manne zu dem Jubeltage darbieten will, auch von Deutschland aus möglichst bereichert werde, so muss ich meinerseits doch verzichten, dazu beizutragen; denn die wenige freie Zeit, welche mir die Amtstätigkeit lässt, brauche ich zur Vorbereitung auf eine Ende Februar anzutretende (5.) wissenschaftliche Reise nach Afrika.

„So kann ich dem Unternehmen nur den besten Erfolg wünschen und Ihnen herzlich danken, dass Sie, wie ich wohl annehmen darf, dasselbe ins Leben gerufen und mir davon Mitteilung gemacht haben."

Letters addressed to Professor Boas were received later from Geheimrat Professor WALDEYER of Berlin, and Professor O. T. MASON of Washington, as follows:—

BERLIN,
15. Januar 1906.

HOCHGEEHRTER HERR KOLLEGE!

Gern würde ich mich an der Ihnen zu widmenden Festschrift beteiligt haben, fand jedoch, durch früher eingegangene Verpflichtungen gebunden, nicht die nötige Musse, um einen Beitrag zu liefern, der Ihrer würdig gewesen wäre. Seien Sie aber dessen gewiss, dass ich zu Ihren aufrichtigsten Verehrern gehöre und Ihre für die Anthropologie in allen ihren Zweigen so ausgezeichnete und wahrhaft förderliche Tätigkeit hoch bewerte und schätze. Dieser Empfindung zum Ausdrucke diene der herzliche Glückwunsch, den ich hiermit Ihnen zu Ihrem heutigen Gedächtnis- und Ehrentage sende. Möchten Ihnen noch viele nur reiche Erfolge auf dem Gebiete unserer Wissenschaft beschieden sein!—Bei Ihrer Frische und Arbeitskraft sehe ich im voraus diesen Wunsch sicher in Erfüllung gehen.

Ihr hochachtungsvoll ergebener

Waldeyer.

WASHINGTON, D.C.,
March 6, 1906.

DEAR DR. BOAS:

It is with especial pleasure that I send you words of good cheer and my blessing on your reaching the twenty-fifth anniversary of your University Doctorate, attained August 9th, 1881. By a happy coincidence, the present National Museum building had just been finished and anthropology received an immense impulse.

The first annual report of the Bureau of Ethnology also made its bow, and the Government Printing Office had just graduated the Introduction of Powell on Languages of the Indians, of Mallery on Sign Language, and of Yarrow on Mortuary Customs. You were yourself not long in enriching the publications of the same Bureau with your splendid monograph on the Central Eskimo. Your interest still continues, and may it long enrich the pages of its ethnological monographs.

I do not forget that you come still nearer to me in the volumes of the Smithsonian Institution and of the National Museum, treating of Houses and House-Life, of Social Organizations and Secret Societies, and of the mind of Primitive Man.

But I should miss a large part of the enjoyment in writing you this letter if I failed to emphasize in this connection the pleasure I myself have had in my studies of culture-history with you and through your published works.

I regret that the state of my health will prevent my being present at your festival and my making a more substantial contribution to the presentation volume that is to mark the twenty-fifth year of your Doctorate.

Yours ever sincerely,

O. T. Mason

Professor F. W. PUTNAM of the Peabody Museum of Archæology and Ethnology, Cambridge, Mass., sent the following message to the editor under date of April 16, 1906:—

As I am forbidden all forms of literary work, for the present, owing to my serious illness in September last, it has been impossible for me to prepare a paper for the Boas Festschrift.

I recall with pleasure my first meeting with Dr. Boas, in 1886, when he came to Buffalo to attend the meeting of the American Association for the Advancement of Science and introduced himself to

me in connection with my position as Permanent Secretary of the Association. At my suggestion the Council of the Association extended
to him the courtesy of making him Foreign Associate Member for the
Buffalo Meeting. Soon after that Mr. Scudder secured him as assistant editor of "Science," and from that time he became one of us.

In 1891, when I was appointed Chief of the Department of Ethnology of the World's Columbian Exposition, Dr. Boas, at my urgent
request, was appointed Chief Assistant of the Department. During
that time of untold trials and difficulties in making the first general
scientific anthropological exhibit in this country, I was supported by
a large corps of loyal and efficient assistants; but to none did I owe
so much as to Dr. Boas for the final success that attended our efforts.
At the close of the Exposition, in 1893, Dr. Boas took charge of the
collections made by our department, as Curator of the Department
of Anthropology of the Field Columbian Museum in Chicago, which
owes its existence primarily to our department of the Exposition.

In 1896, two years after I took charge of the Department of Anthropology of the American Museum of Natural History, I was again
able to secure Dr. Boas's co-operation in the development of this department, giving to him the sections of Ethnology and Somatology.
In this work, his extraordinary energy in planning and collecting soon
made the exhibits from British Columbia and Alaska as well as those
from the Eskimo, both east and west, of the first importance. The
numerous parties sent out to study the languages and myths of our
Indian tribes secured material which added to the unprecedented development of the ethnological collections.

It had been evident for many years that we must have a knowledge
of the ethnology of the Asiatic side of the Northern Pacific in order
to solve several important problems in connection with the study of
the tribes on the western coast of America. The importance of such
a comparative study of the ethnology of both sides of the North Pacific
was brought to the attention of President Jesup, who always showed
an interest in the Department and was ready to give his personal
aid to its development. An elaborate plan, covering a number of
years, was formulated, and Dr. Boas was put in special charge of
the direction of the work, which was carried on under the generous
patronage of Mr. Jesup. The results of this expedition, now being
published under the able editorship of Dr. Boas, are so far-reaching
and so important in all future investigations of the migrations of early
peoples in America and Asia, that Dr. Boas's name will always be indel-
ibly and most honorably associated with the great problems involved

in the comparative ethnology of the two continents. I need not refer to the later ethnological work in China, which Dr. Boas has so largely intrusted to you to execute, except as another instance illustrating his persistence in working for a project on which he has embarked.

Dr. Boas's interest and influence in anthropology have always been wide, and have covered the field to a remarkable extent. His researches relating to the physical characters of the various American peoples were of importance early in his American studies, while his linguistic researches and his study of myths and ceremonials have made him our acknowledged leader in these branches of the science. His present positions in connection with the Jesup Expedition, the Bureau of Ethnology, and Columbia University, are giving him admirable opportunities to develop and make known the results of his studies.

It has always been a pleasure to me to have Dr. Boas associated with me in the many undertakings we have carried on for the advancement of Anthropology in America, and I feel that all workers in the several subjects covered by the term "ethnology," as we now use it, will agree that Dr. Boas's influence has been most helpful and beneficial to American students; while his energetic and critical mind has been of incalculable importance in every phase of anthropological investigation during the past two decades in America.

I beg of you, dear Dr. L., to accept this letter as a brief expression of my appreciation of the great worth of one with whom I have had so many and intimate associations, and to let it be communicated to him with the assurance of my most cordial good-will and sincere wishes for his continued health and prosperity, that he may engage in many more important works for the benefit of all students of anthropology and the advancement of its teachings among men.

I am, with regret that I must withhold myself from saying more,

Cordially yours,

F. W. Putnam

Following is a letter received from Professor W J McGee, Director of the St. Louis Public Museum: —

It is a pleasure to tender a word of appreciation of the eminent contributions made by Professor Franz Boas to American ethnology, and thereby to the general science of man.

The Western Hemisphere has made rich contributions to anthropology. Most of these emanated from North America, many of them chiefly from the United States. Albert Gallatin studied aboriginal tongues as a means of classifying the native tribes for both administrative and scientific purposes; John Wesley Powell extended Gallatin's lines, adding the use of primitive speech as an index to primitive thought as one of the ends of linguistic research; Lewis H. Morgan investigated aboriginal terms as a means of defining social organizations; Daniel G. Brinton applied linguistic research to the tracing of philosophies; and to this eminent group must be added Boas as the foremost investigator of American aboriginal tongues regarded as indices to the development of language itself. Others in numbers have contributed to knowledge of the development of mankind on the Western Hemisphere, and hence toward illumining one of the mòst significant, albeit obscure, of all the long chapters in the development of the human kind. Some have interpreted skeletons, others the artifacts of lowly culture; still others, including the galaxy of linguistic students, primitive activities observed among living tribes by the pioneers. Of all these activities, none are more characteristic of mankind, and none bear more directly on the course of human progress, than the development of that speech by which all other activities are co-ordinated and perpetuated; and it remained for Boas to open and occupy this special field, — the field of language, considered no less as an end than a means of research.

Boas, although the leading occupant of the special field of American linguistics, has done more: he has kept in touch with all other branches and aspects of the science of man; and thereby, to his own credit and to that of the institutions with which he is and has been connected, he takes rank as a general anthropologist, and one of the foremost of those now living.

The tribute so thoughtfully proposed by you and others to Professor Boas is well merited; and I am delighted to join in offering it.

Yours cordially,

WJMcGn

Dr. LIVINGSTON FARRAND, professor of anthropology at Columbia University, expressed himself thus in the following letter, addressed to the editor:—

It is a very welcome opportunity you extend to me to join in the congratulations to Professor Boas on the twenty-fifth anniversary of his Doctorate.

The volume of studies contributed by his colleagues and pupils is a tribute to his influence as a working anthropologist. To me the most striking thought that arises is of Professor Boas's great service to what might be termed the academic side of anthropology in America.

It will be admitted, I think, by every one concerned with the teaching of anthropology in the United States, that the present position of the science as a department of university study is due more to his influence than to any other factor.

When Dr. Boas joined the Faculty of Columbia University, in 1895, a new stimulus was given to the initial efforts to gain recognition of the subject in the curriculum. His broad scholarship, his wide sympathies, and his uncompromising scientific ideals, at once produced their effect, and anthropology was soon placed on a new footing in the University.

It was not long before his work at Columbia, both by his example and through his students, began to show in other institutions. Departments already established were strengthened and broadened, the subject was introduced where it had been unknown before, and to-day, after a lapse of some ten years, the outlook is most promising for the general recognition of anthropology in the university world. I speak advisedly when I repeat that the present situation is due more to Professor Boas than to any other influence.

I am sorry that the necessary limits of a letter prevent my enlarging on this phase of Professor Boas's service; but I suppose I must content myself with giving my hearty approval to the plan of a commemorative volume, and with expressing the wish that his fiftieth anniversary will find him engaged with undiminished power and with all his well-known energy in anthropological research.

Very sincerely yours,

Livingston Farrand

The general plan of arrangement of the papers is, physical anthropology, philology, general anthropology and decorative art, American archæology and ethnology, European and Asiatic subjects. This scheme, however, could not be strictly adhered to because of the late arrival of some papers and of obstacles connected with the printing of others, which for this reason had to be transferred to the end. Grube's paper was made the leading article on account of the symbolical appropriateness of the congratulatory feature of the play which it contains. A bibliography of Professor Boas concludes the volume, which it is hoped will be welcomed by his numerous friends and co-workers. The figure on the back of the title-page represents a Sioux Indian, after a drawing made by Mr. Rudolf Cronau.

The publication of this book has been unduly delayed, owing to manifold typographical and other technical difficulties, much to the regret of the editor, who offers his apologies for its late appearance.

May the papers embodied in this volume be deemed not unworthy of the name of the man in whose honor they have been written; and may this garland of contributions, gathered from both sides of the Atlantic, bespeak again the harmonious collaboration of international science and the peaceful unity of the world in the realm of thought!

B. L.

CONTENTS.

xv

CONTENTS

LIST OF ILLUSTRATIONS.

PLATES.

xviii

TEXT FIGURES.

042

中国和欧洲的鸟型战车

BOAS ANNIVERSARY VOLUME

ANTHROPOLOGICAL PAPERS

WRITTEN IN HONOR OF

FRANZ BOAS

PROFESSOR OF ANTHROPOLOGY IN COLUMBIA UNIVERSITY

PRESENTED TO HIM ON THE TWENTY-FIFTH ANNIVERSARY

OF HIS DOCTORATE

NINTH OF AUGUST

Nineteen Hundred and Six

NEW YORK

G. E. STECHERT & CO.

1906

BRONZE BIRD-CHARIOT, CHINA.

THE BIRD–CHARIOT IN CHINA AND EUROPE.

BY BERTHOLD LAUFER.

IN an interesting paper entitled " A Curious Aino Toy," [1] EDWARD S. MORSE discusses a wooden toy, in the form of a bird on wheels, supposed to be of Ainu origin. Although the idea of wheels, foreign to this tribe, is evidently borrowed from the Japanese, yet Morse had never come across such a toy in Japan. He further figures a similar wooden specimen pertaining to the Yakut in Siberia, and another excavated by Flinders Petrie in the cemetery of Hawara in Egypt, dating back not later than the first century of our era. Morse sets forth the opinion that " this toy might naturally have originated among a civilized people like the Egyptians, who portray wheeled chariots in their early rock sculpture," and concludes that " certainly, unless it can be shown that any kind of an object provided with wheels originated among a savage people, it does not seem an absurd conjecture to suggest the common origin of this toy even among peoples so widely removed in space and time as those above mentioned."

1. These wheeled birds have a much wider dissemination in Eastern Asia than is indicated by Morse, and they form in particular a distinct type among the Chinese antiquities of bronze and nephrite. A number of these have been described and illustrated in the archæological literature of the Chinese. Before reviewing the latter, I will first refer briefly to some actual specimens which have become known to me. Plate XXXIII shows a bronze piece of this kind [2] representing a winged bird,

[1] Bulletin of the Essex Institute, Salem, Mass., Vol. XXV, pp. 1–7.
[2] It is in the possession of my friend, Dr. August Conrady, professor of Chinese at the University of Leipzig, who acquired it at Peking, and to whom I am indebted for his kindness in placing a photograph of it at my disposal. The height of the object is 24.7 cm.; from the bill to the tail-wheel it is 22.3 cm. long; the length of the bird's body is 17.2 cm., and its width is 7.7 cm.

410

with long tail curved downward, resting on two large wheels, and a small wheel attached to the end of the tail.   On its back the bird carries a sacrificial vessel of the type called *ts'un*.   A curious head is brought out in relief on the breast of the bird. Spiral ornaments are engraved on its body, and the graven lines on the wings seem to be intended to indicate plumage.

I do not hazard a conjecture as to the period in which this object may have been made, as I had no opportunity to examine it; but I may say that my general impression would favor a rather recent origin, which I infer chiefly from the modernized formation of the wheel, almost identical with that of the usual North-Chinese travelling-cart of nowadays.   Then we must take into consideration the facts that, of genuine specimens of this type (i.e., such as come down from the periods of the Han and the T'ang), exceedingly few, if any, have survived, and, if such exist, they may be hidden away among the treasures of Chinese private collectors; further, that these very objects are imitated indeed in recent times, of which I had abundant opportunities to convince myself in specimens seen by me in China, the technique of which clearly stamped them as modern productions. These, as a rule, are made on a smaller scale than the antique ones, and easily betray themselves as epigones by the frequent applications of cloisonné enamel, and certainly by their deviation from the standard forms, by their inferior technique, by their plumpness of shape and their crudeness of execution.   It is therefore a matter of some surprise to notice in Dr. S. W. BUSHELL's recent book[1] the figure of such a vessel positively ascribed to the Han dynasty.   The object is much like that illustrated on Plate XXXIII, except that it is lower, and the wings of the bird are entirely concealed behind the wheels.   Dr. Bushell, in his description, remarks that " the curious wheeled wine-vessels commonly called *chiu ch'ê tsun,* or ' dove-chariot vases,' are generally attributed to the Han dynasty (202 B.C.– A.D. 220)."   Though this statement is undeniably correct, it certainly does not justify, without further evidence, the conclusion that the specimen in the South Kensington Museum is

[1] Chinese Art, Vol. I, London, 1904 (Publication of the South Kensington. Museum), Fig. 56 and p. 91.

necessarily also a Han, as asserted in the descriptive matter under the figure. It is naturally impossible to speak positively either for or against its authenticity, without submitting the object in all details of its workmanship to a close inspec-

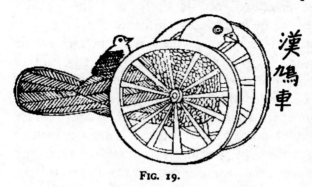

FIG. 19.

tion; nevertheless I cannot suppress the opinion that to me it seems, from the general appearance of its style and technique, to fall rather under the heading of the coarser ware above alluded to. Fur-

ther comments of Dr. Bushell are as follows: " The bird of mythological aspect, which is supposed to represent a dove (*chiu*), has its tail curved downwards, and a trumpet-shaped vase-mouth with scroll-ornament and dragons, and displays on

its breast a grotesque head moulded in relief. Two wheels support it at the side, and a smaller one at the tail, adapting it to circulate on the altar during the performance of the ancestral ritual ceremonies." I do not

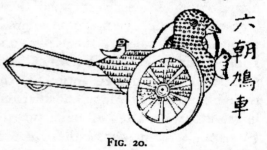

FIG. 20.

know whether the latter statement is the opinion of the author, or whether he derived it from a Chinese writer: I am unable to trace it back to any Chinese source (see end of § 3, p. 419).

2. In the " Po ku t'u," Book 27, pp. 44 a and b, two so-called " dove chariots " are illustrated, the one (Fig. 19) attributed to the Han period,[1] the other (Fig. 20) to the age of the Leu

---

[1] These illustrations are photographed from the edition published in 1753. The cyclopædia San ts'ai t'u hui, section on implements (Book 5, p. 9), gives a reproduction of this engraving with the text of the Po ku t'u, curiously enough grouped in the chapter " Means of Transportation."

ch'ao or Six Dynasties (A.D. 221–589). In the legends over the illustrations, the bird is designated with the one character *chiu*, which is explained as "turtle-dove" or "pigeon." In the accompanying descriptive text, however, the compound *shih* (Giles, No. 9901) *chiu* is used, which is interpreted by Giles as "the cuckoo" (*Cuculus canorus*), while other lexicographers take this compound also in the sense of a pigeon or turtle-dove.[1]

The definition of this object, given by the art-historian WANG FU (first half of the twelfth century), the author of the "Po ku t'u," is to the effect that it consists of two wheels, between which the dove is placed, so that it moves through the motion of the wheels. In both pieces, the large dove is conceived of as the mother-bird, which in the Han specimen carries her young one on the back, and in that of the Leu ch'ao, two young birds, — one on her breast, the other on her back. In the front part of both, there is, according to the description

---

[1] COUVREUR (Dictionnaire classique de la langue chinoise, p. 1047 a), *shih chiu*, "huppe," "pigeon ramier." PALLADIUS (Chinese-Russian Dictionary, p. 528), *gorlitsa*, i.e., "turtle-dove." EITEL (A Chinese Dictionary in the Cantonese Dialect, p. 612 a) explains *shih* by "wood-pigeon," and *shih chiu* by "turtle-dove." Giles's interpretation evidently goes back to O. F. v. MÖLLENDORFF, The Vertebrata of the Province of Chihli (Journal of the North-China Branch of the Royal Asiatic Society, N. S., Vol. XI, Shanghai, 1877, p. 93): "Shih chiu has been explained as a kind of wood-pigeon or turtle-dove. But the description points evidently to the cuckoo; more especially the mention of the habit of that bird of not building nests, but laying its eggs in the nests of other birds, would not admit of any other identification." According to the investigations of T. WATTERS, Chinese Notions about Pigeons and Doves (Journal of the North-China Branch of the Royal Asiatic Society, New Series, Vol. IV, Shanghai, 1868, p. 229), *chiu* is a generic term for doves, *shih chiu* is a wood-pigeon or a dove of some sort (p. 238). The term *shih chiu* occurs in the Shih king (JAMES LEGGE, The Chinese Classics, Vol. IV, Part I, p. 222; Shih king, ed. COUVREUR, p. 157), where Legge translates it by "turtle-dove," Couvreur by "hoopoe." The meaning "cuckoo" would hardly be commensurate with the passage in question, as the *shih chiu* is here introduced as the symbol of filial piety and maternal love, a notion which is attributed by the Chinese just to the dove (see T. WATTERS, l.c., p. 236). The Chinese statement given above, on which Möllendorff's identification with the cuckoo is based, is an idea which seems to go back to the verse in the Shih king: "The nest is the magpie's; the dove dwells in it" (J. LEGGE, l.c., p. 20 and note p. 21); but this does not justify us in assigning to the word *chiu* or *shih chiu* the signification of "cuckoo," even though a confusion of the two birds be admitted. An engraving of this dove (*shih chiu t'u*) from T'u shu chi ch'êng, Vol. 579, ch'in ch'ung tien Book 28, chiu pu hui k'ao, p. 3 a, is reproduced as a vignette at the beginning of this paper.

given, a perforated knob for the passage of a cord, by means of which, apparently, the chariot can be drawn. Of features not mentioned in the text, we notice that in the former (**Fig. 19**) the wheels are larger and provided with twelve spokes, in the latter (**Fig. 20**) with ten; in the first one, the head and neck of the bird are unadorned, its body is decorated with what appears elsewhere as scales on fishes or dragons, and the six hatched portions behind leave no doubt that they are intended to represent the tail-feathers. In the second object, head, neck, and breast are dotted over with small circles; the body is divided into ten rows with vertical hatchings; and instead of the rounded-off tail-feather of the preceding object, we find here a pointed angular piece to which, as is also expressly added in the text, a third small wheel is attached " to strengthen " the chariot. The opinion of Wang Fu is, that these objects served as amusements, playthings for young boys; and since there is nothing that would conflict with their character as toys, I see no reason for rejecting such an interpretation. This is further accounted for by the author with a quotation from TU SHIH with the cognomen *Yu chiu tzǔ*,[1] to the effect that boys at

古
玉
鳩
車

the age of five years play with dove-chariots, while at the age of seven they enjoy the pleasure of the bamboo horse.[2]

Fig. 21 represents a dove-chariot of white nephrite, reproduced af-

FIG. 21.

ter an engraving in the " Ku yü t'u p'u " (" Illustrated Book of Ancient Jades "), Book 47, p. 12, compiled by LUNG

[1] In the Ku yü t'u p'u (Book 47, p. 13 a) the same quotation is given as derived from a book Kin hai (The Golden Sea), with the *varia lectio* that boys at the age of *six* play with the dove-chariot, and at *seven* with the bamboo horse.

[2] Hobby-horses are mentioned as early as the Han time, as is well attested by the bamboo horses on which the boys of Ping chou went out to receive the virtuous Kuo Chi (38 B.C.–A.D. 47), in token of respect and gratitude for his wise administration, on his return to his old magistracy (see GILES, Dictionary, p. 269 b, and Biographical Dictionary, p. 405; STEWART-LOCKHART, A Manual of Chinese Quotations, p. 73; C. PETILLON, Allusions littéraires, p. 288). See, further, STEWART CULIN, Korean Games (Philadelphia, 1895), p. 32; E. CHAVANNES, Documents sur les Tou-kiue (Turcs) occidentaux (St. Pét., 1903), p. 117.

TA-YÜAN in 1176, and published in 1779. The different character of the two birds,—particularly the formation of the
tail, which varies from the previous ones,—and the ornamental treatment of the wheel-spokes, are striking at first glance. This wheel appears almost identical with, and is probably derived from, a wheel-like object of jade pictured in the same work (Book 47, p. 7), and here given in Fig. 22. The text says that this piece is to adorn the upper part of the state carriage (*yü lu*),

FIG. 22.

and an implement of the time of the Three Generations (*san tai*, i.e., the Hsia, Shang, and Chou dynasties). That wheels of this type were employed for ceremonial carriages in

FIG. 23.

times of antiquity, will be seen from Fig. 23, which is meant to illustrate the *ch'ung ti ch'ê* ("cart with pairs of pheasants") mentioned in the "Chou li."[1] As regards the explanation of

[1] T'u shu chi ch'êng Vol. 1114, k'ao kung tien Book 174, ch'ê yü pu hui k'ao, IX, p. 9, whence also the illustration is derived. COUVREUR (Dictionnaire clas-

the bird-chariot in the "Ku yü t'u p'u," the quotation given in the "Po ku t'u" is repeated, and it is further remarked that "this bird-chariot was an object in the palace of the Six Dynasties." Whether this may imply that the piece there figured comes down from this period, according to the author's view, must remain an open question; at least, he makes no other attempt to fix a date for it.

Finally, Fig. 24 is a bronze dove-chariot of the T'ang dynasty, figured after the "Hsi ch'ing ku chien" (Book 38, p. 27).[1] The accompanying note states that the "Po ku t'u" also contains this implement, with the quotation of TU SHIH, which is then reproduced. It concludes with a new sentence not to be found in the other books, saying that this object was not made in the earliest days of antiquity, which seems plainly to hint at the fact that it first sprang up during the Han period. We thus have now an opportunity of viewing specimens of this

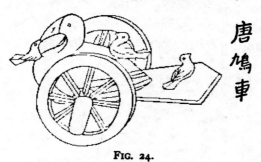

唐
鳩
車

FIG. 24.

type from three ages, — the Han, the Leu ch'ao, and the T'ang. The T'ang piece in Fig. 24 is the simplest of all, void of all decorative elements. It is distinguished from all others by having three young birds,—one on the breast,

another on the back, of the old bird, as in the Leu ch'ao chariot, and a third on the tail, looking in the opposite direction. The form of the tail coincides with that of the Leu ch'ao object.

3. A type widely deviating from the previous ones is found in the eleventh book of the "Hsi ch'ing ku chien." First of all, we meet there, on pp. 27, 28, two vases called *chiu ts'un* ("dove-vases"), each resting on the back of a plastic bird

sique de la langue chinoise, p. 732) translates the term *ch'ung ti ch'ê* by "voiture aux plumes de faisan disposées par paires." The above figure would rather suggest representations of pheasants embroidered on the cart-awnings.

[1] From the quarto edition executed in Japan in 1888, which is an exact facsimile of the original, published at Peking in 1751.

figure, and both attributed to the Han period.   The second of
the two is here reproduced in Fig. 25.[1]   No descriptive text is

FIG. 25.

added, except the measurements and weights of the pieces.
The purely ornamental and conventional style will be readily
observed; and that this bird is by no means a dove, but that
the latter designation is transferred to it merely through seeming analogy, — badly chosen, indeed, — is quite obvious at the
outset.   This feature is still more striking in the former of the
two bird-vases, in which the feet of the bird are set with enormous toes provided with long sharp-pointed claws, so that evidently a bird of prey is there intended.   Our next illustration
(Fig. 26) follows the two vases in the same book of the "Hsi
ch'ing ku chien" (p. 29), and is superscribed as a "dove-
chariot vase of the Han time;" and it will be seen that this

[1] Compare the similar piece in J. LESSING, Chinesische Bronzegefässe, Vorbilder-Hefte aus dem Kunstgewerbe Museum, No. 29, Berlin, 1902, Plate II, Fig. a.
Boas Anniversary Volume. — 27.

wheeled vessel represents the same type of bird as the preceding. It is, besides, the same type as that shown on Plate XXXIII. I can but presume that it is, in fact, a secondary derivation from the former; the feet of the bird being replaced by the two eight-spoked wheels, and a small wheel in the shape of a disk being added to the extreme end of the tail. A clew to the understanding and presumable development of this object is afforded by a

漢
鳩
車
尊

brief explanation in the accompanying text of the "Hsi ch'ing ku chien," in which is this statement: "Compared with the two foregoing vessels [alluded to above], this one is a *plaything,* and that is just the point in which it differs from those sacrificial vessels." If this interpretation is correct, we should have to look upon this object as an adaptation to the sacrificial vase borne by the bird, caused or influenced by the previously described real

FIG. 26.

dove-chariots, which I should like to style genuine or original ones, to distinguish them from the present pseudo-type. Whereas, as we shall see hereafter, the genuine Chinese bird-chariot seems to be derived from a foreign idea, there can be no doubt that the bird-vases, like that illustrated in Fig. 26, are a purely Chinese invention, since there are many analogies to this type which were in existence as early as the Chou dynasty (1122–255 B.C.), when we find the same type of sacrificial vessel standing on the back of elephants and other animals. In the Han period, such vases were placed on birds

called larks (*t'ien chi*) and on so-called auspicious or wonderful animals (*jui shou*); and in the T'ang period, on dragons
and phœnixes combined, and on stags.   I think we may say,
therefore, that the bird-chariot in Fig. 26 is a distinct type,
differing from the others enumerated above, and that it is secondarily derived from a previously existing bird-vase by the
addition of wheels, the analogon to which was found in the then
established dove-chariots, with the same object in view as the
latter implied; i.e., to serve as a toy.   This
deduction is very important, since it implicitly
contains the inference that this wheeled sacrificial vase never was and never could be a
religious or ceremonial object, as Dr. Bushell
concluded (see end of § 1, p. 412), but was
never anything more or less than a simple
plaything.   The occurrence of the vase had
no significance, and was merely incidental in
this toy, a mere grafting of a given favorite
form, serving the purpose of creating a new
variation of this then existing object of play,
—a wholly subsequent and secondary development.

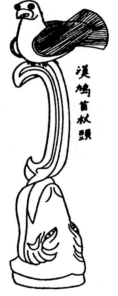

FIG. 27.

4. The utilization of the dove and pigeon
in artistic representations was not frequent
in ancient China.   Besides the dove-chariots,
there are only two kinds of objects known
from the Han time, in which a dove was figured.   The one is a
staff, usually of jade, adorned with the figure of this bird, and
bestowed upon men eighty or ninety years old.   The details
regarding this custom will be found in my forthcoming paper,
"The Pottery of the Han Dynasty."   The engraving Fig. 27 is
from the "Hsi ch'ing ku chien" (Book 38, p. 19).   It represents the handle of such a staff, formed by an inverted animal
(sheep?)-head (presumably to serve as a socket) surmounted
by a dove, the whole apparently made of bronze.   The other
kind of object is a "book-weight" (*shu chên*) of bronze in-

laid with gold and silver (Fig. 28), explained as having the shape of a dove (*chiu chên*), and as originating from the Han time (Hsi ch'ing ku chien, Book 38, p. 39). No description of it is furnished.

漢鳩鎮

FIG. 28.

5. The type of the dove-chariot may raise the question whether there are other ancient vessels extant in the shape of chariots. I have found only one, reproduced in Fig. 29. It is derived from the "Hsi ch'ing ku chien" (Book 38, p. 57), and is entitled *T'ang fang ch'ê hsün lu* ("a censer [lit., 'stove for fragrant herbs'] in the shape of a quadrangular cart of the T'ang period"). Unfortunately, no further discussion of this

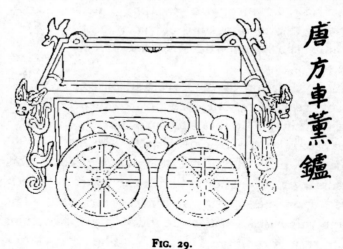

唐方車薰鑪

FIG. 29.

interesting specimen is added. The four monsters with spirally curved tails clinging to the four corners of the four-wheeled vehicle strongly remind one of the type of a hydra (*ch'ih*).

6. Aside from the illustrations in the above-mentioned archæological works, there is no documentary evidence relating to these dove-chariots to be met with in Chinese literature. Neither K'ANG HSI's Dictionary nor the " P'ei wên yün fu " mention this term, and, so far as I know, there is no contemporary record extant in the books of the Han.[1] This silence is very curious and suspicious; and the few unsatisfactory notes which the antiquarians append to this article do not help much in verifying its origin, being of such a nature as hardly to allow of stamping it as a genuine Chinese invention. They admit, on the contrary, that it was unknown in the days of greatest antiquity, and that it did not make its appearance before the era of the Han dynasty, — a period in Chinese art in which large waves of foreign elements burst over the native ideas. It was the time when, as I have tried to show elsewhere, Siberian or Old-Turkish art exercised a far-reaching influence on that of China, and new motives imported from abroad held full sway over the then Chinese artists. Would it not, then, be possible to associate the object under consideration with those other foreign invasions? Would it not be justifiable, under these conditions, to look for analogous phenomena in other spheres of art, which might have been the prototypes of the Chinese idea, and thus afford the foundation for a better explanation of it? We have seen that Morse pointed out the occurrence of a wooden wheeled bird in Egypt, and he is inclined to consider that country as the one where this curious object was first conceived of. But as in Eastern Asia, so in the western part of the Old World, we discover a much wider range and a far more extended geographical distribution of these things than is admitted by Morse. Indeed, almost throughout Europe and Anterior Asia, bird-chariots of bronze occur in large numbers which date from the end of the bronze age. As they have often been described and figured, I will refer the reader to the more important literature regarding these finds: R. VIRCHOW, " Nordische Bronze-Wagen, Bronze-

---

[1] This is particularly confirmed by the great cyclopædias, like Yen chien lei han, T'u shu chi ch'êng, Ko chih ching yüan, which, in mentioning the dove-chariot, are content merely with repeating the one quotation from the Po ku t'u.

Stiere und Bronze-Vögel," in Zeitschrift für Ethnologie, Vol. v,
1873, Verhandlungen, pp. (198)–(207); M. HÖRNES, "Urgeschichte der bildenden Kunst in Europa" (Wien, 1898), pp. 499
et seq., and "Die Urgeschichte des Menschen" (Wien, 1892),
pp. 411, 540–542; E. CHANTRE, "Recherches anthropologiques
dans le Caucase," Vol. II (Text), 1886, pp. 203–205, with 12
figures of such chariots; INGVALD UNDSET, "Antike Wagen-
Gebilde" (Zeitschrift für Ethnologie, Vol. XXII, 1890, pp. 49–75,
particularly pp. 49, 56); JOSEPH HAMPEL, "Altertümer der
Bronzezeit in Ungarn," 2d ed. (Budapest, 1890), Plate LVIII;
SALOMON REINACH, "La sculpture en Europe avant les influences gréco-romaines" (L'Anthropologie, Vol. III, 1896, p. 171);
O. SCHRADER, "Reallexikon der indogermanischen Altertumskunde" (Strassburg, 1901),
p. 930. Virchow remarks
in his paper that groups of
ducks figure on several chariots of bronze, and illustrates one found in Frank-
furt-on-the-Oder, which consists of three wheels connected by an axle; between
the wheels there are four

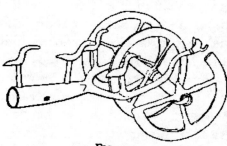

FIG. 30.

birds, two of which stand on the shaft going out from
the axle. This sketch is here repeated in Fig. 30, after
Virchow in the above-quoted paper. Undset (l. c.) discusses,
among others, a bronze chariot from a tomb near Corneto in
Etruria, supposed to belong to the eighth century B.C. On
four four-spoked wheels connected by two axles rests an animal
whose neck, body, and tail betray a bird, that, however, possesses four feet and a head (perhaps that of an ox?) with
horns. In the middle of its back there is a quadrangular opening, and the hollow body thus forms a small vessel. The opening
is closed by a lid formed like the back of a similar animal with
the same bird's tail and neck and horned animal-head. In the
bird-vessels from Italy, he interprets the bird as a duck, and
presumes that these bird-chariots were a sort of sacred objects
which might have some relation to Oriental models. Accord-

ing to him, in northern Europe also, kettle-chariots have been found, representing bronze vases resting on two bent axles.[1] These chariots of European finds are usually conceived of as votive chariots, and assigned to some religious cult. Whether this was so in all cases or not, it is not the place here to discuss; but considering the fact that, according to a well-authenticated tradition, they appear in China from the first as toys, and were there never anything else, I should venture to suggest to our archæologists that such a possibility might be admitted also for a part of the European objects in question. Or might mere ignorance of their true signification on the part of the Chinese have led them to intimate that they were toys?[2] Since European chariots are much older than Chinese ones, since they are found there over a much wider geographical area and in greater numbers, there can be no doubt that this idea must have spread from the west to the east; indeed, if such was the case, it can have found its way to China only by way of Siberia, in the first place through the medium of Scythian tribes, of whom it is now well established that they acted as the mediators also of other motives of art in their transmission from Europe to Siberia, and thence farther to China. Unfortunately, among the antiquities of Siberia, no such bird-chariots have hitherto been discovered. Should chance ever bring one to light, the evidence of the migration of the Chinese dove-chariot, and of the idea underlying it, from Europe to China, would be settled beyond any doubt. After all, the Yakut specimen figured by Morse might be looked upon, if not as a survival of former ages, still as a promising factor pointing in the direction of other possible future finds on Siberian soil.

7. Modern toys set on wheels are not rare in Eastern Asia. Professor Conrady informs me that he once saw in Siam a toy made of straw, representing a bird running on wheels. In a collection of toys made by me in Peking, there is a butterfly set on an axle to which two wheels are attached, the whole made

---

[1] See also R. Virchow, l. c., p. 199.

[2] R. Virchow (Ibid.) mentions a bronze chariot with bull and bird heads which he acquired near Burg-on-the-Spree in 1865, when it was about to be worked into a child's toy.

of tin, and colored. Small models of carts are much in vogue as toys in the capital, and the cart-horse usually moves on four small solid wheels.[1] Although an historical connection between these recent toys and the ancient bird toy-carts cannot be directly demonstrated, there is much reason, after all, for the assumption of such a development.

8. My previous remarks are not by any means exhaustive as regards the archæological importance of these ancient Chinese bird-chariots of bronze: for, owing to their representation of wheels, they offer another source of study, from the view-point of ancient means of transportation. It is well known that carts are pictured in great numbers on the stone bas-reliefs of the Han time with several well-distinguished types, and a few others in relief are on metal mirrors of the same period. An immense amount of material is further stored up in Chinese literature regarding vehicles and modes of transportation in ancient and modern times; and the problem as to the origin, history, and distribution of wheeled vehicles over Asia, is one of no mean importance for the elucidation of oldest history.

[1] See, for instance, the figure in I. T. HEADLAND, The Chinese Boy and Girl (New York, 1901), p. 111.

043

亚洲琥珀历史札记

MEMOIRS

OF THE

AMERICAN ANTHROPOLOGICAL
ASSOCIATION

VOLUME I

LANCASTER, PA., U. S. A.

PUBLISHED FOR

THE AMERICAN ANTHROPOLOGICAL ASSOCIATION

1905–1907

HISTORICAL JOTTINGS ON AMBER IN ASIA

BY

BERTHOLD LAUFER

CONTENTS

213

HISTORICAL JOTTINGS ON AMBER IN ASIA
By BERTHOLD LAUFER

INTRODUCTION

Among the natural products that are of importance in the solution of archeological problems none has attracted wider attention than amber. While the predominating position held by it in prehistoric Europe and in classical antiquity may now be considered as fairly outlined, only sparse material is as yet available for the history of amber and of the trade in it in connection with Asia. A few remarks on amber have been translated from Chinese sources by Wilhelm Schott[1] and St Julien,[2] the latter of whom did not find much interest in this question.[3] Of similar character is that which is to be found in the book of Fred. Porter Smith.[4]

A. Pfizmaier[5] has translated eleven notes relating to amber, seemingly after the cyclopedia *Yen chien lei han*, without entering into a discussion of the subject. When K. G. Jacob[6] attempted to review the subject of amber in China from an historical point of view, he met with but scant material on which to base his studies, with the exception of some occasional communications made to him by Hirth and Arendt. Also in the latest and able work of F. de Mély and M. H. Courel,[7] which

---

[1] "Skizze zu einer Topographie der Producte des chinesischen Reiches," *Abhandlungen der Berliner Akademie der Wissenschaften*, p. 266, 361, 1842.

[2] *Industries anciennes et modernes de l'empire chinois*, p. 228, Paris, 1869.

[3] "Nous ne nous étendrons pas sur cette question qui n'offre du reste qu'un intérêt médiocre."

[4] *Contributions towards the Materia Medica and Natural History of China*, p. 12, Shanghai and London, 1871. Compare also A. G. Vorderman in *T'oung Pao*, I, p. 382, 1891.

[5] "Beiträge zur Geschichte der Edelsteine und des Goldes," *Sitzungsberichte der Wiener Akademie, phil.-hist. Cl.*, LVIII, p. 194–197, 1868.

[6] "Neue Studien, den Bernstein im Orient betreffend," *Zeitschrift der deutschen Morgenländischen Gesellschaft*, XLIII, p. 353 et seq., 1889.

[7] *Les lapidaires de l'antiquité et du moyen âge;* Vol. I, *Les lapidaires chinois*, Paris, 1896.

215

is the most extensive investigation of mineral products from Chinese literature, the problem of amber is only slightly touched on,[1] and the authors are tempted to believe that amber seems to have been little diffused in China.[2]   A brief note, of somewhat uncritical character, on the views of Chinese writers regarding the origin of amber, has been published by a Japanese author.[3] I shall not enter into a criticism of it.

Ever since Fritz Noetling,[4] of the Geological Survey of India, published his important researches regarding the amber mines of Burma, and proved that by far the greater part of this mineral[5] is there purchased by Chinese traders and transported into Yünnan province, it has seemed to me essential to search in Chinese documents for any references to this trade, or for confirmation of the provenience of Chinese amber from Burma. Such research has proved this supposition to be correct; and the confirmation of Noetling's investigation from Chinese sources, and the chronological definition of Burmese amber production and Burmese-Chinese amber trade, seem to me one of the chief

---

[1] Loc. cit., p. li.   "On connaît tous les problèmes soulevés par la question de l'ambre en Occident . . . Eut-il en Chine, dans l'antiquité, la même valeur commerciale qu'en Occident ?   Hirth n'a pas examiné la question à ce point de vue : je ne puis m'empêcher de m'y placer, quand il propose comme origine du mot *hou pe*, qui en cantonnais se prononce *fou pak*, l'ἅρπαξ grec dont le ρ tombait nécessairement en passant en chinois."

[2] Ibid., p. lviii.

[3] K. Minakata, "Chinese Theories of the Origin of Amber," *Nature*, LI, p. 294, 1895.

[4] "Das Vorkommen von Birmit (indischer Bernstein) und dessen Verarbeitung," *Globus*, LXIX, p. 217–220, 239–242, 1896.   English readers may be referred to a similar article by the same author— "On the Occurrence of Burmite, a New Fossil Resin from Upper Burma" — in *Records of the Geological Survey of India*, XXVI, p. 31–40, 1893 ; or to J. G. Scott and J. P. Hardiman, *Gazetteer of Upper Burma and the Shan States*, vol. II, pt. I, p. 289–295, Rangoon, 1900.

[5] According to investigations made by Otto Helm of Danzig, it differs in its physical and chemical qualities from the Baltic amber, or succinite, and is, moreover, distinct from any other known fossil resin.   Helm, therefore, suggested that it should be named *burmite*.   But it is justly remarked in the Gazetteer above quoted, p. 289 : "Since, however, the general outward appearance of the two is similar, there seems no reason why the name burmite should be any more generally adopted than the scientific term succinite has been up to the present."   For historical and archeological purposes I deem it preferable to speak of *Burmese amber*, which gives the well-known generic term and at the same time denotes the place of its origin.

results of the present paper, for which, in addition, I have extracted everything worthy of note regarding amber that is to be found in the Chinese cyclopedias and in other historical and geographical Chinese works.   Where passages from Chinese books without further data are quoted, they are derived from the *T'u shu chi ch'êng*, vol. 570, section on National Economy (*shih huo tien*), book 334, chapter on Amber.   Chinese sources give us accounts not only of the amber of Burma, but of that of many other regions.   The material is arranged geographically, and in each geographical section chronologically.   We treat *seriatim* of the amber in India, Tibet, Persia, the Roman Empire, Burma, Turkistan, and of modern European amber importations into China.   But first we give a complete translation[1] of what is written regarding amber in the *Pên ts'ao kang mu*, the great work on natural history by Li Shih-chên (end of sixteenth century), because it reviews all statements on the subject made by medical authorities since the earliest times, and contains no small amount of important geographical and historical data which will serve as a basis for the investigation that follows.

ANCIENT VIEWS ON AMBER

The account of the *Pên ts'ao kang mu* reads as follows :

Li Shih-chên says :

"When a tiger dies, its soul (spirit) penetrates into the earth, and is a stone.   This object resembles amber, and is therefore called *hu p'o* ('tiger's soul').[2]   The ordinary character is combined with the radical *yü* ('jewel '), since it belongs to the class of jewels.   The Sanskrit books call it *a-shih-mo-chieh-p'o*.[3]

The *Pieh lu* says :

Amber is produced in *Yung-ch'ang* (see p. 235).

T'ao[4] Hung-ching (A.D. 452–536) remarks :

---

[1] With the exception only of some passages expounding the medical properties and prescriptions of amber, which are outside the scope of the present paper.

[2] This interpretation is made *ad hoc* to account for the original writing of the apparently foreign loan-word *hu p'o* ('amber') with the characters *hu* ('tiger') and *p'o* ('soul ').   This way of writing is followed in the two *Han shu*.

[3] W. Schott (loc. cit., p. 361) identifies this word with Sanskrit *açmagarbha* (emerald), and assumes a confusion of these two precious substances.

[4] In the *T'u shu chi ch'êng* this name is written erroneously *Liu*.

There is an old saying that the resin of fir-trees sinks into the earth, and transforms itself [into amber] after a thousand years. When it is then burned it still has the odor of fir-trees. There is also amber, in the midst of which there is a single bee, in shape and color like a living one. The statement of the *Po wu chi*, that the burning of bees' nests effects its make,[1] is, I fear, not true. It may happen that bees are moistened by the fir-resin, and thus, as it falls down to the ground, are completely entrapped. There is also amber made by boiling chicken-eggs with the roe of the "dark" fish, but this is not genuine [that is, it is fictitious]. Only that kind which, when rubbed with the palm of the hand, and thus made warm, attracts mustard-seeds,[2] is genuine.[3] Nowadays amber comes from foreign countries and is produced in those places where *fu ling*[4] grows; on the other hand, however, there is nobody who

---

[1] See p. 235, 238.

[2] Pliny says: "When a vivifying heat has been imparted to it by rubbing it between the fingers, amber will attract chaff, dried leaves, and thin bark, in just the same way that the magnet attracts iron."

[3] It is very remarkable that this observation is quite correct. Our own naturalists also have recourse to the magnetic property of amber to distinguish it from spurious productions. O. C. Farrington, in his *Gems and Gem Minerals* (Chicago, 1903, p. 207), remarks: "Celluloid can be distinguished from amber by the fact that when rubbed it does not become electric, and gives off an odor of a camphor instead of the somewhat aromatic one of amber." K. G. Jacob (loc. cit., p. 355), says: "Arendt communicated to me from the twenty-fifth book of the *Shu wu i ming su* (eighteenth century) another foreign word for amber, *tun mou*." In this work, however, as may be seen from *Ko chih ching yüan*, book 33, p. 7 a, the term *tun mou* (written with the characters no. 12221 and 8044 in Giles's *Dictionary*) is quoted from the *Lun hêng*, a book of the famous philosopher Wang Ch'ung (A. D. 27–97), and thus goes back to the first century, that is, the time when the Chinese became familiar with Burmese amber; it may therefore be supposed that this word was derived from some Shan language. The passage of Wang Ch'ung runs thus: "*Tun mou* picks up mustard-seeds, *tun mou* is identical with *hu p'o* = amber." Next to the *Ch'ien Han shu* (p. 225), this is the oldest literary mention of amber, and the first mention in Chinese literature of the magnetic qualities of amber, with which, as is known, the ancients also were well familiar. The Sanskrit term *trnagrāhin* ('attracting grass') proves the same for India, and in Persian and Arabic we have the word *kahrubā* with a similar meaning; in Chinese, thus, *shih chieh* becomes a designation for amber. The priority in the observation of this natural law as regards Asia is secured to the Chinese, and it is by no means necessary to assume that their knowledge regarding this point was derived from the classical West. A keen observer and deep thinker like Wang Ch'ung, who refuted the popular notions of his time as to immortality with the acumen of a modern psychologist (see A. Forke, "Wang-Chung and Plato on Death and Immortality," in *Jour. China Branch Roy. Asiatic Soc.*, XXXI), may very well have been able to find out such a fact for himself.

[4] A false tuber, *Pachyma cocos* Fries (Bretschneider, *Botanicon Sinicum*, pt. III, p. 532–536). Its cinnamon-brown color, and the fact that it grows on fir-trees, are presumably the causes of its being associated with the evolution of amber.

is not aware of the fact that amber may occur in any place, whether there is *fu ling* there or not.[1]

Li Hsün[2] observes:

Amber is a secretion in the wood of the sea-fir. At first it is like the juice of the peach-tree; later it coagulates, and assumes form. Besides, there is southern amber (*nan p'o*), which, however, does not come to us on sea-going junks.

Han Pao-shêng[3] says:

The resin of the liquidambar-tree[4] penetrates into the earth, changes during a thousand years, and thus becomes amber. It is not only fir-tree resin that thus changes, but, generally speaking, it is tree-resins which penetrate into the earth, and all alter in the course of a millennium, and it is not that liquidambar and fir trees exclusively have resin during many years. When bee-nests are burned, the shapes of bees are inside in addition.

K'ou Tsung-shih[5] says:

Nowadays there is amber also among the Western Jung, the color of which is unevenly pale, or brilliantly clear. The amber of the southern regions is deep in color and cloudy. The people of those countries make objects and figures of it by turning it on a lathe.[6] It is said that, when

---

[1] The above sentence from the *Pieh-lu*, and the quotation from T'ao Hung-ching, have been translated by E. Bretschneider, *Botanicon Sinicum*, pt. III, p. 537. His translation of the latter passage is as follows: "The ancients say that the *hu p'o* is the resin of the fir-tree, which, being embedded in the soil during a thousand years, turns into amber. When burned, it emits an odor like that of resin. It sometimes incloses insects. An imitation of the *hu p'o* is produced by boiling hen's-eggs with fish-roe. The genuine *hu p'o*, when rubbed between the hands till it becomes hot, will attract straw. Now, all the *hu p'o* in China is brought from foreign countries." It will be noted that the last clause is omitted in this translation; for the rest, this is all that is given by Bretschneider regarding amber, from the *Pên ts'ao kang mu*.

[2] Author of the *Hai yao pên ts'ao*, an account of the drugs of southern countries, in six books, second half of the eighth century. E. Bretschneider, *Bot. Sin.*, pt. I, p. 45.

[3] Author of the *Shu pên ts'ao*, the materia medica of Szüch'uan province, compiled about the middle of the tenth century. See E. Bretschneider, loc. cit., p. 46.

[4] *Fêng*. Regarding this tree, see Th. Sampson, *Notes and Queries on China and Japan*, III, p. 4-7, 1866.

[5] A celebrated physician of the Sung dynasty, author of the *Pên ts'ao yen i*, published *ca.* 1115 A.D. E. Bretschneider, loc. cit., p. 48.

[6] *Nien*, originally used for the husking of rice by means of a stone roller. Chao Ju-Kua uses this word for the cut glassware of Bagdad (Hirth, *Die Länder des Islam nach chinesischen Quellen*, p. 42, note 3; p. 48, note 3). In Europe, also, amber is worked by turning it on a lathe or by cutting by hand.

in a thousand years *fu ling* is transmuted [into amber], bees and ants will stick to it as if artificially arranged ; but this is not the case with the majority [of amber-pieces].

The *Ti li chi* says :

Hainan and Lin yi (Champa) produce much amber. The resin of fir-trees, filtering into the earth, changes, and is then amber. Around it there are no plants. It extends underground to a depth of at least from five to six feet, and at most from eight to nine feet. There are large pieces of the size of a *hu*.[1] By cutting off its rind, it becomes perfect. This statement is convincing ; but it must be considered that the soil is either suitable or not, and that accordingly the transformation takes place or not. As to the account of the burned bees, I do not know on what evidence it rests.

Su Sung[2] says :

What all people say regarding *fu ling* coincides, although there are slight discrepancies. All agree in stating that it arises from the transformation of fir-resin into *fu ling*, so that the spirit of the *fu ling* is that of the great pine-tree. When the latter is broken or felled without the root being injured, and thus not decayed, its sap flows down and coagulates. It therefore cures the heart and kidneys by pervading them with juice. Now, amber originates thus : When the branches and joints of the pine-tree are still flourishing, they are scorched, especially under the influence of the hot sun. Then the resin flows out of the trunk of the tree, and thickens in large masses on the outside, where the sun strikes it. Thereupon it sinks into the earth, and the juice, moist in the beginning, trickles into the earth for many years, where finally it is preserved only as a lustrous substance. Now it is capable of attracting mustard ; still, however, it keeps its adhesive properties. This is the reason that all sorts of insects stick to it, which happened before the time it penetrated into the ground. There are, accordingly, two substances which are produced out of the pine-tree, but which are each different in their nature. *Fu ling* arises in the female principle, and is completed in the male principle. Amber arises in the male principle, and is completed in the female principle. Both, therefore, cure, regulate, and tranquilize the heart, and stimulate the water.

Lei Hsiao[3] says :

---

[1] A corn-measure holding five or ten pecks.

[2] A high functionary and distinguished scholar under the Sung dynasty, author of the *T'u ching pên ts'ao*, illustrated materia medica, published between 1057 and 1090. E. Bretschneider, loc. cit., p. 47.

[3] Author of the *P'ao chi lun*, a pharmacology, in the fifth century. E. Bretschneider, loc. cit., p. 41, 42.

Generally, we must follow this division : red fir-tree resin, stone amber, water amber, flower amber, "amber of objects and figures," black (jet) amber (*hsi p'o*), and amber proper (*hu p'o*). Among these, red fir-tree resin is like amber, except that it is dull, in large pieces, and brittle, with streaks extending crosswise. As regards water amber, there are many pieces that are not red, but, rather, light in tinge ; among those which are yellow, there are many with furrowed stripes. Stone amber is heavy like stone, yellow in color, but not fit for use. Flower amber resembles " new horse-tails " [1] and the inner part of the pine-tree : it has streaks alternately red and yellow. The " amber of objects and figures " contains objects in its interior, and enters into wonderful relations with the destiny of man. Jet amber is the most excellent of the " figure " ambers. The amber proper is of blood-red [2] color ; rubbed with a piece of cloth and made warm, it can attract mustard-seeds, which is a proof of its genuineness.

Li Shih-chên says :

Amber attracts mustard-seeds, that is, plants in general, and mustard-seeds and grains. The statement of Lei Shih, that it attracts only mustard-seeds, may be an error. The Annals of the T'ang Dynasty contain the fact that in the western regions there is the Kan river in Soghdiana, where the wood of the fir-tree lies in the water, and changes to stone after one or two years. This exactly coincides with the fact that all wood coming from fir and liquidambar [3] trees discharges a secretion into the ground, which changes into amber. This is a universal law. Nowadays there is also amber in Li chiang, [4] in the country of Chin ch'i (' gold teeth '). The statement that *fu ling*, after a thousand years, changes into amber, is also an erroneous tradition. Ts'ao Chao says in his book, *Ko ku yao lun*, [5] that amber is produced with the Western Barbarians (*Hsi Fan*, Kukunor region, and Tibet) and the Southern Barbarians (*Nan Fan*). It is the secretion of the liquidambar-tree that thus changes after many years. The kind the color of which is yellow and beautifully transparent is called *la*

---

[1] Doubtless the name of some plant, which, however, I have not yet been able to identify.

[2] A color attributed also to Sicilian amber.

[3] The connection of the liquidambar (= liquid amber !) tree with the origin of amber doubtless arose from the semi-fluid resin produced by this tree, identical with the storax or liquid storax of commerce, in Chinese *su ho* (see F. Porter Smith, *Contributions toward the Materia Medica*, etc., p. 187).

[4] Also in the *Ta Ch'ing i t'ung chi*, the Geography of the Chinese Empire of the present dynasty (book 382, p. 5 a), amber is mentioned as a product of *Li chiang fu*.

[5] Written in the beginning of the fifteenth century (see E. Bretschneider, loc. cit., p. 162).

p'o ("wax amber").[1]   If the color resembles the red of the fir-resin, and is, moreover, yellow, its name is *ming p'o* ("bright amber").   There is a fragrant kind called *hsiang p'o* ("fragrant" or "incense amber"). The sort produced in Korea (*Kao li*) and Japan (*Wo kuo*)[2] is of a deep-red color.   There is another kind called "fir-branches of bees and ants" (*fêng i sung chi*), which is still better.

HSI AMBER

A special kind, called *hsi*[3] amber, is then treated in the *Pên ts'ao*, as follows:

---

[1] K. G. Jacob (*Zeitschrift der deutschen Morgenländischen Gesellschaft*, XLIII, p. 356, 1889) remarks: "It is interesting that the Chinese mention wax amber just like the Arabs." I think that there is nothing remarkable about this fact, the term "wax" being used simply as a color-designation of world-wide application. Palladius (*Chinese-Russian Dictionary*, p. 483) translates *la p'o*, "light-yellow amber"; Eitel (*A Chinese Dictionary in the Cantonese Dialect*, p. 510 a) the same expression by "sparkling yellow amber"; and Couvreur (*Dictionnaire classique de la langue chinoise*, p. 580 c) by "ambre jaune." Also Pliny (*Hist. Nat.*, 37, 11) speaks of amber of waxen color (cerinum) found in Scythia, and says that white and wax-colored amber was valueless and used merely for fumigating (Blümner, *Technologie und Terminologie*, II, p. 386). Even the naturalists of our own day employ this attribute in the description of certain kinds of amber. G. F. Kunz, for example, in his *Gems and Precious Stones of North America* (New York, 1890, p. 199), says of a species of amber discovered in America that it is wax- or honey-yellow. This proves sufficiently that the coincidence in the terminology emphasized by Jacob is quite a natural incident.

[2] So far as I can see, there are in Chinese literature two references with regard to tribute of amber sent from Japan. In the fifth year of the period *Yung hui* (A.D. 654), corresponding to the Japanese period *Hakuchi* (in Chinese, *Pai chi*, "white pheasant"), under the Emperor Kao Tsung of the T'ang dynasty, the Japanese Emperor Toyoshi-karu-no-Oji (in Chinese, *Hsiao Têh*, 645–654) despatched a tribute of amber and agate, the piece of amber being defined as being the size of a "peck" (tou). Another gift of amber from Japan is mentioned for the fifth year of the period *Hsi ning*, or A.D. 1072, under the Sung dynasty (*T'u shu chi ch'êng*, vol. 1338, *pien i tien*, bk. 33, *Ji pên pu hui k'ao*, I, p. 9 a, 14 b). Jacob, loc. cit., refers, in regard to Japanese amber, to E. Kämpfer's *Beschreibung von Japan*, II, p. 470, 1779. O. Münsterberg, *Japanische Kunstgeschichte*, II, p. 89 (Braunschweig, 1906), figures a metal mirror of the Nara epoch, A.D. 709–784, on the back of which flowers and leaves are formed of engraved pieces of mother-of-pearl and red amber. In the recent trade statistics of Japan I do not find reference to amber.

[3] Character no. 4174, in Giles's *Dictionary*. Giles defines the term as "a kind of jet, described as a mineral amber of a clear black color." I have no doubt that the species described in the following is identical with the gagate of the ancients, as it strikingly agrees with the description given of this mineral by Pliny, and with that of jet in our own books on mineralogy. On gagate compare H. Blümner,

Lei Hsiao (fifth century) says :

Hsi is the most excelling of all ambers, hence its designation *hsi p'o.*

Li Shih-chên says :

There is also *hsi* that is produced : its color is deep black, hence its name.

Su Kung (seventh century) says :

According to an old tradition, fir-resin alters into *fu ling* in a thousand years. In another thousand years it becomes amber. In another thousand years it becomes *hsi.* When these two substances are burned they emit the odor of firs. In shape they resemble black jade, and are light. They are produced in the country of the Western *Jung*, where there are places in which *fu ling* occurs. Pieces of these substances which are now obtained in a stony desert three hundred *li* south from Hsi chou [Yar-khoto, near Turfan] are large, measuring a foot square, black, moist, and light. When burned they emit a strong odor.[1] The people of Turfan (*Kao ch'ang*) call it *mu hsi* ("wooden *hsi*"),[2] and the black jade they call *shih hsi* ("stone *hsi*"). Among the stones found in the soil of Kung chou [in Honan ?] there are those which, when burned, smell like a fir-tree, in quality equal to that of amber. It is customary to make it into vessels incapable of being burned or broken.[3] I fear that neither these two kinds nor amber are products of fir-resin.

T'ang Shên-wei[4] says :

In the *Liang kung tsü chuan,*[5] Nai Kung remarks : "In the plains and stony deserts in the territory of Kiao ho,[6] the soil is dug a hundred

---

Technologie und Terminologie der Gewerbe und Künste bei Griechen und Römern, III, p. 67, 68, Leipzig, 1884.

[1] Compare Pliny's description of gagate (*Hist. Nat.*, 36, 34) : "Niger est, planus, pumicosus, levis, non multum a ligno differens, fragilis, odore, si teritur, gravis." ("It is black, smooth, porous, and light, differs but little from wood, is brittle, and emits a disagreeable odor when rubbed.")

[2] Compare Pliny's "non multum a ligno differens" and the description of jet by O. C. Farrington, *Gems and Gem Minerals* (Chicago, 1903), p. 210 : "Jet is a variety of coal which, being compact, takes a good polish, and hence can be used in jewelry. It is a kind of brown coal or lignite, and retaining, as it does, some of the original structure of the wood," etc.

[3] Compare Pliny (loc. cit.) : "Fictilia ex eo inscripta non delentur, cum uritur, odorem sulpureum reddit, mirumque, accenditur aqua, oleo restinguitur."

[4] Physician, author of the *Chêng lei pên ts'ao*, compiled in 1108. Bretschneider, loc. cit., p. 47.

[5] Presumably identical with the *Liang ssü kung tsü chi*, Chronicle of the Four Worthies of the Liang Dynasty (502–566), by Chang Yüeh (667–730). Bretschneider, loc. cit., p. 169.

[6] Capital of Turfan at the time of the T'ang. Chavannes, *Documents sur les Toukiue (Turcs) occidentaux*, p. 336 b.

feet deep ; below there is *hsi* amber of an extremely pure black, some pieces being as big as a cart-wheel. A powder made from it serves as a remedy for female complaints of the urinary intestines, obstruction of the bowels, and other diseases.

Li Shih-chên says :

Hsi is, of all ambers, that of blackest color,[1] according to the one, because the tinge of the earth colors it with its odors ; according to others, it is only one kind of wood, emitting an exudation, which then solidifies. But it is not necessarily thousand-year amber which undergoes this transformation.

Wang Ts'ê-ching says :

Resin, after a thousand years, makes *fu ling ; fu ling*, after a thousand years, makes amber ; amber, after a thousand years, makes stone-gall (?) ; stone-gall, after a thousand years, makes *wei hsi* ("sublime happiness" ?). Generally, this is all superstitious[2] talk. One cannot wholly depend on Lei Hsiao, either. As regards its odor and taste, it is evenly sweet, and it is non-poisonous. As regards its medical virtues, the *T'ang pên* (A.D. 650) says : "It re-animates the heart, calms the soul [so as to cause sleep], stops bleeding, produces flesh, and in women cures obstruction of the bowels."

Ch'ên Tsang-ch'i[3] says :

Small boys carry it as an amulet ;[4] ground, and dropped into the eyes, it prevents a cataract.

We thus see that the earliest mention of this black substance occurs in the fifth century, and that it was introduced into the pharmacopœia as early as the seventh century. From the Chinese account we learn the fact, not unimportant for the geography of minerals, which, so far as I know, was hitherto unknown to our mineralogists, that gagate, or jet, is, or at least was at a certain time, found near Turfan in Turkistan, and that the Turkish inhabitants of this region were acquainted with it, and probably dug it and sent it to China. This importation

---

[1] Compare Farrington (loc. cit.) : "Jet is sometimes known as 'black amber,' a name not inappropriate when the similarity in origin of the two minerals is considered."

[2] *Shên i ;* literally, "strange things about spirits," "supernatural wonders."

[3] Author of the *Pên ts'ao shi i,* "Omissions in Previous Works of Materia Medica." He lived in the first half of the eighth century. Bretschneider, loc. cit., p. 45.

[4] According to Pliny, amber is beneficial to infants, attached to the body in the from of an amulet.

seems to have continued down to the twelfth century, as is evident from the observations of T'ang Shên-wei, who recommends the application of jet in certain diseases of women. He is chronologically the last author quoted by Li Shih-chên who does not appear to have known the substance from personal observation, but who merely reasons on the statements of his predecessors. After the time of T'ang Shên-wei, this mineral disappeared from the market, on account of lack of importation, as we may infer from the quotations given in K'ang hsi's Dictionary (*sub voce*) — one, from the dictionary *Chi yün* (middle of the eleventh century), defining *hsi* as a " beautiful stone of black color" (apparently traceable back to Su Kung); the other from the later dictionary, *Chêng yün*, which explains this word as "black jade," with the remark that the old explanation, " a beautiful stone of black color," is wrong — which may be taken as sure signs that the material then no longer existed in the empire. In the modern materia medica of China, jet does not occur. It is true that Doolittle[1] gives two equivalents for the word "jet: (1) *pu huei mu*, which is a mistake, as this term denotes asbestos ; (2) *hei yü*, that is, black nephrite, which seems to be based merely on the above-quoted literary sources, perhaps K'ang hsi.

DISTRIBUTION OF AMBER

INDIA

In the *Ch'ien Han shu* ("Annals of the Former Han Dynasty"), written by Pan Ku, who died A.D. 92, and continued and completed by his sister Pan Chao, we hear of amber for the first time in Chinese literature ; the term never occurs in the classical texts. This historical work contains an account of the western countries (*Ch'ien Han shu Hsi yü chuan*), which relates that the country of *Ki pin* produces amber (there written with the characters for " tiger's soul "). This geographical name denoted Cashmir at the time of the Han and Wei.[2] Relations between China and Cashmir commenced at the time of the Emperor Wu (140–85 B.C.),

---

[1] *Handbook of the Chinese Language*, I, p. 269.

[2] Chavannes, *Documents sur les Tou-kiue (Turcs) occidentaux*, St Petersburg, 1903, p. 336 ; see also Schlegel in *T'oung Pao*, série II, vol. I, p. 329, 1900.

and several embassies were despatched from that country to the Chinese court, the last to the Emperor *Ch'êng* (32–7 B.C.), when intercourse with Cashmir ceased, on account of its inaccessibility, till the Wei dynasty (A.D. 386–532).[1] It cannot be presumed, therefore, that during this short period any trade was begun between China and the remote regions of northwestern India, and, still less, that amber from the latter country, in quantities of any importance, could have reached the Chinese Empire. It is certainly not out of the question that the embassies of Cashmir brought along samples of this product for presentation to the court, though nothing is said definitely in our texts regarding this point. Owing to political and geographical conditions, any extensive trade between the two countries was handicapped at the outset; and I think the view of Palladius,[2] that amber was formerly exported from *Ki pin* to China, cannot be accepted. Still, the Chinese account is of great value, for it proves that amber was known in that part of India during the first century before Christ; and the more so, as ancient Sanskrit literature, at least to my knowledge, seems to be silent on this subject. It is not mentioned in the mineralogical section of the *Nighaṇṭurâja*,[3] although the author of this book, Narahari, was a physician from Cashmir.[4] For the rest, our knowledge of Indian amber is based chiefly on the statements of Pliny, in his *Historia Naturalis*, of which chapter 11 of book 37 is devoted to amber. Pliny there alludes to Indian amber in three passages based on the authority of three different informants. First, he reproduces the statement of Nicias, that amber is found also in India, where it is held as a preferable substitute for frankincense.[5] Second, he says, after Ctesias, that there is in India a river called

---

[1] Regarding details, see O. Franke, "Beiträge aus chinesischen Quellen zur Kenntnis der Türkvölker und Skythen Central-Asiens," *Aus dem Anhang zu den Abhandlungen der Königl. Preuss. Akademie der Wissenschaften*, Berlin, 1904, p. 58, 59, 63, 64.

[2] *Chinese-Russian Dictionary*, I, p. 483 b.

[3] Edited and translated by R. Garbe, *Die indischen Mineralien*, Leipzig, 1882.

[4] Ibid., p. vi.

[5] *The Natural History of Pliny*, translated by Bostock and Riley, VI, p. 399, London, 1857.

Hypobarus, a word which signifies " bearer of all good things " ;
that this river flows from the north into the Eastern ocean, where
it discharges near a mountain covered with trees which produce
electrum ; and that these trees are called *siptachoræ*, the meaning of which is " intense sweetness." [1]   His third testimony is
the most important, as it shows that amber was traded from
India to western Asia.   It reads as follows :

> That amber is found in India too, is a fact well ascertained.   Archelaus,
> who reigned over Cappadocia, says that it is brought from that country
> in the rough state, and with the fine bark still adhering to it, it being the
> custom there to polish it by boiling it in the grease of a sucking-pig.
> One great proof that amber must have been originally in a liquid state is
> the fact that, owing to its transparency, certain objects are to be seen
> within — ants, for example, gnats, and lizards.   These, no doubt, must
> first have adhered to it while liquid, and then, upon its hardening, have
> remained enclosed within. [2]

Archelaus, mentioned in this passage, governed Cappadocia as
a Roman province under the Emperor Tiberius.   The fact that
amber was a native product of ancient India is thus confirmed
beyond doubt.

On the other hand, while the records of the ancients do not
hint at the importation into India of amber from Europe, and
while nothing to this effect can be gained from Indian literary
sources, it is curious and surprising to note that the matter-of-
fact Chinese grant us an opportunity of establishing the fact
that such a commercial relation between the imperium Romanum
and India existed.   If anything, such an example reflects the
highest credit on the wonderfully developed historical sense of
the Chinese.   This notice is found in the *Liang shu*, ("Annals of the Liang Dynasty "), written about A.D. 629 and relating to the period A.D. 502–556.

In A.D. 503 an embassy arrived from India at the court of
the Liang with a tribute of native products.   In the description
of India given in the *Liang shu* it is related that the western
part of India holds intercourse with Syria or the Roman Orient
(*Ta Ts'in*) and Parthia (*An hsi*), that the trade goes by way of

---

[1] Ibid., p. 400.

[2] Ibid., p. 402.   Compare also Chr. Lassen, *Indische Altertumskunde*, III, p. 32.

the sea, and that there are products of Syria in great number there, like precious objects, coral, *amber*, gold, the jade *pi*, pearls, *lang kan*,[1] saffron (*yü kin*),[2] and storax.[3]

From Chinese sources we learn also that amber was employed in the most northern and eastern parts of India. In the description of Nepal given in chapter 221 of the *Old History of the T'ang Dynasty*, it is related that the king of Nepal adorns

---

[1] F. De Mély and H. Courel, loc. cit., p. 56, 258.

[2] Bretschneider (*Botanicon Sinicum*, pt. II, p. 232) left the term *yü kin hsiang* undefined. I am led to the conclusion that *yü kin hsiang*, as well as the simple *yü kin*, whatever its meaning may be with regard to native plants, when applied to foreign and particularly to Indian countries, denotes the true *saffron* of India, for the following reasons :

(1) In the description of India given in the *Liang shu*, it is expressly stated that *yü kin* is produced *solely* in *Kï pin*, that is, Cashmir (*T'u shu chi ch'êng*, vol. 1242, *pien i tien*, bk. 58, *T'ien chu pu hui k'ao*, p. 2 b). Now, Cashmir is the classical land famed for the cultivation of saffron, which was and is thence exported to the rest of India, Tibet, and, farther, to Mongolia and China. The industry is very ancient in Cashmir. See W. R. Lawrence, *The Valley of Kashmir*, p. 342–344, London, 1895 ; also *Mémoires de la société finno-ougrienne*, XI, p. 66–68, 1898, where I have given the history of the distribution of saffron and its various designations.

(2) Hsüan Tsang (St Julien, *Voyages des pèlerins bouddhistes*, II, p. 40, 131) mentions the same plant in the Hindukush and in Cashmir ; and I Tsing (J. Takakusu, *A Record of the Buddhist Religion*, etc., p. 128), its occurrence in North India, where Takakusu correctly identifies it with Sanskrit *kuṅkuma*, that is, "saffron," *Crocus sativus* L., but, strangely enough, again contradicts this statement in a footnote, in which he explains *yü kin hsiang* as Japanese golden turmeric, a species of *Curcuma*. Nearly all Chinese, Japanese, Arabic, and Western authors usually confound these two plants, although they are not only distinct species but even belong to quite different families (*Crocus = Irideæ*, *Curcuma longa = Zingiberaceæ* or *Amomeæ*).

(3) Li Shih-chên, in his *Pên ts'ao kang mu*, gives as the Sanskrit equivalent for *yü kin hsiang* the term *ch'a kü mo = çakama* (Tibetan, *sha-ka-ma*), denoting the saffron of Cashmir. He further refers to Ta Ts'in (western Asia) as the country where this plant is produced, which is quite correct and makes sense only when "saffron" is understood by it, the home of which is in Asia Minor. Compare also Hirth in *Jour. China Branch R. A. Soc.*, XXI, p. 221, 1886. Contrary to the statement of F. Porter Smith (*Contributions to the Materia Medica*, etc., p. 223), that saffron is not procurable in Hankow, I may say that genuine saffron may be obtained [there as well as in Peking, in any drug-store ; to be sure, not under the name *yü kin* or *yü kin hsiang*, restricted to the written language and not understood by the present Chinese, but under the colloquial term *hung hua*, that is, "red flowers." Saffron is highly valued by the Chinese and is very frequently used by women for menstrual disturbances.

[3] Compare also Hirth, *China and the Roman Orient*, p. 47.

himself with genuine pearls, rock-crystal, mother-of-pearl, coral, and *amber*.[1]   And the geographical work, *Ying yai shêng lan*, written A.D. 1416, informs us that amber was a product of Bengal.[2]   Thus we can trace the occurrence of amber in India from the first century before Christ down to the end of the middle ages.

Of amber in modern India we have unfortunately no information,[3] and it seems almost as if the ancient natural resources of amber were long since exhausted.   To discover the sites of the mines remains a task still to be performed, in which archeologists and naturalists are equally interested.   I cannot find any references to the occurrence of amber in Cashmir.   Walter R. Lawrence,[4] who wrote the most comprehensive work on this region, has only the following to say regarding economic mineralogy :

The lapidaries import all their more valuable stones, such as agate, bloodstone, cornelian, cat's-eye, garnet, lapis-lazuli, onyx, opal, rock-crystal, and turquoise, from Badakshán, Bukhárá, and Yarkand.   There are, however, certain local stones for ornaments and buttons.   These stones are soft, and are incapable of a high polish.

Neither does A. Cunningham[5] enumerate amber among the mineral productions of the country.,   Lassen, in his *Indische Altertumskunde*, makes no allusion to amber, except the one passage above quoted, derived from Pliny.   Neither does it seem to figure in the materia medica, as it does not appear in Jolly's *Medicin*,[6] nor did we come across it in our contributions to the medicine of the Tibetans.[7]

---

[1] Sylvain Lévi, "Le Népal, Étude historique d'un royaume hindou," I. *Annales du Musée Guimet, Bibliothèque d'études*, XVII, p. 164, Paris, 1905.

[2] *T'u shu chi ch'êng*, vol. 1242, *pien i tien*, bk. 58, *T'ien chu pu chi shih*, II, p. 1 b.

[3] According to Noetling (*Globus*, loc. cit., p. 241 b), the Indian fossil resins have not yet been investigated.

[4] *The Valley of Kashmir*, pp. 64, 65, London, 1895.

[5] *Ladák, Physical, Statistical, and Historical*, p. 229–237, London, 1854.

[6] *Grundriss der indo-arischen Philologie und Altertumskunde*, III, no. 10, Strassburg, 1901.

[7] Heinrich Laufer, *Beiträge zur Kenntnis der tibetischen Medicin*, two parts, Berlin and Leipzig, 1900.   But in the well-known Chinese-Tibetan pharmacopœia

TIBET

In Tibet amber seems to be an article much in use. According to a Chinese account,[1] high and low all wear one or two strings of prayer-beads around the neck; they are made of coral, lapis-lazuli, mother-of-pearl, or even wood; the wealthy wear amber ones, the beads being sometimes as big as cups. Amber appears further among the gifts of tribute sent by the Dalai Lama to the Chinese Emperor,[2] and is mentioned, in another connection, among presents sent to China in the seventeenth century.[3] The Tibetan tribes seem to be, with the Shan, the only ones in Asia among whom amber plays a more extensive ethnographical rôle. The K'amba women in northeastern K'amdo wear a form of head-ornament consisting of a discoidal piece of amber about two inches and a half in diameter, with a coral bead in the center.[4] These amber disks are transported as far as the western parts of Kansuh province, and Rockhill[5] assumes that they are imported from India through Lhasa. The same explorer found amber in rough pieces procurable in Kumbum.[6] No place, to my knowledge, is known in Tibet where amber is actually found. The native designations are not suggestive of an autochthonous occurrence of the material. The usual name for it, *spos shel* (pronounced *pö-shel* or *pö-she;* in Lepcha,[7] *po-she*), means literally "perfumed crystal" (perhaps from the resinous odor which it emits in burning, or even as an indication of a former utilization of amber in the way of incense),

---

(ibid., p. 11; Bretschneider, *Botanicon Sinicum*, pt. 1, p. 104), published by the pharmaceutical firm WAN I-HAO in Peking, amber is listed as a drug, so that this is very likely due to Chinese influence.

[1] W. W. Rockhill, "Tibet," *Jour. Roy. Asiatic Soc.*, p. 225, 1891.

[2] Ibid., p. 244.

[3] Ibid., p. 204.

[4] W. W. Rockhill, *The Land of the Lamas*, p. 184, London, 1891. Idem, *Diary of a Journey through Mongolia and Tibet*, p. 103, Washington, 1894.

[5] *The Land of the Lamas*, p. 60, 110.

[6] W. W. Rockhill, *Diary of a Journey through Mongolia and Tibet*, p. 103.

[7] J. D. Hooker (*Himalayan Journals*, I, p. 122, London, 1855) remarks: "The Lepcha are fond of ornaments, wearing silver hoops in their ears, necklaces of cornelian, amber and turquoise, brought from Tibet, and pearls and corals from the south."

which name, at all events, seems to hint at an importation of the mineral. Another, a literary, designation is *sbur len*, or *sbur long*, which appears simply as a literal translation from the Sanskrit *tṛṇagrāhin* ("attracting straw"). The Tibetan *se-mo-do* is translated by Sarat Chandra Dás,[1] "necklace of amber"; by Jäschke,[2] however, by "a kind of ornament, for example, made of pearls." This term can by no means be a word of Tibetan origin, and I am inclined to suppose that in it we have the original Turkish form of *mu hsi*, which, according to Su Kung, the author of the *T'ang pên ts'ao* (A.D. 650), was the Uighur designation for jet or black amber (see p. 223); the Turkish equivalent for *mu* in *mu hsi*, being Chinese for "wood," is *modo*, so that turkicized it would yield *simodo*, which seems identical with the above Tibetan loan-word *semodo*.

ROMAN EMPIRE

The Chinese were acquainted with the fact that amber was an article used in the Roman Empire. We learn this from two main sources — first from the *Tien lio*, in the words that the country of *Ta Ts'in* (Syria) possesses much amber; and, second, at a much later period, in the Accounts of Western Countries, in the Annals of the T'ang Dynasty (*T'ang shu Hsi yü chuan*), which say that "the soil of *Fu lin*, which is the ancient *Ta Ts'in*, has much amber." The *Tien lio*, as finally proved by Chavannes,[3] is identical with the *Wei lio*, and was written by Yü Huan between A.D. 239 and 265. These are the only two accounts regarding amber in the Roman Orient quoted in the *T'u shu chi ch'êng*, in the chapter on amber. But we see from Hirth's book[4] that the same fact is recorded also in the *Hou Han shu*, which, however, was compiled only in the fifth century, so that the editors of the cyclopedia may have omitted this passage for the reason that they thought it to be a repetition from the *Wei lio*. The other notice is found in the *Chiu T'ang shu*, as well as in the

---

[1] *A Tibetan-English Dictionary*, p. 1274, Calcutta, 1902.

[2] *A Tibetan-English Dictionary*, p. 575 b, London, 1881.

[3] "Les pays d'occident d'après le Wei lio" (*T'oung Pao*, sér. II, vol. VI, p. 519, 1905).

[4] *China and the Roman Orient*, p. 41.

Hsin T'ang shu.[1] The passages above quoted do not admit of the inference that amber was exported from the Roman Orient to China ;[2] their statements are merely restricted to the fact that amber was one of the characteristic products of these western regions.[3] Even granting that the *Hou Han shu* was compiled from documents actually written under the Later Han dynasty (A.D. 25–220), we must admit that before the existence of amber in the antique world came to the ears of the Chinese, they were acquainted with this mineral, first, by means of accounts received from India,[4] and, second, through actual exportation of it from the region of the Shan States into Yünnan (see p. 233 et seq.). Considering this fact, the mere supposition that a trade in amber might have been carried on from Syria to the Far East must lose much of its probability, as Syrian amber, aside from the costliness of all Syrian products, due chiefly to the expense of transportation over the long overland route, could not have competed with the doubtless cheaper Burmese amber, and could hardly have tempted the Chinese to buy. Nor is there to be found in other than historical sources any reference to amber from western Asia except, in going back to the thirteenth century, the single passage in Chao Ju-Kua, who enumerates amber among the products of the West which were brought to Palembang, Sumatra, for transshipment to the Chinese port of Ch'üan chou fu.[5] These articles are stated in a general way to have come from Arabia, so that amber also should be located there, though the vagueness of the wording gives hardly a clew to the tracing of its origin.

---

[1] Ibid., p. 55, 59.

[2] As, for example, is assumed by H. Nissen, "Der Verkehr zwischen China und dem römischen Reiche," *Bonner Jahrbücher*, xcv, p. 19, 1894.

[3] Taking the sentences in the above accounts in their strictest sense, we cannot fail to notice that the verb *ch'u*, which implies the signification of production as well as of exportation, is not there employed in any case, but that the verb is omitted, so that it seems justifiable to supplement only the copula "there is," "it has," etc.

[4] While the Chinese account of the occurrence of amber in India, particularly in Cashmir, goes back to the first century before Christ, the record regarding the same material in the Roman Orient indubitably refers only to post-Christian times, and in my opinion hardly antedates the third century.

[5] Hirth, *Chinesische Studien*, I, p. 39.

PERSIA

Amber is mentioned also as a production of Persia in the Chinese Annals. This notice occurs first in the *Liang shu*, the Books of the Liang Dynasty (A.D. 502–556), under which an embassy from Persia, with tribute, arrived in 547; second, in the *Pei Wei shu*, or Books of the Northern Wei Dynasty (A.D. 386–532); and, third, in the *Sui shu*, or Annals of the Sui Dynasty (A.D. 589–618).[1] But it is not stated that amber was exported from Persia into China.

BURMA

The fact that the Chinese became acquainted at an early date with what we now call "Burmese amber" is seen from the statement in the Annals of the Later Han Dynasty (*Hou Han shu*),[2] that *Ai lao* produces lustrous pearls and amber. *Ai lao* (probably the same as Laos) is the Chinese name of the ancient Shan kingdom, first appearing in history during the first century of our era, whose dominions once extended far into Ssŭch'uan and Kueichou, embracing nearly the whole of Yünnan and parts of Tonkin and Kuanghsi.[3]

In Parker's translation[4] of the account of the country of the *Ai lao*, from the *Hou Han shu*, the passage is given in extenso, and the production of *amber* there appears together with copper, iron, lead, tin, gold, silver, bright pearls, crystal, oyster-pearls, kingfishers, rhinoceroses, elephants, baboons, and tapirs.

In the *T'ang shu*,[5] it is said regarding the *Nan chao*, the descendants of the *Ai lao*, that the nobles ornament their ears with

---

[1] *T'u shu chi ch'ĕng*, vol. 1242, *pien i tien*, bk. 56, *T'iao chi pu hui k'ao*, p. 2 a, 3 a, 4 a.

[2] In the Account of Eastern Barbarians (*Tung i chuan*).

[3] See E. H. Parker, "The Early Laos and China," *China Review*, XIX, p. 67–106, 1890, and "The Old Thai or Shan Empire of Western Yünnan" (ibid., XX, p. 337–46), where its history is given from Chinese sources. G. Devéria, *La frontière sino-annamite*, p. 117, Paris, 1886; E. Rocher, "Histoire des princes du Yunnan," *T'oung Pao*, X, p. 19, 1899. See also A. v. Rosthorn, *Die Ausbreitung der chinesischen Macht in südwestlicher Richtung*, p. 42 et seq., Wien, 1895.

[4] *China Review*, XIX, p. 70 a.

[5] Ibid., p. 76 b.

pearls, green-stone, and *amber*,[1] which shows that amber was actually used by the Shan.

I think that the first acquaintance of the Chinese with this amber may date from the first century of our era, when their relations with Yünnan and its manifold tribes became more intimate. During the following centuries the references in Chinese literature to the amber of this region become more frequent, and we see that the amber utilized in China was actually supplied from the region named, and is located in such places of southern Yünnan as are near the Burmese frontier, along the ancient trade-route leading from Burma to southwestern China.[2]

The first dictionary that alludes to the location of this amber is the *Kuang ya*,[3] which contains the following passage :

Amber is a pearl.[4] Above and beside it, no plants grow. The least depth [in which it occurs in the soil] amounts to five feet ; the greatest depth is from eight to nine feet. It is as big as a *hu* [a measure holding ten pecks]. By cutting off the rind, the amber is obtained. At first it is like the gum of the peach-tree ;[5] but by being stiffened and hardened it assumes form. The people living in that district work it into head-pillows. It is produced in *Po nan hsien*.[6]

---

[1] Compare also Parker in *China Review*, XX, p. 341 a. This is the first historical reference to the *nadonay*, cylindrical ear-plugs of amber, still worn by the Kachin tribes, and mentioned by Noetling in *Globus*, loc. cit., p. 240 b, 241 a.

[2] Regarding this commercial highway, see C. Ritter, *Erdkunde von Asien*, III, 2d ed., p. 746–51, Berlin, 1834. G. W. Clark, *Kweichow and Yünnan Provinces*, p. 16 et seq., Shanghai, 1894. Other articles traded from Burma to Yünnan are nephrite, rubies, and cotton. Since the British occupancy of lower Burma, the importance of this ancient route is gradually waning, as goods now sent by steamer from Rangoon to Canton, and thence forwarded by boat to Pai sêh, and farther by horse, reach Yün nan fu, the capital of the province, in two months, while the average time for a caravan from the capital to Mandalay, Maulmain, or Rangoon, and back, is about four and a half months.

[3] A dictionary by Chang I (about A.D. 227–240), E. Bretschneider, *Botanicon Sinicum*, pt. I, p. 164 ; according to T. Watters (*Essays on the Chinese Language*, p. 38, Shanghai, 1889), published about A.D. 265. This passage from the *Kuang ya* is quoted in the *P'ei wên yün fu*, bk. 100 b, p. 220 a, as a commentary on the above passage of the *Hou Han shu*.

[4] This idea is presumably based on the ancient and universal custom of fashioning amber into beads.

[5] Compare Pliny (translation of Bostock and Riley, VI, p. 401): "Amber is produced from a marrow discharged by trees belonging to the pine genus, like *gum from the cherry*, and resin from the ordinary pine."

[6] In the Han period, the name of Lan chou or Yung ch'ang fu in Yünnan. The

The introduction into a dictionary of the facts just cited proves that amber was then generally known, and that trade in the product from Yünnan into the remainder of China had been carried on for a certain length of time.

The *Po wu chi*, or Records of Remarkable Objects, in its section on medicines, states that the *Shên hsien chuan* (Biographies of Taoist Immortals),[1] says :

> The resin of the pine-tree and the cypress penetrates into the earth, and changes in a thousand years into *fu ling*. *Fu ling* changes into amber. Amber (*hu p'o*) is the same name as river-pearl (*kiang chu*). Nowadays *fu ling* grows on the T'ai shan,[2] but there is no amber there. In Yi chou, in Yung ch'ang fu (Yünnan province), amber is produced, but there is no *fu ling* there. Some people assert that it is by burning bees' nests that [amber] is made. There is as yet no explanation for these two [divergent] statements.

That amber is produced in Yung ch'ang is further stated in one of the oldest works on materia medica, the *Pieh lu*, as quoted in the *Pên ts'ao kang mu*. This is a book frequently referred to by T'ao Hung-ching (A.D. 452–536), and was the standard work of eminent physicians in the Han and Wei periods, as is indicated chiefly by the geographical names mentioned in it, which are invariably those used during that age.[3] It is quite possible that the *Pieh lu* existed in literary form at an earlier date than the *Kuang ya*, and certainly earlier than the *Hou Han shu*, so that it might claim priority in recording the Burmese-Yünnan amber.

The earliest European account of Yünnan amber is given, so

---

two districts of Ai lao and Po nan, with parts of I chou, were united into the prefecture of Yung ch'ang in A.D. 69.

[1] As the *Po wu chi* was written by Chang Hua (A.D. 232–300) in the latter part of the third century (Wylie, *Notes on Chinese Literature*, p. 192 ; Bretschneider, *Botanicon Sinicum*, part I, p. 181), and as Ko Hung, author of the *Shên hsien chuan*, wrote early in the fourth century (Wylie, loc. cit., p. 219 ; according to Bretschneider, loc. cit., p. 42, he died A.D. 330), the above quotation in the *Po wu chi* from the *Shên hsien chuan* must be considered a later interpolation. This agrees with the remark of Wylie that the *Po wu chi* appears to have been lost during the Sung, and the present work was compiled probably at a later period, from the extracts contained in other publications.

[2] The famous sacred mountain in Shantung, which so early a work as the *Pieh lu* defines as the locality of Pachyma.

[3] E. Bretschneider, *Botanicon Sinicum*, part III, p. 2, 3.

far as I can see, by the Portuguese Jesuit Father Alvarez Semedo, who arrived in China in 1613 and wrote his book on China about 1633 (first published in Rome, 1643). I quote from the English edition,[1] as follows :

Yunnan is a great countrie, but hath little merchandise, I know not any thing is brought from thence, unlesse it bee that matter, whereof they make the beads for chapplets, which in Portugall they call *Alambras;* and in Castile, *Ambares;* and are like Amber, they are counted good against catarre ; it is digged out of mines, and some times in great peices : it is redder than our Amber, but not so cleane.

In another passage of the same work (p. 27), Semedo remarks : "They are very excellent in workes of Ivory, Ebony and Amber."

Old Father DuHalde was familiar also with the fact that the Chinese amber came from Yünnan,[2] and is our witness thereof with regard to the first part of the eighteenth century ; nor did Carl Ritter[3] fail to call attention to it.

The fact that the principal supply of amber in the present Chinese market comes from Yünnan province is clearly seen from the reports of the Chinese Imperial Maritime Customs, in which Yünnan is always mentioned as the source of this product.[4] I was personally given the same information, too, when making a collection of Chinese medicines at Hankow in 1903. The specimens of amber there obtained by me are doubtless of Burmese origin.

If the Chinese sources, from the third century, invariably refer to Yung ch'ang as the source of amber, this does not mean that it was produced also in that region, which must be considered as the mere transit mart from Burma to China, as the staple place from which the Burmese goods were distributed over the rest of the country. That this was so in fact, becomes

---

[1] *History of that Great and Renowned Monarchy of China*, p. 9, London, 1655.

[2] *A Description of the Empire of China* (English transl.), I, p. 122, London, 1738 : "It produces red amber, but no yellow. Some think that the rubies and other precious stones are brought hither from the kingdom of Ava." Likewise, Grosier, *Description générale de la Chine*, p. 86, Paris, 1785.

[3] *Die Erdkunde von Asien*, III, p. 754, Berlin, 1834.

[4] *List of Chinese Medicines, Published by Order of the Inspector General of Customs*, no. 488, p. 450, Shanghai, 1889.

evident from the *Ta ch'ing i t'ung chi*, the official geography of
the empire, wherein it is stated, in the description of Yung
ch'ang (bk. 380, p. 8 b), under the heading of the local products, that the amber of Yung ch'ang "is produced in the soil
of all western barbarians of Burma."

In nearly all Chinese writings on Burma the amber production
of that country is mentioned. In the *Mien fan hsin chi* ("New
Memoirs of the Burmese Frontier"), a brief pamphlet by an
unknown author,[1] the gems of Burma are enumerated as follows :
kingfisher-colored jade (*fei tsui yü*), topaz, large rubies, large
sapphires, cat's-eyes, and various kinds of amber ; likewise in
the *Mien hsün k'ao lio*,[2] by Kung Ch'ai from Ningpo, who,
besides, mentions precious nephrite, ivory, betel-nuts, copper,
iron, lead, tin, and stone-oil ; further in the *Mien hsün so chi*, by
Hsien Shu,[3] etc.

Finally, the fact that Yünnan amber is derived from the
amber mines of Burma is confirmed beyond any doubt by the
researches of F. Noetling. The center of the amber industry
is the village of Maingkhwan, inhabited by Shan, in the district of Myitkyina, which is only about 110 English miles from
the city of Yung ch'ang.[4] According to Noetling,[5] by far the
larger portion of raw burmite is bought up by Chinese traders,
and transported on the route north of Mogung, via Myitkyina,
into Yünnan. He supposes that, as the commercial relations
of the Yünnanese with the northern part of Burma have existed
for a long time, burmite has gone that way for a long time ; if
burmite was ever an object of trade in antiquity, it certainly
found its way to China at that early date. This supposition
meets with strong corroboration from the previous research ;
and we may say that, according to our Chinese sources, the
exploitation of the amber mines of northern Burma must date

---

[1] Reprinted in the Geographical Cyclopedia, *Hsiao fang hu chai yü ti ts'ung
ch'ao*, sec. X, bk. 4, p. 224 b.

[2] Ibid., p. 226 b.

[3] Ibid., p. 252 a.

[4] Compare the map of Noetling, in *Globus*, loc. cit., p. 219, and Bretschneider's
Map of China.

[5] Loc. cit., p. 241 a.

back at least to the first century of our era, if indeed they were not worked in times far anterior to the advent of the Chinese in these regions.

Though there is no doubt that the so-called amber of Yung ch'ang is of Burmese provenience, there are two places in the province of Yünnan to which an indigenous production is ascribed. The one is Li chiang fu, which we found quoted as such in the *Pên ts'ao kang mu*, by Li Shih-chên, and confirmed by the Imperial Geography ; the other is Ning chou, in Lin an fu, as stated in the *Nan Man chi*, or Accounts of the Southern *Man*,[1] which is a work of the time of the Sung,[2] thus :

In the sand of Ning chou[3] there are cliffs full of bees. When the cliffs collapse, the bees come out of the earth. The people burn them, and make them into amber.[4]

This matter requires further investigation on the spot. It is curious to note that in so full a description of Yünnan as that given by E. Rocher,[5] in which all the mineral resources of the province are discussed, no mention whatever is made of amber.

As regards other places of finds for amber in China proper, few trustworthy accounts can be noted. Hirth[6] quotes from the *Man shu*, a book of the ninth century, that a " mountain of amber " (*hu p'o shan*) is situated eighteen days' journey west of the city of Jung ch'ang, in Ssŭch'uan. In the *Ssŭ ch'uan t'ung chi* — the geographical description of this province, in which all products are carefully enumerated for each prefecture — I can find no statement regarding the occurrence of amber. According to Alexander Williamson,[7] amber is found in Lu ngan fu, province of Shansi ; while in Richthofen's *China*,[8] in which the

---

[1] *Man* is a general designation for all non-Chinese tribes of southern and south-western China.

[2] E. Bretschneider, *Botanicon Sinicum*, pt. I, p. 176.

[3] In Lin an fu, Yünnan.

[4] The belief in the production of amber from bees doubtless originated in the finding of insects embedded in pieces of amber.

[5] *La province chinoise du Yün-nan*, 2 vols., Paris, 1879–80.

[6] Quoted by Jacob, loc. cit., p. 356.

[7] *Journeys in North China*, I, p. 160, London, 1870.

[8] Vol. II.

mineral sources of this province are treated in detail, amber is
not mentioned.   In the *Ko chih ching yüan*[1] a passage is quoted
from the *Kuang chi*, a work of the time of the Liang dynasty
(A.D. 502–556), as saying that amber is found in Po p'ing, a
district in the Tung ch'ang prefecture of western Shantung.   As
fossil resins are frequent everywhere, these various notices of
native amber may be credible ; but it must be borne in mind
that these finds need not necessarily refer to true amber, but
may be related amber-like resins, such as retinite and others ;
and, further, that in these cases it is doubtless only a question
of occasional finds, not of a systematic mining process with a
commercial end in view.   If this ever were the case, Chinese
sources would not be silent on this point.   However it may be,
it is safe to assume that from the very beginning of the utilization of amber by the Chinese, the principal supply reached them
from Burma overland by way of Yünnan.

TURKISTAN

During the   middle ages another source for amber was
opened to China through her relations with the Turkish tribes
of Central Asia.   Under the reign of Emperor T'ai Tsu of the
Posterior Chou dynasty (A.D. 951–960), in the second month of
the first year of the period Kuang shun (A.D. 951), the Uighur
(*Hui hu*) of Hsi chou[2] sent as tribute to the Chinese court six
large and small lumps of nephrite, one lump of the nephrite
called *pi*,[3] nine catties of amber (*hu p'o*), and twenty pieces of
great amber (*ta hu p'o*).   In the following year, 952 (third
month of the second year of the period Kuang shun), the same
tribe despatched another tribute, among which were fifty catties
of amber.[4]   Under T'ai Tsu of the Northern Sung dynasty, in
the twelfth month of the period K'ien têh (A.D. 965), the king

---

[1] Book 33, p. 7 a.
[2] Yar-khoto, near Turfan.   See Chavannes, *Documents sur les Tou-Kiue (Turcs)
occidentaux*, p. 357, St Petersburg, 1903.
[3] Giles, *Dictionary*, no. 9009.
[4] *T'u shu chi ch'êng*, vol. 547, *shih huo tien*, bk. 185, *Kung hsien pu hui k'ao*,
v, p. 8 a.

of the country Yü t'ien (Khotan)[1] sent envoys to China with tribute containing five hundred catties of amber.[2] In the geographical work *Huan yü chi*, published A.D. 976–983, amber is enumerated among the products of Khotan and Samarkand.[3] Also in the Geography of the Ming it appears as a product of Samarkand, together with gold, silver, copper, iron, nephrite, coral, and glass.[4] It is not known to me that amber is found in situ in Turkistan or elsewhere in central Asia ; hence we must surmise that it was brought to the Turks from the west.

A clew to the manner in which amber came into the possession of these Turkish tribes is possibly furnished by a passage occurring in the Persian work *Jami ul hikayat*, written by Nur Eddin Mohammed Ufi, of the thirteenth century. He says of the products of China imported thence into Khorassan, at the time following the overthrow of the Samanides :

All sorts of textiles are found with them [the Chinese], some of which are brought to Khorassan, with marvelous curiosities. Their merchandize consists of resin, incense, and *yellow amber* coming from the country of the Slavs. This is a resin thrown out by the sea of the Slavs.[5]

On the ground of this passage I am inclined to assume that in the middle ages a trade in amber was maintained from Russia, by way of Siberia,[6] to Turkistan, whence it was carried to China,

---

[1] The passage in the *Shih i chi*, quoted in *Ko chih ching yüan*, bk. 33, p. 7 b, that a certain Han Fang brought from the kingdom of Khotan, as a present, a phenix of amber six feet in height, is doubtless apocryphal.

[2] *T'u shu chi ch'êng*, loc. cit., bk. 187, VII, p. 2 b.

[3] W. Schott, "Skizze zu einer Topographie der Producte des chinesischen Reiches," *Abhandlungen der Berliner Akademie*, p. 371, 372, 1842.

[4] *T'u shu chi ch'êng*, vol. 1242, *pien i tien*, bk. 53, *Ki pin pu hui k'ao*, II, p. 2 a.

[5] Ch. Schefer, "Notice sur les relations des peuples musulmans avec les Chinois," *Centenaire de l'école des langues orientales vivantes*, p. 8, Paris, 1895.

[6] According to Philipp Johann von Strahlenberg (*Das nord- und östliche Theil von Europa und Asia*, p. 332, Stockholm, 1730), some pieces of amber have been found between the Chatanga and Yenisei rivers, toward the Arctic sea, and also in the sandy desert between Mongolia and China. This account is confirmed by the investigations of Fr. Th. Köppen, "Vorkommen des Bernsteins in Russland" (*Petermanns Mitteilungen*, XXXIX, p. 249–253, 1893); an abstract of this article appears in *Nature*, XLIX, p. 181, 1893. According to Köppen, amber occurs along the arctic shores of Russia and Siberia, over wide tracts, more frequently between the mouths of the Ob and the Yenisei, but also farther inland, a hundred versts east from the Yenisei, on Stefatin river, which falls into the Agano, farther on the Cheta, a side-

and that this product was again sent from China to other countries in the west.

In the chapter on the Uighur (*Hui hui chuan*), in the *Sung shih*, there is a reference to a tribute of amber from Kucha. In A.D. 1010, the King of Kucha, Khagan, despatched an envoy, Li Yen-fu, to bring incense, medicine, *hua jui* ("stamens, pistils"), cloth, renowned horses, a one-humped camel, a sheep with big tail, a saddle and bridle with jade ornaments, amber, and *t'ou shih*.[1]

EUROPEAN AMBER IN CHINA

Finally, it remains to record the importation of amber into China by European nations. S. Wells Williams[2] asserts that the consumption [of amber in China] for court beads and other ornaments is great, and shows that the supply is permanent, for none is brought from Prussia. In a strict sense the last statement may be true in so far as there was hardly ever direct communication between the Baltic coast of Prussia and the shores of China ; but, taken in the sense that no Prussian amber ever found its way to China, it is not correct. From two sources it can be shown that the Portuguese traded amber to China at least as early as the eighteenth century, very probably much earlier; and there can be no doubt that they had shipped this amber from Europe.[3] In the sixty-fourth chapter of the *Chinese Gazetteer of the Province of Kuangtung*, there is a brief summary

---

stream of the Chatanga, where it is dug by the Yakut, and on lakes Ladannach and Tartach. This Siberian species, however, is very likely not genuine amber, but retinite. As to Russia, amber is found over a vast area, almost uninterruptedly from the Baltic to the Black seas and sporadically in two places in the Caucasus. It seems to me that it was not the inferior Siberian material that was made the object of medieval trade to inner Asia and China, but more probably the highly valued Baltic amber, which alone was of sufficient worth to justify so complicated and expensive a traffic.

[1] *T'u shu chi ch'êng*, vol. 1341, *pien i tien*, bk. 51, *K'ui tsú pu hui k'ao*, I, p. 11 b.

[2] *The Middle Kingdom*, II, p. 398, New York, 1901.

[3] L. Riess, "Geschichte der Insel Formosa" (*Mitteilungen der deutschen Ges. für Natur- und Völkerkunde Ostasiens*, VI, p. 427), states that the Dutch of the seventeenth century imported amber into Formosa among some other articles of trade for further export to China; but nothing seems to be known about the origin of this amber.

of the Portuguese settlement of Macao, in which it is said that the Chinese receive from them in trade, ivory, *amber*, coarse and fine woolen cloths, redwood, sandalwood, pepper, and glass.[1] In the Chinese *Chronicle of Macao*,[2] amber of two kinds, "gold" and "water" amber, is mentioned as being in the possession of the Portuguese. On the occasion of his visit to Canton, amber of European provenience was there observed by Nordenskiöld :

Ebenso scheint der Bernstein in hohem Ansehen zu stehen, besonders solche Stücke, welche Insekten enthalten. Bernstein wird in China nicht gefunden, aber von Europa eingeführt ; derselbe ist oft verfälscht und enthält grosse chinesische Käfer mit den Spuren der Nadeln, auf welche dieselben aufgespiesst waren.[3]

Even in Mandalay the succinite, or Prussian amber, is now easier to procure, and cheaper than the Burmese amber.[4] According to Noetling,[5] succinite has been worked in Mandalay to a large extent since the decrease of the importation of burmite. It is therefore not at all impossible that Prussian amber arrives in China overland by way of Burma. That Prussian amber is imported to Chinese ports appears from the reports of the Chinese Maritime Customs. In *Port Catalogues* of the Chinese Customs Collection at the Austro-Hungarian Universal Exhibition, page 267 (Shanghai, 1873), under the port of Shanghai, Europe and Japan are given as places of production of amber ; on page 424, under the port of Canton, the coasts of Prussia, the China sea, and the Indian archipelago are given for the same ; and on page 470, again, "the Archipelago, Prussia," etc. I believe that too great stress should not be laid on these statements. They reflect the opinions or general knowledge of the customs officials concerning the provenience of amber rather than give the result

---

[1] Sir George Th. Staunton, *Miscellaneous Notices Relating to China*, p. 88, London, 1822.

[2] *Ao mĕn chi lio* (bk. 11, p. 39 b), latter part of eighteenth century (Wylie, *Notes on Chinese Literature*, 2d ed., p. 60), but probably earlier, as the preface of one of the two authors, Chang Ju-lin, is dated 1750 (M. Courant, *Catalogue des livres chinois*, 1, p. 104, Paris, 1903).

[3] Adolf Erik Freiherr von Nordenskiöld, *Die Umsegelung Asiens und Europas auf der Vega*, 11, p. 372, Leipzig, 1882.

[4] *Gazetteer of Upper Burma*, 11, pt. 1, p. 294.

[5] *Globus*, loc. cit., p. 242 b.

of a trustworthy inquiry regarding just those ambers which are imported into China. The statement, " China sea, Indian archipelago, Japan," savors much more of the verbal information given by a Chinese clerk on the customs staff than of authenticated facts. In all probability it is quite impossible to determine the exact source of modern ambers ; all we may safely infer is that the ambers of the treaty ports are shipped from abroad and are most likely of German origin.

According to the Returns of Trade of the Chinese Customs (vol. IV, pt. II, p. 410, Shanghai, 1906), 36 piculs of amber from foreign countries, valued at 1033 Haikwan taels, were imported from Hongkong to Canton in 1905.

The amber now used in Korea comes exclusively from Germany. The import from Germany amounted in 1898 to 11,495 yen ; in 1899, to 510 yen ; in 1900, to 2,111 yen ; in 1901, to 5,280 yen.[1]

IMITATIONS OF AMBER

Regarding imitations of amber, S. Wells Williams[2] remarks that " the Chinese have also learned to imitate amber admirably in a variety of articles made of copal, shellac, and colophony." But the most important of these materials is glass, to which, as is well known, the Chinese understand most eminently how to impart, by means of metal oxides, the most beautiful varied colors, imitating those of jade, malachite, lapis-lazuli, and amber. Rosaries of amber-colored glass beads — manufactured, like all glassware, in Po shan, Shantung province — are. frequently used. Very curious is the fact that in Canton, for the manufacture of ornaments, a sort of yellow substance much resembling amber is employed, which is nothing but the upper part of the beak of a crane.[3] A powder of amber, employed in medicine, is much adulterated with colophony and copal.[4]

---

[1] Handelsbericht des Kaiserlichen Konsulats in Sôul für das Jahr 1901, in *Handelsberichte über das In- und Ausland*, Asien, Serie II, no. 36, p. 11, 12, 14, Februarheft, 1903.

[2] *The Middle Kingdom*, II, p. 398, New York, 1901.

[3] *Catalogue of the Chinese Collection of Exhibits at the Paris Exhibition, 1900*, p. 132, Shanghai, 1900.

[4] *Hankow List of Medicines*, p. 18, 1888.

That such fictitious products do not date from yesterday, but are almost as old as amber itself in China, may clearly be seen from the statements of the eminent physicians T'ao Hung-ching and Lei Hsiao, who tested the genuine article by its magnetic properties. The cyclopedia *Ko chih ching yüan*[1] quotes from a treatise *Ts'ai huo yüan liu* ("Sources and History of Wealth"), apparently a treatise on political economy, to the effect that there are many counterfeits of amber, and that only the kind whose color is like that of blood, which, when made warm by being rubbed with cloth, attracts mustard-seeds, is genuine. A curious imitation of amber is described in the *Shên nung pên ts'ao ching*:[2]

Amber can be made from chicken-eggs by the following method: Take an egg, mix the yolk and white of it, and boil it. As long as it is soft, an object can be cut out of it; this must be soaked in bitter wine for several nights, until it hardens; then rice-flour is added to it.

Whether this fabulous composition was really intended to replace amber may be doubted. In the *T'u shu chi ch'êng*, the same passage is given directly after the *Shên nung pên ts'ao ching*, but without the first and last sentences, above quoted, so that the word "amber" does not appear there. According to the *Ch'i wan lin* ("Forest of Extraordinary Curiosities"),[3] whose statements on amber generally agree with those of the *Ko ku yao lun*, as quoted by Li Shih-chên,[4] counterfeit amber is made of dyed sheep's-horn.

NOTE. — Objects made of amber, as described in Chinese sources, and after actual specimens, will be treated in a separate paper.

COLUMBIA UNIVERSITY,
NEW YORK CITY.

---

[1] Book 33, p. 7 a.
[2] *Ko chih ching yüan*, loc. cit.
[3] Ibid.
[4] See p. 221.

044

中国古代青铜器

THE CRAFTSMAN

Volume XII   APRIL, 1907   Number 1

Contents

Published by Gustav Stickley, 29 West 34th St., New York

25 Cents a Copy : *By the Year, $3.00*

THE CRAFTSMAN

GUSTAV STICKLEY, EDITOR AND PUBLISHER
VOLUME XII   APRIL, 1907   NUMBER 1

MARVELOUS BRONZES THREE THOUSAND YEARS OLD FOUND IN ANCIENT GRAVES AND AMONG FAMILY TREASURES IN CHINA: BY DR. BERTHOLD LAUFER

 HINESE bronzes have not yet found the recognition and appreciation due them both from their archæological importance and their value to our own art industries. That the latter could profit by a close study of those works and receive from them new inspiration and ideas in technique and forms of ornaments, is obvious, and has been fully acknowledged by the museums of Industrial Art of Vienna and Berlin, which have issued instructive publications about Chinese bronzes, particularly designed for the purposes of the craftsman. The bronze-workers of this country now have an opportunity of learning from the great examples of Chinese art by a study of the present collection at the Natural History Museum, from which our illustrations are drawn.

In China the archæologist does not share the happy fate of his colleague in Greece, Egypt and other lands, who enjoys the pleasure and privilege of personally bringing to light the costly treasures of bygone ages hidden away in the soil. The Chinese penal code makes special provision for any disturbance of graves, but it is not, as is generally believed, a deep-rooted feeling of reverence and awe for the burying-places of the dead which handicaps the attempts of the foreign investigator in trying his spade on promising spots. Neither ancestor-worship nor superstitious belief has ever deterred the enterprising Chinese treasure-seeker from opening tombs and delving deep in the ground. Ever since the early days of the Han period, this undaunted rifling of graves has been in unchecked operation, partly to satisfy the curiosity of real antiquarian interest, partly from motives of selfish gain. Hardly any people cherish and prize their antiquities more than the Chinese, and a collection of ancient art-trea-

3

BRONZES, THREE THOUSAND YEARS OLD

sures becomes the unrivaled pride, nay, the highest valued property and inheritance of a family, and is handed down from father to son. It is not merely adoration or affection that prompts them to hoard these relics of the past, but also inquisitive and actually "scientific" interest, that love for research which dominates the tendency to store up large collections in the hands of an individual. Numerous are the books written by Chinese collectors on their bronzes and paintings, and many are the inscriptions accompanying them; these publications are usually adorned with fine wood-engravings unsurpassable in softness and delicacy of line. Particularly in their bronzes have Chinese scholars pursued most industrious and ingenious studies. As early as the eleventh century, an Imperial Museum was founded, in which the highest productions available at the time of the arts of casting, sculpturing and painting were hoarded—a collection which every modern art museum might look upon with justifiable envy. The descriptive catalogues then issued at the command of a broad-minded, art-loving monarch now form an indispensable source of information concerning the forms, significance, periods, and ornaments of bronzes and jades.

The uninterrupted demand in the native market for art works has created two unavoidable evils—the development of a special profession of art-dealers, and the wholesale manufacture of countless imitations to meet a demand often far exceeding the supply. Even family heirlooms which fell into the tradesman's greedy hands from time to time were not enough to fill the orders, so that nothing was left but to dig in the ground for the new and unexpected. Noted dealers still keep a host of employees running about the country, treasure-hunting, under cover of night. That their work is detrimental to scientific research is evident. No information can be obtained under such circumstances regarding the exact locality or the particular conditions under which the finds have been made. Neither do these adventurers care for all the treasures found in the grave. All minor and not marketable objects, which to the scientific mind would have great value as revealing former religious customs and worship, are carelessly thrown aside; only profitable pieces being selected from the plunder. As all trades in China are closely allied in guilds and unions, the professional spirit is developed to a marked degree. And it is exactly this commercial monopolization of the art-trade and the effective organization of the art-dealers which are the causes of the foreign

4

From Collection in Natural History Museum, N. Y.

"FLOWER VASE OF A HUNDRED RINGS."
BUT THREE OF THESE BRONZES ARE IN
EXISTENCE: SUNG DYNASTY 960-1126 A. D.

CENSER: RATS STEALING GRAPES. MING DYNASTY 1368-1640 A. D.

LIBATION CUP FOR OFFERING WINE TO DECEASED ANCESTORS: SHANG DYNASTY 1766-1154 B. C.

MUSICAL RATTLE: HAN DYNASTY 200 B. C.-23 A. D.

From Collection in Natural History Museum, N. Y.

CENSER: MING DYNASTY 1368-1640 A. D.
CENSER: MING DYNASTY 1368-1640 A. D.

ORNAMENTAL VASE: HAN DYNASTY 200 B. C.-23 A. D.

VESSEL FOR CARRYING WINE: CHOU DYNASTY.
BATTLE AXE: HAN DYNASTY.
LIBRATION VESSEL: CHOU DYNASTY 1122-255 B. C.

From Collection in Natural History Museum, N. Y.

student being hampered at the outset in any active exploring work. The fears and beliefs of the people with regard to tampering with graves might eventually be overcome by closer personal acquaintance with them, by winning their confidence and sympathy, by tactful procedure in handling coffins and skeletons which, after examination, would be reburied at the discretion and expense of the investigator; even the revengeful spirits of the dead and the raging ire of the offended local gods might be pacified by an equivalent sacrifice in cash value deposited in the *yamen* or with the temple's priesthood. But to oppose the sacred prerogatives of an established trade organization would mean a vain struggle against a superior force, with no possible hope of victory; from the view-point of these traders—exclusive trust magnates, as it were—they would not hesitate to brand all efforts as illegal competition, as a menace to their business, as an impudent encroachment upon their ancient and inherited rights, and to denounce the offender as a dangerous villain, guilty of high treason and sacrilege, who should be punished by the unrelenting hatred and persecution of the populace.

However discouraging and to some extent unfruitful it may be, the student of archæology has no other choice than to take what falls to his lot. But it must not therefore be presumed that his task is by any means easier than that of his fellow-worker who harvests the results of his own excavations. The intricate and mysterious ways of the Chinaman form a harder soil to work upon than that in which he plows. To him, it is comparatively easy to interpret the language of his spoils by a skilful combination of all circumstantial evidence brought out in the exploited field; while the collector of archæological specimens in China is confronted with the single piece only, just offered for sale, on which alone he must exercise all his wits to bring out its period or to judge its historical and artistic merits. His brain must always be vigilant and alert, and his knowledge extending over numerous historical and philological subjects. He must be able to decipher seals and inscriptions in the ancient style of character, which in itself is a complicated study, and he must be familiar with the language and terminology of the dealers, with their queer fashions and customs, with their hundredfold tricks and manipulations, against which he must keep a constant lookout. And no less important is the finding and seizing of the right opportunity and the managing to obtain the services of the proper men.

9

BRONZES, THREE THOUSAND YEARS OLD

WHILE on my mission in China on behalf of the American Museum of Natural History, it became clear to me after careful consideration that the opening I desired was in the ancient capital of Hsi-an-fu, province of Shensi, whither, as is well known, the Empress Dowager and the Court had taken refuge in 1900. There, in the once flourishing center of Chinese civilization—the metropolis of the Han emperors under whom art had attained to a remarkable height, my hope of obtaining genuine material in bronze and clay from the early epochs of Chinese art was finally fulfilled. Nothing could be discovered in Peking or in the large treaty ports of a character to rival the venerable art-treasures of Hsi-an-fu. This city was, and still is, the distributing center of the whole trade in art objects which are shipped from there to the capital, to Hankow and Shanghai. I had the good fortune to meet there Mr. Su, an enlightened, well-educated Mohammedan, whose family had been in the antiquarian's business since the seventeenth century, and who enjoys a reputation all over the country for being an honest, straightforward connoisseur of antiquities.

The way in which the art-trade is carried on in Hsi-an-fu is a matter of curiosity in itself. The shops of the dealers are tiny rooms, dimly lightly and a never-failing source of wonder to the new arrival. Trifling bric-a-brac is heaped up in the front room, some crumpled paper paintings spread over the walls; not a sign that important art objects would ever be forthcoming. The foreigner whose eyes are accustomed to the magnificent, glaringly gilt stores of Shanghai and Peking has not yet learned that the true Chinese antiquarian never exposes his heart-loved treasures to the profane eye. What he displays openly is cheap trash to allure the innocent and ignorant. Woe to him who is trapped in this pitfall; he will never rise to see himself treated to a good genuine piece. It requires patience, proper introduction, personal acquaintance, and the power of wholly adapting one's self to Chinese usages, to be initiated into the sanctum where true art wields the scepter; it is not the possibility that the foreigner may be willing to pay the price—or any price, that induces the Chinese to lift the veil; but the certainty that he possesses a discriminating knowledge and judgment. Only this affords a passport to the hall of adepts and to fair treatment. The shrewd Chinaman is well aware of the fact that he can palm off on the inexperienced

10

VESSEL IN ONE CASTING:
CHOU DYNASTY 1122-225 B. C.

TEMPLE BELL. INLAID WITH GOLD AND SILVER. STRUCK
ON THE KNOBS, EACH PRODUCING A DIFFERENT
MUSICAL SOUND. CHOU DYNASTY· 1123-255 B. C.

foreigner an imitation at the same price as an original. Why therefore should he let him have the genuine article of which he does not recognize the value? Another peculiarity of the art-dealer is that he does not talk about his objects; the buyer of ancient art is expected by him to know all about them as an expert and is responsible for his own failures. If he is disappointed, he must take the blame himself. It has also become an established rule that antiquities must be paid for, cash down, at the very moment of the sale; while on the other hand, there is hardly anything that a Chinaman can not obtain on credit. Another interesting point is that in Hsi-an-fu no discount is allowed on any great work of art, except by small houses which may be in immediate need of cash. All the world knows how dearly a Chinaman loves bargaining and haggling, and how he advances prices to a point he never dreams of realizing, just for the pleasure and excitement of a bargain. But for the real works of art such haggling is not permitted, and where the valuation is thought excessive, a piece may as well be given up at the start. How the prices are made is a mystery; there are no fixed rules and standards, everything depends on chance and circumstance, and on the rarity of a piece; a trade mark with date, or an inscription consisting of a few characters, always commands an additional sum; in lengthy inscriptions the number of characters is carefully counted, and a conscientious estimate is put upon each of them.

There are two sources of supply for the art-dealers of Hsi-an-fu— first, the numerous and practically inexhaustible ancient graves in Shensi Province, many of which belong to the Han period, and second, the transactions with distinguished families residing in the city. Of these, there is a goodly number and many of them are wealthy, as the place is a favorite resort of retired officials. Because of the difficulty people not engaged in actual business encounter in finding a good opening to invest their capital—great real estate openings are lacking in China—they buy up valuable antiquities as an investment on which no losses are liable to be incurred. Many families have a large proportion of their money in such property. If then, for a journey, a marriage, a funeral or other occasion some ready cash is required, an heirloom is disposed of through a middleman who acts as broker for the family. According to all precedent, to deal directly with the owner is impossible. A place and a time are appointed for

13

BRONZES, THREE THOUSAND YEARS OLD

the examination of the piece in question. Wonderful in such cases is the completeness of their departure from the customary Chinese deliberateness; to effect a speedy transaction, the term for the exhibition is limited with rigorous sternness to a few hours, after which the piece is taken away and the meditative customer who could not make up his mind on the instant will never see it again. My own success in bargaining was fair, for the majority of the large pieces of bronze in my collection represent treasured heirlooms from the possession of noted families in old Hsi-an-fu.

LIKE the peoples of Northern and Central Europe, the Chinese passed through a genuine Bronze Age, during which only bronze and copper weapons, implements, and vessels were employed, and iron was entirely unknown. This period terminated at about 500 B. C. The art of casting bronze had reached its greatest perfection before that time, and was in a highly flourishing condition at the period of the earliest dynasties. The process followed was always that known as *à cire perdue,* of which Benvenuto Cellini has left us such a classical description. A great influence in the development of bronze vessels was the worship of ancestors, which culminated in a minutely ritualistic cult that created an epoch of artistic vases. The prescripts of the ancient rituals exactly determined the shape, alloys, measures, capacity, weight, and ornaments for each type of these vessels, and their forms were defined according to the nature of the offerings, which were wine, water, meat, grain, or fruit. The adjustment of the proportions of the single parts is most admirable in the majority of them. The libation cup from which wine was poured in worship of the spirits of the dead, and which, according to the explanation of the Chinese, has the shape of an inverted helmet, is a relic of the Shang dynasty (B. C. 1765-1145), and is the most ancient example of this art in our collection. The bell, the large bowl, the vase with handles formed into animals' heads, and the vessel for carrying wine come down from the time of the Chou dynasty (B. C. 1122-247). The bell is a masterpiece encrusted with gold and silver, proving that the art of inlaying was well understood at this early period. It is remarkable that all these ancient bronzes, despite their colossal dimensions, were executed in one and the same cast, bottom, handles, and decoration included, and rank, even from the view-point

14

BRONZES, THREE THOUSAND YEARS OLD

of the modern bronze-caster, among the greatest works of art ever created in metal.

DURING the Middle Ages, a great renaissance of art arose under the Sung, when bronze vases of most artistic workmanship were turned out. While the deep religious spirit which inspired the creations of the early masters had gradually died away, the worldly element now came more and more to the front, and with it a more human touch. Greater stress was laid by the new artists on elegant forms, on pleasing and harmonious proportions, on delicate treatment of ornamental details. The "Vase with a Hundred Rings," which is actually adorned with that number of movable rings on its four sides, is a good example of the accomplishments of this period. In its shape, it imitates one of the honorific vases of the Chou, which at that time by imperial grace were devoted to the commemoration of exceptionally heroic deeds and bestowed upon worthy officials as a mark of distinction. During the Sung and the later Ming periods, such vases served decorative purposes in the way of flower-vases. The addition of the rings is likewise not an inheritance of the past, but an idea of the Sung artists. The traditions of the latter survived to the Ming dynasty and down to the end of the eighteenth century.

The Ming period excels in number and beauty of incense-burners. Incense proper came to China from India, and incense was burned in religious worship only after the introduction of Buddhism. The censer as a type of vessel is by no means of Indian origin, but is derived from the form of one of the sacred ancestral vessels of the Chou. In no other bronze work has the creative power of the artist shown such great variety of beauty.

EDITOR'S NOTE.—The series of old Chinese bronzes here shown are the result of a recent expedition to China under the auspices of the American Museum of Natural History, New York, made possible through the generosity of Jacob H. Schiff, Esq. As the collection is considered the largest and most representative ever brought out from the Chinese Empire to this country, the foregoing narrative by Dr. Berthold Laufer, Chinese scholar and Oriental explorer, setting forth the peculiar and little known methods of obtaining these ancient master-pieces, together with a general description of the specimens here reproduced, is of timely and instructive interest.

15

045

药学和自然科学历史研究

SCIENCE

A WEEKLY JOURNAL DEVOTED TO THE ADVANCEMENT OF SCIENCE, PUBLISHING THE OFFICIAL NOTICES AND PROCEEDINGS OF THE AMERICAN ASSOCIATION FOR THE ADVANCEMENT OF SCIENCE

FRIDAY, JUNE 7, 1907

CONTENTS

MSS. intended for publication and books, etc., intended for review should be sent to the Editor of SCIENCE, Garrison-on-Hudson, N. Y.

SOME PHYSIOLOGICAL VARIATIONS OF PLANTS, AND THEIR GENERAL SIGNIFICANCE[1]

IN a survey of the domain of the biological sciences in recent years, one of the most significant facts is found in the extent to which physiology has invaded those fields of this domain which, in the earlier stages of development, seemed entirely apart from and independent of physiological relations. When species of plants were supposed to have been created at the beginning just as we find them to-day, and to transmit their original characters unchanged to their remotest possible descendants, there was no physiological question as to the variations within species, and none as to the relation of species to each other nor as to the origin of new species. In that view there could be no origin of new species. They were all created at the beginning, and then the Creator rested.

When botany first began to be a science it was merely an attempt to classify plants, that is, to discover the characters of species as they were originally created, to group together those that were most alike and to separate those that were unlike. The characters used in the first attempts at classification were more or less superficial, and systematic botany was merely a study in formal external morphology.

But a change has come; and this change began with the general acceptance, among biologists, of the view that species are not

[1] Presidential address delivered before the Michigan Academy of Science at Ann Arbor, Mich., March 28, 1907.

any given epidemic there are always certain individuals who never contract the disease. They have a certain natural immunity to that particular disease, and this immunity is due to some physiological peculiarity. So in a field of rusted or mildewed wheat some individual plants show themselves more resistant than their fellows to the species of rust fungus found upon that species of host. By selecting and propagating these immune individuals we may develop an immune race or strain. The problem is not always so simple as here stated. It may happen that a race immune to one disease may be very susceptible to another, or immunity may be accompanied by other qualities altogether undesirable. One might be led to suppose, on reading certain popular articles intended to show how new forms of plants are produced, that it is only necessary to imagine an ideal plant and then set to work to create it. Nothing is farther from the truth than this. Nature does sometimes produce something new, as a stoneless plum, or a nectarine on a peach tree. But man must take the materials furnished by nature, combine them in new ways, or modify them within limits which are usually soon reached. He can not create a wheat plant immune to rust, nor a watermelon resistant to the wilt fungus. But if nature furnishes a few individuals with the desired qualities, man can propagate the individuals possessing those qualities, and by rigid selection maintain the qualities to a high degree. If it is possible to cross the plants with other species or with varieties of the same species, he may be able to combine in the same individual a number of desirable qualities. Having obtained these qualities in one individual, he can best conserve them by vegetative propagation, such as by grafts, cuttings, bulbs or tubers, according to the habit of the plant propagated. He may care nothing

whatever about the limits of species or varieties except in so far as their physiological relations help or hinder his combinations. Following MacDougal's method, it may be possible to produce in plants some new characters. But even if it were possible to produce in this way really new species, it is hardly within the range of possibility that we could choose beforehand the kind of a species we would produce. It would be a case of 'cut and try.' If the result be a form with desirable qualities, let it be preserved, but if it be worthless, let it die. Nature has repeated this experiment ten thousand times. If we would imitate her we must search out her secrets in the physiological realm. She conceals them well, but is not unwilling to reveal them to him who questions her with a hearing ear, a seeing eye, and a thinking brain, tools which she herself has given him.

JAMES B. POLLOCK
UNIVERSITY OF MICHIGAN

A PLEA FOR THE STUDY OF THE HISTORY OF MEDICINE AND NATURAL SCIENCES[1]

FOR a number of years a new current of thought has been gradually coming to the front in the minds of scientific thinkers of the times. The nineteenth century, the mental development of which is now assured, has of late been severely criticized for its unhistorical character, and perhaps not without reason. Over this inheritance from the preceding generation a certain dissatisfaction is being more and more keenly felt in the most diverse branches of science. The main trend of the last century was naturalistic and economic to a marked degree; so much so, that the new methods discovered in natural science, and

[1] Read before the American Anthropological Association, at the meeting of the American Association for the Advancement of Science, December 31, 1906.

the vast progress resulting therefrom, seemed to foreshadow an entirely unprecedented epoch in the history of science, and the generation of that age was only too eager to sever all links connecting it with the accomplishments of former ages.

The inauguration of the twentieth century presents a somewhat contrary aspect. One of its primary tendencies has been towards a restoration of our lost connection with the eighteenth century and with earlier periods, resulting in a movement of such earnest and impressive character that we can not foretell at the present moment whether the eighteenth century will not, at some day not far off, seem nearer to us than the sober prose of the nineteenth.

It is not mere chance that at the dawn of the new age the war-cry 'Historical investigations!' is sounded from all camps, and that in consequence a broader scientific knowledge is obtained through this pursuit of historical research. In nearly all lines, students had become weary of the worn and time-honored ruts, and from the dry atmosphere of specialized specializations yearned for the purer air of loftier heights; and not least among the causes of this reaction was the disappointment due to the misapplications and failures of the evolutionary theory. New ideals were thus created, and found their expression in an extended historical movement, which led to radical changes and to amplifications in literary activity, in academic instruction, and in museum policy—or rather in encouraging prognostics of a new museum era—at least, so far as Germany, Austria and Switzerland are concerned. To give a concise idea of what has been accomplished, and is being proposed to be done in this line, is the object of this paper.

To review even hastily all literary pursuits pertaining to this large field is naturally beyond the scope of my purpose.

The most noteworthy, in my estimation, are the following: the journal *Zoologische Annalen,* founded in the interests of the history of zoology in 1904 by Max Braun, professor of zoology at the University of Königsberg, and the organization of the Deutsche Gesellschaft für Geschichte der Medizin und der Naturwissenschaften, in Hamburg, on September 25, 1901,—a most active and industrious society, which, now under the able leadership of Professor Karl Sudhoff, of Leipzig, has thus far published six volumes of 'Mitteilungen zur Geschichte der Medizin und Naturwissenschaften.' The pages of this journal are full of interesting original contributions and copious reviews concerning the history of anthropology, botany, zoology, geography, geology, mineralogy, chemistry, physics, mathematics, astronomy, technics, and medicine. A distinguished production of German scholarship is the 'Handbuch der Geschichte der Medizin,' established by Theodor Puschmann, the late celebrated medico-historian of Vienna, and edited by Max Neuburger and Julius Pagel. It was recently completed in three volumes, with thirty-one contributors, and embraces the history of medicine in all its departments and epochs, among all peoples of the globe, inclusive of primitive tribes. Despite its very numerous shortcomings—chiefly due to inaccessibility or want of material, especially on Asiatic medicine, but partially also to lack of historical criticism—it remains, nevertheless, a remarkable monument, but more prospective than retrospective. The recent proposed action of the Berlin Academy of Sciences in regard to the publishing of a complete edition of the Greek medical authors may also be mentioned in this connection; and the new epoch-making researches on the life, personality and works of Theophrastus Paracelsus.

In the academic institutions of Germany and Austria, broad and liberal space is now allotted to instruction and research-work in the history of medicine, natural sciences, and particularly in that of cultural plants and domestic animals. It is an officially acknowledged, fostered and encouraged subject of teaching and study; and there is now hardly any German university, however small, where it would not find a competent representative. Only a year ago (1905) an institute for the study of the history of medicine, in connection with a full professorship, was established at the University of Leipzig, the chair being occupied by Professor Sudhoff, who tells me that thus far there has been an average attendance of from fifteen to twenty-five students in his courses. This institute will regularly issue scientific publications from the beginning of next year (1907).

Berlin has two professors for the history of medicine—Pagel and Schweninger—it having been customary for many years for students of medicine to be allowed to choose their theses from this field, which has been done by many of them with evident success. Regular courses are offered there, besides, in the history of epidemic diseases, of anatomy, of chemistry, of astronomy, of cultural plants. In the last-named subject, four courses are tabulated this winter—one of general character, and three special ones relating to the cultural plants of Africa, and to those of the German colonies and the tropics, respectively. The University of Vienna has likewise two representatives of medical history (Neuburger and v. Töply), general courses and a special course on the history of physiology (Kreidl). Innsbruck possesses a specialist in the history of zoology (v. Dalla-Torre). An *extraordinariat* for the history of medicine has been founded at Würzburg (Helfreich); and courses on the subject are provided for at Bonn, Göttingen, Breslau, Heidelberg, Tübingen, Munich, Marburg, Kiel, Rostock (with even three teachers), further at Graz in Austria; and at Basel, Zürich and Bern in Switzerland.

On November 13, 1906, the cornerstone of the German Museum of Masterpieces of Natural Science and Technics, in Munich, was laid,—the last creation born from this young historical spirit. A question much ventilated now, in the circles of Germany interested, is the plan of a comprehensive museum for the history of medicine, illustrating its development, from the times of prehistoric man down to the present day, in anatomy, surgery, hygiene, endemic diseases and other phases.[2] Such a medical museum, fully deserving of the name, as yet exists nowhere. The medical faculty of the University of Paris moved a resolution to this effect some years ago, but the scheme has not yet been carried out. The only institution that has thus far made any attempt in this direction is the Germanic Museum of Nürnberg, whose very beautiful collections, however, are restricted rather to pharmaceutical than to purely medical antiquities. The first temporary exhibition relating to medical history was held in Düsseldorf in 1898, on the occasion of the annual assembly of the German naturalists and physicians; and similar tendencies developed at the Russian congress of physicians at Moscow in 1900, with greatest success.[3]

[2] See a paper by Sudhoff, ' Zur Grundsteinlegung des Deutschen Museums von Meisterwerken der Naturwissenschaft und Technik,' Begrüssende Gedanken und Ausblicke (reprint from *Münchener Medizinische Wochenschrift*, No. 46, 1906).

[3] Compare report on address by B. Reber, ' Über Notwendigkeit und Wert von Sammlungen betreffend die Geschichte der Medizin,' in report on 78. Versammlung Deutscher Naturforscher und Arzte (reprint from *Münchener Medizinische Wochenschrift*, No. 47, 1906, p. 8).

I now venture to suggest that such a museum, representing the development of medicine, natural sciences and technics in their whole range, be established in this country, perhaps here in New York, which seems to be the most appropriate place for it; and I am under the strong impression that such an institution would be of wide and universal benefit to our public at large, and would contribute immensely towards the furtherance of science, both natural and historical, and also considerably aid the cause of anthropology. The temporary tuberculosis exhibit in this city last winter may serve as a technical example of what could be accomplished here. If a sufficient number of notable physicians of New York could be interested in the far more extensive plan just proposed—the carrying out of which would not require an exorbitant capital—its realization would seem to be within easy reach. Nothing would be more welcome to us than the sympathetic cooperation of physicians, to interest whom in the study of anthropology we must make many more and larger efforts, especially when we consider how signally anthropology, in its theoretical and practical bearings, has progressed and been advanced by medical men in Europe. One of the foremost tasks of the future American museum devoted to medical science would certainly be to represent the accomplishments of the hygiene, and technical inventions. In this way we should enlist the interest of physicians in our native population; and students of anthropology might also profit from their mode of viewing the subject or from an active participation in our work. A museum of this type, if developed on the broadest lines, may indeed lead also to new and fruitful anthropological work. I need hardly accentuate here the point that a full historical representation of all endemic and the great epidemic diseases (analogous to

the idea of the tuberculosis museums), in connection with the development of hygiene, would be a matter of great public service—an undertaking which should meet with the support of all philanthropists. It goes without saying that a museum of this kind would be a scientific, social and educational potency of the highest order—an agency of social progress, not inferior in rank to art or ethnographical museums.

At the same time I may be allowed to express the wish that the study of the history of medicine and the other natural sciences be taken up in this country with the same energy as on the other side of the Atlantic. I need not dwell here on a discussion of the manifold advantages of such pursuits, as the development of all science as an emanation of human culture naturally falls under the head of anthropology.

The most obvious gain which could be derived from the carrying out of these suggestions would be closer affiliation and more intimate contact of all the sciences. In the pursuit of historical investigations, we are all on common ground, and the character of the subject necessitates mutual dependence and assistance. It logically leads to a plea for cooperation, through the efficiency of which many of our most important problems are awaiting their final solution. Allow me to recall to you the study of the history of cultural plants and domestic animals, as constituting the framework of all higher forms of human culture. These topics have engaged the attention of anthropologists to a very limited extent only, being mainly worked up by botanists and zoologists, and occasionally by geographers and economists. The leading books on the subject are little satisfactory from the historical point of view, while historical investigations already in existence suffer from the lack of botanical or zoological accuracy. There is an unharmonious dissonance between these various attempts; and a synthetical representation that should seek to reconcile the conflicting standpoints is still a vain hope. The reason is the isolation of the single sciences, each of which, being restricted to its peculiar resources and methods, is intent on solving a problem in which a goodly number of them are involved. Naturally, only one solution to a problem is possible, whether it be attacked through physical or historical research; and if the results obtained by either are mutually contradictory, this is equivalent to saying that the particular science alone is unable to solve it, and that the solution should be undertaken by a concentration of energy of all the sciences concerned in the specific case. To cite a practical example, take the origin and propagation of our cereals, or the long history of the domestication of the ox or the horse—problems around which, finally, the most ancient history of Asia and Europe centers. There is no science which, by the mere exercise of its own limited faculties, could reach a decisive solution of them; but I am fully confident of ultimate success through a cooperative combination of the various sciences involved, which, in this case, are geology, botany, zoology, archeology, history and anthropology. The individual can not master all these sciences; and, instead of dividing our strength by working singly from isolated positions, we should advocate the uniting of all available forces for the best good of the same cause. The identical observation holds for all historical studies of sciences. The students of Oriental fields, for example, whether their work be in the Egyptian, Arabic, Indian or Chinese departments, are almost daily confronted with the wonderfully rich scientific lore of these peoples referring to subjects in which they themselves are not competent; but it is on the shoulders of these

very students that the accumulation of a large portion of the material rests, on which the historian of science can build. One of the most remarkable instances of this sort of cooperation which I have in mind, and which might be extended over many other lines, was the association of the Orientalist Karabacek in Vienna with the naturalist Johann Wiesner, for the investigation of ancient Arabic, Chinese and Turkestan rag-papers, the microscopical and chemical analysis of which confirmed step by step, in minutest details, every result of the history of the invention of rag-paper contributed from Chinese and Arabic sources. The result of their joint labors, carried through many years, I consider one of the greatest triumphs of modern science. But there are many more culture problems of equal importance whose solution must be achieved in a similar manner. Let me refer you only to the history of the invention of gunpowder and of the magnetic compass, both of which are still very obscure in fundamental points, and the working-up of which requires a whole force of well-trained specialists—Arabists, Sanskritists, Sinologues, and men well versed in chemistry, technology, physics and their history.

A study of some of the principal questions in this field is further of profound significance in an interpretation of the methods and results of anthropology. Allow me to exemplify this briefly from the instance of mathematical history. The relation of the concepts of mathematics to the human mind and to the development of culture is still a matter of controversy, and one of burning actuality, just at the present time. A solution on the basis of an historical method is one of the aspects of this problem. The historical position of mathematics, however, is as yet very far from being defined, and no criterions are agreed upon which will admit at the outset of stamping a mathematical thought and theorem as borrowed or independent. The most striking feature in the history of this science is the fact that the same results, even in the highest branches of it, have frequently been obtained by different peoples and at various epochs, with little or no possibility of pointing out an historical connection between such coincidences. The quadrature of the circle, for example, was made the object of correct speculation in China, even in pre-Christian times; or the rule of Horner 'for solving equations of all orders,' established in 1819, was known to the Chinese 520 years earlier, when, in an arithmetical treatise published in 1299, roots were extracted as high as the thirteenth power.[4] Paul Harzer,[5] astronomer at the University of Kiel, last year submitted the mathematical knowledge of ancient Japan to a careful and ingenious examination, and has arrived at the conclusion that the Japanese found spontaneously adequate evaluations of the ratio π, and made the independent discovery of the binomial theorem, which they utilized for obtaining important results. Modern criticism, with its aggressiveness towards the groundwork of human knowledge, towards even that which seems most secure, has recently attacked also the foundations of mathematics, generally looked upon as the most unobjectionable science, and has designated its results, like those of other sciences, as more or less conventional, not necessitated by the nature of the human mind.[6] To us, mathematics is essentially an outcome of human culture; and the question arising from an anthropological view-point

[4] A. Wylie, 'Jottings on the Science of the Chinese Arithmetic,' in his 'Chinese Researches' (Shanghai, 1897), pp. 163, 184, 185.

[5] Paul Harzer, 'Die exakten Wissenschaften im alten Japan,' Kiel, 1905.

[6] Harzer, *ibid.*, p. 26.

is, Are the phenomena of mathematical thoughts to be considered as on an equal footing with those of language, religion or medicine, and, accordingly, capable of methodical anthropological treatment, or are they the particular productions of individual thinkers, and, accordingly, conducive only to an exclusively historical analysis? It is impossible for the present to pronounce a verdict on this intricate problem, though I should like to say tentatively, and with all reserve, that the present state of our knowledge of the mathematics of India, China and Japan would almost seem rather to favor the acceptance of the former theory. At all events, the ventilation of this question is well illustrative of the paramount importance of the study of the history of mathematics and its principal bearings on our views of the intellectual history of man.

The practical proposition which I finally wish to lay before you is, that working committees, cooperative in character, be organized, each consisting of a limited number of members selected equally from students of natural sciences and students of anthropology, especially those in Oriental fields, and pursuing given problems *viribus unitis.* Each of these unions, which need not be of an official character, but may be freely private voluntary alliances of interested students, should be in charge of a particular branch of science. Altogether, seven may be necessary—one for the study of the history of mathematics and astronomy; others for that of cultural plants, domestic animals, physics, chemistry, technology and medicine. Each committee should be so constituted that the united forces of its laborers will represent a consummate systematic knowledge of the subject in question, and take up, suggest, encourage and elaborate pending problems by the concerted action of all its participants. By this method of research, much time and labor would be saved, and more positive and enduring results would be secured.

In concluding, let me call your attention also to the fact that we do not yet possess a history of anthropology, and that broad-minded contributions to the history of our science are an urgent necessity. Goethe has said somewhere that the history of science is the science itself; and I believe, further, that only by a correct appreciation of the development of our science are we able to be just towards our fellow-workers. Now that so many of our prominent leaders, like Brinton, Powell, Cushing, Virchow, Bastian and Ratzel, have passed away, whatever we may personally think of the value of their work and its influence upon future generations, it is our duty to come to an objective understanding of their activity and aspirations, and to write the pragmatic history of anthropology in the life and labors of its most conspicuous representatives.

BERTHOLD LAUFER

SCIENTIFIC BOOKS

THE MISSION FOUREAU

Documents Scientifiques de la Mission Saharienne, d'Alger au Congo par le Tchad. Par F. FOUREAU, chef de la mission. IIme fasc., Orographie, Hydrographie, Topographie, Botanique; IIIme fasc., Geologie, Petrographie, Paléontologie, Esquisse Ethnographique, notes sur la faune, Prehistorique, Aperçu Commercial, Conclusions économiques, Glossaire. Index. Atlas. Paris, Masson et Cie. 1905. 4to, 1210 pp., maps and ills.

While the Mediterranean and mid-African colonies of France have been for some time fairly well known, the efforts to connect them by a line of geographical exploration had been rendered ineffectual by the difficulties and dangers of the route through the desert, and the hostility of the natives. Several expeditions met with disaster and were exterminated by the fanatical population.

Finally the expedition organized and carried out by Foureau in 1898 to 1900 met with success. This happy result had been well earned, because Foureau had already given twenty-three years to Saharan exploration under the auspices of the Ministry of Public Instruction. In 1898 his itineraries already amounted to 21,000 kilometers, of which more than 9,000 km. were in previously unexplored country.

In 1894, M. Renoust des Orgeries left to the Société de Géographie a considerable sum of money, to be devoted to the geographical development of the French colonies. Out of this legacy the society devoted 250,000 francs to the purpose of the Sahara Mission, a sum to which the government added not only funds but an escort of 250 picked soldiers under the command of a most competent African officer, Com. Lamy, who fell during an attack by an African chief, in the very moment when the success of the expedition was assured.

With the concurrence of men of science, the commander of the expedition has prepared this report, which by the assistance of government and various scientific societies, is now published in magnificent style by the Société de Géographie, with a preface by Alfred Grandidier.

Together with the reports indicated by our synopsis of the title, the work is replete with well-executed maps, sketches, plans of towns, views and everything which could be of use to future explorers, including minute notes as to the presence, amount and quality of water, pasturage, cultivated land, wild animals, etc. If one is startled by the frequent notation, along the river, of the presence of 'oyster banks,' hundreds of miles from the sea, reflection recalls the colonies of the fresh-water *Ætheria* to which these notes undoubtedly refer.

The mass of information in this encyclopedic work, it is, of course, impossible to summarize. A few notes may, however, have interest for the reader. While a large portion of the herbarium suffered from termites and the wreck of a canoe, nevertheless a good number of plants are recorded; and numerous

046

中国人与菲律宾群岛的关系

Smithsonian

Miscellaneous Collections

VOLUME L

(QUARTERLY ISSUE, VOLUME IV)

," EVERY MAN IS A VALUABLE MEMBER OF SOCIETY WHO, BY HIS OBSERVATIONS, RESEARCHES,
AND EXPERIMENTS, PROCURES KNOWLEDGE FOR MEN."—SMITHSON.

No. 1789

CITY OF WASHINGTON
PUBLISHED BY THE SMITHSONIAN INSTITUTION
1908

WASHINGTON, D. C.
PRESS OF JUDD & DETWEILER, INC.
1908

THE RELATIONS OF THE CHINESE TO THE PHILIPPINE ISLANDS

By BERTHOLD LAUFER

The history of the Spaniards on the Philippines is an endless chain of frictions and struggles with the Chinese immigrants and settlers, so that the history of the Philippines during the last three centuries is very closely interwoven with an account of the relations between these two peoples. The trade with China was by far the most important business of the Spanish colony—and with it the fortunes of the colony rose and fell. An abundant mass of material has been stored up by Spanish writers since this early contact with the East, from which the political and commercial history of Chinese intercourse might well be compiled; but no attempt has heretofore been made to call to witness coeval Chinese sources, and to compare Spanish accounts with Chinese testimony on the same subject. To advance a step in this direction and do justice to the *audiatur et altera pars* is the prime object of the present paper.

The Chinese have been acute observers of foreign nations and countries, and in their astoundingly vast amount of literature we find many valuable reports on the geography, history, and ethnology of the neighboring peoples. The history of the Malayan Archipelago (particularly, for example, Java and Sumatra) during the precolonial age would be almost shrouded in mystery but for the material regarding these islands hoarded up by the Chinese.[1] The principal Chinese sources of which I have made use are the Annals of the Ming dynasty, or the "Ming shih," which, in chapter 323, furnishes an account of all islands in the eastern Pacific known to the Chinese at that time, and also of the Portuguese, Spanish, and Dutch, who then made their first appearance in the Far East. Furthermore, the annals of the provinces of Kuang-tung and Fuhkien frequently speak of the Philippines, and describe historical and other incidents relating to them, for the natural reason that the traders and seafaring people of those parts of China were most

---

[1] The principal papers on this subject are: L. de Rosny, Les peuples orientaux connus des anciens Chinois (Paris, 1881); W. T. Groeneveldt, Notes on the Malay Archipelago and Malacca, compiled from Chinese sources (Batavia, 1876), and Supplementary Jottings to this paper in *T'oung Pao* (1896), vol. VII, pp. 113-134. Groeneveldt has not dealt with the Philippines.

248

active in transmarine undertakings. The geographical literature of the Chinese also abounds in accounts of the Philippines. The most important of these, frequently alluded to in these pages, is the "Tung hsi yang k'ao" ("Investigations regarding the Eastern and Western Ocean"), published in 1618—a very useful geographical work in twelve books, the descriptions of which, as a rule, refer to the time when Europeans first began to visit the Malayan regions.[1]

The appearance of the Spaniards in Eastern waters was not the first occasion on which the Chinese had taken cognizance of these Western people. From the accounts of the Arabs, they had gained a certain knowledge of Spain as early as the beginning of the thirteenth century. Chao Ju-kua, a member of the imperial family of the Sung dynasty (960-1278) and superintendent and commissioner of customs in Ts'üan-chou-fu, a coast town northward from Amoy, in Fuhkien Province, came in close touch with merchants from India, Persia, Syria, and Arabia, who traded in that port with the Chinese, and availed himself of this opportunity to collect valuable data regarding the countries and peoples of the West. In his book, "Chu fan chi," written between 1209 and 1214, a brief description is given of Spain under the name *Mu-lan-p'i; i. e.*, the Arabian word *Murâbit*, which we find hispanicized as the dynasty of the *Almoravides*. He relates that Spain entertained a lively commerce with the *Ta-shih* (Arabs), and emphasizes the large size of her ships, which could carry several thousand men. Wheat, melons, pomegranates, lemons, rice, and salads are mentioned as the products of the country, and it is curious to find merino sheep mentioned as being several feet high and having tails the size of a fan.[2]

In the "Ming shih" and according to later sources, the name for Spain is *Yü-ssŭ-la* (or *Yü-mi-la*, by confounding the similar characters for *mi* and *ssŭ*), apparently an imitation of the sounds in the name *las Islas*, which the Chinese had heard from the Spaniards on the Philippines. The ordinary designation for Spain and Portugal, however, is simply *Hsi yang* (*i. e.*, "Western Ocean"), with the distinction that Spain is called *Hsiao hsi yang* ("the small Western Ocean"), and Portugal *Ta hsi yang* ("the great Western Ocean"), as the Portuguese were the first of the two to come under the notice of the Chinese. At a later period these names, *Hsi yang* and *Ta hsi yang*, were used in a general way for Europe, the name of which,

---

[1] See A. Wylie, Notes on Chinese Literature, 2d edition, p. 58; Groeneveldt, Notes on the Malay Archipelago, p. VIII.

[2] F. Hirth, Die Länder des Islâm nach chinesischen Quellen (Leiden, 1894), pp. 48-50, 63.

toward the close of the fifteenth century, became known in China also in the transcription *Ou-lo-pa*. In the Portuguese-Chinese vocabulary appended to the "Ao men chi lio," the Chinese chronicle of Macao, we find that the Portuguese name for Ta hsi yang (Portugal) is *Lien-nu*, by which is evidently understood *Lusitania*; while the Portuguese name for Spain is written *Wo-ya*, reading in the Amoy dialect *Nga-ñia* (that is, in Portuguese apparently *Hespanha*). These designations, however, were those of geographical and diplomatic language; the popular term by which the Portuguese and Spanish were both spoken of, and even confounded with each other in literature, was *Fo-lang-ghi; i. e.*, "the Franks."

The main island of the Philippine group, Luzon, was known to the Chinese, long before the Spanish Conquest, under its native name *Luzong*, which appears in the texts in the form *Lü-sung*. This name was also extended to the entire group of islands, and, furthermore, was applied as a tribal name to the native population. At the time when the Spaniards took possession of the Philippines the name Lü-sung designated principally the city of Manila, but it was then transferred also to the Spaniards, who are the "Luzon men" of the Chinese annals, or, officially, *Ta Lü-sung kuo*. At that time a nickname was also invented for the Spaniards in the form *Sung-tsai*, which may be explained as follows: The character "sung" in Lü-sung is identical with that in the name of the Sung dynasty of China, which, like all dynasties, has the adjective *ta* ("great") prefixed to its title. In contrast to the great Sung dynasty, the foreign Lü-sung men were contemptuously called *Sung-tsai; i. e.*, the little Sung. A still more derogatory term under which the Spaniards go in the "Ming shih," in passages of a kind to provoke the criticism of the author, is *Man (i. e.*, savages), originally a name restricted to the primitive aboriginal tribes of southern China.

In modern times the Portuguese retained their old historical name, *Ta hsi yang kuo*, in diplomatic intercourse with the Chinese court, while the Spaniards adopted the transliteration *Ta Jih-ssŭ-pa-* (or *pan*)-*ni-yakuo*, which has come up since the early days of the Jesuits; also Lü-sung is still the Chinese name for Manila, Luzon, and the Philippines generally, and *Lü-sung yen* ("Luzon smoke") is a common term for Manila cigars. In connection with this terminology it might not be without interest to add that the name "America" occurs for the first time in Chinese literature (about a century after its discovery) in the "Ming shih" (chap. 326) as "*A-mo-le-kia*," in connection with a report on the famous Jesuit

Matteo Ricci, who presented to the Emperor a map of the world on which he stated that "there are in the world five parts of the globe. The fourth of these was America."[1]

It is at a comparatively late date that Chinese history makes mention of the Philippine Islands; and this fact is the more striking, since some of the adjacent isles to the south are touched upon much earlier. The Moluccas, for example, are first mentioned, under the name *Mi-li-kü*, in the Annals of the T'ang dynasty (618-906), in determining the site of the island of Bali, although no special description of them is given earlier than the sixteenth century.[2] Puni— that is, Brunei, or the northwest coast of Borneo—appears in the history of the Sung dynasty (960-1279),[3] and we cannot but think that navigators sailing there must have passed the great island of Palawan or some isles of the Sulu Archipelago. However this may be, the Philippines are not actually mentioned by name in literature earlier than the time of the Ming dynasty (*Ming shih*, chap. 323, p. 11 a). In the fifth year of the period Hung-wu (1372) the first embassy from the Philippines arrived in China with tribute. The site of Luzon is stated on this occasion to be in the South Sea very close to Chang-chou in Fuhkien. The Emperor reciprocated the gifts of this embassy by despatching an official with presents of silk gauze woven of gold and colored threads to the king of the country. From this first mention of the Philippines in Chinese history we should not be so narrow-minded as to infer that Chinese intercourse with the Philippines dates from just the year 1372; on the contrary, the fact that there was a Philippine embassy in that year points to a long commercial intercourse between the two peoples, which had escaped the knowledge of the court historiographers at Peking. Although the imperial geography of the Ming, the "Ta Ming i t'ung chi," states expressly that no investigation of Luzon had been made by earlier generations, this is refuted by the fact that we meet with an account of the Philippine tribes in the before-mentioned "Chu fan chi" of Chao Ju-kua in the thirteenth century.[4] Chao Ju-kua describes a country in the north of Borneo which he calls *Ma-yi(t)*, which name Professor Blumentritt thinks is identical with Bay, the

---

[1] E. Bretschneider, Mediæval Researches from Eastern Asiatic Sources, vol. II (London, 1888), p. 324.

[2] Groeneveldt, loc. cit., p. 117.

[3] Ibid., pp. 106, 108.

[4] The passage in question has been translated by Professor Hirth in his book "Chinesische Studien," p. 40.

territory of Manila;[1] and speaks further of a country called *San hsii*[2] ("The Three Islands"—Ka-ma-yen, Pa-lao-yu, and Pa-ki-nung). The sketch of the native population given by him is very interesting. He says:

On each island lives a different tribe. Each tribe consists of about a thousand families. As soon as a foreign ship comes in sight, the natives approach it to barter. They live in rush huts. As there are no springs in the mountains, the women carry two or three jugs at the same time on their heads, in which they fetch water from the springs in the plain, and with this load they ascend the mountains as easily as if they were walking on level ground. In the most hidden valleys live people called *Hai-tan* (the Aëta or Negritos). They are of small stature, have round brown eyes and frizzled hair, and their teeth shine between their lips. They live high up in the tops of trees,[3] where they dwell in families of from three to five individuals. Crawling through the thickets of the forests, they shoot from ambush at passers-by; wherefore they are much dreaded; but if a porcelain cup is thrown towards them they rush on it, shouting with joy, and escape with their spoil.

Then the mode of trading with the merchants of the Chinese ships is related. The native articles traded were cotton, cotton goods, beeswax, cocoanut, and fine mats, while the Chinese exchanged for them silk parasols, porcelain, and baskets plaited of rattan. Even in 1572 the inhabitants of Cagayan told the captain Juan de Salcedo, that their cotton weavings were bought up yearly by Chinese and Japanese traders. Chinese-Philippine trade must therefore have existed early in the thirteenth, and very likely in the latter part of the twelfth century.

Perhaps a still earlier ethnographical allusion to a Philippine tribe

---

[1] I am rather inclined to believe that the island of Mindoro is meant, which, according to Blumentritt (Versuch einer Ethnographie der Philippinen, p. 65), was called *Mait* in oldest times. In all likelihood the Chinese were acquainted with Mindoro at an earlier date than with Luzon. It was on Mindoro that in 1571 Spaniards and Chinese met for the first time. The Three Islands are probably Busuanga, Calamian, and Peñon de Coran. Of other localities mentioned by Chao Ju-Kua, Pai-pu-yen may be identified with the Babuyan north of Luzon; Pu-li-lu with Polillo, eastward from Luzon.

[2] An island group of the same name is mentioned in the History of the Mongol Dynasty as situated near Formosa, with a population of only 200 families. In language these people seem to have been different from the Formosans, for the latter could not understand the speech of an interpreter from there in the service of the Chinese. The group is certainly not the same as that above (see *Yüan shih*, chap. 210, pp. 4-5).

[3] The *Tung hsi yang k'ao* mentions a mountain range on Luzon by the name of *Fou-ting-shan*. It says: "Wild barbarians dwell in nests on the top of these mountains, and shoot from trees at birds and animals, which they eat uncooked. One cannot follow their trails."

is contained in the "History of the Sung Dynasty" (*Sung shih*), in the chapter giving the history of Formosa for that period. After a few remarks on the native tribes of the island, the report goes on to say:

Near them (*i. e.,* the Formosans) is the land of the Pi-sia-ye (Visaya), whose language is not understood [on Formosa]. They go naked, and from the way they stare, one would say they are not like other people. In the period Shun-hsi (A. D. 1174-1189) the chief of that country daringly took some hundreds of his men and suddenly appeared in the Bay of Ts'üan-chou (Fuhkien Province). In Wei-t'ou and other villages they committed outrages and murder. In their plundering they looked chiefly for iron implements, spoons, and chop-sticks. When people shut their doors, they desisted, and only cut off the rings of the door-knockers. When spoons and chop-sticks were thrown to them, they stooped to gather them. When they saw a rider clad in iron, they struggled among themselves to cut off his armor; then, joining forces, slew him mercilessly. In close combat they availed themselves of spears, to which a rope of more than a hundred feet in length was attached with which to handle the weapon, for they save their iron and do not recklessly throw it aside. They had no boats or oars, but rafts made of bamboo poles tied together. Hurriedly they carried these off jointly, set them afloat, and disappeared.

The identification of the Pi-sia-ye mentioned in this text with the Visaya of the Philippines has already been proposed by Terrien de Lacouperie,[1] but has been rejected by G. Schlegel[2] on the ground that it is impossible that those islanders should have been able to make the long passage over sea on rafts, as the Chinese historian says—a feat which, however, was possible from Formosa to Fuhkien. Schlegel accordingly seems to infer, but does not state explicity, that the Pi-sia-ye are a Formosan tribe.[3] His arguments, however, are by no means valid. The Chinese text is not at all ambiguous, and says plainly that it was a country beside or near Formosa, and one with a different language; there is here, consequently, the question of a non-Formosan tribe. In the description of the Formosan tribes, Chinese authors never use the word "country" (*kuo*) as used above in connection with the Pi-sia-ye, but speak only of clans and tribes; furthermore, a tribe of the name Pi-sia-ye has never existed, nor does it exist, on Formosa. The mere linguistic evidence, however (*i. e.,* the phonetic coincidence of Pi-sia-ye and Visaya or Bisaya), is not in itself sufficient proof for assuming the identity of these people with the Philippine tribe of that desig-

---

[1] The Languages of China before the Chinese, p. 127.

[2] *T'oung Pao* (1895), vol. VI, p. 182.

[3] James W. Davidson (The Island of Formosa, p. 3) falls into the same error.

nation. Culture-historical considerations must be added to make the evidence convincing.

There can be no doubt that the aborigines of Formosa form part of the Malayan group of peoples; and from the oldest account which we possess regarding them, which is contained in the Chinese Annals of the Sui dynasty, full evidence of the fact may be obtained, that in the beginning of the seventh century, when the Chinese first discovered the island, its culture was of a thoroughly Malayan character. Moreover, it has been observed that the languages of Formosa are more closely akin to those of Luzon than to any other Malayan stock, a large number of words being in common, even terms expressing relationship, and that striking agreements in the two cultures exist; e. g., in the practice of and ideas concerning head-hunting. I am under the impression that Formosan and Philippine-Malayan cultures are only two variations of one and the same North-Malayan culture-type. The fact that the Formosans are immigrants is self-evident and confirmed by native traditions. Theoretically, there are only two ways possible for this immigration: either the Formosans came from the original seats of the Malayan stock or from the direction of the Philippines. I concur with Prof. H. Kern and P. W. Schmidt in the view that the Malayan home was somewhere off the east coast of Farther India—a theory now splendidly corroborated by the discovery by Schmidt of the relationship of the Malayan with the Mon-Khmêr languages. If the Formosans had taken their starting-point from there, they would doubtless have gradually followed the coast-line of the East-Asiatic mainland, and, touching along the shores of China, have reached their present home. Then, however, we should have expected that they would never have lost contact with the continent, and would have had some idea of the Chinese. The fact is, however, that the Formosans never had any cognizance of China, nor the Chinese of them, before the year 607, and at the first military expedition of the Chinese, in 610, the two cultures suddenly clashed like two alien worlds. The reason for this late mutual acquaintance may be sought partly in natural events, as in the fact that the channel which separates the island from the continent is shallow and perilous to navigation, and in that the whole region is the center of typhoons. On the part of the Formosans, the additional fact comes in that they were not and are not skilled seafarers, in contradistinction to their relatives. No Formosan word referring to boat-gear agrees with any Malayan word; their national vessel is still the raft.[1] We hear nothing of ships and maritime enterprises, and it is strange indeed that they never visited the neighboring Chinese coast. This is one of the chief reasons which incline me to think that the tribe which, according to the "Sung-shih," made a piratical move toward Fuhkien at the end of the twelfth century, can not well have been of Formosan origin. If we now consider the only possible way which emigration to Formosa could have taken place—that is, from the Philippines—there is no longer any reason for wondering why the Pi-sia-ye should not have come from the same direction. Formosa can have been populated only from that region; not, however, as the result of a bold seafaring enterprise, but rather, it would seem, through several accidental adventures. In this connection the following incident may be instructive. In August, 1886, some fishermen in the neighborhood of Anping (now Tainan, southwest Formosa) picked up a castaway canoe in which there were three men, two women, and a child in a starving condition. They proved to be natives of an island to the north of Luzon, who had been blown to sea in a typhoon, and had ultimately drifted to the shores of Formosa, having been thirteen days without food, and dependent on rain water for drink.[2] It seems to me quite conceivable that in times gone by people may have thus drifted repeatedly to Formosa from the Philippines, especially from Luzon, making a series of emigrations which finally led to the settlement of the island. Such casual drifting was perhaps the case also with the Pi-sia-ye, who reached Formosa first,

---

[1] C. Imbault-Huart, L'île Formose (Paris, 1893), p. 273. Some tribes may have formerly possessed canoes also.

[2] The following case, recorded by Davidson (The Island of Formosa, p. 580), also deserves mention in this connection. The Riru tribe of the Kirai district of the northern Ami (in southeast Formosa) state that their forefathers originally lived in an island to the east of Formosa. One man, called Tipots, and his family were out at sea in two canoes when a terrific gale arose, sweeping them away from their home-land and wrecking them on the coast of Formosa, where they built houses and gave life to the present Riru tribe. This tribe possesses an old canoe which they claim is the model of the one used by their forefathers. At present the village people once a year put the canoe into the sea and mimic the landing of their ancestors. After this ceremony they worship spirits of their departed ancestors. A more fanciful tradition is to the effect that their ancestors came from over the sea on the back of a large tortoise. "Thus it would appear," concludes Davidson, "that the traditions of the north Ami describe comparatively recent occurrences and are in the main very possible, if not probable."

and then, driven away by inhospitable natives,[1] turned to the shores of China. I think that in this manner a rational explanation of the event may be given. The account of the "Sung shih" is certainly incomplete and abrupt; but how should the Chinese, then entirely ignorant of the far-off Philippine Archipelago, have obtained a more detailed knowledge of a handful of people who paid only a flying and hostile visit to their coast? But the brief sketch of them showing their craving for iron, their mode of fighting, their bold Viking raid, is an ethnographical document of great impressiveness. If the identification of Pi-sia-ye with Visaya is justified, we have here the oldest historical allusion extant to a tribal movement and event of the Philippines.

A special section in the "Ming shih" is devoted to the Malayan tribe of the *P'ing-ka-shi-lan,* which I identify with the Pangasinan, who inhabit the western and southern shores of the Bay of Lingayen, on Luzon. Before the Conquest their territory extended much farther northward, but they were gradually repulsed by the Ilocanos. Since 1572 they have been subjected to the Spaniards, and at the present time they are all Catholics.[2] According to the Chinese records as preserved in the "Ming shih" (chap. 323, p. 20), they seem to have formed a small realm of their own in the beginning of the fifteenth century. Their first embassy to China mentioned was despatched in 1406 to the court of the Emperor Yung-lo, whom they presented with excellent horses, silver, and other objects. In return they received paper money and silks. Their second embassy falls two years later, in 1408; and a third was sent in 1410. In the former of these last two embassies the chieftain appeared personally with a large retinue, having selected two men from each village subject to his authority, each of whom led a number of his tribal clan to bring tribute to the court. The Emperor bestowed paper money (*ch'ao*) on the two sub-chiefs, and six pieces of an open-work variegated silk fabric for coats and linings for a group of a hundred men. Their followers also received gifts. In the same year, 1410, another embassy from the Philippines is mentioned, the

---

[1] Compare especially chapter IX in Davidson's book, "Wrecks and Outrages on Navigators." It must be also remarked that the communication between Formosa and Luzon had no difficulties. According to Davidson (p. 563), the present plains tribes of Formosa, once in prosperous and powerful circumstances, formerly crossed the Bashee Channel to the south and maintained communication with Luzon. The traveling distance from Formosa to Manila is given by the Chinese to be 60 "watches" (*kêng*), i. e., 6 days and nights (*Ming shih*, chap. 323, p. 18 *b*).

[2] F. Blumentritt, Versuch einer Ethnographie der Philippinen, pp. 21-22.

head of which was a high official called Ko-ch'a-lao.   He brought with him the products of his country, particularly gold.   The natives therefore must be credited with the exploitation of gold before the advent of the Spaniards.[1]   This becomes evident also from a passage in the "Wu hsio pien," a history of the Ming dynasty published in 1575.   It is quoted in the "Tung hsi yang k'ao" (chap. 5, p. 1) as follows: "Luzon produces gold, which is the reason of its wealth; the people are simple-minded, and do not like to go to law."

As to how far the political influence of the Chinese extended over the Philippines in prehispanic times, we have only scanty information.   The "Ming shih" (chap. 323, p. 11 a) relates on this point that in 1405 the Emperor Yung-lo sent a high officer to Luzon, who was to govern the country.   The result of his visit was the embassy from Luzon' under Ko-ch'a-lao in the same year.   How long Yung-lo's delegate remained on the island and of what character his jurisdiction was are not narrated, but it is not at all incredible that the ambitious Yung-lo exercised a kind of supremacy, or at least claimed a prerogative of protection, over the Philippine Islands; for since its establishment the rule of the Ming dynasty has been characterized by a tendency toward expansion, from a desire to extend its fame over land and sea to the farthest extremities of the world.

In Yung-lo's time the Chinese started an extensive exploration of the Indian Ocean.   In 1407 the eunuch Chêng-ho undertook a memorable expedition, accompanied by a fleet of sixty-two large ships, carrying 27,800 soldiers; and on his crusades, repeated several ·times in a space of about thirty years, he visited a number of countries in the Indian Ocean as far as the Arabian Gulf, and obtained the nominal allegiance of their rulers.   For this reason the "Ming shih" abounds in geographical and ethnological descriptions of all Asiatic countries and peoples from Central Asia to Asia Minor.

Then Vasco da Gama had not yet navigated around the Cape of Good Hope; no European sail had yet been visible on the Pacific and Indian oceans, of which the Chinese and the Arabs were the unrestricted masters and the only representatives of an immense trade.   It therefore seems not impossible that in that great age of maritime discoveries the enterprising Emperor had cast his eyes Philippineward and had won a temporary nominal suzerainty over the native tribes of Luzon.

---

[1] Compare M. Sonnerat, Voyage aux Indes Orientales et à la Chine (Paris, 1782), vol. II, p. 114.   Also De Morga mentions native gold-mines.

Some of the older Spanish authors also entertained the view that the Philippines were once subject to Chinese rule; and Father Gaubil relates in the *Lettres édifiantes* that Yung-lo maintained a fleet with thirty thousand men, which sailed to Manila at various times.

It was in 1571 that the Spaniards and Chinese met for the first time at Mindoro, before Legazpi, the conqueror of the Philippines, undertook his expedition to Manila.[1] That there was a colony of Chinese on Luzon before the arrival of the Spaniards, there can be no doubt, as it is clearly stated also in the "Ming shih" (chap. 323, p. 11 b), which says that "formerly the people of Fuhkien lived there because the place was conveniently near. They were traders of abundant means, ten thousand in number, who, as a rule, took up a long residence there, and did not return home until their sons and grandsons had grown up. When, however, the Franks snatched away this country, the Spanish king despatched a chief to suppress the Chinese. As he was concerned lest they might revolt, he expelled many of them. All those remaining had to suffer from his encroachments and insults."

According to the Ming Annals (chap. 323, p. 11 a) it was about the commencement of the Wan-li period (*i. e.*, 1573) that the Franks made their first appearance in Philippine waters. There is a curious tradition reported by the Chinese chronicler in connection with the first settlement of the Spaniards and their foundation of the city of Manila. This tradition runs as follows: "The Spanish Franks surpassed the people of Luzon in strength, and for a long time interchanged commerce with them. When they perceived that· the country was weak and could be occupied, they bestowed rich presents on the king and demanded a plot of land as big as an ox-hide for building houses and living there, The king did not suspect any trickery, and assented. These men thereupon cut the hide of an ox into narrow strips, pieced these together until they extended the length of a thousand fathoms, and in this way encompassed the whole land of Luzon, which they then claimed, in accordance with their agreement. The king was exceedingly taken aback; but, as he had already given his promise, there was no way out of it but to yield to their demand. Thus these men obtained the land, erected houses, and built a city, where they planted firearms and safeguarded it against the attacks of highwaymen. Finally they took advantage of the king's unpreparedness, came upon him unawares, killed him

---

[1] F. Blumentritt, Die Chinesen auf den Philippinen (Leitmeritz, 1879), p. 1.

and his people, and took their country, the name of which was thenceforth 'Luzon-Spanish-Franks.' "[1]

In this tradition a repetition of the classic story of the ruse of Queen Dido in connection with the foundation of Carthage will be recognized at once (see Appendix). That the Chinese tradition regarding the occupation of Manila, however, is not quite without foundation in some details, may readily be seen by a comparison with the Spanish account of Antonio de Morga, whose "Sucesos de las Islas Filipinas" was published in Mexico in 1609.[2] At the Bay of Manila the Spaniards found two fortified towns separated by a large river, each in possession of a chief. The Spaniards entered the town by force of arms, and took it, together with the forts and artillery, on the day of Santa Potenciana (the 19th of May, 1571), upon which the natives and their chiefs gave in and submitted, and many others of the same island of Luzon did the same.[3] Then the commander-in-chief, Legazpi, hurried to the scene from Panay and established a town on the very site of Manila, which the chief presented to the Spaniards for that purpose. In the words of De Morga, *he took what land was sufficient for the city.*

After 1410 no further relations of China with the Islands are recorded until 1576, in which year an imperial army was forced to fight against the corsair *Lin Tao-k'ien* or *Lin-fung.*[4] The inhabitants of the Philippines took an active part in the suppression of the rebels, and, in recognition of the service rendered to them by China, sent an embassy which traveled by way of Fuhkien. The speaker

---

[1] The *Tung hsi yang k'ao* (chap. 5, p. 1 *b*), after relating the same story, has the following in addition: "The King of Yü-ssǔ la (Islas, *i. e.,* Spain) despatched a chieftain to guard the place. After several years a change in the government took place. The Chinese who formerly traded with Luzon now do their business with the Franks. The Chinese go to Manila in great numbers, traveling to and fro. Those who make a long stay and do not return home are called *Ya-tung* (Cantonese, *At-tung;* literally, 'pressing the winter'). They live crowded together in the *Kan* (*i. e.,* the *parian* of the Spaniards). The number of those born there has gradually increased to tens of thousands. Occasionally there are found among the elder sons and grandsons those who cut off their hair." De Morga remarks that the Christians among the Chinese differ only in that they cut their hair short, and wear hats, as do the Spaniards.

[2] English translation by Lord Stanley, published by the Hakluyt Society (London, 1868). It is this edition to which reference is made in this paper. A new translation has just been issued by Blair and Robertson in two volumes (Cleveland, 1907).

[3] De Morga, p. 18.

[4] By Spanish authors he is called *Limahon,* from the Amoy pronunciation, *Lim-hong.*

of this delegation was a Mohammedan, who probably made himself understood in Arabic through Chinese Mohammedan interpreters, as, add the Annals, was also the custom in Corea.

In 1571, three years after its foundation, Manila was attacked and nearly taken by Lin-fung. The city was saved only by the valor of the hero Salcedo. This event was recorded at great length by the Augustinian monk Fray Gaspar, in his "Conquista de las islas Filipinas," which appeared in Madrid in 1698.[1] The "Ming shih" alludes to Lin-fung only once, in the passage above quoted; but the Chinese Annals of the province of Fuhkien, the Chronicle of Chang-chou, and the "Hai kuo t'u chi"[2] give fuller accounts of his piratical enterprises. The Spanish embassy mentioned in the "Ming shih" as having arrived after the expulsion of the corsair is confirmed by the Spanish documents of the time. The governor, Labezares, considered it his principal task to entertain peaceable and amicable relations with an empire whose pirates alone were able to shatter the Spanish possessions in Asia. He was led to such a policy still more by commercial considerations. The commander of an imperial Chinese war vessel, who had been sent out from Chang-chou to look after Limahon and who was charmed with the chivalrous character and the generosity of the Spaniards, offered to take Spanish envoys over to China in his ship. This embassy consisted of two military officers and two Augustinian friars. The instructions given by Labezares to this mission are not without a tinge of modern politics. He declared to the Viceroy of Fuhkien that the Spaniards were animated by the desire to live on friendly terms with the Chinese Empire, and to promote commerce between the two peoples. He requested that missionaries be admitted into the empire, and particularly that a Chinese port be ceded to Spain, whence, like Portugal in Macao, she could trade undisturbed with China; the envoys to pay attention to the customs and manners of the Chinese, and especially to study what articles of merchandise were best suited for interchange between China and the Philippines and what industrial products of Spain and her colonies would promise a fair market in China. This man, Labezares, was evidently more than three centuries ahead of his time. The embassy was unsuccessful in effecting its object, although it humiliated itself so far as to perform the kotow before the viceroy, and returned to Manila in 1575, accompanied by three Chinese captains who had come to bring Limahon

---

[1] F. Blumentritt, loc. cit., pp. 5-16.
[2] A geographical work published in 1844 (Wylie, Notes on Chinese Literature, p. 66).

in chains back to their country. These officers carried rich presents to the governor from the Viceroy of Fuhkien. Meanwhile, however, Dr. Francisco de Sande had succeeded Labezares as "Gobernador." The very learned but also conceited Dr. Sande now claimed these presents for himself, while the Chinese declared they were authorized to deliver them only to Labezares. The pride of Sande was sensibly hurt by this little incident, and from that day he showed such an antipathy toward everything Chinese that he endangered the interests of the Spanish Crown by his narrow-minded policy with regard to China.

Fray Gaspar relates that in 1576 a Chinese war junk arrived at Manila with a despatch from the Viceroy of Fuhkien, in which it was stated that the Emperor had read all the Spanish letters of Labezares, and consented to cede to the Spaniards an island between Canton and Pakian under the same conditions as Macao had been turned over to the Portuguese. *This account meets with no confirmation in the Chinese annals.* Sande did not accept this offer, and offended the Chinese ambassadors by not reciprocating the presents sent to him from the Emperor. The brightest idea that dawned on him was to saddle on the returning embassy two monks, who, however, never saw the shores of China. The Chinese had humor enough to unload this clerical ballast at Bolinao, soon after sailing from Manila.

Sande conceived the daring plan of conquering China by force of arms, and deluged King Philip II with a mass of alluring reports depicting in glowing colors the feasibility of such a scheme. These form fascinating reading matter, and are now easily accessible in the fourth volume of Blair's and Robertson's monumental work, "The Philippine Islands." Philip II flatly rejected this project, and ordered Sande to further amicable relations with China; and since that time Spain has taken no further political action toward China.

The *first* great political event related in the "Ming shih" is the rebellion of the Chinese *P'an Ho-wu* in 1593, who stabbed the then Spanish governor, or, as the Annals call him, chieftain, Don Perez Gomez das Mariñas. His name is preserved in Chinese under the form Lang Lei Pi-li Mi-lao, and that of his son as Lang Lei Mao-lin, which is intended for Don Luis das Mariñas. "Lang" is a term of respect, meaning a "gentleman" generally, and evidently represents a translation of the Spanish "Don," while "Lei" seems to stand for Luiz, "Pi-li" for Perez, and "Mi-lao" or "Mao-lin" for Mariñas. The Chinese account of this incident reads as follows:

In the 8th month of the 21st year of the period Wan-li (1593), when the chieftain Don Perez Gomez das Mariñas undertook a raid on the Moluccas, he employed two hundred and fifty Chinese to assist him in the combat. It was P'an Ho-wu who was their lieutenant. The savages lay down, drowsy, in the daytime and commanded the Chinese to row the ship. As they were somewhat lazy, they were suddenly beaten with a whip, so that several of them died. Ho-wu said, "Let us revolt and die in that way. Should we submit to being flogged to death or suffer any other such ignominious death? Should we not rather die in battle? Let us stab this chieftain to death and save our lives. If we are victorious, let us hoist the sails and return to our country. If we should succumb and be fettered, it will be time enough then to die!" Then all of them at night stabbed the chieftain to death, and, seizing his head, shouted in a loud voice at the savages, who were frightened and arose, not knowing what was going on. They were all killed with the sword. Several fell into the water and died. Ho-wu and the others took possession of their gold, valuables, and military armor. Then they prepared the ship for their return, but lost their way, and proceeded to Annam, where they were robbed by the people of that country. Wei-kuo, Wei-t'ai, and thirty-two other men, being near to another ship, seized it; and when they returned (ashore), the chieftain's son, Lang Lei Mao-lin (Don Luis Perez das Mariñas), who stopped at So-wu (i. e., Cebu) learned of the affair from them. Leading his troops, he passed quickly on (to Manila), and dispatched to China a priest to state the wrong done to his father, with the request that the war junk, gold, and valuables be returned, and that those men who had incurred his enmity be executed, and thus offer retribution for his father's life. The (Chinese) governor, Hsü Fu-yüan, informed the governor-general of the two Kuang provinces of the matter through an official communication and politely sent the priest back. They pardoned Wei-t'ai for having arranged this matter. Ho-wu remained in Annam and did not venture to return. This was the first (Spanish) chieftain who had been slaughtered. Those of his division who came down to Manila expelled the Chinese into the outer part of the city. They demolished their huts; and when Mao-lin (Das Mariñas) returned he ordered them to build houses outside of the city, that they might live there together. It is reported that when some pirates once came from Japan, Mao-lin feared they might join with the Chinese, which he considered would be a calamity, and again decided to drive them out. Fu-yüan sent an envoy (to Manila) to invite the Chinese to come back (to China). The barbarians, however, provided the messengers with food for the voyage, and sent them home; for the Chinese merchants, from their love of profit, did not care to risk their lives, so for a long time they again dwelt together in the city.[1]

Antonio de Morga,[2] after describing Mariñas' plan to conquer the Moluccas, thus narrates the events of the expedition:

The governor and those who accompanied him passed the time playing on the poop till the end of the first watch; and after he had gone into his cabin to rest, the other Spaniards went to their quarters for the same purpose, leav-

---

[1] The same event is briefly alluded to in another passage of the *Ming shih*, where the history of the Moluccas is narrated. In Groeneveldt's translation of this passage (Notes on the Malay Archipelago, p. 118), the erroneous rendering "Portuguese" must in each case be corrected into "Spaniards."

[2] The Philippine Islands (London, Hakluyt Society, 1868). p. 35.

ing the usual guards in the midship gangway and in the bows and stern. The Chinese rowers three days back had agreed to rise up and seize the galley whenever they should find a favorable opportunity, from a desire to save themselves the labor of rowing on this expedition, or from coveting the money, jewels, and other articles of value on board, as it seemed to them ill to lose what was offered to their hands. They had provided themselves with candles and white shirts, and had appointed some of their number as chiefs for the execution of the plan; and they carried it out that same night, in the last watch before dawn, when they perceived that the Spaniards slept. At a signal which one of them gave, at the same moment all put on their shirts and lit their candles, and with their *catans* in their hands they at once attacked the guards and those that slept in the quarters and in the wales ("arrumbadas," planks or frames on which soldiers sleep), and, wounding and killing, they seized upon the galley. But few Spaniards escaped—some by swimming to land, others in the boat which was at the stern. The governor, when he heard the noise in his cabin and perceived that the galley was dragging, and that the rabble was cutting down the awning and was taking to the oars, hurried out carelessly, and his head being unprotected at the hatchway of the cabin, a few Chinese who were watching for him there, split his head with a catan. He fell, wounded, down the stairs into his cabin, and two servants whom he had within carried him to his bed, where he died immediately. The same fate met the servants, who were stabbed through the hatch. The only Spaniards that remained alive in the galley were Juan de Cuellar, secretary of the governor, and Padre Montilla, of the order of Saint Francis, who slept in a cabin amidship; and they stayed there without coming out; and the Chinese did not dare to go in, thinking that there were more Spaniards, until next day, when they took them out, and let them go on the coast of Ylocos, of the island of Luzon itself, in order that the natives might let them take water on shore, of which they were short.

The Spaniards who were in the other vessels, close to land, although they perceived from their ships the lights and the noise in the galley, thought it was some maneuver that was being executed; and when afterwards they knew, after a short space, through those who escaped, swimming, what had happened, they could give no assistance, and remained quiet, as everything was lost, and they were few in number, and not in sufficient force. So they waited till morning, and when it dawned they saw the galley had already set the mainsail, and was sailing wind astern, returning to China, and they could not follow it.

As the wind served, the galley sailed all along the coast of the island until leaving it. It took in some water at the Ylocos, and left there the secretary and the friar. It [the galley] attempted to cross to China, and not being able to fetch it, brought up at the Kingdom of Cochin China, where the King of Tunquin took from them what was in the galley, and two large pieces of artillery which had been embarked for the expedition to Maluco, and the royal standard, and all the jewels, money, and precious things, and left the galley to go ashore on the coast. The Chinese dispersed, and fled to different provinces. The governor, Gomez Perez, met with this disastrous death, with which the enterprise and expedition to Maluco, which he had undertaken, ceased also. Thus his government ended, after he had held it for little more than three years.

* * * In Manila the seizure of the galley and death of the governor became known very shortly, and with this astounding news the townspeople and the men-at-arms, who had remained there, met together in the house of the licentiate, Pedro de Rojas, to treat of what it was fitting to do; and first of all to elect him as governor and captain-general; and then they sent Captain Juan Ronquillo del Castillo, with other captains, in two frigates (for there was no other vessel) in pursuit of the galley; which was fruitless, for they never saw it. In like manner the governor sent to Don Luys Dasmariñas, and to the fleet and army, which was in Pintados waiting for Gomez Perez, advising them of his death, and of what had happened, and of the new election which had fallen upon him for the government, and ordered them to come with all speed to Manila, which was left very much deserted, and without the necessary precautions for anything that might occur.

If the death of the governor, Gomez Perez Dasmariñas, was unfortunate, as much for the loss of him personally as for such a good opportunity having been lost for the conquest of Terrenate, the success of which expedition was held to be certain, the return of the fleet and arrival of the troops in the city was none the less a fortunate event; since not many days later (anticipating the usual time for their navigation), a quantity of ships from China came to Manila with many men on board and little merchandise, and seven mandarins with the insignia of their office. This gave sufficient motive for suspicion that they had had notice of the departure of the fleet to Maluco, and of the city having remained defenceless, and that on this occasion they came to attempt to take the country; from which they desisted when they found the city with more troops than ever, and they returned without showing any particular motive which had brought them, and without any sign of consciousness being given by one side or other. Only the governor, Don Luys, was on the alert and very watchful, and took the proper arrangements, especially with respect to the Chinese and their quarters and parian.

Whilst Don Luys Dasmariñas governed, the suspicions and fear *continued with respect to Japan,* and people lived in anxiety as to that, and on account of the Chinese. The governor sent his cousin, Don Fernando de Castro, to China with letters and dispatches to the Viceroy of Canton and the Viceroy of Chincheo, where it was understood that there were many of the Chinese who had seized upon the galley and killed the governor, Gomez Perez. Supposing that they had gone there with it, a request was made for the guilty to be given up for punishment, and that the royal standard, artillery, and the other things which they had carried off should be restored. This was not obtained, because, as the galley went to Cochin China and the Chinese dispersed in so many directions, it could not be effected, though at the end of a few days a few of the guilty Chinese were brought from Malacca to Manila, whom the captain-major, Francisco de Silva de Meneses, had found there. From these it was known more accurately what had passed with respect to the seizure of the galley and death of the governor, and justice was done upon them.

The fuller account of Antonio de Morga agrees fairly well with the concise Chinese report, except that De Morga neglects to mention the cruel maltreatment of the Chinese sailors, and adduces no other reason for their revolt than their craving for treasure. The

Chinese embassy which he credits with the plan of taking Manila is of course identical with the peaceable envoy of the "Ming shih," whose task it was to bring his countrymen back to China. In this, as in subsequent cases, we find the Spaniards, in their dealings with the Chinese, misinterpreting their motives of action, and in consequence doing them injury and injustice. This was due chiefly to their ignorance of the language and to their lack of well-trained interpreters. From other temporary Spanish records it also becomes evident that Das Mariñas fell a victim to his own rashness and inconsiderateness. He had a large army ready to conquer the Moluccas, but was not able to secure rowers enough for his galleys. He therefore seized by force any Chinese in the parian of Manila he could lay hold of and had them chained to the banks of oars on the galleys. Most of these wretched victims were peaceable merchants and artisans. Besides these, he forced into his service as soldiers a number of Chinese traders and sailors who had just arrived from China. His murder is fixed by the Spanish chroniclers as having taken place on the night of October 25–26, 1593, which tallies exactly with the statement of the "Ming shih;" also the facts there told of the mission of Luis das Mariñas to China to ask indemnity for his slain father are confirmed by the Spanish authors. He returned without having effected his purpose; but the Portuguese gobernador of Malacca sent some of the murderer who had been caught there to Manila, where they were executed. It will be observed that the simple accounts of the Chinese are not valueless either in corroborating or in supplementing the Spanish records, and put in a much clearer and better light the true motives of the Chinese people, which could be but imperfectly understood by the Spaniards of those times.

An instructive example of how myth sometimes develops from history is furnished by Juan de la Concepcion, whose voluminous "Historia General de Filipinas" appeared at Manila in 1788–'92, in fourteen volumes. His account of the Chinese mutiny in 1593 is partial and one-sided. In speaking of the death of the governor, he says the Chinese split his head in two with their *alfanges*. He retired severely wounded, lay down on his bed, took the prayer-book of his order in his hands and an "imagen de Nuestra Señora y con estos consuelos de su piedad, dió su alma al Señor." The older sources relate nothing of such a touching scene, but agree in saying that his head was cut off at a blow.

From an historical point of view, the cruelty of Das Mariñas toward the Chinese, and his death, which resulted from it, form important factors in the long line of relations between China and the West and the opening act in a deplorable series of unjust wars and inhuman outrages. This event, no doubt, must have left a deep and lasting impression on the minds of the Chinese world and furnished good grounds for their prejudices against foreigners. And not only that: the Spanish system of treating the Chinese became the model of the Chinese in their treatment of foreigners. This is expressly stated by an English writer, who remarked seventy years ago,[1] "That the Chinese authorities are not entirely ignorant of the situation of their countrymen at Manila, we infer from the well-attested fact that the system which they have long been endeavoring to impose upon foreigners here [in China] has been borrowed from the Spanish Government. We are informed on the very best authority that Pwankequa, the father of a late well-known senior *Hong* merchant and grandfather of him who bears the same name now, saw there the harsh treatment inflicted on the Chinese in order to keep them in subjection, and marked it as a 'model and motive' to be acted on, after his return to Canton. He was a man of considerable influence in regard to all measures concerning foreigners, and the restriction on their privileges which he caused to be introduced have been gradually becoming more severe since the middle of the last century."

Indeed, if we would fully grasp the innermost causes of the Boxer rebellion, we must go back to the history of the relations of the Spaniards to the Chinese in the Philippines.

When the famous governor, Pedro de Acuña, arrived at Manila, in 1602, trade with China had reached its climax. Yearly thirteen to fourteen thousand merchants assembled at a kind of fair, when with the spring monsoons the large junks came from China. Silks and nankeens, porcelain, copper and iron, besides many other products, were exchanged for Mexican silver. At that time there were, according to Argensola, thirty thousand Chinese settled in Manila.[2] This prosperity was destined not to last, however, for in the following year there appeared in Manila a Chinese mission in search of an El Dorado, an expedition which, though it deserves a place among the wildest and most visionary of quests after gold, yet was fraught with the greatest consequences for the Chinese inhabitants of the country. The story would have a humorous tinge were it not for the fact that the folly of one man cost the lives of twenty-five thousand.

---

[1] *Chinese Repository* (1834), vol. II, p. 350.
[2] F. Blumentritt, Die Chinesen auf den Philippinen, p. 23.

Antonio de Morga, an eye-witness, gives an interesting and graphic account of these events in his temporary records.[1] It reads as follows:

In the month of this year of 1603 there entered into the Bay of Manila a ship from Great China, in which, as the sentinels announced, there came three great mandarins, with their insignia as such, and they came out of the ship and entered the city with their suite. They went straight, in chairs carried on men's shoulders, very curiously made of ivory and fine woods and gilding, to the royal buildings of the High Court, where the governor was waiting for them with a large suite of captains, and soldiers throughout the house and in the streets where they had to pass. When they arrived at the doors of the royal buildings, they were set down from their chairs, and entered on foot, leaving in the street their banners, equipage, lances, and other insignia of much state which they had brought; and went as far as a large hall, well fitted up, where the governor received them standing up, the mandarins making many low bows and courtesies after their fashion, and the governor answering them in his. They told him, by means of the interpreters, that the king had sent them, with a Chinaman whom they had brought with them in chains, to see with their own eyes an island of gold, which he had informed their king was named Cabit,[2] and was close to Manila, which was in the possession of no one; and that he had asked the king for a quantity of ships, and that he would bring them back laden with gold; and if it was not as he had stated, let them punish him with death; and they had come to ascertain the truth of the matter, and to inform their king of it. The governor replied to them in few words beyond giving them a welcome, and inviting them to rest in two houses which had been prepared for them within the city, where they and their people could lodge, and that their business would be talked of later. Upon this they went out again from the royal buildings, and at the door mounted their chairs on the shoulders of their servants, who wore colored clothing, and they were carried to their lodgings, where the governor ordered them to be abundantly provided with whatever they required for their maintenance during the time of their stay.

The arrival of these mandarins seemed suspicious, and [it was thought] that they came with a different intention from that which they announced, because, for people of so much understanding as the Chinese possess, to say that the king sent them on this business seemed to be a fiction. Amongst the Chinese themselves, who came to Manila about the same time with eight merchant ships, and those who were established in the city, it was said that these man-

---

[1] Hakluyt edition, p. 217.

[2] That is, Cavite, called in the writings of the Chinese Chia-i (in Cantonese, Kia-yit), which is the city of Cavite. The *Tung hsi yang k'ao* (chap. 5, p. 3 b) remarks that it was originally only a mountain, and that the Spaniards had founded a city there from fear of the Red-haired (i. e., the Dutch), and concealed gingals behind the walls; in case pirates appeared, they repulsed them by means of these gingals, but did not venture to oppose them in open attack. According to the same passage, the mountain Ki-i shan mentioned by Chang-Yi as the gold mountain is a mistake for Kia-i or Kia-yit, and would therefore be identical with the mountains around Cavite. This agrees perfectly with the statement of De Morga.

darins came to see the country and its condition, because the king wished to
break off relations with the Spaniards, and to send a large fleet before the year
was out, with a hundred thousand men, to take the country.

The governor and High Court were of opinion that they should be watchful
in guarding the city, and that these mandarins should be handsomely treated,
but that they should not go outside of the city, nor be allowed to administer
justice (as they were beginning to do among the Sangley[1] men), at which
they felt some regret: they were desired to treat of their business, and then
return shortly to China, without the Spaniards letting themselves appear conscious or suspicious of anything else than what the mandarins gave out. The
mandarins had another interview with the governor, and he said to them more
clearly, and making rather a joke of their coming, that it caused amazement
that their king should have believed what that Chinaman they had got with
them had said; and that even had there been in truth any such gold in the
Philippines, the Spaniards would let it be carried away, the country belonging as it did to His Majesty. The mandarins replied that they understood
well what the governor explained to them, but that their king had bid them
come, and they were bound to obey him, and bring him an answer, and that,
having done their business, they had fulfilled their duty and would return.
The governor, to shorten the matter, sent the mandarins with their prisoner
and servants to Cabit, which is the port, two leagues from the city, where they
were received with many discharges of artillery, which were fired at the time
they disembarked, at which they showed much fear and timidity; and when
they landed they asked the prisoner if that was the island of which he had
spoken to the king. He answered that it was. They asked him where was the
gold. He replied that all that they saw there was gold, and that he would
make it good with his king. They put other questions to him, and he always
made the same answers, and all was taken down in writing, in the presence
of some Spanish captains who were there with private interpreters; and when
the mandarins had ordered a basketful of earth to be taken from the ground,
to carry it to the King of China; and when they had eaten and rested, they
returned the same day to Manila with the prisoner. The interpreters said
that this prisoner had said, when hard pressed by the mandarins to answer
to the purpose the questions they put to them, that what he had meant to say
to the King of China was, that there was much gold and wealth in the possession of the Spaniards and natives of Manila, and that if a fleet and men
were given him, he offered, as a man who had been in Luzon and knew the

---

[1] The Chinese were called by the Spaniards Sangleyes, derived from a word
of the Amoy-dialect, "seng-li," trade. Each Chinese had to pay a head-tax
"tribute," not to a Spanish official, but to his "capitan," who was a kind
of mayor over the parian, called capitan de sangleyes, or alcalde mayor, and
enjoyed a high authority among his countrymen. The wealthy Chinese
would pay the tribute for their poor fellowmates. It was the principle of the
Spaniards not to meddle with the inner affairs of the parian; the capitan
represented the mediator between the Spanish authorities and the Chinese
population. Sangley means only "trader, merchant," not "class of merchants," as Schott makes out in a note to Jagor's Reisen in den Philippinen
(p. 272), nor "itinerant dealers," as Blumentritt (Chinesen auf den Philippinen, p. 18) explains after Barrantes.

country, to take it, and bring back the ships laden with gold and riches. This, together with what the Chinese had said at first, seemed of much importance, especially so to Don Fray Miguel de Benavides, archbishop-elect of Manila, who knew the language, and that it went much further than what the mandarins had implied. The archbishop, therefore, and other monks, warned the governor and the city, publicly and secretly, to look to its defense, because they held it as certain that a fleet from China would shortly come against it. The governor at once dispatched the mandarins and put them on board their ship with their prisoner, having given them a few presents of silver and other articles, with which they were pleased. Although, according to the opinion of the greater number of the townspeople, the coming of the Chinese against the country was a thing very contrary to reason, yet the governor began in a covered manner to make preparation of ships and other things for the purpose of defense; and he hastened to complete considerable repairs which he had begun to make in the fort of Santiago, at the point of the river, constructing a wall with its buttresses in the inner part of which looks to the parade, of much strength for the defense of the fort.

After the departure of the mandarins, suspicion against the Chinese constantly increased, and an uprising against Spanish rule was imputed to them—a charge heralded, first of all, by the influential clergy, but which was not justified by any plausible arguments. The well-to-do class of the Chinese population had certainly no mind to stake their lives and hard-earned property in a revolution. The preservation of the Spanish possession of Manila was a point of the most vital interest to them, for only under such conditions could they be enabled to amass wealth. If the Philippines should ever come under Chinese sway, trade with the Spaniards would naturally cease, and thus their means of subsistence be cut off. It was only the over-hasty initiative steps and the oppressive measures of the colonial government which incited the Chinese, first of all the proletarian class, to put an end to the unsafe situation by a general riot, into which finally the patricians were also forced, under pressure of a preposterous policy enforced by the mailed fist of the Spaniards.

Since 1598 Manila had also had a colony of Japanese.[1] Acuña summoned the Japanese nobles, and laid before them the question as to what part they would take in case of a Chinese insurrection. Their response was, already known to Acuña, that they would fight by the side of the Spaniards. This secret understanding was promulgated in the parian, where it provoked an indescribable panic. Part of the traders fled, but the majority were ready to kill the

---

[1] An interesting passage extracted from a Japanese work of travel, and relating to the life of Japanese on Luzon, will be found in the *Journal of the China Branch of the Royal Asiatic Society*, new series (1865), vol. II, pp. 79-80.

Spaniards rather than have the hands of the Spaniards laid on them. In vain now were Acuña's efforts to restore peace. It was already an open secret that the Chinese had fixed the uprising for Saint Francis Day (October 4). A Tagal woman had learned this from her Chinese husband, and betrayed it to her father-confessor, who, of course, had nothing more urgent to do than to inform the Gobernador. Fierce combats during eighteen days followed between the Spaniards and the Chinese, which are full of romantic incident and teeming with merciless massacres. The lives of twenty-three thousand Chinese, according to the Spanish accounts, were sacrificed in the name of His Most Catholic Majesty the King of Spain, and twenty-five thousand according to the "Ming shih;" but in 1604 Chinese trade again flourished, and in 1605 six thousand Chinese again inhabited the parian.[1]

Let us now turn to the account of the Ming Annals, which runs thus:

In 1602 two adventurers, Yen Ying-lung and Chang-Yi, came forward with the assertion that there was a mountain, Ki-i-shan, on Luzon containing gold and silver ore. An exploitation of these mines, so they said, might yield yearly ten thousand taels, or ounces, of gold and thirty thousand taels of silver. This rumor reached the ears of the Emperor Wan-li, who issued an edict that a commission be sent to Manila to verify the truth of this startling news. The court was highly amazed at this decree;.and the President of the Imperial Censorate in Peking, Wên Shun-su, was bold enough to memorialize the Throne, and to attempt to dissuade the Emperor from such an erratic act.[2] He clearly set forth the danger of the Emperor's eccentric plan, and pointed out that it would provoke the Spaniards to acts of aggression. "I have heard," he said, "that the city of Hai-ch'êng has a highly developed maritime trade, which amounts to at least thirty thousand taels a year. Its inhabitants make every effort to seek commercial advantages, and it would therefore be utterly unreasonable to sail over the sea to Ki-yi, where I am sure gold and silver are not everywhere to be found, and to employ people there to mine the gold. The disadvantage arising from the carrying out of the imperial

---

[1] F. Blumentritt, Die Chinesen auf den Philippinen, pp. 26-29.

[2] The Censorate is one of the most curious institutions of administration in China. It is, so to say, a substitute for our modern idea of a constitution. The censors exercise a certain supervision over all deeds of court and provincial officials, and freely denounce to the Emperor any defects in their conduct. They receive, for delivery to the Emperor, appeals either of the people against their officials or of officials against their superiors, and they even have the right to accuse the sovereign and to send him warnings and admonitions. They are inviolable, and cannot be called to account for their official doings. Among the memorials of Chinese censors to the Throne, we find a great many documents which breathe a dauntlessness and frankness of speech worthy of a Cato.

decree is extremely great, and calamities and crimes would be sure to follow the dispatch of an army there."[1]

The governor of Fuhkien was not inclined to go himself, but, compelled by the imperial decree, dispatched the assistant district magistrate of Hai-ch'eng (in Chang-chou fu, Fuhkien), named Wang Shih-ho, with a hundred individuals from the same city, to go to Luzon, together with Chang-Yi, to investigate the matter. When the Spaniards heard the news they were terror-stricken. The Chinese, who had a temporary residence there, thus addressed the envoys: "The Imperial Court has really no other intention than that such perverse evil-doers shall breed trouble!" When the governor came to understand a little the intention of their visit, he ordered the clergy to scatter flowers on the road which the imperial envoys would take, and to treat them with respect. He provided a large escort of soldiers to receive them. Shi-ho and his retinue entered the house of the governor, who entertained them with a feast, and after making inquiries, said, "The Imperial Court sends an embassy with the view of exploring our mountains. Each mountain has its owner. How will you explore them? There are mountains in China; could our country go there and open them? Furthermore, you speak of trees on which gold beans grow. Which is the tree that produces them?" Shih-ho could not answer, and looked at Chang-Yi. Chang-Yi replied, "This entire country is gold. Why is it necessary to inquire for beans?" All, without exception, burst out laughing, seized hold of Yi, and wanted to kill him. The Chinese all requested Shih-ho to return to China. He died heartbroken. The governor of Fuhkien was informed of this, and was requested to pass sentence on Yi for his wild speeches. In the meantime the Spaniards were suspicious that the Imperial Court was secretly planning to raid their country, and that the Chinese settlers were treacherously plotting to kill them. The next year the rumor was circulated that troops were to be detailed to take possession of the country. In consequence of this, prices in the iron market rose considerably. The Chinese, in their craving for profit, exhausted their supplies of iron, selling every inch in their possession. The governor issued an order to have the names of the Chinese registered, and divided them into groups of three hundred men, each group to reside in one building. The Spaniards broke into these houses and slew them. As their intentions thus became clear, the Chinese fled in large numbers to the outlying farms. The governor dispatched troops to attack the multitude. As they had no arms, they were killed. A great number took refuge in the mountains of the interior of Luzon (*Ta lun shan*). The savages followed them thither, assaulted them again, and killed a number. As the troops of the savages met with some resistance in the fight, the governor repented, and sent an envoy to deliberate concerning peace with them. The Chinese, suspecting this to be merely a pretext, threw the envoy down and killed him. The governor fell into a great passion, assembled his army, penetrated the city, and set an ambush, so that a great famine broke out among the Chinese near the city. They descended the hills, attacked the city, and suffered a decisive defeat from the division, which fell out of ambush. The total number of those killed in the successive battles amounted to twenty-five thousand. The governor, after holding an in-

---

[1] History was to prove that his prediction was right; but at that moment, when the nation was maddened by a thirst for gold, no one paid any attention to the words of the clear-sighted censor.

quest, ordered that the property of all Chinese be plundered, which the soldiers did sincerely, knowing that treasures had been hoarded up by them. The Spaniards sent a letter to the governor of Fuhkien, saying that the Chinese had plotted a rebellion, but had failed in their plan, and that they had already requested the relatives of the dead to depart with their children. The governor, Hsü Hsio-ch'ü, promptly informed the Emperor of the revolt, who, in dismay and affliction, issued a decree that justice be administered to the instigators.

In 1604 (second month) the Emperor held a council, and said, "Yi and his accomplices have deceived the Imperial Court and bred quarrel beyond the sea, in which they caused the death by sword of twenty thousand wealthy merchants. This is a terrible disgrace to our country, and he must atone for this crime with his life. His head, hung on a pole, shall be sent over the sea to the chieftain of Luzon who dared kill the merchants." Accordingly the officials passed sentence on the criminal and made known the Imperial will to the governor, Hsio-ch'ü, who, in response, transmitted an official dispatch to Manila, censuring the perpetrators of the great slaughter and ordering the burial of the dead and the return of their wives and children. After that time the Chinese gradually flocked back to Manila; and the savages, seeing profit in the commerce with China, did not oppose them. For a long time they continued to gather again in the city.

So runs the account of the "Ming shih." We notice that not the slightest mention is made in it of an intended invasion of the Philippines, which existed merely in the imaginations of the frightened Spaniards. Even enlightened Spanish writers admit that the insurrection of the Chinese must be attributed to a panic on the part of the Spaniards which drove the Chinese into revolt. Several other Chinese books speak of this tremendous massacre. The local Chronicle of Hai-ch'êng states that eighty per cent of the Chinese slaughtered at Manila on the occasion were natives of that city, and the year in which it took place was one of dark foreboding, for in the same month a hurricane swept over Hai-ch'êng, which caused the river to rise so high that it flooded the country around and carried away part of the wall and fortifications of the city and drowned thousands of people, with their cattle and property. The Annals of T'ung-an, a city not far from Amoy, likewise mention this hurricane, and attribute it to the machinations of foreign priests at Manila. As we find that the principal instigator of the massacre was, to all appearances, the archbishop of Manila, Don Fray Miguel de Benavides, the historian of T'ung-an certainly comes very near the truth when he "smells a clergyman at the bottom of the affair."[1]

The history of the Chinese on the Philippines up to modern times

---

[1] G. Phillips, Early Spanish Trade with Chin Cheo (*China Review*, vol. XIX, p. 254).

may now be briefly outlined.[1] In 1639 there was another great
rebellion of the Chinese in Manila, still more obstinate and longer
than that of 1603. In 1662 Chêng Ch'êng-kung, the famous pirate
hero, known to the Spaniards and Portuguese as Kogseng or Koshinga (Koxinga), who drove the Dutch from Formosa and established a kingdom there that he might continue his struggle against
the Manchu, sent a letter to the Gobernador de Lara in which he
accused the Spaniards of suppressing the Chinese, and demanded
that the governor submit to his rule immediately. Upon his failure
to do so, the corsair stated that he would come to Manila with his
entire force and wipe out the city. His threats caused a panic in
Manila, but he died during the preparations for the expedition, and
his son and successor to the throne of Formosa concluded a treaty
of amity with the Spaniards. Their pent-up anger now burst forth
in hatred toward the Sangleys, who were charged with having had
an understanding with Koshinga. The parian was pillaged and its
inhabitants killed or expelled. Nevertheless the Chinese appeared
again, and their settlement was again tolerated. However great the
hatred of the Spaniards and Filipinos toward them was, they were
conscious of the fact that without Chinese trade and industry the
Philippines could not exist. Since the seventeenth century the Philippines have been in decadence, owing to the decline of Spanish
power. The consequence was that Manila lost its attractions for the
big Chinese capitalists, who preferred to invest their money in the
flourishing Dutch colonies, and that after the second half of the
seventeenth and eighteenth centuries the Chinese immigrants came
from the lowest classes of the coast population of Kuangtung and
Fuhkien—"poor devils," whose capital was made up of diligence and
thrift only. In 1709 the Chinese were banished from Manila under
the pretext that they were carrying off the public wealth; but they
did not hesitate to come back again. In the course of the eighteenth
century they settled down also in the smaller places on the island of
Luzon. In 1747 a royal order for their final expulsion arrived from
Madrid, the execution of which was suspended. When the British,
in 1762, captured Manila and demanded the surrender of the Islands,
the Chinese all joined the English. . The governor, Señor Anda,
then gave the order "All Chinese on the island to be hanged!" which
was conscientiously carried into effect. Many Chinese retreated
with the English, after they had returned Manila to the Spaniards
on the conclusion of peace. Nevertheless the parian was populated

---

[1]Compare F. Blumentritt, Die Chinesen auf den Philippinen, pp. 30-33.

again during the next years, though orders were issued from Madrid not to tolerate any settlement of Chinese at Manila. This, like all subsequent ordinances of Spain, was entirely futile in checking Chinese immigration, which continued, in fact, until the end of Spanish rule on the Islands.

That even the present Manchu dynasty still considered the Philippines as one of its tributary States appears from the official work, "Ta Ch'ing hui tien," the rules and regulations of this dynasty, in the section on "Court tribute" (*ch'ao kung*), in which the country of Luzon also figures among the vassals and tribute-bearers of China. It is stated there that it was conquered in the time of the Ming by the Franks, but the name remained unchanged. Trade was interdicted by K'ang-hsi, but resumed again under Yung-chêng.[1]

Toward the middle and end of the eighteenth century a number of small geographical treatises appeared in China which attempt to study the geographical positions and conditions of the islands in the southeastern part of the Pacific by furnishing sailing directions to navigators and describing the peculiar features of the native tribes and foreign colonization. The Philippines were described repeatedly in this period. The most interesting of these little works is the "Hai tao yi chi," by Wang Ta-hai, published about 1791. The author had made a voyage to Batavia in a Chinese junk, and describes many of the Channel Islands from personal observation, and other countries from information gathered from various sources during his travels.[2] As an example of this literature, I will give an abstract of a pamphlet entitled "Records of Manila," written by Huang K'o-ch'ui about 1790. After a brief discussion of the various names under which the Spaniards were known in his time, the author goes on to say that the appearance of these men resembles that of the Chinese. "Their hats," he remarks, "are high and angular, their clothes have narrow sleeves. The articles they make use of in eating and drinking are identical with those of the Dutch. Their silver money, which is current in Fuhkien and Kuangtung, is cast and adorned with the portrait of their sovereign. The island of Luzon is in the southeast of the Fuhkien Sea at a distance of 1,000 li. The number of the native population must be estimated at least at 100,000. The products of this country are gold, tortoise shell,

---

[1] G. Jamieson, The Tributary Nations of China (*China Review*, vol. XII, p. 98).

[2] A. Wylie, Notes on Chinese Literature, 2d edition, p. 65. The *Hai kuo wên kien lu* ("Record of What I Heard and Saw of the Sea Countries"), by Ch'ên Lun-kiung, published in 1744, describes the sea route to Luzon (*T'oung Pao*, vol. IX, p. 296).

Baroos camphor, birds' nests, sea-slugs, ebony, redwood, fish, and salt.   These are all considered the best beyond the sea.   Formerly, at the time of the Ming dynasty, Spain took this country and founded the city of Kuei-tou (Cavite) on the outer lake (*i. e.,* Manila Bay),[1] near the coast of the Western Ocean.   They set a guard on the isle of Kêng-i, west from the city, that they might have this territory far and near under their control.   The winds are extremely severe."

Now follow some curious remarks on the Catholic religion in Manila.   The Spanish monks are designated as the foreign "Buddhist priests" (*fan sêng*)—a term derived from Sanscrit samgha, the Buddhist clergy.   "The foreign priests," comments the author, "have established a church," a word which he expresses by *Pa-li yüan* ("a hall of the padres"), *pa-li* reproducing the sounds of the Spanish *padre.*   "By means of a waterfall they make a clock strike in the church day and night.   At the hours of noon and midnight it strikes the first stroke, and so on until twelve strokes sound, and this is repeated."   To make this explanation clear in his language, the Chinaman had a great difficulty to overcome, as his day is divided into twelve parts, each comprising two hours of our time.   Then he continues:

They do not sacrifice to their ancestors, but worship only their God *Wei-lo,*[2] and, what is still stranger, the padres forgive people their sins.   All the people regard the holy water with great esteem.   The corpse of the king of the padres (probably bishop) is fried and turned into fat.   A father of the religion superintends the work.   If somebody desires to embrace their faith, they order him to take an oath to the effect that his body shall now belong to Wei-lo.   After the oath the padre takes the holy water of the corpse and pours it over his head.   Therefore it is called the "waterfall" water.   At the celebration of a wedding the religious father takes a chain and fastens it around the neck of the man and the woman.[3]

On every seventh day they go to church and beg the padre for forgiveness of sins, and this they call "hearing mass" (*k'an mi-shih*=la misa).   There is also a nunnery especially for the administration of funds with which to defray the needs of the country.   This nunnery is a strict and dignified institution, and is kept locked, while the men who retire into monastic life enjoy an acknowledged authority and are greatly honored.   The daily necessaries of life are transmitted to the nuns by means of a revolving frame like a Chinese

---

[1] In Luzon, according to the view of the Chinese author, there are three lakes—an outer, a middle, and an inner one.

[2] This is doubtless intended for Spanish *cielo.*   The Annals of Kuangtung give a number of Spanish words in Chinese transcriptions, and write *cielo* with the characters *hsi-lo* (Cantonese, *sai-lo*).

[3] My friend Mr. Bandelier explained to me that this custom is still observed also in remote parishes of Spanish South America.

peck-measure, which is on the wall. Among these women there are those who really desire to enter the monastery for the cultivation of moral conduct.

The sailing-ships made in Spain are extremely large, with very strong sails and spars. They carry guns and cannon, which are kept in readiness so that pirates can not come near them. The people of Luzon avail themselves of the sextant, which reflects the surface of the water, shallow stones, and deep-lying rocks. There is nothing that the sextant can not penetrate. This method is more convenient and admirable than the compass. Whenever the people of Luzon are guests of the Chinese they constantly make merry. Their ships are supplied with oars, and it is pleasant to note how clever they are in steering. The large sailing-vessels that come to Manila take three months for their voyage up to the time of landing. When these boats return to their home country, the nature of the water is not the same, and it is neces-sary to reckon five months for the voyage. The Chinese have now for a century been in mutual commercial intercourse and peace with them. In the period K'ien-lung (1736-'95) the red-haired Ying-kuei-li (English) suddenly dispatched over ten ships straightway to oppress Manila. They desired to occupy this country and to convert the people. The padres were willing to pay them off with presents, and thus got free from the English in a courteous manner.[1] The English thereupon turned to China for trading purposes. Such are the records of Manila.

In a Chinese album containing wood engravings of ethnical types, the "Huang ch'ing chih kung t'u" (i. e., "Pictures of the Tribute-Bearing Peoples of the Manchu dynasty"), published in 1752 by order of K'ien-lung, we find in the first book (p. 70), among other types of European nations, the portrait of a Spanish Jesuit and a nun, as well as that of a Spaniard from the Philippines, styled "bar-barian from the country of Luzon," and a woman ("barbarian woman") as his counterpart. These two plates are accompanied with the following flattering explanation:

Luzon is situated in the Southern Sea. It is very near to Chang-chou, in Fuhkien Province. In the commencement of the Ming period it sent tribute to court. In the period Wan-li it was the Franks (Spaniards) who ab-sorbed this country and forthwith gave its name to it. The Franks, being in the southwest of Cambodia, had formerly exterminated Malacca, and then divided the Moluccas with the Dutch (Red-Hairs) until they broke into Luzon. Their wealth and power increased more and more by sojourning in Macao and trading there. The barbarians inhabiting Luzon (i. e., the Spaniards) are of tall stature, and have high noses, pupils like those of cats' eyes, a mouth like that of a hawk, and their clothing is much adorned. They are identical with the people of Spain and Portugal, in Europe. The women coil the hair, in which hairpins are here and there displayed, and wear ear-rings. The neck is bare, and around the breast they wear a short tunic.

[1] The statement is correct in so far as, after the capture of Manila by the British (1762), the private property of the inhabitants was saved from plunder on condition that a ransom of a million pounds be paid, half of which was in money, and the other half in notes on the Spanish Treasury.

They have long petticoats, underneath which they wear a sort of round frame-work of two or three strips of rattan, one above another (probably identical with the old-style hoopskirt). Over the coil of hair they always wear a net.[1]

Two very curious observations with regard to natural history in the Philippines are recorded in a small geographical work, "K'un yü t'u shuo," published (in Chinese) by the Jesuit father Ferdinand Verbiest, about 1673, in which he followed principally a geography of the world written by Pantoja, an Italian Jesuit, in compliance with an imperial order, half a century earlier.[2] The passage reads as follows:

In the southeast of Kuang-chou, Luzon is situated. This country pro-duces falcons. When the king of the falcons flies up, the flock of other falcons follow him to take birds and animals as booty. The king of the falcons first takes the pupils out of the eyes of these animals, and afterwards a covey of hawks devour their flesh. Furthermore, there is a tree there which animals are not able to go near. As soon as they pass it they fall down dead at its foot.

Whether these statements have any foundation in fact, I am not now prepared to say.

After the Spaniards had been unsuccessful in establishing direct commercial relations with China in the port of Amoy, the people of Hai-ch'êng sent their junks to Manila, and extensive trade was car-ried on between the two cities. The bulk of Chinese merchandise, the chief article of which consisted in silk, pottery, and metal-ware, was made over to the ports of New Spain and Peru, which thus became a large market for Chinese manufactures. This trade was a source of immense profit to China. The importation of silver into Manila from Spanish America during two hundred and fifty years of intercourse (1571–1821) is computed by De Comyn at four hun-dred million dollars; and a large share of this, perhaps half, passed over to China.[3]

The entire Spanish colony subsisted until the nineteenth century

---

[1] This is the well-known silk net called by the Spaniards redecilla.

"The women wear no caps, but tie a kind of network silk purse over their hair, with a long tassel behind, and a ribbon tied in a bow-knot over their forehead. This head-dress they call redecilla, and it is worn indiscriminately by both sexes" (Richard Twiss, Travels through Portugal and Spain, in 1772 and 1773 [London, 1775], p. 33).

[2] A. Wylie, Notes on Chinese Literature, 2d edition, p. 58.

[3] Chinese Repository, vol. VIII, p. 173; see also G. Phillips, Two Mediæval Fuhkien Trading Ports (T'oung Pao (1895), vol. VI, p. 456).

10

exclusively on the Chinese trade.[1] Despised, hated, and feared as the Chinese were, they were nevertheless indispensable to the Islands, and were practically their masters and rulers from an economical viewpoint. The boots made by Chinese shoemakers in Manila were so low in price that they could be sold with a large profit in New Spain. As early as 1603 De Morga wrote:

> It is true that the city can neither go on nor maintain itself without these Chinamen, because they are the workmen in all employments. They are very industrious, and work for moderate wages.

After the great massacre of 1603 the Spaniards felt keenly the lack of the Chinese. There was no food to be found to eat, nor shoes to wear, not even for very exorbitant prices. "The native Indians," laments the chronicler, "are very far from fulfilling these offices, and have even forgotten much of husbandry, the rearing of fowls, flocks, cotton, and the weaving of robes, which they used to do in the times of their paganism."

De Morga gives a most extensive account of the manner of Chinese trade, of the articles traded, of their transshipment to America, and of the conditions of the life of the Chinese in the Philippines. To enter into a discussion of this subject is beyond the scope of the present paper; but I cannot refrain from relating a humorous incident which occurred in the history of early Spanish-Chinese trade. It is taken from a tract printed in Mexico in 1638 and embodied in Thevenot's "Voyages Curieux." These Chinese, says our authority, were so eager for gain that if a particular article of merchandise was a success one year, they tried the market again with it the follow-

---

[1] In Pieter Nuyts' (Dutch Governor of Formosa) Report on the Chinese Trade to the Governor-General and Councillors of the United East India Company, written in 1628, it is aptly remarked: "It is, indeed, certain that the only support of the Spaniards and Portuguese in India is the China trade. The wars we [i. e., the Dutch] have everywhere waged against them, with the disgrace they have come to in Japan, have so weakened them, and ruined their trade in other countries, that there is no other place except China where they can make any profits worth mentioning. Accordingly, if we could succeed in depriving them of this trade, or at least in lessening their profits from the same, as we have often done elsewhere, they would be compelled to abandon their best settlements, such as Macao, Manila, Malacca, and Timor; while their factory at Moluccas would lapse of itself. The authorities at Manila clearly see this," etc. (Wm. Campbell, Formosa under the Dutch [London, 1903], p. 53). About the same time, the merchants of Amoy petitioned the authorities, complaining that the Dutch, by their constant attacks on vessels trading with the Spanish, had completely destroyed the lucrative trade formerly carried on beween Amoy and Manila (James W. Davidson, The Island of Formosa, Past and Present [London, 1903], p. 12).

ing year. A Spaniard who had lost his nose got a Chinaman to make him a wooden one to hide his deformity. The artist made such a splendid imitation that it pleased the Spaniard immensely and induced him to pay him the exorbitant sum of twenty dollars for it. The Chinaman, lured by the large sum paid to him, loaded a ship the following year with wooden noses, and returned to Manila with great expectations. Matters, however, did not turn out at all as he had anticipated, and he was only laughed at for his trouble; for, in order to have found a market for this new merchandise, it would have been necessary to have cut off the noses of all the Spaniards in the country.[1]

Regarding the mode of Chinese-Spanish commerce, the "Tung hsi yang k'ao" (chap. 5, p. 6 a) has the following:

As soon as the (Chinese) ships arrived they sent out men to hurry with all dispatch to the chieftain (i. e., the governor of Manila) to bring him presents of silk. The duties which they levied were rather high,[2] but the meshes of their nets were so close that there was no escape. Our people who have intercourse with them remained there without returning home, for the reason that they had the advantage of being but a short distance off and they quickly made money. There was much opportunity for quarrels, but later on they became more cautious. Our people at home were anxious lest the emigrating class might be too numerous there and after their return later on breed rebellion. It was therefore ordered that each junk should carry only two hundred men, and that the number of junks sailing should not exceed a fixed number. Returning home and sailing out again, the number of men was increased to four hundred, the number of ships remaining the same. When our people put to sea many gave a false name and figured only as a number. While their investigation was going on they suddenly escaped in the midst of it and went back to that country. The name of the market is *Kan nei.*[3] Formerly it was within the city; afterwards, when they (i. e., the Spaniards) became suspicious, they transferred it to the outskirts of the city and founded a new *Kan.*[4]

---

[1] *China Review,* vol. XIX, pp. 245-246.

[2] According to De Morga, the duty was 3 per cent.

[3] The term by which the Chinese quarter in Manila is designated, the *parian* of the Spaniards. *Kan* is the Cantonese pronunciation of North Chinese *chien* (Giles' Dictionary, No. 1603), and means "a mountain torrent;" *Kan nei,* "inside of the mountain torrent."

[4] "To Manila, all Chinese wares are openly sent from China in Chinese junks which pay export duty to the Emperor of China; and, in order to attract Chinese merchants and secure a monopoly of trade, the Spaniards were in the habit of advancing large sums of money, but the Chinese often failed to return with the value in goods. This went on for several years, till we settled here and the ravages of the pirates began; whereupon Chinese vessels were first kept at home, and then gradually began to visit us, so that during the last few years very little trade has been carried on at Manila." Thus wrote Pieter Nuyts as early as 1628 (Wm. Campbell, Formosa under the Dutch, London, 1903, p. 52).

The following localities which I am able to identify are mentioned in the "Tung hsi yang k'ao":

Ta-Kiang (*i. e.*, "the great harbor," "the great Manila Bay") is the very first place reached in coming from the Eastern Ocean. A great government board is established there, and a city built of stone. The Franks guard this place under the rule of a chieftain. Rice and grain grow plentifully; but the only other products are objects made of leather and horn. Before the bay is reached, the Pi-kia-shan[1] is visible.

Nan-wang is contiguous to Ta-Kiang. In passing farther along, there are two tiny villages, Wei-mi-yen and Wei-yen-t'ang, which produce leather, horn, and cotton.

Tai-mei Kiang enters with sinuous windings into the configuration of the land, and is therefore called Tortoise-shell Bay (*tai-mei wan*). It is surrounded by a mountain which serves as a land-mark. All ships sailing to Luzon must observe this sign-post and steer towards it. This mountain is thus set up like a guard. Although the name "Tortoise-shell" is given, tortoise-shell is not produced there, but the only product is sappan-wood.

Lü-p'ēng[2] is southward from Luzon, and produces univalve and bivalve shells.

Mo-lao-yang[3] is situated behind Manila, and produces cotton, oil, hemp, and cocoanuts.

There are some other localities mentioned and described in the same work, but as I am still doubtful in regard to their identification, I must leave this for some other occasion.

There are three anthropological problems which must be taken up in considering the relations of the Chinese to the Philippines. The first is a question of physical anthropology, an investigation of which should show what proportion of Chinese blood is contained in the races and tribes at present inhabiting the Islands. Through intermarriage of the Chinese with Malayan women, a class of half-bloods has arisen whom the Spaniards call Mestizos de Sangley, or Mestizos chinos. They are described as people of tall stature, of sturdy build, intelligent, and possessed of the keen commercial abilities of their fathers. The retail trade of the country and the small banking business are largely in their hands. According to the views of many writers, the Igorrotes on Luzon of the present day represent a mixed race, the descendants of wild mountain tribes and those Chinese pirates who escaped the sword of the Spaniards after the expulsion of the great corsair, Limahon, in 1574. This, like many

---

[1] *Pi-kia* is a frame of porcelain, brass, copper, or crystal, on which to rest writing-brushes, usually made in the shape of cragged mountains; mountains, therefore, are again compared with this object. *Shan* means "mountain."

[2] Apparently identical with the Island of Lubang, discovered and conquered by Salcedo in 1569.

[3] I think that the identification of this name with *Morong* would be justifiable.

other problems, should be solved by extensive physical research.   An ethnological question of great importance would be a study of the traces of Chinese material culture, still remaining, in the life of the Philippine tribes.   Such research requires, of course, a deeper knowledge of Philippine ethnology than is available at present, and more extensive and better-classified collections than are now at our disposal.   From a cursory inspection of the Philippine material in the American Museum of Natural History, in New York, it seems to me that Chinese influence is particularly to be observed in connection with the industrial crafts of the Christian peoples, as in agriculture, fishery, navigation, pottery, and weaving.   The types of Philippine footgear almost seem to be derived from China.

Another important problem in connection with the history of Chinese-Spanish-American trade would be to determine what influence objects of Chinese culture may have had on the peoples of Mexico and Peru.   This question has been ventilated by Dr. Walter Hough, in his paper "Oriental Influences in Mexico."[1]   Dr. Hough refers to a number of useful plants which were at that time introduced from the East into Mexico, probably by way of the Philippines, like the cocoanut, the banana, the plantain, the mango, and others.[2]   He mentions, further, some evidences of contact in the industrial arts, as the making of palm-wine, the close resemblance in construction and shape of the rain-coats used in Mexico to those of China, and other items.   To obtain a satisfactory solution of this problem, first of all, the ancient Spanish sources on South America and Mexico should be diligently searched for all references concerning early Chinese trade and imports; secondly, such remains of these as exist should be eagerly sought for and collected, particularly in the line of ceramics and textile manufactures;[3] and, finally, the actual influence, if any, of these on the corresponding industries of American peoples should be investigated.

---

[1] *American Anthropologist*, 1900, pp. 66-74.

[2] See, however, O. F. Cook (The Origin and Distribution of the Cocoa Palm, *Contributions from the U. S. National Herbarium*, vol. VII, No. 2, Washington, 1901, p. 259), who contradicts this view.   The cocoanut-palm is doubtless indigenous in America.

[3] The following notice is interesting in this respect: "Grau y Monfalcon in 1637 reported that there were 14,000 people employed in Mexico in manufacturing the raw silk imported from China.   This industry might be promoted by the relaxation of the restrictions on trade.   It would also be for the advantage of the Indians of Peru to be able to buy for five pence a yard linen from the Philippines, rather than to be compelled to purchase that of Rouen at ten times the price" (from Documentos inéditos del archivo de Indias, in Blair's and Robertson's The Philippine Islands, vol. I, p. 69).

APPENDIX

The Dido Story in Asia

The above Chinese account of the foundation of Manila through the Spaniards (p. 259) contains the well-known ruse of Queen Dido in connection with the founding of Carthage.[1] This is not the only case of its record in Chinese literature. E. Bretschneider[2] refers to Du Halde's La Chine (vol. I, p. 185), where the same tradition is repeated with reference to the settling of the Dutch on the Island of Formosa in 1620. Du Halde's account is drawn from a Chinese source, the "Annals of Formosa" (T'ai-wan fu chi), which imputes the Dido trick to the Dutch. James W. Davidson[3] reproduces the story, and inclines to see in it an actual historical event. It is certainly far from this. In the Dutch sources regarding the history of Formosa, nothing of the kind is to be found. We have here nothing more than a simple tale, which has spread over almost the entire continent of Asia; and it is most curious to note that in nearly all cases the Asiatic peoples with whom the story is found make the tricksters some European nation who were then invading their country. This is sufficient proof to show that this is the case of a comparatively recent story-migration, which is further evidenced by its absence in any Asiatic literary records of earlier date.

The first to call attention to the wide diffusion of the Dido story was Reinhold Köhler.[4] The same subject was taken up by Henri Cordier,[5] Raoul Rosières,[6] René Basset,[7] and N. Katanof.[8] Despite

---

[1] See O. Rossbach, Dido (Pauly's Realencyklopädie, vol. IX [Stuttgart, 1903], pp. 426-433); Meltzer, Dido (Roscher's Lexikon der griechischen und römischen Mythologie [Leipzig, 1885], col. 1012-1018).

[2] *China Review*, vol. IV, p. 386; and Mediæval Researches from Eastern Asiatic Sources (London, 1888), vol. II, p. 319.

[3] The Island of Formosa, Past and Present (London and New York, 1903), pp. 12-13: "The wily Dutchman, with an old trick in mind, proceeded to cut the ox-skin in very long narrow strips, and, after fastening them together, produced a line of sufficient length to surround a vast plot of ground, while the Japanese were struck dumb with astonishment."

[4] Sagen von Landerwerbung durch zerschnittene Ochsenhaut (Th. Benfey's *Orient und Occident*, 1864, vol. III, pp. 185-187).

[5] La légende de Didon (*Revue des Traditions populaires*, 1887, vol. II, pp. 295 and 354); further parallels by Sébillot (*ibid.*, p. 355).

[6] *Ibid.*, vol. VI, pp. 52-54.

[7] *Ibid.*, vol. VI, pp. 335-338.

[8] Türkische Sagen über Besitznahme von Ländern nach Art der Didó (*Revue orientale*, [Budapest, 1902], vol. III, pp. 173-179).

the great zeal of these authors in collecting the material in question, I have found several versions myself not recorded by any of them. Two ways for the migration of the tradition from Europe into Asia are discernible—a land route and a sea route. From Byzance, where it was well known, it seems to have wandered into Russia, and from the Russians to the Ugrians and the Turkish tribes of Siberia. Among both Ugrians and Turks, the tricksters are the Russians. The Syryän tell of the foundation of Moscow in the same way as the Chinese that of Manila, and explain the name of the city by the word "Mösku," which in their language means "a cowhide."[1] The Cheremiss also have it in regard to the Russians, and the Russian farmers themselves relative to a wealthy landowner of their own. Three Turkish versions have been noted by W. Radloff;[2] others are known from among the Kirghiz and Yakut, and from Tashkend and Hami.[3] Through the medium of European nations, the story seems to have spread over the regions around the Indian Ocean in the sixteenth and seventeenth centuries. In India the foundation of Calcutta is connected with it.[4] In Burma, Adolf Bastian[5] has recorded it. In this case the trickster is a female slave of the Burmese king Dwattabong. When the Portuguese penetrated into Cambodia, in 1553, they employed the same trick of cutting a buffalo hide, according to the tradition of the Cambodians.[6] Finally we find it current among the Chinese, as already stated.

There are two points of interest in the dissemination of this story: First, it affords one of the few examples of a Western tale spreading to the extreme East, while as a rule the stream of folk-lore flowed from east to west in the old world; secondly, it shows that the transmission of folk-lore still goes on, even in recent times, by mere oral accounts. While in almost all cases where folk-lore is handed over from Asia to Europe we have been able to trace the fact of migration back to written sources transferred from nation to

[1] J. A. Sjögren, Gesammelte Schriften, vol. 1 (Historisch-ethnographische Abhandlungen über den finnisch-russischen Norden [Petersburg, 1861], p. 301).

[2] Proben der Volkslitteratur der türkischen Stämme Süd-Sibiriens, vol. IV (Petersburg, 1872), pp. 11-12, 139-141, 179-181.

[3] See Katanoff, loc. cit.

[4] J. Todd, Annals and Antiquities of Rajasthan (London, 1832), vol. II, p. 235. Regarding a Tibetan legend containing the same motive see Sylvain Lévi, Le Népal, vol. II (Paris, 1905), p. 7.

[5] Die Völker des östlichen Asiens, vol. V (Die Geschichte der Indochinesen, p. 25).

[6] H. Cordier, loc. cit.

nation, and extant in polyglot translations, there is no such written testimony for the legend of Dido in any Asiatic literature to which, as the starting-point, all the current versions could be reduced. Thus we are led to presume, especially because of the introduction of Europeans into the plot, that its occurrence in southern and eastern Asia is due to the oral stories of European sailors and merchants, who had probably imbibed it during their school-days, while its propagation in Siberia seems to have emanated from the mouths of vagrant Russian adventurers.

It may not be without interest to American readers to repeat here some American parallels of the Dido story once discussed by the great linguist, Pott. In his essay, "Etymologische Legenden bei den Alten" (in the Journal *Philologus*, 1863, Supplementary vol. II, p. 258), he quotes from a work by Kottenkamp (Die ersten Amerikaner im Westen, p. 382) the following: "The Indian reminded us of the fraudulent procedure which had once been practised from Pennsylvania against the Delawares. The whites had purchased a plot of land not larger than they would be able to encompass with a cowhide, and the Delawares had been infatuated by the appearance of the small area. The whites, however, cut up the hide into thin strips and covered a space a thousand times larger than the deceived Delawares had sold." To this, Pott remarks in parentheses. "Whether a white exploited in such a way the tradition of Dido which he had learned in school, by transforming poetry into prose and serious reality, may remain undecided. This matter, however, has been told by Indians on the occasion of the foundations of various establishments by Europeans. Thus this trick of land acquisition on the part of the Dutch at their first settlement in the State of New York has been related by Iroquois to subsequent travelers; likewise the story of the same swindle served for the provocation of the Ohio Indians in those times of which we speak."

047

眼镜的历史

MITTEILUNGEN

ZUR

GESCHICHTE

DER

MEDIZIN UND DER NATURWISSENSCHAFTEN

HERAUSGEGEBEN VON DER

DEUTSCHEN GESELLSCHAFT
FÜR
GESCHICHTE DER MEDIZIN UND DER NATURWISSENSCHAFTEN

UNTER REDAKTION VON

SIEGMUND GÜNTHER und KARL SUDHOFF
MÜNCHEN   LEIPZIG

VI. JAHRGANG.

MIT FÜNF ABBILDUNGEN.

HAMBURG und LEIPZIG.
VERLAG VON LEOPOLD VOSS.
1907.

Mitteilungen
zur
Geschichte der Medizin und der Naturwissenschaften.

| No. 23. | 1907. | VI. Bd. No. 4. |

I. Originalabhandlungen.

Zur Geschichte der Brille.

Von Berthold Laufer.

Mit Interesse habe ich die im letzten Hefte dieser Zeitschrift (Band VI, S. 221—223) veröffentlichten Ausführungen von Prof. J. Hirschberg über die „Geschichte der Erfindung der Brillen" und die sich daran anschliefsende Erörterung von Prof. G. Oppert gelesen. Das darin erwähnte Buch von Hirschberg, in welchem seine Theorie ausführlich dargestellt sein soll, ist mir noch nicht zugänglich geworden. Meine Kritik derselben mufs ich daher nach Möglichkeit einschränken und ziehe es vor, durch neues aus der chinesischen Literatur beigebrachtes Material zu beweisen, dafs die Ansicht von der ursprünglichen Erfindung der Brille in Indien die gröfste Wahrscheinlichkeit für sich hat. A priori ist ja das Ergebnis von Hirschberg im höchsten Grade unwahrscheinlich, da es allen bisherigen Erfahrungen und Analogien in der Kulturgeschichte und in der Geschichte der Erfindungen insbesondere widerspricht; Kristallbrillen treten im europäischen Mittelalter, in Indien und in China auf, und da läfst sich vom historischen Standpunkt von vornherein vermuten, dafs diese kombinatorische Erfindung in jedem dieser drei Kulturkreise nicht unabhängig erfolgt sei, sondern dafs ein geschichtlicher Zusammenhang hier vorliege. Doch ich will nicht vorgreifen, sondern erst die Tatsachen sprechen lassen.

Die folgenden Angaben sollen nur als ein vorläufiges Ergebnis einer ersten Untersuchung dieses Gegenstandes in der chinesischen Literatur gelten, ohne Anspruch auf Erschöpfung und endgültige Behandlung, und nur wirklich Wesentliches und Tatsächliches bringen. Auch verzichte ich mit Rücksicht auf die Leser dieser Zeitschrift auf allen gelehrten sinologischen und philologischen Apparat. Brillen waren in der Zeit des chinesischen Altertums gänzlich unbekannt und werden in der Literatur nicht früher als in Schriften des 13. Jahrhunderts erwähnt und beschrieben, treten also in China in derselben Periode auf wie in Europa, ein hinreichend verdächtiger Umstand, der schon

26

allein dem Historiker zu denken geben muſs. Die chinesischen Aufzeichnungen besitzen nun den einen groſsen Vorzug der Ehrlichkeit,
und es ist den Chinesen niemals in den Sinn gekommen, sich mit
fremden Federn zu schmücken und sich Erfindungen anderer Völker
zu vindizieren; wenn es sich um Erzeugnisse des Auslandes, Kulturpflanzen, Tiere, Mineralien, Gegenstände des Handels und der Industrie,
handelt, versäumen sie fast nie, wenn immer möglich, den Ort der
Herkunft gewissenhaft zu registrieren. Über die ersten Brillen erfahren wir nun, daſs sie aus den Ländern Zentralasiens, d. h. aus Turkistan, kamen. Diese Nachricht wird von CHAO HSI-KU, einem Mitglied
der kaiserlichen Familie der SUNG, gegeben, in seinem Werke *Tung
t'ien ch'ing lu*, mit der Angabe, daſs sie aus einem Buche „Erzählungen
von Leuten der Yüan- (mongolischen) Dynastie" (*Yüan jên siao shuo*)
geschöpft sei. Er erwähnt, daſs alte Leute, die nicht imstande sind,
feine Schriftzeichen zu unterscheiden, mit Hilfe dieser die Augen be-
deckenden Brillen die Schrift klar lesen können; er nennt sie *ai-tai*,[1]
ein poetischer Ausdruck, der von dem Aussehen trüber Wolkenmassen
gebraucht wird. Schon dieses dichterische Attribut deutet auf die
Fremdartigkeit und Wertschätzung des Gegenstandes hin. Der jetzt
gebräuchliche Name *yen-king*, d. i. „Augenspiegel" (entsprechend dem
älteren deutschen Synonym „Augenspiegel" für Brille[2]), trat schon
damals auf, scheint aber erst später in allgemeineren Gebrauch gekommen zu sein, denn in einem Buche des 18. Jahrhunderts (*Shu wu
i ming su*) heiſst es: „Die *ai-tai* nennt man jetzt gewöhnlich *yen-king*."
Die Einführung von Brillen aus Turkistan nach China wird ferner in
zwei weiteren Werken aus derselben Periode berichtet. In dem einen
derselben (*Pai shih lei pien*) wird von zentralasiatischen Gefangenen
erzählt, die Brillen in China verfertigten, um sie „als kostbares Erbe
den Generationen zu hinterlassen". Ein Hauptmann in Ho (einem Orte
in der Provinz Shansi), heiſst es weiter, besaſs eine Brille von einem
Material wie weiſse Glaspaste (*pai liu-li*), die Augen von der Gröſse
eines Geldstückes, mit zusammenlegbaren Stangen aus Knochen. Auf
Befragen, von wo sie hergekommen sei, gab er die Auskunft, daſs er
seit langem einen Barbaren aus der Provinz Kansuh in seinem Dienst
gehabt hätte, der sie ihm zum Geschenk gemacht habe. Die Bezeichnung „Barbar" (*i jên*) weist deutlich darauf hin, daſs es sich um einen
Nicht-Chinesen handelt, also wohl um einen Tanguten oder Türken.

---

[1] Geschrieben mit den Schriftzeichen Nr. 21 und 10561 im chinesischen
Wörterbuch von GILES.

[2] F. KLUGE, Etymologisches Wörterbuch der deutschen Sprache, 6. Auflage, S. 58.

Ein anderes Buch (*Fang chou tsa yen*) enthält die Nachricht, daſs
Brillen in Turkistan im Wege des Austausches mit edlen Pferden
gekauft würden, was uns eine Vorstellung von der anfänglich über-
triebenen Einschätzung des Artikels verleiht, worauf auch die oben
gebrauchte Benennung „Kostbarkeit" (*pao*) hindeutet. Die erste Bekanntschaft mit Brillen darf auf Grund der angeführten Stellen in den
Beginn der Mongolen-Dynastie (1260) versetzt werden und erfolgte
von Turkistan aus. Das Wörterbuch des Kaisers K'ANG-HSI zitiert
ferner eine Stelle aus einem Werke *Fang yü shêng lio*, wonach Brillen
aus Malakka exportiert wurden; das Datum dieses Werkes ist mir
unbekannt; da aber Malakka nicht vor dem Anfang des 15. Jahrhunderts in der chinesischen Literatur erwähnt wird, kommt hier offenbar eine spätere Handelsbeziehung in Frage. Niemand wird aber wohl
glauben, daſs in Turkistan oder in Malakka Brillen selbständig erfunden
worden sind, oder gar in beiden Gebieten unabhängig voneinander.
Die Angabe dieser beiden Herkunftsgebiete und die von Turkistan insbesondere zwingt mit Notwendigkeit zu der Annahme und kann nichts
anderes bedeuten, als daſs diese Brillen in letzter Instanz auf Indien
zurückgehen. Dabei müssen wir uns gegenwärtig halten, daſs die
Länder von Ost-, Zentral- und Westasien im 13. Jahrhundert nach
den Eroberungszügen der Mongolen sich weit näher gerückt waren
und in viel engeren Verbindungen zu einander standen als dies in irgend
einer späteren Periode der Geschichte der Fall war. Die innigen
Kulturbeziehungen zwischen Indien und Turkistan und zwischen Turkistan und China sind so zur Genüge bekannt, daſs es durchaus nichts
Überraschendes hat, wenn wir im Gefolge dieser Kulturwanderungen
auch die Brille von Indien über Zentralasien nach China gelangen
sehen. Später, im Zeitalter der Ming, als die chinesischen Fahrzeuge
den indischen Ozean beherrschten und die Überlandrouten des Handels
gegen die bequemeren Seewege zurücktraten, wurden indische Brillen
auch über Malakka nach China eingeführt. Da nun Brillen in der
zweiten Hälfte des 13. Jahrhunderts in China bekannt waren, so dürfen
wir mindestens eine Generation zurückrechnen, um ihre Einführung
nach Turkistan zu erlauben, und eine weitere Generation, um Zeit
für die Entwickelung der Erfindung in Indien selbst und den Brillen-
handel nach Turkistan zu lassen. Es wird daher kaum zu viel behauptet sein, wenn wir annehmen, daſs dementsprechend die Erfindung
der Brille spätestens im Anfang des 13. oder gegen Ende des 12. Jahrhunderts in Indien gemacht sein muſste. Damit ist der Satz von
Prof. HIRSCHBERG (S. 223): „Es wird sich schwerlich erweisen lassen,
daſs dies schon vor dem Jahre 1300 unserer Zeitrechnung dort üblich
war" hinreichend widerlegt, und es ist vor allem erwiesen, daſs die

26*

Brillen in Indien früher als in Europa bekannt gewesen sein müssen, wo sie nach den Angaben von Hirschberg nicht vor 1270 auftreten. Die zeitliche Priorität Chinas gegenüber Europa im praktischen Gebrauch der Brille scheint mir gleichfalls festzustehen. Es wird nun zunächst Sache der Kenner der Sanskritliteratur sein, hier einzusetzen und die Geschichte der Brille in Indien eingehender zu verfolgen. Ich hoffe, daſs sich Herr Prof. Oppert, dessen Anschauung durch die chinesische Literatur so glänzend gerechtfertigt wird, dieser Untersuchung aufs neue annehmen wird. Prof. E. Wiedemann in Erlangen wäre vielleicht für eine diesbezügliche Durchforschung der arabischen Literatur zu gewinnen, auf die ja bei der Frage des Zusammenhanges der indischen Erfindung mit Europa alles ankommt. Daſs den Arabern, wie Hirschberg sagt, Brillen ganz unbekannt gewesen sein sollen, wäre bei ihrer groſsen Kenntnis der Optik höchst wunderbar;[1] jedenfalls ist aus dem bisherigen negativen Befund nichts Definitives zu schlieſsen, denn wir sind noch nicht ans Ende der arabischen, noch der indischen, noch der chinesischen Literatur angelangt, im Gegenteil, für die kulturhistorische Forschung beginnen erst diese Gebiete sich langsam zu erschlieſsen.

Alles, oder das wenige, was Hirschberg für seine unhistorische Ansicht von der selbständigen Erfindung der Brille in Europa anführt, spricht nicht für, sondern gerade gegen diese Theorie. Der so oft in der Geschichte wiederholte Fall, daſs in Europa mehrere Entdecker einer „Erfindung" fast gleichzeitig auftreten, ist der zuverlässigste psychologische Beweis dafür, daſs keiner von ihnen die „Erfindung" gemacht hat, sondern daſs es sich bestenfalls um eine Wiedererfindung, ein Wiederfinden oder um ein Experimentieren auf Grund in der Literatur gefundener Beschreibungen handelt, — vergleiche Kompaſs und Schieſspulver. Die europäischen Völker haben natürlich in ihrem Dünkel alles erfunden, was jetzt, wie erwiesen, aus dem Orient stammt, und waren bei all ihrer eingebildeten Überlegenheit nicht einmal imstande, die Ziffer Null zu erfinden, die sie samt dem Positionssystem durch die Vermittelung der Araber von den Indern erlernten, während doch das alte zentralamerikanische Kulturvolk der Maya diese Erfindung unabhängig in vorkolumbischer Zeit gemacht hat. Ich fürchte sehr, daſs es mit der angeblichen Erfindung der Brille in Europa nicht besser stehen wird als mit der Reihe der übrigen „Erfindungen", die das ausgehende Mittelalter gemacht haben will. Wenn, wie bisher all-

---

[1] In einigen unserer Encyklopädien wird auf die Optik des Arabers Alhazan († 1038) verwiesen, wo zum ersten Male von Vergröſserungsbrillen die Rede sein soll.

gemein geschehen, und worauf auch Hirschberg hinweist, Roger Bacon als der erste europäische Schriftsteller in Anspruch genommen wird, der von Brillen spricht, so wird niemand vergessen, daſs derselbe Roger Bacon auch als erster in Europa das Rezept für die Bereitung des Schieſspulvers mitteilt, und daſs er auf seinen Reisen in Spanien mit arabischen Gelehrten verkehrte und ihre Schriften studierte. Aus diesem Kreise wird auch zweifelsohne seine Kenntnis der Brille stammen. Nach einer literarischen Beschreibung, wenn einmal die Idee gegeben war, Brillengläser schleifen und sie in ein Gestell einsetzen, erfordert wahrlich keine groſse Kunst, und es ist ja gewiſs möglich, daſs ohne direkten Import einer asiatischen Brille an verschiedenen Orten Europas von verschiedenen Leuten Brillen auf Grund schriftlicher oder mündlicher Nachrichten verfertigt worden sind. Aber die aus solchem Tatbestand gezogene Schluſsfolgerung, daſs darum die originale Erfindung Europa gebühre, ist völlig unberechtigt und kann gegenüber der chronologischen Priorität Indiens und Chinas nicht länger aufrechterhalten werden.

Aus welchem Material die ersten über Turkistan nach China importierten Brillen fabriziert waren, ist nicht deutlich gesagt; nur in dem einen oben erwähnten Falle wird bemerkt, daſs es sich um eine weiſsem *liu-li* (Glaspaste) ähnliche Masse handelte. Sicher aber befanden sich Kristallbrillen darunter, denn aus Kristall haben die Chinesen sehr bald ihre Brillen selbst verfertigt. Es verlohnt sich, in diesem Zusammenhang darauf hinzuweisen, daſs nach den Zeugnissen chinesischer und europäischer Schriftsteller Kristall im ganzen chinesischen Reiche verbreitet ist und seit alter Zeit den Chinesen bekannt war, weil noch immer die trügerische Schluſsfolgerung beliebt ist, daſs, wenn sich ein bestimmtes Material in einem bestimmten Gebiete befindet, auch die daraus gefertigten Gegenstände einheimische Erfindung oder Idee sein müssen. Wir sehen hier wieder einmal den Fall, daſs auch das Gegenteil davon zutreffen kann.

Europäische Brillen wurden in China bereits im Anfang des 18. Jahrhunderts eingeführt, wie ich einem aus dem Jahre 1725 stammenden Zolltarif entnehme, der im siebenten Kapitel der chinesischen Chronik von Amoy (*Hsia mên chi*) abgedruckt ist; daraus geht hervor, daſs aus Europa importierte Brillen damals ebenso hoch besteuert wurden wie Kristallbrillen einheimischen Fabrikats, d. i. $\frac{1}{2}$ Tael oder Unze Silbers für hundert Stück.

Es wäre interessant, chinesische und indische Brillen in bezug auf Form und Technik eingehend zu vergleichen. Die indischen sind mir leider nicht bekannt. In China werden gegenwärtig Brillen aus

Glas und Kristall[1] verfertigt, letztere besonders in Suchou und Canton, wo sich große Kristallschleifereien befinden. Die Gläser, deren es konvexe und konkave gibt, sind im allgemeinen größer als die uns-rigen, nicht oval, sondern kreisrund und mit breiter Einfassung von Schildkrötenschale versehen. Die Stangen sind aus Messing oder Kupfer, werden aber nicht auf die Ohren gelegt, sondern zwischen den Schläfen festgehalten. Eine ganz eigentümliche Methode des Brillentragens, die ich selbst niemals in China beobachtet habe, wird von Sir JOHN FRANCIS DAVIS in seinem Buche „China" (Band II, S. 222, London 1857) beschrieben und abgebildet. Hier sind an den Enden der Brillen-stangen seidene Schnüre befestigt, die über die Ohren geschlungen werden, nach vorne herunterhängen und unten mit Gewichtstücken beschwert und so gehalten werden. Vom ethnographischen Standpunkte stellt somit die chinesische Brille einen von der europäischen ver-schiedenen Typus dar und kann auch aus diesem Grunde nicht auf dieselbe zurückgeführt werden.[2] Brillen werden nicht nur zur Hebung von Kurzsichtigkeit, sondern auch zum Schutz der Augen gegen Sonnen-brand und Staub, besonders auf Reisen in Nord-China, getragen. Dem Kristall wird ein wohltätiger Einfluß auf das Auge zugeschrieben, und er steht natürlich höher im Ansehen und im Preise als das Glas. Unter den Literaten kommt denn auch noch das Bedürfnis hinzu, sich durch die Brille ein würdevolleres und gelehrtes Aussehen zu geben. Merkwürdig ist die chinesische Brillenetikette in ihrer Übereinstimmung mit einer noch in gewissen deutschen Kreisen beobachteten Sitte, in denen bei der ersten Vorstellung Kneifer abgenommen werden. Beim deutschen Militär besteht (oder bestand wenigstens zu meiner Zeit) die Vorschrift, daß bei der Ehrenbezeugung gegenüber Vorgesetzten Kneifer rückhaltlos von der Nase verschwinden müssen, und Schreiber dieses wurde einst während seiner Dienstzeit in den Straßen Leipzigs von einem Unteroffizier zur Rede gestellt und abgekanzelt, weil er diese Höflichkeitsregel außer Acht gelassen hatte. In China besteht ganz ähnliches: bei der ersten Begegnung, Vorstellung oder Begrüßung gilt als gute Form, die Brille abzunehmen, und als grobe Unhöflichkeit, den neuen Bekannten durch die Brille anzustarren. Kein Beamter

---

[1] Auch aus Rauchquarz oder Rauchtopas (*ch'a shui tsing*, d. i. „Tee" [teefarbener] Kristall), siehe A. J. C. GEERTS, Les produits de la nature japonaise et chinoise, p. 248.

[2] Auch DAVIS bemerkt an der angeführten Stelle: "If anything could prove the Chinese spectacles to be original inventions, or not borrowed from Europe, it would be their very singular size and shape, as well as the strange way of putting them on."

darf in Audienz eine Brille in Gegenwart des Kaisers tragen, und
ebenso kein niederer vor einem höheren Bamten. Im Falle, daſs der
wirklich Kurzsichtige bei Anwesenheit seines Vorgesetzten ein Schriftstück lesen muſs oder zu einem anderen Zwecke seiner Brille bedarf,
erbittet er die Erlaubnis, ohne ein Wort zu sprechen, indem er auf
seine Brille zeigt, und setzt sie unverzüglich auf. Ist ihr Gebrauch
aber nicht mehr erforderlich, muſs er sie sogleich wieder abnehmen.[1]

Jedem Besucher Chinas muſs die Tatsache aufgefallen sein, daſs
die Zahl der Brillenträger verhältnismäſsig gering ist, und daſs die
überwiegende Mehrheit sich vorzüglicher Sehschärfe erfreut. Diese
Ansicht wird von Dr. JOHN DUDGEON, einem Arzt und hervorragenden
Kenner der chinesischen Krankheiten und hygienischen Verhältnisse, bestätigt.[2] Zum Schluſs möchte ich darauf hinweisen, daſs Untersuchungen
chinesischer und indischer Brillen angestellt werden sollten, um fest-
zulegen, inwieweit bewuſst oder unbewuſst angewandte optische Gesetze
für das Schleifen der Gläser und Kristalle maſsgebend sind, und inwieweit sie wirklich der Kurzsichtigkeit abzuhelfen vermögen; denn daſs
sie dafür bestimmt waren, geht aus den alten chinesischen Berichten
mit voller Deutlichkeit hervor. Auch könnte eine solche Untersuchung
neues Licht auf den historischen Zusammenhang chinesischer und indischer Brillen und eventueller gemeinsamer optischer Kenntnisse werfen.

New-York, Columbia University.

Beiträge zum Streit zwischen Liebig und Mulder.

Von H. SCHLOSSBERGER.

Daſs LIEBIG bei seiner durchgreifenden reformatorischen Tätigkeit besonders auf dem Gebiete der organischen Chemie öfters auf heftigen Wider-
stand gestoſsen ist, daſs er durch seine Überzeugung genötigt wurde, ihm
falsch dünkende Ansichten anderer Forscher zu bekämpfen, und so teilweise
in sehr schwere und langwierige Kämpfe, die zum gröſsten Teil zu seinen
Gunsten entschieden wurden, verwickelt wurde, ist bekannt. Ich möchte
nun im folgenden durch einige Briefe, die an meinen Groſsvater, den 1860
gestorbenen Professor der Chemie in Tübingen, Dr. JULIUS SCHLOSSBERGER,
gerichtet waren, den Gang des bekannten Streites zwischen LIEBIG und dem
Holländer MULDER illustrieren. Der Grund dieser Zwistigkeit lag bekanntlich

---

[1] Vgl. auch SIMON KIONG, De la politesse chinoise, Shanghai 1906, p. 4.

[2] In der Publikation der International Health Exhibition, London,
1884: China, Public Health, National Education, Diet, Dress, and Dwellings
of the Chinese, London 1885, p. 172: "The eyesight of the people remains
good, and very few, in advanced life, feel the need of spectacles."

048

一种关于汉字起源的理论

AMERICAN ANTHROPOLOGIST

NEW SERIES

ORGAN OF THE AMERICAN ANTHROPOLOGICAL ASSOCIATION
THE ANTHROPOLOGICAL SOCIETY OF WASHINGTON,
AND THE AMERICAN ETHNOLOGICAL
SOCIETY OF NEW YORK

PUBLICATION COMMITTEE

VOLUME 9

LANCASTER, PA., U. S. A.

PUBLISHED FOR

THE AMERICAN ANTHROPOLOGICAL ASSOCIATION

1907

A THEORY OF THE ORIGIN OF CHINESE WRITING[1]

By BERTHOLD LAUFER

It is not my purpose in this paper to initiate the reader into the mysteries of Chinese writing, nor to present a feat of sinological erudition. I merely wish to illustrate the application of a principle derived from the investigation of primitive ornamentation to the question of the origin of ancient Chinese writing.

Every casual observer will be impressed by the decidedly ornamental and picturesque feature of Chinese characters; and this observation coincides perfectly with the view held by the Chinese themselves, that writing is an art — a decorative art — which is as eagerly aspired to, and occupies the same high rank, as painting. The art of painting itself received a strong impetus from that of penmanship, and is still markedly graphic in character. All the famous painters have at the same time been noted calligraphists; and their autographs, one or two words dashed off with a bold stroke of the brush, excite as much admiration and are as greatly prized as their sketches or water-colors. Writing, consequently, offered the first field for the practice of art: it was the beginning of drawing and painting; hence in view of this fact we are justified in questioning its claims, from the anthropological viewpoint, of the development of decorative art.

For such a study we must entirely eliminate the modern forms of characters, which have been in use for two thousand years, and turn to the oldest existing specimens of writing, which are handed down on the bronzes of the Shang dynasty, dating from the third millennium before Christ. At that early age the formation of writing was completed; all further stages in its development are either new combinations or simplifications and changes of form conditional upon the changes in writing implements. The invention of the writing-brush, of ink, and of rag-paper, necessarily produced a

---

[1] Read at the meeting of the New York Academy of Sciences, March 25, 1907.

487

tremendous effect on the shaping of characters, with a tendency toward more rounded, graceful, and pleasing forms; while the oldest writing materials — like bamboo, wood, stone, and bronze, later on also silk — inscribed with a clumsy stylus and varnish, certainly allowed of only rudely executed characters. From this field an abundant supply of examples could be furnished on the question as to how ornaments change under the influence of new technique and material.

Another point that must appeal to the anthropologist is the fact that the Chinese have anticipated us, dissected, analyzed, and interpreted all their characters in numerous philological works commanding high respect. From the results of their painstaking research, foreign scholars have elaborated their system of writing, and usually have adhered to the native interpretations with implicit faith. But these interpretations, however ingenious and convincing they may at first seem, have only a relative value as personal impressions or popular traditions. Chinese scholars began with deliberation to reflect upon the composition and meaning of their characters, and to arrange them in analytical dictionaries, as late as post-Christian times, after writing itself had been in constant use for at least three thousand years; so that practically they could have known nothing about its original growth. What they have to say concerning this point is equivalent to the oral interpretations that we now receive from primitive tribes regarding the signification of their ornamental patterns, and must be regarded in the same critical light. The agreement between the two phenomena is so close that, just as different members of a tribe or of different tribes of the same stock may ascribe to the same ornament a different meaning, various Chinese authors give widely varying and sometimes contradictory explanations of the symbolism underlying their characters; and the traditions crystallizing around them have oscillated and also changed at times.

Chinese writing is not the result of one and the same principle, nor the product of one homogeneous mold; several factors have combined toward its production, and during a period covering many centuries. The most efficient method of construction was by means of a large number of phonetic elements combined with ideographic signs.   Nearly nine-tenths of all the characters now existing are formed on the basis of this principle.   If we eliminate this and other comparatively recent developments, we come upon a group of about six hundred simpler signs, called by the Chinese "pictures of objects," which admit of no reduction into single components.

It is on this limited class of characters that European sinologues have founded the theory of a pictographic origin of Chinese writing, which, for the rest, is merely the reiteration of what the Chinese themselves think on the subject.   It is asserted that these characters, now conventionalized in drawing, abbreviated, and disfigured, were developed from an original realistic picture portraying the object which the character is intended to represent.   It will be readily seen that here we have the same condition of things, and the same theory, as formerly advanced regarding the origin of primitive ornament, when many conventional patterns, through the process of evolution, were traced back to the realistic prototype from which the pattern was named; and I am inclined toward the conviction that, just as we were obliged to dispel that belief, we shall be compelled to abandon the long-cherished theory of the pictographic origin of Chinese writing.   Not that I would transfer merely through analogy the results of research in primitive art to the problem under consideration, but I wish to substantiate my belief with the evidence accruing from this particular field, and thus corroborate what has been ascertained from a study of the ornamentation of modern times.

The proposition that the six hundred primitive symbols were evolved from real pictures is not borne out by the facts, as they are clearly laid down in the ancient bronze inscriptions of the Shang period.   Among the characters there preserved we meet with no expression of realism, with no adequate likeness or full figure, but only with symbols consisting of brief, sketchy, and shadowy outlines — conventional designs in which no sort of development from a natural picture to a state of gradual conventionalization can be traced.   In most cases such a development would be materially impossible and illusory at the outset.   What could it signify in general, and to primitive man in particular, to speak of reproducing a representation true to nature — of water, river, cloud, wind, earth,

metal, fire, and many others that we find among the earliest attempts
of Chinese drawing? He must needs turn, with no other alterna-
tive, to conventional symbols to express the ideas of such objects.
In fact, any realistic representations that could be construed as having preceded writing, and finally resulted in it, do not exist, and
have never existed. They do not even exist as survivals in art, and
if they ever did we should justly expect there to discover them.
Ancient art, however, is in perfect harmony with ancient writing.
As all primeval characters represent conventional designs, so is all
early Chinese art as decidedly conventional and traditional as any
art can be; and I may go a step farther by making bold to say
that in the art prior to our era, illustrative of a development extending over three thousand years, there is not a trace of realism or of
naturalism apparent in any artistic production. All patterns are
either strictly geometrical or consist of animals and monsters conventionalized to extremes, while the human figure plays hardly any
conspicuous rôle. Realism appeared in Chinese art only a few
centuries after the beginning of the Christian era, in the works of
prominent individual artists, as though it were the result of a reaction
directed against the monotonous traditionalism of the older national
art. Not one natural bird, not a single natural tree or flower, do
we discover in the archaic period, the Han dynasty included, until,
in the seventh century, the great painters of lifelike birds and
flowers arise in the time of the T'ang.

The opinion that conventional forms are evolved from realistic
representations is without substantial foundation, and is refuted, so
far as China is concerned, by historical evidences such as these. If
realism in art proves to be the product of such recent times, it is
difficult to imagine how it could have existed during the epoch of
the embryonic formation of writing, whose beginnings must be conjectured to have been at least in the fourth millenium B. C. So
that there is nothing left for us but to conclude that the oldest forms
extant are also identical with the earliest primeval forms, which of
course had no predecessors. These forms, if we analyze them fairly,
are composed of a certain number of lines, strokes, dots, combinations of these, and simple ornamental figures which are variously
interpreted as certain objects or are named after them. Rows of

dots, for example, according to the different ways in which they are surrounded by lines, are identified with raindrops in the one case, with grain and rice in two other cases, and, in still other combinations, with sparks of fire or nuggets of metal. It will be recognized that it was not the picture of an object, or any attempt to draw a life-like design, that was the primary agency in the formation of writing, but a group of conventional ornamental forms. These received individual names by which to distinguish them one from another, the name being suggested by a process of association, in the primitive mind, of the design with the object to which the name referred. Thus, naturally, a vertical stroke would suggest the stem of a tree or a piece of wood ; a curved line, a snake or a river ; a zigzag line, the top of a mountain. This designation adhered to the ornament traditionally, and name and design finally became so thoroughly yoked together that the symbol called to mind the name, and the name the symbol, until they became inseparably united. I will not dwell at length on the final process that led to the conception of ornaments as true writing, in which the design was fixed at last as a character, and its name was substituted by the word conveying the idea of the object that this name implied. This was by no means an abstract process of intentional rationalism, but a development as purely emotional as the original creation of ornaments. It was doubtless prompted by the early existence of an elaborate system of ritual symbolism and by the facts that ornamental combinations and compositions are treated as legible rebuses which have dominated the art and religious customs of China from the days of antiquity until the present time. Whatever the psychical basis of this concluding step may have been, I think we may say now that the beginnings of Chinese writing are not pictographic, but ornamental and symbolic.

This theory receives strong corroboration from two other ideographic systems of writing occurring in eastern Asia — that of the Lolo and that of the Miaotse. Of the latter, we have a single specimen preserved in a Chinese book of the year 1683, giving two short songs in the original script, with an interlinear version in Chinese. The Lolo writing, consisting of about three thousand characters, has become better known through the investigations of

Father Vïal, who sees in it one of the oldest forms of Chinese writing ; while other scholars consider it as adaptations to and reconstructions of ancient Chinese characters. Although traditionally its invention is attributed to a Chinese who lived about the year 550, there is no resemblance whatever between Lolo and Chinese or between Lolo and Miaotse characters. The Lolo and Miaotse symbols are quite independent and original in their outward structure, and no doubt originally represented indigenous ornaments of those particular tribes. The stimulus of adapting these ornamental designs to the purposes of writing was unquestionably received from the Chinese, while the forms themselves were autochthonous. This supposition accounts as well for the above tradition as for the facts as we find them at present, and in my opinion there is no other possible way of explaining them.

COLUMBIA UNIVERSITY,
NEW YORK CITY.

049

卡尔施《同性恋研究》书评

BOOK REVIEWS

Forschungen über gleichgeschlechtliche Liebe. Von F. KARSCH-HAACK. I.
Band. *Das gleichgeschlechtliche Leben der Ostasiaten : Chinesen Japaner
Koreer.* München : Seitz & Schauer, 1906. 8°, ix, 134 pp.

This is a most scholarly production by an assiduous worker, a deep
thinker, and a genial philanthropist. In the *Jahrbuch für sexuelle Zwi-
schenstufen* (vol. III, 1901, pp. 72–201) the author, who is a Privatdocent in the University of Berlin, discussed the occurrence of pederasty
and tribady among primitive tribes, pointing out the existence of homosexual individuals among the Negroes, Malayans, American Indians, and
Arctic peoples. In this new treatise the Chinese, Japanese, and Koreans
are dealt with from the same point of view ; in a second volume he pro-
poses to treat of the Hamites, Semites, and the culture nations of
America, while the Aryans will occupy the third and fourth volumes.

The leading thought of these investigations is, as stated in the preface
(p. ix), that the above-named phenomena as effects of sexual impulse are
not "vices," but manifestations always and everywhere appearing which
are deserving neither of contempt nor social ostracism or brutal persecution by law, and that accordingly among single races and peoples they do
not differ essentially or in principle, but in the characteristic forms of
their occurrences there are variations corresponding to the ethnic traits
of the peoples. Students of East-Asiatic cultures will feel greatly indebted
to the author for the present volume, which represents a new and most
interesting contribution to our knowledge of the culture of the Chinese
and Japanese, with much new light on their innermost thoughts. It is
undoubtedly a valuable character study of these peoples. The sources
available for such a study are utilized with remarkable completeness,
with conscientiousness and sound critical acumen. With regard to
Chinese historical data which are quoted from sources that are now anti-
quated, and the spelling of proper names, the author would have done
well to consult a sinologue ; it is impossible to determine, for example,
what person the emperor "Qua-Tschesi" (p. 11) is.

The reviewer, who essays an appreciation of this book merely in the
attitude of a student of culture, openly admits that its subject proper, in
its physiological and medical aspects, is entirely foreign to him ; with that
reserve becoming his ignorance of the matter, he ventures to say that in

390

the chapter on China a clear distinction seems to him not to have always been drawn between really homosexual persons and occasional homosexual actions of otherwise normal individuals, such as was doubtless the case, for example, of the Emperor K'ien-lung, for one can by no means stamp him as a homosexual, as it is known from history that he left five sons.

From his consideration of homosexual life in China, which is organized in all forms, developed in all degrees, and spread over all classes of society, the author formulates the conclusion that pederasty cannot palpably weaken the vitality of an otherwise healthy nation nor check the progressive increase of the population — that it cannot be the expression of the decadence of a people. The vital force, the power of resistance, and tenacity of Chinese culture, and the extent of the population would speak eloquently against any assumption to the contrary. What, from our prejudiced and narrow point of view, we call prostitution, in China and Japan is a fundamentally different institution, and a juster understanding of it is attempted by Karsch (p. 69).

The history of the sexual relations of the Japanese is the more interesting portion of the book, as in the treatment of this many more sources are available; indeed, the Japanese themselves have revealed to us so many features of their sexual life. The author believes he is able to prove, by the testimony of history, law, literature, and art, that in Japan there was a period of natural, naïve, and unscrupulous practice and cultivation of mutual men-love which has been artificially suppressed only since the latter part of the nineteenth century under the influence of Occidental ideas. No law ever stood in the way of pederasty. In the famous codification of the Hundred Laws of Iyeyasu (seventeenth century) by the first Shōgun of the Tokugawa family (doubtless the greatest personality whom Japan has ever produced) the intercourse of men and women is set forth as the fundamental law of human society, and marriage is recommended to all who have transgressed the sixteenth year of age. This common sense in natural things, however, did not shield the great legislator from the sober and objective judgment of others who deviated from the norm established by him. Article 86 of his code runs : "Male and female prostitutes, dancing-girls and persons roving about at night are unavoidable in towns and flourishing places of the country. Although the habits of men are often impaired by this, yet greater vileness would come forth if severe prohibitions were issued. But games at dice, intoxication, and sexual debauchery must be strictly forbidden." From the tenor of this it is unambiguously evident (according to Karsch) that the legislator regarded intercourse with boys and sexual dissolution as entirely distinct things, and wanted them viewed in a different light. Japanese fiction is replete in descriptions of homosexual relations, the most prominent work being the "Great Mirror of Man-Affection," by the novelist Ibara Saikaku (1687), which is said to be an unvarnished realistic production not devoid of deeper sentiments nor of poetic beauty, and in all events a mine for the culture study of the Japanese people. About 1830 there appeared a catalogue enumerating no fewer than 177 Japanese works on pederastic subjects (p. 118). I fully concur with the author in his judgment on that branch of Japanese painting branded as " obscene " by the ordinary philistine spirit (p. 106), on that art of the nude which is certainly nothing but an outlet for the overflowing joy of life and sound sensuality unfettered by disguise and hypocrisy.

The Samurai, the military nobility, were in the habit of keeping fine young boys or youths in addition to their wives. Now, it is a curious fact that Satsuma was anciently and still is the center of pederasty, and it is also true that the bravest and most warlike people come from this province and clan of Satsuma. Lovers of boys are said there to be manlier than lovers of women. Until 1868 there was in Satsuma a law forbidding, under penalty of death, young men under 30 years of age to touch a woman. This law, remarks Karsch, was due to the fact that the population of Satsuma forms an exceedingly warlike tribe, ten to twenty thousand men of which were permanently at war and must have been concerned about the fidelity of their wives at home, had not the importunity of the youthful male progeny thus been checked. This can hardly be the true reason, but is merely the subsequent reflection of the Japanese on the subject. The actual Samurai idea which endeavored to deter young men from seeking women under this formidable threat was rather to drive them intentionally to homosexual intercourse. On this point and these conditions in general on the island of Kiushu the present writer has direct information from Japanese who lived there, and he may thus, for the rest, confirm the report of the author. Eye-witnesses assert that pederasty is still widely prevalent in the army and navy, being an inheritance from the Samurai ; and it is said to have contributed not a little to the successes in the war against Russia. Though this may seem to be asserting too much, it cannot be denied that the military spirit of Japan was an essential factor in the cultivation of specific forms of manly relations ; certainly it was not the cause of them, which remains as mysterious to Japan as to all other countries.

Considering the investigations of Karsch, there can be no doubt that homosexuality is an ethnological problem worthy the attention and reflection of the student of anthropology, though it is from the anthropological point of view that it is difficult for the reviewer to subscribe to all the opinions and judgments of the author. First of all, one is not inclined to believe that he has succeeded in entirely proving that these phenomena were ever regarded by the Japanese as perfectly natural up to the period of the restoration. This is such a far-reaching statement, of such paramount anthropological and psychological importance, and it would represent such an extraordinary case, that it deserves some discussion. Strangely enough, Karsch himself furnishes the material from which just the reverse of his thesis may be deduced. He thinks (p. 77) that the first allusion to pederasty in Japanese literature is found in the *Nihongi* (completed A.D. 720), in the annals of the empress Jingō, under the designation "atsunahi no tsumi," which he translates by "Vergehen der Männerliebe," referring to Hepburn's *Japanese-English Dictionary* as giving the meanings "crime, trespass," etc., for *tsumi*, but unfortunately, as he remarks, no information regarding *atsunahi*. But on what authority his own translation rests, the author does not state, although he quotes the whole passage in which this expression occurs from Aston's excellent and well-known version of the *Nihongi*, in which the correct interpretation is given. To make the whole case intelligible to the reader, and by reason of the importance of this alleged first historical reference to pederasty in Japan, we quote literally this interesting story from Aston's *Nihongi* (I, 238):

"Prince Oshikuma, again withdrawing his troops, retreated as far as Uji, where he encamped. The Empress proceeded southwards to the land of Kiï, and met the Prince Imperial at Hitaka. Having consulted with her ministers, she at length desired to attack Prince Oshikuma, and removed to the Palace of Shinu. It so happened that at this time the day was dark like night. Many days passed in this manner, and the men of that time said :— 'This is the Eternal Night.' The Empress inquired of Toyomimi, the ancestor of the Atahe of Ki, saying : 'Wherefore is this omen?' Then there was an old man who said : 'I have heard by tradition that this kind of omen is called Atsunahi no tsumi [Aston's note : "The calamity of there being no sun"].' She inquired : 'What does it mean?' He answered and said :— 'The priests (hafuri) of the two shrines have been buried together.' Therefore she made strict investigation in the village. There was a man who said :— 'The priest of Shinu and the priest of Amano were good friends. The priest of Shinu fell ill, and died. The priest of Amano wept and wailed, saying :— 'We have been friends together since our birth. Why in our death should there not be the same grave for both?' So he lay down beside the corpse and died of himself, so that they were buried together. This is perhaps the reason.' So they opened the tomb, and on examination found that it was true. Therefore they again changed their coffins and interred them separately, upon which the sunlight shone forth, and there was a difference between day and night."

Atsunahi, or *atsunai*, is an archaic Japanese term, *atsu* meaning 'hot' and poetically used for 'sun' in compounds only, *nai* being the negative copula ('not to be'). Aston's explanation, "the calamity of there being no sun," or plainly a solar eclipse, is quite appropriate, while that of Karsch is arbitrary. But, assuming the latter to be correct, he has placed himself in the position of sawing off the very branch of the tree on which he sits, for if in this tradition intercourse between men be considered a crime—a crime of such an extent as to cause the sun to darken—it shatters his theory of an original natural concept of homosexual acts in Japan and would prove that in ancient Japan such acts were condemned. I should even go so far as to say that an unbiased mind could not find in this tradition a hint at those relations which our author infers from it. The plain words of the text do not bear out his interpretation. All that is said is that the two priests had been good friends from childhood, and it is only in their burial in a common grave that the abnormity of the case comes to cause its connection with a contemporaneous eclipse of the sun. Surely if Karsch's conception of a sexual intercourse and his reading into the text "Vergehen der Männerliebe" were correct, the whole story would be inconsistent. Why, if there is here the question of the "crime of man love," is not the sun made to disappear during the lifetime of the men, as would be most logical, instead of so doing only after their death? It is quite evident that it is only the unusual entombment of the two men that forms the keynote of the tradition. In this case it is not conducive to the evidence of homosexuality in ancient Japan.

Yet again (p. 97) we are told that in the *Norito*, the ancient rituals of Shintō, homosexual intercourse is not mentioned as a crime or sin, although sodomy is expressly named, which seems most noteworthy to our author, who thinks it would be inconsiderate to infer from this that pederasty had then been unknown. The passage to which he alludes may now be conveniently read in Aston's recent book on Shintō (London, 1905, p. 300). There is no evidence to show that ancient Shintō, either in an official or an unofficial form, ever sanctioned or tolerated pederasty, and if it did not condemn it, nothing can be followed from this regarding the existence or non-existence of such a custom. Shintō had very little, if any, concern with sexual relations; nor did it pronounce a verdict on adultery (see Aston, p. 91), although this does not prove that it was in silent sympathy with it.

As this is all the evidence gathered by Karsch from the ancient Japanese sources, it cannot be said that what he seeks to prove is valid for this early period ; and I am inclined to think that it did not then exist, at least not so manifestly as to attract public attention. And here an *argumentum ex silentio* seems to be somewhat conclusive, as all sexual relations are spoken of otherwise with unveiled naïveté and play an important part in the *Kōjiki*, the most ancient records of Japan. Now, if Karsch will make one believe that pederasty is inborn, so to speak, and hence natural to the Japanese, why does it not manifest itself in some form in the most natural productions of the *Kōjiki* ? I am far from disbelieving that at a certain period and among certain classes of people it was practised as a thing seemingly and perhaps effectively natural to them : all that we hear and read about it in regard to the class of Samurai makes indeed the striking, not to say appalling, impression of naturalness and ingenuity. This state of naturalness however is apparently a secondary development, and not by any means the original idea, as emphasized by our author ; it is a subsequent thought gradually bred and traditionally taught and handed down by the Samurai, and , we may admit, also by the celibate Buddhist priests. Even from the law of Iyeyasu it follows that the legislator only tolerated the practice, not that he approved of it. It is not too much to say that there is hardly a country under the sun that follows such sound principles and enjoys such wholesome conditions in matters of sexual intercourse as Japan, from which the hypocritical white world could learn many a lesson looking to the regeneration of its rusty morals, and that it is just this art of conforming to matter-of-fact living that the unique genius and exceptional greatness of Japan is due.

We do not deny any facts conscientiously recorded by Karsch concerning homosexual life ; we fully believe in them, but we desire to accentuate that which he utterly neglects to state, that also in Japan they form the exception to the rule, and, offset by normal sexual conditions, they lose much of the magnification to which they appear to be subjected when viewed individually, and when severed from a universal consideration of the ruling ties of love.

It further seems to me that we are not justified in saying, with Karsch, that the sudden reaction and legal measures taken by the Japanese government against pederasty in recent times are due solely to the influence of Western methods. It is true that these clauses of the Japanese penal code breathe the same spirit as corresponding ones in our criminal law and follow almost the same tenor; but it would mean to dispossess the Japanese lawyers of the freedom of the psychological motive by imputing to them the "forcible suppression of native genius," as Karsch puts it, through the imposing of a merely foreign law upon their people. There are many sections in our penal code that did not find an echo in that of the Japanese, owing to the entire lack of an actual basis for them in their environment. But the adoption of the clause against "unnatural offenses" sufficiently shows that the modern legislators of Japan were guided, and could not but have been guided, by a psychological motive in the reception of this law, which is to say that they were *not* led by the idea of that naïve and natural feeling toward this matter which our author tends to insinuate was the case with the mass of the Japanese. . And this is further strong ground for our view that this natural concept of homosexuality was not general, but was restricted to certain classes to whom it was secondarily instilled by tradition and education.

Here we must touch upon another weak side of the book. In his laudable attempt to do justice to a widely misunderstood question, Karsch looks disdainfully on all tendencies and powers opposing homosexuality; but he does not try to analyze or explain this antagonism. It is true that the homosexual individual has a claim to justice and to objective, impartial judgment. The phenomenon itself is an inexplicable enigma, and its world-wide propagation in ancient and modern times renders it all the more difficult of solution. Aside from this universality we can not, by way of purely scientific reasoning, attribute to it any other descriptive term than that it is *abnormal*, according to our present knowledge. To say that it is unnatural is certainly a fallacy, first, because everything occurring in natural, i. e., in human or nature, life, is implicitly natural, and, secondly, because the favorite conclusion, "it is against my nature, consequently against nature," is illusionary and deceptive of one's self. But these intellectual deductions cannot blind our eyes to the existence of certain emotions which dominate the soul of the individual as well as the life of the peoples of the globe. It is evident beyond cavil that all men and all women of normal sexual sentiment have an innate aversion to all abnormal sexual practice, and particularly to homosexuality, and as certain as the existence of the latter is, so certain also is the psychological abyss separating heterosexuals and homosexuals. This is not only a psychological but also an anthropological fact, and accordingly an anthropological problem for investigation, as it pervades all mankind; for it cannot be mere coincidence that the laws of primitive and of civilized peoples alike make provisions against abnormal intercourse.[1] The general animosity of law toward homosexuals is the crystallization of social and ethnic sentiment, and to study the foundations and reasons of this sentiment among peoples is one of the great requirements of anthropology. Certainly the question whether this sentiment is objectively justified or not, does not concern us as anthropologists, but moves along an entirely different line. This is also the reason why I believe that Mr Karsch, despite his noble efforts, will convince or convert few readers to his beliefs, which seem to culminate in the idea that homosexuality has the same privilege of existence as heterosexuality, a deduction which the majority cannot accept by reason of just those uncanny elementary ethnic emotional thoughts that haunt us common normal individuals, and which Mr Karsch, not being an anthropologist, is prone to stamp with such commonplace terms as prejudice and ignorance.

However all this may be, and how far our opinions may differ, it does not belittle the great value of Karsch's serious and thorough work, which deserves the widest attention of all thinking anthropologists.

B. LAUFER.

Sex and Society. Studies in the Social Psychology of Sex. BY WILLIAM I. THOMAS. Chicago: University of Chicago Press, 1907. 12°, 366 pp.

This book is chiefly a collection of special articles published from time to time in periodicals. The chapter headings are: Organic Differences in the Sexes, Sex and Primitive Social Control, Sex and Social Feeling, Sex and Primitive Industry, Sex and Primitive Morality, The Psychology of Exogamy, The Psychology of Modesty and Clothing, The Adventitious Character of Woman, The Mind of Woman and the Lower Races.

The general anabolic and katabolic conception of the sexes is accepted by the author at the start as the organic basis of society. While this is now the traditional view in biology and sociology, the author presents arguments in support of this sex antithesis as expressed in psychic and social activities. On the social side the male is considered as unsocial, or disposed to wander about detached, while the female because of her association with children forms the nucleus of a social group. In a general way the theory of maternal descent is accepted, but the author rejects the idea that promiscuity is implied in such a condition for the tie binding the woman and the children is a real, if not the real, social bond. However, the ever prevailing tendency toward male social authority is considered

---

[1] See, e. g., Post, *Grundriss der ethnologischen Jurisprudenz*, II, pp. 390–392.

050

纽厄尔和李白诗集

Dr Weiser was buried on the spot, and his grave temporarily marked, according to a letter from Colonel McPhail, "by three picket pins in a triangle, 12 inches apart, set at six feet south from the spot of burial, and extending four inches above the ground. "Subsequently, after 35 years of neglect, in 1898, search was made for Dr Weiser's grave by Mr Brower in order to mark it more permanently. He did not find the picket pins, nor any spot resembling the grave ; but he "gathered up a quantity of large and small bowlders in the northwest corner of the camp [Goodell] and at the point indicated by Colonel McPhail by blue cross [on a plot submitted by Brower], erected a small mound of earth and stone and placed a marble slab at a long rifle pit." A field-sketch of this by Mr Brower gives dimensions of the mound covered by bowlders, as 8 feet by 6 feet and 3 feet high, elongated east and west, a small marble slab lying flat in the center on the top, on which were engraved the words "DR. JOSIAH S. WEISER 1863."

The accompanying photographic view, by Mr Brower, was labeled by him : *Dr. Weiser's Grave, Kidder Co., North Dakota.* In the view the marble slab is invisible, indicating that it was small. By this time it may have been removed, and the group of stones might be considered the work of the aborigines. Numerous stone cairns, well known to be of aboriginal origin, at the present time are mere groups of stones that show little evidence of the purpose for which they were gathered.

N. H. WINCHELL.

W. W. Newell and the Lyrics of Li-T'ai-Po. — In Dr Chamberlain's bibliography of the late W. W. Newell, given in the last number of the *American Anthropologist*, I miss one of Mr Newell's last and most interesting works which, however, has unfortunately not been given to the public. This little volume bears the title "Lyrics of Li-T'ai-Po [Chinese Poet of the Eighth Century] by Michitaro Hisa and William Wells Newell (Printed Not Published)," xiv, 62 pp. The preface is dated "Wayland, Mass., August, 1905," and in it the origin of the book is set forth. Michitaro Hisa, a Japanese student at Harvard from 1891 to 1895, later professor of economics at Kioto (died 1902), became a close friend and frequent visitor in the family of Mr Newell, and, in response to inquiries concerning Chinese poetry, brought him translations from several authors ; among these, Mr Newell was especially interested in versions of Li-T'ai-Po, the greatest and most original poetical genius of China. The literal prose renderings of Hisa, following character by character the Chinese text, were brought into metrical form by Newell

who made it his first object to reproduce sentiment and language as closely as possible. "The results," Mr Newell says, "were shown to Hisa who furnished advice and suggestions; in this manner came into being the verses here printed, not for circulation or public notice, but for the sake of record, and as memorial of a friend whose delicate perception and deeply poetical spirit are mainly responsible for their existence, but who will never look upon their permanent form." Last Christmas, when Mr Newell attended the meeting of the American Anthropological Association at New York, he was good enough to present me with a copy of this book and to ask my judgment of it. I was just going to submit to him a plan for its publication, when the sad news of his death came. I have compared with the original text several of the twenty-six poems here selected, and in my estimation the translation is admirable and even unique. The few existing translations of some of Li-T'ai-Po's poetry in French and English give at least a mere circumstantial paraphrase of the text, while Mr Newell's rendering, in the epigrammatic terseness of its style, gives an excellent reproduction of the true spirit of the original. If there are in existence more copies of the book, which I am told Mr Newell printed with his own hands, they should certainly be circulated.

                                        BERTHOLD LAUFER.

Archeological Collections from San Miguel Island, California. — A series of specimens of bone, stone, and shell artifacts, obtained from ancient graves on the island of San Miguel, off the coast of Santa Barbara county, California, is shown in the accompanying plates. The data and photographs were furnished by the late Horatio N. Rust of Pasadena, California.

Plate XXXI, nos. 6, 7, and 8, illustrate small stone picks used in roughing-out the objects of shell and especially in making the perforations which were afterward to be enlarged and rendered symmetrical by the sandstone drills shown in nos. 1–5. Nos. 9–12 are supposed to be abrading stones and to have been used in giving the final shape to the various implements of shell and bone.

No. 1 of the lower half of the same plate represents a piece of shell formed by the stone pick referred to above and is ready for perforation. Nos. 2 and 3 are of shell, and show the use of the pick and drill. No. 4 is of stone. Nos. 5 and 6 indicate the use both of the drill and the abrading implement. Nos. 9–13 have been further elaborated with the latter implements. These hook-like objects have generally been classed as fish-hooks, but were regarded by Mr Rust as ornaments. He conceived that they may have served as a means of holding or attaching strings of beads or other pendant objects.

051

玉米传入东亚考

Congrès International

des

AMÉRICANISTES

XVᵉ SESSION

TENUE À QUÉBEC EN 1906

QUÉBEC

DUSSAULT & PROULX, IMPRIMEURS

1907

THE INTRODUCTION OF MAIZE
INTO EASTERN ASIA[1].

BY BERTHOLD LAUFER.

1. Much has been written about the introduction of maize from America into Europe and its further dissemination over the Old World; and it might seem almost superfluous to take up this question again, were it not for the fact that as yet no satisfactory investigation of the subject exists. The history of maize, in accordance with pragmatic methods based on all the available documentary evidence, remains still to be written. The interesting point of view to be pursued in connection with this question seems to me the following.

Of all problems of ancient cultural history, none is more obscure, and less enlightened by literary records, than the origin and propagation of our cereals, as only a cursory glance at the famous books of De Candolle and Hehn will convince one. Now, maize belongs to the few cereals which were distributed from one centre in recent historical times, — times which we are able to check from written documents. It is certainly of value to determine within which time maize spread from the New into the Old World, from Europe to the Far East; and through the medium of which nations, and by which commercial or geographical routes, it was transplanted. The history of maize may thus prove to be an object-lesson by affording us an adequate judgment as to the relative degree of rapidity with which cereals may be carried from country to country, and as to what agencies are efficient in their dissemination. What may happen with regard to products in historical times is liable to occur also in a

---

1. The parenthetical references relate to the corresponding numbers in the list of Chinese characters at the end of the paper.

prehistoric period, if ethnic and economic conditions may be supposed to have then been on the same or a similar basis. The geographical area which is of chief interest to us in this paper includes India, Central Asia, and China. In the sixteenth century, when maize was introduced into this region, travelling and transportation facilities there were not easier than in earliest historic and prehistoric ages ; and if we were to succeed in pointing out a definite course which maize may have taken inside of this geographical province, and in marking the boundaries of this route by means of certain chronological data, we should have a clew or an analogon as to the possibility of kindred occurrences in the same region in most ancient times. In this idea, then, lies the chief object of the present paper ; and by this, as well as by the utilization of new material, it is plainly distinguished from its predecessors. It has nothing to do with the chimerical attempts of some former writers who believed in a pre-American existence of maize in Eastern Asia [1], but its chief purport is to show through which channels maize has entered China. To anticipate its main results, I shall try to demonstrate that maize, introduced into India probably by the Portuguese, spread north-ward to Sikhim, Bhutan, and Thibet, finally from there into Ssech'uan, the province of western China bordering on Tibet, and then from the west into the other parts of China, without any interference of European nations, like Spaniards or Portu-guese. That in those days this route from India into Western China was a high-road of great importance for just the migration of cultural plants, we shall try to show by the example of another cereal, sorghum, by way of introduction inte the subject proper.

2. The accounts on the varieties of sorghum are united in

---

1. Like v. SIEBOLD and MAYERS. — The question, if it can now be called a question, whether there is an indigenous species of maize in Eastern Asia, is mainly one of botany, but not of sinology. If maize were indigenous to Asia, we should expect to find there either a wild form, from which the cultivated species was derived, or the Asiatic species to be differentiated from that of America, neither of which is found. The close identity ot the Asiatic with the American maize proves that both are one and the same botanically ; that is, the former is derived from the latter, and the botanical agrees with the historical result. All historical proofs adduced heretofore for the pre-American existence of maize in Asia are based on misinterpretations of sources, limited knowledge of the whole subject, and strange absence ot log c.

the cyclopædia " T'u shu chi ch'êng ", Vol. 468, po wu hui
pien, section on plants (ts'ao mu tien), Book 30, shu pu hui
k'ao, pp. 14 a-15 b. Sorghum was not known in the period of
Chinese antiquity, and is not mentioned either in classical or in
other early literature. It first occurs under the name *shu shu* ([1])
in the " Ch'i min yao shu " of CHIA SSE NIU ([2]), who is said to
have lived in the fifth century A. D. [1]. This notice is as follows :
" The spring month is most suitable for burying the seeds [of
the sorghum] in the earth. The stalk is over ten feet high.
The ears are big like brooms, the grains black like lacquer or
like frog's eyes. When it is ripe, it is harvested by mowing and
gathering it in sheaves, which are set up. The fruit yields a
grain which is hulled and eaten. Oxen and horses may be fed
with the refuse, and even the waste material may be utilized.
The stalks can be made into brooms for cleaning pots ; the blades
can be plaited into door-screens, mats, and fences. Besides, it
is served at table, so that there is nothing that need be thrown
aside. Thus it is one of the most serviceable grains, and indispensable to the farmer " ([3]).

The further texts in the " T'u shu chi ch'êng " are as
follows. The master YüAN HU [2] says, " *Shu shu* (*Sorghum vulgare*) did not exist in times of old. Later generations, I think,
must have obtained the seeds from another country. As its
glutinous properties come near to those of *shu* [3], this name *shu*
has been borrowed and transferred to it. The people of the
present time, whenever they point it out, call it simply *shu*,
ignorant of the fact that there is also the kind of *shu* called *liang
shu* (*Setaria italica*), wherein they make a mistake. Different
from this one is another kind, *yü mi* (' jade rice ', that is, maize,)
also called *yü mai* (' jade wheat ') or *yü shu shu* (' jade sor-

---

1. BRETSCHNEIDER, Botanicon Sinicum, Part I, p. 77 : " This work
contains many interesting particulars regarding the cultivation of the cereals,
vegetables, fruits, trees, etc., then grown in China. It is also of interest on
account of its numerous quotations from previous ancient writings now lost. "
Bretschneider (ibid., p. 78) enumerates also *Sorghum vulgare* among the
plants treated of in this book which can be identified.

2. Literary name of Hsü KUANG-CH'I (1562-1634), author of the
treatise on agriculture *Nung chêng ch'üan shu.*

3. That is, the glutinous variety of *Setaria italica.*

ghum '). Also, as regards this kind, the seeds were obtained from another country, and its designations *mi, mai,* and *shu shu,* are all borrowed names " (⁴).

He further says, " In the northern parts of China the soil is not favorable to wheat and other grains. The seeds of sorghum are much more suitable to it. On the fifth day after the Commencement of Autumn (Aug. 11), although sunk in the rain-water to a depth of ten feet, it cannot be spoiled. Only before the Commencement of Autumn (Aug. 7), is it spoiled by rain-water. Therefore in the regions of the north, they build embankments two to three feet [high] to shield it from the violence of the water. By these efforts to ward it off for several days, no harm is done, however big the rain may be " (⁵).

He further says, " Where in the country of Ts'in [i. e. North China] there is salt soil, sorghum is planted in the ground, for it is especially suitable to the sowing of sorghum. It is necessary to plough early, from the first to the last in the solar term *Ch'ing ming* [that is, from April 5 to 19]" (⁶).

WANG YING ¹ says, " Sorghum (*Shu shu*) is sown in the northern regions to provide for the lack of grain. The refuse is fed to oxen and horses. It is the most excellent of all cereals. The people in the south call it *lu tsi*" (⁷).

LI SHIH-CHÊN says, " Sorghum is convenient to sow : in the spring month, the seeds are scattered ; in the autumn month, it is gathered. The stalk is over ten feet high ², in shape resembling the *lu ti* ; also the fruit inside, and the leaves, are like the *lu* ³. The ears are big, like brooms. The grains (*li*) are big, like pepper, of red and black color, and as hard as hulled rice. The fruit (*shih*) is yellow and red in color. There are these two kinds, — a glutinous kind, of which, with glutinous rice and glutinous *Setaria italica*, fermented wine can be made, and cakes ; a non-glutinous kind, which can be made into

1. Author of the *Shih wu pên ts'ao,* a treatise on eatables, published in the beginning of the sixteenth century (BRETSCHNEIDER, Botanicon Sinicum, Part I, p. 53).

2. According to BRETSCHNEIDER (Die Pekinger Ebene, Pet. Mitt. Ergaenzungsheft, No. 46, p. 17), the sorghum-plants near Peking reach a height of twelve feet or more.

3. A kind of reed (*Phragmites Roxburghii* and *P. communis*).

dumplings and into congee. Also the waste material can be utilized : cattle can be nourished with it. Brooms can be made of the small tips. From the blades door-screens and mats can be plaited. It contributes to our table, and is of extreme advantage to the nation. The people of the present time use it wrongly for their offerings in place of millet (*Panicum milia-ceum*). The husks of this grain, when soaked in water, assume a red color, and red wine can be made of it. The " Po wu chi " says, Where sorghum is sown, there will be many snakes after a number of years" (8) [1].

Li SHIH-CHÊN says, " It is not so very long ago that *Shu shu* (*Sorghum vulgare*) made its appearance, but now there is a great quantity of it in the northern part of the country. The ' Kuang ya ' mentions *ti liang* [2] and *mu chi* (' wooden millet '). The sorghum too is a kind of millet, and is as high and big as the reeds *lu* and *ti*, wherefore in common language it has all these names. It was first cultivated in *Shu* (Ssech'uan), and is therefore called *Shu shu* ; that is, millet of Ssech'uan " (9).

If we analyze the preceding records, it is easily recognizable that the different varieties of sorghum are treated indiscrimi-nately [3]. The most striking fact, from an historical point of view, is that both Li Shih-chên and Hsü Kuang-ch'i agree in the statement that sorghum can be only a recent introduction, the former saying that it did not date so far back in the past, but grew plentiful in the north of China in his time (that is, the second half of the sixteenth century), the latter positively denying

---

1. No value whatever is to be attributed to this quotation. The *Po wu chi*, alleged to have been written in the third century A. D., was lost during the Sung ; and the present work which goes under this name was drawn up at a much later period, compiled from extracts in other publications, and padded with numerous anachronisms and marvellous inventions.

2. *ti*, a kind of reed with a pithy stem.

3. Also F. PORTER SMITH, Contributions towards the Materia Medica and Natural History of China, p. 202 (Shanghai, 1871), justly remarks, " The Sorgo, or Chinese Northern Sugar Cane, is described in the Pen Ts'au along with the Sugar-cane and the *Holcus Sorghum* [now *Sorghum vulgare*], or Barbadoes Millet. " — A. DE CANDOLLE, Origin of Cultivated Plants, pp. 381, 382, confounds the two Chinese species by identifying the Chinese *kao liang* with *Sorghum saccharatum*. — Already BRETSCHNEIDER (Chinese Recorder, Vol. III, p. 289 a) referred to the fact that the glutinous kind of Li Shih-chên is *S. saccharatum*, and his non-glutinous kind, *S. vulgare*

its occurrence in times of antiquity, and referring to an introduction from a foreign country. Neither of them — according to
the general experience in the history of the dissemination of
cereals, which so suddenly appear and then spread with such
wonderful rapidity — is able to assign a definite date to the
introduction ; but Li Shih-chên affords a most valuable clew for
unravelling the mystery by his interpretation of *Shu shu*, the
name for *Sorghum vulgare*, as millet (*shu*) of Ssech'uan (*Shu*),
in which province, according to him, it was first grown. Thus
far, matters would be easy but for the fact that the mention of
sorghum is ascribed to two much older works, the " Kuang ya "
and the " Ch'i min yao shu ". How can the opinions of Li Shih-
chên and Hsü Kuang-ch'i regarding a recent importation be
reconciled with this condition of affairs ?

The " Kuang ya " is a dictionary by CHANG I, published
about A. D. 265 [1]. Now, Li Shih-chên quotes from this
dictionary the two terms *ti liang* and *mu chi;* but there is no
evidence whatever that these two terms, which went out of use
long ago and seem solely restricted to the work in question, ever
denoted sorghum or related plants. The word *ti* refers to a kind
of reed with a pithy stem [2] ; and the great importance of reeds
in the economical life, which form an industry along the Yangtse
only second to bamboo, and a certain similarity between reeds
and sorghum [3], may be held responsible for the simile employed
by Li Shih-chên. More serious and more difficult is the passage
in the " Ch'i min yao shu " of the fifth century, in which a
variety of sorghum is undeniably described. I think, however,
that a way out of this difficulty is possible. The variety described
in the " Ch'i min yao shu " is, in my opinion, *Sorghum saccharatum ;* the variety of recent introduction, mentioned by Li
Shih-chên and Hsü Kuang-ch'i, is *Sorghum vulgare* [4]. This

---

1. T. WATTERS, Essays on the Chinese Language, p. 38 (Shanghai,
1889).

2. Notes and Queries on China and Japan, Vol. III, 1869, p. 97 ;
BRETSCHNEIDER, Botanicon Sinicum, Part II, p. 272.

3. Compare the name *lu su* (' reed millet ') for *Sorghum saccharatum.*

4. English names for this cereal are, Indian millet, African millet,
Barbadoes millet, great millet, Guinea-corn, Kafir-corn, broom-corn, durra
or doura, coffee or chocolate-corn (the latter U. S.).

decision rests mainly on the fact that the grains of the sorghum are described as black (and as black as lacquer) in the " Ch'i min yao shu ", which is indeed the case with *Sorghum saccharatum*, while Li Shih-chên speaks of red and black grains, thus comprising the two varieties.

To which of the two varieties of sorghum Wang Ying, who wrote about half a century before Li Shih-chên, alludes in the passage quoted above, must remain undecided, but in all likelihood he means *Sorghum vulgare*, possibly both. However this may be, it is perfectly safe to assume that *Sorghum vulgare* was introduced from abroad into China not long before the time of Li Shih-chên, possibly a century or so, say about the end of the fifteenth.

The question as to the original habitat of the two sorghum species — the views of botanists still seem to be divided between tropical Africa and India [1] — does not here concern us. Suffice it to bear in mind that they must have been at home in India, where it is still the staple food, for a long time, the Indian *Sorghum vulgare* being doubtless alluded to by Pliny [2]. If this sorghum, as regards China, was first planted in Ssech'uan, I see no other possibility than to assume that it had migrated from India northward into Tibet, and from Tibet had reached western China. We find it eagerly cultivated by the Lepcha in Sikhim [3], who live with Tibetans and in the immediate neighborhood of Tibet.

---

1. A. DE CANDOLLE, Origin of Cultivated Plants, pp. 381, 382, seeks their origin in tropical Africa, whence they were introduced into Egypt, afterward into India, and finally into China. A. ENGLER, in HEHN's Kulturpflanzen und Haustiere (7th ed., p. 504, Berlin, 1902), leaves it an open question.

2. HEHN, l.c., p. 502.—" Taking India as a whole, it may be broadly affirmed that the staple food-grain is neither rice nor wheat, but millet. Excluding special rice-tracts, varieties of millet are grown more extensively than any other crop, from Madras in the south, to at least as far as Rajputána in the north. The two most common kinds are great millet *(Sorghum vulgare)*, known as joar or jawǎri in the languages derived from the Sanskrit, as jonna in Telugu, and as cholam in Tamil; and spiked millet *(Pennisetum typhoideum)*, called bájra in the north, and kambu in the south. " W. W. HUNTER, The Indian Empire, 3d ed., p. 582 (London, 1892). See further W. CROOKE, Things Indian, p. 226 (New York, 1906).

3. MAINWARING and GRUENWEDEL, Dictionary of the Lepcha Language, p. 533 (Berlin, 1898). H. H. RISLEY, The Gazetteer of Sikhim, pp. 74 et seq. (Calcutta, 1894).

In the great Dictionary in four languages, published by order of the Emperor Ch'ien lung, the term *Shu shu* is reproduced in Manju as *shushu* ; in Mongol, *shishi* ; and in Tibetan, *sa lu*. The original meaning of the latter word is " rice ", derived from Sanskrit çâlî, which goes to show that the Tibetan language also (in the same way as Chinese, Manju and Mongol) possessed no indigenous word for " sorghum ", and conveyed this idea to a Sanskrit word which had already been adopted. This expediency, however, would not have been resorted to but for the derivation of the plant itself from India.

3. The more important notices in regard to maize from Chinese sources are those of W. F. MAYERS [1], and S. W. BUSHELL [2]. Mayers stands principally on the passage of the " Pên ts'ao kang mu ", where he reads that " the seed of the *yü shu shu* [maize] came from the lands beyond the Western Frontiers (i. e., from Central Asia) ". We shall see hereafter that this translation is wrong. He then refers to the " Nung chêng ch'üan shu ", written by Hsü Kuang-ch'i in 1619, who says that the plant was anciently obtained from the territories on the west of China [3]. From the fact that, within less than three quarters of a century after the discovery of America, Li Shih-chên describes maize as a plant perfectly well known in China, but is at the same time unable to assign a date for its introduction into the country, he concludes that maize must have been cultivated in China for a considerable period before the

---

1. Maize in China. Notes and Queries on China and Japan, Vol. I, pp. 89-90. (Hongkong, 1867).

2. Maize in China, Ibid., Vol. IV, p. 87.

3. I cannot find the passage, thus translated by Mayers, in the *Nung chêng ch'üan shu*. Hsü Kuang-ch'i speaks of maize only in the passage, translated above, where he treats of sorghum, and says of maize that it came from another (foreign) country. This passage will be found in his work *Nung chêng ch'üan shu*, Book 25, p. 9 b. Among the plants enumerated and illustrated in this work, maize is not included. In the agricultural cyclopædia *Shou shih t'ung k'ao* (Book 24, p. 4), in which all important matter of the *Nung chêng ch'üan shu* is embodied, a woodcut of maize with the text of the *Pên ts'ao kang mu* only is given. In the *T'u shu chi ch'êng* (l.c.), where the section on maize follows that on sorghum, the latter text also is given, and curiously enough, accompanied with the description of the " wild millet " (*yeh shu*) of the *Nung chêng ch'üan shu*. I do not see what the latter species has to do with maize. In neither of these books are any extracts given from the work of Hsü Kuang-ch'i regarding maize, so that I am at a loss to explain what the foundation of Mayers's statement may be.

arrival of Europeans.   " Botanists must decide ", he continues,
" whether or no the theory of its introduction from Central Asia
is tenable, but for my own part I incline to think that it very
probably reached China from Japan ".   This lack of logic is very
strange.   From his sources and his way of interpreting them,
Mayers could have drawn no other conclusion than that maize
arrived in China from the west : he found nothing to indicate a
trace of an importation from Japan, and there is in fact nothing
whatever to justify such a view.   The consequence of his oscil-
lation was that he was misunderstood by many subsequent
writers, some of whom seem to quote him without having had
access to his original article.   Thus, for example, I. REIN [1]
remarks, " In 1869 Mayers proved that maize came to China
through Portuguese and Spaniards, so that the Japanese names
which refer to China possess no greater value than the names
' Welschkorn ' and ' Turkish wheat ' among us ".   Mayers,
however, denied point-blank the possibility of its having been
brought to China by the Portuguese, and he did so justly, since
there is no evidence for such a statement.   A. DE CANDOLLE [2]
credits Mayers with the statement " that early Chinese authors
assert that maize was imported from Sifan (Lower Mongolia,
to the west of China) long before the end of the fifteenth
century, at an unknown period ; the importation through
Mongolia is improbable to such a degree that it is hardly worth
speaking of it, and as for the principal assertion of the Chinese
author, the dates are uncertain and late ".   Mayers does not use
the term " Sifan " in his paper on maize.   This entire passage
was then copied, with Candolle's other views of a Portuguese
introduction of maize into China, by JOHN W. HARSHBERGER [3].

The brief notice of Bushell quoted above is directed against
Mayers.   He refers to the fact that maize is not described in any
Chinese work anterior to LI SHIH-CHÊN's " Pên ts'ao kang
mu " ; and the time which elapsed between the discovery of

---

1. Zur Geschichte der Verbreitung des Tabaks und Mais in Ostasien
(Petermanns Mitteilungen, Vol. xxiv, 1878, p. 215).

2. Origin of Cultivated Plants'(New York, 1885), p. 392.

3. Maize : A Botanical and Economical Study, Contributions from the
Botanical Laboratory of the University of Pennsylvania, Vol. i, No. 2,
Philadelphia, 1893, p. 157.

—17

America and the date of this work seems to him amply sufficient for its introduction into China.   Bushell quotes two further passages from Chinese books dealing with maize [1].   He does not, however, take any standpoint with regard to the western origin of maize held by Chinese authors, and his only refuge are the Portuguese [2].

4. The two most important passages not considered heretofore, and suggestive of an introduction of maize into China from Tibet, are contained in the cyclopædia " Ko chih ching yüan ", [3] published in 1735.   The one is derived from a book, " Liu ch'ing ji cha ", of an author, T'IEN I-HÊNG ([10]) ; the other, from the " Hsio pu tsa shu " of WANG SHIH-MOU ([11]).   The latter is known as a botanical author who left many treatises on plants, and died in 1591 [4].   The former work is not quoted by either Bretschneider or Wylie ; but its author, T'ien I-hêng, is given by WYLIE [5] as having edited, during the Ming dynasty, an anthology of the productions of celebrated poetesses.   As the quotations in the cyclopædias are usually well arranged in chronological order, and as the passage of Wang Shih-mou is apparently based on, if not actually derived from, T'ien I-hêng, we may infer that he was the former's predecessor, and wrote, say, about the middle of the sixteenth century.   His remarks are as follows :

" Maize (*yü mai*, 'imperial wheat') is produced in Tibet (*Hsi*

---

1. There is a bibliographical mistake in this paper, which deserves correction.   Bushell refers to the first volume of Ramusio, published in 1550, as containing a very accurate woodcut of an ear of maize.   The various editions of Ramusio being in the Lenox Library of New York, I may state that this woodcut is inserted in the second edition of 1554, but not in the first of 1550.   The text to this illustration runs thus (Vol. 1, p. 385) : " La Mirabile et famosa semenza detta maiz nell' Indie occidentali, della quale si nutrisce la metà del mondo, i Portughesi la chiaman miglio zaburto, del quale n'è venuto gia I Italia di colore bianco e rosso e sopra il Polesene de Borgo, e Villa Bona seminano i campi interi de ambedui i colori. "   In China the white and yellow varieties of maize are also found.

2. " At this time the Portuguese were in constant communication with India and China, and it is highly probable that maize was introduced into Asia by them, either from Europe or directly from America. "

3. Book 61, p. 10.

4. BRETSCHNEIDER, Botanicon Sinicum, Part 1, p. 150.

5. Notes on Chinese Literature, 2d ed., p. 243.

fan). Its former name is 'Tibetan wheat' (*Fan mai*). Since it was formerly brought as tribute to the court, it received for this reason the name ' imperial wheat ' (*yü mai*) [1]. The stems and leaves are of the same kind as those of the panicled millet (*Panicum miliaceum*), the blossoms like the ears of the rice-plant. Its husks are like a fist, and long. Its awns are like red velvet. Its grains are as big as the fruit of the water-plant *ch'ien* (*Euryale ferox*), and lustrous white. The blossoms open in the crown, and the fruit appears at the joints " ([12]).

The passage of WANG SHIH-MOU, in his " Hsio pu tsa shu", runs thus : " The maize (*Hsi fan mai*, 'Tibetan wheat') resembles in shape the panicled millet. Branches and leaves form extraordinarily large knots closely joined together. People boil and eat it. Its taste is very like that of the fruit of *Euryale ferox* " ([13]).

The expression used for Tibet in these two passages is *Hsi fan* [2], which designates first the Tibetan tribes inhabiting parts of Western China [3] ; second, the nomadic Tibetan tribes of the Kansuh borderland or the Kukunor region [4], and in a wider sense Tibet and Tibetans in general [5]. The names *Hsi fan mai* or *Fan mai* for " maize ", have been perpetuated in literature : they are employed in the botanical treatise, " Kuang ch'ün fang pu ", revised and republished in 1708 [6].

5. We must now go back to Mayers's assertion of a Central Asiatic origin of Chinese maize. This was based on the following sentence in the " Pên ts'ao kang mu : " — " Yü shu shu chung ch'u hsi t'u " ([14]); that is, the seeds of maize came from the western territory (*hsi t'u*). The whole question naturally pivots

---

1. The same view is expressed in the *Kuang ch'ün fang pu* (see BUSHELL, l.c.).

2. Usually translated " western barbarians ," but *fan* was doubtless intended in the beginning as a transliteration of the designation by the Tibetans of their own nation as *Bod* (pronounced B'ö').

3. G. DEVÉRIA, La frontière sino annamite (Paris, 1886), p. 167.

4. W. W. ROCKHILL, The Land of the Lamas (London, 1891), p. 72.

5. Already in the Annals of the T'ang dynasty (see ROCKHILL in Journal of the Royal Asiatic Society, 1891, p. 280 ; also BUSHELL, The Early History of Tibet, Ibid., 1880, p. 2).

6. See BUSHELL in Notes and Queries, Vol. IV, p. 87.

around this latter term. Mayers translates it by " the lands beyond the Western Frontiers (that is, Central Asia) " [1]. I am not aware on what ground he does so ; nor can I find any authority for it, nor any passage in Chinese literature where *hsi t'u* plainly refers to any countries of Central Asia, which, as well known, are called *hsi yü*. The word *t'u* (" earth, land, region ") invariably relates to the native soil of China [2]; and *pei t'u*, for example, most frequently used by Li Shih-chên and other botanical authors, never means anything else but the northern territories of China. Thus, the meaning of *hsi t'u* is " Western China ", comprising the provinces Yünnan, Ssech'uan and Kansuh, and Ssech'uan in particular is understood by this expression [3]. Compare, for example, the following passage in the " Nung chêng ch'üan shu, " where *Rubia cordifolia* (*ch'ien* ([15])), a plant indigenous in China, is in question : " Hsi t'u ch'u chê chia, chin pei t'u ch'u ch'u yu chih ([16]) " ; i. e., " That kind growing in the western regions of China is the best, now it occurs everywhere also in the northern regions. "

Li Shih-chên must accordingly be understood in the passage above as saying, " The seeds of maize are produced in (or come from) the western part of China. " We thus find him alluding to the same region as the first place of production of maize as he did in regard to sorghum. This testimony of Li Shih-chên, the greatest authority on subjects of natural history in and long after his time, is of great importance. Though he makes no direct reference to Central Asia or to Tibet, but only to Ssech'uan or western China generally, we are allowed to draw from this statement the same conclusion as in the case of sorghum, that maize also came to China from India by way of Tibet. We have thus brought together three direct literary evidences for the derivation of Chinese maize from Tibet. A fourth argument could be found in the term *Jung shu*, which means " pulse of the western barbarians *Jung* ", among whom

---

1. Also BRETSCHNEIDER, Chinese Recorder, Vol. III, p. 225, understands the passage in the sense that maize was introduced from Central Asia.

2. Sometimes adopting the meaning " native, indigenous. "

3. In modern Su hua, *hsit'u* means " opium from Shansi Province " (*hsi*, " Shansi ; " *t'u*, " opium ").

the Tibetans are also included. This expression occurs in the
" Kuang ch'ün fang pu ¹."

6. Two early European authors confirm the existence of
maize in China in the sixteenth and seventeenth centuries. The
one is the Augustinian monk I. GONZALEZ DE MENDOZA, in his
" History of the Great and Mighty Kingdom of China, " which,
written in Spanish, appeared in Rome in 1585. He informs us
in this book that the Chinese — besides wheat, barley, millet —
cultivate also the same *maize* which constitutes the principal
food of the Indians in Mexico ². As Mendoza's work is made
up from the reports of some friars of his order, — chiefly Martin
de Herrada, who had visited the port of Ch'üan chou fu, Fuhkien
Province, in 1577 ³, for three months, — we are led to assume
that in that year maize was an object of cultivation in Fuhkien,
and that it must have been brought there before that time. This
date is very important, since it affords a *terminus ad quem*, and
allows us to conclude that maize, after having traversed China
from west to east, had reached her easternmost parts by 1560,
or at least 1570.

Bretschneider mentions only this one allusion ⁴ to maize in
Mendoza ; but there is another still more interesting passage to
be found in his reports. In Chapter 3 of Book III, in which he
treats of the taxes in kind which the Emperor of China derives

---

1. BUSHELL in Notes and Queries, Vol. IV, p. 87.

2. BRETSCHNEIDER, Early European Researches into the Flora of China
(Journal of the North-China Branch of the Royal Asiatic Society, N.S., No.
XV, Shanghai, 1880, p. 4) ; and History of European Botanical Discoveries
in China, London, 1898, p. 8. Bretschneider adds the remark, " This latter
statement, made at so early a date, has a peculiar interest for us, for it is
now a well established fact that maize is not indigenous to China, but has
been introduced since the discovery of America. "

3. Not 1575, as Bretschneider twice has it. Compare RICHTHOFEN,
China, Vol. I, p. 649.

4. In the English edition, The Historie of the Great and Mightie Kingdome of China, London, 1588, republished by GEORGE STAUNTON for the
Hakluyt Society, London, 1853, it is to be found in Vol. I, p. 15 (B. I,
Ch. 3) : " On their high grounds, that are not good to be sowne, there is
great store of pine trees, which yeelde fruite very savorie : chestnuts greater,
and of better tast, then commonly you shall finde in Spaine : and yet betwixt
these trees they do sow maiz, which is the ordinarie foode of the Indians of
Mexico and Peru, and great store of panizo [panic grass], so that they doe
not leave one foote of grounde unsowen. "

from his province, he states that " of wheat called Mayz, twentie millions two hundred and fifty hanegs [1]" are obtained. This seems to represent an enormous quantity, especially if we compare with it the other items given by him : of millet, 24 millions of hanegs ; of panizo, 14,200,000 ; of wheat, 33,120,200 hanegs, etc. It is very hard to see how the Augustine monks, who had paid only a flying visit to Kuangtung and Fuhkien, should have obtained such accurate statistical material for the total number of provinces, which, even in the present day, it is very difficult to secure, and then only approximately. At all events, it cannot fail to show that a most extensive cultivation of maize was then carried on in China, — a cultivation of such dimensions as to forcibly lead to the presumption that it must have been in progress at least for the period of a generation ; that is, since about the year 1540.

There is another interesting passage in Mendoza which may be discussed in this connection, and which deserves some critical consideration. In the second volume of the Hakluyt edition (London, 1854), p. 57, we read, " In this province [Fuhkien], and all the rest of the fifteene in that kingdome, they gather much wheate, and excellent good barley, peese, borona, millo, frysoles, lantesas, chiches, and other kindes of graines and seedes, whereof is great abundance, and good cheape. " The cereal " borona " is explained in a note of Staunton as " a sort of grain resembling maize or Indian corn. " In the Dictionary of the Spanish Academy (p. 162 c), this word is derived from Celtic bron or bara (" bread "), and interpreted (1) as millet (mijo), and (2) as maize. The main question is, first of all, whether the word " borona " is used in the original Spanish edition, but, not having access to it, I cannot say. In the Italian edition of 1586 [2], which immediately followed the Spanish version, and was published two years earlier than the English,

---

1. From Spanish *fanega*, on an average equal to an English bushel and three fifths.

2. Dell' Historia della China descritta dal P. M. Gio. Gonzalez di Mendozza dell' Ord. di S. Agost. nella lingua Spagnuola. Et tradotta nell' Italiana dal Magn. M. Francesco Avanzo, cittadino originario di Venetia... Alla Santità di N. S. Papa Sisto V. In Roma. Appresso Bartolomeo Grassi 1586. P. 179. A copy of this now very rare edition is in the Library of Columbia University.

this passage runs as follows : " In tutta questa Provincia, come anco nell' altre di quel Regno, si raccoglie molto formento buono, orzo, spelta, miglio, fagiuoli, cece, lente, e altri grani, e legumi, e'l tutto in gran copia, e per poco prezzo, nondimeno la biada più famigliare à tutto il Regno, e più commune à i naturali, e à i vicini, è il riso. " It thence follows that *borona* corresponds here to the word *spelta* ; that is, spelt (*triticum spelta*). Again, in the English edition (vol. 1, p. 15) we read, " They doo sowe wheate, barlie, rye, and oates, and manie other kindes of graine ", which corresponds to the fo lowing sentence in the Italian text (p. 8) : " Ricolgono gran quantità di cotone [which is missing in the English], di formento, d'orzo, di spelta, e d'avena, e di altre diverse sorti di grani che moltiplican notabilmente nel frutto. " In this case, the Italian *spelta* corresponds to English " rye ", so that I am inclined to believe that *spelta* represents the true original reading, which seemed to embarrass the English translator, so that in one case he tried to render it by *borona*, in another by *rye*. This critical comparison is not without value, for the mention of rye should have struck Mendoza's English editor, who had been travelling in China himself. As is well known, rye is a cereal which does not exist [1] in China, and with which the Chinese became acquainted only in the Amur region through the Russians. On the other hand, the mention of *spelt* is also striking, as no definite proofs of the occurrence of this cereal in China seem to exist : at least, it is not mentioned in the writings of Bretschneider, nor is it, to my knowledge, identified with any Chinese designation, unless one of the many compositions formed with *mai* (" wheat "), of which the Chinese recognize many varieties, is identical with it. I know of only one modern author who ascribes spelt to China ; and, as he is or was a recognized authority on the agriculture of that country, I think that Mendoza's testimony deserves consideration, and is very valuable in connection with his statement. This is S. SYRSKI, who in his excellent sketches of Chinese economy [2], remarks with regard to rye, that he has

---

1. BRETSCHNEIDER in Chinese Recorder, Vol. III, p. 225 a, 286 b.

2. In Fachmännische Berichte über die Oesterreichisch-ungarische Expedition nach Siam, China, und Japan, herausgegeben von K. v. SCHERZER, Stuttgart, 1872, Anhang, p. 94.

not seen it, nor could he learn anything about it from the Chinese ; but about spelt (*Iriticum spelta*), he observes that it is regarded by the Chinese as a kind of wheat, and is, in spite of an opposite view held by several, sown in November or December, and cut in June [1].

The other allusion to maize is found in the « Relatione della Grande Monarchia della Cina », of the Portuguese Jesuit, ALVAREZ SEMEDO, who arrived in China in 1613, and wrote his account about 1633 (first published in Rome, 1643). He mentions, with regard to the province of Peking, that it produces *maize*, wheat, and some rice for the use of the Emperor's court, the mandarins, and the soldiers [2]. From this passage we may safely infer that maize was cultivated in the environment of the capital at least from the beginning of the seventeenth century. The later Jesuit authors on China do not speak of maize, and nothing can be discovered in European literature of the sixteenth and seventeenth centuries regarding an importation of maize into China by any western nation.

Bretschneider quotes only this one passage from Semedo, but it is very important to note that in fact it is mentioned there in several places, and in such a way that we may safely assume that maize was generally known throughout northern China in the beginning of the seventeenth century, and particularly in the provinces of Shensi, Shansi, and Chihli. Semedo speaks of maize in the following passages (I quote from the English edition of « The History of That Great and Renowned Monarchy of China, » London, 1655) : —

---

1. A new critical edition of Mendoza, made on the basis of the Spanish original, is an urgent desideratum. Wherever terms of cultural history come up and are to be laid stress upon, the English version leaves in the dark or in doubt, in many cases, as illustrated above, which is, however, only one of many. By the way, I wish to remark that in the same volume of Mendoza the maize of Mexico is also spoken of in the journey of the Franciscan Father Martino Egnatio, who travelled from Sevilla to Mexico, Manila, and China. On p. 390 of the Italian edition he says, " In Mexico Vi si semina, e raccoglie quasi tutto l'anno, e in ogni luoco, così il formento, di che quei terreni son fecondissimi, come il maiz, ch'è l'ordinario sostegno, non sol de gl'Indiani, e de i Negri, ma anco de i cavalli, che ci sono in gran copia, buoni, e belli, etc. " And again, on p. 311, " Mangiando tutto l'anno l'herba verde, e'l maiz, ch'è il formento de gl'Indiani. "

2. BRETSCHNEIDER, l. c., p. 6.

(1) " The Northern Provinces use for their proper sustenance Wheate, Barly, and Maiz ; eating Rice but seldome, as we doe in Europe; leaving it for the Southern Provinces; which although they have Wheat in great plenty, make use of it with the same moderation, as we doe of Rice, or any other sort of fruit. " (p. 4.)

(2) " The second [of the northern provinces] is Xemsi [Shensi] ; it lyeth in 36 degrees and more, to the West : it is very large, but dry for want of water, as also are the three neighbouring Provinces : notwithstanding it doth abound in Wheate, Barly and Maize, of Rice they have but little. " (pp. 15, 16.)

(3) " Riansi'¹ [Shansi] is the third of these six Northern Provinces... ; it hath many mountaines, which makes their Harvest but poore : there is little wheate, lesse Rice, but most Maiz. " (p. 19.)

(4) " Pekin is the fift Province... The soyle is very drie, and favourable for health, but barren of fruits for the common sustenance : But this want is supplyed by that general prerogative of courts which draw all to them, and overcome in this the proper nature of the place.   It hath Maiz, Wheate, and little Rice, only for ² the use of the people of the Palace, which is very numerous, the Mandarines and Souldiers being many thousands. "

7. The introduction of maize by an overland route from India to China in the first part of the sixteenth century is corroborated by the following indirect proofs : —

(1) There is no historical evidence to show that maize was brought to China by way of the sea, either by Spanish or Portuguese, or by Chinese mariners themselves.   It is impossible to presume that maize came to China from the Philippines ; for when Martino de Herrada made his trip from Luzon to China in 1577, he found the plant already cultivated there to a large extent, and this was at a time when the Spaniards had just com-

---

1. In the first sentence of *this* chapter 3 ; p. 15, printed *Kiansi*.
2. This clause refers only to rice; but not to the previous cereals.

menced to take a footing in the Philippines and to settle at
Manila.   It was only in 1571 that the Spaniards and Chinese
met for the first time, at Mindoro, before Legazpi, the conqueror
of the islands, undertook his expedition to Manila.   It cannot,
therefore, have been the Spaniards who brought maize to China.
On the contrary, maize must have been in China long before it
was planted in the Philippines ; and the proposition whether
the Chinese of the province of Fuhkien — who were early settlers
on Luzon [1], who kept up a lively trade with Manila in the
sixteenth and seventeenth centuries, and who cultivated maize,
according to the testimony of Herrada in 1577 — were not the
very people to take along from their home seeds of maize, and
transplant them in the Philippines, should be considered and
investigated.   Of the character and all single items of Spanish-
Chinese trade, we have the excellent accounts of De Morga and
other Spanish writers, who do not waste a word over maize in
Spanish relations with China.

(2) The same remark holds good with regard to the Portuguese, who have not been the transplanters of maize into China ;
and here I readily concur with Mayers in all arguments brought
forward by him against the Portuguese [2].   Further, nothing of
the kind is to be ascertained from Portuguese sources nor from
Chinese sources dealing with the latter ; for example, in the
" Chronicle of Macao " (" Ao mên chi lio "), where all products
of the Portuguese market are enumerated and described, but
maize is completely ignored.

(3) While, with regard to the American potato and tobacco,
the Chinese are well aware of their foreign origin, concerning
maize they are still ignorant of this fact, and look upon it as a

---

1. Compare the Annals of the Ming dynasty (*Ming shih*), Book 326,
where the following is on record under the heading of Luzon *(Lü sung)* :
" Formerly the people of Fuhkien lived there, because the place was handy.
They were traders of abundant means, ten thousand in number, who, as a
rule, took up a long residence there, and did not return home until their
sons and grandsons had grown up.   When, however, the Franks (that is,
the Spaniards), snatched away this country, the Spanish king despatched a
chief to suppress the Chinese.   He was concerned lest they might revolt,
and expelled many of them.   All those remaining had to suffer from his
encroachments and insults. "

2. Notes and Queries, Vol. 1, p. 90 a.

domestic cereal [1]. The introduction of the potato and of tobacco took place at a period posterior to that of maize ; and these are recorded in Chinese literature as having taken their origin from the Philippines, whereas nothing to this effect is stated regarding Indian-corn.

8. After this historical proof, I shall endeavor to show that the geographical distribution of maize agrees with this state of things. I am unable to say when and by whom maize was introduced into India ; and instead of repeating the hackneyed phrase, unaccompanied by any documentary evidence, that it is due to the Portuguese (why not to the Arabs? [2]), I propose, as the next necessary task, to pursue this matter in all its historical sources in its relation to India of the sixteenth century.

There is no doubt, that for a very long time maize has held an important place in the agricultural life of North India [3]. Already LASSEN [4] makes mention of it, saying that it is not yet much propagated. W. W. HUNTER [5] observes that Indian-corn is cultivated to a limited extent in all parts of India. The entire area cultivated with maize is estimated at 3, 156,842 acres [6]. In Kooloo [7] it belongs to the main crops with rice, opium, tobacco, wheat, barley, and amaranth. By some authors it is considered, next to rice, the most important crop, and is believed

---

1. Compare the tradition communicated by MAYERS (Notes and Queries, Vol. I, p. 90) as prevalent in Kueichou Province, that maize was introduced there by General Ma Yüan (14 B.C. A.D. 49), and the notes of BRETSCH-NEIDER (Chinese Recorder, Vol. III, p. 225) to the effect that the oldest men in Peking whom he asked about maize agreed in stating, that, as long as they could remember, maize had been cultivated there. Maize has passed also into the stock of proverbial sayings (see A. H. SMITH, Proverbs and Common Sayings from the Chinese, Shanghai, 1902, p. 223).

2. Which might not be impossible, in view of the fact that the Persian name of maize is *ghendum i Mekkā* ("wheat from Mecca"). This seems to prove, according to BRETSCHNEIDER (Chinese Recorder, Vol. III, p. 225), that maize, after having been brought to Europe, spread over Asia from west to east.

3. Regarding the American plants generally in India, see Statistical Atlas of India, 2d ed., Calcutta, 1895, p. 25.

4. Indische Altertumskunde (Bonn, 1847), Vol. I, p. 247.

5. The Indian Empire, 3d ed., p. 583.

6. Statistical Atlas of India, p. 26 b.

7. HARCOURT, Himalayan District of Kooloo, Lahoul, and Spiti (London, 1871), p. 164.

to have a greater range of temperature than any other of the
cereals.   In Behar its culture is carried on to a large extent, and
it forms, with the various kinds of millet, the staple article of
food for the bulk of the inhabitants [1].   As to the northwestern
provinces of India, the people to the east eat rice and pulse ;
those to the west, wheat, barley, and millets.   The farmer to
the west pays his rent out of his wheat and cotton, and grows a
patch of maize or millets as food for his family, and as fodder
for his cattle [2].

The most notable circumstance for the end we have in view
is the fact that maize is cultivated in Cashmir [3], Nepal [4], and
Sikhim [5], the three border lands of Tibet, from each of which
countries the plant could have crossed to Tibet.   In the remarkable culture of the small mountainous tribe of the Lepcha
in Sikhim, a people closely related in language to the Tibetans,
maize plays a significant rôle, and it is surprising to note what
a rich terminology they have developed with regard to its
economy.   There are four words for maize, two for the flowers
of maize [6] ; and no less than eighteen varieties, by means of
attributes, are distinguished [7].   Further, the head of maize
bears several names, according to its growth.   There is a
special expression for the young head when first appearing, for
the head when seed commences to appear, when the grain

---

1. H. Drury, The Useful Plants of India, 2d ed. (London, 1873),
p. 453.

2. W. Crooke, The North-Western Provinces of India (London, 1897),
p. 39.

3. W. R. Lawrence, The Valley of Kashmir (London, 1895), pp. 336,
337.

4. " Besides rice and wheat -- the latter particularly for purposes of
distillation — maize and murva (a sort of millet) are grown in Nepal, which
come more and more into general favor, owing to the increasing expensive-
ness of life ; maize is harvested in the rainy season. " See Sylvain Lévi,
Le Népal, Vol. I, pp. 303, 304 ; Annales du Musée Guimet, Bibliothèque
d'études (Paris, 1905), Vol. xvii.

5. H. H. Risley, The Gazetteer of Sikhim (Calcutta, 1894), p. 75.
T. D. Hooker, Himalayan Journals (Vol. I, p. 148, London, 1855), remarks
that the flowers of maize are occasionally hermaphrodite in Sikhim, where
they form a large drooping panicle, and ripen small grains. This is, however, a rare occurrence, and the specimens are highly valued by the people.

6. Mainwaring and Gruenwedel., Dictionary of the Lepcha Language,
p. 508 b.

7. Ibid., p. 19 a.

begins to get a little larger, when the grain begins to get a little firm, when the grain has acquired firmness, and for the head when ripe. Considering that all these terms are autochthonous, the ethnologist is confronted with the interesting fact that a small primitive tribe has exercised no small amount of linguistic ability concerning an imported product, — imported not quite four hundred years ago. This development of maize terminology is doubtless proof that the culture of the plant is old, relatively old of course (dating from the beginning of the sixteenth century), and that it has gained a deep-rooted influence on the minds of the people. It is therefore not improbable that it was the Lepcha who transmitted Indian-corn to the Tibetans, though it is not out of the question that it came to Tibet through different channels, from various regions of India ; for example, from Cashmir, especially as the Ladâkhi of Cashmir are acquainted with the plant, and use the same designation for it as in the written language of Tibet [1]. Maize seems to be grown chiefly in the eastern parts of Tibet [2]. It thrives also in the region of the Kukunor [3].

9. In China, the provinces of Ssech'uan and Yünnan appear as the chief centres of maize production. " Throughout North Szech'uan the maize harvest is the great event of the year. At harvest time the châlet-like cottages of the hill people present a very bright appearance, as the local custom is to tie the golden corncobs in festoons all round the house, varied here and there

---

1. *Marmospeylothok* (H. Ramsay, Western Tibet, p. 95 b), *mamoipe* (G. Sandberg, Handbook of Colloquial Tibetan, p. 152 b) = *mar me bai lo tòg*, given in the Dictionary of four languages (Yü ch'i sse t'i Ch'ing wên chien, Book 28, p. 44) as the equivalent of Chinese *yü shu*, Manju *aiha shushu* (" glass-bead sorghum "), and Mongol *erdeni shishi* (" precious sorghum "). The signification of the Tibetan term for maize is somewhat puzzling : *lo tog* means " a crop, harvest ; " *mar* means " butter ; " *mar me*, " a lamp " (lit., " butter-fire ") ; but *mar me* with the following suffix *ba* is otherwise an unknown formation, which in this combination could hardly mean " lamps. " Other Tibetan terms for maize are *abras mo spos shel* (" rice amber ") and *abras mo dkar adsom* (Jaeschke, A Tibetan-English Dictionary, p. 400 a) ; *bachi* is a colloquial word for cakes made of maize-meal (Sandberg, l. c., p. 150 a).

2. See W. W. Rockhill, Diary of a Journey through Mongolia and Tibet (Washington, 1894), p. 364.

3. W. Filchner, Das Kloster Kumbum in Tibet (Berlin, 1906), p. 24.

with bunches of red pepper" [1].   Between Ta chien lu and Wa
sze k'ou, remarks ROCKHILL [2], the soil is cultivated wherever
possible, Indian-corn being the principal crop.   It may generally be said that the ordinary food of the Szech'uanese consists
of maize, bread, bean-curd, rice, vegetables, and a little boiled
pork [3].   The French consul, P. BONS d'ANTY [4], found everywhere in Yünnan fields of maize.   " The valley of Mai-Tzu
Ping [in Yünnan] is closely cultivated, the principal crops
being buckwheat, maize, barley, and oats [5]. "

It is interesting to note that among all so called aboriginal
tribes of western and southern China, maize forms the favorite
and principal food, and is more highly appreciated by these
tribes than by the Chinese.   The Lolo, who live in the hills,
grow wheat, *maize*, oats, beans, buckwheat, rice, potatoes, and
poppies [6].   Maize is prepared by them by pounding and steam-
ing, in the same way as rice [7].

In the " Gazetteer of Upper Burma and the Shan States" [8],
there is a reference to a small group of Miao tze, who eat nothing
but Indian-corn, of which they cultivate about a hundred acres,
with twice that area of poppies.   They live in the village Taping-
hsö, in the Ko Kang circle of the northern Shan State of Hsen
Wi ; the inhabitants call themselves *Mung*, and came twenty
years ago from Shun ling, or Shun ning, east of Yung ch'ang
and south of Ta li.   Most of the immigrants have gone back
to Yünnan, five households, numbering thirty-two in 1892,
remaining in the above village.

Not far from the city of Pi chieh, in Kuei chou Province,
in the midst of well-cultivated hills, there are Miao tze, who
subsist on fruit, buckwheat, maize, and potatoes ; they do not

---

1. G. T. L. LITTON, Report of a Journey to North Szechuen (Shanghai,
1898), p. 8 a.

2. The Land of the Lamas, p. 299.

3. Ibid., p. 307 ; see also pp. 301, 304, 309, 314.

4. Excursions dans le pays Chan chinois et dans les montagnes de thé,
p. 61.   Série d'Orient, No. 3, Shanghai, 1900.

5. R. LOGAN JACK, The Back Blocks of China (London, 1904), p. 120.

6. Gazetteer of Upper Burma, Part I, Vol. I, p. 614.   E. ROCHER, La
province chinoise de Yünnan, Vol. I, p. 125 ; Vol. II, p. 11.

7. ROCHER, l. c., Vol. II, p. 12.

8. Part II, Vol. III, p. 223.

harvest rice in sufficient quantity to make it the basis of their nourishment, and they prefer to sell it to the towns people [1]. The same author, speaking in general of the Miao tze, — who keep themselves outside of the pale of the Chinese population, and build their villages on the summits of mountains, surrounded by palisades or earth-walls, — remarks that around their villages they raise maize (*blé de Turquie*), much buckwheat, and rice, when the soil is favorable to them [2].

With regard to the cultivation of maize on the Island of Hainan by the aboriginal tribe Li, B. C. HENRY [3] makes the following remarks: « They also grow a superior quality of maize, of which we saw a good supply in almost every house. The ears were large and well-developed, the corn white and of an excellent flavor. From the rafters of the house hung hundreds of fine ears with the husk partly stripped off, the whole thoroughly dried and preserved by the constant heat and smoke from the fire. The extensive cultivation of maize by these aboriginal tribes and also by the Ius [= Yü] in the Lien-chow district touches a very interesting question in botany ; namely, the source whence this grain was introduced into China. [Now follows a brief résumé of Mayers's paper.] From whatever source it came, it soon gained favor, and is now very extensively cultivated by the Chinese, but especially by the aboriginal peoples, among whom it seems to be almost as great a favorite as among the American Indians. It forms a main portion of the sustenance of both the aborigines in the north-west corner of Kwangtung and of those in Hainan. The ease with which it can be cultivated in the hilly country they inhabit and the rich return of fruit it yields for the labor expended are probably the reasons for its extensive cultivation. »

A great centre of maize-cultivation is found in Upper Burma, which appears as the natural continuance of that in northern India and Yünnan, accomplished prior to the advent

---

1. E. ROCHER, La province chinoise de Yünnan (Paris, 1879), Vol. I, p. 59.

2. Ibid., Vol. II, p. 4.

3. Glimpses of Hainan (The Chinese Recorder, Vol. XIV, Shanghai, 1883, p. 343).

of Europeans in these regions. This cultivation is carried on by Kachin, Shan, and Chinese in the same way, by the latter and Red Karen usually for distilling spirit flavored with stramonium [1]. The Wa tribes also make,. besides rice-spirit, a beverage out of fermented maize, and are particularly fond of eating the barm from which the liquor has been strained off [2]. The Burmese name for " maize " is *pyaung-bu*, derived from *pyaung* (" millet "). It is grown, together with millet, sesamum, and the numerous varieties of peas and beans, wherever there is not enough water for paddy cultivation [3].

It is astonishing to read of the high altitudes in which maize occurs. The Tingpan Yao or Yao Mien, an agricultural people, cultivate it only in the hills, and generally at an altitude not lower than four thousand feet above sea-level, growing paddy, cotton, maize, and poppies [4]. In La chang chai, a Chinese village in the Ko Kang trans-Salween circle of the northern Shan State of Hen Wi, situated at a height of fifty-four hundred feet on the ridge above the Ching Pwi stream, where the slopes above and below the village are so steep that it is nearly impossible to walk straight down them, the inhabitants cultivate large fields of opium, hill-rice, and maize [5]. Another Chinese village, Ling keo tsai, in the same region, where the same plants are grown, stands on a steep slope at a height of forty-five hundred feet [6].

According to CARL RITTER [7], maize is largely cultivated in Siam (*kaopot* of the Siamese), particularly in the mountainous districts, without being an object of exportation, as is nowhere the case in Asia, because its cheapness does not counterbalance the cost of transportation.. In one of the latest and best accounts

1. Gazetteer of Upper Burma, Part II, Vol. I, pp. 316, 467 ; Vol. II, p. 613.

2. Ibid., Part I, Vol I, p. 507 ; Part II, Vol. II, p. 57.

3. Ibid., Part I, Vol. II, p. 341.

4. Ibid., Part I, Vol. I, p. 602.

5. Ibid., Part II, Vol. II, p. 1.

6. Ibid., Part II, Vol. II, p. 51. References to maize may be found in this Gazetteer under numerous districts. Compare Part. II, Vol. III, pp. 16, 49, 50, 106, 133, 323, 346, 360, and many others.

7. Die Erdkunde von Asien (Berlin, 1834), Vol. III, p. 1093. ,

of Siam [1], however, it is said that maize and millet are grown in small plots in the plains and in fields on the higher lands. As they do not require much water, two crops can often be raised in a year ; but the amount grown is small, and is not increasing.

10. The names for maize in China are the following : *Fan mai* and *Hsi Fan mai*, " Tibetan wheat ; " *yü mai*, " imperial wheat ; " *yü mai*, " jade wheat " (in " Nung chêng ch'üan shu "); *yü shu shu*, " jade sorghum " (in " Pên ts'ao ") ; *yü kao liang*, the same ; *t'ien hua tou*, " small-pox beans " (from the similarity of the grains to pustules) [2] ; *yü mi*, " jade rice " (in the " Nung chêng ch'üan shu " and in Peking colloquial) ; *pao êrh mi* ([18]), " sheath rice " (northern colloquial) [3] ; *pao suk* ([19]), " sheath millet " (Cantonese) [4] ; "*pang tze*, lit., " club, drumstick [5] ; *chên chu mi* ([20]), " pearl rice " (in Shanghai) [6] ; *su mi* ([21]), " millet rice " (in Canton) [7] ; *pao ku* ([22]), " sheath grain " (colloquial, sometimes also literary) [8] ; *Jung shu* ([17]), " pulse of the *Jung*, (western) barbarians (in " Kuang ch'ün fang pu ").

I further find a singular expression, *yü chiao tze* ([23]) [9], used by the native population of Fan chih hsien ([24]) in Tai chou, Shansi Province [10]. The following colloquial terms for maize are quoted in the " Yin shih pu " ([25]) (" Book on Eatables and Drinkables "), p. 9 b : *pao lu*, " sheath reed ; " *yü su*, " twisted (?) millet ; " *leu ku* ([26]), " the sixth grain, " the standard number of recognized grains being five ".

---

1. The Kingdom of Siam, Ministry of Agriculture, Louisiana Purchase Exhibition, edited by A. CECIL CARTER (New York and London, 1904), p. 163.
2. GILES, Dictionary, p. 509 c.
3. According to Ch'êng téh fu chi, Book 28, p. 8. Also *pao mi* occurs.
4. L. AUBAZAC, Dictionnaire français-cantonnais, p. 192.
5. GILES, Dictionary, p. 859 c.
6. WILLIAMS in Notes and Queries, Vol. I, p. 78.
7. Ibid.
8. Tsun hua chou chi, Book II, p. 7 b.
9. *Chiao* is a kind of aquatic grass, cultivated for its sweet stalks (BRETSCHNEIDER, Botanicon Sinicum, Part III, p. 352), *yü* means " jade. "
10. In the Fan chih hsien chi, Book 3, p. 1 b.
11. See further, on terminology of maize, PARKER in China Review, Vol. XIX, 1890, p. 192 a.

—18

11. Maize found early favor in China as an article of food. Li Shih-chên remarks that the grains may be eaten baked or roasted ; in case they are roasted, they burst open into a white flower of an appearance similar to that of glutinous rice when roasted. The " Kuang ch'ün fang pu ", first published in 1630, mentions the making of flour from maize, from which cakes are produced. The native population in the district of Jehol makes a gruel of it [1]. The manufacture of wine and bread from maize is also noted in literary records [2].

Maize as an article of food is thus described in a small book, " Yin shih pu (26) " (" Book on Eatables and Drinkables "), published in 1861 : —

" The maize must be gathered as long as it is young. Husk and beard are then removed, and it is eaten roasted. Its taste is very sweet and fine. When it is old, the grains are as hard as stone. By pounding and grinding, it is made into food [3]. Also the waste material can be saved for important objects. Only sorts of poor quality are dry by nature, and when eaten afford only half satiation. As the following foods are all easy to digest, — tung ch'iang [4], finger-millet [5], huang ching [6], and jade-bamboo [7], — each kind may be mixed with the grain of maize, and the whole, prepared as food, will satisfy hunger. Also wine is made of it, as one of the utilizations of waste materials which are not completely recorded here (29). "

A singular utilization of maize is made in Pauk, a township in the Pakok-ku district in the Minbu division of Upper Burma.

---

1. Ch'êng têh fu chi, Book 28, p. 8. There the above passage from the Kuang ch'ün fang pu is also quoted.

2. San shêng pien fang pei lan (27) (ed. 1830), Book 9, p. 16 ; Yang shêng shih chien (or Shih wu pên ts'ao (28), Shanghai, 1898, Book 1, p. 5 b.

3. This, as does also practical experience in China, refutes the curious statement of WILLIAMS (The Middle Kingdom, Vol. 1, p. 772) : " Maize, buckwheat, oats, and barley are not ground, but the grain is cooked in various ways, alone or mixed with other dishes. " See also A. HOSIE, Manchuria (London, 1901), p. 179.

4. Nitraria Schoberi, Willd.

5. Or Coracan (Eleusine coracana, Gaertn.), Chinese shan tse.

6. Polygonatum macropodum, (BRETSCHNEIDER, Botanicon Sinicum, Part III, No. 7).

7. Yü chu, rhizome of Polygonatum officinale, L. (Ibid., No. 8).

People there grow it, not so much for the sake of the grains as for the innner sheath of the cob, the leaves of which are used as wrappers for Burmese cheroots. This product, when ready for the market, is called *hpet*. There is a great demand for it, and at least a million pounds are exported annually at a value of from six to seven lakhs of rupees [1].

12. The Island of Formosa seems to occupy a peculiar position in the history of maize, in the opinion of Mayers. In " Notes and Queries on China and Japan [2]", attention had been called to a passage occurring in the " T'ai wan fu chi [3] ", Chronicle of Formosa, wherein maize is designated as *fan mai* (" foreign corn ") ; and the same writer added that the Chinese, both at Formosa and in Amoy, consider maize to be a foreign introduction of modern times, and in the colloquial also call it " foreign corn. " With reference to this statement, MAYERS [4] commented that " it is no doubt very probable that maize in Formosa may have a European origin, inasmuch as the Dutch and Spaniards made extensive settlements on the Island during the seventeenth century, and may well have been the first to introduce the plant there, but this does not in any way affect the question of its introduction in the Chinese Empire. " Though it may well be possible that the Dutch, who encouraged and promoted agriculture on Formosa, planted or even re-introduced maize there, it is hardly credible that maize was not introduced there before by the Chinese from Fuhkien. The term *fan mai* need not be referred to the Dutch, but existed, as we have seen, in the sixteenth century, derived from *Hsi fan* (" Tibet "). CARL RITTER [5] (after Klaproth), designating Formosa as the " Kornkammer " of the province of Fuhkien, says that the island yields abundant crops of rice, but also corn, millet, *maize*, vegetables of all descriptions, and truffles. It is strange that C. IMBAULT-HUART, in his excellent book " L'île Formose, histoire et

---

1. Gazetteer of Upper Burma and the Shan States, Part II, Vol. II, p. 758.
2. Vol. I, p. 74.
3. First edition in 1694 (WYLIE, Notes on Chinese literature, p. 47).
4. Ibid., p. 90.
5. Die Erdkunde von Asien (Berlin, 1834), Vol. III, p. 871.

description " (Paris, 1893), does not make mention of maize (chapter " agriculture et produits, " p. 203) ; nor does JAMES W. DAVIDSON [1], who very carefully lists and describes all the manifold products and industries of the island.

13. As regards Japan, we must still acquiesce in the view of old KAEMPFER, who notices maize in his " Amœnitates exoticæ " (published 1712), p. 835, under the name of ' Kjokuso, vulgo Nan ban kiwi, id est Milium populi septentrionalis [but nan means " south "], quia a Lusitanis ex India primum invectum [2] '. The Japanese name Namban kiwi would agree with this view, by Namban the Portuguese or Dutch being understood ; while another name for maize — tô-morokoshi, the latter part written with the Chinese characters yü shu shu — would allude to an introduction from China, as tô corresponds to the Chinese character T'ang meaning " China. "   The significance of the whole term is " Chinese sorghum " (see especially REIN, The Industries of Japan, " pp. 52-55, (New York 1889).   The plant is also grown on the Luchuan Islands [3] ; and in the Luchuan name, tô-nu-ching, the first syllable is identical with that of the Japanese term [4] .

14. The results of this paper may be summed up as follows : —

(1) No share is due either to Portuguese or to Spaniards in connection with the introduction of maize into China, and probably, also into Further India and other parts of Eastern Asia. With regard to India, the question has not yet been investigated, but the introduction there through the Portuguese has some probability.

(2) Maize did not reach China from the seacoast, but came

---

1. The Island of Formosa (Shanghai, 1903).

2. E. Bretschneider, History of European Botanical Discoveries in China, p. 28.

3. B. H. CHAMBERLAIN, The Luchu Islands and their Inhabitants (The Geographical Journal, 1895, p. 301).

4. B. H. CHAMBERLAIN, Essay in Aid of a Grammar and Dictionary of the Luchuan Language.   Transactions Asiatic Society of Japan, Vol. XXIII Supplement, 1895, p. 249.

overland from Tibet, first into Ssech'uan and other parts of western China, whence it rapidly spread to the north, south, and east. The year 1540 might well be conjectured as that of the first introduction, and from 1560 to 1570 maize had reached the eastern parts of China in the province of Fuhkien.

In concluding these notes, I may be allowed to come back to the proposition advanced in the introductory remarks, that the history of maize is an instructive historical example which might be fruitfully applied to the prehistoric dissemination of ancient cereals, giving an idea, at least, of how cereals might have travelled in prehistoric days. Of all the manifold gifts of the New World, maize spread the most rapidly ; and the most interesting result of the previous investigation seems to be the fact that maize travelled with much greater speed than the ships of the European nations which then shared in the universal trade, for, long before the arrival of Europeans in China, maize was known there as an overland arrival, so that the idea of a European origin of it never struck the Chinese ; and than the other cultural plants of America, like the potato, tobacco, ground-nut, pine-apple, custard-apple [1], etc. ; and, last but not least, it is worth while adding, that maize travelled even faster than syphilis, which, after the discovery of America, so quickly spread in Europe. This latter circumstance is also remarkable as showing that maize and syphilis, which seem to have started from America at about the same time, were not each other's equal in rapidity of movement, in which maize was doubtless superior, although, *a priori*, the reverse, perhaps, might be expected. If it is allowable to draw a general conclusion or law from the preceding, I should venture to say something like the following : —

It seems that the rapidity with which cereals are disseminated vies with that of all other objects connected with human

---

1. The custard-apple (*Anona squamosa* L. or *reticulata*), a native of the West Indies, called *fan li chi* (" foreign *lichi* "), was first presented to the Emperor by the brigadier-general of Hangchou in 1699. This notice is given in the Chinese Chronicle of Macao (Ao men chi lio, Book 11, p. 34 b) Presumably this fruit had entered southern China some decades earlier, as it is mentioned by MICHAEL BOYM in his Flora Sinensis, published at Vienna in 1656 (see BRETSCHNEIDER, Early European Researches into the Flora of China, p. 23).

culture ; that a land route is preferred over a sea route as their way of propagation, and that overland propagation is effected in a shorter space of time than marine propagation ; and that cereals spread more rapidly than all other cultural plants, or even, perhaps, than infectious diseases.   Counting a generation as, on an average, thirty years, we might well say that, during the first generation after the discovery of America, maize became known and planted in Europe ; at the end of this period it must have reached India ; and during the second generation it spread over all China, so that, after about seventy or eighty years, its wanderings to the farthest East were completed.

Mayers had labored under the belief that in the sixteenth century, maize was " a grain only dimly known as a new curiosity by botanists in Europe " ; and to refute him, Bushell referred to the first volume of Ramusio, published in 1550, where maize is pictured and described.   But Ramusio cannot by any means be called an early author to speak of maize, for the first mention of maize in literature occurs as early as only one year after the discovery of America, in the famous " Decades " of PIETRO MARTIRE AB ANGHIERA ; that is, in the first book of the first decade, which was written in November, 1493, or at least then completed [1].

Mr. BANDELIER, further, calls attention to a passage in the " Historia Natural y General de las Indias, " by GONZALO FERNANDEZ DE OVIEDO Y VALDÉS [2], which was written after

---

[1]. This quotation, as well as the following ones, is due to the erudition of Mr. A. F. BANDELIER, who kindly consented to their publication in this paper.   In the English translation of the above work by RICHARD EDEN, under the title The Decades of the Newe Worlde or India (1555), the passage is as follows : " They make also an other kynde of breade of a certayne pulse, called PANICUM, muche lyke vnto wheate, whereof is great plentie in the dukedome of Mylane, Spayne, and Granatum.   But that of this countrey is longer by a spanne, somewhat sharpe towarde the ende, and as bygge as a mannes arme in the brawne.   The graynes whereof are sette in a maruelous order, and are in forme somwhat lyke a pease.   While they be soure and vnripe, they are white ; but when they are ripe they be very blacke. When they are broken they be whyter then snowe.   This kynde of grayne they call MAIZIUM. " Compare also E. O. v. LIPPMANN, Abhandlungen und Vortraege, Leipzig, 1906, pp. 335-338.

[2]. Vol. I, edition of 1851, p. 268 (Lib. VII, Cap. I) : " Como soy amigo de la leçion de Plinio, diré aqui lu que diçe del mijo de la India, y pienso yo que es lo mismo que en estas nuestras Indias llamamos mahiz, el qual auctor diçe aquestas palabras : ' De diez años acá es venido mijo de la India, de color negro de grande grano : el tallo como cañas, cresçe siete

1530, but not later than 1557, and to the " Catalogus Plantarum" of CONRAD GESSNER (Zürich, 1542), fol. 63 ¹, who refers to Ruellius as his authority.

HARSHBERGER ² mentions that maize is quoted by Hieronymus Bock in his " Neu Kreüterbuchs, " Strassburg, 1539.

These various sources prove that maize was well known in Europe in the first half of the sixteenth century ; and it is therefore not surprising, but appears as a natural consequence of history, that about and from the middle of that century it entered the horizon of Chinese authors.

For the rest, a thorough critical investigation remains to be made of the immediate introduction of maize from America into Europe, and of its gradual importation into the single European countries, into Africa, and particularly into India.

---

piés : es dicho lobas, é es fertílíssimo sobre todas las cevadas : de un grano nasçen tres sextarios : siembrase en lugares húmidos. ' Por estas señas que este auctor nos da, yo lo avria por mahiz, porque si diçe que es negro, por la mayor parte el mahiz de Tierra firme es morado escuro ó colorado, I tambien hay blanco, é mucho dello amarillo. Podria ser que Plinio no lo vido de todas estas colores, sino de lo morado escuro que pareçe negro. El tallo que diçe que es como cañas, assi lo tiene el mahiz, y quien no lo conosçiesse I lo viesse en el campo, quando está alto, penssará que es un cañaveral. "

1. " Milium Indicum in Italiam Neronis principatu conuectum, nigrum colore, Amplum grano, arundineum culmo. Adolescit ad pedes altitudine septem, pregrandibus culmis, lobas uocant, omnium frugum fertilissimum, ex uno grano terni sextarij gignuntur. Seri debet in humidis. " This, I think, is derived from Pliny. On fol. 64 recto appears " Ich meyn es sye dz man Türckisch, Heydisch, oder Indisch korn nennet : etliche kórner werdent braunschwartz, die andern gál."--Beside this : " Hodie Galli in hortis ostentationis gratia serunt, grano pisum aequante, atro, stipula arundinea, quinum apud senumûe pedum proceritate, qua milium saracenicum quasi peregrinum nominant, nec ante quindecim annos hunc aduectum. "

2. L.c., p. 156.

CHINESE CHARACTERS

(1) 蜀秫　(2) 賈思勰思齊民要術　(3) 春月種宜用下土。莖高丈餘。穗大如帚。其粒黑如漆。如蛤眼。熟時收刈成束。攢而立之。其子作米可食。餘及牛馬。又可濟荒。其莖可作洗帚。稭稈可以織箔編蓆夾籬。供爨。無有棄者。亦濟世之一穀。農家不可闕也。

(4) 元扈先生曰。蜀秫古無有也。後世或從他方得種。其黏者近秫。故借名為秫。今人但指此為秫。而不知有粱秫之秫誤矣。別有一種玉米或稱玉麥或稱玉蜀秫蓋亦從他方得種。其曰米麥蜀秫皆借名之也。

(5) 又曰。北方地不宜麥禾者乃種此尤宜下地。立秋後五日。雖水潦至一丈深。而不

能土壞之．但立秋前水至
即壞．故北土築堤二三
尺以禦暴水．但求隄防
數日即客水大至亦無
害也．(6)又曰．秦中鹵地
則種蜀秫．下地種蜀秫
特宜早．須清明前後耩．
(7)汪穎曰．蜀黍北地種之
以備缺糧．餘及牛馬．穀之
最長者．南人呼爲蘆穄．
(8)李時珍曰．蜀黍宜下地．
春月布種．秋月收之．莖
高丈許．狀似蘆荻而內
實葉亦似蘆．穗大如帚．
粒大如椒紅黑色米性
堅．實黃赤色．有二種．黏
者可和糯秫釀酒作餌．
不黏者可以作糕煮粥．
可以濟荒．可以養畜．梢
可作箒．莖可織箔蓆編
籬．供爨最有利於民者．今
人祭祀用以代稷者誤矣．

其穀殼浸水色紅.可以紅
酒.博物志云.地種蜀黍
年久多蛇.　(9)李時珍曰.
蜀黍不甚經見.而今北方
最多.按廣雅.荻粱木稷
也.蓋此亦黍稷之類.而高
大如蘆荻者.故俗有諸
名.種始自蜀.故謂之蜀黍.
(10)田藝衡留青日札.(11)王世
懋學圃雜疏(12)御麥出
於西番.舊名番麥.以其
曾經進御故名御麥.幹其
葉類稷.花类稻穗.其紅瑩
苞如拳而長.其鬚如而節.
絨.其粒如茨實大而枝葉.
白.花開於頂.實結於節.
(13)西番麥形似稷.而枝葉.
奇大結子纍纍煮食之.味
亞茨實.(14)玉蜀黍種出
西土.(15)苕(16)西土出者
佳.今北土處處有之.(17)戎
菽(18)包兒米(19)包粟(20)珍珠米

(21)粟米 (22)包穀 (23)玉荿子
(24)繁峙縣 (25)飲食譜 (26)苞
蘆。秈粟。六穀。(27)三省
邊防備覽 (28)養生食鑑。
食物本草 (29)嫩時采得。
去苞鬚煮食。味甚甜美。
老則粒堅如石。舂磨爲
糧。亦爲救荒要物。但
粗糲性燥。食宜牛飽。
庶易消化。至東廧稷子。
各種雜糧。及黃精玉竹
之類。並可充飢作食。造
酒濟荒。茲不備載。

052

落花生传入中国考

NOTE ON THE

Introduction of the Ground-Nut
into China

BY BERTHOLD LAUFER

The ground-nut or peanut (*Arachis hypogæa*) is generally believed to be a native of Brazil, where several species of it are indigenous, and to have spread into the Old World after the discovery of South America.   Historical dates, stepping-stones in the migration of this plant, remain to be determined.

As regards the date of the introduction of the ground-nut into China, BRETSCHNEIDER [1] gave his opinion in the following words : —

" I think, this plant has been introduced into China in the last [eighteenth] century, for the Pên-ts'ao does not mention it. It is first described and represented in the Chi wu ming shih t'u k'ao [published in 1848], under the names *lo hua shêng* [2] and *fan tou* (" foreign bean "). In the descriptive part of the work it is stated that the *lo hua shêng* is not an indigenous plant, but came by way of sea from southern countries.   There it is said that at the time of the Sung (960-1280) or the Yüan (1260-1368) *lo hua shêng*, with cotton and a number of other plants, were first brought from the sea countries to Canton. "

In a footnote to this quotation he remarks, " The author may be right, that all the above-mentioned plants were introduced into China, but he errs regarding the time of their introduction. "

1. Chinese Recorder, Vol. III, p. 224 b.
2. This name means " born from blossoms fallen to the ground, " which is an exact description of the peculiar way of growth of this plant.   Another name is *ch'ang shêng kuo*, " fruit of long life. "

There are, however, much earlier references to the ground-nut in Chinese literature than in the botanical work of 1848 indicated by Bretschneider. A goodly number of notes regarding this plant, are to be found in a book, " Pên ts'ao kang mu shih i " ("Omissions in the Pên ts'ao [materia medica] of Li Shihchên "), by CHAO NU-HSIEN, of Hangchow, published in 1765, Book 7, pp. 72b-76b. This valuable work, not made use of by Bretschneider, is full of interesting notes, especially regarding the newly introduced plants of the sixteenth and seventeenth centuries. The authors of the various quotations in regard to the peanut, all agree in the statements that it did not exist in China in times of old, that it came from abroad, and that it was recently (most of the passages come down from the seventeenth century) introduced into China. The most important passage is extracted from the " Chronicle of Hsien chü " (meaning " Fairies' Abode "), a city in the prefecture of T 'ai chou, Chehkiang Province ; and it is the edition of this book issued in the Wan li period (1573-1619), which is here quoted (" Wan li Hsien chü hsien chi ") as saying that peanuts came originally from, or were grown in, Fuhkien, and that seeds of these were but recently obtained and planted in the district under consideration. If, accordingly, in the period 1573-1619, ground-nuts were brought from the province of Fuhkien into the adjoining northern province of Chehkiang, we may be right in assuming that before the year 1573 peanuts had been planted in Fuhkien ; and this is for the time being the earliest date which I can make out for the existence of this fruit as regards China.

In a book, " Hua ching " (" Mirror of Flowers "), published in 1688, it is stated that peanuts are grown in great quantity in the southern part of Chehkiang, which shows that the cultivation of the plant must have made progress in that part of China during the seventeenth century.

The fact that the ground-nut must have arrived in China at least during the end of the sixteenth century, as shown above, is further corroborated by the full description of it in the " Wu li hsiao shih ", a work compiled early in the seventeenth century, and printed in 1664. It is called here also by the name *fan tou* (" foreign bean "). In another book, entitled " Hui shu ", the resemblance of the stalks, leaves, and blossoms of the plant to

the kinds of beans, is emphasized. This latter book, also makes
the origin of the nut in Fuhkien Province [1].

The following is contained in the same work, " Pên ts'ao
kang mu shih i " :

" In the Chronicle of Fu ch'ing [1] (Fu ch'ing hsien chi) it is
stated : It comes from abroad (foreign countries) ; in former
years it did not exist [in China]. Spreading over the ground,
it grows in orchards ; when the blossoms have faded, threads
come out of the middle of them and droop into the ground,
where the fruits are formed : thence its name. One plant can
produce two or three seeds. They are roasted and eaten, being
of very fragrant and fine taste. In the first year of the period
K'ang hsi (1662), a Buddhist monk, Ying Yüan, travelled to
Fu-sang [2] to search for seeds, which he sent to China, and on
his return, one could still press oil out of them. Now it is grown
in the province of Min (Fuhkien), and that produced in Hsing
hua is of the first quality. It is called *huang t'u* ('yellow
earth '), is of sweet taste and full of nuts. The sort produced
on T'ai wan (Formosa) [3] is called *pai t'u* (' white earth '), is acrid
to the taste, and the nuts are thin. They are simmered in oil,
and eaten, if not well done, serviceable as a purgative ".

The records of the Chinese allow of the inference of the

---

1. Fuhkien and Chehkiang are still the most flourishing districts for the
culture of ground-nuts, which are, according to R. FORTUNE, a staple summer production in the hilly country of Chehkiang. They are also grown
most extensively during the summer season in the southern provinces, more
particularly in Fuhkien. The very sandy soil near the river (between Yen
chow fu in Chehkiang and the border of Anhui) yields good crops of the
ground-nut. They are also plentiful in the light sandy soil near Kü chou
fu, Chehkiang (see BRETSCHNEIDER, History of European Botanical Discoveries in China, p. 451). C. ABEL (1816-17) first met with *Arachis hypogœa*
cultivated in fields on the banks of the Yangtse, and continued to observe
it through the whole of the provinces of Kiangsi and Kuangtung. The
seeds are roasted before they are eaten. The Arachis in China, according
to him, bears from two to four seeds in each capsule (Ibid, p. 233). I myself
found peanuts grown plentifully in the province of Shensi, in whose capital,
Hsi an fu, the itinerant venders of the roasted nut form one of the indispensable street-pictures.

1. In Fu chou fu, Fuhkien Province.

2. A fabulous country in the sea, east from China ; by Klaproth and
Schlegel identified with Saghalin, which is quite unfounded. In this passage
it seems to stand for Formosa.

3. On peanut cultivation in Formosa see JAMES W. DAVIDSON, The
Island of Formosa, p. 547.

first cultivation of the peanut in Fuhkien, particularly in the seaports Hsing hua and Fu ch'ing. Besides, they assert that the plant came from abroad. It cannot be otherwise, therefore, as the trade of Fuhkien was largely directed towards the south and east, than that it arrived either from the Malayan Archipelago or the Philippines. No Chinese source makes a definite statement regarding the name of the country. This seems to prove that the introduction is not due to Spaniards, Portuguese, or Dutch, but that it was effected by Chinese sailors or traders of Fuhkien.

The first European botanists to mention *Arachis hypogœa* as a plant cultivated in China are LINNÉ [1], his specimen being brought from the Canton region by his pupil Peter Osbeck, who gives as his Chinese name *fy shin* (Cantonese : *fa shang* ; *fa = hua* 'flower'), and LOUREIRO in his " Flora Cochinchinensis " (1789) [2], possibly also PALLAS in his " Reisen durch verschiedene Provinzen des russischen Reiches " (1768-73) [3].

---

1. BRETSCHNEIDER, Early European Researches into the Flora of China (Journal North-China Branch Royal Asiatic Society, N.S., Shanghai, 1880, No. xv, p. 96.
2. Ibid., p. 145.
3. Ibid., p. 128.

053

论维吾尔语的佛教文学

T'OUNG PAO

通報

OU

ARCHIVES

CONCERNANT L'HISTOIRE, LES LANGUES,
LA GÉOGRAPHIE ET L'ETHNOGRAPHIE
DE
L'ASIE ORIENTALE

Revue dirigée par

Henri CORDIER
Professeur à l'Ecole spéciale des Langues orientales vivantes

ET

Edouard CHAVANNES
Membre de l'Institut, Professeur au Collège de France.

SÉRIE II. VOL. VIII.

LIBRAIRIE ET IMPRIMERIE
CI-DEVANT
E. J. BRILL
LEIDE — 1907.

ZUR BUDDHISTISCHEN LITTERATUR DER UIGUREN

VON

BERTHOLD LAUFER.

———→✳←———

„Die Verfolgung des Buddhismus durch die Mohamedaner erstreckte sich von den Ufern des Oxus bis Lob-nor, den westlichen Grenzen des chinesischen Reiches und dauerte eine lange Reihe von Jahrhunderten. Noch unter den ersten Nachkommen von Tchingis-Khan und Kublai sehen wir den Buddhismus unerschüttert in Kashgar und noch weiter östlich, und er erlischt erst nach dem Fall der Yüan-Dynastie. Allein die Verfolgung der Mohamedaner war verheerender, als die durch die Tırthika's hervorgebrachte Umwälzung. Jene vertilgten alles, was den Typus des Heidentums trug: Überlieferungen sowohl als Schriften, und so auch die reiche Literatur, welche man bei den uigurischen Buddhisten voraussetzen darf, da Gelehrte aus deren Kreise unter der Dynastie Yüan nach Peking gerufen wurden, um an dem Gelehrten-Ausschuss Teil zu nehmen, welcher mit der Vergleichung der tibetischen und chinesischen Bücher des Buddhismus beauftragt war" [1].

Bei der auf Veranlassung des Kaisers Kubilai veranstalteten Revision und Übersetzung der buddhistischen Schriften sollen unter den Gelehrten, die schliesslich den Druck besorgten, ausser Mönchen,

---

[1] W. Wassiljew, Der Buddhismus, S. 79. Über die Zerstörung buddhistischer Tempel in Indien durch die Mohammedaner gibt eine anschauliche Schilderung Tāranātha in seinem Werke bKa-babs-bdun-ldan (ed. by Sarat Chandra Das, Darjeeling, 1895), p. 70.

die Sanskrit, Tibetisch, Mongolisch und Chinesisch beherrschten, auch solche gewesen sein, die das Uigurische verstanden [1]). Die Tatsache, dass eine uigurische buddhistische Litteratur bestanden hat, wurde durch die russische Turfan-Expedition von 1898 geliefert, die in den Höhlentempeln Turfans Schriftstücke in türkischer Sprache entdeckte, die sich nach der Untersuchung von W. RADLOFF als Fragmente buddhistischer Bücher herausstellten [2]).

Da nun bei der in Turkistan energisch fortgesetzten Forscher-arbeit alle Aussicht vorhanden zu sein scheint, dass sich uns eines Tages der buddhistische Kanon, oder wenigstens Teile desselben, in uigurischer Sprache erschliessen wird, so möchte ich auf die aus dem colophon eines tibetischen Sūtras erschlossene Tatsache hinweisen, dass chinesische Sūtras ins Uigurische übersetzt worden und uigurische Sūtras in Peking gedruckt worden sind. Da das in Rede stehende uigurische Sūtra im Jahre 1330 in einer Auflage von tausend Exemplaren hergestellt worden ist, so scheint es durchaus im Bereich der Möglichkeit zu liegen, dass das eine oder andere Exemplar erhalten geblieben ist und dereinst ans Tageslicht kommen wird, wobei denn zur Festsetzung des Textes und der Übersetzung die tibetische Version von grosser Bedeutung sein wird.

Das tibetische Sūtra führt den Titel *sme-bdun žes-pa skar-mai mdo* „Sūtra von den ‚Grosser Bär‘ genannten Gestirnen". Der Titel ist ferner auf chinesisch und mongolisch in tibetischer Transcription angeführt. Aus der Umschrift *bī-du tsʻid zin gin* lässt sich der chinesi-sche Titel 北斗七星經 [3]) reconstruiren. Der mongolische Titel ist so umschrieben: *do-lo-an ₂e-bu-gan ne-re-tʻu ho-don-nu su-dur*,

---

1) A. GRÜNWEDEL, Mythologie des Buddhismus in Tibet und der Mongolei, S. 66.

2) Nachrichten über die von der Kaiserlichen Akademie der Wissenschaften zu St. Petersburg im Jahre 1898 ausgerüsteten Expedition nach Turfan, Heft I, St. Pet. 1899, S. 68 u. flgde.

3) Nicht in BUNYIU NANJIO's Catalogue verzeichnet.

woraus sich das mongolische *Dologhan äbüghän* [1]) *närätü odon-u sudur* ergibt.

Die tibetische Ausgabe dieses kurzen Sūtras gehört nicht zu den Seltenheiten. Ein handschriftliches Exemplar befindet sich in der Universitätsbibliothek von Cambridge, und ein gedrucktes in der von Oxford in England, und vor kurzem hatte ich Gelegenheit, einen guten Holzdruck desselben in einer umfangreichen Sammlung von Sūtras zu finden, die im Besitz des Herrn WILBERFORCE EAMES, Direktor der Lenox Library von New York, ist [2]). Da die tibetische Übersetzung erst im Jahre 1336 verfertigt worden ist, so ist es selbstverständlich, dass sie nicht in die damals abgeschlossen vorliegenden kanonischen Sammlungen des Kanjur und Tanjur aufgenommen ist; das Fehlen der Schrift im chinesischen Tripiṭaka scheint darauf hinzudeuten, dass sie unkanonisch ist.

Das Colophon, dessen Text ich anhangsweise mitteile, besteht aus zwei Teilen, die ich als A und B bezeichne, einem Teil von 36 neunsilbigen Versen und einer darauffolgenden Partie in Prosa. Wie sich ergeben wird, scheinen die beiden Teile zwei verschiedenen Redaktionen derselben Schrift anzugehören oder auf zwei verschiedenen Traditionen zu beruhen. Was die folgende Übersetzung betrifft, so möchte ich bemerken, dass die Lesung von Colophons zu den mühsamsten Aufgaben der tibetischen Philologie gehört, und dass ich mir nicht schmeichle, alle Schwierigkeiten des Textes gelöst zu haben.

Colophon A lautet:

---

1) Wörtlich „die sieben Grossväter oder Alten'. Der gewöhnliche Ausdruck für das Siebengestirn ist *dologhan odon*. Die Bezeichnung *dologhan äbüghän* findet sich nicht bei KOWALEWSKI, wohl aber in dem vortrefflichen mongolischen Wörterbuch von GOLSTUNSKI, Vol. III, p. 145 b.

2) Aus einer Vergleichung der drei Exemplare hat sich mir ergeben, dass der New Yorker Holzdruck genau mit dem von Oxford übereinstimmt, während das Manuscript von Cambridge einige unbedeutende, meist auf Schreibfehlern beruhende Abweichungen zeigt.

Wer das von dem Lehrer, dem vollendeten Buddha, vorgetragene „Sutra von den ‚Grosser Bär' genannten Gestirnen" standhaften Sinnes im Gedächtnis behält und ehrt, dem erwächst Heil [1]): in dieser Erkenntnis hat der Zu-gur-c°e [2]), mit Namen U-rug-bo-ga [3]), der von Kindheit an beständig an diese Schrift glaubte, sie immerwährend gelesen und geehrt. Kraft seines Strebens nach einer Würde und kraft seiner Gebete hat der friedlich regirende Herr, der verdienstreiche, der eine Verwandlung des die Erlösung bewirkenden Buddha ist, der Prinz TEMUR [4]), den Wunsch gehegt, ein dereinst die Gefilde des langen Lebens erreichender [5]) grosser Fürst zu werden.

---

1) Dies wird im Verlaufe des Sutra selbst des näheren ausgeführt.

2) Dieser und der folgende Name Urugboga, die sich nach dem Zusammenhang nur auf den weiter genannten Kaiser Tob Temur beziehen können, sind aus den mir zur Verfügung stehenden Quellen nicht festzustellen, auch nicht im *Yüan-shih* enthalten. Ganz rätselhaft ist der tibetische Name *T'iu-kvan-t'iñ-mur*, den SARAT CHANDRA DAS in seinem tibetischen Wörterbuch (p. 577) demselben Herrscher zuschreibt. Über *Zu-gur-c'e* wage ich eine Vermutung zu äussern. Das in unseren Wörterbüchern in dieser Bedeutung nicht belegte Wort *gur* habe ich wiederholt in Verbindung mit den Namen der chinesischen Dynastieen gelesen, wo es also den Sinn von „Dynastie" hat. In einer tibetischen Lebensbeschreibung des Kriegsgottes Kuan-ti wird letzterer unter dem Namen Yuu (羽 in 關羽) als „Minister des Kaisers Hsien-ti (126—220) der Han-Dynastie in dem grossen Lande Mahācīna" (*Ma-hā-tsi-nai yul-gru c'en-por Han-gur-gyi rgyal-po Ṣyan-dhī žes-bya-ba-žig-gi blon-po*) eingeführt, wo also *Han-gur* in der Bedeutung „Han-Dynastie" über allem Zweifel steht. Ebenso kommt im Colophon einer dhāraṇī im Kanjur (Kanjur-Index, ed. SCHMIDT, no. 502, p. 76) die Verbindung *T'añ-gur-gyi dus* vor, d. i. zur Zeit der T'ang-Dynastie, und ebenda *c'en-po C'iñ-gur*, d. i. die grosse Ts'ing-Dynastie. So werde ich zu der Vermutung geführt, dass das im obigen Texte gebrauchte *Zu-gur-c'e* „die grosse Yüan-Dynastie" bedeutet (resp. „der aus der grossen Yüan-Dynastie"). *Zu* ist freilich in diesem Sinne nicht belegt, wäre aber entweder als Schreibfehler aus dem üblichen *Yvan* (= Yüan) entstanden oder als noch unbekannter tibetischer Name für die Dynastie denkbar.

3) Der Name klingt mongolisch, ist aber weder aus chinesischen noch mongolischen Quellen als Name des Kaisers Tob Temur zu eruiren. Mong. *uruk* bedeutet „Familie, Verwandte auf Seiten der Frau, Stamm".

4) Wie aus dem im Folgenden genannten chinesischen Datum der Periode *T'ien-li* hervorgeht, handelt es sich wohl um den Kaiser Tob Temur (1330—1332); in der mongolischen Geschichte wird er Jiyaghatu, in der tibetischen auch Goyugan genannt.

5) Ich verhehle mir durchaus nicht die Schwierigkeiten dieser Stelle, deren obige Übersetzung ich nur mit allem Vorbehalt gebe. Das tibetische *ts'e-riñ-žiñ* würde bei dieser Auffassung dem chinesischen 壽域 entsprechen, worüber CHAVANNES, Dix inscriptions

Nach der Ankunft des bLo-ldan Byaṅ-cʿub-sems-dpa bdag-po [1]) bestieg er den Tron des Kaisers Se-cʿen (d. i. Kublai) und sagte: „Meines Herzens Wünsche sind nun befriedigt; ohne dass ich Zweifel habe, ist mir sicheres Wissen aus dieses Schrift (nämlich dem obigen Sūtra) entstanden. Dieses in uigurischen Lettern vorhandene buddhistische Sūtra soll, da es bisher von keinem anderen übersetzt worden ist, damit die zahlreichen Mongolen ihm gläubig ihre Verehrung bezeigen, in die Sprache meiner Mongolen übersetzt werden. Alle Wünsche der zehntausend mal tausend Wesen sollen, ebenso wie es bei mir der Fall war, dadurch befriedigt werden". Mit diesen Worten liess er tausend Exemplare davon drucken und unter alle verteilen. Kraft der Wirkung der Gnade dieses Verdienstes verlängerten sich der Kaiser und die Kaiserin das Leben samt ihrer Nachkommenschaft und vermehrten ihre Verdienste. Möchten sie die Würde des das Ende erreichenden Buddha erlangen! Mögen sich im Reiche Feinde und Aufruhr beruhigen, und möge es sich wohl befinden, von bösen Geistern und Unfällen verschont! Mögen Regen und Wind zur rechten Zeit kommen und keine Hungersnot sein! Mögen meine Worte und mein Herzenswunsch erfüllt werden! Möge ich mit meinen Eltern, Söhnen und übrigen Verwandten, samt meinen noch lebenden geistlichen Brüdern und vielen Wesen schon auf dieser Welt durch die Religion Befriedigung der Wünsche und das dauernde Erreichen der Gefilde der Seligen [2]) erlangen.

---

chinoises de l'Asie centrale (Paris, 1902, p. 84) zu vergleichen ist. Das in Rede stehende Sūtra verspricht demjenigen, der seine Lehren in sich aufnimmt, Verlängerung des Lebens, und darin ist sicher der Hauptgrund für seine Übersetzung in verschiedene Sprachen, seine weite Verbreitung und Popularität und besonders seine Beliebtheit beim Kaiser zu suchen. Hier klingt ein Nachhall jener alten alchemistischen Ideen durch, die seit den Tagen des Tsʿin Shih-huang-ti die chinesischen Kaiser auf die Jagd nach dem Elixir der Unsterblichkeit getrieben haben. Diese Ideen im Zusammenhang mit dem Buddhismus hat CHAVANNES in der Einleitung seines Buches Voyages des pèlerins bouddhistes (Paris, 1894, pp. xv et seq.) trefflich geschildert.

1) Vermutlich ein Beiname des γYuṅ-ston rDo-rje-dpal (1284—1376), des Hierarchen von bKra-šis-lhun-po, der den Kaiser im Jahre 1330 besuchte (s. Journal Asiatic Soc. Bengal, Part I, No. I, 1882, p. 21).

2) Tib. bde ba can žiṅ = Sanskrit sukhāvatī.

Colophon B lautet:

Am ersten Tage des zehnten Monats, eines Drachenmonats, des ersten Jahres der Periode T'ien-li [1]) (1330) ist das Buch gedruckt worden. Dieses Sūtra ist aus Indien von einem indischen Paṇḍita und Hsüan Tsang [2]) mitgebracht und in China übersetzt worden. Während es sich so in China mehr und mehr verbreitete, wurde es auf Aufforderung des Ministers des grossen Kaisers, des im Geschlecht der Bodhisattva geborenen, von Weisheit und Meditation erfüllten Gim-rtse-goṅ-lu tai-hui γyui-ši T'ai ,u-rug-po [3]) von dem Meister

[1]) Tib. t'en-li (= 天曆). Die Handschrift von Cambridge hat t'in-li. Das Datum kann sich naturgemäss nur auf die mongolische und uigurische Version beziehen, da die tibetische erst 1336 zustande kam.

[2]) Tib. T'aṅ zam ts'aṅ, d. i. Transcription des chinesischen 唐三藏 T'ang San Tsang, San Tsang (Beiname des Hsüan Tsang, 602—664) der T'ang Dynastie. Die Richtigkeit dieser Identifikation wird mir aus dem Colophon einer tibetischen aus dem Chinesischen übersetzten Schrift br.Jed t'oq yaṅ-ṭii za-ma-tog in der Library of Congress in Washington bestätigt. W. W. ROCKHILL, der dieselbe erworben hat, beschreibt sie in seiner Abhandlung "Tibet" (Journal Royal Asiatic Society, 1891, p. 235) als "a book on divination containing most of the Chinese methods, and which is probably a translation of some Chinese work". Das Colophon lautet: „Dies sind die von dem grossen Meister T'ang Zan-tsaṅ (Hsüan-Tsang) gesammelten Erklärungen über die Wirkungen der Gestirne. Am 15. Tage des ersten Monats des ersten Jahres der Periode Chên-kuan (tib. ṭin-kvan, = 627) der T'ang-Dynastie stellte der Kaiser T'ai-Tsung der T'ang an die grossen Minister die Frage, woher es komme, dass Werke der Tugend übenden Männern, obwohl sie drei oder vier Tage lang ohne Unterbrechung die Götter ehrten und Gaben spendeten, kein Vorteil noch Segen erwüchse, worauf der grosse Meister T'ang Zan-tsaṅ bei eben dieser Gelegenheit seine Fragen beantwortete". In diesem Texte sind nun die chinesischen Schriftzeichen unter der tibetischen Zeile glossirt, und zwar T'ang Zan-ts'aṅ zweimal 唐三藏, T'ang T'ai-Tsun (sic!) 唐太宗 und T'ang Ṭin-kvan 唐貞觀. Unter dem Namen T'ang Sêng wird Hsüan-Tsang in der tibetischen Geographie des Mincbul Chutuktu erwähnt, der eine Geschichte aus dem Hsi yü ki citirt (s. WASSILJEW's russische Übersetzung, St. Petersburg, 1895, p. 66). Die Existenz einer tibetischen Übersetzung des Hsi-yü-ki ist sehr wahrscheinlich. Schliesslich möchte ich darauf hinweisen, dass der bei SCHIEFNER (Eine tibetische Lebensbeschreibung Çâkjamuni's, S. 80 und 101) erwähnte chinesische Gelehrte ,De-snodgsum-pa (= S. tripiṭaka = chin. san tsang) niemand anders als eben unser Hsüan-Tsang ist.

[3]) Dieser Name ist bis auf den letzten Bestandteil ,u-rug-po, der mongolisch zu sein scheint (vergl. oben U-rug-bo-ga) aus chinesischen Schriftzeichen zusammengesetzt, deren Identifikation mit grossen Schwierigkeiten verknüpft ist. Ich vermute, dass lu chinesisch 路 wiedergeben soll; in Gim-rtse-goṅ (letzteres vielleicht = 江) müsste demnach ein

der Lehre der Uiguren, PRAJÑĀÇRI, iu mongolische Sprache und Schrift übersetzt und in zwei tausend Exemplaren gedruckt. Die von ALIN-TEMUR tai-se-du [1]) verfertigte uigurische Übersetzung wurde in tausend Exemplaren gedruckt. Die Schrift wurde verschenkt und unter Mongolen und Uiguren verbreitet. Selbst diejenigen Staatsbeamte (tib. *tai-hu* = 大夫), die früher der Religion der Mongolen angehangen hatten, traten durch den Segen dieser Schrift zur Religion des Buddha über, empfingen die Weihen und genossen deren Vortrefflichkeit. Später im Feuer-Rind Jahre (d. i. 1336) haben der Übersetzer MAHĀPHALA und ÇRI-ĀNANDAVAJRA im Kloster Guṅ-t'aṅ die Schrift ins Tibetische übersetzt und redigirt.

Der Gegensatz zwischen den beiden in Versen und Prosa geschriebenen Colophons ist augenfällig. Der Hauptwiderspruch, der sie unvereinbar macht, liegt in der Angabe, dass nach A tausend, nach B zwei tausend Exemplare der mongolischen Version gedruckt worden sind, und dass nach A die Veranlassung zu derselben in einem Befehl des Kaisers Top Temur, nach B in dem Wunsche eines Ministers angegeben wird, wenn man nicht zu dem Compromiss seine Zuflucht nehmen will, dass der Minister lediglich als

---

geographischer Name stecken (aber welcher?). *Tai-hu*, das weiter unten wieder begegnet, ist offenbar Umschreibung von 大夫. Der Laut f, der dem Tibetischen fehlt, wird bei Umschreibung chinesischer Wörter in jüngeren Texten durch *ap‘* oder *p‘h* (mit untergeschriebenem *h*) wiedergegeben, in älteren Texten dagegen durch *h*. In einer tibetischen Schrift aus dem 13. Jahrhundert finde ich den Namen der Stadt Ch‘ĕng-tu fu durch *Šin-tu hu* transcribirt. In *γyui-ši* glaube ich 御史 und in *T‘ai* den Familiennamen (vermutlich 臺) erkennen zu sollen.

1) Offenbar identisch mit dem Uiguren 阿隣帖木兒, dessen Biographie im *Yüan-shih*, Cap. 124, p. 5, kurz skizzirt wird, und dessen auch in der Geschichte des Kaisers Top-Temur (*Yüan-shih*, Cap. 32—36) häufig Erwähnung geschieht. Er war Präsident des Finanzministeriums, 大司徒, womit die tibetische Transcription tai-se-du übereinstimmt. Die Worte in seiner Biographie 善行翻譯諸經 dürfen als willkommene Bestätigung der ihm von dem tibetischen Texte zugeschriebenen Übersetzung gelten.

26

Exekutive des kaiserlichen Willens gehaudelt habe. Ein weiterer Widerspruch ist darin gegeben, dass es nach A den Anschein hat, als wenn die uigurische Übersetzung bereits vor der Zeit des Kaisers Top Temur existirt habe, während iu B die Drucklegung derselben nach dem Druck der mongolischen berichtet wird. Ich glaube, dass wir iu der nüchternen Prosa von B die Wirklichkeit und iu der Poesie von A Anklänge frommer Legendenbildung zu erblicken haben, die ja bei der grossen Frömmigkeit des Kaisers nicht wunderzunehmen ist uud immerhin sich um den Keim eines Faktums gruppiren mag. B enthält jedenfalls die historischen Tatsachen, und die augenblicklich für uns wichtigste, dass im Jahre 1330 eine uigurische Übersetzung des Sūtras vom Gestirn des Grossen Bären in einer Auflage vou tausend Exemplaren in Peking gedruckt worden ist.

Die von W. Radloff [1]) übersetzte Stelle aus dem Fragment eines uigurischen Sūtras kommt übrigens in unserem Sūtra nicht vor.

Iuhaltlich bietet das Sūtra wenig Interessantes, aber es gewinnt ein gewisses kulturhistorisches Interesse dadurch, dass darin der türkische Cyklus der zwölf Tiere aufgezählt wird, und zwar in der Weise, dass die in einem bestimmten Jahre geborenen Menschen zu eiuem der sieben Sterne des Siebengestirns in Beziehung gesetzt und unter den Schutz des betreffenden Sternes gestellt werden, dem zu diesem Zwecke eine bestimmte Getreideart geopfert werdeu muss. Die zwölf Tiere sind demnach auf die sieben Sterne künstlich verteilt, und zwar in folgender Weise:

---

1) L. c., S. 78.

Name der Sterne [1].	Name der Tiere.	Nummer des Tieres im Cyklus.
tam-lań	Ratte	1
kun-min	Rind	2
	Schwein	12
lu-sun	Tiger	3
	Hund	11
un-k'u	Hase	4
	Hahn	10
lim-c'im	Drache	5
	Affe	9
wu-gu (hu-gu)	Schaf	8
	Schlange	6
bu-gur (p'o-gun)	Pferd	7

Man sieht, dass bei dieser Verteilung ein gewisses System an-
gewandt ist. Da nach Colophon B das Original des Textes in Sanskrit
abgefasst war und angeblich von einem indischen Paṇḍita und Hsüan
Tsang aus Indien nach China gebracht und dort übersetzt worden
ist, so wären wir vor die Frage gestellt, ob der Cyklus der Tiere
bereits im Sanskrittext vorhanden und demnach vor der Zeit des
Hsüan Tsang in Indien bekannt war. CHAVANNES hat in seiner
scharfsinnigen Studie: Le Cycle turc des douze animaux [2] diese Frage
gründlich untersucht und ist inbezug auf Indien zu dem Ergebnis

1) Ich vermag die tibetischen Namen, die in den Wörterbüchern nicht enthalten sind,
nicht zu erklären. Tibetisch sind sie jedenfalls nicht, sondern sehen wie Transcriptionen
aus dem Chinesischen aus. Mit den von SCHLEGEL, Uranographie chinoise, p. 859, gegebenen
Namen der Sternbilder des Grossen Bären lassen sie sich aber nicht identificiren.

2) T'oung Pao, 1906, pp. 51—122.

gelangt, dass der Tiercyklus dort unbekannt gewesen zu sein scheint: das Mahāsaṃnipāta-sūtra, der älteste buddhistische Text, in welchem derselbe erwähnt wird, ist, wie aus der darin entfalteten Kenntnis der Geographie Centralasiens hervorgeht, im östlichen Turkistan entstanden oder jedenfalls stark überarbeitet worden [1]). Dass der türkische Kreis der zwölf Tiere im alten Indien und in der älteren buddhistischen Litteratur unbekannt war, scheint gewiss zu sein, aber es scheint mir kein Widerspruch zu der Ansicht Chavannes', vielmehr eine Bestätigung seiner Gesamttheorie über die Entstehung und Verbreitung des Cyklus zu sein, wenn Anzeichen vorhanden sind, dass in einer späten Periode, die sich annähernd zwischen das 7. und 11. nachchristliche Jahrhundert datiren lässt, der türkische Tiercyklus in Indien bekannt geworden ist, und zwar von den Gebieten Centralasiens her. Diese Anzeichen sind:

1) Der in Rede stehende Text des Bärengestirnsūtras, in welchem ich nach eingehender Prüfung nichts habe finden können, was gegen eine Entstehung der Schrift in Indien zu sprechen vermöchte. Ob der Text wirklich von Hsüan Tsang mitgebracht wurde, bleibt ja zweifelhaft, da er in der Liste der von dem Pilger erworbenen und übersetzten Bücher nicht erwähnt zu sein scheint; aber wann auch immer der Sanskrittext entstanden sein mag, die Erwähnung des Tiercyklus in seiner Beziehung zu den sieben Sternen muss darin vorhanden gewesen sein, da diese Idee den Anfang und die Grundlage des ganzen Sūtras bildet.

2) Im 19. Kapitel der tibetischen Lebensbeschreibung des Padmasambhava, des Stifters des Lamaismus im 8. Jahrhundert, erlernt Padmasambhava von einem indischen Lehrer in der Stadt Guhya in Indien die Astrologie und wird also über den Tiercyklus im Zusammenhang mit den zwölf Nidāna's [2]) unterwiesen: „Wenn der

---

1) T'oung Pao, 1906, p. 93.
2) Vergl. CHAVANNES, l c , p. 86.

Gott der aukti den Segen gibt, so steht dies im Zusammenhang mit dem *Mäuse*jahr der avidyā. Wenn der aschenfarbene Elephant im Mutterleibe sich sechsfach verwandelt, so hängt es zusammen mit dem *Stier*jahre der saṁskāra's; wenn im Mutterleibe das Innere in regenbogenfarbene Kreise sich verwandelt, so steht es im Zusammenhange mit dem *Tiger*jahre des vijñāna; wenn im Moment des Hervorkommens das Ohr aufrechtstehend herauskommt, so hängt dies zusammen mit dem *Hasen*jahre des nāmarūpa; wenn zur Zeit der Geburt aus den Himmeln eine Stimme ertönt, so steht dies im Zusammenhang mit dem *Drachen*jahre der sechs âyatana's; wenn dem Verkörperten der Nāgakönig das Bad gibt, so hängt dies zusammen mit dem *Schlangen*jahre des sparça; wenn auf einem isabellfarbigen Pferde der Goldene reitet, so kommt dies von dem *Pferde*-jahre der vedanā; wenn ein Gott Schafsmilch vorsetzt, so kommt dies von dem *Schafs*jahre der tṛṣṇā. Wenn der Affe Hanumān Honig bringt, so kommt dies von dem *Affen*jahre des upādāna. Wenn der Fürst der Gefiederten, der Garuḍa, Weihrauch herbeibringt, so kommt dies von dem *Vogel*jahr des bhava; wenn der Hund Taudiya die Lehre Buddhas hört, so hängt dies zusammen mit dem *Hunde*jahr der jāti. Wenn die neun eisernen Schweine eingebohrt in Kraftanstrengung wetteifern, so hängt dies zusammen mit dem *Schweine*jahre des Alterns und Sterbens: so stehen im Zusammenhang zwölf Vorgänge im Leben eines Buddha und die zwölf Nidāna's, welche in einem indischen Jahre umlaufen" [1]. Es ist offenbar, dass die hier gegebene Einkleidung des Cyklus nur auf indischem Boden entstanden sein kann.

3) Eine Belegstelle für das Vorkommen dieses Tiercyklus in Indien gibt F. K. Ginzel [2] nach Erard Mollien, doch bin ich nicht in der Lage, diese Quelle zu prüfen.

---

1) Nach der Übersetzung von A. Grünwedel, Ein Kapitel des Tä-še-sun, Abdruck aus Bastian-Festschrift, S. 13—14.

2) Handbuch der mathematischen und technischen Chronologie, Band I, Leipzig, 1906, . 87.

4) Es muss daran erinnert werden, dass die Tibeter ihre gegenwärtige Zeitrechnung gleichzeitig mit der Einführung des sogenannten
Kālacakra Systems aus Centralasien erhalten haben, und zwar aus
dem Reiche Šambhala. Einem der Hauptvertreter des Kālacakra,
Atīça, wird die Gestaltung des jetzigen Kalenders und das Rech-
nen nach sechzigjährigen Cyklen zugeschrieben [1]). Das Jahr 1026
ist der Anfang des ersten tibetischen Cyklus, und das voraufgehende
Jahr 1025 gilt als das der officiellen Reception des Kālacakra Systems.
Bemerkenswert ist auch die Tatsache, dass in Indien nach dem
Sūrya Siddhānta in demselben Jahre 1026 ein neuer Cyklus der
Bṛihaspati-Jahre begann [2]). Es ist nun seltsamer Weise bisher immer
die Tatsache übersehen worden, dass die Grundlage der tibetischen
Chronologie von dem türkischen Cyklus der zwölf Tiere gebildet
wird, und dass die tibetische Tradition selbst astronomische und
chronologische Kenntnisse von den Türken herleitet. Es ist unrich-
tig, wie E. Schlagintweit [3]) getan hat, dabei ausschliesslich auf Indien
und China zu verweisen, und wenn Ginzel [4]) sein Kapitel „Zeitrechnung in Tibet" mit den Worten eröffnet: „Am nächsten mit
der indischen Zeitrechnung verwandt ist die der Tibetaner; in derselben zeigt sich neben einer gewissen Ursprünglichkeit, indischer
und chinesischer Einfluss", so will das soviel besagen, dass diese
„gewisse Ursprünglichkeit" eben türkisch ist [5]), und dass sich der

---

1) A. Grünwedel, Mythologie des Buddhismus, S. 58.
2) Ginzel, l. c., S. 407.
3) Buddhism in Tibet, p. 273.
4) L. c., S. 403.
5) Grünwedel (Die orientalischen Religionen, in Kultur der Gegenwart, S. 140) ist
durchaus im Recht, wenn er von einer gewissen alten Kulturgemeinschaft der Tibeter mit
ihren früheren türkischen und mongolischen Nachbarn spricht. Der wirtschaftliche Typus
der alten tibetischen Kultur stimmt durchaus mit dem der alttürkischen überein und steht
am nächsten dem der Hsiung-nu, mit denen sie auch auffallende Züge in der sozialen Organisation gemeinsam haben. — Rockhill (Journal Royal Asiatic Soc., 1891, p. 207) sagt
anlässlich der Erörterung der tibetischen Chronologie in einer Note: "The Chinese and also
Father Desgodins state that the Tibetans follow the Mohammedan (Turkestan?) system of
calculating time. See Peking Gazette, Nov. 19, 1887, and C. H. Desgodins, Le Thibet, p. 369.
I have been unable to learn anything of this". Diese Angaben erklären sich jetzt von selbst.

indische und chinesische Einfluss im wesentlichen auf Punkte der Terminologie beschränkt. Wie die Chinesen und die Mongolen, so haben auch die Tibeter den Cyklus der zwölf Tiere von den Türken erlernt und übernommen, die, wie CHAVANNES trefflich gezeigt hat, denselben in originaler Weise ausgebildet und chronologischen Zwecken dienstbar gemacht haben, und dieser mit dem Kālacakra System übernommene Cyklus ist das Fundament der ganzen tibetischen Chronologie, und mit dem Kālacakra wird er wohl auch nach Indien gelangt sein. Ich vermute, dass die Bezeichnung Kālacakra, „das Rad der Zeit", (tib. *dus-kyi ak'or-lo*) ursprünglich nichts anderes bedeutete als eben den türkischen Cyklus der zwölf Tiere, und dass der Cyklus endlich dem ganzen System, in dem er offenbar eine grosse Rolle gespielt haben muss, den Namen geliehen hat. Damit ist auch implicite gesagt, dass das Kālacakrasystem selbst türkischen Ursprungs sein muss. CSOMA [1]), welcher zuerst über diese jüngste Phase des Buddhismus berichtete, hat das Land Šambhala, dem sein Ursprung zugeschrieben wird, am Jaxartes localisirt. Ich habe dagegen den Eindruck gewonnen, dass jenes Reich mit der Gegend von Khotan zu identificiren ist, und erlaube mir am Schluss einen Text zur Rechtfertigung meiner Ansicht anzufügen. Freilich wird sich erst völlige Klarheit in dieser Angelegenheit gewinnen lassen, wenn die ganze Litteratur über Kālacakra im Tanjur bearbeit sein wird, was ein dringendes Erfordernis ist, um dies kulturgeschichtlich wie religionsphilosophisch gleich interessante System zu erschliessen.

Der im Folgenden mitgeteilte Text ist einer Handschrift mit dem Titel *Šambha-lai lam yig*, d. i. „Reiseführer nach Šambhala", entnommen. Unter diesem Titel gibt es eine ganze Reihe verschiedener

---

1) Note on the Origin of the Kála-Chakra and Adi-Buddha Systems. Journal Asiatic Soc. of Bengal, Vol. II, 1833, pp. 57—59, und A Grammar of the Tibetan Language, Calcutta, 1834, pp. 192—193.

Werke ¹). Nach Sarat Chandra Das ²) wäre ein Werk dieses Namens
von dem *Paṇ-c'en bLo-bzaṅ dPal-ldan Ye-šes* von *bKra-šis-lhun-po*
geschrieben worden, der von 1737 bis 1779 gelebt hat. In meiner
Handschrift ist kein Verfassername angegeben, aus einer Reihe von
Umständen aber, unter denen am meisten ins Gewicht fällt, dass
darin eine Beschreibung von Peking und dem dortigen Kaiserpalast
mitgeteilt wird, die nur auf die Mongolenzeit passt, glaube ich schliessen
zu dürfen, dass die Version meines Werkes im 13. Jahrhundert
entstanden ist.

Das von Šambhala handelnde vierte Kapitel lautet darin folgendermassen:

„Wenn man auf eine weite Entfernung nach Norden reist, wo
die Königin K'om-k'om haust, da sollen sich prächtige Wälder und
Gewässer befinden. Von dort weiter nordwärts ist unter den sechs
dort befindlichen Gebieten das Königreich von Khotan (*Li-yul*) ³)
das grösste. Dem viereckigen Eisenberg entlang muss man auf der
Handelsstrasse sechs Tagereisen zurücklegen. Da ist der Fluss Sita,
der von Westen nach Osten fliesst ⁴), und dem entlang die Hor ⁵)
leben: gewöhnlich haben sie keine Häuser, sondern wohnen lediglich
in Filzzelten. Die Zelte der Vornehmen haben zwei- und sogar

---

1) Eines derselben ist im Tanjur enthalten (Asiatic Researches, Vol. XX, p. 584); s.
auch Schiefner in Mélanges asiatiques, Vol. I, p 405.

2) A Tibetan-English Dictionary, p. 1231.

3) Nach Wassiljew, Der Buddhismus, S. 80, bezeichnet Li-yul „buddhistische Land-
striche im Norden von Tibet und insbesondere Khotan"; ebenso in seiner Geographie Tibets
(russ , St. Petersburg, 1895, p. 57); siehe ferner Schiefner, Tibetische Lebensbeschreibung
Çâkyamuni's, S 60, 97; Rockhill, The Life of the Buddha, p. 230.

4) Daraus scheint hervorzugehen, dass Sita nicht, wie bisher angenommen, mit dem
Oxus oder Jaxartes identisch ist, sondern eher der Tarim zu sein scheint.

5) Über die Namen Hor und Sog s. besonders Hodgson, Essays on the Languages etc.
of Nepál and Tibet, London, 1874, pp 65 et seq. In der späteren tibetischen Litteratur
werden Hor und Sog fast ohne Unterschied zur Bezeichnung der Mongolen gebraucht; ursprünglich bezog sich aber Hor auf Türkstämme und insbesondere auf die Uiguren, wie das
Wort denn auch noch gegenwärtig in West-Tibet zur Benennung der Türken gebraucht wird.

dreifache Dächer. Für Transportzwecke bedienen sie sich der Kamele, von denen je 110 eine Ladung ausmachen. Am Fusse der Südseite der Schneeberge, welche Šambhala von aussen einschliessen, liegt eine grosse Stadt, in der alle Menschen, Männer sowohl als Frauen, sich in folgender Weise geschlechtlich vereinigen: der Penis befindet sich an der Innenseite des rechten Oberschenkelmuskels, während sich die weiblichen Geschlechtsteile an der entgegengesetzten Seite, am linken Oberschenkel, befinden. Der Fötus verbleibt drei Monate im linken Oberschenkel, und dann soll die Geburt stattfinden.

Was die Ausdehnung des Landes Šambhala betrifft, so ist sie im Süden etwa die Hälfte eines kleinen Jambudvīpa, auf der Nordseite aber ist es ein grosses Land, das *Yo-gsum*, mit dem gewöhnlichen Namen aber Šambhala heisst. Auf allen Seiten ist es von 500 Yojana langen Schneebergen umgeben. Drei innere Bergketten sind in acht Blätter von der Form eines Lotus zerschnitten, so dass sich die Ränder der äusseren und inneren Berge berühren. Im Südwesten befindet sich Wasser und ein für Menschen gangbarer Weg. Mongolen (*Sog-po*), Kaufleute, jung verheiratete Frauen u. s. w. sollen häufig diese Strasse hinauf und herab ziehen. Auf der Ostseite gibt es nur einen Fluss und Vögel, doch auch einen Pass, der früher niemals von Menschen begangen sein soll. In kurzer Entfernung mündet dieser Fluss in einen äusseren See. In dem leeren Raume bildet das ganz von Wasser umgebene Land eine Insel, eben jene, welche gemäss der im Kālacakra gegebenen Erklärung eine der elf Inseln ist, die vom Wasser des äusseren Sees umgeben sich einzeln von einander abgelöst haben.

Was das Centrum des Landes betrifft, so erreicht man dasselbe, wenn man oben von den erwähnten Lotusblättern aus zwei oder drei Tagereise hinansteigt; dort befindet sich der Königspalast,

genannt Ka-la-lha [1]), dessen vier Seiten jede elf Yojana lang ist. Auf
dessen Südseite befindet sich der Park *Ma-la-ya*, in welchem das
von dem König Sucandra (*Zla-bzaṅ*) errichtete grosse Glos-sloṅ (?)-
Maṇḍala des Kālacakra und das von dem Kulika [2]) Puṇḍarīka (Rigs-
idan Pad-ma dkar-po) errichtete kleine Glos-sloṅ-Maṇḍala beide im
Innern eines Vajra-Zelthauses aufgestellt sind. Auf der Nordseite des
Palastes liegt eine grosse Stadt, wo die Sklavinnen des Palastes
leben. Im Osten und Westen davon ist ein grosser See von der
Gestalt des Mondes am achten Tage des Monats. Jener Königspalast
enthält vier grosse Gebäude mit zwei oder drei übereinander gebauten
Dächern aus Ton, deren Spitzen mit vergoldeten Zinnen gekrönt
sind. Die übrigen Gebäude haben in der Regel nur zwei Dächer,
gefällig wie der Stil der tibetischen. In den beiden Palästen wendet
man die bei uns in Tibet in früherer Zeit allgemein gebrauchte
ạByam Schrift [3]) an. Es sind dies die beiden von dem grossen König
als Behausungen für die grossen Geistlichen errichteten beiden Gebäude, nämlich das On-ạdab gsal-k'aṅ-rtse und das Gra-mda ya-k'a
k'aṅ-gsar, wo man die auf Baumwollpapier geschriebenen beiden
Klassen der Sūtra und Tantra gründlich versteht.

Was die äussere Erscheinung des Königs jenes Landes anbelangt, so ist sein Haupthaar geteilt und in einen Knoten gebunden;
er trägt eine buntfarbige Mütze mit fünf Zipfeln, welche die fünf
Kasten symbolisiren; er sitzt auf einem sehr hohen Tron. Einstmals,
als er zahlreichen Personen religiöse Vorträge hielt, habe auch ich
sein Antlitz geschaut und seine Worte gehört. Zu den Kleinodien
jenes Königs gehören folgende: er besitzt den Wunschbaum (kal-
padruma = tib. *dpag-bsam šiṅ*) in Gestalt von hölzernen Buchdeckeln,

---

1) Wahrscheinlich Schreibfehler für *Kalapa*.

2) Siehe A. GRÜNWEDEL, Mythologie des Buddhismus, S. 41.

3) Der Name dieser Schrift fehlt in den Wörterbüchern, wenn nicht *ạbyam* Schreibfehler für *ạbam* ist; das *ạbam-yig* Alphabet ist in CSOMA's Grammar auf den Tafeln 31—35
dargestellt.

der ihm jeden Gedanken erfüllt, und die Nāga haben ihm ein Juwel verliehen, das, sobald er auf seinen Tron gelangt, alle Wünsche des Herzens erfüllen soll. Unter anderen wunderbaren Dingen, die er besitzt, ist folgendes. Wenn er einen Boten aussendet, so überreicht er ihm Zauberkräfte besitzende Gegenstände, und zwar ein Schwert für den König der Mitte und je eine Salbe für jeden der 96 Vasallenfürsten des Lotusblattgebirges; der Bote braucht sich nun nur des Landes zu erinnern, in das er zu gehen hat, um in einem Augenblick dort angelangt zu sein. Die gewöhnlichen Nahrungsmittel sind Reis, verschiedene Arten Baumfrüchte, Weizen und anderes. Die Menschen tragen grossenteils weisse Kleider und rote Mützen mit fünf emporragenden Spitzen, doch ohne Flügel. Die Frauen haben gewöhnlich sowohl rote Kleider als Mützen und tragen weiche Shawls und Edelsteinschmuck. Die Mönche tragen in der Regel eine Mütze, aber keine Stiefel [1]), und es gibt viele unter ihnen, die bis zu drei Mönchsgewändern besitzen. Bei den Versammlungen in ihren Schulen liegt das Hauptgewicht darauf, dass sie keine Worte äussern, sich in der Tat mit wenigen Erläuterungen begnügen und sich auf die Erkenntnis (*ñams-rtogs*) verlassen".

Der Geschichte des Buddhismus unter den Türken in Centralasien könnte kein grösserer Dienst erwiesen werden als durch eine Übersetzung aller auf das Kālacakra bezüglichen Texte des Tanjur.

---

[1]) Während die tibetischen Lama Stiefel tragen.

Tibetischer Text des Colophons des Sūtra
sMe-bdun žes-pai skar-mai mdo.

A.

ston-pa rdzogs-pai Saṅs-rgyas-kyis gsuṅs-pai
sme-bdun žes-pai skar-mai mdo-sde ạdi
brtan-pai sems-kyis dran-žiṅ gaṅ mcᶜod-pa
de-la pᶜan-pa ạbyuṅ-žes rab-šes-nas
5 ₍U-rug-bo-gai miṅ-can Zu-gur-cᶜe
cᶜuṅ-ṅu dus-nas rtag-par cᶜos ạdi-la
yid-cᶜes-ldan-pas rgyun-du klog-ciṅ mcᶜod
raṅ-gi go-ạpᶜaṅ tsᶜol-žiṅ gsol-ạdebs-pas
mtᶜuṅ-par skyob-pai bdag-po bsod-nams-can
10 grol-mdzad ston-pa Saṅs-rgyas sprul-pa gaṅ
Tᶜe-mur rgyal-bu yun-du tsᶜe-riṅ-žiṅ
tᶜug-can rgyal-bu cᶜen-por ạgyur-bar ạdod
bLo-ldan Byaṅ-cᶜub-sems-dpa bdag-po de
bsleb-nas Se-cᶜen rgyal-poi gdan-sar bžugs
15 bdag-gi yid-la ạdod-pa tsᶜim gyur-pas
tᶜe-tsᶜom med-par cᶜos ạdir ṅes-šes skyes
Yu-gur yi-ger cᶜos-kyi mdo-sde ạdi
sṅon-cᶜad gžan-gyis bsgyur-ba med-pas-na
maṅ-poi Hor-rnams dad-pas mcᶜod-gyur ces
20 bdag-gis Hor-gyi skad-du bsgyur-ba yin
ṅa-yis ji-ltar bsam-pa ạgrub-ạgyur-na
sems-can kᶜri pᶜrag stoṅ pᶜrag ạdod-pa kun
bdag ñid ji-bžin tsᶜim-par gyur-cig ces
stoṅ pᶜrag par-du btab-nas kun-la bkye
25 ạdi-yi bsod-nams driṅ-gyi ạbras-bui mtᶜus
bdag-po rgyal-po dpon-mo brgyud-par bcas
yun-du sku-tsᶜe riṅ-žiṅ bsod-nams ạpᶜel

mtᶜar-tᶜug Saṅs-rgyas go-ạpᶜaṅ rñed gyur-cig

rgyal-kᶜams dgra daṅ ạkrug-pa ži-ba daṅ

30 gdon daṅ bar-cᶜad rims-med bde gyur-cig

cᶜar rluṅ dus babs mu-ge med ạgyur-žiṅ

bdag-gis smras daṅ bsam-don ạgrub-gyur-cig

bdag daṅ pᶜa ma bu sogs gñen rnams daṅ

ạtsᶜo-bai spun daṅ sems-can maṅ-por bcas

35 ạjig-rten ạdir yaṅ cᶜos-kyis ạdod-pa tsᶜim

bde-ba-can žiṅ rtag-par pᶜyin-gyur-cig

B.

Tᶜen-li daṅ·poi-lo ạbrug-gi-zla-ba bcu-pai-tsᶜes-gcig-la par-du btab-pa-yin. mdo ạdi rGya-gar-gyi-yul-nas rGya-gar-gyi paṇḍita gcig daṅ, Tᶜaṅ Zam-Tsᶜaṅ-gis kᶜyer-te rGyai-yul-du bsgyur-ro. rGya-cᶜen-poi-yul-du rgyas-par gyur-ciṅ, gnas-pa-las rgyal-po cᶜen-poi blon-po byaṅ-cᶜub-sems-dpai rigs-su byuṅ-ba dad-pa daṅ, šes-rab daṅ, tiṅ-ñe-ạdzin-daṅ-ldan-pa Gim-rtse-goṅ-lu tai-hui-γyui-šī Tᶜai ₃u·rug-po-yis bskul-te Yu-gur-gyi bstan-pai bdag-po Prajñā-çrī-s Hor-gyi skad daṅ yi-ger bsgyur-nas stoṅ pᶜrag gñis par-du btab. ₃A-lin Tᶜi-mur tai-se-du·s Yu-gur-gyi skad-du bsgyur-te, stoṅ pᶜrag gcig par-du btab-nas, cᶜos-kyi sbyin-pa byas-te Hor daṅ Yu-gur-la rgyas-par byas-šiṅ, tai-hu ñid kyaṅ sṅan-cᶜad Sog-poi cᶜos-lugs ạdzin-pa-la, cᶜos ạdii byin-rlabs-kyis Saṅs·rgyas-kyi cᶜos-la žugs-šiṅ rab-tu byuṅ-ste ạdii yon-tan-rnams ñams-su myoṅ-bar gyur-pao.

slad-kyis me mo glaṅ-gi lo-la lo-tstsᶜa-ba Ma-hā-pᶜa-la daṅ, Çrī ᶜĀ-nan-da-va-jra-s Guṅ-tᶜaṅ-gi gtsug-lag-kᶜaṅ-du Bod-kyi skad daṅ yi-ger bsgyur-ciṅ žus-te gtan-la pᶜab-pao.

054

日本春宫图

ἈΝΘΡΩΠΟΦΥΤΕΙΑ

Jahrbücher

für

Folkloristische Erhebungen und Forschungen

zur

Entwicklunggeschichte der geschlechtlichen Moral

unter redaktioneller Mitwirkung und Mitarbeiterschaft von

Prof. Dr. Thomas Achelis, Gymnasialdirektor in Bremen, Dr. Iwan Bloch, Arzt für Haut- und Sexualleiden in Berlin, Prof. Dr. Franz Boas, an der Columbia-Universität in New-York V. S. N., Dr. med. und phil. Georg Buschan, Herausgeber des Zentralblattes für Anthropologie in Stettin, Geh. Medizinalrat Prof. Dr. Albert Eulenburg in Berlin, Prof. Dr. Anton Herrmann, Herausgeber der Ethnologischen Mitteilungen aus Ungarn, in Budapest, Prof. Dr. Juljan Jaworskij in Kiew, Dr. Alexander Mitrovió, Rechtsanwalt in Knin, Dr. Giuseppe Pitrè, Herausgeber des Archivio per lo studio delle tradizioni popolari in Palermo, Dr. med. Isak Robinsohn in Wien, Prof. Dr. Karl von den Steinen in Berlin u. anderen Gelehrten

gegründet im Verein mit

Prof. Dr. med. Bernhard Hermann Obst,
weiland Direktor des Museums für Völkerkunde in Leipzig

herausgegeben

von

Dr. Friedrich S. Krauss
In Wien VII/2, Neustiftgasse 12

IV. Band.

Leipzig
Deutsche Verlagsactiengesellschaft
1907

Bezugpreis für jeden Band 30 Mk.

Dä treibt a d Stei[1] eini Schaf und a d Widder,
Aba do heilign Zeitn kimt a und stehlts wieda.
s Stehln des tragt zweni und Preisroß san gar,
Drum macht er und sei Bua an Pfarra an Narr.

14.

Da Pfara vo Reischbeurn des is a schlaucha Mo,
Bei den greift d Reiffeißn und Kreditbank a nimma o.
Sei Vieh hot a vosteigert und dabei glacht,
Wei äm dö duma Bauern an Haufn Geld ham as Haus zuwi[1] bracht.
An Knecht hot a furtgschickt weis mit da Ökonomie niks mehr is
Aba Dirn hot a gholtn[2] für d Nachzucht ganz gwiß.

Zum Schluß ein dreifaches Hoch auf den Dekan von Gaißach.

Ein japanisches Frühlingbild.
Von Berthold Laufer, New York.

China und Japan sind unendlich reich an volktümlichen Kunst-
darstellungen, die, was Japan betrifft, erst zu einem Teil bekannt,
was China anbelangt, überhaupt noch nicht zugänglich geworden
sind. Es ist bewußt populäre Kunst, von Leuten des Volkes, ihrer
sozialen Stellung nach Kunsthandwerkern, nicht Künstlern, geschaffen,
für die breiten Schichten des Volkes bestimmt. Die allgemeine Anschauung geht dahin, daß die Schule der Ukiyoye (wörtlich ,Bilder
der dahinfließenden, vergänglichen Welt') Japans, die im schwarz-
weißen und buntfarbigen Holzschnitt das Höchste geleistet hat, eine
durchaus einheimische, echt japanische Kunstrichtung sei. Dies
trifft für die späteren Phasen in der Entwicklung der xylographischen
Technik und des wesentlichen Inhalts gewiß zu, auf keinem anderen
Gebiete hat sich Japan auch japanischer bewährt als in dem der
Holzschneidekunst, und doch muß daran erinnert werden, daß das
Nachbarreich, die Quelle all seiner Kultur, eine gleiche Kunstrichtung derselben Form und desselben Inhalts besessen hat und noch
besitzt. Freilich, niemand hat ihr bisher Aufmerksamkeit geschenkt,
unsere Spezialliteratur über chinesische Kunst enthält kein Wort

---

[1] zu, [2] zurückbehalten.

darüber, von den allgemeinen Darstellungen der Kunstgeschichte
ganz zu schweigen, und unsere Museen schweigen sich ebenso gründlich darüber aus. Und doch kann sich jeder, der irgend eine chinesische Stadt mit sehenden Augen durchwandert oder irgend ein einfaches Bürger- oder Bauernhaus betritt, täglich und stündlich von
ihrer Existenz überzeugen, von ihrer großen Wertschätzung beim
Volke und ihrer Bedeutung für das gesellige Leben. In ihrer Technik
sind diese Bilderbogen meist roh, obwohl sich neuerdings in Shanghai
eine Schule gebildet hat, die der in Europa üblichen Durchschnitt-
ware kaum nachstehende Farbendrucke herstellt. Aber was auch
immer der künstlerische Wert dieser Erzeugnisse sein möge — Ästhetik
hat mich stets herzlich wenig gekümmert — sie sind das Ent-
zücken des Ethnographen und eine unerschöpfliche lebenswahre
Quelle der Anregung und Belehrung für das Studium des Volkslebens.
Da ist China, wie es leibt und lebt, wie es trinkt und zecht, wie es
spielt und lacht, wie es feiert und hochzeitet und das Leben fröhlich genießt, das ausgelassene Treiben der Kinder, die munteren
Spiele der Knaben, die beschaulich-geschäftige Tätigkeit der Mädchen
und alle Phasen im Leben der Frau. Natürlich, sie steht im Mittelpunkt dieser ganzen Kunst, ebenso wie in Japan, die Frau, von der
man nicht spricht, und in noch höherem Grade, die Frau, von der
man spricht. Die berühmten Schönheiten und Sängerinnen von
Shanghai und Peking werden immer und immer wieder porträtiert,
kahnfahrend, Lotosblüten im See pflückend, in einem Gartenpavillon
oder in ihrer Häuslichkeit. Darstellungen von Szenen populärer
Bühnenstücke und beliebter Schauspieler sind ungemein häufig. Der
Humor kommt nie zu kurz, und Folgeszenen lustiger Bilder auf
einem Blatt vereinigt sind ganz nach Art der Münchener Bilderbogen. Idyllen aus dem Leben von Pantoffelhelden, die in China
ebenso florieren als bei uns, gehören dabei zu den geschätztesten
Sujets. In anbetracht der Tatsache, daß uns das Familienleben der
Chinesen und vor allem das äußere und innere Leben ihrer Frauen
verschlossen bleibt, sind diese fliegenden Blätter, die allenthalben
auf den Straßen feilgehalten und in großen Auflagen über das ganze
Land verstreut werden, ein willkommener Ersatz für den Mangel
direkter Beobachtung und erschließen uns tiefe Einsichten in ihr
innerstes Fühlen. Ich habe daher von diesem Hilfsmittel reichlich
Gebrauch gemacht und auf meinen Reisen in China keine Gelegenheit vorübergehen lassen, solche ethnographische Dokumente
zu sammeln, die im Laufe der Zeit auf viele Hunderte angewachsen

sind. Ob und wie sie sich werden veröffentlichen lassen, ist mir vorläufig noch ein Rätsel; wären es prähistorische Topfscherben, so hätte sie längst ein Museum auf würdigen Tafeln publiziert, aber es ist ja pulsierendes Leben der Gegenwart. Bei der Engherzigkeit und fossilen Verdummung, mit der gegenwärtig unsere amerikanischen Museen verwaltet werden, ist ohnehin an solche Publikationen nicht zu denken.

Diese Volkskunst steht in bewußtem Gegensatz zu der ernsten, gleichsam ‚akademischen‘ Kunstmalerei, die dem eigentlichen Volksleben fernsteht. Wir sind schulmäßig gewöhnt, in dem Chinesen den ernsten und nüchternen Realphilosophen zu sehen; gewiß, der Chinese ist ernst und muß ernst genommen werden, viel ernster noch in Zukunft als bisher geschehen. Aber mit dem Ernst, der Wirkung uralter moralischer Erziehung und ritualer Einrichtungen, ist das Wesen seiner Psyche noch lange nicht erschöpft. Neben dem offiziellen Menschen kommt auch der natürliche Mensch zu seinem Recht. Im allgemeinen ist der Chinese, nicht nur Bürgers- und Bauersmann, sondern auch der strenge Konfuzianer, Beamter oder Gelehrte, ein heiterer lebensfroher Genußmensch, nicht einer, der dem Genuß sinnlos und bedingunglos fröhnt, sondern der die Freuden des Daseins mit Maß und Weisheit zu genießen versteht. Kaum ein Volk hält so viel auf die Bewahrung des Decorums und aller äußeren Regeln des Anstands und guter Sitten, nicht als einer rein formellen Äußerlichkeit, sondern wurzelnd in einem stark ausgeprägten Moralitätbewußtsein. Das Ritual des Konfuzius ist der Ausfluß seiner praktischen Ethik. Ihre Literatur ist ungewöhnlich frei von dem, was unsere Moralisten ‚Schmutz‘ nennen, und ist selbst von Missionaren als eine der ‚reinsten‘ gepriesen worden. Dabei darf aber nicht vergessen werden, daß es sich hier nur um die offizielle oder anerkannte Literatur handelt; es gibt eben eine ungeheure Masse anderer Literatur, die darum, weil sie anders ist, nicht zur Literatur gezählt wird. Und diese Literatur ist gerade die volktümliche, die von der großen Masse gierig verschlungen wird. So gibt es Romane von zynischstem Naturalismus, gegen die sich die Versuche der modernen Franzosen wie das erste Erröten des erwachenden jungen Mädchens ausnehmen, bürgerliche Lustspiele mit aktuell gegebenen Situationen, vor denen die freieste Bühne Europas auf immer zurückschrecken würde. Das eheliche, oft genug zum unehelichen gemachte Bett und der Nachtstuhl spielen in diesen Stücken eine sichtbare Rolle auf der Bühne. Für die auf ihr üb-

liche Freiheit der Rede will ich nur ein ganz zahmes Beispiel an-
führen, das gleichzeitig charakteristisch für die Art und Weise ist,
wie der chinesische Schauspieler das Publikum mitspielen läßt. Der
Held des Dramas ist so sehr von Liebe zu einer Schönheit ent-
flammt, daß er gleich auf offener Szene seine Leidenschaft stillen
will. Sie wehrt ihn ab. Er: ‚Aber warum denn nicht? Das ist
doch die natürlichste Sache von der Welt, das tun doch alle
Menschen‘. Sie: ‚Sehr schön! Aber es geht doch nicht hier in der
Öffentlichkeit vor dem ganzen Publikum‘. Er: ‚Ach, das macht
doch nichts!‘ Sie: ‚Nun sieh Dir bitte diesen ehrwürdigen alten
Herrn mit grauem Haar in der ersten Reihe des Parketts an‘ (sie
zeigt wirklich auf den Betreffenden); ‚wolltest Du es verantworten,
ihm die Schamröte ins Gesicht zu treiben?‘ Er: ‚Nun ja, dann
warten wir bis bis später!‘ Ebenso gelangt in den Volksliedern die
erotische Seite des Liebelebens in den stärksten Tönen zum Ausdruck, und ein besonderer Zweig der oben geschilderten Volkskunst
ist die Darstellung erotischer Szenen.

Diese Bilder heißen euphemistisch ‚Frühlingbilder‘ (ch‘un hua,
in japanischer Aussprache shungwa); die Bezeichnung ‚Frühling‘
wird vielfach ganz passend für die Regungen des Geschlechtstriebes
gebraucht. ‚Frühlingsmedikamente‘ sind Aphrodisiaca. In Japan ist
ferner der Ausdruck warai-ye d. i. Bilder zum Lachen, gebräuch-
lich, sodann makura-ye d. i. Kissenbilder; Bücher mit solchen Abbildungen heißen makura-zōshi. Aus China sind mir auch gefaltete
Albums mit Malereien bekannt, die ganze Zyklen von Coitusszenen
darstellen, oder die Geschichte eines Liebepaares in der wechselnden Entwicklung der Ereignisse; manche darunter sind von technischer Vollendung der Ausführung und bei der bekannten Begabung
der Ostasiaten für die Auffassung und Darstellung der körperlichen
Bewegung meisterlich naturwahr. In Japan scheint der buntfarbige
Holzschnitt in diesem Fache zu überwiegen. Zu den künstlerischen
Leistungen der Chinesen gehören auch Karikaturen von Europäern,
die in solchen Szenen dargestellt werden. In China sah ich ein
Album, vortrefflich gemalt, in dem Jesuiten auf diese Weise verspottet werden; ferner erwarb ich eine Serie bis in das Detail der
Kostümornamentik sehr fein ausgeführter Malereien, die wahrscheinlich aus dem 18. Jahrhundert stammen und die sexuelle Geschichte
eines Europäers und einer Europäerin in der Tracht der Rokokozeit
illustrieren. Die chinesischen Typen überwiegen naturgemäß, und
ihnen haftet nicht der Stempel der Satire an. Die Mehrzahl der

von mir gesehenen Darstellungen kann auch nicht einfach als obszön bezeichnet werden, wenn auch noch weniger als naiv; sie sind künstlerisch veredelt und der Ausdruck einer überschäumenden Lebenslust. Manche nehmen sich wie Anleitungen für angehende Liebende aus in erschöpfender Vorführung aller Stellungmöglichkeiten und mit besonderer Berücksichtigung der Terrainschwierigkeiten, z. B. in freier Natur, im Garten, auf dem Stuhl etc. Daneben gibt es natürlich auch viele rohe Darstellungen, besonders in Holzschnitten, die in Peking von hausierenden Spielzeughändlern in den Strassen verkauft werden. Es sei bemerkt, daß nach dem chinesischen Strafgesetz die Verkäufer ‚unmoralischer Publikationen‘ eine Strafe von hundert Stockschlägen und Transportation auf drei Jahre verwirken; die Käufer erhalten hundert Stockschläge, die Urheber dieselbe Zahl und außerdem lebenslängliche Transportation bis zu einer Entfernung von 3000 li. Trotz dieser Strafandrohungen scheint aber dieser Kunstzweig eifrig zu blühen, denn auf Verlangen kann man solche Bücher überall leicht erlangen, und aus persönlicher Erfahrung weiß ich, daß sie auch von Beamten mit Vorliebe gekauft werden.

In Japan haben die Frühlingbilder eine noch viel tiefere Bedeutung für das Volksleben als in China gehabt, denn sie sind jetzt von der nach europäisch-amerikanischem Muster prüden Regierung strengstens verboten und unterdrückt worden, wie auch der Phalluskultus. Besonders die Illustration von Romanen mit geschlechtlichen Szenen war in Japan bis zur Zeit der Restauration in vollem Schwang, jedenfalls wird sie auch jetzt noch im verborgenen betrieben. Ich habe eine nach etwa hundert Bänden zählende Sammlung dieser Art im Besitz eines alten deutschen Residenten in Yokohama inspiziert, der sie auf heimlichnächtlichen Streifzügen im Laufe vieler Jahre zusammengebracht hatte. In Japan sollen solche Bücher als eine Art Instruktionhefte zur Brautausstattung in die Ehe tretender Mädchen gehört haben. Sicher dienten sie alt und jung zur Unterhaltung und Belustigung. Charakteristisch für die japanischen Frühlingbilder ist die phantasiereiche Mannigfaltigkeit der Positionen, und zwei auf den entsprechenden chinesischen Bildern nie vorkommende Züge, die Anwesenheit von Zuschauern im Hintergrund, die durch Luken und Schiebetürspalten neugierig hereingucken, und sehr häufig neben dem menschlichen Begattungakt eine Parallele aus dem Tierleben, besonders rammelnde Katzen. Auch die Darstellung von Massenpaarungen in einem Raume ist nichts ungewöhnliches. Ein Beispiel der letzteren Gattung ist auf unserer Tafel nach

einem japanischen Originalholzschnitt reproduziert. Die Situation
läßt an Deutlichkeit nichts zu wünschen übrig und spricht für sich
selbst. Aber dies, und darin liegt der besondere Wert dieses Bunt-
drucks, ist das einzige Frühlingbild, das ich gefunden, dem anscheinend eine mythologische Bedeutung zu Grunde liegt. Diese
geht aus dem großen Ungetüm mit dem Fischkopf hervor, das die
lustige Gesellschaft plötzlich überrascht und in wirrer Hast ausein-
andersprengt. Das Bild hat leider keinerlei Beischriften, die das
Sujet erklären würden, und ich muß von vornherein bemerken, daß
mir der eigentliche Sinn der Darstellung unklar ist. Ich habe sie
bisher verschiedenen gebildeten Japanern vorgelegt, die gleichfalls
nicht imstande waren, eine befriedigende Erklärung darüber zu geben.
Vielleicht wird diese Veröffentlichung dazu beitragen, diese Frage
zu klären. Die drei das Monster begleitenden Männer sind Hand-
werker, der eine, der drohend seine Säge zum Angriff auf die Fest-
teilnehmer schwingt, ist ein Zimmermann; sein Nachbar scheint einen
Bohrer oder anderes Instrument zu halten. Man könnte so vermuten,
daß die Idee, die der Darstellung zu Grunde liegt, ein Kampf der
ehrbaren Zünftigkeit gegen die Ausschweifung sei; ich kann mich
aber in dieser Deutung auch irren. Vielleicht handelt es sich um
die Illustration einer uns unbekannten Lokalsage, deren es ja in Japan
so viele gibt. Unzweifelhaft ist jedenfalls, daß das Bild eine tiefere
mit dem Wesen der Phallusverehrung in Verbindung zu bringende
Symbolik besitzt. Darauf deuten zunächst die drei großen weißen
Kalebassen, die auf dem violetten Rock des Ungetüms angebracht
sind. Die Kalebasse ist in Ostasien ein Symbol des Phallus. Ferner
schwingt das Frauenzimmer unten rechts einen großen Penis in der
Rechten, den sie anscheinend dem mit dem Kopf nach vorn auf
dem Boden liegenden, um seinen Verlust klagenden Manne ausge-
rissen hat. Hier handelt es sich vermutlich um die magische Verwendung des Phallus, über die jüngst W. G. Aston in seinem trefflichen Buche ‚Shinto‘ (London 1905), p. 196, gehandelt hat. Ebenda,
pp. 186—198 und p. 363, findet man eine gute Auseinandersetzung
über den Phalluskultus in Japan [1]), der in den östlichen Teilen des
Landes auch jetzt noch nicht ausgestorben sein soll.

---

[1]) Ältere Literatur darüber: W. E. Griffis, The Religions of Japan, New York
1896, pp. 27—32, 49—52, 88, 380—384, und Edmund Buckley, Phallicism in Japan,
publiziert von der Universität Chicago. — Zu beachten ist die neueste, die einschlägige
ethnologische Literatur fast erschöpfende Monographie von Dr. Friedrich S. Krauss:
Das Geschlechtleben in Glauben, Sitte und Brauch der Japaner. Ein Beitrag zur Erforschung der Anthropophyteia, 1907.

Tafel X.

Dr. Berthold Laufer: Ein japanisches Frühlingbild.

Ein Geist mit einem Fischkopf stört strafend eine geschlechtlicher Ausschweifung frönende Gesellschaft.

055

蒙古文文献概说

VIII. évfolyam. 1907. 2—3. szám.

KELETI SZEMLE.

KÖZLEMÉNYEK AZ URAL-ALTAJI NÉP- ÉS NYELVTUDOMÁNY KÖRÉBŐL

A M. TUD. AKADÉMIA TÁMOGATÁSÁVAL

A NEMZETKÖZI KÖZÉP- ÉS KELETÁZSIAI TÁRSASÁG MAGYAR
BIZOTTSÁGÁNAK ÉRTESITŐJE.

REVUE ORIENTALE

POUR LES ÉTUDES OURALO-ALTAÏQUES.

SUBVENTIONNÉE PAR L'ACADÉMIE HONGROISE DES SCIENCES.

JOURNAL DU COMITÉ HONGROIS DE L'ASSOCIATION INTER-
NATIONALE POUR L'EXPLORATION DE L'ASIE CENTRALE ET
DE L'EXTRÊME-ORIENT.

SZERKESZTIK ÉS KIADJÁK
Rédigée par
Dᴿ KÚNOS IGNÁCZ ✶ Dᴿ MUNKÁCSI BERNÁT.

BUDAPEST.
En commission chez **Otto Harrassowitz**
Leipsic.

SKIZZE DER MONGOLISCHEN LITERATUR.

— Von Berthold Laufer. —

Die nachstehende Zusammenstellung habe ich auf Wunsch des Herausgebers dieser Zeitschrift, Dr. Bernhard Munkácsi, verfasst. Sie soll denjenigen als ein Hülfsmittel dienen, die eine Einführung in das Gebiet der mongolischen Literatur wünschen, sich mit dem bisher Geleisteten und Erreichten vertraut machen und den Weg zu eigener Arbeit gewiesen werden wollen. Eine Geschichte der mongolischen Literatur kann noch nicht geschrieben werden, dazu fehlt es noch gänzlich an soliden Vorarbeiten, dazu sind noch zu viele ungehobene Schätze da. Prof. A. Pozdnějev hat eine lithographirte Ausgabe seiner «Vorlesungen über die Geschichte der mongolischen Literatur»[1] in zwei Heften, 1896—1897 veröffentlicht; diese Arbeit ist sehr kritisch und gründlich, doch keineswegs eine Literaturgeschichte im eigentlichen Sinne, vielmehr eine literarische Studie der über die mongolischen Denkmäler bestehenden europäischen Literatur. Mehr kann indessen vorläufig auch nicht geboten werden. Meine Bemerkungen sind daher im wesentlichen ein Referat. Einiges Neue habe ich bei der Besprechung der Quadratinschriften hinzugefügt; ferner habe ich von den in der Kgl. Bibliothek zu Dresden vorhandenen kalmükischen Handschriften, von denen ich einen eigens angefertigten Katalog besitze, Gebrauch gemacht, und vor allem von meiner eigenen an tausend Bände zählenden Bibliothek mongolischer und tibeti-

---

[1] Im folgenden als «Pozdnějev, Lit.» citirt. Mir steht leider nur das erste Heft zur Verfügung; das zweite soll die Quadratinschriften behandeln.

scher Drucke.[1]) Leider konnte ich dieselbe hier noch nicht in
dem vollen Umfange benutzen, wie ich gewünscht hätte, da die
Katalogisirung derselben noch lange nicht abgeschlossen ist.
Ich hoffe auf die Nachsicht derer rechnen zu dürfen, denen
meine Anführung der Literatur, besonders der russischen, nicht
vollständig genug erscheint; die ältere und neuere russische
Literatur auf diesem Gebiete ist sehr schwer zugänglich, zum
grossen Teil vergriffen, und hier in den Bibliotheken von New-
York ist so gut wie nichts davon vorhanden.

Die für Zeitschriften gebrauchten Abkürzungen sind dieselben wie die in der Orientalischen Bibliographie üblichen.
«Kasan», mit folgender Nr. und Zahl citirt, bezieht sich auf den
Katalog «Каталогъ Санскритскимъ, Монгольскимъ, Тибетскимъ,
Маньджурскимъ и Китайскимъ книгамъ и рукописямъ, въ
Библіотекѣ Имп. Казанскаго Университета хранящимся», Kasan, 1834, 30 p. (enthält die Liste der von Kowalewski auf seinen Reisen erworbenen Bücher und Handschriften). Die Ab-
kürzung «Göttingen» bedeutet «Verzeichnis der Handschriften
im Preussischen Staate. I Hannover 3 Göttingen» (d. i. Band
III), Berlin, 1894.

Hülfsmittel zum Studium der mongolischen Sprache.

1. **Grammatik.** — Als beste praktische Grammatik für
die mongolische Schriftsprache ist die von A. BOBROVNIKOV[2]) zu
empfehlen; die von SCHMIDT[3]) ist wohl als veraltet zu betrachten und vielfach uncorrekt in der Bezeichnung der Aussprache.

---

[1]) Zur Förderung der tibetisch-mongolischen Studien bin ich jeder
Zeit gern bereit, Fachgenossen Werke aus dieser Sammlung zur freien
Benutzung zu übersenden.

[2]) А. Бобровниковъ, Грамматика монгольско-калмыцкаго языка,
Kasan, 1849, XI und 400 p.

[3]) I. J. SCHMIDT, Grammatik der mongolischen Sprache, St. Petersburg, 1831, 4°, XII u. 179 p. Auch in russischer und französischer Übersetzung. Das Buch von CARLO PUINI, Elementi della grammatica mongolica (Florenz, 1878) ist ein danach gearbeiteter kurzer Abriss. LÉON FEER,
Tableau de la grammaire mongole (Paris, 1866, 8 lithographirte Seiten,
4°) gibt ein Schema der Grammatik.

Gut ist auch Kowalewski's[1]) kurze Grammatik. Neuere Arbeiten sind die Vorlesungen von Kotvič[2]) und Rudnev,[3]) in denen vor allem die Phonetik eine fortgeschrittene wissenschaftliche Behandlung erfährt. Interessante Beiträge zu letzterem Kapitel findet man vor allem in W. Radloff's «Vergleichende Grammatik der Nördlichen Türksprachen» und in P. Schmidt's wertvoller Abhandlung «Der Lautwandel im Mandschu und Mongolischen».[4]) Sehr tüchtige Leistungen sind auch die Arbeiten von G. J. Ramstedt,[5]) der sich um die Erforschung der modernen mongolischen Dialekte grosse Verdienste erworben hat.

Für das Kalmükische besitzen wir die treffliche Arbeit von A. Popov.[6]) Vergleichend mit dem Mongolischen ist das Kalmükische in der schon erwähnten Grammatik von Bobrovnikov behandelt. Dagegen ist die lithographirte Schrift von H. A. Zwick[7]) gänzlich wertlos, da sie nichts weiter als ein Abklatsch

---

[1]) Осипъ Ковалевскій, Краткая грамматика монг. книжнаго языка, Kasan, 1835. 197 p.

[2]) Вл. Л. Котвичъ, Лекцій по грам. монг. языка, St. Pet., 1902.

[3]) А. Д. Рудневъ, Лекцій по грам. монг. письменнаго языка, Выпускъ I, St. Pet., 1905.

[4]) Journal of the Peking Oriental Society, Vol. IV, Peking, 1898, pp. 29—78.

[5]) Das Schriftmongolische und die Urgamundart phonetisch verglichen (Journal de la Société finno-ougrienne, Vol. XXI, 2, Helsingfors, 1902, 55 p.). — Über die Konjugation des Khalkha-Mongolischen (Sonderabdruck aus den Mémoires de la Société finno-ougrienne, Vol. XIX, Helsingfors, 1902, XV und 119 p.). — Mogholica, Beiträge zur Kenntnis der Moghol-Sprache in Afghanistan (Aus Journal de la Société finno-ougrienne, Vol. XXIII, 4, Helsingfors, 1905, III und 60 p., 1 Tafel). — Über mongolische Pronomina (Aus Journal de la Société finno-ougrienne, Vol. XXIII, 3, Helsingfors, 1904, 20 p.). [Manche der hier mitgeteilten Beobachtungen sind nicht neu, sondern bereits von W. Schott gemacht worden.] — Vergl. А. Д. Рудневъ, Защита докторской диссертаціи Г. І. Рамстедта и два его труда въ области монгольской грамматики, St. Pet., 1904, 26 p. (Sonderabdruck aus Труды Троицкосавско-Кяхтинскаго Отдѣленія Имп. Русск. Геогр. Общества, Vol. VI, No. 2, 1903, pp. 98—123).

[6]) Грамматика калмыцкаго языка сочиненная Александромъ Поповымъ, Kasan, 1847. IX und 390 p.

[7]) Grammatik der west-mongolischen, das ist Oirad oder Kalmükischen Sprache, Königsfeld, 1851. 147 p.

12*

der Grammatik von Schmidt mit Übertragung der mongolischen
in kalmükische Schrift ist.

Für das Burjatische kommen die Werke von M. A. CASTRÉN,[1])
G. BÁLINT,[2]) A. ORLOV,[3]) N. A. VOLOŠINOV[4]) und A. D. RUDNEV[5])
in Betracht. Den Dialekt der Khalkha haben VITALE und DE
SERCEY[6]) kurz behandelt, den des Ordos M. G. SOULIÉ[7]) (aber
unzuverlässig). P. MELIORANSKI[8]) hat eine interessante Arbeit
eines arabischen Philologen über die mongolische Sprache veröffentlicht.

Sprachwissenschaftlich ist das Mongolische von W. SCHOTT[9])
betrieben worden, dessen Abhandlungen auch jetzt noch von
Wert sind und auf jeden Fall eine fesselnde Lektüre bilden.

---

[1]) Versuch einer burjätischen Sprachlehre nebst kurzem Wörterverzeichnis, herausgegeben von A. SCHIEFNER, St. Pet., 1857. XV und 244 p.

[2]) Az éjszaki burját-mongol nyelvjárás rövid ismertetése (Nyelv-
tudományi Közlemények, Vol. XIII, Budapest, 1877). Von magyarischen
Arbeiten über das Mongolische sind ferner zu nennen J. BUDENZ, Rövid
mongol nyelvtan (Ibid., Vol. XXI, 1890), und G. BÁLINT, Párhuzam a
magyar és mongol nyelv terén, Budapest, 1877.

[3]) Грамматика монголо-бурятскаго разговорнаго языка, Kasan,
1878. X und 265 p.

[4]) Русско-монголо-бурятскій переводчикъ. 2. Ausgabe, St. Pet.,
1898. XVIII und 108 p.

[5]) Матеріалы для грам. монг. разговорнаго языка. Zap., Vol. XIV,
1902, pp. 036—042.

[6]) Grammaire et vocabulaire de la langue mongole (Dialecte des
Khalkhas), Peking, 1897. VIII und 68 p.

[7]) Éléments de grammaire mongole (Dialecte Ordoss), Paris, 1903.
VII und 87 p.

[8]) Арабъ филологъ о монгольскомъ языкѣ (Arabischer Text, Übersetzung und Erläuterungen). Zap., Vol. XV, 1904, pp. 75—171.

[9]) Besonders zu nennen sind (ausser zahlreichen kleineren Schriften)
die grösseren Abhandlungen Über das Alta'ische oder Finnisch-Tatarische
Sprachengeschlecht, Berlin, 1849; Altajische Studien oder Untersuchungen
auf dem Gebiete der Altai-Sprachen (Abhandlungen der Berliner Akademie,
1860, pp. 587—621); dasselbe, zweites Heft, ibid., 1861, pp. 153—176;
dasselbe, drittes Heft, ibid., 1867, pp. 89—153; dasselbe, viertes Heft,
ibid., 1870, pp. 267—307 (auch unter dem Titel: Die fürwörtlichen An-
hänge in den tungusischen Sprachen und im Mongolischen); dasselbe,
fünftes Heft, ibid., 1872, pp. 1—46. Das Zahlwort in der tschudischen
Sprachenclasse, wie auch im Türkischen, Tungusischen und Mongolischen
(Abhandlungen der Berliner Akademie, 1853, pp. 1—29).

Dagegen haben die vielen älteren allgemein-sprachwissenschaftlichen Werke (MAX MÜLLER, WHITNEY, STEINTHAL, FR. MÜLLER u. a.), in denen das Mongolische gelegentlich einmal gestreift wird, jetzt kaum mehr als eine historische Bedeutung. Das Mongolische hatte das eigentümliche Schicksal, von dieser schematisirenden Schule dazu verurteilt zu sein, als abschreckendes Beispiel einer ewig-starren, unwandelbaren Sprache geschulmeistert zu werden, während nicht nur im Verhältnis der Schriftsprache zu den lebenden Sprachformen, sondern sogar innerhalb der Perioden der Schriftsprache selbst bemerkenswerte Entwicklungsprocesse vor sich gegangen sind. Interessant sind die verschiedenen Schriften von HEINRICH WINKLER,[1] die sich besonders auf die Syntax der altaischen Sprachen beziehen, aber leider von grosser Hast und Flüchtigkeit zeugen. Bemerkenswert ist die vergleichende Grammatik der altaischen Sprachen von JOSEF GRUNZEL.[2]

2. **Lexikographie.** — Sehen wir von den älteren Versuchen kleiner Vokabulare ab,[3] so war in Europa die erste wissenschaftliche Leistung auf dem Gebiet der mongolischen Lexikographie das Wörterbuch von I. J. SCHMIDT.[4] Es lag nicht in dem Plan des Verfassers, wie er im Vorwort bemerkt, ein den ganzen mongolischen Sprachschatz umfassendes Wörterbuch herauszugeben, sondern sich auf eine Auswahl der nö-

---

[1] Besonders Uralaltaische Völker und Sprachen, Berlin, 1884.

[2] Entwurf einer vergleichenden Grammatik der altaischen Sprachen nebst einem vergleichenden Wörterbuch, Leipzig, 1895.

[3] Das älteste Werk derart ist das Vocabularium Calmucko-Mungalicum des Schweden PHILIPP JOHANN VON STRAHLENBERG (Das Nord- und Östliche Theil von Europa und Asia, Stockholm, 1730, S. 137—156). — Das Handwörterbuch der westmongolischen [d. i. kalmükischen] Sprache (Donaueschingen, 1853) von H. A. ZWICK hat nur geringen Wert. Ein kurzes russisch-kalmükisches Wörterbuch (Краткій русско-кальмыцкій словарь) von P. SMIRNOV wurde in Kasan, 1857, (127 p.) veröffentlicht, ein anderes Русско-кальмыцкій словарь von C. GOLSTUNSKI, St. Petersburg, 1860, 136 p.

[4] Mongolisch-Deutsch-Russisches Wörterbuch, nebst einem deutschen und einem russischen Wortregister, von I. J. SCHMIDT, herausgegeben von der Kaiserlichen Akademie der Wissenschaften, St. Petersburg, 1835. 613 p., 4°.

tigsten und gebräuchlichsten Wörter und Wortformen zu beschränken. Die Zahl der von ihm aufgenommenen Wörter beläuft sich auf 13,000, reicht aber dennoch nicht aus, um Werke wie z. B. den von ihm herausgegebenen Sanang Setsen zu lesen. Dagegen dürfte es für die Lektüre leichterer Erzählungen, Jātaka etc., genügen. Die deutschen und russischen Indices am Ende mit Verweisen auf den mongolischen Teil machen die Arbeit auch jetzt noch wertvoll. Einen wesentlichen Fortschritt bedeutet das gross angelegte Wörterbuch von KOWALEWSKI,[1] der viele Jahre unter den Mongolen zugebracht und so Gelegenheit gehabt hatte, einen grossen Teil seiner Materialien an Ort und Stelle zu sammeln und zu controliren. Ausserdem besass er eine grosse Belesenheit in der mongolischen Literatur; die von ihm ausgezogenen Werke sind im Vorwort zusammengestellt und werden häufig unter den Stichwörtern citirt. Er gibt auch Vergleiche aus den tungusischen und türkischen Sprachen sowie Sanskrit, tibetische und chinesische Äquivalente; die tibetischen sind freilich nicht immer correkt, und trotz der grossen Vortrefflichkeit im allgemeinen sind zahlreiche Mängel in Einzelheiten vorhanden. Überholt ist dieses Werk jetzt durch die gediegene Arbeit von GOLSTUNKI,[2] die eine erschöpfende Darstellung des mongolischen Sprachschatzes gewährt und viel neues, bei KOWALEWSKI nicht vorhandenes Material enthält, altes natürlich in verbesserter Form.

3. Chrestomathien. — An europäischen Ausgaben mongolischer Texte, die grösstenteils in Russland gedruckt worden sind, ist kein Mangel. In Peking gibt es zwei Pressen, die sich

---

[1]) Dictionnaire mongol-russe-français, dedié à Sa Majesté l'Empereur de toutes les Russies, par JOSEPH ÉTIENNE KOWALEWSKI, Kasan, 1844, 1846, 1849. 3 Vols., 2690 p., 4°. Auf dem Titelblatt des zweiten und dritten Bandes ist hinzugefügt: couronné par l'Académie impériale des sciences.

[2]) Монгольско-русскій словарь, составленный К. Ѳ. Голстунскимъ, St. Petersburg, Vol. I, 1895, 268 p. Vol. II, 1894, 462 p. Vol. III, 1893, 491 p. Дополненія къ монгольско-русскому словарю, 1896, 89 p., 4°, autographirt. Dass dieses Wörterbuch nicht im Druck erschienen ist, kann nicht genug bedauert werden; die autographirte russische Currentschrift ist für die Augen eines nicht-russischen Lesers keineswegs eine Freude.

ungefähr vom **Anfang des achtzehnten Jahrhunderts** bis in die neuere Zeit mit der Herstellung mongolischer (wie tibetischer) Holzdrucke in guten, lesbaren und billigen Ausgaben befasst haben; dieselben sind verhältnismässig leicht zu beschaffen, wiewohl viele jetzt vergriffen sind. Für das erste Studium empfehlen sich die folgenden drei Chrestomathien, die eine treffliche Gesamtübersicht über das Gebiet der mongolischen (mit Ausschluss der kalmükischen) Literatur gewähren und in die mannigfaltigen Stilarten der einzelnen Literaturgebiete einführen. Die Chrestomathie von Popov[1] gibt eine Auswahl von zehn Stücken erzählenden und historischen Inhalts mit Commentar und hat den Vorzug, mit einem kurzen, aber recht übersichtlichen Glossar ausgestattet zu sein, das dem Anfänger viel Zeit und Mühe erspart. Auf einer weit breiteren Basis ist das ausgezeichnete zweibändige Werk von Kowalewski[2] angelegt, das gleichzeitig sehr geeignet ist, in die Gedankenwelt und Geschichte des Buddhismus einzuführen. Die Erläuterungen sind sehr eingehend und instruktiv, in einem guten Index registrirt und auch jetzt noch brauchbar. Die neueste Arbeit auf diesem Gebiete ist die von A. Pozdnějev,[3] der seine Vorgänger an Mannigfaltigkeit und Reichhaltigkeit des Materials überflügelt und die Fülle der mongolischen Literatur in allen ihren Zweigen vor Augen führt. So dankenswert diese Auslese, besonders neu gebotenen Stoffes, ist, so lässt doch die Art und Weise der Edition viel zu wünschen übrig und steht hinter dem Werke von Kowalewski ganz entschieden zurück. Die Texte sind ohne alle Noten abgedruckt, und das Buch hat seltsamer

---

[1] Монгольская Христоматія для начинающихъ обучаться монгольскому языку, изданная Александромъ Поповымъ, Kasan, 1836, zwei Teile. I. Texte, pp. 1—144; Commentar, pp. 145—318. II. Glossar, 199 p.

[2] Монгольская Хрестоматія, изданная Осипомъ Ковалевскимъ, Kasan, 1836, 1837, 2 Vols. Vol. I, XVI, Texte, pp. 1—243; Commentar, pp. 247—588. Vol. II, Texte, pp. 1—207; Commentar, pp. 211—532; Index, pp. 533—594.

[3] Монгольская Христоматія для первоначальнаго преподаванія составленная А. Позднѣевымъ, съ предисловіемъ Профессора Н. И. Веселовскаго. Изданія Факультета Восточныхъ Языковъ Имп. С.-Пет. Университета, No. 7, St. Pet., 1900. XVIII und 416 p. Texte, 4°.

Weise nicht einmal ein Inhaltsverzeichnis, von einem Index
ganz zu schweigen. Die Bemerkung auf dem Titelblatt, dass es
für den «Anfängerunterricht» bestimmt sei, bedarf daher wohl
einer gewissen Einschränkung; unter Anleitung eines sachkun-
digen Lehrers vermag es der Anfänger natürlich zu benutzen,
aber wie soll der auf sich und seine Hülfsmittel allein ange-
wiesene — und das ist doch gewiss die Mehrzahl der ausserhalb
Russlands das Mongolische Studirenden — damit fertig werden? Manche Texte sind ohne Kenntnis des Chinesischen oder
Tibetischen gar nicht verständlich; wie soll z. B. der Anfänger
ohne Anweisung die vielen Transkriptionen tibetischer Namen
auf p. 379 (z. B. smon-lam rab abyams-pa bstan-adzin grags-pa,
u. a.) verstehen? Selbst in mongolischen Drucken werden solche Fremdnamen stets an der Seite links in tibetischer Schrift
glossirt.

Die mongolische Sprache in China und Korea, und einheimi-
sche Lexikographie.

Für die Chinesen hat die mongolische Sprache seit den Tagen
der Yüan-Dynastie stets eine grosse Bedeutung gehabt, die sich
im Zeitalter der Ming und der Ts'ing eher gesteigert als verringert
hat; denn nach der Eingliederung der Mongolei in das Verwaltungs-
system des Reiches war die Unterhaltung eines administrativen
Amtes erforderlich, zu dessen Obliegenheiten naturgemäss die
Übersetzung chinesischer Edikte und Gesetze in die Sprache
der Mongolen und Abfassung mongolischer Dokumente gehörte.
Die Chinesen haben daher auch nicht die Mühe gescheut, Beiträge zur mongolischen Grammatik und Lexikographie zu liefern,
und an der in Peking zur Ausbildung von Dolmetschern in
fremden Sprachen bestehenden Schule nahm das Mongolische
die vornehmste Stellung ein.[1]) Diese Schule wurde im Jahre

---

[1]) Vergl. F. HIRTH, The Chinese Oriental College (Journal of the
China Branch of the Royal Asiatic Society, New Series, Vol. XXII, Shanghai, 1888, pp. 203—219). G. DEVÉRIA, Histoire du collège des interprètes
de Péking (Fragment), Mélanges Harlez, Leiden, 1896, pp. 94—102. S.
auch ABEL-RÉMUSAT, De l'étude des langues étrangères chez les Chinois,
Mélanges asiatiques, Vol. II, Paris, 1826, pp. 242—265.

1407 unter dem Kaiser Yung-lo aufs neue organisirt und in acht Klassen eingeteilt: die erste derselben war die mongolische. An der Spitze der Anstalt standen 38 Direktoren oder Lehrer, Examina wurden jährlich abgehalten, und Belohnungen in Form von Promotionen an die Studenten verliehen. Im Jahre 1644 wurde die Schule unter der neuen Dynastie officiell als ein Staats-Institut wieder eröffnet, und unter den zehn damals gelehrten Sprachen stand das Mongolische wieder an der Spitze; 1658 soll es als Unterrichtsgegenstand aufgegeben worden sein, eine Nachricht, die uns seltsam anmutet, da gerade unter den folgenden Kaisern K'ang-hsi und K'ien-lung das Mongolische am Kaiserhofe mit Vorliebe literarisch gepflegt wurde und während des achtzehnten Jahrhunderts, wie auch noch gegenwärtig, für amtliche Zwecke in stetem Gebrauch war. ABEL-RÉMUSAT,[1] dem wir diese Nachrichten verdanken, sucht diesen Umstand dadurch zu erklären, dass er sagt, das Mongolische sei zu verbreitet gewesen, als dass man noch Dolmetscher nötig gehabt hätte; dies ist indessen nicht ganz plausibel, da doch früher, als die Dolmetscher bestanden, die Sprache mindestens ebenso weit verbreitet war, und von einer grösseren Ausdehnung derselben nach jener Zeit auch kaum die Rede sein kann. Prof. HIRTH entdeckte eine handschriftliche Ausgabe des Werkes *Hua i yi yü* in 24 Bänden, in welchem die in der Dolmetscherschule getriebenen Sprachen für die Bedürfnisse von Chinesen dargestellt sind. Der erste Band enthält ein sachlich geordnetes mongolisches Vokabular,[2] die chinesische Wortbedeutung in der Mitte, das mongolische Wort in Originalschrift rechts davon, und die Aussprache in chinesischen Schriftzeichen zur Linken. Im achtzehnten Bande befindet sich eine Sammlung mongolischer Aktenstücke mit chinesischer Übersetzung.

---

[1] Recherches sur les langues tatares, Paris 1820, p. 220.

[2] A. WYLIE (Notes on Chinese Literature, 2. Ausg., Shanghai, 1901. p. XX) bemerkt, dass noch ein Vokabular der mongolischen Sprache vorhanden sei, unter dem Titel *Hua i yi yü*, im Jahre 1382 von einer kaiserlichen Commission ausgearbeitet, 15 Jahre nach der Unterdrückung der Yüan Dynastie. Es ist nicht unwahrscheinlich, ja, recht denkbar, dass das Hirth'sche Vokabular in seiner Grundlage auf diese oder eine ähnliche Quelle aus der Yüan-Zeit zurückgeht.

Lange bevor sie Herren von China wurden, fanden es die Manju bei ihrem diplomatischen Verkehr mit den Mongolen nützlich, eine gewisse Fertigkeit in ihrer Literatur zu erlangen, und sie sollen sogar junge Leute im Studium der mongolischen Sprache geschult haben.[1]) Von diesem Einfluss legt vor allem die Bildung der manjurischen Schrift beredtes Zeugnis ab, die nach dem Vorbild der mongolischen gemodelt ist.

Von grösstem Interesse ist die Frage, inwieweit das Mongolische die chinesische Sprache zu beeinflussen vermocht hat. Die Annalen der Yüan-Dynastie wimmeln naturgemäss von mongolischen Eigennamen, doch gleichfalls von Lehnwörtern aus dem Mongolischen; aber auch stilistische und phraseologische Eigentümlichkeiten jener Epoche mögen auf mongolischen Einfluss zurückzuführen sein. Ein umfangreiches Material zur Beurteilung dieser Frage hat CHAVANNES[2]) gesammelt, der in der chinesischen Kanzleisprache des dreizehnten Jahrhunderts eigentümliche Wendungen und Formeln findet, deren Einfluss sich noch gegenwärtig sowohl im officiellen Stil als in der Umgangssprache fühlbar macht. CHAVANNES gelangt indessen zu der Ansicht, dass sich nicht alle diese Erscheinungen aus dem Mongolischen erklären lassen, und dass die Ursachen, welche das Chinesische damals so stark modificirt haben, noch nicht genau bestimmt werden können.[3])

Unter der Ming-Dynastie scheint das Mongolische in China wesentlich unter dem Gesichtspunkt des von der Politik diktirten Nutzens gewürdigt worden zu sein (s. oben), doch erwähnt RÉMUSAT[4]) ein sachlich geordnetes mongolisch-chinesisches Wörterbuch eines Han-lin Ho-yüan-kiei (?).

Aus dem Zeitalter der Ming besitzen wir sonst nur wenige

---

[1]) RÉMUSAT, Recherches sur les langues tatares, Paris, 1820, p. 219, und WYLIE, Chinese Researches, Shanghai, 1897, Part IV, p. 261.

[2]) Inscriptions et pièces de chancellerie chinoises de l'époque mongole, T'oung Pao, 1904, pp. 357—447; 1905, pp. 1—42.

[3]) T'oung Pao, 1904, pp. 417—418. Über eine dem Mongolischen entlehnte interessante grammatische Bildung vergl. ibid., p. 389, Note 1.

[4]) Recherches sur les langues tatares, Paris, 1820, p. 218; doch die Angabe, dass das Mongolische in uigurischer (Kao-ch'ang) Schrift geschrieben war, muss Zweifel erwecken.

mongolische Dokumente. Dazu gehört ein merkwürdiger Brief, den POZDNÉJEV von einem Lama im Tempel *Yung-ho kung* in Peking 1893 erlangte. Dieser Brief wurde im Jahre 1580 von Altan Khan (1504—1582) an den Kaiser Wan-li gerichtet, um ihm die Absendung des gewöhnlichen Tributs zu melden. Das Schriftstück ist chinesisch und mongolisch abgefasst, letzteres eine recht unvollkommene Übersetzung des chinesischen Textes, die wohl nur von einem Chinesen angefertigt worden sein kann. Der Übersetzer hat nämlich jedes chinesische Wort in einem Wörterbuch aufgesucht und es gerade in der Form übernommen, wie er es dort vorfand; so hat er Verbalformen im Imperativ abgeschrieben, selbst wenn der Sinn des Textes einen anderen Modus verlangte; versagte das Wörterbuch, so transkribirte er das chinesische Wort auf mongolisch. Der mongolische Text hat also kaum einen Wert und verdient gar nicht zur mongolischen Literatur gezählt zu werden. Noch viel weniger aber kann man ihn, wie CHAVANNES[1] versucht hat, als ein Beweisstück für den unglaublichen Verfall der mongolischen Literatur nach den Yüan in Anspruch nehmen und der Meinung sein, Altan Khan habe niemand in seiner Umgebung gehabt, der fähig gewesen wäre, einen correkten mongolischen Brief zu schreiben.[2]

Unter der gegenwärtigen Dynastie, namentlich im achtzehnten Jahrhundert, stand das Studium des Mongolischen in China in grosser Blüte. Im Jahre 1780 erschien unter dem Titel *San ho pien lan* (12 Vols.) eine Übersicht der chinesischen, manjurischen und mongolischen Sprache, die im ersten Kapitel eine Manju Grammatik mit chinesischer Erklärung, im zweiten eine vergleichende Zusammenstellung der manjurischen und mongolischen Sprachformen enthält, nebst einem mongolischen

---

[1] Journal Asiatique, Janv.-Févr., 1896, p. 179.

[2] Der Brief ist veröffentlicht und übersetzt von A. POZDNÉJEV, Ein neu entdecktes Denkmal der mongolischen Literatur aus der Zeit der Ming-Dynastie (in Восточныя Замѣтки. Сборникъ статей и изслѣдованій профессоровъ факультета восточныхъ языковъ С. П. Унив., St. Pet., 1895, pp. 367–386, 1 Tafel). CHAVANNES (Journal Asiatique, Janv.-Févr., 1896, pp. 173–179) hat eine ausführliche Analyse der Arbeit mit Übersetzung des Briefes gegeben.

Syllabar und einigen Regeln über die Aussprache der Casus-
endungen. Das letztere Kapitel ist von H. C. v. D. GABELENTZ[1])
bearbeitet worden. Der wertvollste Bestandteil dieses Werkes
sind aber die Bände 2—10, die ein vergleichendes Wörterbuch
enthalten, in der ersten Columne Manju in alphabetischer Ordnung, in der zweiten chinesisch (Umgangssprache), in der dritten
mongolisch in Originalschrift, in der vierten Umschrift des
Mongolischen in Manju.[2]) Da in der Manju Schrift jeder Laut
durch ein eigenes Schriftzeichen ausgedrückt wird, lässt sich
vermittelst dieses Vokabulars die Aussprache der mongolischen
Wörter leicht erlernen; ja, es mag demselben ein gewisser historischer Wert zukommen, da darin offenbar die Aussprache jener
Zeit, des Endes des achtzehnten Jahrhunderts, festgelegt ist.

Die grosse Bedeutung, welche das Mongolische für die
Manju in der damaligen Zeit besass, zeigt sich aber vor allem
in einer Reihe auf kaiserlichen Befehl verfasster grosser Wörter-
spiegel. Dazu gehört der auf Befehl des Kaisers K'ang-hsi im
Jahre 1708 von einem Stabe von Gelehrten herausgegebene
«Spiegel der Manju und mongolischen Sprache» in 21 Bänden,[3])

---

[1]) Mandschu-mongolische Grammatik aus dem Sân-ho-pián-lân
übersetzt. Zeitschrift für die Kunde des Morgenlandes, Band I, Heft 3,
Göttingen, 1837, S. 255—286. Vergl. P. G. v. MÖLLENDORFF, Essay on
Manchu Literature (Journal of the China Branch of the Royal Asiatic
Society, Vol. XXIV, 1889) p. 7, No. 2. — Nach GABELENTZ ist das Buch
1772 gedruckt worden, nach MÖLLENDORFF 1780, nach J. ZACHAROV (Пол-
ный маньчжурско-русскій словарь, St. Pet., 1875, p. XVIII) 1792. Es mag
ja verschiedene Ausgaben zu jener Zeit gegeben haben; ein Exemplar,
das ich in Peking erlangte, hat ein Vorwort datirt 1780 (chinesisch K'ien-
lung kéng-tse; manju-mongolisch: weisses Rattenjahr der Periode K'ien-
lung), und daher gebe ich einstweilen diesem Datum als dem der Verfasserschaft des Werkes den Vorzug. Verfasser desselben ist ein Manju
namens FUGIYÔN (chinesisch Fu-tsung).

[2]) Band 11 und 12 enthalten ein Ergänzungsvokabular, besonders
von Composita, gleichfalls alphabetisch nach dem Manju geordnet, nicht
aber, wie MÖLLENDORFF sagt, kurze Sätze in drei Sprachen.

[3]) P. G. v. MÖLLENDORFF, Essay on Manchu Literature, p. 12, No.
34. Vergl. M. ABEL-RÉMUSAT, Notice sur le dictionnaire intitulé: Miroir
des langues mandchoue et mongole (Notices et Extraits des Manuscrits
de la Bibliothèque du Roi, Vol. XIII, 1-re partie, 1838, pp. 1—41), der
die Manju und mongolischen Texte der beiden Vorreden mit Übersetzung
gibt; HIMLY, T'oung Pao, Vol. VI, p. 258.

in dem das Manju durch das Mongolische erklärt wird. Unter den von K'ien-lung publicirten Wörterbüchern kommen für das Mongolische drei Ausgaben in Betracht, ein Spiegel des Manju mit mongolischer und chinesischer Übersetzung in 24 Bänden, herausgegeben 1780 oder etwas später, ein viersprachiges Wörterbuch in 10 Bänden, dem noch das Tibetische hinzugefügt ist,[1] und das grosse viersprachige Wörterbuch in 36 Bänden.[2] Diese verschiedenen Ausgaben sind alle nach demselben Schema, einer vom chinesischen Standpunkt gemachten Einteilung in Sachen, angelegt.[3] Die Benutzung des Mongolischen ist daher für uns sehr erschwert, aber die Chinesen selbst haben diesem Übelstand durch eine andere Arbeit abgeholfen.

Zu ihren vorzüglichsten Leistungen nämlich in dem Felde der mongolischen Philologie gehört das grosse mongolisch-chinesisch-manjurische Wörterbuch (in 16 Bänden, 4°, nebst einem Bande die Einleitung enthaltend), von dem auf Antrag des *Li-fan-yüan* im Jahre 1891 eine vorzügliche Neuausgabe gedruckt worden ist *(Mongol-un üsük churiyāksan bičik)*. Dies

---

[1] P. G. v. MÖLLENDORFF, l. c., p. 13, No. 43, und p. 14, No. 44.

[2] Ohne Vorrede und Datum, bei v. MÖLLENDORFF nicht erwähnt. Der chinesische Titel ist *Yü ch'i ssŭ t'i Ts'ing wên kien*. In dieser glänzend gedruckten Ausgabe befinden sich je vier Wörter auf jeder Seite, erste Zeile Manju, zweite tibetisch, dritte mongolisch, vierte chinesisch, ohne Hinzufügung von Transkriptionen. — Das British Museum besitzt ferner ein Unicum, ein fünfsprachiges Wörterbuch (36 Vols. in Gross-Folio) von hervorragend schöner Handschrift, die nur zum kaiserlichen Gebrauch bestimmt gewesen sein kann, worauf auch die Einbände in gelber Seide hinweisen. Inhaltlich stimmt dieser Thesaurus genau mit dem vorher genannten viersprachigen Wörterbuch überein: die fünfte Sprache ist Djagatai-Türkisch. Das Djagataische ist zuerst in Originalschrift und dann in manjurischer Transkription gegeben; das Tibetische ist sogar jedesmal mit zwei Manju Transkriptionen versehen, die eine wörtlich der tibetischen Schrift folgend, die andere die Aussprache reproducirend. Dieses Werk scheint niemals gedruckt worden zu sein; ich habe bei meinem Aufenthalt in Peking eifrige Nachforschungen angestellt, es ist aber kein weiteres Exemplar aufzutreiben.

[3] Wer keines der Originale zur Hand hat, kann sich über das Anordnungsprincip und den Inhalt aus J. KLAPROTH, Verzeichnis der chinesischen und mandschuischen Bücher und Handschriften der Kgl. Bibliothek zu Berlin (Paris, 1822, S. 97 108) orientiren.

zeigt, dass auch noch gegenwärtig bei den zuständigen Behörden das Interesse am Mongolischen aufrechterhalten und gepflegt wird. Das Wörterbuch ist nach dem mongolischen Alphabet geordnet und bei seinem schönen Druck ein sehr bequemes Nachschlagewerk, das stets befragt werden sollte, wenn Kowalewski oder Golstunski versagen oder im Zweifel lassen.[1]

Dass Chinesen und Manju sich auch noch in neuerer Zeit dem Studium des Mongolischen hingeben, geht unter anderen Anzeichen aus der Tatsache hervor, dass im Jahre 1830 ein Konversationsbuch zum praktischen Gebrauch in den drei Sprachen gedruckt worden ist; dies hat den Brigadegeneral des roten Banners und Prinzen des Aimak der Barin, *Dämäk*, zum Verfasser, der zu diesem Zweck einen zur Zeit K'ien-lung's verfassten Manju Text in seine Muttersprache verdolmetschte, sich aber bei der Niederschrift des Manju Alphabets bediente, um den grösseren Lautreichtum der Umgangssprache anschaulicher darzustellen, wozu die mongolische Schrift bei der vielfach vorkommenden Mehrdeutigkeit mancher Buchstaben kaum ausgereicht hätte. Aus diesem Grunde sind die Texte von grossem Interesse, indem sie einmal die Phonetik und dann den ganzen Stil der Umgangssprache zu veranschaulichen suchen. Prof. W. GRUBE hat sich daher ein schätzenswertes Verdienst durch eine kritische Ausgabe des mongolischen Textes in genauer Umschrift erworben.[2] Der chinesische Text ist mit Übersetzung in WADE's

---

[1] Ein von mir in Peking erworbenes Exemplar befindet sich jetzt im Chinese Department der Columbia University, New-York. Prof. A. POZDNÉJEV hat begonnen, dieses Wörterbuch herauszugeben und als Beilage in den Nachrichten des Orientalischen Instituts von Wladiwostok erscheinen zu lassen (unter dem Titel Монгольско-Китайско-Маньчжурскій словарь въ русско-французскомъ переводѣ). Bisher habe ich davon die Seiten 1—48 gesehen (ohne Vorwort), woraus hervorgeht, dass es sich um einen genauen Abdruck des obigen mongolisch-chinesischen Wörterbuchs handelt.

[2] Proben der mongolischen Umgangssprache, WZKM, Band XVIII, 1904, S. 343—378; Band XIX, 1905, S. 29—61. Der Text dieses oder eines ganz ähnlichen Buches muss übrigens bereits vor 1830 publicirt worden sein, denn schon TIMKOWSKI (Reise nach China durch die Mongoley in den Jahren 1820 und 1821, Band II, Leipzig, 1825, S. 359) erwähnt in der Liste seiner in Peking erworbenen Bücher «Hundert Gespräche in mongolischer Sprache, mit chinesischer Übersetzung». Auch werden die

«Tzŭ êrh chi» abgedruckt;[1]) mit noch grösserem Nutzen wird man aber den Manju Text zu Rate ziehen, den P. G. v. Möllendorff[2]) in Originalschrift mit Noten und Übersetzung veröffentlicht hat.

Im Jahre 1848 veröffentlichte der Minister Saišangga, ein Manju, ein Compendium des Mongolischen in vier Bänden, das Manju Wörter und Redensarten mit chinesischer und mongolischer Übersetzung enthält.[3]) Fast aus derselben Zeit, der diese Lehrbücher angehören, stammt eine Bemerkung von Jakinf Bičurin, die hier einen Platz verdient:

«Zur Bildung der Jugend in der mongolischen Literatur sind von der chinesischen Regierung Schulen in Kalgan und Peking errichtet. In der ersteren werden čakharische Kinder in der manjurischen und mongolischen Sprache unterrichtet. In Peking ist für die ihre Literatur studirenden Mongolen eine Prüfung festgesetzt, bei welcher Aufgaben zum Übersetzen von Werken in Prosa und Versen gegeben werden. Die zu Prüfenden sind in drei Classen abgeteilt, nämlich in solche, die Studenten, solche, die Candidaten, und solche, die Magister werden.»[4])

Das Studium der mongolischen Sprache ist auch bis nach Korea gedrungen. Die regirende Dynastie dieses Landes hat das schon früher bestehende Amt der Dolmetscher in vier Sek-

mongolische und die dreisprachige Ausgabe im Katalog von Kasan vom Jahre 1834 unter Nr. 17 und 18 angeführt.

[1]) Ist aber leider in der 3. Auflage (Shanghai, 1904) fortgelassen.

[2]) A Manchu Grammar with analysed Texts, Shanghai, 1892, pp. 15—50. Vergl. G. v. d. Gabelentz, ZDMG, Band XVI, 1862, S. 539.

[3]) P. G. v. Möllendorff, Essay on Manchu Literature, l. c., p. 7, Nr. 3. Der Name Saišangga bedeutet nicht, wie Möllendorff übersetzt, the praiser, sondern der Lobenswerte. Der dreisprachigen Vorrede zu seinem Buche, von dem ich ein Exemplar aus Peking mitgebracht habe, entnehme ich die Angabe, dass der Verfasser besonders günstige Gelegenheit zum Studium des Mongolischen gehabt hatte. Im Jahre 1811, als er nur vierzehn Jahre alt war, begleitete er seinen Vater, der im Auftrag des *Li-fan-yüan* (der Behörde in Peking für die Verwaltung der auswärtigen Provinzen) als Postdirektor nach Sairusu geschickt wurde, in die Mongolei, und gab sich dort dem Studium der mongolischen Literatur hin.

[4]) Denkwürdigkeiten über die Mongolei von dem Mönch Hyakinth. Aus dem Russischen übersetzt von K. F. v. d. Borg, Berlin (G. Reimer), 1832, S. 152.

tionen für das Studium des Chinesischen, Mongolischen, Japanischen und Niüchen eingeteilt (letzteres später durch das Manju ersetzt). Im Jahre 1469 waren diese vier Abteilungen vollständig constituirt, was aus den zu dieser Zeit veröffentlichten Regierungsstatuten hervorgeht, welche die Liste der in jeder Abteilung studirten Bücher mitteilen. Damals war natürlich die Macht der Mongolen schon seit einem Jahrhundert gebrochen, und wenn trotzdem das Studium des Mongolischen beibehalten wurde, so zeigt dies, dass die Beziehungen der Koreaner zu den Mongolen auch damals noch eine gewisse Bedeutung hatten. Von den in den Statuten von 1469 angeführten Büchern ist eine grosse Zahl verloren gegangen, einige existiren noch in Ausgaben aus dem achtzehnten Jahrhundert, die sich in der Bibliothek der Ecole des Langues Orientales in Paris befinden; die koreanischen Nachrichten über diese Bücher stammen vom Ende des siebzehnten Jahrhunderts. MAURICE COURANT[1]) zählt zwanzig derartige Werke auf, darunter Sprachlehrbücher, eines enthaltend mongolischen Text mit koreanischer Transkription und Übersetzung der Redensarten am Schlusse, gedruckt 1741, ein anderes ähnlichen Inhalts mit dem Titel «Erklärung der mongolischen Sprache», 1737 gedruckt und 1744 bei den mongolischen Examina gebraucht. In der königlichen Bibliothek zu Söul existiren auch eine Geschichte des Studiums des Mongolischen und ein nach Materien geordnetes mongolisches Wörterbuch in mehreren Bänden.

Sehr umfangreich und bedeutend ist die lexikographische und grammatische Literatur der lamaischen Völker. In Tibet wurzelnd, ist weitaus der grösste Teil derselben ins Mongolische übertragen worden und in tibetisch-mongolischen Paralleldrucken zugänglich. Ich berühre sie hier nur insoweit, als das Mongolische in Betracht kommt. Obenan steht das kolossale Werk des Kun·dga rgya-mts'o («Der Ocean der Namen», *min-gi rgya-mts'o*), das 1718 publicirt wurde und seinen etwas pompösen Titel nicht so ganz zu Unrecht trägt; denn es ist in der Tat eine bewun-

---

[1]) Bibliographie coréenne, tableau littéraire de la Corée. Vol. I, Paris, 1895, pp. CLXXV—CLXXVI, 93—99. Supplément à la bibliographie coréenne, Paris, 1901, p. 4.

dernswerte philologische Leistung ersten Ranges. Es gibt davon zwei verschiedene Holzdruckausgaben mit gleichlautendem Gesamttitel, von denen die eine in vier Büchern 367 grosse Folioblätter umfasst.

Ein beachtenswertes und recht brauchbares Werk ist auch das unter dem Namen *Bod-kyi brda-yig rtogs-par sla-ba* («Die tibetischen Sprachregeln in leicht fasslicher Darstellung») gehende tibetisch-mongolische Wörterbuch, von dem I. J. Schmidt zur Abfassung seines tibetischen Lexikons ausgiebigen Gebrauch gemacht hat.[1] Abel-Rémusat hat den mongolischen Text der Einleitung abgedruckt und übersetzt.[2] Das Werk ist nach Art chinesischer Bücher in Folio gedruckt und besteht aus drei Teilen (im ganzen 190 fols.). Der erste Teil enthält ein Wörterbuch nach dem tibetischen Alphabet geordnet und einen Abschnitt über synonymische Glossen, verfasst auf Befehl des Kaisers K'ang-hsi (1662--1722) und des Dalai Lama auf Grund früherer Quellenwerke (aufgezählt fol. 173); darauf folgt ein ergänzender Teil von 12 fols. ebenfalls in Anordnung des tibetischen Alphabets, verfasst 1737 von einem Oirat Gūši Biligun Dalai und einem Khalkha Gūši bLo-bzaṅ bzod-pa, endlich ein kleiner Nachtrag von 4 fols.

Der lCaṅ-skya Lalitavajra (Rol-pai rdo-rje) und bLo-bzaṅ bstan-pai ñi-ma verfassten in Gemeinschaft, um die Übersetzung des Tanjur ins Mongolische zu erleichtern, ein grosses vergleichendes Wörterbuch der tibetischen und mongolischen Sprache *(Bod Sog-gi skad gñis šan sbyar)*, in welchem sie nicht nur die rein religiöse Literatur, sondern auch Werke über Logik, Sprachwissenschaft, alte und neue Orthographie, Kunsttechnik und Medicin verarbeiteten.[3] Dieser Thesaurus diente dann als Grundlage für die mongolische Tanjur-Übersetzung.

---

[1] Ein Exemplar befindet sich unter der Signatur Zx **2904** fol. in der Kgl. Bibliothek zu Berlin, mit dem Vermerk «sehr selten». In Peking ist jedoch das Werk noch aufzutreiben, und ich selbst habe dort ein Exemplar davon erlangt.

[2] In seiner Abhandlung Notice sur le Dictionnaire intitulé: Miroir des langues mandchoue et mongole (Notices et Extraits des Manuscrits de la Bibliothèque du Roi, Vol. XIII, première partie, 1838, pp. 42—61).

[3] G. Huth, Geschichte des Buddhismus in der Mongolei, Band II,

Ein fünfsprachiges Wörterbuch in zwei Bänden (98 und 96 fols.) zur Zeit K'ien-lung's verfasst und 1783 von Amiot in zwei Exemplaren an die Bibliothèque du Roi in Paris geschickt,[1]) ist von Rémusat[2]) besprochen und herausgegeben worden; es ist buddhistischen Inhalts[3]) und enthält unter anderem die Namen der Buddhas, der 32 Kennzeichen und der 80 Schönheitsmerkmale eines Buddha, soweit es Rémusat behandelt hat. Das ganze Wörterbuch ist eine Bearbeitung der Mahāvyutpatti durch den lCaṅ-skya Khutuktu von Peking.[4])

Für buddhistische Studien und die Lektüre buddhistischer Schriften ist das von A. Schiefner herausgegebene dreisprachige Wörterbuch sehr nützlich.[5]) Es enthält in 71 sachlich geordneten Abschnitten die buddhistische Terminologie nach dem grossen im Tanjur (Sūtra, Vol. 123) abgedruckten Wörterbuch Mahāvyutpatti.

S. 292. Zweifellos ist jene Arbeit auf der Mahāvyutpatti basirt und vermutlich identisch mit dem von Wasiljef (Mélanges asiatiques, Vol. II, St. Pet., 1856, p. 383) erwähnten Werke *sGra sbyor bam gñis*, einem als klassisch geltenden Commentar zur Mahāvyutpatti, welcher hauptsächlich der Übersetzung aus dem Tibetischen ins Mongolische dienen sollte; zu demselben Zweck verfasste nach Wasiljef der lCaṅ-skya Khutuktu ein Werk *mK'as-pai ʿbyuṅ-gnas* («Fundgrube der Weisen»), von dem nach Schiefner (Mélanges asiatiques, Vol. I., St. Pet., 1852, p. 411) auch eine Übersetzung existirt.

[1]) Mémoires concernant les Chinois, Vol. XI, p. 616.

[2]) Abel-Rémusat, San, si-fan, man, meng, han tsi yao ou Recueil nécessaire des mots Sanskrits, Tangoutains, Mandchous, Mongols et Chinois. Fundgruben des Orients (Mines de l'Orient), Band IV, Wien, 1814, S. 183—201, 1 Tafel. Derselbe, Sur un vocabulaire philosophique en cinq langues, in seinen Mélanges asiatiques, Vol. I, Paris, 1825, pp. 153–183 (vergl. auch pp. 452–454; dies ist ein etwas erweiterter Abdruck der früheren Arbeit). Später hat C. de Harlez den Sanskrit-tibetischen Text desselben Wörterbuchs herausgegeben und übersetzt (s. Le Muséon, Vol. X, Louvain, 1891, p. 139).

[3]) Der erste Band enthält 35, der zweite 36 Sektionen.

[4]) Vergl. A. Schiefner's Vorwort zur Ausgabe der Buddhistischen Triglotte, St. Pet., 1859.

[5]) Buddhistische Triglotte, d. h. Sanskrit-Tibetisch-Mongolisches Wörterverzeichnis, gedruckt mit den aus dem Nachlass des Barons Schilling von Canstadt stammenden Holztafeln und mit einem kurzen Vorwort versehen von A. Schiefner. St. Petersburg, 1859. Im Langformat tibetischer Bücher, 37 fols.

Schrift und Druck.

Schrift. — Vor dem Anfang des dreizehnten Jahrhunderts ist bei den mongolischen Stämmen von dem Gebrauch einer Schrift keine Rede. Beides, Schrift und Literatur, ist ihnen durch die civilisatorische Mission des Buddhismus zuteil geworden. Im Laufe eines Jahrhunderts haben sie in der Hauptsache drei Schriftsysteme gepflegt, die Schrift der Uiguren seit etwa 1204, ‚P‘ags-pa oder die sogenannte Quadratschrift seit 1269, und die eigentliche mongolische Schrift seit 1311. Die praktische Verwendung der Schrift beginnt langsam bereits unter Činggis Khan; im Verkehr mit der Kin-Dynastie von China bediente man sich damals der chinesischen Charaktere, im amtlichen Austausch von Dokumenten mit anderen Nationen aber der Schrift der Uiguren,[1] die bekanntlich von dem syrischen Estranghelo Alphabet abgeleitet ist und durch nestorianische Missionare in Centralasien verbreitet worden war. Ein genaues Datum für die Einführung der chinesischen Schrift lässt sich nicht aufstellen; im allgemeinen mag die Zeit von 1210 bis 1220 dafür gelten; das Jahr 1219 lässt sich mit Sicherheit für den Gebrauch der chinesischen Schrift in Anspruch nehmen.[2]

Der taoistische Mönch Ch‘ang Ch‘un freilich, der auf die Aufforderung von Činggis eine Reise an dessen Hof unternahm, welche ihn in den Jahren 1221 bis 1224 durch Centralasien nach Persien und an die Grenzen Indiens brachte, erzählt, dass die Mongolen keine Schrift besässen, dass sie ihre Angelegenheiten durch mündliche Vereinbarung regelten und gewisse Zeichen in Holz einschnitten (d. h. sich der Kerbhölzer bedienten) an Stelle von Contrakten. Nur wenige Jahre nach dem

---

[1] S. Pozdnéjev, Lit., pp. 8—9. Über die Geschichte der uigurischen Schrift s. W. Radloff, Das Kudatku Bilik, Teil I, St. Pet., 1891, pp. LXXXIII et seq.

[2] Ein schönes Beispiel für den Gebrauch des Chinesischen zu jener Zeit hat Chavannes in seiner fesselnden Arbeit Inscriptions et pièces de chancellerie chinoises de l’époque mongole, T‘oung Pao, 1904, pp. 368—372, in einem Edikt von Činggis vom 11. April 1223 gegeben, das Ch‘ang Ch‘un von Samarkand nach China mitnahm.

13*

Besuch des Reisenden in der Mongolei wurde die uigurische
Schrift von Činggis bei seinem Volke eingeführt (nach PALLA-
DIUS),[1] doch sie muss früher in Gebrauch gewesen sein. denn
Ch῾ang Ch῾un selbst berichtet, dass Činggis seine Gesandten
mit goldenen Täfelchen (p῾ai-tse) als Zeichen ihrer Machtbefugnis
ausstattete, die mit Inschriften versehen waren, dass er Gesetze
erliess, Briefe, Dokumente und offenbar auch tagebuchartige
Aufzeichnungen schrieb. Ja, als Činggis einst mit Ch῾ang Ch῾un's
Antworten besonders zufrieden war, sagte er, wie Ch῾ang Ch῾un
erzählt: «Deine Worte liegen mir am Herzen», und liess das
Gespräch in uigurischer Schrift aufzeichnen.[2] Die obigen Worte
des Autors sind daher wohl nur auf die grosse Masse des Volkes
zu beziehen, dem naturgemäss in der Periode des Činggis die
Schrift noch fremd war, während sie sich am Hofe und für
amtliche Zwecke Eingang verschafft hatte. Die mongolische
Chronik *Altan Tobči* und das chinesische Werk *Yüan shih lei
pien* wie das *Yüan-shih* (Kap. 121) verknüpfen übereinstimmend
die Einführung der uigurischen Schrift bei den Mongolen mit
dem Namen des Uiguren Tatatungga, Sekretär des Fürsten der
Naiman, die Činggis im Jahre 1204 unterwarf. Dieser unter-
richtete auf Befehl des Herrschers seine beiden Söhne in der
uigurischen Schrift, so dass sich der Gebrauch derselben bei
den Mongolen ungefähr von dieser Zeit ab datiren lässt.[3] Sie
erhielt sich unter ihnen bis zum Jahre 1278, als ein kaiserliches
Edikt des Inhalts erlassen wurde, dass die militärischen Patente,
die bisher mit uigurischen Charakteren beschrieben waren, von nun
an die nationale Schrift tragen sollten, die seit 1269 aufgekommen
war. Es folgte ein Edikt im Jahre 1284, in dem die uigurische
Schrift in den Aktenstücken der inneren Verwaltung verboten
und befohlen wurde, dass alle Regierungspublikationen von einer

---

[1] E. BRETSCHNEIDER, Mediæval Researches from Eastern Asiatic
Sources. Vol. I, London, 1888, p. 53.

[2] POZDNÉJEV, Lit., p. 11.

[3] POZDNÉJEV, Lit., pp. 14—16. Vergl. auch GABELENTZ, Zeitschrift
für die Kunde des Morgenlandes, Band II, Göttingen, 1839, S. 19—21;
ABEL-RÉMUSAT, Nouveaux mélanges asiatiques, Vol. II, Paris, 1829, pp.
61—63; WYLIE, Chinese Researches, Shanghai, 1897, Part IV, pp. 257—258.

Umschrift in mongolischen Charakteren begleitet sein sollten.[1])
Wie wir indessen aus den Briefen des Argun und Öljäitü ersehen, hat sich die uigurische Schrift unter den Mongolen Per-
siens noch etwas länger erhalten.

Die sogenannte ‚P‛ags-pa Schrift ist das Werk des Gross-
lamas von Sa-skya, ‚P‛ags-pa (d. i. der Ehrwürdige) bLo-gros
rgyal-mts‛an (1234—1279),[2]) der vom Kaiser Kubilai im Jahre
1261 nach China berufen wurde, und ist eng mit der Bekehrung
des mongolischen Kaiserhofes zu der tibetischen Form des
Buddhismus verknüpft. Auf Wunsch des Kaisers construirte er
auf Grund des tibetischen Alphabets eine Schrift für die Mongolen, in der die tibetischen Buchstaben von oben nach unten
laufend verbunden und in vertikalen Columnen von links nach
rechts angeordnet waren. Durch ein kaiserliches Edikt[3]) wurde
diese Schrift 1269 officiell eingeführt, mit dem Befehl, dass sie
in allen amtlichen Dokumenten zur Anwendung kommen sollte.
Fünf Monate später wurden in demselben Jahre in jeder Provinz Schulen für das Studium der neuen Schrift errichtet. Drei
weitere Edikte zu ihren Gunsten wurden Ende 1272, 1273 und
1275 bekannt gemacht: alle chinesischen Aktenstücke sollten
von einer mongolischen Übersetzung in diesen Charakteren begleitet sein, und eine Abteilung der kaiserlichen Akademie in
Peking *(Han-lin yüan)* sollte sich besonders der mongolischen
Sprache und Literatur widmen; der erste Präsident dieser Sektion war ein Gelehrter namens Sa-ti-mi-ti-li.[4])

---

[1]) WYLIE, Transactions of the China Branch of the Royal Asiatic
Society, Part V, Hongkong, 1855, p. 69.

[2]) Seine Biographie im *Yüan-shih*, Kap. 202; bei Sanang Setsen
(I. J. SCHMIDT, Geschichte der Ost-Mongolen, S. 115—117, vergl. auch S.
392—398); nach *Grub-mt‛a šel-kyi me-loṅ* (verfasst 1740, JASB, 1882, Part
I, No. I, pp. 67—68); bei ‚Jigs-med Nam-mk‛a (G. HUTH, Geschichte des
Buddhismus in der Mongolei, Band II, S. 139—159). Vergl. KÖPPEN, Die
lamaische Hierarchie und Kirche, S. 97—99; A. GRÜNWEDEL, Mythologie
des Buddhismus in Tibet und der Mongolei, S. 63—66.

[3]) Übersetzt von A. WYLIE, l. c., p. 68.

[4]) Im allgemeinen vergl. über die ‚P‛ags-pa Schrift G. PAUTHIER,
De l'alphabet de Pa'-Sse-Pa (Journal Asiatique, 1862, pp. 1—47), im besonderen die im nächsten Abschnitt zu den einzelnen Denkmälern ange-
führte Literatur. — Der mongolische Name *Dörbäljin* («viereckig») hat

Während sich diese Schrift als wirkungsvolle Lapidarschrift und auch als Kanzleischrift bis zum Ende der Yüan-Dynastie zu erhalten gewusst hat, erwies sie sich bald zu umständlich und schwerfällig für den täglichen Gebrauch, vor allem für die Herausgabe ganzer Bücher, wie sie die Übersetzung der buddhistischen Texte erheischte. Man nahm daher wieder seine Zuflucht zu dem bequemeren uigurischen Alphabet, und C'os-kyi Od-zer (mongolisch Nomun Gäräl, «Licht der Religion») von Sa-skya[1]) schuf zwischen 1307 und 1311 auf Grundlage desselben eine mongolische Schrift, die auch die noch jetzt gebräuchliche ist. Auch diese Schrift wird in senkrechten, von links nach rechts laufenden Reihen geschrieben.

Die Kalmüken haben sich eine eigene Schriftform geschaffen (verfasst von Zaya Paṇḍita 1648) offenbar in Anlehnung an die runderen und gefälligeren Typen der Manju Schrift; sie ist natürlich nur eine Variante des ostmongolischen Alphabets, von dem auch das der Manju abgeleitet ist, mit dem wesentlichen Unterschied jedoch, dass jedem einzelnen Laut ein bestimmtes graphisches Zeichen entspricht, während im Ostmongolischen gewisse Lautpaare, wie z. B. *o* und *u*, *a* und *ä* im Inlaut, *d* und *t*, *y* und *ds* durch ein und dasselbe Schriftbild ausgedrückt werden; auch bezeichnet das Kalmükische die Länge der Vokale durch einen Querstrich.

Buchdruck. — Von den Chinesen haben die Mongolen das Verfahren des Holztafeldrucks erlernt, das sich für ihre

---

ihr die Bezeichnung Quadratschrift verschafft; auf tibetisch heisst sie *Hor-yig* («Mongolenschrift»).

[1]) Seine Biographie bei G. HUTH, Geschichte des Buddhismus in der Mongolei, Band II, S. 160 164. Vergl. I. J. SCHMIDT, Forschungen im Gebiete der älteren religiösen, politischen und literärischen Bildungsgeschichte der Völker Mittel-Asiens, St. Pet., 1824. S. 128—129; KÖPPEN, l. c., S. 100 -101; WYLIE, Chinese Researches, Part IV, pp. 260—261. — Eine kurze Darstellung der Geschichte der mongolischen Schriften bei T. DE LACOUPERIE, Beginnings of Writing in Central and Eastern Asia, London, 1894, pp. 76, 89, 176, wo auch die ältere Literatur ziemlich vollständig citirt ist. Von neuerer ist noch zu nennen I. EUTING, Tabula scripturæ Uiguricæ, Mongolicæ, Mandshuricæ, Strassburg, 1891; FRIEDRICH MÜLLER, Zur Frage über den Ursprung der uigurisch-mongolisch-mandžuischen Schrift. WZKM. Band V, 1891, S. 182—184.

Schrift als ebenso bequem und praktisch erweist als für das
Chinesische und Tibetische. Druckfehler werden dadurch ganz
ausgeschlossen, und etwaige Irrtümer sind meist leichte Schreibfehler des Copisten, der dem Drucker die Handschrift liefert.
Die ältere im uigurischen und ‚P῾ags-pa Alphabet geschriebene
mongolische Literatur scheint ausschliesslich im Manuskript geblieben zu sein, was wohl der Hauptgrund dafür ist, dass sie
so leicht der Vernichtung anheimgefallen ist. Dagegen kam der
Druck sehr bald nach der Bildung der nationalen Schrift durch
C῾os-skyi Od-zer auf, und wir besitzen Nachrichten über frühe
Drucke besonders buddhistischer Schriften aus der chinesischen
und tibetischen Literatur (s. den Paragraphen **Buddhistische
Literatur**). Leider ist von diesen mongolischen Inkunabeln des
vierzehnten Jahrhunderts noch nichts ans Licht gekommen, und
wie es scheint, auch in China nichts derartiges erhalten worden;
doch dürfen wir uns einstweilen noch mit der Illusion trösten,
dass die Bibliotheken der grossen Klöster in Tibet und der
Mongolei solche Schätze bergen mögen. Soweit ich zu urteilen
vermag, gehören alle jetzt vorhandenen mongolischen Buch-
drucke dem Zeitalter der gegenwärtigen Manju Dynastie an, und
keiner scheint über die Periode des Kaisers K῾ang-hsi (1662—
1722) hinauszugehen; wenigstens ist mir noch kein älterer Druck
in die Hand gekommen. Den beiden erleuchteten Herrschern
K῾ang-hsi und K῾ien-lung ist ein grosser Anteil an der Förderung mongolischen Buckdrucks und Literatur zuzuschreiben.
Ebenso wie auf dem Gebiete der chinesischen, manjurischen
und tibetischen Sprache, wurden auch mongolische typogra-
phische Meisterwerke während des achtzehnten Jahrhunderts
hergestellt, die, nur als Proben der Buchdruckerkunst genommen, jeder grossen Bibliothek Ehre machen würden. In Peking
sind seit dem Anfang des achtzehnten Jahrhunderts zwei grosse
Druckereien im Felde der lamaischen Literaturen tätig gewesen,
die, wie ich mich selbst überzeugt habe, auch jetzt noch in
Betrieb sind, und von denen die eine mit grosser Energie weiter-
arbeitet. Diese Anstalt liegt beim Tempel *Sung-chu ssŭ* im östlichen Teile der Kaiserstadt, die andere an der Strasse gegenüber dem grossen Lamatempel *Yung-ho kung.* Beide befinden
sich (und waren wohl seit jeher) im Besitz von Chinesen, wie

auch die ganze Produktion der Bücher von chinesischen Arbeitern besorgt wird, da ja natürlich der Blockschnitzer das Manuskript gar nicht zu lesen braucht, auch gar keine Chance dazu hat — denn er klebt das Blatt des Manuskripts mit der Schriftseite nach unten auf den Holzblock — und bei seiner rein mechanischen Arbeit ebenso gut ein Analphabet sein kann. Die Inhaber dieser Anstalten sind Drucker, Verleger und Buchhändler in einer Person. Ihre Kundschaften finden sie an den mönchischen Insassen der Lamatempel in und um Peking und an den zahlreichen, die Hauptstadt im Winter besuchenden mongolischen Händlern. Die vielen und besonders schön (meist schöner als die tibetischen) hergestellten mongolischen Ausgaben und Auflagen beweisen, dass letztere eine gern bereite und gut zahlende Abnehmerschaft sind. Erwähnenswert ist die Tatsache, dass medicinische Werke (in erster Reihe solche kanonischen Ansehens) höher im Preise stehen als andere und unter keinen Umständen einen Rabatt vertragen, ja jedes Handeln um den geforderten Preis ausschliessen. Die in einem Tempel des Kaiserpalastes aus der Zeit K'ien-lung's aufbewahrten Druckplatten zum Kanjur und Tanjur, wahrhafte Meisterwerke der Holzschneidekunst, sind im Jahre 1900 ein Opfer der europäischen «Civilisation» geworden und teils von Soldaten als willkommenes Brennholz beim Kochen verwendet, teils als «Andenken» und Beutestücke in alle Winde verstreut mitgenommen worden.

Die Technik und innere Einrichtung mongolischer Bücher ist identisch mit der der tibetischen. Wie diese, haben sie das den indischen Handschriften nachgeahmte Langformat und bestehen aus einzelnen losen, zweiseitig bedruckten Blättern von sehr dickem und starkem Papier. Umfangreiche Bände werden in einzelnen Teilen vermittelst durch den Rand gezogener, aus Papier gedrehter Schnur zusammengehalten. In Peking ist das in Tibet bestehende Verfahren, die Blätter zwischen Holzdeckeln, die mit Schnüren oder Seidenbändern umwickelt werden, aufzubewahren, nicht üblich. Die obere Seite des ersten und die Rückseite des letzten Blattes sind gewöhnlich gelb gefärbt. In den meisten mongolischen Werken sind die beiden ersten Seiten in Rotdruck (rot als der heiligeren, in religiösem Sinne verdienstlicheren Farbe), der Rest aber in Schwarzdruck herge-

stellt. Der Titel findet seinen Platz in der Mitte der Vorderseite
des ersten Blattes, von einem Rechteck mit schwarzen Rand-
linien eingerahmt, das sich oft von einem besonders gefärbten
Untergrunde abhebt. Im Eingang wird gewöhnlich der Titel
wiederholt, oder wenn es sich um eine Übersetzung handelt,
zuerst der Sanskrit, dann der tibetische, dann der mongolische
Titel mitgeteilt. Die Mehrzahl der mongolischen Peking-Drucke
beginnt aber noch vor dem Titel mit der bekannten buddhistischen Zufluchtsformel in Sanskrit, transkribirt in mongolischen
Schriftzeichen. Auf den beiden ersten Seiten sind in der Regel
am Rande Miniaturen, Holzschnitte von Göttern, Buddha, Heiligen oder Lamen angebracht, meist mit Bezug auf den Inhalt
des Werkes ausgewählt. Am Rande links läuft die mongolische
Pagination, häufig mit einer verkürzten Fassung des Titels verbunden; am Rande rechts sind dieselben Blattbezeichnungen
chinesisch, in der Regel mit einem oder mehreren chinesischen
Schriftzeichen, die dann auch hinter dem Titel auf dem Titel-
blatte erscheinen. Diese Zeichen haben gewöhnlich keinen Inhalt,
sondern nur die Bedeutung einer Signatur, als Merkmal für
den Drucker, wie für den Buchhändler, die ja, wie gesagt, als
Chinesen das Buch nicht lesen können. Unter dieser Signatur
trägt der Buchhändler das betreffende Werk in seinen Katalog
ein, aus dem es ein sprachkundiger Käufer auswählt, und findet
es für ihn eben mit Hülfe dieser Signatur. In einigen aus dem
Chinesischen übersetzten mongolischen Sūtra wird auch der
volle Titel des chinesischen Originals durchgehend am Rande
bemerkt, in derselben Weise, wie wir das jüngst an Handschriften-
funden aus Turkistan beobachtet haben, was also wohl auf eine
alte Praxis zurückgeht. Am Schluss der meisten mongolischen
Werke ist, wie in indischen und tibetischen, ein Colophon angehängt, in dem die Namen von Verfassern, Übersetzern, Bear-
beitern, Redakteuren, Druckort, Druckjahr[1]) u. s. w. angegeben
werden. Manche dieser Schlussworte sind von grossem Umfange,
recht mitteilsam über die Geschichte des Werkes, seine ver-

---

[1]) In den Peking Ausgaben gewöhnlich durch die *Nien-hao* der
chinesischen Kaiser bestimmt, auch mit Zusatz des Jahres im tibetisch-
mongolischen Tiercyklus.

schiedenen Schicksale und Editionen und daher von grösstem Wert als literarhistorische Dokumente. Das Studium eines jeden Werkes sollte daher von dem des Colophons seinen Anfang. nehmen. Die Schlussseite ist gewöhnlich zu beiden Seiten wieder mit bildlichen Darstellungen verziert. Die meisten mongolischen Drucke von Peking weisen ausserdem noch ein besonderes Schlussblatt auf, das ganz mit den Figuren der vier Himmelskönige (mahārāja) bedeckt ist.

Das Format und der Umfang der mongolischen Bücher ist im Durchschnitt weit grösser als das der entsprechenden tibetischen, die dasselbe Werk enthalten. Dem vielsilbigen und formenreichen Charakter der Sprache gemäss beansprucht ein mongolischer Satz einen grösseren Raum als ein denselben Gedanken ausdrückender tibetischer Satz. Dann wird auch durch die vertikalen Columnen der Schrift eine grössere Breite des Blattes technisch bedingt.

Es gibt auch zahlreiche tibetisch-mongolische Paralleldrucke, in welchen das Mongolische als Interlinearversion des tibetischen Textes erscheint und jedes mongolische Wort genau unter das entsprechende tibetische der horizontalen tibetischen Reihe gesetzt ist; in diesem Falle geht natürlich der Eigencharakter des mongolischen Satzes verloren, indem auf eine Reihe höchstens zwei, meist jedoch nur ein Wort kommt. Neben den bilinguen kommen denn auch drei- und viersprachige Texte vor, in den vier Hauptsprachen des nördlichen Buddhismus (chinesich, manjurisch, tibetisch, mongolisch). Letztere Bücher sind indessen in der Regel in der Form und Technik chinesischer Bücher gedruckt; dazu gehören vor allem die kaiserlichen Ausgaben der mehrsprachigen Wörterbücher, Kalender, medicinische Handbücher u. a.

Merkwürdig sind kleine Ausgaben von Gebeten und Sūtra in der Form von gefalteten Albums; das Buch besteht aus einem einzigen langen, nur auf einer Seite bedruckten Blatt Papier, das zwischen zwei Pappdeckeln gefaltet und in seiner ganzen Länge entrollt werden kann. Dies ist eine Nachahmung chinesischer buddhistischer Bücher.

Auffallend ist, dass die Mongolen fast gar keine ihrer historischen und volkstümlichen Werke gedruckt haben, die lediglich in Handschriften verbreitet sind, während Tibet gerade an gedruckten Büchern dieser Gattungen Überfluss hat.

Als Druckorte mongolischer Bücher sind ausser Peking die Städte Urga und Kükä Khota («Blaustadt»)[1]) bekannt geworden; doch gibt es jedenfalls viele Klöster in der Mongolei, die sich mit der Herstellung von Drucken befassen. Die Burjaten werden von den früheren englischen Missionaren als geschickte Graveure von Drucktafeln gerühmt. Dagegen scheinen die Kalmüken niemals den Buchdruck betrieben zu haben;[2]) mir ist kein kalmükischer Druck je zu Gesicht gekommen, sondern nur Handschriften.

Literatur.

Alte Literaturdenkmäler. — Die ältesten Denkmäler der mongolischen Literatur sind nicht in der gegenwärtig allgemein üblichen Schrift, sondern teils in uigurischer, teils in der sogenannten Quadratschrift abgefasst.

Von den Denkmälern in uigurischer Schrift, die aus dem dreizehnten und vierzehnten Jahrhundert stammen, sind die folgenden auf uns gekommen:

1. Das älteste bisher bekannte ist eine im Kreise Nerchinsk nahe beim Flusse Onon gefundene, in eine Granitplatte gehauene Inschrift von fünf Zeilen aus der Zeit des Činggis Khan, die sich jetzt im Asiatischen Museum zu St. Petersburg befindet; sie wurde publicirt und übersetzt von I. J. SCHMIDT,[3])

---

[1]) Mélanges asiatiques, Vol. I, St. Pet., 1852, pp. 408, 415. — Kükä Khota liegt nördlich der grossen Mauer in der chinesischen Provinz Shansi (chinesisch *Sui-yüan*, in *So-p'ing fu*). Die Stadt ist Sitz eines Khutuktu und hat fünf Lamatempel (KÖPPEN, Die lamaische Hierarchie und Kirche, S. 381).

[2]) «Bei den Kalmüken habe ich keine Künstler gefunden, die Schrift zu schneiden imstande wären, und bei den Mongolen sind sie nur selten; desto gemeiner ist diese Kunst in Tibet und China.» P. S. PALLAS, Sammlungen historischer Nachrichten über die mongolischen Völkerschaften, Band II, S. 370.

[3]) Bericht über eine Inschrift aus der ältesten Zeit der Mongolen-Herrschaft, Mémoires de l'Académie des Sciences de St.-Pétersbourg, VI-ième série, Vol. II, 1834, pp. 243–256. Vergl. H. C. v. D. GABELENTZ in Zeitschrift für die Kunde des Morgenlandes, Band II, 1839, S. 18—21.

dessen Arbeit von Dorji Banzarov[1]) wesentlich verbessert wurde. Schmidt versetzt die nicht datirte kurze Inschrift in das Jahr 1219 oder 1220, Banzarov zwischen 1224 und 1230. Die Bearbeitung der Inschrift ist noch nicht befriedigend, und sie erfordert noch weiteres Studium.

2. Die interessantesten mongolischen Dokumente in uigurischer Schrift sind zwei Briefe, welche Argun (1284—1291), der Sohn des Ilkhan Hulagu (1258—1265) im Jahre 1289, und dessen Sohn Öljäitü im Mai des Jahres 1305 an den König Philipp den Schönen von Frankreich gerichtet haben, um ihm ihre Bundesgenossenschaft anzubieten. Die Briefe werden noch im Staatsarchiv von Paris (Archives nationales de France) aufbewahrt und wurden zuerst von Abel-Rémusat[2]) bekannt gemacht und übersetzt. Eine verbesserte Transkription in modern mongolischer Schrift und correkte Übersetzung wurde danach von I. J. Schmidt[3]) publicirt. Beide Briefe sind auf Rollen ko-

---

Eine schöne Reproduktion der Inschrift ist in W. Radloff's Atlas der Altertümer der Mongolei, Lief. 1, St. Pet., 1892, Tafel XLIX. Nr. 3.

[1]) Объясненіе монгольской надписи на памятникѣ князя Исункé, племянника 'Іингисъ-Хана, in der Sammlung seiner Schriften 'Іерная Вѣра etc.. St. Pet.. 1891, pp. 88—105. Vergl. auch Pozdnéjev. Lit., pp. 51—79.

[2]) Mémoires sur les relations politiques des princes chrétiens, et particulièrement des rois de France, avec les empereurs mongols. Premier mémoire: Rapports des princes chrétiens avec le grand Empire des Mongols, depuis sa fondation sous Tchinggis-Khan, jusqu'à sa division sous Khoubilaï (Mémoires de l'Académie des incriptions et belles-lettres, Vol. VI, Paris, 1823. pp. 396—469). Second mémoire: Relations diplomatiques des princes chrétiens avec les rois de Perse de la race de Tchinggis, depuis Houlagou, jusqu'au règne d'Abusaïd (Ibid.. Vol. VII. 1824, pp. 335—438, 7 Tafeln). Abgekürzter Bericht über diese Arbeit in seinen Mélanges asiatiques, Vol. I, Paris, 1825, pp. 401—412.

[3]) Philologisch-kritische Zugabe zu den zwei mongolischen Original-Briefen der Könige von Persien Argun und Öldshäitu. St. Petersburg. 1824. 31 p., mit besonderem Nachtrag von 4 p. über fünf mongolische Münzen von Ghasan. (Diese Notiz mit Zusätzen von Jacquet in Nouveau Journal Asiatique, Vol. VIII, 1831, pp. 344—348). Vergl. ferner Pozdnéjev, Lit., pp. 79—123, der auch Facsimiles der Briefe gibt. Die beste Facsimile Reproduktion derselben befindet sich im Atlas von Prince Roland Bonaparte, Documents de l'époque mongole, Paris, 1895, Tafel XIV. Auch G. Pauthier (Le livre de Marc Pol, Appendice Nr. 5, pp. 775—781) hat die

reanischen Papiers geschrieben: Argun's Rolle ist 1·8 m lang
und 0·25 m breit; die Öljäitü's 3 m lang und 0·50 m breit.
Argun, König von Persien, verspricht in seinem Schreiben, im
nächsten Jahre ein Lager vor Damas zu beziehen, um seine
Truppen mit denen der Kreuzfahrer zu vereinigen, und im Falle
er Jerusalem nähme, die Stadt dem König von Frankreich zu
überweisen. Öljäitü, König von Persien, kündigt in seinem
Schreiben die Versöhnung der Fürsten vom Stamme Činggis nach
45jährigen inneren Kämpfen und die Absendung zweier Gesandten an.

3. Ein Paizah,[1] Silbertäfelchen, mit mongolischer Inschrift
in uigurischen Charakteren, des Abdulla Khan der goldenen
Horde, 1848 von einem Bauern im Dorfe Grušovka am Dnjepr
im Gouvernement Jekaterinoslav in Süd-Russland gefunden und
von Baron von Stieglitz an die Akademie der Wissenschaften
in St. Petersburg gesandt, wo die Inschrift von DORJI BANZAROV[2]
entziffert und übersetzt wurde. Sie lautet: «Durch die Kraft des
ewigen Himmels, durch die Hülfe seines grossen Schutzes und
seiner Macht:[3] wer dem Befehl des Abdulla nicht gehorcht, ist
strafwürdig und soll sterben.» Der hier erwähnte Abdulla regirte nach den Münzfunden von 1362 (oder 1363) bis 1368
(oder 1369).

4. Die Befehle *(yarlik)* der Khane der Goldenen Horde,

---

beiden Briefe in modern mongolischer Schrift, Transkription und Übersetzung abgedruckt. Auf die Briefe bezieht sich der Aufsatz von A. SCHIEF-
NER, Ein kleiner Beitrag zur mongolischen Paläographie, Mélanges asiatiques, Vol. II, St. Pet., 1856, pp. 487—489.

[1] Chinesisch *p'ai-tse*, s. darüber YULE, Marco Polo, 3. Aufl., Vol. I,
pp. 350—354.

[2] Erklärung einer mongolischen Inschrift aus einer im Jekaterino-
slawschen Gouvernement ausgegrabenen Silberplatte. Bulletin de la classe
hist.-phil. de l'Acad. de St. Pét., Vol. V, Nr. 9, 1848 (mit Nachtrag von
O. BÖHTLINGK in Nr. 12). Vergl. ferner D. BANZAROV, in der Sammlung
seiner Schriften Черная Вѣра etc., St. Pet., 1891, pp. 52—53; GRIGORJEF,
Journal Asiatique, 1861, p. 544; POZDNÉJEV, Lit., pp. 124—127, wo auch
die Facsimiles der Inschrift reproducirt sind (letztere auch in YULE'S
Marco Polo, 3. Aufl., Vol. I, p. 355).

[3] Über diese typische Formel in mongolischen Edikten vergl. E.
CHAVANNES, T'oung Pao, 1904, pp. 395—396.

der des Tokhtamyš Khan, 1393 an den Grossfürsten von Lithauen Jagaila geschrieben,[1]) und der des Kutlugh Timur, an einen gewissen Muhamed-Beg mit der Würde des Tarkhan 1397 verliehen.[2])

5. Zu diesen Denkmälern kann man auch einige im Kreise Minusinsk in Sibirien befindliche Felsen- nnd Höhleninschriften rechnen; vor allem ist die von Abakansk bemerkenswert, die Pozdnějev[3]) facsimilirt und ausführlich erörtert hat. Die Inschrift selbst ist nicht mehr vorhanden, sondern soll 1806 von Klaproth bei seiner Rückkehr aus China mit einem Säbel ausgekratzt worden sein, weil sie nach seinen Worten eine unvorteilhafte Äusserung gegen die Russen enthielte.[4]) Bereits Pallas[5]) hatte die Aufmerksamkeit auf dieses Denkmal gelenkt, und 1847 fertigte Castrén[6]) eine Copie desselben an.[7])

Die früher gehegte Meinung, dass die Quadratschrift nur eine vorübergehende Bedeutung gebabt habe, und wenige Denkmäler in derselben abgefasst worden seien, ist durch die gründliche Untersuchung von A. Pozdnějev als hinreichend widerlegt zu betrachten. Pozdnějev[8]) hat durch zahlreiche historische Zeug-

---

[1]) D. Banzarov, l. c., pp. 122—123.

[2]) Pozdnějev, Lit., pp. 50, 145. — Jos. v. Hammer, Uigurisches Diplom Kutlugh Timur's vom Jahre 800 (1397), Fundgruben des Orients, Band VI, Wien, 1818, pp. 359—362. Vergl. ZDMG, Band IV, 1850, S. 518.

[3]) Lit., pp. 127—144.

[4]) Ibid., p. 131.

[5]) Reise durch verschiedene Provinzen des russischen Reiches, Band II, S. 688—689.

[6]) Reiseberichte und Briefe aus den Jahren 1845—1849, St. Pet., 1856, S. 375.

[7]) Die Namen des Kaisers Münkä und vieler mongolischer Khane aus dem Hause des Juči und Hulagu sind auf den Münzen in uigurischer Schrift dargestellt, doch übergehe ich dieselben, da sie nicht zur Literatur gehören (s. Pozdnějev, Lit., p. 49). Besondere Erwähnung verdient die von Mongolisten nicht beachtete, sehr interessante Abhandlung von Karabacek, Der unmittelbare Einfluss der mongolischen Invasion (1241—1242) auf die Münzverhältnisse Ungarns (Separatabdruck aus dem VI. und VII. Bande der Wiener Numismatischen Zeitschrift, 1874—1875, 9 p., 1 Tafel). Das Hauptergebnis ist der Nachweis des Vorkommens einzelner uigurisch-mongolischer Buchstaben auf Münzen ungarischen Gepräges aus der angezeigten Periode.

[8]) Lit., s. vor allem sein Schlussergebnis p. 240.

nisse aus der chinesischen Literatur dargetan, dass die Quadratschrift, unter dem Namen «mongolische Staatsschrift» *(Mong-ku
kuo tzŭ)* bekannt, während des ganzen Zeitraums der Dynastie
Yüan, als die allgemeine officielle Schrift der Verwaltung verwendet wurde, und zwar die Quadratschrift ausschliesslich; auch
die mongolische Sprache spielte eine sichtbare Rolle im Regie-
rungsmechanismus bis zum Sturze der Dynastie. Die uigurische
Schrift, die gleichzeitig in derselben Periode existirte, fand bei
den Mongolen hauptsächlich Verwendung in der buddhistischen
Literatur.

Das beste Hülfsmittel zum Studium der mongolischen
Quadratinschriften ist der glänzende, auf Kosten des Prinzen
ROLAND BONAPARTE herausgegebene Atlas.[1]) Ein undatirtes sil-
bernes P'ai-tse mit mongolischer Quadratschrift ist im Jahre
1846 von einem russischen Kaufmann Ananjin im Kreise Minusinsk gefunden worden. Dasselbe haben AVVAKUM[2]) und dann
I. J. SCHMIDT[3]) bekannt gemacht und übersetzt. Der Inhalt ist
ähnlich der oben erwähnten mongolischen P'ai-tse Inschrift in
uigurischer Schrift und wird von Avvakum interpretirt: «Durch
die Kraft des Himmels: der Name des Münkä-Khan sei gehei-
ligt; wer ihn nicht achtet, soll untergehen und sterben!» Eine
Silberplatte mit ganz ähnlicher Inschrift wurde 1853 im Dorfe
Njuksk (Kreis Verchne-Udinsk, Provinz Transbaikalien) gefunden
und von Saveljev beschrieben;[4]) sie befindet sich jetzt in der

---

[1]) Documents de l'époque mongole des XIII-e et XIV-e siècles. Inscriptions en six langues de la porte de Kiu-yong Koan, près Pékin,
lettres, stèles et monnaies en écritures ouïgoures et 'Phags-pa, dont les
originaux ou les estampages existent en France. Paris, 1895, folio, 15
Tafeln und 5 Seiten Text. Vergl. die Anzeige von W. BANG, WZKM, Band
X, 1896, S. 59–66.

[2]) Das Citat der Abhandlung desselben s. bei D. BANZAROV, Черная
Вѣра etc., p. 50; ebenda auch über die diese Inschrift betreffende Contro-
verse. V. GRIGORJEF, Lettre sur l'origine et les monuments de l'écriture
carrée (Journal Asiatique, 1861, pp. 529—558).

[3]) Über eine mongolische Quadrat-Inschrift aus der Regierungszeit
der mongolischen Dynastie Jüan in China, Bull. hist.-philol. de l'Acad.
de St.-Pét., Vol. IV, Nr. 9, 1846. Danach ist das P'ai-tse in YULE's Marco
Polo, 3. Aufl., Vol. I, Tafel gegenüber p. 352, reproducirt.

[4]) Труды Вост. Отдѣла Имп. Археолог. Общества, Vol. V, pp. 160—165.

Eremitage zu St. Petersburg. Ein seiner Form wegen merkwürdiges Stück (nach Art einer Medaille) aus Gusseisen, die Schriftzeichen in Silber eingelegt, ist von POZDNÈJEV[1]) abgebildet und erläutert worden. Die Inschrift lautet: «Durch die Kraft des ewigen Himmels: wer den Befehlen des Khans nicht gehorcht, der soll getötet werden!» Das Objekt wurde im Bezirk Bogomilsk (Kreis Marinsk, Gouvernement Tomsk) von einem Bauern im Boden gefunden.[2])

Ich lasse nun eine kurze Aufzählung der wichtigeren historischen Quadratinschriften in chronologischer Ordnung folgen:

1. Die älteste bisher bekannt gewordene stammt aus dem Jahre 1283,[3]) also 14 Jahre nach der officiellen Einführung der P'ags-pa Schrift (1269), und ist eine chinesisch-mongolische Bilingue, die einen Befehl des kaiserlichen mongolischen Prinzen Ananda, König von Ngan-si, enthält, anlässlich einer Landabtretung in Yung-shou hsien in der Provinz Shensi.[4])

2. Inschrift vom Jahre 1288 (Herkunftsort nicht genau bestimmt, wahrscheinlich von einem Tempel des Confucius in Mittel-China); den unteren Teil des Steines nimmt die chinesische Inschrift ein, den oberen Teil die mongolische, die aber nur eine phonetische Umschrift des chinesischen Textes dar-

---

[1]) Объясненіе древней монгольской надписи на чугунной дощечкѣ, доставленной въ Императорскую Академію Наукъ Г. Винокуровымъ. Записки Академіи Наукъ, Vol. XXXIX. St. Pet.. 1881, pp. 1—13, 1 Tafel.

[2]) Ausser Inschriften haben die Mongolen-Kaiser auch Münzen und Medaillen in P'ags-pa Schrift hinterlassen (s. Documents de l'époque mongole, Tafel XV; E. DROUIN, Notice sur les monnaies mongoles, Journal Asiatique, Mai-Juin, 1896, pp. 486—544; E. CHAVANNES, Note sur une amulette avec inscriptions en caractères Pa-Se-Pa, Journal Asiatique, Janv.-Févr., 1897, pp. 148—149).

[3]) Doch siehe weiter unten über eine ältere Inschrift aus dem Jahre 1275.

[4]) Reproducirt in Documents de l'époque mongole, Tafel XII, Nr. 1. Übersetzung des chinesischen Textes von G. DEVÉRIA, Notes d'épigraphie mongole-chinoise avec une notice de W. BANG (Extrait du Journal Asiatique, 1896), Paris, 1897, pp. 7—18; Übersetzung des mongolischen Textes von W. BANG, ibid., pp. 18—30. Vergl. CHAVANNES, T'oung Pao, 1904, pp. 404—413, der eine Inschrift aus dem Jahre 1276 mitteilt, in der eine Zeile chinesisch in Quadratschrift transkribirt vorkommt (p. 409 und Tafel VIII).

stellt.[1]) Es werden darin gewisse Privilegien an die confucianischen Literaten erteilt.

3. Inschrift vom Jahre 1294, ehemals im Tempel des Confucius in Shanghai, aus dem sie jetzt merkwürdiger Weise verschwunden ist, gleichfalls chinesisch und mongolische Transkription des chinesischen Textes, zuerst publicirt und übersetzt, mit einer eingehenden und wertvollen Geschichte der Quadratschrift von ALEXANDER WYLIE.[2]) Dies ist ein Edikt des Kaisers Temur zu Gunsten der Lehre des Confucius, welches durch das ganze Reich verbreitet wurde. Die Inschrift wurde infolgedessen in verschiedenen Teilen Chinas in Stein gemeisselt, und mehrere derselben sind noch vorhanden. Im Confucius-Tempel zu *Sungkiang* (Provinz Kiang-su) ist das genaue Gegenstück zu der Inschrift von Shanghai in einer Steintafel vorhanden.[3]) WYLIE bemerkt am Schlusse seiner Abhandlung,[4]) dass nach Angabe des Buches «Bericht über die Steininschriften von Shantung» sich eine Copie der Inschrift ausschliesslich in mongolischen Charakteren im Confucius-Tempel von *K'ü-fu* befinde. Diese Nachricht stimmt nicht ganz genau. Da ich im Januar 1903 die Stadt K'ü-fu besuchte, hatte ich Gelegenheit, Abklatsche der dort befindlichen drei mongolischen Inschriften zu nehmen; es sind dort zwei Steine vorhanden, einer, der ausschliesslich die mongolische Transkription der vorher erwähnten Inschrift enthält, und ein anderer, mit chinesisch-mongolischem Text, 1294 datirt, in der aber die Schlusspartie der Shanghai Inschrift fehlt,

---

[1]) Reproducirt in Documents, Tafel XII, No. 2. Übersetzt von G. DEVÉRIA. l. c., pp. 30—39.

[2]) Ancient Inscription in Chinese and Mongol (Transactions of the China Branch of the Royal Asiatic Society, Part V, Hongkong, 1855, pp. 65—81). Vergl. auch WYLIE, Chinese Researches, Shanghai, 1897, Part IV, p. 259. G. PAUTHIER, Le livre de Marc Pol, Paris, 1865, Appendice Nr. 3, pp. 768—771 (Von Appendices Nrs. 3—6 gibt es unter dem Titel Inscriptions mongoles einen Sonderabdruck, 15 p.). Über die Schicksale des Steines s. WYLIE in JRAS, Vol. V, 1870, p. 29, Note 5.

[3]) Ein Abklatsch dieses Steines reproducirt auf Tafel XXXII des Werkes des P. LOUIS GAILLARD, Nankin d'alors et aujourd'hui, aperçu historique et géographique, Shanghai, 1903 (Variétés sinologiques, No. 23); derselbe gibt auch eine Übersetzung der Inschrift, pp. 299—302.

[4]) L. c., p. 77.

und auch andere Varianten vorkommen.[1]) Ich werde darüber, sowie über die dritte mongolische Inschrift, gelegentlich an anderer Stelle berichten.

4. Edikt vom Jahre 1309 der Wittwe des Kaisers Dharmapāla (mong. Tarmabala), welches die Geistlichen der Zahlung aller Abgaben entbindet und Achtung vor ihrem Eigentum gebietet. Der inschriftliche mongolische Text dieses Edikts wurde 1830 in einem buddhistischen Tempel von *Pao-ting fu* (Provinz Chihli) von dem Archimandrit Avvakum von der russischen Mission in Peking entdeckt und von A. A. Bobrovnikov[2]) publicirt.

5. Edikt vom Jahre 1314, mongolischer Text mit chinesischer Übersetzung auf einer Steinsäule in *Chu-chi hsien*, Hsi-an fu, Provinz Shensi,[3]) der ein besonderes Interesse durch Erwähnung der Christen[4]) hat. Devéria[5]) hat eine französische Übertragung der russischen Übersetzung dieses Denkmals durch Grigorjev[6]) gegeben. Mit derselben Inschrift hatte sich schon früher Hans Conon v. d. Gabelentz[7]) beschäftigt auf Grund eines Textabdrucks in einer chinesischen epigraphischen Sammlung vom Jahre 1618. Dies war die erste sich an die Quadratinschriften knüpfende wissenschaftliche Arbeit überhaupt und macht dem Scharfsinn dieses eminenten Forschers alle Ehre. Dieselbe Inschrift hat ferner A. Wylie[8]) behandelt.

---

[1]) Vergl. *K'ü-fu hsien chi* (Chronik von *K'ü-fu*), Buch 52, p. 19.

[2]) Памятники монгольскаго квадратнаго письма, St. Pet., 1870, pp. 15—30. Auch unter dem Titel Грамоты вдовы Дарма-баловой и Буянту-хана, писанныя квадратнымъ письмомъ, in Труды Восточнаго Отд. Имп. Русскаго Арх. Общества, Vol. XVI.

[3]) Reproduktion des Abklatsches der Inschrift in Documents, Tafel XII, Nr. 3.

[4]) Über die mongolische Bezeichnung der Christen s. jetzt Chavannes, T'oung Pao, 1904, p. 420, Note 7.

[5]) L. c., pp. 40—42.

[6]) Bei Bobrovnikov, l. c., pp. 30—50.

[7]) Versuch über eine alte mongolische Inschrift. Zeitschrift für die Kunde des Morgenlandes, Band II, Heft 1, Göttingen, 1839, S. 1—21, mit drei Tafeln. Nachtrag zur Erklärung der altmongolischen Inschrift, Ibid., Band III, 1840, S. 225—227.

[8]) Sur une inscription mongole en caractères P'a-sse-pa, Journal Asiatique, 1862, pp. 461—471 (1 Tafel); danach ist der mongolische Text

6. Edikt vom Jahre 1316, chinesisch und Umschrift des chinesischen Textes in Quadratschrift, das den Eltern des Philosophen Mong-tse posthume Ehrentitel verleiht.[1]) DEVÉRIA hatte den Abklatsch dieser Inschrift in Peking käuflich erworben, und daher den Ort, wo sich der Stein befand, nicht in Erfahrung gebracht. Auf meinen Reisen in der Provinz Shantung fand ich den Stein im Tempel des Mong-tse in *Tsou hsien* (südlich von *K'ü-fu*, in *Yen-chou fu*, Shantung).

7. Dort befindet sich ausserdem eine chinesisch-mongolische Inschrift vom Jahre 1331,[2]) deren Herkunft im Atlas von ROLAND BONAPARTE gleichfalls unbestimmt gelassen war. Das Mongolische ist auch hier wiederum blosse Transkription. Es ist ein Edikt, in welchem dem Mong-tse ein posthumer Ehrentitel verliehen wird.

8. Inschrift vom Jahre 1345 im Tore *Kiü-yung kuan* im Nan-k'ou Pass auf dem Wege von Peking nach Kalgan, wo sich Inschriften in Sanskrit, tibetisch, mongolisch, uigurisch, chinesisch und in Hsi-hsia (tangutisch) befinden.[3])

---

mit WYLIE's Übersetzung in G. PAUTHIER, Le livre de Marc Pol, pp. 772—774, reproducirt. Vergl. CHAVANNES (T'oung Pao, 1904, pp. 414—415 und 422—426), der eine Übersetzung des chinesischen Ediktes der Inschrift gibt, das bisher eine Crux der Sinologen gebildet hatte, und zeigt, •dass der chinesische Text selbst vollkommen verständlich ist, und dass er in dem gewöhnlichen Kanzleistil der mongolischen Khane von China geschrieben ist».

[1]) Reproducirt in Documents, Tafel XIII, Nr. 1. Übersetzt von DEVÉRIA, l. c., pp. 83—85.

[2]) Reproducirt in Documents, Tafel XIII, Nr. 2. Übersetzt von DEVÉRIA, l. c., pp. 85—87. Beide Inschriften werden im *Tsou hsien chi* (Chronik von *Tsou*, Buch X) angeführt, wo im ganzen 43 Inschriften aus der Yüan-Dynastie aufgezählt werden.

[3]) Sämtliche Inschriften reproducirt in Documents, Taf. II—XI, die mongolische auf Tafel VIII; die letztere übersetzt von GEORG HUTH, Note préliminaire sur l'inscription de Kiu-yong Koan, quatrième partie: Les inscriptions mongoles (Journal Asiatique, Mars-Avril, 1895, pp. 351—360). Die Inschriften sind zuerst teilweise studirt worden von A. WYLIE, On an Ancient Buddhist Inscription at Keu-yung kwan, in North China (JRAS, New Series, Vol. V, 1870, pp. 14—44). Vergl. ferner C. IMBAULT-HUART, Note sur l'inscription bouddhique et la passe de Kiu-Young-kouan (Revue de l'Extrême Orient, Vol. II, 1883, pp. 486—493). Die chinesische

14*

Zu der oben mitgeteilten Liste von Quadratinschriften kann ich noch einige weitere hinzufügen, teils auf Grund chinesischer Quellen, teils solche, die ich selbst gesehen und abgeklatscht habe. Am reichsten scheint die Provinz Shantung an mongolischen Inschriften gewesen zu sein, die aber leider zum grossen Teil spurlos verschwunden sind; die von *K'ü-fu* und *Tsou hsien* habe ich bereits erwähnt. Die Chronik der Stadt *Po-shan* in Shantung (*Po-shan hsien chi*, Buch V a, p. 2. b) gibt die Nachricht, dass sich im dortigen Tempel *Ling-ch'üan-miao* zwei Steine mit mongolischen Inschriften befänden, die eine vom Kaiser Ogodai im Jahre 1235, die andere 1291 errichtet. Ich habe den betreffenden Tempel selbst besucht und erfahren müssen, dass die Inschriften leider nicht mehr vorhanden sind; auch officielle Erkundigungen bei dem dortigen Kreisvorsteher ergaben kein Resultat über ihren Verbleib. Das ist um so mehr zu bedauern, als die Inschrift von 1235 ja wohl nicht in Quadratschrift, sondern mutmasslich in uigurischer Schrift abgefasst war, und bisher keine mongolischen Inschriften von dieser Schrift in China gefunden worden sind. Über den Inhalt einer dieser Inschriften (welcher von beiden, ist nicht gesagt) enthält die Chronik von *Po-shan* eine schwache Andeutung an einer anderen Stelle (Buch IV b, p. 6 b), wo im Kapitel der Katastrophen, welche die Stadt betroffen haben, erzählt wird, dass in der Periode *Chêng-yu* (1213—1216) des Kaisers Hsüan Tsung der Kin-Dynastie das Wasser des Hsiao Flusses sich plötzlich gelb färbte und für einige Tage auf eine Ausdehnung von vierzig Meilen *(li)* zu seiner Quelle zurückkehrte. An dieser Stelle bemerkt nun der Verfasser der Chronik: «Siehe die mongolische Inschrift im *Ling-ch'üan* Tempel», die also eine Schilderung dieses Ereignisses enthalten haben muss.

　　In den Tempeln des *T'ai-shan* sind vier chinesisch-mon-

Inschrift ist von E. CHAVANNES (Journal Asiatique, Sept.-Oct., 1894, pp. 355—368), die tibetische von S. LÉVI (Ibid., pp. 369—373), die uigurische von W. RADLOFF (Ibid., Nov.-Déc., pp. 546—550) übersetzt worden. Die Inschriften auf der westlichen Wand enthalten die abgekürzte Redaktion eines buddhistischen Sūtras, das CHAVANNES nach der Version des chinesischen Tripiṭaka in den Mélanges Harlez (Leiden, 1896, pp. 60 81) übersetzt hat.

golische Inschriften vorhanden, 1302, 1306, 1307 und 1310 datirt, deren chinesische Texte im *T'ai-shan chi* (Buch XVIII, pp. 44—49) mitgeteilt sind.

Über eine chinesisch-mongolische Inschrift aus dem Jahre 1275 (12. Jahr der Periode *Chih-yüan*), welche somit die älteste bekannte Quadratinschrift sein würde, berichtet die Chronik von *T'ung-chou fu* (*T'ung-chou fu chi*, Buch XXVI, p. 49). Der Stein befindet sich in *Han-ch'êng hsien* (Provinz Shensi) und enthält ein Edikt[1] anlässlich der Erbauung eines kaiserlichen Palastes. Das *Kuan-chung kin shih chi* (eine Sammlung epigraphischer Denkmäler der Provinz Shensi, publicirt 1782, Buch VIII, p. 11 b) erwähnt ein inschriftliches Edikt vom Jahre 1319 mit einem mongolischen Text in der oberen und einem chinesischen in der unteren Partie: der Stein befindet sich im Tempel *Kuang kuo (se)* in der Stadt *Ho-yang* (Präfektur *T'ung-chou*, Provinz Shensi).

Für die Anwendung der ‚P'ags-pa Schrift in Büchern hat bereits WYLIE[2] ein Beispiel gegeben. Im 81. Kapitel der Encyklopädie *King ch'uan pai pien* (gedruckt 1581) ist das kleine populäre Werk über die Familiennamen *(Po kia hsing)* in mongolischen Charakteren transkribirt, die jedoch eine sehr entstellte Form zeigen. Im Jahre 1307 überreichte Po-lo Temur, ein hoher Würdenträger, dem Kaiser ein Exemplar des *Hsiao-king* («Buch der Kindesliebe») in ‚P'ags-pa geschrieben, das der Kaiser drucken und verbreiten liess.[3] Von selbständigen mongolischen Werken in dieser Schrift hat sich leider nichts erhalten, wenigstens ist bisher keines ans Licht gekommen.[4]

---

[1] Der chinesische Text im *Kuan-chung kin shih chi*, Buch VIII, pp. 4 b—5 a.

[2] Ancient Inscription in Chinese and Mongol, l. c., pp. 77, 78, und Notes on Chinese Literature (neue Ausgabe, Shanghai, 1901) p. 187. Facsimile-Reproduktion bei J. EDKINS, Sanscrit and Mongolian Characters (Transactions of the China Branch of the Royal Asiatic Society, Part V, Hongkong, 1855, pp. 101—108, 4 Tafeln).

[3] WYLIE, l. c., p. 69, und JRAS, New Series, Vol. V, 1870, p. 28; RÉMUSAT, Nouveaux mélanges asiatiques, Vol. II, p. 3.

[4] Listen solcher Werke auf Grund chinesischer Nachrichten geben ABEL-RÉMUSAT, Recherches sur les langues tatares, Paris, 1820, pp. 197—

Neuere Inschriften. — Im Anschluss an die altmongolischen Inschriften in ‚P‘ags-pa wird es passend sein, hier ein Wort über die neueren, in der gewöhnlichen Schrift abgefassten Inschriften anzuschliessen. Die Mongolei und das nordöstliche China sind reich an epigraphischen Denkmälern in mongolischer Sprache, doch ist erst ein sehr geringer Teil davon bekannt gemacht worden. Was zunächst China betrifft, so befinden sich in allen Lamatempeln in und um Peking sowie in Jehol zahlreiche grosse Steininschriften, darunter viele von ungewöhnlichen Dimensionen, und der Mehrzahl nach in den vier Sprachen chinesisch, tibetisch, mongolisch und manjurisch abgefasst; die meisten stammen aus den Zeiten der Kaiser K‘ang-hsi und K‘ien-lung und sind teilweise von den Kaisern selbst verfasst und geschrieben; sie sind von grossem historischen Interesse für die Entwicklung des Lamaismus und seiner Beziehungen zum chinesischen Reiche, besonders im siebzehnten und achtzehnten Jahrhundert.[1]) Über den wesentlichen Inhalt der Jehol-Inschriften kann man sich vorläufig aus dem trefflichen Buche von O. Franke[2]) orientiren.

Ausser den auf den Lamaismus bezüglichen gibt es in der Provinz Chihli nur wenige mongolische Inschriften, die ein anderes Gebiet streifen. Dahin gehört z. B. eine im Tempel der Stadtgottheit (*Ch‘êng huang miao*) von Jehol von K‘ien-lung 1773 errichtete chinesisch-manjurische und türkisch-mongolische Inschrift, dahin ferner die auf die Geschichte des Islams bezügliche viersprachige Inschrift (chinesisch, Manju, djagatai-türkisch und mongolisch) vom Jahre 1764 in der Moschee gegenüber dem Kaiserpalast in Peking.[3])

214; G. Pauthier, Journal Asiatique, Janv.-Février. 1862. pp. 18--19; Pozdnéjev, Lit., pp. 158—240.

[1]) Während meines Aufenthalts in Peking habe ich von allen diesen Inschriften Abklatsche herstellen lassen, die ich zu einer Epigraphia Lamaica zu vereinigen gedenke.

[2]) Beschreibung des Jehol-Gebietes in der Provinz Chihli, Leipzig, 1902.

[3]) Reproducirt und übersetzt von G. Devéria, Musulmans et Manichéens chinois (Journal Asiatique, Nov.-Déc., 1897, pp. 445—454); der darin erwähnte von Devéria nicht identificirte Tempel *Fu-ning* befindet sich in

Eine der frühesten Inschriften der Manju Dynastie ist die vom Jahre 1639 in Sam-jön-do, Korea, zur Erinnerung an die Unterwerfung dieses Landes im Jahre 1637. Der chinesische Text ist zuerst von W. R. Carles,[1] dann besser von W. W. Rockhill[2] übersetzt worden; ferner hat Pozdnějev[3] die manju-mongolische Version derselben Inschrift behandelt.

Eine viersprachige Inschrift des Kaisers K'ien-lung, zur Erinnerung an die Besiegung der Kalmüken im Jahre 1757 errichtet, ist von M. Amiot[4] übersetzt worden, ebenso eine gleichfalls viersprachige Inschrift desselben Herrschers im Ili-Gebiet zur Feier der Rückkehr der Torgoten aus Russland im Jahre 1771.[5] N. Pantusov[6] hat eine viersprachige Inschrift vom Jahre 1755, die sich auf dem Berge Gadyn am linken Ufer des Flusses

Jehol. W. Bang (Keleti Szemle, Vol. III, 1902, pp. 94—103) hat die Manju Version der Inschrift (mit einem Facsimile derselben) in Transkription und Übersetzung publicirt und die mongolische Recension sowie den historischen Teil (p. 95, Note) versprochen, die meines Wissens bisher nicht erschienen sind.

[1] A Corean Monument to Manchu Clemency (Journal of the China Branch of the Royal Asiatic Society, New Series, Vol. XXIII, 1888, pp. 1—8); gibt eine Copie des chinesischen, und auch des manju-mongolischen Textes (auf Tafel gegenüber p. 284 desselben Bandes).

[2] China's Intercourse with Korea from the XVth Century to 1895. London, 1905, pp. 39—44.

[3] Каменописный памятникъ подчиненія Маньчжурами Кореи. Zap., Vol. V, 1890, pp. 37—55 (gibt den Manju und mongolischen Text der Inschrift).

[4] Monument de la conquête des Eleuths, Mémoires concernant les Chinois, Vol. I, Paris, 1776, pp. 329—400.

[5] Monument de la transmigration des Tourgouths, ibid., pp. 405—418 (vergl. auch pp. 419—427). Vergl. Köppen, Die lamaische Hierarchie und Kirche, S. 214; Abel-Rémusat, Nouveaux mélanges asiatiques, Vol. II, Paris, 1829, pp. 102—105.

[6] Nach dem Referat von W. Barthold in Mitteilungen des Seminars für orientalische Sprachen, Band I, Abt. 1, Berlin, 1898, S. 206. Die Abhandlung selbst war mir nicht zugänglich. — N. Pantusov entdeckte ferner in Arasan am Flusse Kegen im Kreise Jarkent sechs Steine mit tibetischen und mongolischen Inschriften, ferner vier Steine mit mongolischen Inschriften im Kreise Kopalsk in Semiréčensk (s. Orientalische Bibliographie, Band XV, 1902, Nr. 1219, 1337; die beiden Abhandlungen sind mir leider nicht zugänglich).

Sumbe, Nebenflusses des Tekes, befindet, besprochen und die
chinesische Inschrift abgebildet; sie soll zur Erinnerung an den
Sieg dreier chinesischer Officiere und 22 Soldaten über ein Heer
von 6500 Dsungaren gesetzt worden sein.

In seinem «Atlas der Altertümer der Mongolei» (1. Lieferung, St. Petersburg, 1892) hat WILHELM RADLOFF auch einige
mongolische Inschriften aus der Mongolei reproducirt, darunter
eine vom Tempel Erdenidsu[1]) (Tafel LI) und eine von den Ruinen
eines ehemaligen Tempels von Tsaghan-Baišing («Weisses Haus»)
am Flusse Tola (Tafel LX). Letztere ist samt der dazu gehörigen tibetischen Inschrift von GEORG HUTH[2]) herausgegeben und
übersetzt worden; sie stammt aus dem Jahre 1601 und berichtet
die Gründung des an der Stelle eines ehemaligen Khans-Palastes
errichteten Tempels der Rotmützen-Sekte. RADLOFF's Atlas gewährt ein prächtiges Bild von der vielseitigen Fülle der Altertümer der Mongolei, die in vier Gruppen zerfallen, prähistorische Denkmäler (Grabhügel, Felszeichnungen u. s. w.), alt-
türkische Inschriften aus der Periode der Tu-küeh, die bis zur
Mitte des achten Jahrhunderts den grössten Teil der nördlichen
Mongolei beherrschten, die Denkmäler der Uiguren, die von der
Mitte des achten bis zu Ende des neunten Jahrhunderts in der
Mongolei herrschten (Ruinen von Kara Balghasun am Orkhon),
und endlich die Denkmäler der Mongolenzeit seit dem dreizehnten Jahrhundert.[3])

---

[1]) Gegründet 1585 auf der Stelle des alten Karakorum, der vom
Kaiser Ogotai 1235 erbauten Hauptstadt.

[2]) Die Inschriften von Tsaghan Baišin. Tibetisch-mongolischer Text
mit einer Übersetzung sowie sprachlichen und historischen Erläuterungen, Leipzig, 1894. Derselbe, Sur les inscriptions en langue tibétaine et
mongole de Tsaghan Baisching et sur le rapport de ces monuments avec
«l'histoire du Bouddhisme en Mongolie» composée en tibétain par 'Jigs-
med nam-mk'a (Actes du X-e Congrès international des Orientalistes,
session du Genève, 1894), Leide, 1896, pp. 175—180. Vergl. die Anzeige
von W. BANG, WZKM, Band X, 1896, S. 255—262, der einige Verbesse-
rungen zur Übersetzung der mongolischen Inschrift vorschlägt.

[3]) Erwähnung verdient noch die Tatsache, dass es chinesische
Porzellane mit mongolischen Aufschriften gibt: S. W. BUSHELL (China
Review, Vol. XI, p. 331, und Chinese Art, Vol. II, London, 1906, p. 43,
mit Abbildung der Stücke) hat einen bemalten Teller und Weinschale

Historische Literatur. — Pozdnějev[1]) unterscheidet drei Arten historischer Werke: 1. originale Erzeugnisse mit den vorherrschenden charakteristischen Zügen des Epos; 2. Werke unter dem Einfluss chinesischer Autoren verfasst, in denen sich Räsonnements historischer Kritik und sogar historischer Philosophie vorfinden; 3. Werke, in denen man den Einfluss der Manju annehmen muss, deren Verwaltungsformalismus sich auch in der mongolischen Geschichtschreibung widerspiegeln soll. Vergessen ist in dieser Aufzählung, dass es auch ins Mongolische übersetzte tibetische Geschichtswerke gibt.

Das vornehmste und best bekannte historische Werk der mongolischen Literatur ist die Geschichte des Sanang Setsen in zehn Büchern.[2]) Ihre Bearbeitung durch den Akademiker Schmidt bezeichnete eine Meisterleistung für die damalige Zeit, die aber auch jetzt noch unübertroffen dasteht. Obwohl wir vieles darin gegenwärtig in anderem Lichte sehen, und sich sehr vieles nach dem heutigen Stande der Wissenschaft besser übersetzen und kritischer verarbeiten liesse, so fragt es sich doch, ob je eine gleich Schmidt geduldige und ausdauernde Arbeitskraft eine Revision seines Werkes wagen würde.[3]) Sanang Setsen gibt einen kurzen Abriss der Geschichte des Buddhismus in Indien und Tibet (Buch I—III) und erzählt dann die Geschichte der Mon-

---

besprochen, die auf der unteren Seite in mongolischer Schrift den Namen *Baragon Tümäd* «Westliche Tümäd», tragen, womit wohl ein Prinz dieses Stammes gemeint ist, der eine Tochter des chinesischen Kaisers Taokuang (1821—1850) heiratete, und für den das Service jedenfalls fabricirt worden ist. Solche Stücke mit lamaistischen Darstellungen sind in Peking und in New-Yorker Privatsammlungen gar nicht selten anzutreffen.

[1]) Mongolische Chrestomathie, p. IX.

[2]) Geschichte der Ost-Mongolen und ihres Fürstenhauses, verfasst von Ssanang Ssetsen Chungtaidschi der Ordus; aus dem Mongolischen übersetzt, und mit dem Originaltexte, nebst Anmerkungen, Erläuterungen und Citaten aus andern unedirten Originalwerken herausgegeben von Isaac Jakob Schmidt. St. Petersburg und Leipzig, 1829. 4°. XXIV und 509 p.

[3]) Für das Lesen des Originaltextes wird man gut tun, die in Kowalewski's und Pozdnějev's Chrestomathien aufgenommenen Abschnitte zu Rate zu ziehen, in denen die Druckfehler des Textes von Schmidt verbessert sind.

golen von ihren Anfängen bis zum Jahre 1662.[1]) Für die zwischen dem Sturz der Yüan-Dynastie und dem Beginn der Manju-
Dynastie liegende Periode ist er auch jetzt noch eine unschätz-
bare Quelle. Die Mängel seiner Chronologie, Topographie und
Ethnographie sind oft genug ins gebührende Licht gerückt
worden und können nicht fortgeleugnet werden, aber manche
Kritiker sind doch mit dem alten Herrn zu streng ins Gericht
gegangen. Ganz unbegreiflich ist, wie ein so sorgsamer Forscher
vom Schlage EMIL BRETSCHNEIDER's[2]) zu der Behauptung kommt,
dass seine mongolische Geschichte hauptsächlich auf Überlieferungen und nicht auf officiellen Dokumenten beruhe, während
doch unser Autor am Schlusse seines Werkes sieben seiner hauptsächlichen Quellenwerke namhaft macht[3]) und auch im Verlauf
desselben aus anderen Werken citirt.[4]) Es kann, nach dem
ganzen Charakter seiner Aufzeichnungen und nach seinem Verhältnis zu seinen Vorgängern zu urteilen, kaum davon die Rede
sein, dass er irgend welche mündliche Traditionen direkt aufgezeichnet, vielmehr das, was an solche anklingt, älteren schriftlichen Quellen entlehnt hat. Gerade aber in der Aufbewahrung
der alten mongolischen Tradition liegt die eminente und unver-
gängliche Bedeutung seines Werkes, und wenn BEREZIN einer
Angabe BRETSCHNEIDER's zufolge sagt, dass das Nichtvorhanden-
sein der Geschichte des Sanang Setsen für die Geschichtswissen-
schaft keinen Verlust bedeuten würde, so ist das nichts weiter
als öde Schulmeisterei, die ihre Censur auf Grund einer pedan-
tischen Fehlerschnüffelei erteilt, aber nicht auf die Tiefe des
Kerngehalts vorzudringen vermag. Mit Recht hat WILHELM
SCHOTT[5]) geurteilt: «Wer Sanang Setsen's Werk nur vom Stand-

---

[1]) Eine Analyse und Kritik des Werkes gibt ABEL-RÉMUSAT, Observations sur l'histoire des Mongols orientaux, de Sanang-Setsen (Extrait
du Nouveau Journal Asiatique, Paris, 1832, 88 p.); wiederabgedruckt in
seinen Mélanges posthumes, Paris, 1843, pp. 373—458.

[2]) Mediæval Researches from Eastern Asiatic Sources, London,
1888, Vol. I, p. 194.

[3]) SCHMIDT, l. c., S. 299.

[4]) Zum Beispiel S. 21, wo er das *Burkhan maktagal-un tailburi* des
Bakši Biligun-Chuyak anführt.

[5]) Älteste Nachrichten von Mongolen und Tataren, historisch-kriti-

punkte der Vollständigkeit und äusseren Wahrheit beurteilt, den
befriedigt es weniger als die trockenste Chronik, deren Verfasser
sich zum Gesetz gemacht hätte, nur Glaubwürdiges aufzuneh-
men, das Netz der Jahre gut auszufüllen und alles genau zu
datiren ... Wer aber nach innerer Wahrheit sucht, das Sonst
und Jetzt der Mongolen lebendig erfassen, die Gegensätze
lebendig anschauen will: der findet bei Sanang Setsen gewiss
mehr und Besseres als bei vielen anderen.» Sanang Setsen hat
eben nicht für die Historiker Europas geschrieben, sondern als
Mongole, der sich für die Heldenzeit des Činggis begeisterte.
«Die vaterländische Heldenzeit vor dem Wendepunkte zur
allmählichen Verdummung, in deren Gefolge immer die Ein-
knechtung kommt, hat für Sanang Setsen, obschon auch sein
Geist der Hierarchie verfallen und mit Legenden aus Indien
und Tibet angefüllt ist, einen solchen Zauber, dass er, während
er diese Zeit in lose verknüpften Sagen uns schildert, über sich
selbst zu stehen scheint. Da ist keine Spur von Frömmelei oder
geistlicher Überwachung» (SCHOTT). Er hat uns die alten Überlieferungen und Sagen von Činggis aufbewahrt und uns die
ritterlichen Gefühle dieses Volkes kennen gelehrt, das «nicht
blosser Sucht nach Beute, nicht blossem Hang zu roher Ver-
wüstung jene beispiellosen Erfolge dankte.» Man lese die Schilderung vom Ende des grossen Welteroberers, seine Abschieds-
rede, sein Begräbnis und die ergreifende Totenklage,[1]) um Sanang Setsen's Bedeutung als Schriftsteller zu würdigen. Die Aufzeichnungen der chinesischen und mohammedanischen Schriftsteller mögen vom historischen Standpunkt zuverlässiger sein
(vom psychologischen sind sie es gewiss nicht), dennoch können
sie uns den Sanang Setsen nicht ersetzen mit seiner Überlieferung der correkten Schreibung unzähliger mongolischer Personen-
und Ortsnamen.[2]) Dass er wesentlich als gläubiger (nach KÖPPEN

---

sche Abhandlung. Abhandlungen der Berliner Akademie, 1846. S. 2 des
Sonderabdrucks.

[1]) SCHMIDT, l. c., S. 103—109.

[2]) CARL RITTER (Die Erdkunde von Asien, Band II, Berlin, 1833, S.
390) bemerkt ausdrücklich, dass er die Geschichte des Sanang Setsen
«von allen ihren geographischen Seiten gewissenhaft benutzt zu haben
glaubt».

«überfrommer») Buddhist geschrieben, ist ihm nicht weiter zu verargen, gestehen wir doch auch unseren eigenen Historikern das Recht zu, vom Parteistandpunkte eines religiösen oder politischen Bekenntnisses Geschichte zu behandeln. [1]

Was wir von Sanang Setsen's Lebensumständen wissen, erfahren wir aus seinem eigenen Werke, in dem er einigemale von sich in der dritten Person spricht. Danach wurde er im Jahre 1604 aus fürstlichem Geschlecht im Ordos geboren, als Sohn des Batu Hung-taiji [2] und Urenkel des Khutuktai Setsen Hung-taiji von den Baragon Gar. Ursprünglich Sanang Taiji genannt, erhielt er in seinem elften Jahre den Titel seines Grossvaters und wurde Sanang Setsen Hung-taiji genannt. Siebzehn Jahre alt, wurde er von Bušuktu Jinong unter die Zahl seiner hohen Beamten aufgenommen und erhielt die gerichtliche Verwaltung übertragen, wobei er die Gnade und das Zutrauen des Jinong in hohem Grade genoss. Nach dessen Tode rief er 1627 den zweiten Sohn desselben, Rinčen Äyäči Daičing, zum Khagan aus. Darauf unternahm er zur Unterdrückung eines Aufruhrs einen Feldzug in die Gobi, von dem er im Jahre 1634 zurückkehrte. Er gewann auch die Čakhar und andere Stämme für seinen Herrscher, der sich in demselben Jahre noch einmal krönen liess; dem Sanang Setsen, «der von Anfang an sein Gefährte gewesen und seine Feinde ihm geneigt gemacht und hergebracht hatte, verlieh er den Titel Ärkä Sätsän Hung und den Oberbefehl über das Vordertreffen des Kriegsheers, wie auch des Centrums bei grossen Treibjagden.»

In seinem 59. Lebensjahre (1662) verfasste der alte Herr seine Chronik «auf den Wunsch und das Verlangen vieler Wissbegierigen». «Die Fehler und Irrungen in derselben», schliesst er sein Buch, «bitte ich, mit Nachsicht und Geduld zu behandeln, und ihr, mit Vorzügen und Verdiensten ausgerüstete Weise

---

[1] Über den epischen Charakter des Werkes von Sanang Setsen s. den Paragraphen «Volksliteratur».

[2] Die früher beliebte Übersetzung dieses Titels durch «Schwanenfürst» ist wohl irrtümlich; er ist offenbar identisch mit chinesisch *hung tai-tse* «Rotgürtel», den die Seitenverwandten des kaiserlichen Hauses tragen (s. W. F. MAYERS. The Chinese Government, 2. Aufl., Shanghai, 1886, p. 4, Nr. 30).

und Gelehrte, prüft und verbessert das Fehlerhafte! Wer aber
ohne Vorurteil dieses Werk liest und ein wenig Gutes zu seiner
Belehrung darin findet, der möge wie in einen Spiegel hinein-
blicken, in welchem er, denen gleich, welche vermittelst des
himmlischen Cintāmaṇi[1]) das Verborgene zu ergründen wünschen,
die Lingchoa[2])-Blume der ewigen Weisheit aufblühen sieht.»

Dem Werke des Sanang Setsen ıst auch die Ehre einer
Übersetzung in das Chinesische zu teil geworden, die unter
dem Titel *Mong-ku yüan-liu* «Ursprung und Geschichte der Mongolen» zuerst im Jahre 1777 erschien. Dieses Werk war schon
ABEL-RÉMUSAT[3]) bekannt, der bereits das Datum festgestellt und
die darauf bezügliche Stelle aus dem Katalog der K'ien-lung'-
schen Bibliothek (Kap. V, p. 29) wiedergegeben hat, ebenso
W. SCHOTT.[4])

Auf Sanang Setsen und andere mongolische Geschichtswerke ist auch die von Jigs-med Nam-mk'a im Jahre 1818
tibetisch verfasste und von G. HUTH herausgegebene und übersetzte «Geschichte des Buddhismus in der Mongolei» basirt.
Eine Übersetzung dieses Werkes ins Mongolische wurde von
dem lamaischen Grosswürdenträger Zam-ts'a laut Angabe im
Colophon abgefasst;[5]) es ist aber nicht gesagt, ob dieser Befehl
zur Ausführung gelangt ist.

Die officielle Geschichte der Mongolen *Altan Däbtär* («Das
Goldene Buch») ist nicht auf uns gekommen, jedoch in der

---

[1]) Wunschedelstein.

[2]) Lotus (chinesisch *lien-hua*).

[3]) Observations sur l'histoire des Mongols orientaux, Paris, 1832.
pp. 85—86.

[4]) Älteste Nachrichten von Mongolen und Tataren, l. c., p. 3, Note.
Vergl. ferner eine Notiz von FR. HIRTH, Über eine chinesische Bearbeitung der Geschichte der Ost-Mongolen von Ssanang Ssetsen (Sitzungsberichte der Bayer. Akademie, München, 1900, Heft II, S. 195—198). —
E. HAENISCH, Die chinesische Redaktion des Sanang Setsen im Vergleiche
mit dem mongolischen Urtexte (Mitteilungen des Seminars für orientalische Sprachen, Band VII, Abt. 1, 1904, S. 173–199).

[5]) G. HUTH, Geschichte des Buddhismus in der Mongolei, Strass-
burg, 1896, Band II, S. 446. Vergl. die gehaltreiche Recension von A.
GRÜNWEDEL, WZKM, Band XII, 1898, S. 70—74; Anzeige von L. FEER,
Journal Asiatique, Janv.-Févr., 1897, pp. 159—165.

Form erhalten, dass sie in die persische Geschichte (Djami ut Tevarikh) des Rašid-eddin (1247—1318) und in die chinesischen Annalen der Yüan-Dynastie *(Yüan-shih)* übergegangen ist. Rašid erwähnt sie ausdrücklich, mit anderen historischen mongolischen Dokumenten, als im Schatzhaus des Ghazan Khan aufbewahrt und der Aufsicht des Seniors der Begs anvertraut.[1]

Eine andere vollständige Geschichte der Mongolen in mongolischer Sprache, in 32 Heften, soll im Kaiserpalast zu Peking aufbewahrt werden und aus der Mongolei für den Tronfolger gebracht worden sein.[2]

Eines der merkwürdigsten und frühesten Erzeugnisse der mongolischen historischen Literatur ist das unter dem chinesischen Titel gehende Werk *Yüan ch'ao pi (mi) shi*, «die Geheime Geschichte der Yüan-Dynastie». Der Archimandrit PALLADIUS, einer der vorzüglichsten Kenner der chinesischen und mongolischen Literatur, hat davon eine Übersetzung in Band IV, pp. 1—258, (1866) der «Arbeiten der Russischen Mission in Peking» geliefert, auf Grund eines Auszuges aus der chinesischen Übersetzung des mongolischen Originals. Letzteres war im Jahre 1242 beendet zur Zeit einer grossen Versammlung am Flusse Kerulen und repräsentirt somit das älteste vorhandene Werk der mongolischen Literatur. Es behandelt die früheste Geschichte der Mongolen, die Regierung von Činggis Khan und den Anfang der Regierung von Ogotai. Zu Beginn der Ming-Dynastie wurde eine chinesische Übersetzung davon verfertigt. Unter dem Jahre 1382 wird dieselbe im *Hung-wu shih lu* «Ausführliche Berichte über die Regierung des Kaisers Hung-wu» erwähnt. Es wird

---

[1] BRETSCHNEIDER, Mediæval Researches from Eastern Asiatic Sources, Vol. I, London, 1888, p. 197. — Die mongolischen Dokumente, deren sich Rašid bediente, waren in uigurischer Schrift abgefasst von sechs Schreibern unter Leitung des Pulād Beg, der besser als irgendeiner die Genealogie der Mongolen kannte, mongolisch sprach und die mongolischen Werke lesen konnte. — Abul Ghazi, der drei Jahrhunderte später schrieb, bediente sich für seine Geschichte der Mongolen und Tataren ausser dem Werke seines Vorgängers siebzehn anderer Geschichten des Činggis in mongolischer Sprache (DESMAISONS, Histoire des Mogols et des Tatares par Aboul-Ghâzi Bêhâdour Khan, Vol. II, St. Pet., 1874, p. 35).

[2] KOWALEWSKI, Mongolische Chrestomathie, Vol. II, p. 331.

dort gesagt, dass der mongolische Text der chinesischen Übersetzung angehangen wurde, aber nicht in der Originalschrift,
sondern indem man die mongolischen Laute durch chinesische
Schriftzeichen wiedergab. Es gelang PALLADIUS, eine handschriftliche Copie der Ming Ausgabe des Werkes, begleitet von dem
mongolischen Text in chinesischen Charakteren, aufzufinden, die
auch EMIL BRETSCHNEIDER[1]) benutzt hat. Er behauptet, dass trotz
vieler Archaismen und Schreibfehler, die in diesem Texte vorkommen, die Wiederherstellung des mongolischen Originals wenig
Schwierigkeiten für den Kenner beider Sprachen biete. Für alle,
welche die Geschichte der Mongolen studiren, gewährt nach
seinem Urteil dieses seltene Dokument alter mongolischer Literatur ein hohes Interesse. Es bestätigt im allgemeinen Rašid
eddin's Aufzeichnungen, und gelegentlich finden sich Stellen
darin, die wie eine wörtliche Übersetzung der Äusserungen des
persischen Historiographen klingen. Dies beweist, nicht nur, dass
Rašid dieselbe Informationsquelle wie der unbekannte Verfasser
des *Yüan ch'ao pi shih* benutzt hat, sondern auch die Güte der
altmongolischen Tradition. Im allgemeinen sind die Daten des
mongolischen Werkes mit denen der mohammedanischen Schriftsteller in Übereinstimmung, wiewohl in einigen wenigen Fällen
grosse chronologische Schnitzer und Verschiebungen von Ereignissen vorkommen.

PALLADIUS' Handschrift ist in den Besitz der Universitätsbibliothek von St. Petersburg übergegangen, und Prof. POZDNĚ-
JEV[2]) hat ein Facsimile des Textes mit Übersetzung und Noten
publicirt.

---

[1]) Siehe Journal of the North-China Branch of the Royal Asiatic
Society, N. S., Vol. X, 1876, pp. 88—89, oder E. BRETSCHNEIDER, Mediæval
Researches from Eastern Asiatic Sources, London, 1888, Vol. I, pp. 192—193.

[2]) Транскрипція палеографическаго текста Юань-чао-ми-ши [nicht
gesehen]. — Derselbe, О древнемъ китайско-монгольскомъ историческомъ
памятникѣ «Юань-чао-ми-ши», St. Pet., 1882, 22 p., 1 Tafel. — Seit vielen Jahren habe ich mich um ein Exemplar dieses wichtigen Werkes bemüht und Prof. POZDNĚJEV persönlich in Wladiwostok um Beschaffung
eines solchen gebeten, ohne es trotz seines Versprechens zu erhalten. Die
russischen Gelehrten sind immer kampffähig bereit, über die Vernachlässigung ihrer Schriften von seiten ihrer westlichen Kollegen Klage zu
führen, aber was tun sie selbst für die Verbreitung ihrer Schriften im

Eine interessante und besonders für die alte Sagenkunde
wertvolle Chronik ist das *Altan Tobči* («Goldener Knopf») oder
Ärdänin Tobči («Kostbarer Knopf»), verfasst 1604, von dem die
russischen Missionare in Peking zwei handschriftliche Exemplare
beschafft hatten. Der gelehrte Lama GALSANG GOMBOJEV[1]) veranstaltete danach eine Ausgabe des Textes mit russischer Übersetzung. BRETSCHNEIDER[2]) bezeichnet ihn etwas übertrieben als
einen sehr verworrenen Bericht der Geschichte der Mongolen
bis zum sechzehnten Jahrhundert und im allgemeinen schwer
verständlich. «Wir finden darin», bemerkt er, «eine grosse Fülle
von Namen von Männern, Orten und Ereignissen unzusammen-
hängend aufgezeichnet, und in den meisten Fällen sind wir in
Verlegenheit, was wir aus diesen Geschichten machen sollen,
und zu entscheiden, auf welche Periode sie sich beziehen sollen;
ein Vergleich des Altan Tobči mit den chinesischen historischen
Berichten jedoch befähigt uns, darin einen Kern von Echtheit
zu erkennen.» Der Stil dieser Chronik ist schwieriger als der
des Sanang Setsen, weit kürzer und gedrängter und voll von
Idiotismen und auch veralteten Ausdrücken. Manche Sagen sind
hier in einer ursprünglicheren und reineren Form überliefert
als bei Sanang Setsen, der manche altertümliche Züge nicht
mehr recht verstanden zu haben scheint, wie ich bereits früher
an einem Beispiele gezeigt habe.[3]) Die Chronik nimmt eine

Auslande? Die meisten russischen Publikationen haben nicht einmal einen
Verleger, und die Organisation des russischen und gar erst des sibirischen
Buchhandels ist ein wahres Mysterium. Aus Ägypten, Syrien, Indien,
Indochina, Australien, Java, Japan, China, ja aus Tibet ist es jetzt leichter
(und durchaus keine Schwierigkeit) Bücher zu erlangen denn aus Russland und Sibirien. Ich möchte daher den russischen Kollegen zurufen:
es ist durchaus nicht unsere Schuld, wenn ihr hier unbekannt bleibt;
macht euere Arbeiten der Welt zugänglich, und wir werden sie lesen und
studieren.

[1]) Altan Tobči, mongolische Chronik, in den Arbeiten der orientalischen Abteilung der Kaiserl. archäologischen Gesellschaft, Teil VI. St.
Petersburg. 1858. XIV und 234 p. Der Text weist leider viele Druckfehler
auf. Die Geschichte der Vorfahren des Činggis und dessen Regierung ist
daraus in POZDNĚJEV's Chrestomathie. pp. 105—126. abgedruckt.

[2]) Mediæval Researches, Vol. II. p. 159.

[3]) Der Urquell, Neue Folge, Band II, 1898, S. 148—149.

Mittelstellung zwischen *Yüan ch'ao pi shih* und *Sanang Setsen* ein.

Unter demselben Namen *Altan Tobči* gibt es auch eine andere Chronik, die mit der vorhergenannten nichts als den Titel gemeinsam hat. Den handschriftlichen Text derselben erlangte POZDNĚJEV in der nördlichen Mongolei; nach seinem Urteil stammt er wahrscheinlich von den Kharačinskischen Khošunen, denn zwei darin vorkommende Legenden, welche die Mongolen gewöhnlich auf Činggis Khan beziehen, werden auf Khasar, den Stammvater der Kharačinen, übertragen. Dieser Stamm war mehr als alle anderen mongolischen Geschlechter dem Einfluss der Chinesen unterworfen, weshalb auch der Autor des Altan Tobči die Erzählungen chinesischer Historiker benutzen konnte. In seinem Texte lassen sich unschwer fast wörtliche Übersetzungen aus dem *Yüan shih lei pien*,[1]) *Ming shih* und anderen chinesischen Quellen erkennen; auch sein Trachten nach historischer Kritik und hier und da eingestreute Reflexionen werden sicher auf diesen Einfluss zurückgehen. Ein Bruchstück aus diesem Werke ist in POZDNĚJEV's Chrestomathie, pp. 126—148, abgedruckt.

In den Erläuterungen zu seiner Ausgabe des Sanang Setsen hat SCHMIDT reichhaltige Auszüge aus einer kalmükischen Chronik, *Nom garkhai todorkhai toli* («Der die Ausbreitung oder Geschichte der Religion erklärende Spiegel»)[2]) oder *Bodhi-mör* («Pfad der Erleuchtung»), mitgeteilt. Dies ist eine Bearbeitung der tibetischen Chronik *rGyal-rabs gsal-bai me-lon* «Der die Genealogie der Könige [d. i. von Tibet] erklärende Spiegel», eine Geschichte Tibets, die im Jahre 1327 von dem Sa-skya-pa bSod-nams rgyal-mts'an im Kloster bSam-yas verfasst wurde. Dieses tibetische Werk verdient eine vollständige Übersetzung, zumal da gerade die wichtigsten und interessantesten Partien in SCHMIDT's Auszuge fehlen, und die meisten Namen darin aus dem Tibetischen

---

[1]) Eine abgekürzte chinesische Geschichte der Yüan-Dynastie von *Kiai Shan*, herausgegeben 1699, in 42 Kapiteln (BRETSCHNEIDER. Mediaeval Researches. Vol. I, p. 191).

[2]) Auch *Gägän Toli*, «Glänzender, erhellender Spiegel», Übersetzung von tib. *gsal-bai me-lon*.

ins Mongolische übersetzt erscheinen. Eine mongolische Ausgabe desselben ist in Peking gedruckt worden;[1] eine kalmükische Recension ist in Dresden (Eb 494b, 158 fols.) vorhanden.

Popov[2] erwähnt ein Werk *Jirükän-ü Tolta* («Schmuck des Herzens»), vorwiegend historischen Inhalts, aus dem er einen Abschnitt über die Herkunft der mongolischen Buchstaben mitteilt; darin wird auch die erste Einführung und Ausbreitung des Buddhismus bei den Mongolen erzählt. Der vollständige Text ist in Pozdnějev's Chrestomathie (pp. 360—379) zugänglich gemacht, der dasselbe (p. XV) für eine der besten original-mongolischen Grammatiken erklärt: es sollen darin Regeln zur Abfassung eines musterhaften literarischen Stils niedergelegt sein, auch die Beziehungen der Schrift- zur Umgangssprache berührt werden.[3] Es handelt sich aber darin in der Hauptsache um Regeln über Rechtschreibung und Aussprache. Schon Bobrovnikov hat den richtigen Gebrauch davon gemacht und wiederholt daraus in seiner mongolisch-kalmükischen Grammatik citirt. Die Verfasserschaft dieser Schrift wird von der Tradition dem Urheber des gegenwärtigen mongolischen Alphabets, C′os-kyi Od-zer, zugeschrieben; doch ist der Beweis dafür noch zu erbringen.

Eine neuere mongolische Chronik ist das *Ärdäni-yin Ärikä* («Kostbarer Kranz»), die ausser einer allgemeinen Geschichte der Mongolen bis 1849 Materialien zur Geschichte der Khalkha von 1636 bis 1736 enthält, und die A. Pozdnějev[4] in Text und Übersetzung veröffentlicht hat.

---

[1] Das Kapitel von der Erscheinung des Avalokiteçvara mongolisch in Kowalewski's Chrestomathie. Vol. II. pp. 32—37; dasselbe deutsch bei I. J. Schmidt. Forschungen im Gebiete der älteren religiösen, politischen und literärischen Bildungsgeschichte der Völker Mittel-Asiens, St. Petersburg, 1824. S. 193—206.

[2] Mongolische Chrestomathie, p. VIII. Dieser Abschnitt ist übersetzt von I. J. Schmidt. Geschichte der Ost-Mongolen. S. 392—394, 397—398, und Pallas, Sammlungen historischer Nachrichten etc., Band II, S. 356—369.

[3] Es ist seltsam, dass Pozdnějev in der Einleitung zu seiner Chrestomathie sich in keiner Weise auf eine Erörterung der Verfasserschaft und Zeit der von ihm behandelten Werke einlässt.

[4] Монгольская лѣтопись «Эрдэнійнъ-эрикэ». Подлинный текстъ съ переводомъ и объясненіями, St. Pet., 1883, 8°, 421 p.

Die Kalmüken besitzen eine ganze Anzahl historischer Schriften. Auf der Kgl. Bibliothek zu Dresden ist die «Erzählung, wie die Dörbön-Oirad von den Mongolen besiegt wurden.»[1] Den kalmükischen Text der Geschichte des Ubaši-Hung-Taiji hat C. GOLSTUNSKI[2] herausgegeben, ebenfalls GALSANG GOMBOJEV[3] mit einer russischen Übersetzung. Andere kalmükische Chroniken hat A. POZDNĚJEV[4] bearbeitet.

Ein modernes mongolisches Geschichtswerk, *Bolor Toli* («Krystallspiegel»), das nach 1820 geschrieben sein muss, ist kurz von A. RUDNEV[5] analysirt worden. Im ersten Buche wird ein Abriss der buddhistischen Kosmologie und des Buddhismus in Indien entworfen, im zweiten Buch werden China und Tibet behandelt, mit Biographien der Lamen, im dritten die Mongolen. Diesem Schema nach zu urteilen, lehnt sich das Buch an gewisse bekannte tibetische Geschichtswerke an, enger vermutlich an das *Grub-mtʻa šel-kyi me-loṅ* («Krystallspiegel der Siddhānta»), mit dem es ja das Stichwort des Titels gemeinsam hat.

Sicher gibt es auch aus dem Chinesischen ins Mongolische übersetzte historische Werke. Dafür halte ich z. B. das im Katalog von Kasan, Nr. 178, erwähnte mongolische Buch in hundert Kapiteln, das nach der dort gegebenen Beschreibung eine ausführliche Schilderung des Krieges der Chinesen gegen die Dsungaren (von 1754 an) und das östliche Turkistan enthalten soll.

---

[1] *Dörbön Oirad Monggoli daruksan tūji*, 10 fols., MS. Dresd., Eb 404i. Die Dörbön Oirad sind die vier Stämme der Kalmüken: Sungar, Torgod, Khošod und Dörböd.

[2] Убаши Хунъ-Тайджійнъ тӯджи (zusammen mit Djanggar und Siddhi Kür), St. Pet., 1864, lithographirt, pp. 1—7.

[3] In seiner Ausgabe des Altan Tobči, St. Pet., 1858, pp. 198—224.

[4] Памятники исторической литературы Астраханскихъ Калмыковъ, St. Pet., 1885, 130 p., 8°. Vergl. auch von demselben, Астраханскіе Калмыки и ихъ отношеніе къ Россіи до начала нынѣшняго столѣтія, St. Pet., 1886, 32 p., 8°, und seine Beiträge zur Geschichte der sungarischen Kalmüken in dem Buche von N. I. VESELOVSKI, Посольство къ вюнгарскому Хунъ-Тайчжи Цэванъ Рабтану капитана отъ артиллеріи Ивана Унковскаго и путевой журналъ его за 1722—1724 годы, St. Pet., 1887, pp. 239—264 (nebst sechs Seiten kalmükischen Texts).

[5] Zap., Vol. XV, 1904, pp. 032—034.

15*

Geographische Literatur. — Beschreibungen sind in der mongolischen Literatur selten und den älteren Denkmälern gänzlich fremd; doch gibt es Schilderungen von China und Heiligtümern jüngeren Datums. Als Muster des beschreibenden Stils hat Pozdnějev in seiner Mongolischen Chrestomathie (pp. 15—21) aus der unlängst publicirten ausführlichen Biographie des lCaṅ-skya Khutuktu zwei Stücke mitgeteilt, eine Schilderung von Peking und eine von Jehol. Darin wird ein Bild der Strassen, Paläste und Läden, sowie des Handels der Hauptstadt entworfen, in der anderen Skizze besonders die Parks und Seen des dortigen Kaiserpalasts beschrieben. Ziemlich reich dagegen ist das mongolische Schrifttum an Reisebeschreibungen, die bisher leider sehr wenig bekannt geworden sind. Der Khubilgan Jebtsun-damba Khutuktu hat seine Reise nach Erdeni-tsu in Form eines Tagebuchs geschildert, und der burjatische Khanbo-Lama Dsayayi seine Fahrt nach Tibet mit einer systematischen Beschreibung der von ihm besuchten Orte und besonderen Skizzen ganzer Gegenden. Es gibt auch noch mehrere andere nicht veröffentlichte Beschreibungen von Tibet-Reisen burjatischer und kalmükischer Lamas. Ein recht interessantes Beispiel dieser Art, die Reisebeschreibung eines Kalmüken, der mit zwei Gefährten über Urga nach Lhasa und zurück über See nach Russland reiste (1891—1894), hat Pozdnějev[1]) in Text und Übersetzung mitgeteilt.

Schamanische Literatur. — Von der alteinheimischen Religion der Mongolen galt bisher allgemein die Annahme, dass sie keine Literatur besässe, sondern ausschliesslich auf mündlichen Überlieferungen beruhe. Es gelang jedoch Pozdnějev, bei seinem Aufenthalt unter den Burjaten 1879 von dortigen Schamanen eine Handschrift zu erwerben, die eine Darstellung des Schamanenglaubens der Burjaten enthält und in seiner Mongolischen Chrestomathie, pp. 293—311, abgedruckt ist. Hier

---

[1]) Сказаніе о хожденіи въ Тибетскую страну Мало-Дёрбётскаго Бāза Бакши. Калмыцкій Текстъ съ переводомъ и примѣчаніямъ. Издан. Фак. Вост. Языковъ С. П. Ун., St. Pet., 1897. VIII und 260 p. Analyse des Inhalts von W. BARTHOLD, Mitteilungen des Seminars für orientalische Sprachen, Band I, 1. Abt., Berlin, 1898, S. 187—190.

werden die schamanischen Götter und Schutzgeister, die privaten und lokalen Ongone, die Schamanen und Schamaninnen,
die bei der Weihe zum Schamanen zu vollziehenden Ceremonien,
ihre Kleider und Geräte, ihre Gebete bei verschiedenen Gelegenheiten, die Herstellung der Ongone, die bei Krankheiten zu
beobachtenden Riten und Opfer u. s. w. behandelt. Dieses Werk
würde gewiss eine Übersetzung verdienen. Einen Ersatz dafür
hat man einstweilen in den trefflichen Schriften von Changalov,
der aus dem Munde burjatischer Schamanen aufgezeichnete
Texte mit Übersetzungen bietet.[1]) Wertvoll ist auch noch die
Arbeit von Banzarov.[2])

Buddhistische Literatur. — Das Übersetzungswerk buddhistischer Schriften scheint frühzeitig begonnen zu haben.
Schon Cʻos-kyi (oder Cʻos-sku) Od-zer von Sa-skya, der Schöpfer
des eigentlichen mongolischen Alphabets, befasste sich mit der
Übertragung von Dhāraṇī und Çāstra ins Mongolische.[3]) Der

---

[1]) М. Н. Хангаловъ и Агапитовъ, О шаманствѣ у Бурятъ Иркут-
ской губерніи. Извѣстія Восточно-Сибирскаго Отдѣла Императорскаго
Русск. Географическаго Общества, Vol. XIV, 1883, Nr. 1—2. Хангаловъ, Новые матеріалы о шаманствѣ у Бурятъ, Записки Восточно-Сиб.
Отдѣла Имп. Русск. Геогр. Общ. по Этнографіи, Vol. II, Nr. 1, Irkutsk,
1890. Danach H. Kern, Over de godsdienstleer der Burjaten, Amsterdam,
1893 (K. Akademie van Wetenschappen). Шаманскія повѣрія инородцевъ
восточной Сибири. Ibid., Vol. II, Nr. 2, Irkutsk, 1890, mit 10 Tafeln
(Zeichnungen von Schamanentrommeln).

[2]) Дорджи Банзаровъ, Чечная Вѣра или шаманство у Монголовъ
и другія статьи, St. Petersburg, 1891.

[3]) Köppen, Die lamaische Hierarchie und Kirche, S. 101; Huth,
Geschichte des Buddhismus in der Mongolei, Band II, S. 164. Nach Kowalewski (Mongolische Chrestomathie, Vol. II, p. 459) ist er der Übersetzer des Bodhisattvacaryāvatāra von Çāntideva (im Tanjur, Vol. 23 der
Sūtra; s. Wasiljev, Der Buddhismus, S. 228, und Schiefner's Übersetzung
des Tāranātha, S. 163—168). Diese Übersetzung ist jedoch damals nicht
gedruckt worden, sondern wurde in fehlerhaften Abschriften überliefert,
bis sie mit der Zeit unverständlich wurde. Die Lamas vom Tempel *Sung-
chu-ssŭ* in Peking trieben drei mongolische Manuskripte auf und drei gedruckte tibetische Exemplare mit einigen Commentaren. Da sie zwischen
den einzelnen Versionen keine Ähnlichkeit fanden, entschlossen sie sich
mit Hülfe des lČangča Khutuktu zu einer Verbesserung der Übersetzung
zu allgemeinem Gebrauch. Ein Exemplar dieser verbesserten mongolischen
Übersetzung befand sich im Besitz von Kowalewski.

Kaiser Yesun Temur (1324—1327) der Yüan-Dynastie berief den Lama dGa-ba bsod-nams von Sa-skya und den mongolischen Übersetzer (Lo-tsā-ba) Šes-rab seṅ-ge und liess viele Pravacana durch sie übersetzen.[1] Unter dem Kaiser Tub Temur wurde im Jahre 1330 das «Sūtra vom Gestirn des grossen Bären» *(dolo-gan äbügän närätü odonu sudur)* ins Mongolische übersetzt und in 2000 Exemplaren gedruckt, wohl das älteste beglaubigte Beispiel eines mongolischen Druckes.[2]

Es sei daran erinnert, dass Europa die ersten eingehenderen Nachrichten über den Buddhismus aus der mongolischen Literatur erhalten hat. PALLAS[3] hat in seinem berühmten klassischen Werke nach den Angaben seiner Dolmetscher treffliche Mitteilungen über die Kosmologie und Mythologie, das Göttersystem und die geistliche Hierarchie, wie den Cultus der mongolischen Buddhisten gemacht und auch auszugsweise Übersetzungen aus mongolisch-buddhistischen Schriften gegeben; vor allem hat er schon eine Biographie des Buddha nach drei Originalschriften bekannt gemacht.[4] Auch B. BERGMANN[5] hat die Übersetzung einer buddhistischen Kosmologie geliefert, und TIMKOWSKI[6] ein kurzes Leben Buddhas nach mongolischen Berichten.

Der Kanjur, die hundertbändige tibetische Version des Tripiṭaka, wurde im Anfang des XVII. Jahrhunderts vollständig ins Mongolische übersetzt unter dem Legs-ldan Khutuktu Khagan der Čakhar, der von 1603 bis 1634 regirte; die Übersetzung war 1623 beendet.[7] «Seitdem gediehen bis auf die

---

[1] HUTH, l. c., S. 166.

[2] LAUFER, T'oung Pao, 1907, pp. 395, 397.

[3] Sammlungen historischer Nachrichten über die mongolischen Völkerschaften. Zweiter Teil, St. Pet., 1801.

[4] L. c., S. 410—419.

[5] Nomadische Streifereien unter den Kalmüken in den Jahren 1802 und 1803, Riga, 1804—1805, Band III, S. 185—230. Das kalmükische Original in Dresden (Eb 404k, 19 fols.).

[6] Reise nach China durch die Mongoley in den Jahren 1820 und 1821, Band III, Leipzig, 1826, S. 387—408.

[7] G. HUTH, Geschichte des Buddhismus in der Mongolei, Bd. II, S. 248. I. J. SCHMIDT, Sanang Setsen, S. 418. Auf WADDELL's (The Buddhism of Tibet, p. 38) durch keine Quellenangabe gestützte Behauptung,

Gegenwart», bemerkt der tibetische Chronist, der uns diese Nachricht überliefert, «die Übersetzung der Schriften, die Anhörung der Erklärungen und die übrigen Religionsübungen vortrefflich, und infolgedessen begannen alle Gegenden der Mongolei auf dem Pfade lauterer Tugend zu verharren und sich eifrig der Gläubigkeit, der Freigebigkeit und einer sonstigen vornehmen Lebensweise zu befleissigen.» Die Übersetzung wurde von mehreren Schriftgelehrten unter Leitung des Kun-dga Odzer ins Werk gesetzt. Schon früher waren zur Zeit des Kaisers Gulug oder Wu-Tsung (1308—1311) Teile des Kanjur und Tanjur übersetzt worden.[1] Jene Übersetzung wurde auf Befehl des Kaisers K'ang-hsi (1662—1722) noch einmal revidirt und dann gedruckt.[2]

Der Kaiser K'ien-lung (1736—1795) beauftragte den lČanskya Rol-pai rdo-rje und bLo-bzaṅ bstan-pai ñi-ma, auch den Tanjur, den Commentar zum Kanjur in 225 Bänden, aus dem Tibetischen ins Mongolische zu übersetzen. Dieser Arbeit aber standen weit grössere Schwierigkeiten entgegen als der viel leichteren Übersetzung des Kanjur, da sich im Tanjur eine grosse Reihe wissenschaftlicher Abhandlungen mit einer reich entwickelten eigentümlichen Terminologie finden, darunter Werke über Sprachwissenschaft, Metrik, Rhetorik, Lexikographie, Medicin, Kunsttechnik. Zu diesem Zwecke mussten die beiden Lamen erst ein technologisches Wörterbuch in tibetischer und mongolischer Sprache schaffen, das ihrer Arbeit als Grundlage

---

Sa-skya Paṇḍita habe bereits den Kanjur ins Mongolische übersetzt und mit den chinesischen Texten collationirt, ist gar kein Wert zu legen.

[1] Sanang Setsen (S. 121) sagt, dass es der grösste Teil der Sûtra und Dhāraṇī war. Das tibetische Werk *Grub-mt'a śel-kyi me-loṅ* (verfasst 1740), Kap. 11: berühmte Dhāraṇī und andere Religionsschriften des Kanjur und Tanjur (s. JASB, 1882, p. 61, Z. 8; die Übersetzung dieser Stelle von CHANDRA DAS, p. 68, ist nicht ganz correkt). Vergl. POPOV, Mongolische Chrestomathie, p. 94.

[2] HUTH, l. c., S. 291. CHANDRA DAS (l. c., p. 69) übersetzt «teilweise gedruckt», was durch den tibetischen Text nicht gerechtfertigt wird. KOWALEWSKI (Mongolische Chrestomathie, Vol. I, p. 264) sagt, dass die Revision unter Yung-chêng in Peking durch den lČangča Khutuktu, und der Druck in der ersten Hälfte des achtzehnten Jahrhunderts auf kaiserliche Kosten stattfand.

dienen sollte. Mit Hülfe eines Stabes vieler Gelehrten, «welche
die Quellen studirt hatten, und beide Sprachen sprechender
Lochāva» begannen die beiden das Übersetzungswerk im Oktober-
November 1740 und sollen mit der ganzen Übersetzung im No-
vember-December 1741, also in Jahresfrist, vollständig fertig
geworden sein, zufolge dem tibetischen Historiker ‚Jigs-med
Nam·mk´a.[1]) Diese Angabe ist ganz unglaublich, auch wenn das
Heer der Mitarbeiter noch so gross gewesen ist. Es kann sich
dabei höchstens um einen ganz geringen Bruchteil des Tanjur
gehandelt haben. Das geht auch aus den weiteren Ausführungen
hervor. «Danach überreichten sie die Übersetzung dem grossen
Kaiser zur Einsicht und Prüfung. Der nun liess ihnen voller
Freude in reichem Masse Lobeserhebungen zuteil werden und
liess den Übersetzern Dankgeschenke und unfassbare andere
Gnadenbezeugungen zukommen. Die Übersetzung liessen diese
auf Kosten des Kaisers drucken und in alle Landesteile der
grossen Mongolei verbreiten. Dies war die Hauptursache des
Fortbestehens des Edelsteins der Lehre.» Hier ist von einer
grossen Druckauflage die Rede, und in diesem Falle hätte sich
im Laufe der Zeit das eine oder andere Exemplar auffinden
lassen, bisher ist keines aufgetrieben worden. Nun wird zwar
auch im *Grub-mt´a šel-kyi me-lon* positiv berichtet, dass sowohl
der Kanjur als der Tanjur in mongolischer Sprache zur Zeit
von K´ien-lung gedruckt worden seien.[2]) Diese Angabe ist indessen sehr verdächtig, da der Verfasser 1740 starb, nachdem
er sein Werk etwa eine Woche vor seinem Tode beendet hatte.[3])
Er konnte also schlechterdings nichts von der angeblich erst
1741 abgeschlossenen Tanjur-Übersetzung und noch weniger
über die erst später erfolgte Drucklegung derselben wissen. Die
Stelle muss daher als nachträgliche Interpolation betrachtet
werden.

Unter dem Titel «Buddhistischer Katechismus» hat Kowalewski in seiner Chrestomathie (Vol. II, pp. 99—158) nach
einer Handschrift den Text eines Werkes *Tonilkhu-yin čimäk*

---

[1]) Huth, l. c., S. 291—293.

[2]) JASB, 1882, pp. 62, 69.

[3]) JASB, 1881, pp. 187. 188.

kämägdäkü šastir («Çāstra genannt Schmuck der Erlösung») publicirt, das ein Auszug oder, wahrscheinlicher, Übersetzung eines tibetischen Buches ist. In POZDNĚJEV's Chrestomathie (pp. 200—228) ist dieselbe Schrift in einer vollständigeren und berichtigten Fassung reproducirt. Sie ist im wesentlichen dogmatischen Inhalts und erläutert die Grundbegriffe der buddhistischen Lehre.

Das Sūtra der zweiundvierzig Artikel, das eine Übersicht der hauptsächlichen buddhistischen Lehren enthält, ist in dreifachem Text von L. FEER[1]) herausgegeben worden. Es wurde ins Mongolische von Prajñodayavyāsa im Zeitalter des Kʻienlung (wahrscheinlich 1781) übersetzt.[2])

In der Entwickelung des Mahāyāna Buddhismus hat sich ein System ausgebildet, in dessen Mittelpunkt der Lichtgott Buddha Amitābha (mongolisch *Amida*) steht und die Lehre von einem im Westen Indiens gelegenen Paradiese Sukhāvatī. Die Vorstellung davon hat auf alle Völker des nördlichen Buddhismus einen tiefen Eindruck gemacht, und die Sehnsucht, inmitten dieses himmlischen Freudenortes wiedergeboren zu werden, hat zu zahlreichen Schilderungen desselben in Wort und Bild Veranlassung geboten. Eine mongolische Beschreibung, vermutlich im fünfzehnten Jahrhundert aus dem Tibetischen übersetzt, hat J. PODGORBUNSKI[3]) ins Russische übertragen.

Ein eigentümliches Erzeugnis der tibetischen Literatur ist das *Māṇi bka-ạbum* («Die hunderttausend kostbaren Vorschriften»). Von der Tradition wird die Verfasserschaft dem tibetischen König

---

[1]) Le Sūtra en quarante-deux articles. Textes chinois, tibétain et mongol, autographiés par LÉON FEER, d'après l'exemplaire polyglotte rapporté par l'abbé Huc. Paris, 1868 (Maisonneuve), 39 p.

[2]) LÉON FEER, Le Sūtra en 42 articles traduit du tibétain avec introduction et notes, Paris, 1878, pp. 49, 76, 77. In der trefflichen Einleitung findet man auch die früheren Übersetzungen besprochen. CH. DE HARLEZ (Les quarante-deux leçons de Bouddha, texte chinois avec traduction, introduction et notes, Bruxelles, 1899) hat eine andere chinesische Version desselben Sūtras bekannt gemacht.

[3]) Зерцало мудрости, которое, разсказавъ о происхожденіи царства Сукавади, ясно представитъ достоинства этого свѣтого царства (переводъ съ монгольскаго). Извѣстія Восточно-Сибирскаго Отдѣла Имп. Русскаго Общества, Vol. XXVI, Irkutsk, 1895, pp. 1—31.

Sroṅ-btsan sgam-po zugeschrieben, was unhaltbar ist und wohl
nur daher rührt, dass sich das Werk hauptsächlich mit diesem
Könige und seinen gesetzlichen Bestimmungen befasst (ausserdem dient es der Verherrlichung des Gottes Avalokiteçvara);
aber es liegt kein Grund vor, mit WASILJEF anzunehmen, dass
das Werk «zweifellos modern und auf Befehl der Dalai Lamas
zur Aufrechthaltung ihrer Autorität geschrieben worden sei.»[1]
Es enthält unzweifelhaft viele alte Bestandteile, darunter besonders Gesetze und Aussprüche des ersten historischen Königs
von Tibet, die auf eine gegründete Tradition aus jener Zeit
zurückgehen. Eine tibetische Ausgabe (329 fols.) wurde 1730 in
Peking gedruckt, eine sehr schöne mongolische ebenda 1713 in
zwei Bänden (291 und 251 fols.); von der kalmükischen Bearbeitung befindet sich ein Bruchstück von 58 fols. in Dresden
(Eb 404a).

Inhaltlich scheint mit dem *Māṇi bka-abum* das von KOWA-
LEWSKI[2]) erwähnte Werk *Dürbän aimak T'übät ulus irgändür üli-
gär khaoli baiguluksan šastir* verwandt zu sein, in welchem
religiös-moralische Vorschriften zur Leitung der Bewohner in
den vier Provinzen von Tibet niedergelegt sind.

Das beliebteste Werk der tibetischen Literatur, die Lieder
des Milaraspa, eines wandernden Bettelmönchs des elften Jahrhunderts, und seine Biographie, ist von Širägätü Gušri Čos-rje
ins Mongolische übersetzt worden,[3]) der eine reiche literarische
Tätigkeit entfaltet hat. Ebenso verdient besondere Erwähnung

---

[1]) Mitteilung an W. W. ROCKHILL (The Life of the Buddha. London,
1884, p. 213). Eine Analyse des Werkes bei E. SCHLAGINTWEIT, Buddhism
in Tibet, London, 1866, p. 84. Der Anfang ist unter dem Titel «Auszug
eines grossen tangutischen Werks Mani-Gambo, welches die Legenden
vor den grossen Burchanen Abida, Chondschin-boddi-saddo und Schak-
tschamunih enthält» bei PALLAS, Sammlungen historischer Nachrichten
über die mongolischen Völkerschaften. Band II, S. 396—409. übersetzt.
Das zweite Kapitel übersetzt von ROCKHILL, The Land of the Lamas,
London, 1891, pp. 326—334. Vergl. KÖPPEN, Die lamaische Hierarchie
und Kirche, S. 58.

[2]) Mongolische Chrestomathie, Vol. II, pp. 330—331.

[3]) HUTH, l. c., S. 248; derselbe, Die Inschriften von Tsaghan Baišiṅ, Leipzig, 1894, S. 28. Der erwähnte Paṇḍit ist Verfasser der tibetischen Inschrift.

die mongolische Übertragung des interessanten tibetischen Legendenwerkes von Padmasambhava, der in der Religionsgeschichte Tibets eine bedeutende Rolle spielt. Die in Peking hergestellten mongolischen und tibetischen Ausgaben gehören zu den schönsten Erzeugnissen der lamaischen Presse. Der mongolische Holzdruck umfasst 292 Blätter grössten Formats mit dreissig Zeilen auf jeder Seite; als Übersetzer nennt sich Sakya Türüb Kälämürči.

Das Hauptwerk des Reformators bTsoṅ-kʿa-pa, 1356—1418, (*Byaṅ-cʿub lam-gyi rim-pa*, kurzweg *Lam-rim* [mongolisch *mür-ün tsärgä*] genannt, «der Stufenweg zur Vollkommenheit»)[1]) ist von den Mongolen mit grossem Eifer studirt worden, besonders im achtzehnten Jahrhundert, als zwei grosse Klöster Porhantu und Tala mit einer eigenen *Lam-rim*-Fakultät zum Studium dieses Systems gegründet wurden.[2])

Von den zahlreichen Abteilungen der buddhistischen Literatur sind die *Jātaka* und *Avadāna* weitaus die beliebtesten und populärsten. Unter den Mongolen haben besonders zwei Sammlungen, das *Üligärün Dalai* («Meer der Gleichnisse») und das *Altan Gäräl* («Goldglanz») grösste Verbreitung gefunden. Jenes Werk ist eine Bearbeitung der aus dem Chinesischen ins Tibetische übersetzten, im Kanjur enthaltenen bekannten Sammlung *mDzaṅs-blun* («Der Weise und der Tor») von 51 Erzählungen, die I. J. Schmidt herausgegeben und übersetzt hat.[3]) Im Vorwort zu dieser Ausgabe bespricht er auch das Verhältnis der tibetischen zu der mongolischen Version.[4]) Während sie dem Hauptinhalt nach gleich sind, ist die Erzählung im mongolischen Text häufig breiter gehalten und paraphrasirt, öfters auch mit kleinen Zusätzen versehen, die im tibetischen fehlen; doch

---

[1]) Köppen, Die lamaische Hierarchie und Kirche, S. 112. H. Wenzel, Tsonkhapa (Short Notice on his Works), JRAS, 1892, pp. 141—142.

[2]) Huth, Geschichte des Buddhismus in der Mongolei, Band II, S. 375; s. ferner S. 378, 379, 382, 383, 387, 388, 392. Das Werk wurde von Ṅag-dbaṅ blo-gros ins Mongolische übersetzt (ibid., S. 367).

[3]) Der Weise und der Thor. Aus dem Tibetischen übersetzt und mit dem Originaltexte herausgegeben von I. J. Schmidt. 2 Vols., St. Petersburg, 1843.

[4]) Vol. I, pp. XVI, XVII.

finden sich auch in letzterem bisweilen kurze Stellen, die der mongolische nicht hat. Der mongolische Dzanglun besteht aus 52, der tibetische aus nur 51 Kapiteln.[1] SCHMIDT schloss daher in Anbetracht dieser Varianten, dass die mongolische Version nach einer anderen tibetischen Edition als der im Petersburger Kanjur vorhandenen übersetzt sei, eine Vermutung, die ihm noch dadurch bestärkt wurde, dass ein von ihm eingesehenes Exemplar der kalmükischen Übersetzung des Dzanglun mit jener tibetischen Kanjur-Ausgabe in allen Stücken übereinstimmte. In Petersburg sind verschiedene Ausgaben des Üligärün Dalai vorhanden; einzelne Legenden sind edirt worden, vier in KOWALEWSKI's Chrestomathie (Vol. I, pp. 5—41), eine in der von POPOV (pp. 41—53, entsprechend dem 6. Kapitel des Dzanglun), eine in der Grammatik der mongolischen Sprache von I. J. SCHMIDT (St. Petersburg, 1831, S. 129—142). Die Pekinger Holzdruckausgabe (230 fols.) ist im Jahre 1714 publicirt worden.

Eine kalmükische Version des Werkes von 287 fols. ist in Dresden (Eb 404c) vorhanden. Am Schlusse derselben wird in kalmükischer Umschrift sowohl der Sanskrit Titel *Damamūkonāmasūtra*, als auch der tibetische Titel *aDzaṅs blun žesbya-ba mdo* gegeben, und als kalmükischer Titel *Mädä-täi* (ostmong. *mädäghä-täi*) *mädä-ûghäigi ilagukči kämäkü sudur* «Sūtra genannt Besieger der Verständigen und der Unverständigen.» Die Übersetzung wird dem Zaya Paṇḍita zugeschrieben (siebzehntes Jahrhundert).

Eine dem *Üligärün Dalai* ähnliche Sammlung ist das *Üligärün Nom* («Gesetz oder Religionsbuch der Gleichnisse»), in welchem sich teils aus jenem Werke entlehnte Erzählungen in abgekürzter Form finden, teils viele andere hinzugefügt sind, zur Einpflanzung der vom Buddhismus gelehrten Tugenden.

---

[1] Diese besondere Erzählung (das siebente Kapitel) ist bei SCHMIDT (Vol. I, pp. XVIII—XXXI) im mongolischen Text mit Übersetzung abgedruckt. Manche der von SCHMIDT hervorgehobenen Varianten werden darauf zurückgehen, dass der von ihm benutzte tibetische Text eine nicht sehr correcte Abschrift war (s. A. SCHIEFNER, Ergänzungen und Berichtigungen zu Schmidt's Ausgabe des Dsanglun, St. Petersburg, 1852, S. 1—2). SCHIEFNER hat zur Herstellung der richtigen Lesarten von der mongolischen Version reichlich Gebrauch gemacht.

Der Stil dieses Werkes ist etwas schwieriger als der des vorhergenannten, kurz, aber klar und verständlich. Vier Kapitel daraus
sind in KOWALEWSKI's Chrestomathie (Vol. I, pp. 41—93) aufgenommen, das erste bei POZDNĚJEV (pp. 228—265). [1]

Von grossem Interesse ist das Werk *Čindamani Ärikä*
(tibetisch *Nor-bu p'reṅ-ba*, «Kranz der Edelsteine»), die Bear-
beitung einer tibetischen Sammlung religiöser Legenden, die der
berühmte Lama Jū (Jo-bo) Atīça (983—1055) über die früheren
Taten des Avalokiteçvara und Brom Bakši erzählt hat. KOWA-
LEWSKI (Chrestomathie, Vol. I, pp. 93—182) hat den Text zweier
Abschnitte daraus abgedruckt, und bemerkt (p. 394), dass der
Stil leicht und anziehend sei, und dass viele poetische Stellen
in Versen eingestreut seien. Einen langen Abschnitt daraus hat
I. J. SCHMIDT[2]) übersetzt. Ein guter Holzdruck des mongolischen
Werkes (344 fols.) ist unter der Regierung K'ang-hsi's in Peking
herausgegeben worden.

Das *Altan Gäräl*, wie es mit seinem abgekürzten Titel
genannt wird, ist die Übersetzung eines im Kanjur enthaltenen
Mahāyāna Sūtra (*Suvarṇaprabhāsottama-sūtrendrarāja*, Kanjur-
Index, ed. SCHMIDT, Nr. 556, p. 81). Es ist ein bei den Mongolen sehr beliebtes Werk, wie die zahlreichen und gut gedruckten Ausgaben desselben beweisen; das war bereits im
sechzehnten Jahrhundert der Fall, wie wir aus einer interessanten Stelle bei Jigs-med Nam-mk'a ersehen, wo ein Oirat dem
dritten Dalai Lama gegenüber (bSod-nams rgya-mts'o, 1543—
1586) von diesem Buche mit Ehrfurcht spricht.[3] Ebenda wird
berichtet, dass Gušri Khan (geb. 1581) diese nebst vielen anderen Schriften ins Mongolische übersetzen liess, so dass die

---

[1]) In einem Exemplar dieses Werkes im British Museum findet
sich der folgende offenbar aus der Feder eines englischen Missionars
stammende erheiternde Erguss: Much bad stuff in this book, especially
for young people; some of the stories are fascinating and might lead the
young mind astray.

[2]) Geschichte der Ost-Mongolen, S. 424—488. Näheres über diese
zwanzig Jātaka s. Journal of the Buddhist Text Society of India, Vol. I,
1893, p. 32, wo auch das erste derselben übersetzt ist.

[3]) G. HUTH, Geschichte des Buddhismus in der Mongolei, Band II,
S. 250.

mongolische Version Ende des sechzehnten oder Anfang des siebzehnten Jahrhunderts vorhanden gewesen sein dürfte. Sanang Setsen (1662) hat sie gekannt und citirt daraus.[1]

Eines der schönsten mongolischen Jātaka ist die «Geschichte von dem Knaben, der ohne Sattel auf einem schwarzen Ochsen reitet». Hier treten Buddha in Gestalt eines armen Knaben und ein von seiner Gelehrsamkeit aufgeblasener Brahmane auf; sie beginnen eine Disputation und legen einander Fragen vor, wobei der Brahmane arg ins Gedränge gerät und dem Knaben nicht widersprechen kann, dessen scharfsinnige Antworten seinen Stolz dämpfen.[2] Der Lehrer fragte zum Beispiel: «O Knabe, was ist rechts, was links?» Der Knabe antwortete: «Der Westen liegt rechts, der Osten links, der Norden hinten, der Süden vorn.» Der Lehrer fragte weiter: «Stehen dir Vater und Mutter nahe?» «Mein Vater steht mir nahe, meine Mutter nicht.» «Wenn deine Mutter, ehe du auf die Welt gekommen wärest, mit dir gestorben wäre, so wäret ihr in ein Grab gelegt worden; wie sollte sie dir also nicht nahe stehen?» «Vater und Mutter sind den Wurzeln des Baumes zu vergleichen. Gäbe es keine Wurzeln, wie sollte der Baum entstehen? Sollten wir Menschen keine Eltern haben, wie sollten wir unseren Ursprung nehmen? Wie es leicht ist, den abgesprungenen Reif eines Wagenrades zu ersetzen, so schwer ist es, Vater und Mutter zu finden, wenn sie einmal tot sind. Da sie mich ins Dasein gerufen, wird es schwer sein, alle die Mühen zu vergessen, die sie um meine Erziehung gehabt haben.»

Von manchen ihrer buddhistischen Lieblingsgeschichten lassen die Kalmüken besondere Abschriften cirkuliren. So gibt es eine in vielen Exemplaren verbreitete Ausgabe der Legende von Manoharā unter dem Titel *Kündü bilik arilgakči Manuhari*

---

[1] I. J. SCHMIDT, Geschichte der Ost-Mongolen, S. 11, 307. Das 26. Kapitel ist in Text und Übersetzung in SCHMIDT's Grammatik der mongolischen Sprache, S. 143—176; der Text des 20. Kapitels in POPOV's Chrestomathie, pp. 138—144.

[2] Der Text in POPOV's Chrestomathie, pp. 19—39. Eine deutsche Übersetzung dieser Erzählung von ANTON SCHIEFNER in der St. Petersburger Zeitung, 1849, Nr. 79; ein Sonderabdruck dieses Aufsatzes befindet sich in der Bibliothek der Deutschen Morgenländischen Gesellschaft in Halle.

okin tänggäri-yin tūji «Die von schweren Sünden reinigende Geschichte von der Devī Manuhari».[1]) Dies ist eine freie Bearbeitung des von A. SCHIEFNER[2]) nach dem tibetischen Kanjur übersetzten Sudhana Avadāna (auch in der Sammlung Divyāvadāna, Nr. 30, und Avadānakalpalatā, Nr. 64). Ich habe bereits bei einer früheren Gelegenheit[3]) gezeigt, dass in dieser Erzählung die Nachbildung einer Scene aus dem vierten Akt von Kālidāsa's Drama Vikramorvaçī eingeflochten ist, die in der kalmükischen Version mit grosser dichterischer Freiheit und Schönheit behandelt ist. Eine kalmükische Bearbeitung des Viçvantara Jātaka ist gleichfalls in Göttingen[4]) vorhanden.

Zu den unterhaltendsten Werken der ostmongolischen und kalmükischen Literatur gehört die Märchensammlung *Siddhi-Kūr*, eine Bearbeitung der indischen *Vetālapañcaviṃçati*[5]) (*kūr* =

---

[1]) Handschrift in Dresden Eb 404h (21 fols.) und in Göttingen, Asch 110 (23 fols.). Eine Collationirung beider MS. zeigt nur geringfügige Abweichungen. Beide tragen am Schluss den Vermerk: «Auf Befehl des Dsasak-tu Hung-taiji vom Paṇḍita Guši ins Mongolische übersetzt und von einem Schüler namens Buddhakala niedergeschrieben.» Die Übersetzung ist also gegen Ende des siebzehnten Jahrhunderts verfertigt worden.

[2]) Tibetan Tales derived from Indian Sources. London, 1906 (nur unveränderter Neudruck der früheren Auflage), pp. 44—74.

[3]) Der Urquell, Neue Folge. Band II, 1898, S. 156—157.

[4]) Asch 113 (15 fols.).

[5]) BENJAMIN BERGMANN, Nomadische Streifereien unter den Kalmüken, Band I. Riga, 1804, S. 247—351. — BERNHARD JÜLG, Die Märchen des Siddhi-Kūr, kalmükisch. X. Erzählung (als Probe einer Gesamtausgabe). Festgruss aus Österreich an die XX. Versammlung deutscher Philologen. Wien, 1861. 6 Seiten. 4°, 2 Seiten Text. — B. JÜLG, Die Märchen des Siddhi-Kūr. Kalmükischer Text mit deutscher Übersetzung und einem kalmükisch-deutschen Wörterbuch. Leipzig, 1866. Diese Arbeit ist vielfach überschätzt worden; der Text ist nach einer schlechten Handschrift gedruckt und ganz kritiklos behandelt; die Kritik, die sich JÜLG daher von GOLSTUNSKI zugezogen hat, war völlig berechtigt. — B. JÜLG, Kalmükische Märchen. Die Märchen des Siddhi-Kür oder Erzählungen eines verzauberten Toten. Ein Beitrag zur Sagenkunde auf buddhistischem Gebiete. Aus dem Kalmükischen übersetzt Leipzig, 1866 (dasselbe wie oben ohne Text und Vokabular). — B. JÜLG, Mongolische Märchen. Erzählung aus der Sammlung Ardschi Bordschi. Ein Seitenstück zum Gottesgericht in Tristan und Isolde. Mongolisch und deutsch nebst dem Bruchstück aus

ostmong. *kägür* «Leichnam»; *siddhi-kūr* = *vetālasiddhi;*. Die mongolische Redaktion enthält jedoch nur 23 Erzählungen. Die Identität beider Werke ist zuerst von THEODOR BENFEY[1] erkannt worden.

Die von Jülg erwähnte mongolische Bearbeitung der *Simhāsanadvātrimçati* («Die 32 Erzählungen des Löwentrones») ist von H. C. V. D. GABELENTZ ins Deutsche übersetzt worden. Die Handschrift seiner Übersetzung wird auf der Königl. Bibliothek zu Berlin aufbewahrt und ist leider nie publicirt worden. Ich habe daraus fünf indische Fabeln veröffentlicht.[2] Die Rahmenerzählung dieser Sammlung ist kurz folgende: König Kasna lässt in einem Hügel, der die Wirkung hervorbringt, dass eine Schar auf demselben spielender Knaben mit wunderbarer Weisheit erfüllt wird, einen prächtigen Tron ausgraben, den zweiunddreissig mit menschlicher Rede begabte Holzfiguren zieren;

Tristan und Isolde. Innsbruck, 1867. 37 p. Der Vermerk auf der Rückseite des Umschlags «Erster mongolischer Druck im ausserrussischen Europa» ist eine völlig grundlose Renommage: ABEL-RÉMUSAT in Paris und die Britische Bibel-Gesellschaft in London haben ein Menschenalter früher mongolische Texte mit guten Originaltypen gedruckt. — B. JÜLG, Mongolische Märchensammlung. Die neun Märchen des Siddhi-Kür nach der ausführlichen Redaction und die Geschichte des Ardschi-Bordschi Chan. Innsbruck, 1868 (Text und Übersetzung, dasselbe auch nur in Übersetzung, ibid., 1868). — Sagas from the Far East; or, Kalmouk and Mongolian Traditionary Tales. With historical preface and explanatory notes. By the author of «Patrañas» (anonym). London. 1873. Dies Buch enthält die 23 Erzählungen des Siddhi-Kür und die Geschichte von Ardši-Bordši nach JÜLG, nebst einem Index. — Eine dem Text von JÜLG weit vorzuziehende Recension in echter kalmükischer Volkssprache hat C. GOLSTUNSKI, St. Petersburg, 1864, lithographirt herausgegeben. Der verdiente Lama GALSANG GOMBOJEV (Шидди-Куръ, собраніе монгольскихъ сказокъ, Этнографическій сборникъ Имп. Русск. Геогр. Общ., Vol. VI, St. Pet., 1864, pp. 1—102) hat eine vollständige Übersetzung der mongolischen Version verfertigt, die A. SCHIEFNER nach seinem Tode mit einer Einleitung veröffentlicht hat. Nach diesen Vorarbeiten und der deutschen Übersetzung von BERGMANN kann man der Leistung von JÜLG keine grosse Originalität nachrühmen, noch auch darin einen wesentlichen Fortschritt gegenüber seinen Vorgängern erkennen.

[1] Mélanges asiatiques de l'Acad. de St.-Pétersbourg, Vol. III, 1859, pp. 170—203; vergl. SCHIEFNER, ibid., pp. 204—218.

[2] ZDMG, Vol. LII, 1898, pp. 283—288.

diese halten nach einander den Herrscher, der sich auf den Tron setzen will und den Anspruch darauf mit der Erzählung seiner ruhmvollen Taten geltend macht, von seinem Vorhaben zurück und rechtfertigen jedesmal ihre Abweisung mit einer Erzählung aus dem Leben ihres nach ihrer Ansicht würdigeren Königs. Es entstehen auf diese Weise, da den 32 Geschichten der Holzmenschen immer je eine des Kasna gegenübersteht, im ganzen vierundsechzig Erzählungen, von denen die zweiunddreissig auf jeder Seite in fortlaufendem Zusammenhang stehen. die eine Serie dem König, die andere den Holzfiguren in den Mund gelegt, unter denen man übrigens nichts anderes als ein agirendes Puppentheater zu verstehen hat. Die Erzählungen der Holzfiguren scheinen fast alle indischen Stoffen ihren Ursprung zu verdanken, sind aber keineswegs wörtlich übersetzt, sondern stark im mongolischen Sinne überarbeitet. Die Geschichte des Königs Kasna dagegen scheint mir grossenteils mongolische Erfindung zu sein und einen mit den Sagen von Geser Khan verwandten epischen Charakter zu tragen. Kasna ist wie Geser der von den Göttern eingesetzte Vertreter des Buddhismus auf Erden und hat die Bestimmung, als solcher die am heidnischen Schamanentum festhaltenden Völker Centralasiens zu bekämpfen, welche die Volkssage treffend in Drachen- und Riesenkönigen verkörpert; es ist die Schilderung eines Conflikts, in dem eine alte Welt zusammenbricht und eine neue Zeit aufleuchtet.[1]

Einen breiten Raum nehmen in der mongolischen, wie in der tibetischen Literatur, die Biographien der Heiligen und Lamen ein, von denen viele einen riesigen Umfang besitzen; besonders erfreuen sich die Lebensbeschreibungen des bTson-k'a-pa,[2] des Reformators des Lamaismus, und der Dalai Lamas grosser Beliebtheit und sind oft in eleganten Drucken aufgelegt worden.

---

[1] Eine ausführlichere Analyse dieser Geschichte teilt A. Rudnev, der den König Gesen nennt, in seinem interessanten Aufsatze mit Замѣтки по Монгольской Литературѣ. I. Три сборника разсказовъ о Бікарміцітѣ. Zap.. Vol. XV. 1904. pp. 026—032.

[2] Mélanges asiatiques, Vol. II. St. Pet., 1856, p. 362. Von den 21 Bänden der gesammelten Schriften (gsuṅ-qbum) des bTson-k'a-pa sind 18 Bände der mongolischen Übersetzung in Kasan vorhanden (ibid.. p. 353). Tibetisch-mongolische Biographie in Kasan, Nr. 163.

So umfasst z. B. die in Peking gedruckte mongolische Biographie des siebten Dalai Lama bLo-bzaṅ bskal-bzaṅ rgya-mts'o (1705— 1758) 346 grosse Folioblätter.

Eine der berühmtesten lamaischen Kultstätten ist der *Wu-tai shan* (tibetisch *Ri-bo rtse-lṅa*, mongolisch *Tabun üdsügürtü aġūla* «fünfgipfliger Berg») in der Provinz Shansi, China, der Verehrung des Mañjuçrī geweiht. Der älteste der zahlreichen Tempel *(Ts'ing liang se* «Tempel der reinen Kühle», tibetisch *Draṅs-bsil)* war bereits Ende des fünften Jahrhunderts gegründet und wurde von den Mongolen 1265 erneuert. Der Berg wird besonders von mongolischen Pilgern besucht, und es existiren zahlreiche Beschreibungen von seinen Heiligtümern in tibetischer und mongolischer Sprache. Eine mongolische Schilderung wurde 1668 in Peking gedruckt, und eine noch weit ausführlichere in zehn Büchern im Jahre 1702 auf Chinesisch, Tibetisch, Mongolisch und Manju.[1]) Ausser einer Reihe tibetischer Schriften über denselben Gegenstand besitze ich eine mongolische aus dem Jahre 1667.[2]) In 49 poetischen Strophen, tibetisch und mongolisch, hat der lCaṅ-skya Latitavajra die Schönheiten des *Wu-tai shan* besungen.[3]) POZDNÉJEV[4]) hat ein anderes elf-strophiges mongolisches Gedicht auf den Berg (herausgegeben Peking, 1794) bekannt gemacht.

Merkwürdig ist die Tatsache, dass auch das von A. SCHIEF-

---

[1]) KOWALEWSKI, Mongolische Chrestomathie. Vol. II, p. 303. Mélanges asiatiques, Vol. I, St. Pet., 1852, pp. 414–415. MÖLLENDORFF (Essay on Manchu Literature, p. 33, Nr. 163) erwähnt eine Manju Beschreibung der Chin-lien Berge mit Zusatz eines Fragezeichens. Offenbar ist der *Wu-tai shan* gemeint. Die chinesische Bearbeitung führt den Titel *Ts'ing liang hsin chi* (4 Vols.) und wurde 1785 auf Befehl K'ien-lung's stark vermehrt und mit Karten und Plänen versehen in acht Bänden wieder herausgegeben.

[2]) Betitelt *Uda* (für *Udaishan)-yin tabun aġūlan-u orosil süsük-tän-ü čikin čimäk* «Führer zu den fünf Gipfeln des Wu-tai, Ohrschmuck der Gläubigen» (mit dem chinesischen Stichwort *Wên-shu chi* «Geschichte des Mañjuçrī», verfasst von Gūšri bLo-bzaṅ bstan-ạdzin. Auch in Kasan, Nr. 164 (vergl. ferner 165, 166).

[3]) Bilinguer Rotdruck in Falteform, gedruckt in Peking, 1767.

[4]) Образцы народной литературы монгольскихъ племенъ, St. Pet., 1880, pp. 41, 309—312.

NER unter dem Titel «Über das Bonpo-Sūtra: Das weisse Nāga-Hunderttausend»[1]) übersetzte tibetische Werk der Bon Religion auch ins Mongolische übersetzt worden ist,[2]) was als eine Bestätigung des schon früher bekannten Umstandes gelten mag, dass sich diese Schrift ebenso gut in den Händen der buddhistischen Gläubigen befindet.

Das buddhistische Schrifttum macht überhaupt den weitaus grössten Bestandteil der mongolischen Literatur aus und ist durch eine Masse vorzüglich gedruckter Peking-Ausgaben am leichtesten zugänglich; von fast allen bedeutenderen Sūtra, Tantra und Dhāraṇī sind gute Einzelausgaben vorhanden; auch an Buddha-Biographien ist kein Mangel.[3]) Die Literatur des nördlichen Buddhismus kann aus dem Mongolischen mit ebenso grossem Nutzen als aus dem Chinesischen studirt werden. Aber in literarhistorischer Hinsicht bleibt hier noch alles zu tun. Von den Namen der zahllosen Übersetzer wissen wir bisher so gut wie nichts; die Daten der Übersetzungen, die Lebenszeit der Übersetzer und Bearbeiter (wie der Name ihrer Klöster und Sekten), die Jahreszahlen der Drucke müssen bestimmt, die verschiedenen Ausgaben und Recensionen collationirt, das bisher

---

[1]) Mémoires de l'Académie de St.-Pét., VII-e série, Vol. XXVIII, Nr. 1, St. Pet., 1880.

[2]) Gedruckt in Peking, 1766, unter dem Titel *Ünän ügätü ärdäni khubilgan Bon-bo-yin arigun abum tsagan lus-un yäkä külgän sudur* (was genau dem tibetischen Titel *gTsaṅ-ma klu abum dkar-po Bon rin-po-c'e ap'rul-dag bden-pa t'eg-pa c'en-poi mdo* entspricht). In einem Exemplar der mongolischen Ausgabe im British Museum ist folgende wunderbare Übersetzung des Titels von einem englischen Missionar gegeben: The true worded book giving description of what men ought to do in to among (sic!) the gods and be in favor with men. Ferner wird von demselben über das Werk bemerkt: This is called the true word book giving directions how to attain to the station of the gods, and favour with men; and has some very moral precepts, but many heathen ideas concerning gods, but useful in contending with the heathen.

[3]) Eine sehr umfangreiche in 24 Büchern, Übersetzung des chinesischen *Shih-kia-mu-ni Fu yüan-liu king*, ist in Peking 1871 in einem prachtvollen Druck wieder aufgelegt worden. — Zu erwähnen ist noch die Schrift von А. О. Ивановскій, Буддійская покаянная молитва (Mongolischer und Tibetischer Text). Zap., Vol. XII, 1899, pp. 07—013.

16*

in den Bibliotheken aufgespeicherte Material systematisch kata-
logisirt werden. Damit würden wir nicht nur ein wertvolles Kapitel
zur Geschichte des Buddhismus, sondern auch einen interessanten Beitrag zur Geschichte des Buchdrucks und Buchwesens
erhalten.

Die vielfach geäusserte Theorie, dass es der Buddhismus
war, welcher die Mongolen aus wilden Eroberern und Welten-
stürmern zu zahmen, friedlich die Steppe begrasenden Lämmern
gemacht habe, gehört zu den vielen legendären Ansichten, die
sich einmal in der Wissenschaft bilden und dann unausrottbar
wie ein Dogma fortschleppen. Aber wer hat sie je ernstlich erwiesen? Tibeter und Chinesen haben, wann immer es ihr poli-
tisches Interesse erheischte, trotz allen Buddhismus ihre Kriege
ruhig weitergeführt, und was soll man erst von den Japanern
sagen? Warum hat der Buddhismus, der sie seelisch wahrlich
doch viel tiefer afficirt hat als die Mongolen, ihre kriegerischen
Neigungen nicht gebrochen? Die Vorliebe unserer Zeit, unzäh-
lige Erscheinungen auf religiöse Ursachen zurückzuführen, und
die Neigung, alles aus gerade *einer* Ursache abzuleiten, tragen
die Schuld an der Verkennung des wahren Sachverhalts. Gewiss hat der Buddhismus wie überall verdienstlich und heilsam
auch auf die Mongolen eingewirkt und ihre Sitten gemildert;
er ist gewiss einer der ihr Leben umgestaltenden Faktoren gewesen, aber er kann nicht ausschliesslich für die Erscheinung
des radikalen Umwandlungsprocesses verantwortlich gemacht
werden. Hier müssen vor allem historisch-politische und wirtschaftliche Ursachen mit im Spiele gewesen sein. Der grosse
materielle Bankrott der Yüan-Dynastie war mehr als alle Philosophie und Religion dazu angetan, die Mongolen von ihrer
Höhe herabzustürzen und recht klein zu machen, und die besonders auf Centralasien gerichtete Expansionspolitik der Ming-
und Manju-Dynastie trug weiter wesentlich mit zu dieser Wirkung bei. Dazu kam und kommt die Aussaugung der Steppensöhne durch gewissenlose chinesische Händler, die sie mit einem
einseitig auf ihren Vorteil berechneten Creditsystem, wie Yama
mit der Todesschlinge, gefangen halten; dazu kommen die un-
vermeidlich wechselnden Geschicke des auf der Viehhaltung gegründeten Wirtschaftssystems. Nicht gering anzuschlagen ist auch

die mongolische Unfähigkeit zur Organisation, zur Bildung eines Staatswesens, die nicht zum mindesten in der eigentümlichen Richtung des Wirtschaftslebens eine ihrer Hauptursachen hat; unter der straffen Leitung einiger überlegenen Persönlichkeiten vermochten sie zeitweise eine politische Einheit zu bilden und fremde Völker zu beherrschen, um wieder auseinander zu fallen, sobald der führenden Hand die Zügel entfielen.

Aber für den Kulturhistoriker bleiben die Mongolen ein instruktiver Fall. Von der Höhe ihrer alten politischen Macht herabgesunken wie die modernen Griechen und Italiener, aller Symptome des ehemaligen Glanzes bar, machen sie gegenwärtig in ihrer halb sesshaft-feldwirtschaftlichen, halb nomadisirenden Lebensweise den Eindruck des irrtümlich und missbräuchlich sogenannten «Naturvolks». Aber inmitten dieser harmlosen Natur- kinder erheben sich lebhafte Handelsstädte, industrielle Centren, vielbereiste Verkehrsstrassen, und vor allem prächtige Tempel und Klöster, die Sitze scholastisch-religiöser Gelehrsamkeit und eifrigen Bücherstudiums, mit ihren reichen Bibliotheken und ständig arbeitenden Druckereien, mit ihren Sammlungen an Skulpturen und Malereien, mit ihren verschiedenen Fakultäten, Lehrstühlen und Vorträgen. Das heilige Wasser des Gangges hat die öde Steppe befruchtet, und wer darf sich unterfangen zu sagen, dass der Buddhismus keine Kulturmacht sei, und wer weiss, welch erneute Lebenskraft er noch in Zukunft für die Völker Asiens gewinnen wird?

Chinesische Literatur. — Auch confucianische Ideen haben durch Übersetzungen chinesischer Schriften Eingang in die Mongolei gefunden. Sie werden besonders zur Erziehung der Beamtenschaft als Lehrbücher in den Schulen gebraucht; der Stil dieser Bücher macht naturgemäss starke Concessionen an die chinesische Construktion. Obenan unter diesen Erziehungs- mitteln steht das *San tzŭ king* «Der Canon der drei Schrift- zeichen» (da jede Zeile des chinesischen Originals aus drei Charakteren besteht), das bekannte von der ganzen chinesischen Jugend wörtlich auswendig gelernte Schulbuch, das kurz und bündig die wichtigsten confucianischen Lehren und allerlei Wis- senswertes aus Geographie und Geschichte mitteilt. Die mon- golische Version findet man in POZDNĚJEV's Chrestomathie, pp.

266—286.[1]) Auch das sogenannte «Heilige Edikt» der Kaiser K´ang-hsi und Yung-chêng ist ins Mongolische übersetzt worden.[2])

Bereits ABEL-RÉMUSAT[3]) hatte eine mongolische Übersetzung dieses Werkes besprochen, deren Handschrift er von Baron Schilling v. Canstadt erhalten hatte. Dieselbe war einer am Schlusse in manjurischer Sprache gemachten Angabe zufolge im Jahre 1806 von einem Russen Vakhili Nobokhiyelub, d. i. Vasili Novozilov, einem Dolmetscher in Peking, verfertigt worden, unter Anleitung eines mongolischen Lehrers. RÉMUSAT hat die drei ersten Abschnitte nach dem Mongolischen übertragen und den mongolischen und manjurischen Text derselben in Parallelreihen (das Manju durch Rotschrift unterschieden) abgedruckt, eine für die Vergleichung beider Sprachen sehr nützliche Arbeit. — KOWALEWSKI führt im Vorwort zu seinem Wörterbuch eine handschriftliche Übersetzung des *Ta hsio* an.

Der chinesische historische Roman von der Geschichte der drei Reiche *(San kuo chi)* hat eine Übersetzung ins Mongolische erfahren; der Text eines Abschnitts ist daraus in POZDNÉJEV's Chrestomathie (pp. 321—343) mitgeteilt. Prof. W. GRUBE[4]) erwähnt eine sehr seltene handschriftliche mongolische Übersetzung des chinesischen Romans *Fan T´ang yen-chuan* «Erzählung von der Empörung gegen das Haus T´ang», von der er ein Exemplar in Peking erworben hat. Über weitere chine-

---

[1]) Vergl. auch v. MÖLLENDORFF, Essay on Manchu Literature, p. 20, Nr. 80.

[2]) Zwei Kapitel daraus bei POZDNÉJEV, pp. 287—292; zwei andere in POPOV's Chrestomathie, pp. 124—137; vier in der von KOWALEWSKI, Vol. I, pp. 229—243. Nach POPOV wäre die Übersetzung aus dem Manju verfertigt und soll sich durch Reinheit des Stils auszeichnen. Nach KOWALEWSKI (l. c., p. 565) ist die mongolische Übersetzung erst im Jahre 1830 im Auftrag eines Ministers Fu-Tsung herausgegeben worden. Gute Ausgabe in den drei Sprachen, 4 Vols., Peking, 1873 (Exemplar in Chinese Department, Columbia University, New-York).

[3]) Notice sur un manuscrit mongol qui contient une traduction de la Sainte Instruction de Khang Hi, avec les amplifications de Young Tching. Notices et Extraits des Manuscrits de la Bibliothèque du Roi, Tome XIII, première partie, Paris, 1838, pp. 62—125.

[4]) Zur Pekinger Volkskunde, Veröffentlichungen aus dem Königl. Museum für Völkerkunde, Band VII, Berlin, 1901, S. 133, Note 2.

sische Werke s. den folgenden Paragraphen und «Juristische Literatur».[1])

Didaktische Literatur. — Das Werk *Oyun Tulkigur* («Schlüssel des Verstandes») enthält Belehrungen und Lebensregeln für die Beamten des Volkes in der Art einer praktischen Philosophie, z. B.: «Wenn du das Volk auf deine Seite bringen willst, streue Wohltaten aus; wenn du stark und mächtig zu sein gedenkst, begünstige dein Heer; wenn deine Angelegenheiten ohne Säumnis ausgeführt werden sollen, ermuntere deine Beamten; wenn du Ruhe geniessen willst, bemühe dich um den Frieden.»[2]) Die mongolische Literatur ist überhaupt reich an

---

[1]) Treffend bemerkt CARL RITTER (Die Erdkunde von Asien, Band II, Berlin, 1833, S. 390—391): «Durch die fortgesetzte Übertragung chinesischer Literatur ins Mongolische und Manjurische wurde ein gemeinsames Band dieser drei Literaturen gefunden, das nicht ohne einigen Einfluss auf Assimilation der Ideen und Kulturen dieser drei so wesentlich verschiedenen Völker, zumal in den Kulminationspunkten, oder den höheren Ständen bleiben konnte, und wenn auch nicht eine Verschmelzung dieser Völker, ihrer Verfassungen, Sitten, Gebräuche herbeiführte, doch eine Annäherung und wenigstens ein gegenseitiges Verständnis so grosser Differenzen anbahnte, wie sich dies in dem Religionscultus, in dem Militärwesen, in der Beamtenwelt, bei dem Hofceremoniell u. s. w. überall seit der Manjuherrschaft gezeigt hat. Es hat hierdurch ein merkwürdiger gegenseitiger Austausch körperlicher und geistiger Kräfte dieser Nationen Ostasiens stattfinden können, dem das kolossale Herrscherreich, bei so verschiedenen Elementen, offenbar einen Teil seiner Dauer in der gegenwärtigen Gestalt zu verdanken hat. Bei geringer nationaleinheimischer literarischer Entwicklung ist hierdurch doch dem Mongolenstamme eine Masse von aussen her eingeimpfter Wissenschaften durch die Übersetzungsliteratur zuteil geworden, die sich, gleich einer Tradition, mit seinem einheimischen Wissen, nach und nach amalgamiren mussten, wovon auch die mongolischen Annalen nicht unwichtige Belege geben. Es ist dies der Sieg jeder geistigen, entwickelteren Nation, dass ihre Literatur von ihrer gewonnenen Höhe hinabgleitet zu dem tieferen Niveau der roheren Nachbarn, gleich quellenden Gewässern in unzähligen Bächen und Strömen nach allen Richtungen hin.»

[2]) Ein Abschnitt aus diesem Buche ist in POPOV's Mongolischer Chrestomathie, pp. 1—17, abgedruckt; einzelne Sprüche in POZDNÉJEV's Chrestomathie, pp. 1—3, der bemerkt, dass die Mongolen diesem Werke ein hohes Altertum vindiciren, was dadurch gerechtfertigt sei, dass kein lamaistischer Einfluss darin sichtbar ist, vielmehr der kriegerische Geist der Zeit des Činggis sich darin widerspiegelt; doch zeigt sich der Autor

didaktischen Spruchsammlungen, meistenteils in Anlehnung an
tibetische, aber auch an solchen von freier Erfindung. Nachahmung eines tibetischen Vorbildes[1]) ist die mongolische Schrift
Tōba-yin gägän-ü togōji (oder *tūji*). Çākyamuni ist auf einer
Wanderung begriffen, um Almosen zu sammeln, und begegnet
einem Verkäufer von Papageien. Dieser setzt seine Vögel in
Freiheit, nachdem er eine Predigt Buddhas über das Mitleid
angehört hat. Einer der losgelassenen Papageien lässt sich auf
den Ast eines Baumes nieder und spricht nun eine Reihe von
lehrhaften Sentenzen aus über gute und schlechte Lamen, Be-
amte, gewöhnliche Männer und Frauen, Jünglinge und Mädchen,
und über allgemeine menschliche Beziehungen.[2])

KOWALEWSKI[3]) hat eine Auswahl von Sprüchen aus verschiedenen Werken, hauptsächlich unter Benutzung eines von
ihm *Rinčen Bumbi* genannten, mitgeteilt. LOUIS ROCHET[4]) hat
170 Sprüche (darunter aber auch einfache Sätze), wie es scheint,
aus verschiedenen Quellen, die er aber nicht angibt, gesammelt; eine grosse Anzahl daraus ist der classischen chinesischen Literatur entlehnt. Die bekannte Spruchsammlung des
Sa-skya Paṇḍita[5]) ist auch ins Mongolische übersetzt worden

mit chinesischer Philosophie vertraut, wie z. B. die oben mitgeteilten
Sätze dartun, so dass sein Buch wohl viel später als die Epoche des
Činggis abgefasst sein muss. Seine Sprache weist einen musterhaften,
reinen Stil auf.

[1]) POZDNÉJEV (Chrestomathie, p. VI) nennt es nach seinem mongo-
lischen Titel *Usun Däbiskärtü Khān-u tūji*, «Geschichte vom Könige Usun
Däbiskärtü», der nach ihm von 866 bis 901 in Tibet regirt haben soll.
Dies ist ein Irrtum, denn der mongolische Name ist wörtliche Übersetzung
des tibetischen Ral-pa-can («der mit wallendem Haar»), Beiname des
Königs K'ri-lde sroṅ-btsan, der von 816 bis 838 regirte. Eine kalmükische
Handschrift jenes Werkes ist in Göttingen (Asch 112, Nr. 2).

[2]) Der Text dieses Werkes ist in POZDNÉJEV's Chrestomathie (pp. 3—
15) abgedruckt.

[3]) Mongolische Chrestomathie, Vol. I, pp. 1—4.

[4]) Sentences, maximes et proverbes mantchoux et mongols, accom-
pagnés d'une traduction française, des alphabets et d'un vocabulaire de
tous les mots contenus dans le texte de ces deux langues. Paris, 1875,
IV und 166 p.

[5]) PH. E. FOUCAUX, Le trésor des belles paroles, Paris, 1858 (Vergl.
G. HUTH, Geschichte des Buddhismus in der Mongolei, Band II, p. XXVIII).

(Sain ügä-tü ärdäni-yin sang); Sanang Setsen citirt sie S. 131 der Übersetzung von SCHMIDT. H. C. V. D. GABELENTZ[1]) hat zweiundzwanzig Sprüchwörter aus diesem Schriftsteller und der Geser-Sage gesammelt, welche die eigenen Anschauungen der Mongolen charakterisiren. Eine Anzahl echt mongolischer Sprüchwörter hat GALSANG GOMBOJEV seinem Aufsatz «Randbemerkungen zu Plano Carpini»[2]) einverleibt.

Mehrere chinesische Spruchsammlungen haben ihren Weg ins Mongolische gefunden und sind in trilinguen Ausgaben vorhanden, so das *San ho ming hsien chi* (2 Vols., 1879)[3]) «Aussprüche von berühmten Weisen in drei Sprachen», ferner das *Lien chu chi* (1728) «Sammlung von Perlenschnüren».[4])

Volksliteratur. — Die Erzeugnisse der volkstümlichen Literatur der mongolischen Stämme setzen sich aus Heldensagen, Liedern, Märchen, Erzählungen und Rätseln zusammen. Russische Forscher haben bereits ein grosses Material aus dem Volksmunde gesammelt, dessen Umfang sich schwer überblicken lässt.

Heldensagen liegen uns in drei verschiedenen Quellen vor: 1. in Fragmenten epischen Charakters, die in den alten Chroniken enthalten sind, 2. in literarischen Buchredaktionen von Heldenmärchen, 3. in den Vorträgen epischer Lieder moderner Erzähler, die zum Teil auf mündlicher Überlieferung beruhen.

HANS CONON V. D. GABELENTZ war der erste, der die dichterischen Bestandteile in der Chronik des Sanang Setsen klar erkannt und in seiner inhaltreichen Abhandlung «Einiges über mongolische Poesie»[5]) dargelegt hat. «Man findet nämlich bei

---

[1]) Zeitschrift für die Kunde des Morgenlandes, Band I, Göttingen, 1837, S. 34—37.

[2]) Mélanges asiatiques, Vol. II, St. Pet., 1856, pp. 650—666.

[3]) G. P. v. MÖLLENDORFF, Essay on Manchu Literature, p. 24, Nr. 109 (Exemplar in Columbia University, New-York). Der Manju Text ist in J. KLAPROTH's Chrestomathie mandchoue, Paris, 1828, pp. 5—23, abgedruckt; Übersetzung, ibid., pp. 195—210.

[4]) G. P. v. MÖLLENDORFF, l. c., p. 22, Nr. 94. Einige Bemerkungen über dieses Werk und Übersetzung des Anfanges desselben bei G. V. D. GABELENTZ, ZDMG, Band XVI, 1862, S. 540—541.

[5]) Zeitschrift für die Kunde des Morgenlandes, Band I, Göttingen, 1837, S. 20—37.

genauerer Betrachtung», sagt er, «dass dieser Schriftsteller an
vielen Stellen seines Geschichtswerkes grössere oder kleinere
Bruchstücke von Gedichten eingeflochten hat, welche offenbar
einem epischen Cyklus der Mongolen angehören und vielleicht
noch jetzt im Munde des Volkes fortleben, wie dies wahrscheinlich auch zu Sanang Setsen's Zeit (1662) der Fall war. Ob sie
einem der sieben Sudur's entlehnt sind, welche er selbst als
Quellen seines Geschichtswerkes angibt, ist wenigstens aus deren
Titel nicht zu bestimmen.» Diese von v. D. GABELENTZ nachge-
wiesenen 25 Gedichte beziehen sich, mit Ausnahme des ersten,[1])
sämtlich auf die Geschichte des Činggis Khan, also auf die
Glanzperiode, und auf die Zeit der Anarchie, also auf die Un-
glücksperiode der mongolischen Geschichte; sie sind Lieder
lyrischen Inhalts, in Form und Gedanken den modernen Volksliedern nicht unähnlich (Klagelieder, Trostlieder, Gebete u. a.).
In der Beurteilung dieser Frage, scheint es mir nun, müssen
wir noch über die von v. D. GABELENTZ formulirte Ansicht hinaus-
gehen: nicht nur diese versificirten Bestandteile selbst scheinen
einem alten epischen Cyklus entlehnt zu sein, sondern auch ein
grosser Teil der Prosapartien, in welchen sie eingestreut sind.
In der Tat macht z. B. das ganze vierte Buch des Sanang Setsen
fast durchweg den Eindruck eines alten epischen Berichts,[2])
den der Chronist natürlich überarbeitet und abgeschwächt haben
mag. Eine Stütze erfährt diese Anschauung aus der Tatsache,
dass die unter dem wirklichen Namen epischer Erzählungen
gehenden Produkte der Mongolen in Prosa abgefasst sind, die
von eingeflochtenen lyrischen Liedern abgewechselt wird; so
z. B. in der Gesersage.[3]) Es kann jetzt gar keinem Zweifel mehr

---

[1]) Das dem Lama in den Mund gelegt ist, der den tibetischen
König gLaṅ-dar-ma tötet, und das jedenfalls einem tibetischen Vorbilde
nachgebildet ist.

[2]) Vergl. die vielen Heldenabenteuer, Reden, Zwiegespräche und
den poetisch gehobenen Stil; selbst alle Rosse erhalten Eigennamen wie
in arischen Epen.

[3]) Ein Lied aus dieser s. bei W. SCHOTT, Über die Sage von Geser-
chan (Abhandlungen der Berliner Akademie, 1851, S. 275); Spottlieder
(Urquell, Band II, 1898, S. 147). In der tibetischen Fassung der Geser-
Sage wechseln beständig Lieder in Versen mit erzählender Prosa ab

unterliegen, dass die Mongolen, sicher bereits in den Tagen des Činggis, uralte Heldenlieder besessen, und dass sich um die grosse Persönlichkeit des Welteroberers bald nach seinem Tode (1227) verschiedene epische Sagenkreise gebildet haben, die zum Teil nicht nur schriftlich fixirt und von den späteren Chronisten wie Sanang Setsen benutzt worden sind,[1]) sondern sich auch in Bruchstücken, und teilweise ganz unabhängig von literarischer Tradition, bis auf unsere Tage mündlich unter dem Volke fortgepflanzt haben. Dieses Faktum geht unwiderleglich aus der Untersuchung hervor, die G. N. POTANIN in seinem vortrefflichen Werke «Die östlichen Motive im mittelalterlichen europäischen Epos» über den Sagenkreis von Činggis Khan veranstaltet hat.[2]) Danach wird es auch völlig klar, warum die Nachrichten unseres Sanang Setsen, der eben zwischen rein historischen und epischen Berichten nicht unterschieden hat, so weit von den nüchternen Darstellungen der übrigen Geschichtsschreiber derselben Ereignisse abweichen, und danach mag sich füglich das Gezeter der Herren Historiker von Fach über ihn endlich zur Ruhe begeben; ich für mein Teil weiss nicht, ob wir dem biederen Mongolen nicht mehr für die Erhaltung seiner

---

Ebenso tragen die alten tibetischen Annalen denselben epischen Charakter wie die mongolischen. Von meinem Gebiete aus gelange ich betreffs Entwicklung des Epos mehr und mehr zu der Ansicht, die E. WINDISCH (Māra und Buddha, Leipzig, 1895, S. 222 ff.) für das indische dargelegt hat.

[1]) Auch im *Yüan ch'ao pi shih* (1242), das W. BARTHOLD (Mitteilungen des Seminars für orientalische Sprachen, Band I, Abt. 1, Berlin, 1898, S. 196) als «ein in aristokratischen Kreisen entstandenes Epos» bezeichnet, kommen zahlreiche Partien in Versen vor (POZDNÉJEV, Образцы народной литературы монгольскихъ племенъ, St. Pet., 1880, p. 320), die weit ausführlicher sind als die entsprechenden im Altan Tobči und Sanang Setsen (ibid., p. 304).

[2]) Г. Н. Потанинъ, Восточныя мотивы въ средневѣковомъ европейскомъ эпосѣ, Moskau, 1899, pp. 73—136. Hier liegt noch ein sehr dankbares Feld für weitere, tiefer eindringende Forschungen vor, die aber erst strenge philologische Vorarbeit verlangen. Die einschlägigen Texte müssten kritisch bearbeitet und neu übersetzt (der altväterische Stil der Übersetzung des alten Isaac Jacob ist für solche Zwecke natürlich nicht zu verwerten), die Sagenmotive sorgfältig herausgeschält, in ihren Beziehungen und Verbindungen zu einander und zu den alten wie den modernen mongolischen Sagenkreisen aufgedeckt werden.

nationalen epischen Überlieferungen dankbar sein sollen, als für eine trockene Liste mathematisch correkter Jahreszahlen, die er nach der überlegeneren Meinung der Schulmeister hätte liefern sollen. Er wollte eben keine Geschichte schreiben, sondern die national-mongolische Tradition der Nachwelt übermitteln.

Von den literarisch fixirten Heldensagen der Mongolen ist die von Geser Khan die bedeutendste. Dieser Held, von dem auch die Tibeter und die türkischen Stämme singen, lebt gleichfalls in den mündlichen Sagen der Mongolen fort, um deren Sammlung sich POTANIN[1]) und POZDNÉJEV[2]) verdient gemacht haben. Die mongolische Version wurde 1716 in Peking (177 fols.) auf Befehl des Kaisers K'ang-hsi gedruckt und danach von I. J. SCHMIDT (St. Petersburg, 1836, 191 p.) herausgegeben, in deutscher Übersetzung 1839.[3]) Dies ist ohne Zweifel das interessanteste Erzeugnis der ganzen mongolischen Literatur, in dem Heldisches, Humorvolles und Poetisches mit Groteskem und Trivialem bunt gemischt ist; leider entzieht es sich vorläufig der wissenschaftlichen Analyse, die ohne die uns noch fehlende Kenntnis des grossen tibetischen Gesar Epos voreilig und verfrüht wäre.

---

[1]) In seinen Werken Очерки сѣверо-западной Монголіи, Выпускъ IV. Матеріалы этнографическіе, St. Pet., 1883, pp. 250—257, 817—820; Тангутско-Тибетская окраина Китая и центральной Монголіи, Vol. II, St. Pet., 1893, p. 120 et passim; Былина о Добрынѣ и монгольское сказаніе о Гэсерѣ (Вѣстникъ Европы, 1890, pp. 121—158). Von seinen zahlreichen Aufsätzen in Этнографическое Обозрѣніе über diesen Gegenstand hebe ich hervor Ордынскія параллели къ поэмамъ ланбардскаго цикла (Vol. XVIII, Nr. 3, 1893) und Греческій Эпосъ и ордынскій фольклоръ (Vol. XXI, Nr. 2, 1894). Ferner sein oben citirtes Werk Восточныя мотивы (s. Geser im Index).

[2]) Сказка про сраженіе Гэсэръ-хана съ Андалмой (Калмыцкія сказки VII), Zap., Vol. IX, 1896, pp. 41—58.

[3]) Eine neue mehr kritische Übersetzung in zeitgemässer Sprache wäre sehr am Platze; im rechten Gewande würde das Werk gewiss einen weiteren Leserkreis anziehen. Die Literatur zur Geser-Sage s. LAUFER, WZKM, Band XV, 1901, S. 78—79. In Ermangelung des Werkes von SCHMIDT halte man sich an die treffliche Analyse der Sage von W. SCHOTT in den Abhandlungen der Berliner Akademie, 1851, S. 263—295.

Von den mündlich vorgetragenen Heldenliedern der Ost-
mongolen wissen wir bisher wenig.[1]

Die Kalmüken besitzen eine Reihe eigener Heldenlieder
von stark rhetorischer Sprache und einer oft ins Ungeheuer-
liche gesteigerten Phantasie. Die Lieder werden nach ihrem
Helden *Janggar*[2] genannt und von Sängern vorgetragen, die
janggarči heissen. BENJAMIN BERGMANN war der erste, der sie
gehört und eine begeisterte Schilderung davon entworfen hat.[3]
Dann gab A. A. BOBROVNIKOV im Jahre 1854 die russische Übersetzung eines Janggar heraus, die F. v. ERDMANN[4] ins Deutsche
übertragen hat. Ferner hat C. GOLSTUNSKI[5] den Text eines solchen
Janggar publicirt.

Das Hauptcharakteristikum der mongolischen Poesie ist die
Verwendung der Alliteration und Assonanz im Verse. Diese Erscheinung ist nicht nur den Mongolen, sondern auch den Manju,
den Türkstämmen, den Wogulen, den Finnen, Esthen und alten
Magyaren eigentümlich; sie erstreckt sich somit über das ganze
weite Gebiet der altaischen und finno-ugrischen Völkerfamilie
und muss daher als ein uraltes gemeinsames Erbgut der gesamten grossen Gruppe in Anspruch genommen werden, ein
Phänomen, das aufs deutlichste den geistigen und historischen
Zusammenhang ihrer einzelnen Glieder offenbart. Die früher
gehegte Ansicht, als man nur Gelegenheit hatte, die Erschei-

---

[1] RAMSTEDT, Über die mongolischen Bylinen (Orientalische Bibliographie, Band XVIII, 1905, Nr. 1378) ist mir leider nicht zugänglich.

[2] Der Name ist wahrscheinlich aus persisch Jehāngīr entstanden.
Die Gesänge sollen der Zeit der Flucht der Kalmüken aus Russland
(1771) angehören. Persischer Einfluss scheint darin vorzuliegen.

[3] Nomadische Streifereien unter den Kalmüken, Band II, Riga,
1804, S. 205—214; Probe eines solchen Heldengesangs (nicht ganz vollendet), Band IV, Riga, 1805, S. 181—214.

[4] Kalmükischer Dschanggar, Erzählung der Heldentaten des erhabenen Bogdo-Chan Dschanggar. ZDMG. Band XI, 1857, S. 708—730.
Bobrovnikov's Arbeit in Вѣстникъ Имп. Русск. Геогр. Общ., Teil XII,
1854, pp. 19—128.

[5] Убаши Хунъ-Тайджійнъ туджи, народная калмыцкая поэма
Джангара и Сиддиту кюрыйнъ-туľи, изданныя на калмыцкомъ языкѣ,
St. Pet., 1864, lithographirt. Janggar: pp. 7—74.

nung an der finnischen Kalewala zu beobachten,[1] dass sie auf
skandinavischen Einfluss zurückzuführen sei,[2] kann gegenüber
ihrer jetzt constatirten weiten Verbreitung über die verwandten
Stämme nicht länger aufrecht erhalten werden. HANS CONON v.
D. GABELENTZ[3] gebührt das Verdienst, sie als erster für das
Mongolische nachgewiesen zu haben. «Das unerlässlichste Erfordernis scheint bei allen Gedichten der Parallelismus der Glieder zu sein, der sich oft durch Wiederkehr derselben Endungen
(Reim) oder derselben Worte *(Refrain)* kundgibt. Beides findet
sich gewöhnlich noch durch mehr oder weniger regelmässige
Alliteration der Versanfänge verstärkt. Dagegen ist ein eigent-
liches Versmass nicht zu entdecken, und selbst die Zahl der
Silben willkürlich, insoweit nicht der Parallelismus gewisse Grenzen festsetzt.» Er bespricht dann die Formen des Verses, Stro-
phenbau etc. und erwähnt auch dieselbe Eigentümlichkeit bei
den Manju, von der er früher mit RÉMUSAT angenommen hatte,
dass sie eine Erfindung des Kaisers Kao-tsung sei, zieht aber
nun die durchaus richtige Schlussfolgerung, dass die Analogie,
welche zwischen den Sprachen beider Völkerschaften stattfindet,
sich auch auf die Form ihrer Poesie erstrecke, und dass die
Manju lange vor Kao-tsung, ja wohl noch vor Eroberung des
chinesischen Reiches, so gut ihre Volkspoesie hatten wie die
Mongolen und andere auf gleicher, ja auf noch niedrigerer Stufe
der Kultur stehende Völker.[4] Dieselbe Frage hat W. RADLOFF[5]
für die Türkstämme behandelt.

---

[1] D. COMPARETTI, Die Kalewala oder die traditionelle Poesie der
Finnen. Halle, 1892, S. 30, 32, 33.

[2] Wohl zuerst und schlagend widerlegt von PAUL HUNFALVY (s.
W. SCHOTT, Altaische Studien, Fünftes Heft. Abhandlungen der Berliner
Akademie, 1872, S. 41—44).

[3] In der bereits oben angeführten Abhandlung Einiges über mongolische Poesie (Zeitschrift für die Kunde des Morgenlandes. Band I,
Göttingen, 1837, S. 22—33).

[4] Ein gutes Beispiel eines manjurischen Strophengedichts mit
Alliteration und Assonanz bei E. v. ZACH, Lexicographische Beiträge,
Band III, Peking, 1905, S. 104—107.

[5] Über die Formen der gebundenen Rede bei den altaischen Tataren (Zeitschrift für Völkerpsychologie und Sprachwissenschaft, Band IV,
S. 85—114).

Für das Mongolische darf man, möchte ich glauben, aus dem Vorhandensein der Alliteration einen wichtigen Schluss auf die Geschichte des Accents ziehen. Während gegenwärtig der Accent im allgemeinen auf der letzten Silbe ruht, muss er zur Zeit der Bildung des alliterirenden Princips offenbar auf die Anfangssilbe wie im Finnischen und Magyarischen gefallen sein, da sonst die Vernehmlichkeit und das Zustandekommen der alliterirenden Anlaute kaum denkbar wäre. Eine erschöpfende Darstellung der mongolischen Verskunst kann im Rahmen dieser kleinen Skizze nicht gegeben werden. Als allgemeine Regel kann aufgestellt werden, dass die Lieder gewöhnlich in vierzeilige Strophen abgeteilt werden, und dass die Anlaute des ersten Wortes in den vier Verszeilen (zuweilen in den drei ersten) übereinstimmen (mitunter auch Konsonant + Vokal).[1] Vokale können, auch wenn sie nicht identisch sind, wie in der alt-deutschen Dichtkunst, zur Bildung der Assonanz verwendet wer-den. Von einer strengen Silbenzählung ist dem Charakter der Sprache gemäss keine Rede, dagegen sind die Verse im Durch-schnitt von annähernd gleichmässiger Länge und werden von einem mehr gefühlsmässigen als bewussten Rhythmus beherrscht. Refrains kommen am Schlusse, am Anfang und auch innerhalb der Strophe vor; überhaupt ist zu beachten, dass die mongo-lische Poesie formell und inhaltlich durchaus keinen primitiven (im ethnographischen Sinne) Charakter trägt, sondern eine ver-hältnismässig weit fortgeschrittene Kunstpoesie repräsentirt. Die zahlreichste und beste Sammlung mongolischer Lieder hat A. POZDNĚJEV[2] zusammengebracht, auf dessen Werk ich auch für die Fragen der Poetik hinweise.

---

[1] Dagegen findet sich auch, besonders in den alten Liedern der Chroniken, Alliteration innerhalb desselben Verses wie im Finnischen.

[2] Образцы народной литературы монгольскихъ племенъ. Выпускъ I (mehr ist mir nicht bekannt geworden), Народныя пѣсни монголовъ, St. Pet., 1880. Texte von 65 Liedern (pp. 1—43), Transkription mit Über-setzung und ausführlichem Commentar (pp. 45—346). — Vergl. auch LAUFER, Über eine Gattung mongolischer Volkslieder und ihre Verwandt-schaft mit türkischen Liedern (Der Urquell, Neue Folge, Band II, 1898, S. 145—157). — Ältere Aufzeichnungen von Liedern bei P. S. PALLAS, Sammlungen historischer Nachrichten über die mongolischen Völker-

Eine wertvolle Arbeit über mongolische Lieder ist von
Prof. C. STUMPF[1]) in Berlin nach den Aufzeichnungen von J. S.
STALLYBRASS veröffentlicht worden. Hier sind auch die Melodien
hinzugefügt, die einzig zuverlässigen Notirungen mongolischer
Musik, die wir haben. Die Gesänge entstammen den Burjaten
am Baikalsee. Leider aber sind die Texte nicht vollständig
wiedergegeben, dagegen zu einigen Liedern, was sehr dankbar
ist, die Entstehungsgeschichte. So wird als Veranlassung zu
dem ersten Liede, dessen Motiv ist «Die weissen Klippen von
Kuderi kann keines Menschen Macht erschüttern; was des
Menschen Herz erschüttert, ist der Abschied von Vater und
Mutter», angegeben, dass es von Badma (wohl Padma), Sohn
des Morkhon, verfasst wurde, als er mit einem anderen jungen
Schüler um 1815—16 nach St. Petersburg berufen wurde, um
I. J. SCHMIDT bei der Übersetzung des Neuen Testaments ins
Mongolische zu helfen; sein Vater fand bei einer Wohnungs-
veränderung die Abschrift unter dem Dache versteckt. Die Weise
des Liedes aber ist viel älter, und es sollen oft neue Texte zu
alten Weisen gemacht werden. Der zwanzigstrophige Text des
dritten Liedes, von dem leider nur die erste Strophe mitgeteilt ist —

---

schaften, Erster Teil, St. Pet., 1776, S. 152 157. Die kalmükische Originalschrift der beiden auf S. 153 154 mitgeteilten Liebeslieder, mit
Interlinearversion versehen, von der Hand von JUST. FRIEDR. MALSCH. in
Göttingen (Asch 143, pp. 63b 64b), mit der Bemerkung: «Diese zwey
gedichte hab ich nicht darum ab geschrieben um sie den Challmüken
vor zu lesen, weil sie sehr leicht auf gedanken fallen würden als ob wir
auch noch dergleichen liebten und bey uns in gebrauch währen, sondern
nur ihre Reth und denkungs art darauss zu sehen und noch viele un-
bekante Wörtter drauss zu erlernen». PALLAS' Lieder sind auch von FR.
MAJER in KLAPROTH's Asiatisches Magazin, Band I, Weimar, 1802, S.
547 554, abgedruckt. Ferner fünf Lieder (drei mit Texten) bei G. TIM-
KOWSKI, Reise nach China durch die Mongoley in den Jahren 1820 und
1821, Band III, Leipzig, 1826, S. 292 298. A. BASTIAN, Geographische
und ethnologische Bilder, Jena, 1873, S. 355 (Übersetzung eines kalmükischen Liedes).

[1]) Mongolische Gesänge. Vierteljahrsschrift für Musikwissenschaft.
Band III, Leipzig, 1887, S. 297 304.

Im Hintergrund des schilfbewachsenen Sees
Stehen gelbe Gänse beisammen;
Unter zwanzigtausend[1]) Männern
Ist Khori's Väterchen erster — —

entstand im Jahre 1836 bei der Anstellung Dorji's als Häupt-
ling der Khori, eines Stammes im Westen des Selenga Flusses,
nach bitteren Kämpfen mit Rivalen. Das schönste der hier mitgeteilten zehn Lieder ist Nr. 6, der Sang eines sterbenden
Helden, der sehr alt sein und aus der Zeit des Činggis Khan
stammen soll. Er lautet:

Der ellenlange schwarze Zopf mein
Mit seinen Flechten liegt nun hier;
Das bunt schimmernde Auge mein
Mit seinem (letzten) Blick liegt nun hier.
Dieser arme Körper mein,
Der über Felsen schritt, liegt nun hier.
Lieber melonenfarbener[2]) Läufer mein,
Nach Haus gesund gelangst du wohl!
Melde doch dem Vater mein:
Getroffen durch den Rücken lag er da.
Melde doch der Mutter mein:
Volle fünfzehn Wunden trug er.

Die Melodie dieses Liedes ist von ergreifender Trauer.
Die Mehrzahl der Gesänge schliessen auf der Quinte, einige
auch auf der Sekunde.

Ein mongolisches Liebeslied ist jüngst von A. D. RUDNEV[3])
in phonetischer Transkription mit Übersetzung und ausführlichen Erläuterungen mitgeteilt worden. Der Anfang desselben
lautet:

In der Stadt Peking ist glückbringend die bunte Lerche;
für das zärtliche Liebespaar ist die Nacht voller Wonne. Mitten

---

[1]) Die Zahl 20 *(khorin)* ist wegen der Alliteration mit dem Namen
Khori gewählt.

[2]) Der Herausgeber übersetzt «rahmfarben»; das Wort *ghuwa* (kalmükisch: *ghū*) des Textes, indessen, ist Transkription des chinesischen
kua «Melone».

[3]) Extrait du Journal de la Société finno-ougrienne, Vol. XXIII,
18 (8 p.).

auf dem See tummelt sich eine buntgefiederte Ente; mögen die Gefühle von uns beiden, die wir uns heimlich zusammengefunden haben, übereinstimmen oder nicht, wir werden immer wieder zusammentreffen.

Man sieht, dass hier bereits chinesischer Einfluss vorliegt in dem Vergleich des Liebesverhältnisses mit dem Leben der Ente, die in China als Symbol der ehelichen Treue gilt. Interessant ist der in diesem Gedicht vorkommende negative Ausdruck: Was über dem Felsenufer kreist, ist nicht ein Habicht, nein, ein Schmetterling. Rudnev spricht, im Gegensatz zu Pozdnějev, die Ansicht aus, dass diese poetische Form sehr häufig vorkomme und für die mongolischen, epischen und lyrischen Gedichte charakteristisch sei. Obwohl das Lied ein typisches lyrisches Volkslied ist, wird ihm ein rein buddhistischer Schluss angehängt mit den Worten: «Mit dem Segen des Dalai Lama wollen wir allesamt uns freuen und fröhlich sein!», was nach den heissen Liebesergüssen der vorhergehenden Strophen wie ein kalter Wasserstrahl wirkt.[1]

Was die aus mündlicher Tradition aufgezeichneten Märchen und Erzählungen betrifft, muss ich mich auf eine Anführung der wichtigeren Sammlungen beschränken. In erster Reihe ist Pozdnějev's schöne Sammlung «Kalmükische Erzählungen» zu nennen, die zehn umfangreiche Texte in Originalschrift mit Übersetzungen bietet.[2] Die ergiebigste Ausbeute haben bisher die Burjaten geliefert.[3]

---

[1] Überhaupt hat der Buddhismus die mongolische (wie auch die tibetische) Volkspoesie nachhaltig afficirt, und man kann ruhig hinzusetzen, recht nachteilig. Er hat diesen derb lebenslustigen, sanges- und festesfrohen Völkern das Lämpchen der Lebensfreude auszublasen gesucht und sonst gesunde Menschenkinder in maskirte Pessimisten und querköpfige Grübler verwandelt; wir haben in Centralasien denselben Effekt der Pfaffenwirtschaft wie in Europa, die das natürliche Empfinden der Menschen knebelt und den freudigen Ausdruck desselben in Lied und Sage mit frevler Hand würgt. Wenn die Mongolen trotz ihrer pfäffischen Servilität sich so viel ihres Eigenen für bessere Stunden des Daseins hinübergerettet haben, so spricht dies genug für die zähe Energie ihres Charakters, und man gewahrt hoffnungsvoll, dass auch dort Luft, Licht und Leben noch nicht ganz ertötet und verloren sind.

[2] Калмыцкія сказки. Zap., Vol. III, 1888, pp. 307–364; IV, 1889,

Die Mongolen besitzen auch kleine lustige Schnurren, von
denen Prof. GABRIEL BÁLINT in Klausenburg[1]) einige gute Proben
mitgeteilt hat. Besonders werden darin die Chinesen verspottet.
Ein Mongole und ein Chinese reisten einst zusammen, als es
zu donnern und zu regnen begann. Der Chinese ritt vorne, der
Mongole hintendrein; im Augenblick, als es donnerte, hieb der
Mongole mit seiner Peitsche so auf den Kopf des Chinesen ein,
dass dieser vom Pferde herabtaumelte. Hierauf stieg der Mongole vom Pferde, verdrehte die Augen, öffnete den Mund und
blieb wie tot liegen. Der Chinese raffte sich auf zu ihm hin
und sagte: «Der Mongole kann nicht einen Blitzschlag ertragen,
mich haben viele, viele — drei Schläge getroffen. Der Mongole
ist doch ein elendes Ding! So steh doch auf, steh auf!» Auf
diese Worte erhob sich der Mongole, als wenn er halb zu sich
käme. «Schau», sprach wieder der Chinese, «ich bin der wackere
Mann, drei Blitzschläge spalteten nicht meinen Kopf. Elender

---

pp. 321—374; VI, 1891, pp. 1—68; VII, 1892, pp. 1 38; IX, 1895, pp.
1 58; X, 1896, pp. 139 185. Die Texte I—VI (ohne die Übersetzung)
sind auch separat unter dem Titel Калмыцкія народныя сказки собран-
ныя въ калмыцкихъ степяхъ Астраханской Губерніи I., St. Pet., 1892
(150 p.) erschienen.

³) H. M. Хангаловъ и H. Затопляевъ, Бурятскія сказки и повѣрья
(Записки Восточно-сиб. Отдѣла Имп. Русск. Геогр. Общ., по отдѣленію
этнографіи, Vol. I, Nr. 1, Irkutsk, 1889, 159 p.). Сказанія Бурятъ, за-
писанныя разными собирателями (Ibid., Vol. I, Nr. 2, Irkutsk, 1890, 159
p.). Reiches Material findet man ferner in den schon erwähnten Werken
von POTANIN. A. D. RUDNEV's Erzählung der Bargu Burjaten, und J. S.
SMOLEV's Burjatische Legenden und Erzählungen (Orientalische Bibliographie, Band XVIII, 1905, Nr. 1380, 1382) sind mir leider nicht zugänglich. Beiträge zum Folklore der Mongolen auch in den Büchern von JAMES
GILMOUR, Among the Mongols (Kap. 28 und 29), More about the Mongols
(London, 1893) und R. LOVETT, James Gilmour of Mongolia (London,
1892); doch sind diese Materialien nicht ganz einwandfrei. Die ältere die
Kalmüken betreffende Erzählungsliteratur, die aber in ausserrussischen
Bibliotheken kaum zugänglich sein dürfte, ist bei V. J. MEŽOV, Bibliogra-
phia Sibirica, Vol. I, St. Pet., 1891, pp. 424 -429 verzeichnet. — Eine
gute Charakteristik der wichtigeren Züge und Motive des mongolisch-
türkischen Folklore gibt W. JOCHELSON in seiner Studie The Koryak (Memoirs of the American Museum of Natural History, Vol. VI, Part I, Leiden und New-York, 1905, pp. 344—352).

¹) Ethnologische Mitteilungen aus Ungarn, Band IV, 1895, S. 70—71.

17*

Mongole, reisen wir weiter!» Und damit setzten sie die Reise fort. Eine andere Geschichte erzählt, wie ein Chinese und ein Mongole, die Reisegefährten waren, einst bei einem zur Winterfeuerung angesammelten Torfhaufen übernachten mussten. Während der Nacht begann der Mongole, wie ein Wolf knurrend, mit den Zähnen am Anzuge des Chinesen zu zerren und ihn in die Waden zu beissen. Der erschrockene Chinese verhüllte seinen Kopf und flehte zum Wolfe: «Dort ist der elende Mongole, den magst du zuerst fressen, und wenn du auch dann noch hungrig bist, gibt er selbst auch sein blosses Gesäss hin.» Am Morgen schwur er bei Himmel und Erde, dass er, während der schlechte Mongole gut geschlafen, neun Wölfe halb tot geschlagen habe.

Interessant ist auch die Tatsache, dass die Mongolen kleine Anekdoten oder Geschichten über chinesische Helden wiederholen, die sie aus dem Munde von Chinesen gehört oder durch Vermittlung der chinesischen Übersetzungsliteratur sich angeeignet haben mögen. So hat z. B. G. TIMKOWSKI[1]) zwei mongolische Geschichten über den sagenumwobenen General Chu-ko Liang (181—234 n. Chr.)[2]) vernommen.

BENJAMIN BERGMANN[3]) hat zehn kalmükische Schwänke, besonders über Streiche von Dieben, mitgeteilt.

Die mongolischen Stämme haben auch ihre Freude an Rätseln und Rätselspielen, die nur an langen Winterabenden erlaubt sind. Es werden zwei gleich starke Parteien gebildet, die sich darüber einigen, wie oft jede Partei die Lösung des von der andern aufgegebenen Rätsels versuchen darf. Löst die befragte Partei das Rätsel nicht, so muss sie eines ihrer Mitglieder an die siegende Partei ausliefern, der dann das Recht zusteht, ein neues Rätsel aufzugeben. GALSANG GOMBOJEV[4]) hat sechzig bur-

---

[1]) Reise nach China durch die Mongoley in den Jahren 1820 und 1821, Band I, Leipzig. 1825, S. 174 175. TIMKOWSKI nennt ihn Kumin, was für K'ung-ming steht, einen seiner Beinamen.

[2]) S. GILES, A Chinese Biographical Dictionary, London, 1898, pp. 180 182.

[3]) Nomadische Streifereien unter den Kalmüken, Band II. Riga, 1804, S. 343 352.

[4]) In M. A. CASTRÉN's Versuch einer burjätischen Sprachlehre, St. Pet., 1857, S. 228 233.

jatische Rätsel in der selenginschen Mundart in Text und Übersetzung mitgeteilt; eine andere von BAZAROV veranstaltete Sammlung hat RUDNEV herausgegeben.[1]) VL. KOTVIČ[2]) hat eine Sammlung kalmükischer Rätsel und Sprüchwörter publicirt.

Juristische Literatur. — Das Recht der Mongolen wird
dadurch zu einer complicirten Materie, als sich drei verschiedene Kulturschichten darin abgelagert haben, — einmal ursprünglich mongolisches Recht, sodann indisch-buddhistische Vorstellungen, die mit dem Lamaismus aus Tibet eingedrungen sind
und kurz als lamaisches Recht bezeichnet werden können, und
drittens chinesisch-manjurisches Recht, das die einheimischen
Volksrechte fast ganz verdrängt hat, und nach dem die weitaus
grösste Zahl der Mongolen noch jetzt regirt werden. Von den
alten Gesetzbüchern, welche zur Zeit ihrer politischen Macht-
stellung in Kraft waren, ist uns leider keines erhalten geblieben,
doch liegen uns einige Andeutungen darüber und Auszüge daraus in der historischen (besonders mohammedanischen) Literatur
vor, die, durch Ausnutzung der neueren traditionellen Volksrechte einigermassen erweitert, zu einer Reconstruktion des alten
mongolischen Rechts verwertet werden können. Um zunächst
das chinesische Recht vorwegzunehmen, so waren die Manju
Kaiser Chinas nach Unterwerfung der Mongolei, die ihnen durch
die Zersplitterung und Uneinigkeit der zahlreichen Stämme
wesentlich erleichtert wurde, von jeher bemüht, ihre eigene Ge-
setzgebung dahin zu verpflanzen. Der Kaiser K'ien-lung (1736—
1795) war der erste, der eine Codifikation aller früher in Sachen

---

[1]) Образцы монгольскаго народнаго творчества. Монгольскій
текстъ и русскій переводъ загадокъ, собранныхъ Базаровымъ среди
баргу- и агинскихъ бурятъ, а также отчасти въ хошунахъ Узумучін
и Ару-Хорчін, подъ редакціей А. Д. Руднева. Zap., Vol. XIV, 1901, pp.
092—0106. (Vergl. ferner Orientalische Bibliographie, Band XVIII, 1905,
Nr. 1372 [nicht gesehen].)

[2]) Калмыцкія загадки и пословицы, издалъ Вл. Котвичъ. Изданія
Факультета Восточныхъ Языковъ, Nr. 16, St. Pet., 1905. — Vier kalmü-
kische Rätsel bei A. BASTIAN, Geographische und ethnologische Bilder,
Jena, 1873, S. 356. — Zwanzig Rätsel der Burjaten nördlich vom Baikal
(Text und Übersetzung) von I. Чистохинъ, Инородческія загадки Тун-
кинскаго края, Извѣстія Восточно-сибирскаго Отдѣла Имп. Русскаго
Геогр. Общества, Vol. XXVI, Irkutsk, 1895, pp. 37—39.

der Mongolen erlassenen Bestimmungen befahl. Diese erste Sammlung umfasste 209 Paragraphen auf zwölf Kapitel verteilt und wurde 1789 in vier nicht sehr umfangreichen Bänden gedruckt, die HYAKINTH BIČURIN ins Russische übersetzt hat.[1]) Doch bald machte sich bei wichtigeren Gesetzen in ihrer praktischen Anwendung auf lokale Zustände und hinsichtlich der Beziehungen zum Pekinger Hofe ein Mangel fühlbar. Im Verlauf von zwanzig Jahren erschienen in den Ministerialberichten eine nicht unbedeutende Zahl von Erlassen, welche das frühere Gesetzbuch verbesserten und vervollständigten. Auf Antrag des die auswärtigen Angelegenheiten leitenden Ministeriums Li-fan-yüan wurde im Jahre 1811 eine besondere Kommission zur Abfassung eines neuen Gesetzbuches eingesetzt, das bereits 1815 vollendet war. In dieser Sammlung wurden von den früheren 209 Paragraphen zwanzig gestrichen, 178 verändert, die übrigen elf wiederaufgenommen, und 526 neue Artikel verfasst. Die manjurische Version dieses Codex wurde von LIPOVTSOV ins Russische übersetzt.[2]) In den folgenden Jahren vermehrte sich so das mongolische Recht noch weiter. Im Jahre 1826 wurde eine Ausgabe in 63 Büchern, und schon 1832 eine Neuausgabe veranstaltet. Es ist eine mongolische Übersetzung davon vorhanden, aus der KOWALEWSKI in seiner Chrestomathie (Vol. I, pp. 183—201) den Text von fünf Abschnitten mitgeteilt hat, — nämlich über die Kopfzählung der Mongolen, über die Ankunft bei Hofe, über Aufnahme von Gastfreunden, über die Treibjagd, und über besondere Gnadenbezeigungen und Wohltätigkeit. Dieses Gesetzbuch enthält auch reiches statistisches Material.

A. POZDNĚJEV hat in seiner Chrestomathie (pp. 379—416) die Texte von 32 officiellen Aktenstücken — Gesetze, Anträge,

---

[1]) О. Iакинфъ, Записки о Монголіи, St. Pet., 1828, Vol. II, pp. 203—339. Deutsch von K. F. v. D. BORG, Denkwürdigkeiten über die Mongolei, Berlin, 1832, S. 320—426. Auszüge bei TIMKOWSKI, Reise nach China durch die Mongoley in den Jahren 1820 und 1821, Band III, Leipzig, 1826, S. 327—342 (vergl. auch Band II, S. 15—16).

[2]) Липовцовъ, Уложеніе Китайской Палаты внѣшнихъ сношеній, съ Маньджурскаго, St. Pet., 1828, 2 Vols., 4°. Mongolische Ausgabe in Kasan, Nr. 139.

amtliche Briefe, kaiserliche Edikte etc. — gegeben, leider aber ohne Anführung der Quellen.

Das älteste Gesetzbuch der Mongolen ist unter dem Namen «Yasa des Činggis» (Khaoli yosun-u bičik) bekannt und rührt von dem grossen Eroberer selbst her; leider ist es uns nicht erhalten, doch erfahren wir von seiner Existenz aus mohammedanischen und mongolischen Quellen.[1]) Es war in mongolischer Sprache abgefasst und in uigurischer Schrift geschrieben; war ja doch der Wunsch, seine Verordnungen und Gesetze schriftlich zu überliefern, die Haupttriebfeder für Činggis zur Annahme und Verbreitung der Schrift. Sein Gesetzbuch war sicher das früheste Erzeugnis der mongolischen Literatur.[2])

Ausser dem Gesetzbuch des Činggis scheint zu seiner Zeit auch ein Werk existirt zu haben, das vorzüglich die Processordnung der Volksgerichte behandelte und unter dem Namen *Kudatku Bilik* des Činggis bekannt war. Wenigstens ist dies die Definition dieses Werkes, zu der neuerdings P. MELIORANSKI[3]) im Gegensatz zu HAMMER-PURGSTALL gelangt, der in diesem Werke eine «Sammlung von mündlichen Befehlen» des Khans sehen zu müssen glaubte. Der Titel wird wohl unter dem Einfluss des uigurischen *Kudatku Bilik* gewählt worden sein, ist aber nicht mit dem *Bilik*, einer Sammlung von Aussprüchen des Khans, identisch, die teilweise bei Rašid Eddin erhalten sind.

Ein interessantes Streiflicht fällt auf die alten Einrichtungen der Mongolen und die Veränderungen ihrer Lebensanschauungen durch den Einfluss des Lamaismus in jener Zeit, als der dritte Dalai Lama bSod-nams rgya-mts'o (1543—1586) den Altan Khan (1504—1582) der Tümäd besuchte (1577) und die Mongolen aufs engste mit den Interessen der Kirche von

---

[1]) Vergl. POZDNÉJEV, Lit., pp. 37—50.

[2]) Bruchstücke und Auszüge bei Mirchavend, Ibn Baṭūṭa, Bartak, Plano Carpini, Rubruk, Marco Polo: der Hauptinhalt ist bei dem arabischen Historiker Makrizi dargelegt (SILVESTRE DE SACY, Chrestomathie arabe, Vol. II, Paris, 1826, pp. 160—165). Vergl. YULE, Cathay, London, 1866, Vol. II, p. 507.

[3]) О Кудатку Биликѣ Чингизъ хана, Zap., Vol. XIII, 1901, pp. 015—023. Vergl. Mitteilungen des Seminars für orientalische Sprachen, Band IV, 1. Abt., Berlin, 1901, S. 254—255.

Lhasa verband. Lama und Khan einigten sich auf die Annahme der Gebote des buddhistischen Dekalogs von seiten der Mongolen und schlugen folgende radikale Umgestaltung der heimischen Gesetze vor: «War früher ein Mongole gestorben, gleichviel ob einer hohen oder niederen Klasse angehörig, so wurden sein Weib, seine Sklaven, seine Pferde, sein Vieh u. s. weiter getötet;[1]) von nun an hingegen sollen so viel Güter, als man ideell als gleichwertig den zum Töten bestimmten Pferden, Vieh u. s. w. schätzt, der Geistlichkeit und dem Lama dargebracht werden; diese sollen um ein gemeinsames Gebet ersucht werden; die fingirte Tötung soll gar nicht vorgenommen werden. Wie früher, verliert, wer einen Menschen getötet hat, nach dem Gesetze Leib und Leben; wer ein Pferd oder ein Stück Vieh getötet hat, geht nach dem Gesetze seiner ganzen Habe verlustig; wenn jemand einen Lama und Träger geistlicher Tracht gestossen, geschlagen oder sonstwie Hand an ihn gelegt hat, so wird des Täters Wohnsitz zerstört.»[2]) Ferner wurden die alten schamanischen Götterbilder (onggot) verboten, die vorhandenen sollten verbrannt, und die Wohnungen Zuwiderhandelnder zerstört werden; blutige Opfer sollten aufhören; statt dessen sollte man sich der Tugend befleissigen und kurz und gut alles so einrichten, wie es bereits in Tibet der Fall war.[3]) Hier ist klar ausgesprochen, dass es sich damals um eine Reception tibetischlamaischen Rechts in der Mongolei handelte.[4])

---

[1]) Vergl. YULE's Marco Polo, 3. Aufl., Vol. I, pp. 250—251.

[2]) HUTH, Geschichte des Buddhismus in der Mongolei. Band II, S. 219—220. Vergl. KÖPPEN, Die lamaische Hierarchie und Kirche. S. 139.

[3]) HUTH, l. c., S. 221.

[4]) «Es gibt bei den Kalmüken und Mongolen nicht nur viele Gewohnheitsgesetze, sondern die Fürsten haben auch unter sich zu verschiedenen Zeiten, seit Činggis Khans Regirung, geschriebene Gesetze errichtet, welche (eine monarchische Verfassung vorausgesetzt) der natürlichen Billigkeit, wo nicht ganz gemäss sind, doch gewiss sehr nahe kommen. Das alleräteste Gesetzbuch (Zaatschin Bitschik [= Tsaghādsayin bičik «Gesetzbuch»]), wonach zwar nicht mehr gerichtet wird, das aber doch, soviel mir davon bekannt geworden, manche Merkwürdigkeit enthalten muss, habe ich mir nicht verschaffen können. Nach demselben war die Unzucht, die mit Beischläferinnen der Pfaffen (dergleichen sich die mongolische Geistlichkeit am meisten bedienet) getrieben wurde,

Das neuere Gesetzbuch der Kalmüken, des Bātur Hung-
Taiji, aus der ersten Hälfte des siebzehnten Jahrhunderts ist
vollständig von PALLAS mitgeteilt worden.[1]) POPOV hat daraus in
seiner kalmükischen Grammatik (pp. 359—365) einen Abschnitt
in Text und Übersetzung mitgeteilt. Doch ist PALLAS' Arbeit
längst in den Schatten gestellt durch die treffliche kritische
Ausgabe und Übersetzung der kalmükischen Gesetze durch
GOLSTUNSKI,[2]) dessen Werk allen Studien über das kalmükische,
wie das ostmongolische Recht zu Grunde gelegt werden muss.

M. CHANGALOV[3]) hat die Rechtssitten bei den Burjaten dargestellt; KOWALEWSKI erwähnt im Vorwort zu seinem Wörterbuch (p. IX a) einen handschriftlichen «Gesetzescodex der
Burjaten.»[4])

---

völlig ungestraft. Wer in Ehebruch mit einer Fürstin betreten wurde
hatte nur eine Ziege und ein Böcklein zur Busse zu erlegen; denn das
Gesetzbuch stellte voraus, dass ein Gemeiner sich nie an eine Fürstin
wagen würde, wann er nicht dazu gereizt worden. Für gemeinen Ehebruch musste der Täter dem Hörnerträger ein vierjähriges Pferd, und
die Ehebrecherin dem Richter ein dreijähriges stellen. Wer einen Fremden bei seiner Sklavin ertappte, konnte demselben alles ausziehen, Pferd,
Geld, und was er bei sich hatte, nehmen und ihn nackend fortjagen; die
Sklavin aber blieb ungestraft. Aus eben diesem alten Codex scheinen
sich auch verschiedene noch heutzutage rechtskräftige Gewohnheitsgesetze
herzuschreiben.» (P. S. PALLAS, Sammlungen historischer Nachrichten
über die mongolischen Völkerschaften, Erster Teil, St. Pet., 1776, S.
193—194.)

[1]) Sammlungen etc., Erster Teil, S. 194—218.

[2]) Монголо-ойратскіе законы 1640 года, дополнительные указы
Галданъ-Хунъ-тайджія и законы, составленные для волжскихъ калмыковъ при калмыцкомъ ханѣ Дондукъ-Дашп, St. Pet., 1880. Einleitung, pp.
1–16; Text, pp. 1—33; Übersetzung, pp. 35–72; Anmerkungen, pp. 73
.143. — CARL KOEHNE, Das Recht der Kalmüken (Zeitschrift für vergleichende Rechtswissenschaft, Band IX, Stuttgart, 1891, S. 445—475) hat
eine kurze auf PALLAS und BERGMANN basirte Compilation des Gegenstandes gegeben, ohne GOLSTUNSKI's Mongolisch-oiratische Gesetze zu benutzen; von Činggis Khans Gesetzbuch ist gar keine Rede, und historische Gesichtspunkte sind überhaupt nicht vertreten.

[3]) Этнографическое обозрѣніе, Moskau, 1895, pp. 100—142.

[4]) Über die gegenwärtige Verwaltung der Mongolei vergl. W. F.
MAYERS, The Chinese Government, 2. Aufl., Shanghai, 1886, pp. 80—92.
Über die Stammeseinteilung der Mongolen s. I. J. SCHMIDT, Die Volks-

　　　Die politischen und commerciellen Beziehungen zwischen
Russland und China begannen um die Mitte des siebzehnten
Jahrhunderts. Die zwischen beiden Ländern abgeschlossenen
Verträge wurden in der Regel aus dem Chinesischen ins Man-
jurische, und einige, wie es scheint, auch ins Mongolische über-
setzt. Die mongolische Version eines solchen Vertrags aus dem
Jahre 1803 befindet sich in Popov's Mongolischer Chrestomathie,
pp. 96—123, wiedergegeben.

　　　Medicinische Literatur. — Obwohl es den mongolischen
Stämmen nicht an einer einheimischen Volksmedicin zu fehlen
scheint, zu der Pallas und Bergmann interessante Beiträge geliefert haben, so scheint doch die bestehende Literatur über die
Heilkunde ausschliesslich auf Übersetzungen aus dem Tibetischen und Chinesischen gegründet zu sein. Dafür ist die Quan-
tität dieser Literatur um so gewichtiger; fast alle umfangreichen
Werke der Tibeter über Medicin haben Übertragungen ins
Mongolische erfahren, vor allem das *Lhan t'abs*, von dessen In-
halt Pozdnějev[1]) eine kurze Analyse entworfen hat, und von
dessen mongolischer Edition er eine Übersetzung in Aussicht
gestellt hat. Bemerkenswert ist die Tatsache, dass hervorragende
Ärzte in der Mongolei Tagebücher führen, in welche sie die
Geschichte der von ihnen behandelten Krankheiten samt der
Heilmethode eintragen. Diese Denkschriften werden nach dem
Tode des Arztes seinem Lieblingsschüler erblich hinterlassen.[2])

stämme der Mongolen, als Beitrag zur Geschichte dieses Volkes und
seines Fürstenhauses (Mémoires de l'Acad. des Sciences de St.-Pét., VI-e
Série, sc. polit., Vol. II, 1834, pp. 409—477); W. Schott, Bevölkerung,
Verfassung und Verwaltung der heutigen Mongolei (nach Hyakinth
Bičurin's Statistischer Beschreibung des Chinesischen Reiches), Erman's
Archiv für wissenschaftliche Kunde von Russland, Band IV, 1845, S.
534—547; П. С. Поповъ, Записки о монгольскихъ кочевьяхъ, переводъ
съ китайскаго (Зап. Имп. Русск. Геогр. Общ., Vol. XXIV, St. Pet., 1895).

　　　[1]) Очерки быта буддійскихъ монастырей и буддійскаго духовен-
ства въ Монголіи (Зап. Имп. Русск. Геогр. Общ., Vol. XVI) St. Pet.,
1887, p. 163, Note.

　　　[2]) Näheres hierüber und die mongolische Literatur der Medicin s.
bei Heinrich Laufer, Beiträge zur Kenntnis der tibetischen Medicin,
I. Teil, Berlin, 1900, S. 11—17.

Astronomische Literatur. — Von dieser, die wohl ausschliesslich in Übersetzungen aus dem Tibetischen und Chinesischen besteht,[1]) ist bisher wenig bekannt geworden.[2]) In Peking wird noch alljährlich ein mongolischer Kalender mit astronomisch-astrologischen Angaben publicirt und in die Mongolei gesandt. Das grosse Interesse, welches die mongolischen Kaiser der Himmelsbeobachtung entgegenbrachten, ist bekannt: das berühmte Observatorium, das ehemals auf der Stadtmauer von Peking stand, wurde 1279 unter Kubilai gegründet.[3]) Die Zeitrechnung der Mongolen ist im wesentlichen mit der der Tibeter identisch.[4])

Bibelübersetzungen. — Die früheste mongolische Version der Schrift, von der wir einige Kenntnis haben, ist die von dem Franziskaner JOHANNES DE MONTE CORVINO (1247—1328) verfasste,

---

[1]) Vergl. z. B. HUTH. Geschichte des Buddhismus in der Mongolei, Band II, S. 333, 335, 343, 379. KOWALEWSKI, Mongolische Chrestomathie, Vol. II, p. 306.

[2]) I. P. ABEL-RÉMUSAT, Odonu närä inu, Uranographia Mongolica sive nomenclatura siderum, quæ ab astronomis mongolis agnoscuntur et describuntur (Excerptum ex opere, Mongolica lingua conscripto, quod in Bibl. Imp. Paris. conservatur). Fundgruben des Orients (Mines de l'Orient). Band III, Wien, 1813, S. 179—196. Dieselbe Arbeit französisch unter dem Titel Uranographie mongole in seinen Mélanges asiatiques, Vol. I, Paris, 1825, pp. 212—239. Die hier bearbeitete Liste der Constellationen ist der mongolischen Übersetzung einer zur Zeit K'ang-hsi's von Jesuiten verfassten astronomischen Abhandlung entnommen (s. auch RÉMUSAT, Recherches sur les langues tatares, p. 223), die mit Tafeln, die Bewegung des Mondes und der Planeten darstellend, versehen ist. Das ganze Werk soll keinen umfassenden Titel haben; der erste Band, mit dem sich RÉMUSAT beschäftigt, ist *Tägri-yin utkha-yin alkhum-un domok* («Verse über die Umdrehung des Himmels») betitelt. Der von RÉMUSAT gegebene chinesische Titel stimmt mit Nr. 4893 *(T'ien wén pu t'ien ko)* im Catalogue des livres chinois von M. COURANT (Vol. II, Paris, 1903, p. 40) überein, ein Werk, das indessen nicht von Jesuiten verfasst ist.

[3]) Vergl. A. WYLIE, The Mongol Astronomical Instruments in Peking (Chinese Researches. Shanghai, 1897, Part III, pp. 1—20); YULE, Marco Polo, 3. Aufl., Vol. I, pp. 446—456.

[4]) Vergl. PALLAS, Sammlungen etc., Zweiter Teil, S. 218—234; B. BERGMANN, Nomadische Streifereien unter den Kalmüken, Band II, Riga, 1804, S. 333—340; SCHMIDT, Geschichte der Ost-Mongolen, pp. XIX—XXI. F. K. GINZEL (Handbuch der mathematischen und technischen Chronologie) hat leider die Zeitrechnung der Mongolen nicht behandelt.

der von Papst Nikolaus IV. als Gesandter an den Hof des Kaisers Kubilai geschickt wurde (1291). In einem Briefe vom 8. Januar 1305 schrieb er: «Seit zwölf Jahren habe ich nunmehr keine Nachrichten aus dem Westen erhalten. Ich bin alt und grauköpfig geworden, doch eher von Arbeit und Leiden als vom Alter, denn ich bin erst 58 Jahre alt. Ich habe eine sachverständige Kenntnis der tatarischen Sprache und Schrift, in die ich das ganze Neue Testament und die Psalmen übersetzt habe, und habe sie mit grösster Sorgfalt abschreiben lassen.»[1] Es ist nicht bekannt, ob diese Übersetzung je veröffentlicht worden ist, und es bleibt zweifelhaft, ob sie überhaupt gedruckt worden ist. Monte Corvino erwähnt auch ebenda, dass er die Absicht gehabt habe, das ganze lateinische Ritual zu übersetzen, und dass er bei der Celebrirung der Messe die Worte der Præfatio und des Canons in mongolischer Sprache gelesen habe.

I. J. SCHMIDT vollendete 1809 die Übersetzung des Evangeliums Matthäi ins Kalmükische, doch sein Manuskript ging im Brande von Moskau zu Grunde, so dass der Druck erst 1815 stattfand.[2] Bis 1817 war die ganze Auflage ausverkauft. Die Übersetzung der Vier Evangelien und der Apostelgeschichte ins Kalmükische wurde 1822 fertiggestellt;[3] eine neue Übersetzung des Neuen Testaments ins Kalmükische wurde von A. POZDNÈJEV[4] mit Hülfe einiger Assistenten unternommen, dessen Werk im Jahre 1896 auf photolithographischem Wege verkleinert in Shanghai reproducirt wurde.

EDWARD STALLYBRASS und W. SWAN wurden 1817 von der London Missionary Society in die Mongolei gesandt, um die

[1] H. YULE, Cathay and the Way thither. Vol. I, p. 202; ABEL-RÉMUSAT, Jean de Montecorvino, Nouveaux mélanges asiatiques, Vol. II. Paris. 1829, pp. 193 198.

[2] Evangelium St. Matthæi in linguam Calmucco-Mongolicam translatum ab ISAACO IACOBO SCHMIDT. Cura et studio Societatis Biblicæ Ruthenicæ typis impressum. Petropoli, apud Fridericum Drechslerum, 1815. 4°.

[3] Eine Sonderausgabe des kalmükischen Evangeliums Johannis ist neuerdings von der Britischen und Ausländischen Bibelgesellschaft veranstaltet worden, ist aber ohne Jahreszahl. Matthäus wurde mongolisch in St. Petersburg, 1819, Markus und Lukas 1821, publicirt.

[4] Vol. I, St. Petersburg. English and Foreign Bible Society, 1887, 586 p. Vol. II, ibid., 1894, 688 p., 8°.

Bibel in literarisches Mongolisch zu übersetzen. Ihre Arbeit wurde von I. J. SCHMIDT revidirt. Das ganze Alte Testament wurde 1840 in Selenginsk und Khodon publicirt. Um dieselbe Zeit liess die Russische Heilige Synode diese Mission durch kaiserlichen Ukas unterdrücken. SWAN und STALLYBRASS kehrten daher nach England zurück, und die British and Foreign Bible Society traf 1843 Anordnungen für die Übersetzung und Publikation des Neuen Testaments, für welche die Arbeit der russischen Bibelgesellschaft als Grundlage diente. Da die in Sibirien gebrauchten Typen verkauft worden und keine mongolischen Typen in England vorhanden waren, wurde das Buch mit Manju Typen gedruckt, welche die British Bible Society schon besass. Im Jahre 1877 wurden für eine Neuausgabe des Neuen Testaments mongolische Typen besonders gegossen. ANTON SCHIEFNER überwachte den ersten Teil des Druckes, nach seinem Tode A. POZDNÉJEV; 1880 war der Druck beendet.[1]) Diese Ausgabe wurde

---

[1]) Die letzte mir zugängliche Ausgabe führt den Titel: The New Testament of our Lord and Saviour Jesus Christ: Translated out of the Original Greek into the Mongolian Language by EDWARD STALLYBRASS and WILLIAM SWAN, many years Missionaries residing in Siberia; for, and at the expense of, the British and Foreign Bible Society. London. printed by William Watts, 1896, 925 p. B. JÜLG (On the Present State of Mongolian Researches, JRAS, 1882, p. 23) sagt, dass das Alte Testament in Sibirien von 1836 bis 1840 übersetzt worden sei. Doch ich habe einen Druck der Übersetzung der Genesis eingesehen, aus dem hervorgeht, dass diese Übersetzung spätestens Ende 1831 beendet und 1833 herausgekommen sein musste. Das in Rede stehende Exemplar enthält das folgende interessante handschriftliche Vorwort:

The Directors of the London Missionary Society will be pleased to accept of this copy of the Book of Genesis, Translated into the Mongolian Language, by the Rev. Edward Stallybrass, as the First Fruits of the Mission Press at Selenginsk, Siberia.

The greater Number of the Larger Types used in Printing this First Edition, were cast from Matrices made at this station, by the Boriate workmen, as were also the Matrices for the Cipher Types, used for the Verses; and the whole work of composing, and almost the whole of the Printing, and Binding of this Edition was Executed by tens of the Young Boriate students of this Station.

The Book of Genesis has the Honour to be the First in the Mongolian Language, in which Ciphers has been used to mark the divisions!

1885 in Shanghai photolithographisch reproducirt und in einzelnen Teilen ausgegeben (neue Ausgabe, Shanghai, 1900).

In die Volkssprache der Khalkha ist das Evangelium Matthäi übersetzt worden, unter Leitung von JOSEPH EDKINS und Bischof SCHERESCHEWSKY, die einen Lama engagirten, um diese Version nach SWAN's und STALLYBRASS' literarischer Bearbeitung zu verfertigen. Die Arbeit wurde 1873 publicirt und 1894 wieder photolithographisch vervielfältigt.[1] EDKINS gab auch einen Christlichen Katechismus in mongolischer Sprache (26 Blätter, Peking, 1866) heraus.[2] Einige christliche Traktate von I. J. SCHMIDT und russischen Missionaren werden von B. JÜLG[3] aufgezählt. TIMKOWSKI[4] erwähnt in der Liste seiner in Peking erworbenen Bücher das «Gespräch eines Christen mit einem Chinesen über den Glauben» in mongolischer Sprache, in zwei Teilen. Dies muss die Übersetzung einer chinesischen Schrift des Jesuiten

The getting up of this Volume in its present form, and the labours attending it, were wholly from the Plan, and exertions of your

Humble Servant

Dec. 1831, Selenginsk. Robert Yuille.

Soli Deo honor et gloria.

Auf der Innenseite des nächsten Blattes findet sich die folgende Censurvermerkung gedruckt:

По порученію Г. Г. Министровъ Внутреннихъ дѣлъ, и Народнаго Просвѣщенія, Я читалъ Монгольскій переводъ книги Бытія, и нашелъ его согласнымъ съ разными переводами протестантскихъ исповѣданій, почему и одобряю его къ напечатанію: Санктпетербургъ, 11 Декабря 1833 года. Академикъ Коллежскій Совѣтникъ и Кавалеръ, Яковъ Шмитъ.

[1] Im allgemeinen vergl. über die mongolischen Bibelversionen M. BROOMHALL, The Chinese Empire, a General and Missionary Survey, London, 1907, pp. 410 414.

[2] [A. WYLIE], Memorials of Protestant Missionaries to the Chinese, Shanghai, 1867, p. 283.

[3] On the Present State of Mongolian Researches (JRAS, 1882, p. 22 des Separatabdrucks). Vergl. auch A. POZDNÉJEV, Nochmals zur Frage über die letzten Veröffentlichungen der Rechtgläubigen Missionsgesellschaft in kalmükischer Sprache, in Journal des Ministeriums für Volksaufklärung 302, pp. 158—172, 1900.

[4] Reise nach China durch die Mongoley in den Jahren 1820 und 1821, Band II, Leipzig, 1825, S. 357.

Matteo Ricci sein,[1] vermutlich des *Ki-jên shih pien*.[2] Ein anderes Werk von Ricci in mongolischer Version wird im Katalog von Kasan unter Nr. 120 angeführt als *Tägri-yin ädsän-ü ünängi jirum* «Wahre Lehren von Gott.» Dies ist das bekannte *T'ien-chu shih i* von Ricci, 1601 verfasst.[3]

Schlusswort. — Im allgemeinen lässt sich die mongolische Literatur als eine receptive und reproducirende charakterisiren. Sie hat das buddhistische Schrifttum Tibets und einen kleinen Bruchteil chinesischer Literatur in sich aufgenommen, ohne sich zu einer originalen Verarbeitung der so empfangenen Ideen aufzuschwingen. Gute Erzähler, wie sie stets waren, wurden die Mongolen eifrige Verbreiter indischer Geschichten, von deren unerschöpflichem Vorrat sie ihren westlichen Nachbarn, Türken und Slaven, abgegeben haben, und von diesem Gesichtspunkt der kulturellen Vermittelung ordnet sich auch ihre literarische Tätigkeit in den Rahmen der Weltliteratur ein. Historischen Sinn haben sie unzweifelhaft besessen, wenn vielleicht auch erst von den Chinesen erworben, und ihre Chroniken, auf die sie gewiss viel Fleiss und Ausdauer verwandt haben, stellen ihrem guten Willen, wie ihrer Bildungsfähigkeit, ein schönes Zeugnis aus. Wo sie, von fremden Einflüssen ungestört, ihre Eigenart frei entfalten konnten, hat es ihnen nie an Originalität und schöpferischen Leistungen des Volksgeistes gefehlt, die von Gemüt und Humor beseelt sind. Sie haben die schlichte empfindende Volksweise und das epische Heldenlied gepflegt. Auf letzterem Gebiet zeigen sie sich den Kirgisen, den Wogulen, den Irtysch-Ostjaken, den Esthen und Finnen geistesverwandt, und Antipoden der Chinesen und jener ganzen ostasiatischen Kulturwelt, die nie einen Heldensang hervorgebracht hat. Das Epos übt seinen Zauber auf uns alle aus, und die Frage seiner Ent-

---

[1] Vergl. den in der Orientalischen Bibliographie, Band IV, 1891, unter Nr. 403 angeführten Titel.

[2] H. Cordier, L'imprimerie sino-européenne en Chine, Paris, 1901, p. 40, Nr. 233.

[3] H. Cordier, l. c., p. 39, Nr. 225. Diese mongolischen Übersetzungen der Schriften von Ricci sind nicht von Cordier notirt und scheinen den Jesuiten selbst, in deren Werken sie nicht erwähnt werden, unbekannt geblieben zu sein.

stehung und Verbreitung ist für jeden Gebildeten von unerschöpflichem Interesse. Auch darin kann uns die mongolische
Literatur führen und wichtige Beiträge zur Lösung sowohl allgemeiner Fragen als insbesondere der noch ungelösten Frage
der Geschichte der epischen Lieder auf dem Boden des mittleren und nördlichen Asiens liefern. Auch hier gehen die Erzeugnisse der Mongolen über die rein literarische Bedeutung
weit hinaus; sie werden zu wertvollen kulturhistorischen und
ethnographischen Urkunden, indem sie uns die Zusammenhänge
mit den Völkern der fernsten Thule enthüllen und die geschichtlichen Fäden, die alles menschliche Denken verbinden, vor uns
aufdecken. Die centrale Stellung, welche die Mongolen in Asien
einnehmen, hat sie auch sozusagen zu einem geistigen Mittel-
punkte gemacht, zu einem Herde, von dem die verschiedensten
Stämme sich einen Feuerbrand mitgenommen haben. Zwei Haupt-
strömungen sind bis jetzt erkennbar, die von den Mongolen in
fremdes und fernes Gebiet hineinführen. In nordöstlicher Richtung sind ihre Sagen zu tungusischen Völkern gedrungen, die
sie weiter zu Korjaken und Tschuktschen bis an die Behrings-
strasse gebracht haben, und selbst darüber hinaus finden wir
Anklänge an der nordpacifischen Küste Amerikas.[1] Auf einem
anderen Wege sind mongolische Heldenmärchen ostwärts in das
Tal des Amur zu den Golden und anderen Tungusen gewan-
dert,[2] ja, sie sind über die Meerenge nach Sachalin zu den Ainu
hinübergegangen. Dieses ganze Phänomen ist nicht weniger bemerkenswert als die Wanderung indischer Märchen nach dem
Westen, und auf einem Gebiet, wo der geschichtlichen Dokumente so wenige sind, muss es eben zur Reconstruktion der
Geschichte helfen. Und dieses Moment allein wäre bedeutsam
genug, um das Studium der mongolischen Literatur zu ermun-
tern und fruchtbar zu machen.

---

[1] Nur nebenher kann ich hier auf die grundlegenden Arbeiten von
FRANZ BOAS (Indianische Sagen von der Nord-Pacifischen Küste Amerikas,
Berlin, 1895), W. JOCHELSON (The Koryak, l. c.) und P. EHRENREICH (Die
Mythen und Legenden der südamerikanischen Urvölker und ihre Bezie-
hungen zu denen Nordamerikas und der alten Welt, Berlin, 1905) hin-
weisen.

[2] LAUFER, American Anthropologist, 1900, pp. 330—331.

Aber noch ein anderer Gesichtspunkt kommt hinzu. Die weltgeschichtliche Bedeutung der Mongolen ist so gross, dass uns alles an dem Geistesleben eines Volkes, das einst Europa zu erschüttern vermochte und seine Könige und Päpste in seinen Bann zog, lebhaft interessiren muss. In der Geschichte unseres Vaterlandes sind sie als barbarische Horden gebrandmarkt, auf der Schulbank werden sie uns schwärzer als chinesische Tusche und als die Vorläufer jener berühmten gelben Gefahr gemalt. Um so mehr hat die Wissenschaft die Pflicht, solchem gro tesken Farbenauftrag entgegenzutreten und dieses Volk auch unserem Herzen menschlich näher zu bringen. In seinen Liedern tritt es uns entgegen in seinen Freuden und Leiden, in seiner innigen Liebe zur Heimat, zur weiten Steppe und zum feurigen Steppenross, in rührender Anhänglichkeit zur Mutter, zur Geliebten, zum Weibe. Wir erkennen die menschlichen und auch die edlen und ritterlichen Züge dieser ungebundenen Steppensöhne, die einst mit so mächtigem Schwertschlag an die Tore unseres Vaterlandes gepocht haben.

TÜRKISCHES VOLKSSCHAUSPIEL.

Orta ojnu.

— Von Dr. Ignaz Kúnos. —

(Zweite Mitteilung.*)

II.

Während der Zusammenstellung meiner auf die *Orta* bezüglichen Daten, erschien in den Spalten eines türkischen Blattes eine auf diese Spiele Bezug habende türkische Mitteilung.**) Verfasser derselben ist Ahmed Rāsim. Obgleich diese kurze Notiz

*) S. Keleti Szemle p. 1—95.
**) S. Ikdam, Journal politique, économique et littéraire. Constantinople, jeudi 11 Juillet 1907.